HANDBOOK OF COGNITIVE TASK DESIGN

Edited by

Erik Hollnagel
University of Linköping, Sweden

LEA
LAWRENCE ERLBAUM ASSOCIATES, PUBLISHERS
2003   **Mahwah, New Jersey**   **London**

Senior Acquisitions Editor: Anne Duffy
Editorial Assistant: Kristin Duch
Cover Design: Kathryn Houghtaling Lacey
Textbook Production Manager: Paul Smolenski
Full-Service Compositor: TechBooks
Text and Cover Printer: Hamilton Printing Company

This book was typeset in 10/12 pt. Times, Italic, Bold, Bold Italic. The heads were typeset in Helvetica Bold, and Helvetica Bold Italic.

Lawrence Erlbaum Associates, Inc., Publishers
10 Industrial Avenue
Mahwah, New Jersey 07430

The editor, authors, and the publisher have made every effort to provide accurate and complete information in this handbook but the handbook is not intended to serve as a replacement for professional advice. Any use of this information is at the reader's discretion. The editor, authors, and the publisher specifically, disclaim any and all liability arising directly or indirectly from the use or application of any information contained in this handbook. An appropriate professional should be consulted regarding your specific situation.

Library of Congress Cataloging-in-Publication Data

Handbook of cognitive task design / edited by Erik Hollnagel.
     p.   cm.—(Human factors and ergonomics)
    Includes bibliographical references and index.
    ISBN 0-8058-4003-6 (alk. paper)
    1. Human-machine systems—Handbooks, manuals, etc. 2. Cognition—Handbooks,
manuals, etc.   3. Task analysis—Handbooks, manuals, etc. I. Hollnagel, Erik, 1941–
II. Series.

TA167.H35   2003
620.8′2—dc21   2003042401

Printed in the United States of America
10   9   8   7   6   5   4   3   2   1

HANDBOOK OF COGNITIVE TASK DESIGN

HUMAN FACTORS AND ERGONOMICS
Gavriel Salvendy, Series Editor

Hendrick, H., and Kleiner, B. (Eds.): *Macroergonomics: Theory, Methods, and Applications.*

Hollnagel, E. (Ed.): *Handbook of Cognitive Task Design.*

Jacko, J. A., and Sears, A. (Eds.): *The Human-Computer Interaction Handbook: Fundamentals, Evolving Technologies and Emerging Applications.*

Meister, D. (Au.): *Conceptual Foundations of Human Factors Measurement.*

Meister, D., and Enderwick, T. (Eds.): *Human Factors in System Design, Development, and Testing.*

Stanney, Kay M. (Ed.): *Handbook of Virtual Environments: Design, Implementation, and Applications.*

Stephanidis, C. (Ed.): *User Interfaces for All: Concepts, Methods, and Tools.*

Ye, Nong (Ed.): *The Handbook of Data Mining.*

Also in this Series

HCI 1999 Proceedings   2-Volume Set

- **Bullinger, H.-J., and Ziegler, J.** (Eds.): *Human-Computer Interaction: Ergonomics and User Interfaces.*
- **Bullinger, H.-J., and Ziegler, J.** (Eds.): *Human-Computer Interaction: Communication, Cooperation, and Application Design.*

HCI 2001 Proceedings   3-Volume Set

- **Smith, M. J., Salvendy, G., Harris, D., and Koubek, R. J.** (Eds.): *Usability Evaluation and Interface Design: Cognitive Engineering, Intelligent Agents and Virtual Reality.*
- **Smith, M. J., and Salvendy, G.** (Eds.): *Systems, Social and Internationalization Design Aspects of Human-Computer Interaction.*
- **Stephanidis, C.** (Ed.): *Universal Access in HCI: Towards an Information Society for All.*

For more information on LEA titles, please contact Lawrence Erlbaum Associates, Publishers, at www.erlbaum.com.

To Agnes

Contents

Series Foreword

With the rapid introduction of highly sophisticated computers, (tele)communication, service, and manufacturing systems, a major shift has occurred in the way people use technology and work with it. The objective of this book series on human factors and ergonomics is to provide researchers and practitioners a platform where important issues related to these changes can be discussed, and methods and recommendations can be presented for ensuring that emerging technologies provide increased productivity, quality, satisfaction, safety, and health in the new workplace and the "Information Society."

Erik Hollnagel, a leading international cognitive ergonomics scientist, assembled some of the best and brightest cognitive ergonomics scientists from around the world to author the 31 chapters of this handbook. This is reflected in the depth and breadth of the discipline coverage.

The 190 figures and 101 tables of the handbook illustrate, in a most effective way, the concept, methods, and tools of cognitive ergonomics. The 1387 references provide a road map for further in-depth study of many of the cognitive ergonomics concepts and methods. Ten of the chapters deal with the theory of cognitive task design, 12 of them deal with its methods, and 9 describe field studies in the area.

The handbook should be of use to both developers and users of cognitively based tasks, systems, tools, and equipment, and of special use to researchers and practitioners in the broader discipline of human factors and ergonomics.

—Gavriel Salvendy
Series Editor

Acknowledgments

This handbook would not have been possible without the dedicated efforts of a great many people, including, of course, the authors. From the very beginning I was supported by the members of the editorial board, who contributed with discussions and viewpoints on the scope of the handbook and who were further instrumental in soliciting potential authors and reviewers for the draft chapters. The members of the Editorial Board are, in alphabetical order:

Rene Amalberti, IMASSA CERMA, Paris, France
Liam Bannon, University of Limerick, Limerick, Ireland
John Flach, Wright State University, Dayton OH, USA
Gudela Grote, Technical University, Zurich, Switzerland
Edwin Hutchins, University of California at San Diego, CA, USA
Kenji Itoh, Tokyo Institute of Technology, Tokyo, Japan
Antonio Rizzo, University of Siena, Siena, Italy
Tjerk van der Schaaf, Technical University, Eindhoven, Netherlands
John Wilson, University of Nottingham, Nottingham, UK
David D. Woods, Ohio State University, Columbus OH, USA
Wan Chul Yoon, KAIST, Taejon, South Korea

A special word of thanks is due to Nalini Suparamaniam from the University of Linköping, who served as my editorial assistant throughout the process, at the same time as she pursued her own doctoral degree. She was extremely helpful in keeping track of the many communications to authors and by gently reminding people of impending deadlines. She performed this sometimes delicate task with great skill, and is part of the reason that the progress stayed reasonably close to the schedule.

Last but not least I would like to thank the many people from all around the world who so willingly accepted the task of reviewing the various chapters. Reviewing is always an additional task, and people who are sought as reviewers are, as a rule, busy with many other things. Reviewing a chapter for a book is also in many ways more difficult than reviewing a paper, because there is the overall context of the handbook to be taken into account. Altogether, 76 people reviewed and commented on the chapters and provided views and comments that were essential not only for the authors, but also for the editorial process as a whole. The reviewers were as follows:

Anthony Adamski	Sue Baker	Jose Cañas
Motoyuki Akamatsu	Kevin Bennet	Woo C. Cha
Jim Alty	Ann Bisantz	Sidney Dekker
John Annett	Alan Blackwell	Francoise Detienne
Kazuo Aoki	Nathalie Bonnardel	Eileen Entin
Nikolaos Avouris	Pietro Carlo Cacciabue	Jonathan Grudin
Sebastiano Bagnara	Todd J. Callantine	Michael Harrison

Brian Hazlehurst
Martin G. Helander
Jean-Michel Hoc
Barbara E. Holder
Kenji Itoh
Denis Javaux
Anker Helms Jørgensen
Rainer Kluwe
Akinori Komatsubara
Tom Kontogiannis
Ryuuji Kubota
Todd Kustra
John Lee
Debbie Lucas
Holger Luczak
Jane Malin
Chris Miller
Dae H. Min
Randall Mumaw

S. Narayanan
Mark Neerincx
John O'Hara
Mieko Ohsuga
Noriyuki Onihashi
Rainer Oppermann
Raja Parasuraman
Jennifer Perotti
Marielle Plat
Scot Potter
Bill Rankin
Bernard Riera
Antonio Rizzo
Janine Rogalski
Ragnar Rosness
Emilie Roth
Bengt Sandblad
Nadine Sarter
Jan Maarten Schraagen

Valerie Shalin
Larry Shattuck
Tom Sheridan
Jonathan Smalley
Kip Smith
Tamara Sumner
Gunilla Sundström
Alistair Sutcliffe
Ola Svenson
Gerrit van der Veer
Yvonne Waern
Toni Wäfler
Bernhard Wilpert
John Wilson
Wan Chul Yoon
Jiajie Zhang
Bernhard Zimolong

Contributors

Anthony J. Adamski
Eastern Michigan University, USA

James L. Alty
Loughborough University, UK

Matthijs Amelink
Delft University of Technology,
The Netherlands

Henning Boje Andersen
Risø National Laboratory, Denmark

John Annett
University of Warwick, UK

Nikolaos M. Avouris
University of Patras, Greece

Ann M. Bisantz
University at Buffalo, USA

Alan F. Blackwell
University of Cambridge, UK

Heiner Bubb
University of Technology Munich,
Germany

Kevin M. Corker
San Jose State University, USA

Sarah Jackson
University of Nottingham, UK

Serge Debernard
Universite de Valenciennes, France

Françoise Decortis
Université de Liège, Belgique

James R. Easter
ManTech Aegis Research Corporation, USA

William C. Elm
ManTech Aegis Research Corporation, USA

John M. Flach
Wright State University, USA

Thomas R. G. Green
University of Leeds, UK

Gudela Grote
Swiss Federal Institute of Technology,
Switzerland

James W. Gualtieri
ManTech Aegis Research Corporation, USA

Michael D. Harrison
University of York, UK

Rachel L. Hewson
University of Cambridge, UK

Robert J. B. Hutton
Klein Associates Inc., USA

Toshiyuki Inagaki
University of Tsukuba, Japan

Kenji Itoh
Tokyo Institute of Technology, Japan

Paul F. Jacques
Wright State University, USA

Philip D. Johnson
University of York, UK

Holger Luczak
University of Technology at Aachen, Germany

Patrizia Marti
University of Siena, Italy

John McCarthy
University College Cork, Ireland

Pamela M. McCraw
Johnson Space Center, USA

Thomas E. Miller
Klein Associates Inc., USA

Claudio Moderini
Domus Academy, Italy

Manfred Mühlfelder
University of Technology at Aachen,
Germany

Max Mulder
Delft University of Technology,
The Netherlands

Mark A. Neerincx
TNO Human Factors, The Netherlands

Sarah Nichols
University of Nottingham, UK

Jennifer J. Ockerman
Georgia Institute of Technology, USA

M. M. (Rene) van Paassen
Delft University of Technology,
The Netherlands

Darby L. Patrick
Wright State University, USA

Marielle Plat
Airbus France, S.A.S., France

Scott S. Potter
ManTech Aegis Research Corporation,
USA

Peter Rasker
TNO Human Factors, The Netherlands

Bernard Riera
University of Reims, France

Antonio Rizzo
University of Siena, Italy

Emilie M. Roth
Roth Cognitive Engineering, USA

Job Rutgers
Philips Design, The Netherlands

Cornelia Ryser
Swiss Federal Institute of Technology,
Switzerland

Ludger Schmidt
University of Technology at Aachen,
Germany

Jan Maarten Schraagen
TNO Human Factors, The Netherlands

Masaki Seki
Central Japan Railway Company, Japan

Valerie L. Shalin
Wright State University, USA

Jonathan Smalley
QinetiQ, UK

Oliver Sträter
Eurocontrol, Belgium

Alistair Sutcliffe
University of Manchester Institute of Science
and Technology, UK

Jacques Theureau
Université de Technologie de Compiègne,
France

Marvin L. Thordsen
Klein Associates Inc., USA

Nikolaos K. Tselios
University of Patras, Greece

Martijn Van Welie
Vrije Universiteit, The Netherlands

Gerrit C. Van Der Veer
Vrije Universiteit, The Netherlands

Ron Westrum
Eastern Michigan University, USA

John R. Wilson
University of Nottingham, UK

Anna Windischer
Swiss Federal Institute of Technology,
Switzerland

David D. Woods
The Ohio State University, USA

Peter C. Wright
University of York, UK

Toni Wäfler
Swiss Federal Institute of Technology,
Switzerland

Acronyms

3D	Three-dimensional
A/THR	Autothruttle
A2/C2	Adaptive architectures and command and control
ACWA	Applied cognitive work analysis
AGARD	Advisory group for aerospace research and development
AGIP	Agent-based intelligent collaborative process construction kit
AHASS	Advanced human adapted supervisory systems
AIO	Action Information Organisation
AM	Attitude-maneuver options on trajectory propagation
AMANDA	Automation and man-machine delegation of action
AMEBICA	Autoadaptive multimedia environment based on intelligent collaborating agents
AOA	Angle of attack
AP	Autopilot
API	Application programmer interfaces
APU	Auxiliary power unit
ART	Action regulation theory
AST	Adaptive structuration theory
ATC	Air traffic controller
ATL	Attitude time line
ATM	Air traffic management
ATOMOS	Advanced technology to optimize manpower on board ships
AVID	Anchor vector ID
AWACS	Airborne warning and control system
BAS	Basic action sequence
BDC	Bed-Down Critic
C.AR.ME	Conservation of area and its measurement environment
C2	Command and control
CAA	Civil Aviation Authority
CAG	Carrier air group
CAHR	Connectionism assessment of human reliability
CARE	Cooperative actions of R&D in Eurocontrol
CAS	Compound action sequence
CBD	Cargo bay doors
CCD	Charge-coupled device
CCO	Combat control officer
CD/s	Cognitive dimension/s (of notations). Commonly used among cognitive dimensions users as a shorthand term to mark references to the theory.
CDM	Critical Decision Method
CDS	Cook decision system

CFM	Cognitive function model
CIC	Combat information center
CMTool	Cognitive modeling tool
COA	Course of action
CPT	Captain
CREA!	Creative research environment for ATM
CREAM	Cognitive reliability and error analysis method
CRO	Character read out
CSCW	Computer supported cooperative work
CSE	Cognitive systems engineering
CTA	Cognitive task analysis
CTD	Cognitive task design
CWA	Cognitive work analysis
CWR	Cognitive work requirements
CWS	Common work space
DCD	Decision-centered design
DDD	Distributed dynamic decision
DG	Diagnosis guide
DI	Direct input
DISC	Demonstrator integrated ship control
DMT	Detailed maneuver table
DTA	Dynamic task allocation
DTM	Designer task model
DV	Digital video
ΔV	Delta velocity
DYN	Dynamics officer
EEG	Electroencephalogram
EGPWS	Enhanced ground proximity warning system
EID	Ecological interface design
ENM	Electrical network management
EPHL	Ephemeris length
ET	Elapsed time
FAA	Federal Aviation Administration
FAN	Functional abstraction network
FAR	Federal aviation regulation
FCU	Flight control unit
FD	Flight director
FDO	Flight dynamics officer
FFS	Full flight simulator
FMA	Flight mode annunciator
FMAQ	Flight Management Attitudes Questionnaire
FO	First officer
FPA	Flight path angle (vertical navigation mode in degrees)
GMTV	Greenwich Mean Time for vector
GOMS	Goals, operators, methods and selection rules
GPWS	Ground proximity warning system
GTA	Groupware task analysis
HAZOP	Hazard and operability analysis
HCI	Human–computer interface or Human–computer interaction
HDG	Heading mode (lateral automated navigation mode)

HEP	Human error probability
HMC	Human–machine cooperation
HRA	Human reliability assessment or human reliability analysis
HSI	Human system interaction or human system integration
HTA	Hierarchical task analysis
HTA(T)	Hierarchical task analysis for team tasks
IAEA	International Atomic Energy Agency
ID	Identification
IDA-S	Template for partially automated functions based on information, decision, supervision, and action
IFF	Identify friend or foe
IH	Information handler
IMCS	Integrated monitoring and control system
IO	Information operations
IRR	Information relationship requirements
JCS	Joint cognitive system
JR	Japan Railway
KAT	Knowledge analysis of tasks
KB	Knowledge-based
KCON	Constant value for the trajectory propagator
KE	Kinetic energy
KVAR	Variable value for the trajectory propagator
LNAV	Lateral navigation
LOA	Level of automation
LOFT	Line oriented flight training
LSO	Landing signal officer
MAGTIF	Marine Air Ground Task Force
MCT	Mission constants table
MDD	Mass data display
MET	Mission elapsed time
MIDAS	Man–machine integrated design and analysis system
MOC	Mission operations computer
MPEG4	Moving Picture Experts Group version 4
MUSE	Method for usability engineering
NAM	Narrative activity models
NASA	National Aeronautics and Space Administration
NASA-TLX	NASA task load index
NAV	Navigation officer
NDM	Naturalistic decision making
NG	New guys
OECD	Organization for Economic Cooperation and Development
OFM	Operator function model
OMI	Operator–machine interface
OMP	Orbital maneuver processor
OMS	Orbiter maneuvering system (type of engine)
ORBB	Orbit number at the beginning of an ephemeris
OTA	Overview task analysis
P&ID	Piping and instrumentation diagram
PC	Planning controller
PDC	Presentation design concept

PE	Potential energy
PF	Pilot flying
PFD	Primary flight display
PLAF	Module's name–planning of allocation
PNF	Pilot not flying
PRA	Probabilistic risk assessment
QRH	Quick reference handbook
RB	Rule-based
RC	Radar controller
RCS	Reaction control system (type of engine)
RCT	Rendezvous constraints table
RDR	Representation design requirements
RELMO	Relative motion analysis
RET	Rendezvous evaluation table
REZ	Road effect zone
RP	Rule provider
RPD	Recognition-primed decision
S&C System	Supervisory and control system
SA	Situation awareness
SAINTEX	Module's name (in reference to Saint-Exupéry, (*Vol de Nuit,* 1931)
SB	Skill-based
SC	Scheduler
SD method	Semantic differential method
SGT	Sub-goal template
SGTR	Steam generator tube rupture
SMAQ	Ship Management Attitudes Questionnaire
SME	Subject-matter expert
SOPs	Standard operating procedures
SPECTRA	Experimental dynamic task allocation system for air traffic control
SSD	State space diagram
STAR	Tactical system for assistance to resolution
STDN	Spacecraft tracking and data network processing
STM	Student task model
TA	Tactical assessor
TAD	Target audience description
TAFEI	Task analysis for error identification
TAKD	Task analysis for knowledge description
TCAS	Traffic alert and collision avoidance system
TDRS	Tracking and data relay system
TE	Total energy
TEAS	Training equipment advisory system
THERP	Technique for human error rate prediction
TIDE	Team integrated design environment
TIE	Technology integration experiment
TKS	Task knowledge structure
TMAQ	Train Management Attitudes Questionnaire
TOD	Temporal overview display
TOTE	Test-operate-test-exit
TPAB	Team performance assessment battery
TPS	Trajectory profile status

Traj	Trajectory officer
TRK	Track mode (lateral automated navigation mode)
TSU	Trajectory software upgrade
TUP	Trajectory update
UAtool	Usability analyzer tool
UI	User interface
UIMS	User interface management systems
UML	Unified Modeling Language
UTM	User task model
UVM	User virtual machine
V/S	Vertical speed (vertical navigation mode in feet)
VAM	Vent-attitude-maneuver options on trajectory processing
VAT	Vector administration table
VNAV	Vertical navigation
VRML	Virtual reality modeling language
VTL	Vent time line
WTS	Weights

HANDBOOK OF COGNITIVE TASK DESIGN

PART I
Theories

1

Prolegomenon to Cognitive Task Design

Erik Hollnagel
University of Linköping, Sweden

Abstract

The design of artefacts, and in particular artefacts involving computing technology, is usually focused on how the artefact should be used. The aim of cognitive task design (CTD) is to go beyond that focus by emphasising the need to consider not only how an artefact is used, but also how the use of the artefact changes the way we think about it and work with it. This is similar to the envisioned world problem, that is, the paradox that the artefacts we design change the very assumptions on which they were designed. The ambition is not to make CTD a new discipline or methodology, but rather to offer a unified perspective on existing models, theories, and methods that can be instrumental in developing improved systems. In this context, cognition is not defined as a psychological process unique to humans, but as a characteristic of system performance, namely the ability to maintain control. The focus of CTD is therefore the joint cognitive system, rather than the individual user. CTD has the same roots as cognitive task analysis, but the focus is on macrocognition rather than microcognition; that is, the focus is on the requisite variety of the joint system, rather than the knowledge, thought processes, and goal structures of the humans in the system.

WHAT IS CTD?

In a handbook of CTD, it is reasonable to begin by defining what CTD is. It is useful first to make clear what it is not! CTD is not a new scientific discipline or academic field; nor is it a unique methodology.

CTD Is not a New Scientific or Academic Field

Since the mid-1970s, the terms *cognitive* and *cognition* have come to be used in so many different ways that they have nearly lost their meaning. In particular, there has been an abundance of more or less formal proposals for lines of activity or directions of study that as a common feature have included the words *cognitive* and *cognition* in one way or another. These range from cognitive ergonomics, cognitive systems engineering, and cognitive work analysis to cognitive tools, cognitive task analysis, cognitive function analysis, cognitive technologies, cognitive agents, and cognitive reliability—to mention just a few.

Some of these have thrived and, by their survival, justified their coming into the world as well as established an apparent consensus about what the word *cognitive* means. Others have been less successful and have in some cases languished uneasily between survival and demise for years. The situation does not in any way seem to improve, because the words *cognitive* and *cognition* are used with ever-increasing frequency and, inevitably, diminishing precision— although at times with considerably success in obtaining funding.

CTD is not put forward in the hope that it will become a new scientific field. Indeed, if that were the case it would signify failure rather than success. As this handbook should make clear, CTD is something that we already do—or should be doing—although we may lack a precise name for it.

CTD Is not a Unique Methodology

The widespread use of the terms *cognitive* and *cognition* is, of course, not the result of a whim but does indicate that there is an actual problem. Indeed, many people are genuinely concerned about issues such as, for instance, cognitive tasks and cognitive reliability, because these are salient aspects of human behaviour in situations of work, as well as of human behaviour in general.

It is because of such concerns that a number of specific methodologies have emerged over the years. A case in point is the development of cognitive task analysis as an "extension of traditional task analysis techniques to yield information about the knowledge, thought processes, and goal structures that underlie observable task performance" (Chipman, Schraagen, & Shalin, 2000, p. 3). Even though cognitive task analysis is used as a common name for a body of methods that may vary considerably in their aim and scope, partially because of the imprecision of the term *cognitive,* they do represent a common thrust and as such a distinct methodology.

Whereas cognitive task analysis is aimed at analysing cognitive tasks, that is, knowledge, thought processes, and goal structures, CTD is not aimed at designing cognitive tasks as such, at least not in the precise sense implied by Chipman et al. (2000). Instead of being a unique design methodology, CTD is proposed as a unified perspective or point of view that can be combined with a number of different methods. This is explained in the following section.

CTD Is a Unified Perspective on Design

By saying that CTD is a perspective, I mean that it designates a specific way of looking at or thinking about design. More particularly, it is a specific way of thinking about the design of systems in which humans and technology collaborate to achieve a common goal. These systems are called *joint cognitive systems,* and this concept is explained later in this prolegomenon. The way of thinking that is represented by CTD can be expressed by a number of principles:

- Every artefact that we design and build has consequences for how it is used. This goes for technological artefacts (gadgets, devices, machines, interfaces, and complex processes) as well as social artefacts (rules, rituals, procedures, social structures, and organisations).
- The consequences for use can be seen both in the direct and concrete (physical) interaction with the artefact (predominantly manual work) and in how the use or interaction with the artefact is planned and organised (predominantly cognitive work). Thus, introducing a new "tool" not

only affects how work is done, but also how it is conceived of and organised. In most cases, this will have consequences for other parts of work, and it may lead to unforeseen changes with either manifest or latent effects.

• The primary target of design is often the direct interaction with or use of the artefact, as in human–computer interaction and human–machine interaction. Interface design, instruction manuals, and procedures typically describe how an artefact should be used, but not how we should plan or organise the use of it. However, the realization of the artefact may affect the latter as much, or even more, than the former.

• As a definition, the aim of CTD is to focus on the consequences that artefacts have for how they are used, and how this use changes the way we think about them and work with them, both on the individual and the organisational level. The ambition is to ensure that CTD is an explicit part of the design activity, rather than something that is done fortuitously and in an unsystematic manner.

CTD is not new, because many of the principles have been recognised and used before. Some may be found among the roots of cognitive task analysis, as described later. Two more recent examples are the concept of "design for quality-in-use" (Ehn & Löwgren, 1997) and the notion of task tailoring (Cook & Woods, 1996). It is also a postulate that this perspective is unified. Perhaps it would be more correct to say that CTD should *become* a unified perspective on design. The argument of this prolegomenon, and of many of the chapters that follow, is that the design of artefacts and systems has very often neglected the aspect of cognitive tasks in this sense, that is, as the way we think about artefacts rather than as the way we use them. A considerable amount of work has been done on issues of usability and usefulness, and several schools of thought have established themselves. However, both usability and usefulness focus more on the direct use of artefacts than on how that use by itself changes how we come to see them.

THE NEED OF CTD

The 20th century has witnessed dramatic changes to the nature of human work, particularly since the 1950s. These changes led to the emergence of human factors engineering, ergonomics, and later developments such as cognitive ergonomics and cognitive systems engineering. The extent of the changes is amply demonstrated by tomes such as the *Handbook of Human–Computer Interaction* (Helander, Landauer, & Prabhu, 1997) and the *Handbook of Human Factors and Ergonomics* (Salvendy, 1997). In relation to CTD, the following two changes were particularly important.

First, work went from being predominantly manual work—or work with the body—to being predominantly cognitive work—or work with the mind. Many manual tasks have disappeared while new cognitive tasks have emerged. For those manual tasks that remain, technology has often changed them considerably.

This change to the nature of human work is the logical continuation of the development that gained speed with the industrial revolution, the essence of which is that machines or technology are used to amplify or replace human functions. Modern information technology has significantly increased the speed by which this development takes place and the types of work that are affected. Even for the kinds of work that are still largely manual, technology or machinery is usually involved in some way. (A checkout counter at a supermarket is a good example of that.) Today there are, indeed, very few types of work that depend on human physical power and abilities alone. Machines and technology have effectively become part of everything we do, whether at work or at home. This means that cognitive tasks are everywhere, and work design is therefore effectively CTD.

Second, cognitive tasks, even loosely defined, are no longer the prerogative of humans, but can also be carried out by a growing number of technological artefacts. Machines and information technology devices have, in fact, for many years been capable of taking over not just manual but also mental work. In other words, the ability to amplify human functions has migrated from the physical (motor) to the cognitive parts of work. It is not just that work has become more cognitive, but also that humans have lost their monopoly of doing cognitive work or cognitive tasks. When we conceive of and build machines, we must therefore also in many cases consider CTD.

The changes to the nature of human work require corresponding changes to the methods by which work is described, analysed and designed. The cognitive aspects of work have traditionally been addressed by cognitive task analysis, as an extension of classical task analysis. However, cognitive tasks are more than just an addition to manual tasks, and the changes affect work as a whole rather than what humans do as part of work. To be able to build efficient and safe sociotechnical systems, we must be concerned about CTD from the point of view of the individual who carries out the work as well as of the work system. CTD thus comprises the study of cognitive work—by humans, by machines, and by human–machine ensembles—and covers the whole life cycle of work from preanalysis, specification, design, and risk assessment to implementation, training, daily operation, fault finding, maintenance, and upgrading. CTD is about the nature of work as it is now and as we want it to be in the future.

THE MEANING OF COGNITION

As previously argued, the words *cognitive* and *cognition* have generally been used without much precision, and Wilson, Crawford, and Nichols make the same point in chapter 5 in this handbook. It is therefore necessary to provide a more precise definition of what cognition means. This is not an easy task, because etymology, semantics, and practice may be at odds with each other. Although it is tempting to accept Jim Reason's definition of cognition as that which goes on "between the ears," it behooves an editor to sound more scientific. This leads to the following line of reasoning.

Cognition is used to describe the psychological processes involved in the acquisition, organisation, and use of knowledge—with the emphasis on rational rather than emotional characteristics. Etymologically it is derived from the Latin word *cognoscere,* to learn, which in turn is based on *gnoscere,* to know. Following the example of Mr. Pott in the *Pickwick Papers* (Dickens, 1837), we could then define cognitive tasks as those that require or include cognition.

This easily leads us to an axiomatic position, which starts from the fact that humans are cognitive beings (or that humans have cognition); hence human performance has a cognitive component. However, following the same line of reasoning, we could also argue that human actions are driven by motives and emotions, and that human performance therefore has a motivational and an emotional component—which indeed it has. Although it is evidently true that humans have cognition, the axiomatic position makes it difficult to extend the notion of cognition to other entities, such as technological artefacts and organisations. It also begs the question of what cognition really is.

An alternative is to use a more pragmatic definition, which is based on the characteristics of certain types of performance. Human performance is typically both orderly (systematic and organised) and goal directed. This can be used as a provisional definition of cognition, and it can be extended to require cognitive tasks to have the following characteristics:

- Cognitive tasks are driven by goals (purposes and intentions) rather than by events. They include cause-based (feedforward) control as well as error-based (feedback) control. Cognitive tasks are therefore not merely responses based on algorithmic combinations of predefined elements, but require thinking ahead or planning over and above complex reactions.
- Cognitive tasks are not limited to humans; cognitive tasks can be found in the functioning of organisations, of certain artefacts (a growing number, but still not many), and of animals.

Polemically, the issue is whether the definition of cognitive tasks is based on an axiomatic definition of cognition, or a pragmatic characterisation of performance. In other words, cognitive tasks could be defined as tasks performed by a system that has cognition—which presumably only humans have (?). Alternatively, cognition could be defined as a quality of any system that has certain performance characteristics, and that therefore can be said to do cognitive tasks. This would base the definition on the characteristics of the tasks and of system performance, rather than on the possible constituents and explanations of internal mechanisms.

From this perspective, cognitive tasks are characteristic of humans, organisations, and some artefacts. (I leave out animals in this context.) CTD is consequently concerned with how functions and structures—of a cognitive system proper and of its environment—can be designed to further the system's ability to perform in a purposeful manner and to let it keep control of what it is doing. Looked at in this way, CTD refers to (joint) cognitive systems as a whole, whether they are biological individuals, artificial intelligences, or organisations. CTD clearly also goes beyond cognitive tasks analysis, as the emphasis is on the potential (future) rather than the actual (past and present) performance.

The importance of CTD stems from the fact that any change to a system—such as the introduction of new technology, improved functionality, or organisational changes—inevitably changes the working conditions for the people in the system, and hence their cognitive tasks. All design is therefore implicitly or explicitly CTD. This is obviously the case for technological artefacts and information devices, because these directly affect user tasks. A little thought makes it clear that the same is true for any kind of design or intentional change to a system, because the use of the system, that is, the way in which functions are accomplished and tasks carried out, will be affected. CTD comprises the study of how intentional changes to system functions and structures affect the conditions for work, and hence the cognitive tasks, and the development of concepts and methods that can be used to improve design practices.

The outcome of this line of reasoning is that cognition is not defined as a psychological process, unique to humans, but as a characteristic of system performance, namely the ability to maintain control. Any system that can maintain control is therefore potentially cognitive or has cognition. In this way the focus of CTD is not just on the characteristics of putative human information processing or capabilities such as recognition, discrimination, and decision making that normally are seen as components of cognitive work. The focus is rather on descriptions of the performance of cognitive systems in the complex sociotechnical networks that provide the foundation of our societies, and how this performance must change to enable the systems to stay in control. CTD therefore struggles with the dilemma known as the envisioned word problem (Woods, 1998), which is how the results of a cognitive task analysis that characterises cognitive and cooperative activities in a field of practice can be applied to the design process, because the introduction of new technology will transform the nature of practice! Or, put more directly, the paradox of CTD is that the artefacts we design change the very assumptions on which they were designed.

COGNITIVE SYSTEMS AND JOINT COGNITIVE SYSTEMS

The concept of a joint cognitive system (JCS) has already been mentioned several times. Although classical ergonomics and human factors engineering have often emphasised the necessity of viewing humans and machines as parts of a larger system, the analysis nevertheless remains focused on the level of elements of the larger system, specifically humans and machines. A consequence of this seemingly innocuous and "natural" decomposition is that the interaction between human and machine becomes the most important thing to study. Cognitive systems engineering (CSE; Hollnagel & Woods, 1983) instead argues that the focus should be shifted from the internal functions of either humans or machines to the external functions of the JCS. This change is consistent with the idea that humans and machines are "equal" partners, and that they therefore should be described on equal terms. Humans should not be described as if they were machines; neither should machines be described as if they were humans. Whereas the former has been the case so often that it practically has become the norm, the latter has only rarely been suggested or practised. (Some examples would be anthropomorphism and animism. Thus most people tend to ascribe psychological qualities to machines.)

An important consequence of focusing the description on the JCS is that the boundaries must be made explicit—both the boundaries between the system and its context or environment, and the boundaries between the elements or parts of the system. A system can be generally defined as "a set of objects together with relationships between the objects and between their attributes" (Hall & Fagen, 1969, p. 81), or even as anything that consists of parts connected together. In this definition, the nature of the whole is arbitrary, and the boundary of the system is therefore also arbitrary. This is illustrated by the following delightful quote from Beer (1959):

> It is legitimate to call a pair of scissors a system. But the expanded system of a woman cutting with a pair of scissors is also itself a genuine system. In turn, however, the woman-with-scissors system is part of a larger manufacturing system—and so on. The universe seems to be made up of sets of systems, each contained within a somewhat bigger, like a set of hollow building blocks. (p. 9)

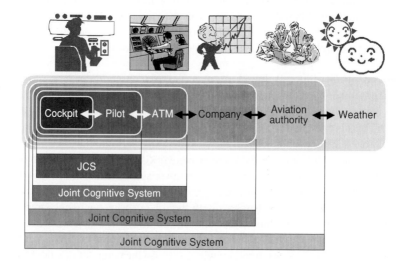

FIG. 1.1. Levels of JCSs.

TABLE 1.1
Pragmatic Definition of the JCS Boundary

	Objects or Factors that may Affect Artefact Use	*Objects or Factors that do not Affect Artefact Use*
Objects or factors that are changed by artefact use	must be considered by CTD	should be considered by CTD
Objects or factors that are not changed by artefact use	should be considered by CTD	need not be considered by CTD

As this examples shows, it is insufficient to describe only one level of the system, for instance that of the woman using the pair of scissors. The woman-*cum*-scissors is in turn a part of the local manufacturing system, which may be a part of a larger production system and so on. If for some reason the people cutting the cloth are given a better pair of scissors, or even a laser cutting tool, this will clearly have effects beyond the cutting itself. The design or improvement of the cutting artefacts must acknowledge this and include it into the design as far as necessary. Failing to do so will render the design basis invalid to some degree, and hence possibly jeopardise the outcome.

For a JCS, the boundary clearly depends on the purpose of the analysis and thereby also on the purpose of the JCS. Consider, for instance, a pilot flying an airplane (Fig. 1.1). The traditional ergonomic approach would focus on the pilot in the cockpit, and the way in which the interface is designed. This approach addresses issues such as information presentation, information access, automation, communication, and the manual and cognitive tasks required by flying as such. The ergonomic approach is based on decomposing the system into its basic elements, that is, the pilot and the cockpit, taking certain physical boundaries for granted.

In a CSE-based approach, the pilot–plane ensemble clearly constitutes a JCS, which exists within the larger air traffic environment. It makes sense to consider the pilot–plane ensemble as a JCS and to analyse, for example, how well it is able to attain its goals, such as arriving at the destination at the scheduled time. Alternatively, the pilots in the cockpit can also be considered as a JCS that is studied separately, such as in cockpit or crew resource management (Helmreich et al., 1997). The environment of the pilot–plane or crew–plane ensemble is, however, not just the atmosphere and the weather, but also the air traffic management (ATM) system, specifically the air traffic controllers. It may therefore in many cases make sense to consider the pilot–plane–ATM ensemble as a JCS and modify the boundaries to correspond to that. In this case the context may be the airline companies. The whole of aviation transportation may also be considered a JCS, and in this case the context could be the national aviation authority and the regulators, and so on.

From the general systems point of view, the environment is that which provides inputs to the system and which (one would hope) reacts to outputs from the system. A more formal definition is as follows: "For a given system, the environment is the set of all objects a change in whose attributes affect the system and also those objects whose attributes are changed by the behavior of the system" (Hall & Fagen, 1969, p. 83). Translated into CTD terms, this means that CTD should consider all objects or factors that may affect the use of the artefact as well as all objects and factors that are changed by the functions and use of the artefact. These principles are summarised in Table 1.1.

COGNITIVE TASK ANALYSIS AND CTD

Even though cognitive task analysis (CTA) and CTD are fundamentally different in their objectives, they do have several things in common. Most importantly, CTA is an important stepping-stone for and complement to CTD.

The label *cognitive task analysis* was coined in 1981 to 1982 to describe the endeavor to understand the cognitive activities required of a man–machine system. It was used in print in a technical report in February 1982 (Hollnagel & Woods, 1982), which later appeared in journal form as Hollnagel and Woods (1983), and it was a major explicit part of the discussions at the workshop in August 1982 on cognitive modeling of nuclear power plant control room operators (Abbott, 1982).

Looking back at the developments then, we can see five threads or approaches to CTA as emerging roughly in parallel in the early 1980s. (The following list is borrowed with kind permission from Woods, Tinapple, Roesler, and Feil, 2002.)

1. *Functionalism.* The need to aid human cognition and performance in complex high-consequence settings gave rise to what might be labeled a *functionalist cognitive engineering thread*. The goal of CTA here was to discover how the behaviour of practitioners adapted to the constraints imposed by the domain, organizational goals and pressures, and characteristics of the artefacts available (Hollnagel & Woods, 1983; Rasmussen, 1986; Rasmussen & Lind, 1981).

2. *The basis for expertise.* The need to expand expertise by means of new forms of training led to studies of the basis for expertise and the knowledge organization of people at different stages of acquiring expertise—what might be labeled a *cognitive learning thread* (e.g., Chi, Feltovich, & Glaser, 1981; McKeithen, Reitman, Reuter, & Hirtle, 1981), later transforming into knowledge engineering.

3. *Cognitive architectures.* The need to understand work through a computer and to support the design of human–computer interfaces led to work that mapped mechanisms of the individual's microcognition onto specific tasks that centered on interaction with computers and computerized devices—what also might be called a *cognitive simulation thread* (e.g., Card, Moran, & Newell, 1983; Kieras, 1988).

4. *Ethnography of workplaces.* The need to understand work culture as a result of the consequences of technology change led to field observation of practitioners at work in their world and ethnographies of work—what might be called an *ethnography of work thread* (e.g., Hutchins, 1980, 1990, 1995; Jordan & Henderson, 1995; Suchman, 1987).

5. *Naturalistic decision making.* Attempts to apply decision making to complex settings led to observations of people at work and critical incident studies. Observing people at work challenged assumptions about human decision making, leading to new models (Klein, Orasanu, Calderwood, & Zsambok, 1993).

Each of these threads included an approach to bring concepts and methods about cognition to work. In one way or another, they all tried to account for operational systems (people, technology, policies, and procedures) in terms of cognitive concepts. All of them also used CTA, or similar labels, as a pointer to develop a research base of cognition at work—even though that term had not itself become in vogue. Finally, they all wrestled with how to apply results from CTA to design.

Of the five threads just mentioned, only some are important for CTD. Because the purpose of CTD is to account for how the use of artefacts affects the ways in which they are understood, neither the acquisition of experience nor the match to human information processing is essential. What remains important are the functional approach and the strong anchoring in practice, that is, the link to macrocognition, rather than microcognition. The concern has also changed from that of yielding information about knowledge and thought processes to that of providing a comprehensive foundation for the design of artefacts and how they are used. In a historical context, CTD brings to the fore the struggle to apply CTA to design that was already present in the early 1980s. One ambition of this handbook is indeed to demonstrate how CTA can be used for CTD, and to argue that the design of any system, or of any detail of a system, be it hardware, software, or "liveware," inevitably also affects the tasks of the people involved (the chapter by Kevin Corker is a good example of that).

DESIGN AND CTD

Design is the deliberate and thoughtful creation of something to fulfil a specific function. Design is therefore usually focused on the actual purpose and use of the artefact, and every effort is made to ensure that the design objectives are achieved. A number of competing design schools exist, ranging from usability engineering and problem-oriented practice to user-centered, participatory, and situated design (e.g., Carroll, 2000; Dowell & Long, 1989; Greenbaum & Kyng, 1991; Nielsen, 1993; Norman & Draper, 1986; Winograd & Flores, 1986). Common to them all is a focus on the artefact in use.

The design of technological and social artefacts—ranging from simple tools used in the home, to complex machines used at work, to rules and regulations governing communication and cooperation—must consider both how the artefact is to be used and how this use affects the established practice of work. As a simple example of the latter, just think of how the introduction of a new and faster photocopier may change the work routines of the people who use it.

It is important for CTD to acknowledge that the use of an artefact may affect the systems in ways that were not among the primary concerns of the designer. Any change to a system will affect the established practice of work. Some of the consequences are desired, and indeed constitute the rationale for the design. Other consequences may be unwanted and represent unanticipated and adverse—or even harmful—side effects. If the artefact is well designed, the changes will be minimal and the unanticipated side effects may be negligible. However, if the design is incomplete or inadequate in one or more ways, the unanticipated side effects may be considerable.

In most cases people adapt to the artefact after a while, and they develop new working practices. This involves learning how to use the artefact, possibly by means of some system tailoring (Cook & Woods, 1996), agreeing on norms and criteria for its usefulness, developing new practices, task tailoring, and the like. As pointed out by the envisioned world dilemma, these responses transform the nature of practice, which may render invalid the criteria for quality-in-use that were the basis for the design. This is illustrated by Fig. 1.2, which shows a progression of steady or stable states of use, in contrast to a simple change from an "old" to a "new" system configuration. These issues become even more pronounced when the design concerns the organisation itself, such as role assignments, rules for communication and control, administrative and work procedures, and so on.

The design methodology must be able to anticipate and account for these changes. In the words of David Woods, "design is telling stories about the future," where the future may change as a result of the users' coping with complexity. To achieve that is obviously easier said than done, because predicting the future is notably a difficult business. CTD does not provide the final methodology to do that, but it does propose a new perspective on system design as a necessary first step.

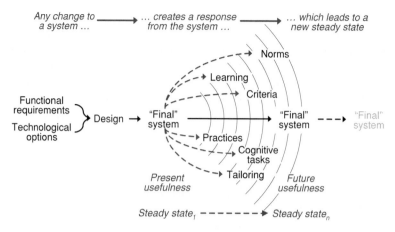

FIG. 1.2.   Context of CTD.

EXAMPLES

So far this prolegomenon has tried to describe what CTD is in an analytical manner. To complement that, a few examples may be of use.

An Analogy: The Road Effect Zone

One way of understanding the meaning of CTD is to consider the analogy of the road effect zone (REZ). A road is usually built to meet specific objectives, that is, to allow a certain volume of traffic to move efficiently from point A to point B. However, whenever a road is built, there are also consequences for the existing wildlife, that is, the ecology. A road thus has an influence that goes beyond its pavement (Aschwanden, 2001).

This is described by the term *road effect zone*. The REZ is defined as the total area over which a road exerts its ecological influence. The REZ is irregularly shaped, and it is determined by factors such as topography, landscape, land use, and waterways. Roads have normally been built without consideration of the REZ, and the result has often been devastating for the ecology and fauna. However, when the road is planned carefully and the REZ is considered, ecological costs can be reduced.

When we design an artefact—such as hardware, software, or rules for work and collaboration—it is not sufficient to ensure that the objectives for designing the artefact are achieved. Since the artefact will have consequences for the structure of work itself—in a sense the "social ecology"—it is necessary to consider this as part of the design. In other words, we need something akin to a *work effect zone,* defined as the total range of work over which the artefact has an influence. Finding effective ways of doing this is a primary objective for CTD.

Digital Pictures as Cognitive Artefacts

One of the more recent examples of how information technology affects daily life is the use of digital photography, which broke into the mass consumer market in 1997. From a technological point of view this can be seen as no more than replacing photographic film with a charge-coupled device (CCD). The primary advantage is that there is no need to develop the film and make copies, which saves both time and money. The main disadvantage was—and is—that the resolution of the picture is considerably poorer than that for a conventional camera.

Digital photography is, however, not just the replacement of one technology with another, but a complete revolution in the way we use pictures. For professional news photographers (if one dare still use that word), the change to digital pictures was accompanied by a wide-ranging change in how pictures are used, affecting not only how the picture is taken, but also how it is distributed, stored, and edited. For other professions, such as dentists, doctors, and real estate agents, there are equally large changes, although they are taking place more slowly. For private consumers, digital photography brings about a new way of thinking about pictures; because there is no need to worry about cost, one can afford to take many pictures and throw away the bad ones either on the spot or at a later time. This essentially amounts to a completely new practice of picture taking, as well as of storing.

Yet another effect is that pictures from digital cameras can be used as a complement to symbols in communication for people with cognitive disabilities (see, e.g., Jönsson & Svensk, 1995; Svensk, 2001). Pictures can, for instance, replace verbal descriptions both in remembering the past and in communicating with others and thereby help people with aphasia or other disabilities.

As these few examples show, the introduction of a new artefact—the digital camera—has wide-ranging consequences for how we think about the function that it provides (taking pictures), and it may lead to innovative applications. The extensive use of digital photography also causes major changes in organisations of many types and on many levels, from camera and film producers to news agencies. Although some of these changes may have been anticipated, it is a fair bet that most of them were not and that even now we do not appreciate the full impact of digital photography on our daily lives. This example illustrates that the design of the artefact is woefully incomplete if it is focused just on taking the pictures. Digital cameras affect the ways in which we use pictures, and thereby change some of the basis for designing the cameras in the first place.

Electronic Mail as Bookkeeping

A different and more straightforward example is the use of electronic mail. From a bumpy beginning in the mid-1980s, e-mail has now become the practical standard for communication—business as well as private. The design of e-mail systems has also converged toward a common style of interface, which usually includes some kind of list or overview of the e-mails that remain in the inbox.

Although the primary purpose of such a list is to show the e-mails that have arrived recently or since the service was last used, the e-mails that have been received usually remain on the list until the user actively deletes them. In this way the list can be used to keep track of e-mails that have been received but not yet acted on. Many people, including myself, use the inbox list as a bookkeeping device and as a way of managing correspondence and activities. In this way the artefact—the e-mail system—changes the way in which I work over and above what was intended. It not only makes it faster and easier to receive and send messages, but also serves as a reasonably effective to-do list.

Other, and perhaps more serious, consequences of the use e-mail are to the way information is archived and events are documented. A number of court cases around the world have shown that this is not an easy problem to solve. This goes way beyond the aim of the example here—e-mail as bookkeeping—but it illustrates in a dramatic sense that the change from one way of doing things to another, in this case the change from physical to electronic media, has far-reaching consequences. Designing a good e-mail system therefore requires more than understanding the knowledge and thought processes of individual users; it requires sufficient imagination to anticipate how the artefact will be used, how it will be brought into established practices, and how the whole fabric of activity will be changed as a result of that.

ABOUT THE HANDBOOK

Although it is common to end an introduction by providing a summary of the contents of the rest of the work, I have chosen not to do that. One reason is that each chapter has its own abstract, which the reader can use quickly to get an impression of what the chapter is about. Another reason is that the chapters are so diverse that they are difficult to put into well-defined groups.

As a consequence of that, the handbook is simply organised in three major parts, called Theory, Methods, and Field Studies, respectively. The various chapters have been put into one of these parts based on whether their main thrust is on theory, methods, or applications. Although several chapters with some justification could put in a different category, the need to achieve some kind of order has dictated the final choice. In this, as in many other cases, the linear ordering imposed by a printed document cannot do justice to the multifariousness of the contents. In any case, there is no single ordering that can satisfy the needs of all readers; so, rather than trying to solve an impossible problem, this prolegomenon has been provided to offer a basic outline of what CTD is. Although the reasoning presented here has developed with the handbook, the outline has in an earlier version been the basis for selecting and reviewing the individual contributions; it is my hope that it will also enable readers to find those chapters that are particularly relevant to their own problems and to make use of the knowledge and experience that these pages offer.

Epilogue: Mr. Pott's Approach

Mr. Pott looked dubiously at Bob Sawyer for some seconds, and, turning to Mr. Pickwick, said:

"You have seen the literary articles which have appeared at intervals in the Eatanswill Gazette in the course of the last three months, and which have excited such general—I may say such universal—attention and admiration?"

"Why," replied Mr. Pickwick, slightly embarrassed by the question, "the fact is, I have been so much engaged in other ways, that I really have not had an opportunity of perusing them."

"You should do so, sir," said Pott, with a severe countenance. "I will," said Mr. Pickwick.

"They appeared in the form of a copious review of a work on Chinese metaphysics, sir," said Pott.

"Oh," observed Mr. Pickwick; "from your pen, I hope?"

"From the pen of my critic, sir," rejoined Pott with dignity. "An abstruse subject I should conceive," said Mr. Pickwick.

"Very, sir," responded Pott, looking intensely sage. "He 'crammed' for it, to use a technical but expressive term; he read up for the subject, at my desire in the 'Encyclopaedia Britannica'."

"Indeed!" said Mr. Pickwick; "I was not aware that that valuable work contained any information respecting Chinese metaphysics."

"He read, sir," rejoined Pott, laying his hand on Mr. Pickwick's knee, and looking round with a smile of intellectual superiority, "he read for metaphysics under the letter M, and for China under the letter C, and combined his information, sir?" (Dickens, 1838, p. 815)

REFERENCES

Abbott, L. S. (Ed.). (1982). *Proceedings of workshop on cognitive modelling of nuclear plant control room operators* (Report No. NUREG/CR-3114, ORNL/TM-8614). Washington, DC: U.S. Nuclear Regulatory Commission.

Aschwanden, C. (2001, February 3). Tread softly. *New Scientist, 2276,* 32–36.

Beer, S. (1959). *Cybernetics and management.* New York: Science Editions.

Card, S. K., Moran, T. P., & Newell, A. (1983). *The psychology of human-computer interaction.* Hillsdale, NJ: Lawrence Erlbaum Associates.

Carroll J. M. (2000). *Making use: Scenario-based design of human-computer interactions*. Cambridge, MA: MIT Press.

Chi, M. T. H., Feltovich, P. J., & Glaser, R. (1981). Categorization and representation of physics problems by experts and novices. *Cognitive Science, 5,* 121–152.

Chipman, S. F., Schraagen, J. M., & Shalin, V. L. (2000). Introduction to cognitive task analysis. In J. M. Schraagen, S. F. Chipman, & V. L. Shalin (Eds.), *Cognitive task analysis* (pp. 3–23). Mahwah, NJ: Lawrence Erlbaum Associates.

Cook, R. I., & Woods, D. D. (1996). Adapting to new technology in the operating room. *Human Factors, 38*(4), 593–613.

Dickens, C. (1837). *The Pickwick Papers*. London: Chapman & Hall.

Dowell, J., & Long, J. (1989). Towards a conception for an engineering discipline of human factors. *Ergonomics, 32,* 1513–1535.

Ehn, P., & Löwgren, J. (1997). Design for quality-in-use: Human-computer interaction meets information systems development. In M. Helander, T. K. Landauer, & P. Prabhu (Eds.), *Handbook of human-computer interaction* (2nd ed., pp. 299–323). Amsterdam, The Netherlands: Elsevier Science.

Glaser, R. (Ed.). (1978). *Advances in instructional psychology*. Hillsdale, NJ: Lawrence Erlbaum Associates.

Greenbaum, J., & Kyng, M. (Eds.). (1991). *Design at work: Cooperative design of computer systems*. Hillsdale, NJ: Lawrence Erlbaum Associates.

Hall, A. D., & Fagen, R. E. (1969). Definition of system. In W. Buckley (Ed.), *Modern systems research for the behavioural scientist*. Chicago, IL: Aldine.

Helander, M., Landauer T. K., & Prabhu, P. (Eds.). (1997). *Handbook of human–computer interaction* (2nd ed.). Amsterdam, The Netherlands: Elsevier Science.

Helmreich, R. L. (1997). Managing human error in aviation. *Scientific American, May 1997,* 40–45.

Hollnagel, E., & Woods, D. D. (1982). *Cognitive systems engineering: New wine in new bottles* (Tech. Rep. No. Risø-M-2330). Roskilde, Denmark: Risø National Laboratory.

Hollnagel, E., & Woods, D. D. (1983). Cognitive systems engineering: New wine in new bottles. *International Journal of Man-Machine Studies, 18,* 583–600.

Hutchins, E. (1980). *Culture and inference: A trobriand case study*. Cambridge, MA: Harvard University Press.

Hutchins, E. (1990). The technology of team navigation. In J. Galegher, R. Kraut, & C. Egido (Eds.), *Intellectual teamwork: Social and technical bases of cooperative work*. Hillsdale, NJ: Lawrence Erlbaum Associates.

Hutchins, E. (1995). *Cognition in the wild*. Cambridge, MA: MIT Press.

Jordan, B., & Henderson, A. (1995). Interaction analysis: Foundations and practice. *The Journal for the Learning Sciences, 4*(1), 39–103.

Jönsson, B., & Svensk, A. (1995). Isaac—a personal digital assistant for the differently abled. In *Proceedings of the 2nd TIDE Congress* (pp. 356–361). Amsterdam: IOS Press.

Kieras, D. (1988). Towards a practical GOMS model methodology for user interface design. In M. Helander (Ed.), *Handbook of human-computer interaction*. New York: North-Holland.

Klein, G. A., Orasanu, J., Calderwood, R., & Zsambok, C. E. (Eds.). (1993). *Decision making in action: Models and methods*. Norwood, NJ: Ablex.

McKeithen, K. B., Reitman, J. S., Reuter, H. H., & Hirtle, S. C. (1981). Knowledge organization and skill differences in computer programmers. *Cognitive Psychology, 13,* 307–325.

Nielsen, J. (1993). *Usability engineering*. Boston MA: Academic Press.

Norman, D. A., & Draper, S. W. (1986). *User centered system design*. Hillsdale, NJ: Lawrence Erlbaum Associates.

Rasmussen, J. (1986). *Information processing and human-machine interaction: An approach to Cognitive Engineering*. New York: North-Holland.

Rasmussen, J., & Lind, M. (1981). Coping with complexity. In H. G. Stassen (Ed.), *First European annual conference on human decision making and manual control*. New York: Plenum.

Salvendy, G. (Ed.). (1997). *Handbook of human factors and ergonomics* (2nd ed.). New York: Wiley.

Suchman, L. (1987). *Plans and situated actions: The problem of human-machine communication*. Cambridge, MA: Cambridge University Press.

Svensk, A. (2001). *Design for cognitive assistance*. Lund, Sweden: Lund Institute of Technology, Centre for Rehabilitation Engineering Research. Available: http://www.certec.lth.se/doc/designforcognitive/index.html

Winograd, T., & Flores, F. (1986). *Understanding computers and cognition: A new foundation for design*. Norwod, NJ: Ablex.

Woods, D. D. (1998). Commentary: Designs are hypotheses about how artefacts shape cognition and collaboration. *Ergonomics, 41*(2), 169–173.

Woods, D. D., Tinapple, D., Roesler, A., & Feil, M. (2002). *Studying cognitive work in context: Facilitating insight at the intersection of people, technology and work*. Columbus, OH: Cognitive Systems Engineering Laboratory, Institute for Ergonomics. Available: url: http://csel.eng.ohio-state.edu/woodscta

2

Hierarchical Task Analysis

John Annett
University of Warwick, United Kingdom

Abstract

This chapter begins by placing Hierarchical Task Analysis in its historical context. The method is presented as a general analytic strategy for providing solutions to initially specified performance problems. The unit of analysis is an *operation* specified by a *goal* and activated by an *input*, attained by an *action*, and terminated by *feedback*. The method is based on the systematic decomposition of goals and subgoals and operations and suboperations to any desired level of detail until the source of performance failure, physical or cognitive, is identified and a solution can be hypothesised. A seven-step procedure is described and illustrated. A variety of uses and adaptations of the method are outlined, including cognitive task design; hazard and operability assessment; training needs in contexts such as power generation, air traffic control, and military command and control tasks. Finally, Hierarchical Task Analysis is realistically evaluated as requiring both time and skill to yield useful results.

INTRODUCTION

Hierarchical Task Analysis (HTA) was first developed in the 1960s (Annett & Duncan, 1967; Annett, Duncan, Stammers, & Gray, 1971; Cunningham & Duncan, 1967) in order to overcome the limitations of classical time-and-motion methods in analysing complex nonrepetitive cognitively loaded tasks. Originally developed for training process control tasks in the steel and petrochemical industries, HTA is now widely used in a variety of contexts, including interface design and error analysis in both individual and team tasks in power generation and command and control systems, as well as many others (Ainsworth & Marshall, 1998; Kirwan & Ainsworth, 1992; Shepherd, 2001).

The process of HTA is to decompose tasks into subtasks to any desired level of detail. Each subtask, or *operation,* is specified by a *goal,* the *input conditions* under which the goal is activated, the *actions* required to attain the goal, and the *feedback* indicating goal attainment. The relationship between a set of subtasks and the superordinate task is termed a *plan,* and several plan types can be distinguished, including *procedures, selective rules,* and *time-sharing* tasks. The overall aim of the analysis is to identify actual or possible sources of performance failure and to propose suitable remedies, which may include modifying the task design and/or providing appropriate training. HTA is probably best seen as a systematic search strategy that is adaptable for use in a variety of different contexts and purposes within the human factors enterprise (Shepherd, 1998).

Origins of HTA

During the 1950s, new technology was introducing major changes in the design of industrial and military tasks. In particular, "automation," while reducing physical effort, was increasing the mental effort required to monitor and control complex machinery. The classical methods of analysing tasks, such as Gilbreth's motion study (Gilbreth, 1911), which had been in use for half a century, were proving inadequate to describe tasks of increasing complexity and ever more important mental content. R. B. Miller (1953, 1962) developed a method for man–machine task analysis that decomposed the main task functions into subtasks. For each subtask specified the display, the relevant control and action to be taken, feedback indicating response adequacy and objective criteria of accuracy and characteristic malfunctions. In short Miller's method focused on performance outcomes and the data-processing characteristics of operator tasks.

During this period, information-processing theories of human performance were extended from the basic concept of limited capacity in the execution of simple perceptual-motor tasks (Fitts, 1954; Hick, 1952) to more general accounts of the acquisition of skill (Annett & Kay, 1956, 1957), selective attention and memory (Broadbent, 1958), and the principle of feedback control (Annett, 1969). Miller, Galanter, and Pribram (1960) articulated a wide-ranging theory of goal-directed behaviour that was based on the concept of a nested hierarchy of feedback loops, referred to as TOTE (Test, Operate, Test, Exit) units, and it is this general conception that underlies the *operation* as the chosen unit of analysis in HTA.

A further influence on the early development of HTA was the concept of *systems analysis,* especially as articulated by Chapanis (1951). The effective output of a complex human–machine system may be characterised in terms of deviations from its designed output, or its error variance. The total error variance σ_T^2 is the sum of the variances of the various system components $\sigma_a^2 + \sigma_b^2 + \sigma_c^2 \cdots$ (provided they are uncorrelated). Because the root mean square error contributed to the total by each component increases quadratically with its size, any steps taken to reduce the major sources of error will contribute disproportionately to the reduction of total error. Putting this principle together with the concept of a nested goal hierarchy, then, we see that the obvious strategy for any human factors analysis is the successive decomposition of goals and subgoals in order to identify significant sources of error variance associated with each. We can then use this information as a basis for proposing modifications that will yield maximum benefit in optimising performance outcomes. With this brief introduction to the historical background, we now take a closer look at the underlying principles before turning to practical advice on the conduct of HTA. For a more extensive account of the early background to the development of HTA, see Annett (2000).

DEFINITIONS AND PRINCIPLES

Analysis Versus Description

In this section the principles on which HTA is based are stated succinctly. These principles were all present or implicit in the original statement of HTA (Annett et al., 1971) but are elaborated here for clarification. Being clear about the underlying principles and the definition of technical terms will make it easier to understand when and how to use HTA. Analysis is not just a matter of listing the actions or the physical or cognitive processes involved in carrying out a task, although it is likely to refer to either or both. Analysis, as opposed to description, is a procedure aimed at identifying performance problems, that is, sources of error, and proposing solutions. This distinction between *task description* and *task analysis* was made by R. B. Miller (1962) and emphasises the purpose of the analysis as providing solutions to initially specified problems. The problem might be to design a suitable interface or perhaps to decide what kind of training to provide, and in each case the course of the analysis might be different depending on what kinds of information are most relevant to the question being asked.

Tasks and Goals

A task is defined in the *Shorter Oxford Dictionary* as "any piece of work that has to be done" and, as such, every task has a goal. The figurative definition of a goal is "the object of effort or ambition." HTA differs radically from earlier methods of analysis by beginning, not with a list of activities, but by identifying the goals of the task. In routine repetitive tasks, actions vary little, while ever the environment and purpose remain constant. In complex tasks, the same goals may be pursued by different routes and different means, depending on circumstances peculiar to each occasion. Hence, simply to list actions without understanding what they are for can be misleading. Complex systems are designed with goals in mind, and understanding how a system attains or fails to attain its designated goal is the primary purpose of the analysis. A goal is best stated as a specific state of affairs, formally a *goal state*. The goal state can be an event or some physically observable value of one or more variables that act as criteria of goal attainment. At any one time a goal may be *active* or *latent*. Active goals are those being currently pursued; latent goals are those that may be pursued under conditions that might arise. The importance of this distinction will become apparent when we consider the concept of a *plan*.

Decomposition and Redescription

Goals are often complex; that is, they are defined by more than one event or by values of more than one variable. When these can be individually identified, the analysis should specify these component goal states by the process of *decomposition*. HTA envisages two kinds of decomposition. The first comprises identifying those goal states specified by multiple criteria, for example to arrive at a destination (an event) having expended minimum effort and with no injury. The second kind of decomposition comprises the identification of subgoals in any routes that may be taken to attain the overall goal state. Goals may be successively unpacked to reveal a nested hierarchy of goals and subgoals. This process of decomposition, also referred to as *redescription,* has the benefit, according to the general principle proposed by Chapanis (1951), of comprising an economical way of locating sources of general system error (actual or potential) in failure to attain specific subgoals.

Operations: Input, Action, and Feedback

An *operation* is the fundamental unit of analysis. An operation is specified by a goal, the circumstances in which the goal is activated (the *I*nput), the activities (*A*ction) that contribute to goal attainment, and the conditions indicating goal attainment (*F*eedback); hence operations are sometimes referred to as *IAF units*. Operations are synonymous with TOTE units (Miller et al., 1960) and are feedback-controlled servomechanisms. Just as goals may be decomposed into constituent subgoals, so operations may be decomposed into constituent suboperations arranged in a *nested hierarchy*. Suboperations are included within higher order (or superordinate) operations, with the attainment of each subgoal making a unique contribution to the attainment of superordinate goals. The suboperations making up a superordinate operation should be mutually exclusive and collectively comprise an exhaustive statement of the subgoals and superordinate goals.

An *action* can be understood as an injunction (or instruction) to do something under specified circumstances, as illustrated by the classic example of a TOTE for hammering a nail into a piece of wood—"Is the nail flush?/No → Hammer! → Is the nail flush?/Yes → Exit!" (Miller et al., 1960). Input and feedback both represent *states* or *tests* in the formulation by Miller et al. These states register either error, therefore requiring action, or the cancellation of error, signalling the cessation of that action. An action can be understood formally as a *transformation rule* (Annett, 1969, pp. 165–169), which is a specification of how a servo responds to an error signal and its cancellation. For example, in a manual tracking task, the transformation rule can be specified by an equation, known as a *transfer function,* which quantifies the control output required to correct for an error signal of given direction and magnitude (McRuer & Krendel, 1959). However, this is a special case, and normally, for example in self-assembly kits, computer software handbooks, and cookbooks, instructions are specified in terms of commonly understood verbs. Some verbs (such as *chamfer, defragment,* and *marinate*) form part of a technical vocabulary that may need redescription in simpler terms (how *does* one chamfer wood, defragment a computer disk, or marinate meat?). These redescriptions comprise a set of suboperations. As already indicated, suboperations collectively redescribe their superordinate operation, but typically we need to know not only the constituent suboperations but also the order, if any, in which they should be carried out (e.g., "to chamfer, first secure the piece to be chamfered, then obtain a suitable file," and so on).

Plans

The specification of the rule, or rules, governing the order in which suboperations should be carried out is called a *plan*. Plans can be of various types; the most common is simply a *fixed sequence* or *routine procedure,* such as "do this, then this, and then this," and so on. Another common type of plan specifies a *selective rule* or *decision*—"if *x* is the case, do this; if *y* is the case, do that." These two types of plan are significant because they imply knowledge on the part of the operator. It may be simple procedural knowledge, or the plan may require extensive declarative knowledge of the environment, the limits and capabilities of the machine, safety rules, and much else besides. In this respect HTA anticipated the requirement for what is now known as cognitive task analysis (Schraagen, Chipman, & Shalin, 2000).

A third distinct type of plan requires two or more operations to be pursued in parallel; that is, the superordinate goal cannot be attained unless two or more subordinate goals are attained at the same time. This is known as a *time-sharing* or *dual task* plan, and this type also has significant cognitive implications in terms of the division of attention or, in the case of team operations, the distribution of information between team members acting together.

When a goal becomes active, its subordinate goals become active according to the nature of the plan. For example, in a fixed sequence the goal of each suboperation becomes active as the previous subgoal is attained. When the plan involves a selective rule, only those goals become active that are specified by the application of the rule; the rest remain latent, and in a time-sharing plan two or more goals are simultaneously active.

Stop Rules

The decomposition of goal hierarchies and the redescription of operations and suboperations might continue indefinitely without the use of a stop rule, which specifies the level of detail beyond which no further redescription is of use. The ultimate stop rule is just that: "stop when you have all the information you need to meet the purposes of the analysis." However, because a general purpose of HTA is to identify sources of actual or potential performance failure, a general stop rule is "stop when the product of the probability of failure and the cost of failure is judged acceptable." This is known as the $p \times c$ criterion (Annett & Duncan, 1967), and its prime benefit is that it keeps the analytical work down to a minimum and it focuses the attention of the analyst on those aspects of the task that are critical to overall system success. In practice, lack of empirical data may mean that p and c can only be estimated, but it is the *product* of the two that is crucial to the decision to stop or continue decomposition. The obvious reason for stopping is that the source of error has been identified and the analyst can propose a plausible remedy in terms of either system design, operating procedures, or operator training, that is, by redesigning the cognitive task. A point that is often overlooked is that performance is a function not only of the system but of the operator. Thus a great deal more detail may be required if the aim of the analysis is to create a training program for complete novices than if the aim is to examine the possible effects of a redesigned equipment or operating procedure on the performance of experienced operators.

HTA may be used for design purposes and need only be carried down to the level of equipment specificity. Team training often follows training in individual operator skills, and the analysis of team tasks may be usefully taken down to the level at which operations involve communication and collaboration between individuals. For example, in the case of interface design, the analysis might proceed only to the degree of detail prior to the physical specification of devices (Ormerod, Richardson, & Shepherd, 1998). In the HTA for teams-HTA(T)-the analysis was specifically aimed at identifying operations that were especially dependent on the exercise of team skills such as communication and collaboration (Annett, Cunningham, & Mathias-Jones, 2000); another variation on the hierarchical decomposition approach, Task Analysis for Knowledge Description (TAKD), was initially aimed at identifying *generic* skills common to a wide range of Information Technology tasks (Diaper, 1989, 2001). In these cases, both the depth of analysis and the most relevant types of data are heavily influenced by the purpose and hoped-for product of the analysis.

HOW TO CARRY OUT AN HTA

HTA is a flexible tool that can be adapted to a variety of situations and needs. Data may be derived from any number of different sources; the analysis can be continued to any desired level of detail and there is no rigid prescription of how the results may be used. Shepherd (1998) refers to HTA as "a framework for analysis" but it is arguable whether the intrinsic adaptability of HTA is necessarily advantageous. A strictly specified procedure may be preferred to an esoteric craft; for one thing, the former is easier to learn (Patrick, Gregov, & Halliday, 2000), and, as Diaper (2001) points out, the process of automating or proceduralising a method

TABLE 2.1
Summary of the Principal Steps in Carrying Out HTA

Step Number	Notes and Examples
1. Decide the purpose(s) of the analysis.	1. Workload, manning, error assessment. 2. Design new system or interface. 3. Determine training content or method.
2. Get agreement between stakeholders on the definition of task goals and criterion measures.	1. Stakeholders may include designers, managers, supervisors, instructors, or operators. 2. Concentrate on system values and outputs. 3. Agree on performance indicators and criteria.
3. Identify sources of task information and select means of data acquisition.	1. Consult as many sources as are available; direct observation, walk-through, protocols, expert interviews, operating procedures and manuals, performance records, accident data, or simulations.
4. Acquire data and draft decomposition table or diagram.	1. Account for each operation in terms of input, action, feedback and goal attainment criteria and identify plans. 2. Suboperations should be (a) mutually exclusive & (b) exhaustive. 3. Ask not only what *should* happen but what *might* happen. Estimate probability and cost of failures.
5. Recheck validity of decomposition with stakeholders.	1. Stakeholders invited to confirm analysis, especially identified goals, and performance criteria. 2. Revert to step 4 until misinterpretations and omissions have been rectified.
6. Identify significant operations in light of purpose of analysis.	1. Identify operations failing $p \times c$ criterion. 2. Identify operations having special characteristics, e.g., complex plans, high workload, dependent on teamwork, or specialist knowledge.
7. Generate and, if possible, test hypotheses concerning factors affecting learning and performance.	1. Consider sources of failure attributable to skills, rules, and knowledge. 2. Refer to current theory or best practice to provide plausible solutions. 3. Confirm validity of proposed solutions whenever possible.

guarantees reliability and often leads to a deeper understanding of the underlying process. However, it might equally be argued that there are few procedures capable of dealing with all eventualities, and the unthinking application of a rigidly proceduralised tool may sometimes lead to a gross misinterpretation of the task. HTA can nevertheless be carried out in a number of different ways that may involve greater or lesser attention to individual steps in the fundamental procedure, which is outlined in the paragraphs that follow. In general, the benefits of HTA, and its reliability and validity, are proportional to the effort that goes into following this procedure. The analyst is nevertheless entitled to trade off effort for value-added by shortening or adapting the procedure to suit specific needs. Some ways of doing this are mentioned in the following steps, which are summarised in Table 2.1.

Step 1: Decide the Purpose(s) of the Analysis

The purpose of the analysis has important implications for the way in which it is carried out, including the preferred data collection procedures, the depth of the analysis, and the kinds of solutions (results) that can be offered. Typical purposes are designing a new system, troubleshooting and modification of an existing system, and operator training, all of which involve the design or redesign of the cognitive tasks required of the operators. When used for system design, the primary source of information is the design team, but few designs are totally novel and, as suggested by Lim and Long (1994), an analysis of a comparable extant system may prove useful in identifying difficulties to be avoided in the new design. In this case relevant data may be collected from records of performance, errors, and accidents; the views of expert users, supervisors, and managers; and by direct observations. Depending on the observed, reported, or even anticipated symptoms of failure, the analyst's attention may well focus on particular aspects of the task, such as displays, communications, complex decision rules, or heuristics to be employed for successful performance. Particular attention may be paid to operations where these play a critical role, and reference to published ergonomic design standards may be valuable.

The intended product of the analysis is also important in determining the appropriate stop rule. For a fully documented training program for novices to be produced, the level of detail may have to be able to generate very specific, plain language, "how to do it" instructions. If the purpose is to identify the type of training required, the analysis should identify operations and plans of particular types that are thought to respond to particular training methods. For example, where inputs are perceptually or conceptually complex, special recognition training exercises may be required, or where procedures are especially critical, operating rules and heuristics and system knowledge are to be learned (Annett, 1991). In summary, the analysis should anticipate the kinds of results that would provide answers to the original questions, such as design recommendations, training syllabi, and the like.

Step 2: Get Agreement Between Stakeholders on the Definition of Task Goals

Task performance, by definition, is goal-directed behaviour and it is therefore crucial to the analysis to establish what the performance goals are and how one would know whether or not these goals have been attained. A common problem is to interpret this question as about observed operator behaviour, such as using a particular method, rather than performance out-comes, such as frequency of errors and out-of-tolerance products. Bear in mind that the effort of analysis is ultimately justified by evidence of the outcomes of performance; this issue is taken up again in the later section on validity.

Different stakeholders (designers, trainers, supervisors, or operators) can sometimes have subtly different goals. It is better to identify problems of this kind early in the analysis by thorough discussion with all relevant stakeholders. If goals appear to be incompatible, the analyst can sometimes act as a catalyst in resolving these issues but should not impose a solution without thorough discussion. As the decomposition proceeds, more detailed goals and more specific criterion measures are identified, and it can emerge that different operators with ostensibly the same overall purpose have slightly different plans (ways of doing things), which may imply different subgoals. Objective performance measures provide the opportunity to compare methods in terms of superordinate criteria.

The key questions, which may be asked in many different forms, are, first, What objective evidence will show that this goal has been attained? and, second, What are the consequences of failure to attain this goal? Answers to the first question can form the basis of objective performance measures that may subsequently be used in evaluating any design modifications or training procedures proposed on the basis of the analysis. Answers to the second question may be used to evaluate the $p \times c$ criterion and hence the degree of detail of the analysis. Answers to both questions form the essential basis for agreement about the system goals. If goals cannot be stated in these objective terms, then the sponsors and stakeholders are unclear about the purposes of the system, and the validity of the entire analysis is called in question.

Step 3: Identify Sources of Task Information and Select Means of Data Acquisition

If the purpose is to look for ways to improve operator performance on an existing system, then records of actual operator performance, including both the methods used by operators and the measures of success or failure will be important. Errors may be rare, but critical incident data can provide useful insights into the origins of performance failure. If the purpose is to make recommendations on a new design, then data relating to comparable (e.g., precursor) tasks may be helpful, but the designer's intentions are critical. In the absence of actual performance data, the analyst should challenge the designer with "what if" questions to estimate the consequences of performance failure. Sometimes data concerning performance on preexisting systems or comparable tasks may provide useful information.

Preferred sources of data will clearly vary considerably between analyses. Interviews with experts are often the best way to begin, particularly if they focus on system goals, failures, and shortcomings. Direct observation may provide confirmatory information but, especially in nonroutine tasks, may yield relatively little information concerning uncommon events that may be critical. Formal performance records such as logs and flight recorders may be available, especially in safety-critical systems, but these are often designed primarily with engineering objectives in mind and the human contribution to system performance can sometimes be difficult to determine from the mass of recorded data. Focused interview with recognised experts aimed at identifying performance problems is often the only practicable method of obtaining estimates of the frequency and criticality of key behaviours. In some cases, in which the informants are unclear about what would happen in certain circumstances and what would be the most effective operator strategy, it may be helpful to run experimental trials or simulations. In at least one (personally observed) case in which even skilled operators were not clear about the cues used in reaching an important decision, an experiment was run in which the effects of blocking certain sources of information were observed. In this instance it was determined that the operator was, unconsciously, using the sound of the machinery rather than the sight of the product to reach a key decision.

Step 4: Acquire Data and Draft a Decomposition Table or Diagram

In general, the more independent sources of data consulted, the greater the guarantee of validity of the analysis. It is all too easy to think of operations simply as actions. The critical feature of HTA is always be to be able to relate *what* operators do (or are recommended to do) and *why* they do it and what the consequences are if it is not done correctly. Only when this is thoroughly understood is it possible to create a meaningful table or diagram. A useful logical check on the validity of a proposed decomposition table is that all suboperations must be (a) mutually exclusive and (b) exhaustive, that is, completely define the superordinate operation.

Notation.   The use of standard notation generally helps in the construction of tables and diagrams as well as the interpretation of results, and it aids communication between analysts and stakeholders. The recommended notation system provides a unique number for each identified operation that may be used in both diagrams and tables. The notation should also specify plans and indicate stops. Stops represent the most detailed level of the analysis, typically the level that is most relevant to the results of the search, and recommendations for further action.

Both tabular and diagrammatic formats have their advantages, and typically both are used. The diagrammatic format often helps to make clear the functional structure of the task, whereas

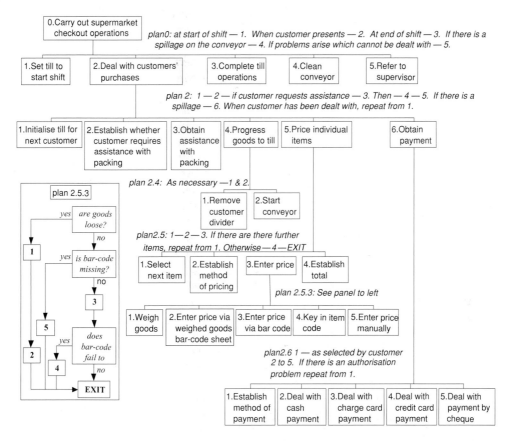

FIG. 2.1.   Analysis of a supermarket checkout task in diagrammatic form (from Shepherd, 2001).

the tabular format is more economical of space and facilitates the recording of supplementary notes, queries, and recommendations. Individual operations are normally numbered with 0 standing for the top-level goal, in effect the title of the task, and with suboperations being numbered in the order of description (which is not necessarily the same as the order of execution). Thus operations 1, 2, and 3 would be the three principal suboperations of task 0; and 1.1, 1.2, and 1.3 would be the three suboperations into which operation 1 is decomposed and 2.1 and 2.2 would be the two suboperations into which operation 2 is decomposed. Each additional digit indicates a new level of decomposition. In the diagrammatic form, a vertical line descends from the superordinate operation to a horizontal line covering the set of suboperations into which it is expanded.

Plans are implicit in the decomposition structure, but they can be made more explicit by adding the appropriate algorithm to the diagram as in the example shown in Fig. 2.1. A succinct way of presenting four basic types of plan in the diagram was employed by Annett et al. (2000), using the symbols > to indicate a sequence, / to represent an either/or decision, + to represent dual or parallel operations, and : to represent multiple operations in which timing and order are not critical. This is shown in Fig. 2.2.

The lowest or most detailed level of decomposition, often being the most important level of description, is typically indicated in the diagram by an underline or some other feature of the box and may be similarly indicated in the table by a typographic variant such as boldface. The tabular layout can be facilitated by the use of the Outline system in Word™, and I have

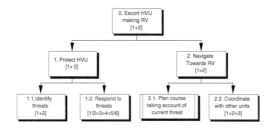

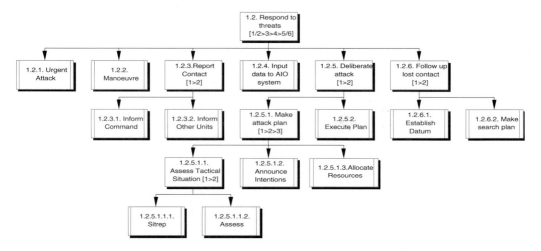

FIG. 2.2.   Principal tasks of an ASW team (top section) and the further analysis of 1.2, Respond to Threats (lower section) (from Annett et al., 2000).

used another program, Inspiration™ (see Fig. 2.2), which is effectively a graphical form of an outline system, as a convenient, easily modifiable form of recording both diagrams and tables with ample scope for additional plain language notes.

Illustrative Examples.   The first example shows a supermarket checkout task from Shepherd (2001), whose book is illustrated with a number of examples, and the second example (from Annett et al., 2000), using the alternative notation already referred to, is based on the analysis of a naval command and control team task. Shepherd's analysis identifies training needs and indicates where special training might be needed. The analysis from Annett et al. identifies operations in which teamwork is particularly critical and was used to develop an objective method of assessing teamwork.

The diagram in Fig. 2.1 shows the supermarket checkout task as comprising five principal sub-operations, one of which (#2) is further broken down into six more of which #4 is further decomposed into two more and #5 into four and #6 into five sub-operations one of which (#3) is redescribed in five sub-operations. Note that the box for each operation contains no more than a brief title, which stands for both the goal and the principal action involved. Table 2.2 gives a verbal description of selected operations together with notes concerning training. The table gives full numerical labels such that operation 2.5.3 (enter price) is identified as at the third level of decomposition and is itself finally redescribed as five suboperations that are listed at the bottom of the table. These can also be identified in the diagram as operations 2.5.3.1 to 2.5.3.5.

TABLE 2.2
Analysis of the Supermarket Checkout Task in Tabular Form (From Shepherd, 2001)

Operations and Plans	Exp.	Notes
0. Carry out supermarket checkout operations		
Plan 0: At start of shift—1. When customer presents—2. At end of shift—3. If there is a spillage on the conveyor—4. If problems arise which cannot be dealt with—5. Refer to supervisor.		
1. Set till to start shift	No	Demonstration as per induction programme.
2. Deal with customers' purchases	Yes	Ditto
3. Complete till operations	No	Ditto
4. Clean conveyor	No	Ditto
5. Refer to supervisor	No	Note that supervisors are not always at their desk. Review methods of communication.
2. Deal with customers' purchases		
Plan 2: 1–2—if customer requests assistance—3. Then—4—5. If there is a spillage—6. When customer has been dealt with, repeat from 1.		
1. Initialise till for next customer	No	
2. Establish whether customer requires assistance with packing	No	Some operators lack skill in asking customers whether they require help.
3. Obtain assistance with packing	No	
4. Progress goods to till	Yes	
5. Price individual items	Yes	
6. Obtain payment	Yes	
2.4. Progress goods to till		
Plan 2.4: As necessary—1 & 2.		
1. Remove customer divider	No	
2. Start conveyor	No	
2.5. Price individual items		
Plan 2.5: 1—2—3. If there are there further items, repeat from 1. Otherwise—4—EXIT.		
1. Select next item	No	
2. Establish method of pricing	No	Training required for more rapid identification on nonbarcoded items.
3. Enter price	Yes	
4. Establish total	No	
2.6. Obtain payment		
Plan 2.6: 1—as selected by customer 2 to 5. If there is an authorization problem, repeat from 1.		
1. Establish method of payment	No	Training required to help operators distinguish between different types of card.
2. Deal with cash payment	No	More training required to identify forged notes and coins.
3. Deal with charge card payment	No	
4. Deal with credit card payment	No	
5. Deal with payment by check	No	
2.5.3. Enter price		
Plan 2.5. See panel at left.		
1. Weigh goods	No	
2. Enter price by means of weighted goods barcode sheet	No	
3. Enter price by bar code	No	
4. Key in item code	No	
5. Enter price manually	No	

Note. From Shepherd (2001).

Figure 2.1 also shows the plans. The plan for 0 comprises a simple procedure, operations 1, 2, and 3 with two conditional operations, 4 and 6. The plan for operation 2.5.3 is described in the form of an algorithm that contains three tests (Are goods loose? Is bar code missing? Does bar code fail to read?). A tabular version can contain the same information with the addition of notes relating to features of the task of special interest. Although the tabular version is in some respects easier to read, the diagrammatic version shows the hierarchical structure, and this can be helpful in understanding the interdependencies of sections on the overall task.

Figure 2.2 shows selected parts of the diagrammatic form and Table 2.3 shows selected parts of the tabular form of a task carried out by the anti-submarine warfare (ASW) team in the operations room of a warship. The purpose of the analysis was to be able to identify and provide objective measures of key team skills.

The team comprises the Principal Warfare Officer (PWO), the Active Sonar Director (ASD), and the Action Picture Supervisor (AcPS), who are also in touch with other individuals and teams, notably the Officer of the Watch (OOW) and the Missile Director (MD). The upper section of Fig. 2.2 shows the overall aim of the team, which is to escort a highly valued unit (HVU; e.g., a troopship) to a rendezvous (RV; operation 0) while identifying (operation 1.1) and responding appropriately to (operation 1.2) submarine threats. The lower section of Fig. 2.2 shows that to identify threats (1.1) the team must, at the same time, scan for possible threats (1.1.1) and classify threats (1.1.2), of which more than one could be present at any given time. The plan for 1.1 is indicated by the notation $[1 + 2]$, where 1 and 2 identify the two suboperations and the $+$ symbol indicates they must be done in parallel. Classification (1.1.2) requires a decision as to the apparent immediacy of the threat followed by a process of checking and investigation.

Table 2.3 shows extracts from the tabular form, including three operations, 1.1, 1.2, and 1.1.2.2. Because the aim is to identify and provide a way of measuring critical teamwork skills, the analysis contains detailed notes on the goal, how goal attainment might be measured, and what kinds of teamwork activities are believed to be involved in attaining the goal.

Step 5: Recheck Validity of Decomposition with Stakeholders

Stakeholders, especially those unfamiliar with HTA, sometimes change their minds as the analysis proceeds. By thinking about the task they may realise that events do not always occur in a standard way or that something that has never been known to happen just could. For this reason an iterative process is recommended wherever possible, with the full analysis being developed over a series of interviews and by cross-checking between sources such as training manuals and experienced operators. The process of cross-checking not only provides the best guarantee of the reliability of the completed analysis but encourages the stakeholders to develop a sense of ownership of the analysis and consequently to share responsibility for recommendations arising from the results.

Step 6: Identify Significant Operations in Light of Purpose of Analysis

In any decomposition method, a decision must be made about the level of greatest detail required, that is, when to stop the analysis. It is common in decomposition methods used by large organisations, such as the military, to adopt a fixed number of levels, such as *job, task,* and *sub-task,* but this means that some parts of the task may be described in excessive detail whereas others require more. As a general rule the recommended criterion (the stop rule) is to stop the analysis of any given operation when the probability of performance failure multiplied by the cost to the system of performance failure ($p \times c$) is acceptably low. Where no data exist,

TABLE 2.3

Extracts From the Tabular Form of the ASW Team Task Analysis

1.1. Identify Threats

Goal	Identify and classify all ASW contacts.
Measure	Contacts identified and classified as quickly and accurately as possible.
Teamwork	Team compiles information from various sources. PWO monitors and directs team.
Plan	$[1 + 2]$ Scanning all sources for information on potential threats is continuous (1.1.1). Classification procedures (1.1.2) follow identification as soon as possible.

1.2. Respond to Threats

Goal	To respond appropriately to an identified threat according to classification.
Measure	Response according to plan for force, without hesitation once relevant information is available.
Teamwork	PWO selects appropriate response based on information provided by team in accordance with standard operating procedures (team plan) plus range and bearing.
Plan	If threat is immediate (e.g., torpedo), go to urgent attack (1.2.1) otherwise execute (1.2.3) to (1.2.5). If contact is lost go to (1.2.6).

1.1.2.2. Chart Check

Goal	To establish whether contact classified as possub lo 1 represents a known feature such as rock, wreck, or pipeline.
Measure	Procedures correctly executed, conflicts resolved, and other units informed.
Teamwork	Sonar operator passes information to ASD, who confers with PWO. PWO calls for "chart check poss. sub. BRG/RG." OOW plots position to agreed margin of error. ASD directs SC to investigate location; AcPS inputs data to system; EW and radar teams check returns on that bearing; OOW and MD check bearing visually. All report results of checks.
Plan	If chart check negative go to (1.2). If information is inconsistent go to (1.1.2.3).

1.1.2.3. Investigate Possub

Goal	To confirm possub classification
Measure	Correct procedures and priority to search; optimal use of assets.
Teamwork	Correct procedures according to team plan.
Plan	If possub confirmed go to (1.2).

Note. From Annett et al. (2000).

values of p and c may be estimated. The rationale for this general rule is that the analysis is essentially a search process aimed at identifying significant sources of actual or potential system failure, so when no more sources can be identified then clearly the analysis should cease. However, modifications to this general rule may be made in certain circumstances. For example, when the analysis is part of the design process, it may be desirable to stop at a level that is device-independent, that is, at a purely functional specification of the component operations, leaving open specific implementation in terms of equipment or language (see Lim & Long, 1994; Ormerod et al., 1998). Another stopping rationale is when the aim of the analysis is to identify certain features of the task. For example, Crawley (1982) used a form of HTA referred to as Overview Task Analysis (OTA) in a study of air traffic control. The analysis was stopped at tasks that could be readily identified and judged by controllers as being particularly demanding or especially satisfying, thus providing a basis for deciding whether automation should be considered. As shown in the previous section, Annett et al. (2000) analysed naval ASW command team tasks in sufficient detail to identify certain types of team work such as operations that critically depended on intra-team communication or discussion, active collaboration, or the synchronisation of actions. The analysis formed the basis for a behaviourally based scoring system, thus enabling instructors to supplement their comments and advice by reference to objectively measurable features of teamwork.

Step 7: Generate and, if Possible, Test Hypotheses Concerning Task Performance

HTA is principally carried out on the assumption that it is a tool to be used by a specialist who is looking for particular classes of problem and has a range of optional recommendations available. The analysis provides a means of generating hypotheses concerning the likely sources of actual or potential failure to meet overall task goals and to propose appropriate solutions. It must be clearly understood that HTA as a method includes neither a set of diagnostic categories nor a set of acceptable solutions to the problems identified. These will depend on the predilections and technical capabilities of the analyst. HTA simply provides an efficient procedure for identifying sources of actual or potential performance failure. However, as a useful source of guidance, Reason's (1990) classification of human error, based on Rasmussen's (1983) taxonomy of skill-based, rule-based and knowledge-based performance, may be helpful in this context. The analyst is prompted to develop hypotheses to account for failures and to propose practical solutions, but these are to be regarded as hypotheses that should be put to the test because this is the only guarantee that the analysis is valid. Several broad classes of specialist interest may be distinguished. These include system design and operator training.

Task Design.   The traditional concerns of human factors and ergonomics are questions such as allocation of function, manning, and interface design. The basic concept of an operation, being essentially that of a *function,* can be applied equally to a human or machine operator and to constraints such as the physical effort and information processing demands of the task. The analysis specifies the design of both the physical and the cognitive tasks, the latter being our primary concern in the present context. Consideration of the input, action, and feedback components of critical operations often provides useful clues about actual or potential problems and hence of possible solutions. Input problems may range from sheer legibility at Rasmussen's skill level, through to the recognition and interpretation of complex patterns at the knowledge level, which are the essential precursors of decision making.

Action problems likewise may range from the accessibility and effort requirements of controls through fixed routine procedures to the use of complex strategies, some following closely specified plans and others relying on sophisticated heuristics. Feedback problems are commonly underestimated, and this may be partially due to the use of action language ("do x") rather than goal-directed language ("achieve state x") in describing tasks. Too often the feedback that indicates successful goal attainment is ignored, perhaps because of the implicit assumption that all statable actions are simple and can be executed without error. The operator's failure to appreciate or to learn the significance of feedback can often be a major source of error that can be exacerbated if, as in so many process control type tasks, there is a significant temporal lag between the control action, such as reducing the temperature of a vessel, and feedback, such as change in chemical composition of the product as shown in a subsequent laboratory test. The problem may become even more severe in cases in which a control action has multiple effects that may be registered at different times and locations and may even interact with other operations.

The goal structure, essentially the plans, of a task also provide important clues to potential or actual performance failures. Consider the three main classes of plan identified in an earlier section. A simple sequence or routine procedure may create few problems if it is short and frequently practised but may otherwise be liable to memory failure, one of the commonest forms of human fallibility (Reason, 1990). The second type, selective rule or decision, clearly implies the existence of a repertoire of rules or strategies to deal with possible contingencies. The analysis should specify what these are if rule-based errors (Reason, 1990) are to be avoided. Examples include both applying the wrong rule and applying the correct rule to the wrong object or just using an inadequate rule. The third category of plan, dual task or time-sharing, provides opportunities for errors of inattention, distraction, and sheer physical or mental overload. Identification of these types of error, some of which ought to be apparent during the initial data collection phase, is probably the most important step to their elimination through techniques for providing various kinds of support, such as checklists, decision aids, equipment redesign, and automation.

Training. Questions relevant to training may include course content, training method, and often both. Training is often considered only when a new system is about to become operational. In a project for the British Army, Annett (1991) recommended the use of HTA at an early stage of system development as a means of choosing between types of training equipment, such as simulators, part-task trainers, embedded trainers, and weapons effects simulation as alternatives to classroom work and on-the-job experience. The choice of trainer was based principally on the characteristics of critical operations and plans identified by the analysis but also took into account the context in which training could be implemented. The recommendations were embodied in an algorithm that proposed the optimal training medium for each operation. For example, critical recognition and control skills typically require practice in high-fidelity simulations but can also be learned in a part-task trainer, whereas investigation and planning skills may be practised in a simple mock-up or, if more complex, in a simulation that provides a realistic range of problems.

USES OF HTA

HTA has been used in a wide variety of contexts for the full range of problems that confront human factors practitioners. Shepherd (2001) cites a range of examples, including simple procedural tasks, such as changing a printer cartridge, using a word processor, and carrying out the supermarket checkout task, through fine motor skills of minimal access (keyhole) surgery to air traffic control and management tasks. In a survey of 30 task analysis studies in the defense industry, Ainsworth and Marshall (1998) found two cases of its use in system procurement, seven for manpower analysis, nine for interface design, five for operability assessment, and two instances of its use in specifying training.

The final (seventh) step in the HTA procedure is to formulate and test hypotheses about possible solutions to the critical performance problems emerging from the analysis. As indicated, different kinds of solutions may be possible, such as changing the design or the task procedure or selecting operators or providing them with special training. As part of the original Hull research project, Duncan (1972) described the use of HTA in the design of training for process control operators in the petrochemical industry. His account is particularly valuable for the thorough discussion of the process of forming hypotheses and proposing solutions based on the analysis and current instructional theory. Other uses include assessing workload and manning requirements. Penington, Joy, and Kirwan (1992) used HTA to determine staffing levels in the control room of a nuclear plant. Fewins, Mitchell, and Williams (1992) described the use of a version of HTA to identify 56 critical tasks in the operation of a newly commissioned nuclear power station for the purpose of assessing the workload. The cognitive task problems identified by the analysis were dealt with by a combination of interface design, staffing levels, and automation. Crawley (1982) used an abbreviated version of HTA to identify critical air traffic control tasks to assess their suitability for automation. Command and control tasks typically involve team work, and HTA-T has been used to identify critical team functions (Annett et al., 2000; Annett & Cunningham, 2000). Shepherd (2001) also outlines HTA applications to production teams, supervision of an automatic railway, and collaboration between members of medical teams.

HTA has been used for the purpose of hazard assessment and error prediction. Baber and Stanton (1994) describe a method, Task Analysis For Error Identification (TAFEI), for designing error-tolerant consumer products that combines HTA with State-Space Diagrams (SSDs). HTA is used to describe human activities and SSDs to describe how the product will behave, that is, how it will move from one state to another; the aim is being to ensure that the design does not involve transitions to undesirable or hazardous states. Reed (1992) also describes the use of HTA in nuclear power plant operation specifically to assess the safety of the operator–machine interface in the handling of hazardous materials. Penington (1992) used a hierarchical analysis in a hazard and operability (HAZOP) study of oil rig drilling operations.

HTA has been recommended as a precursor to systematic design by Lim and Long (1994) in their Method for Usability Engineering (MUSE). Because most new designs begin as alternatives to extant designs, Lim and Long suggest that the HTA of an extant system (or systems) is a convenient way of generating a generalised task model containing the essential functional requirements of the proposed new system and demonstrating their implications for the human user. Ormerod (2000) also advocates a modified version of HTA, called the Sub-Goal Template (SGT) method, as part of the design process. The decomposition proceeds down to the level at which the specification of an operation is independent of any specific equipment or interface design. This approach frees the designer to consider a range of design possibilities.

The principle of hierarchical goal decomposition has become widely used in a number of well-known HCI methods. Some, like GOMS (Goals, Operators, Methods, and Selection rules) by Card, Moran, and Newell (1983), appear to have been developed for the specification of human–computer interaction tasks quite independently of HTA. Others, such as KAT (Knowledge Analysis of Tasks; Johnson & Johnson, 1991) and TAKD (Diaper 2001; Diaper & Johnson, 1991) explicitly incorporate the essentials of HTA into an extended methodology. In these latter cases the theory of knowledge structures is exploited as a means of determining the transfer potential (or "generification") of certain classes of tasks. Extensions of the principles underlying HTA raise an interesting general point.

This chapter has presented HTA as a "complete" route for proceeding systematically from problem to solution, but others have viewed HTA as a "front end" procedure, or the first stage in a number of distinct methods targeted at specific types of problem. In a broad view, most human factors problems span three domains—the machine or external system, the human agent, and the set of functional relationships between these two, that is, the task itself. HTA is essentially a specification of the functionality of a goal-directed system, or, to put it very simply, it provides the recipe for getting something done. However, getting something done depends not just on the recipe but the variable properties of the environment within which the "thing" has to be done and the often limited, but also variable, capabilities of the agent. To get at the heart of a problem, whether it be of task design or remediation, each of these aspects has to be modelled. If HTA is a procedure for modelling the functionality of the task, additional information and constructs are needed to model the nature of the machine environment and the capabilities of the agent. TAFEI does this by using SSDs to model the properties of the machine environment, and KAT provides a way of modelling the knowledge structures of the agent. Significant developments of the original principle underlying HTA such as these surely deserve their unique designation, and no doubt the future will bring more.

USABILITY OF HTA

It is reasonable to expect that the same standards of usability should apply to the methods used by human factors specialists as they apply to the objects of their studies. Ainsworth and Marshall's (1998) survey found a good deal of variability in the application of HTA. For example, only half the studies specified a stop rule, a third did not use diagrams, and only a minority of cases reported their recommendations and how they were derived from the analysis. Ainsworth and Marshall suggest that training in the use of HTA is somewhat variable, and it appears that some practitioners take shortcuts, neglecting some of the important steps outlined in this chapter. They also note that "it appeared that the most insightful analyses were undertaken by analysts who had the most human factors experience" (p. 1617/p. 89).

Stanton and Young (1998) surveyed the use of 27 methods used by members of the Ergonomics Society who were also on the Society's professional register. These included methods such as checklists, questionnaires, link analysis, predictive human error analysis, repertory grids, keystroke level models, and HTA. Although the response rate was low, the consensus view was that HTA was useful but requires more training and practice than most other methods. HTA is sometimes seen as very time consuming, which is certainly true when applied thoroughly to seriously complex tasks. In a study by Stanton and Stevenage (1998), 9 engineering students were given up to 4 hours training and practice in 11 of the most commonly used methods, including HTA. A week later they were required to use each method to evaluate the design of a car radio cassette player and were asked to judge each method on its perceived acceptability, auditability, comprehensiveness, consistency, theoretical validity, use of resources, and usefulness. Time to complete each evaluation was also recorded. The main finding was that HTA was, both objectively and subjectively, the most time intensive of the methods. The students also showed a preference for observation and interview over the more structured methods such as HTA.

Patrick, Gregov & Halliday (2000) carried out a study in which a small sample of students received training in four of the main features of HTA, that is, the decomposition of operations and sub-operations, the definition of operations in terms of objectives, the $p \times c$ stopping rule, and the construction of hierarchical diagrams and tables. They were then required to draw up an analysis of either painting a door or making a cup of tea. Their analyses were then scored on 13 criteria dealing with the principal features of the method. Overall performance was found to be poor, particularly in respect to the students' ability to construct an adequate hierarchy. A second study using the same population and tasks but with enhanced training generated analyses of higher quality, although still not without problems. These results confirm the conclusions reached by Ainsworth and Marshall (1998) and Stanton and Young (1998) that HTA is far from simple and takes both expertise and practice to administer effectively.

Evidence from these studies is suggestive rather than conclusive. However, careful attention to the basic steps summarised in Table 2.1 is recommended as the best guarantee of validity and reliability. In particular, keeping the purpose of the study in sight throughout is crucial to the validity of the analysis, and this is supported by continuing to insist that the stakeholders are clear about the system values and outputs by which performance is to be judged. The ultimate test of validity lies in step 7, the empirical test of the hypotheses on which the recommendations are based. Sadly, such results are rarely, if ever, reported. Reliability rests principally on the skills of the analyst in extracting and cross-checking data from various sources (step 3) and on consultation with stakeholders in step 5. A good analyst will always pursue and try to resolve apparent disagreement between informants or inconsistencies in the data. In this way reliability can be maximised. There are many practical reasons why direct evidence of the validity and reliability of HTA, in common with other analytical methods, is scarce, but perhaps the best evidence that HTA has been found a valuable tool lies in its continued use in a wide variety of contexts over the past 30 years.

REFERENCES

Ainsworth, L., & Marshall, E. (1998). Issues of quality and practicability in task analysis: Preliminary results from two surveys. *Ergonomics, 41*(11), 1607–1617. Also in Annett and Stanton (2000) op. cit. pp. 79–89.

Annett, J. (1969). *Feedback and human behaviour.* Harmondsworth, Penguin.

Annett, J. (1991, November). *A training equipment advisory system* (Report No. D/ER1/9/4/2187/031/APRE). Army Personnel Research Establishment.

Annett, J., (2000). Theoretical and pragmatic influences on task analysis methods. In J.-M. Schraagen, S. F. Chipman, and V. L. Shalin (Eds.), *Cognitive task analysis* (pp. 25–37). Mahwah, NJ: Lawrence Erlbaum Associates.

Annett, J. & Cunningham, D. (2000). Analyzing command team skills. In J.-M. Schraagen, S. F. Chipman, V. L. Stalin. *Cognitive task analysis* (pp. 401–415). Mahwah, NJ: Lawrence Erlbaum Associates.

Annett, J., Cunningham, D., & Mathias-Jones, P. (2000). A method for measuring team skills. *Ergonomics, 43*(8), 1076–1094.

Annett, J., & Duncan, K. D. (1967). Task analysis and training design. *Occupational Psychology, 41,* 211–221.

Annett, J., Duncan, K. D., Stammers, R. B., & Gray, M. J. (1971). *Task analysis.* London: Her Majesty's Stationery Office.

Annett, J., & Kay, H. (1956). 'Skilled performance.' *Occupational Psychology, 30,* 112–117.

Annett, J., & Kay, H. (1957). Knowledge of results and "skilled performance". *Occupational Psychology, 31,* 69–79.

Baber, C., & Stanton, N. A. (1994). Task analysis for error identification: A methodology for designing error-tolerant consumer products. *Ergonomics, 37*(11), 1923–1941.

Broadbent, D. E. (1958). *Perception and communication.* London: Pergamon.

Card, S., Moran, T. P., & Newell, A. (1983). *The psychology of human-computer interaction.* Hillsdale, NJ: Lawrence Erlbaum Associates.

Chapanis, A. (1951). Theory and methods for analyzing errors in man-machine systems. *Annals of the New York Academy of Sciences, 51*(6), 1179–1203.

Crawley, R. (1982, February). *Predicting air traffic controller reaction to computer assistance: A follow-up study* (AP Rep. No. 105). Birmingham: University of Aston. Applied Psychology Department.

Cunningham, D. J., & Duncan, K. D. (1967). Describing non-repetitive tasks for training purposes. *Occupational Psychology, 41,* 203–210.

Diaper, D. (1989). Task analysis for knowledge descriptions (TAKD): the method and an example. In D. Diaper (Ed.), *Task analysis for human-computer interaction* (pp. 108–159). Chichester: Ellis Horwood.

Diaper, D. (2001). Task analysis for knowledge descriptions (TAKD): A requiem for a method. *Behaviour & Information Technology, 20*(3), 199–212.

Diaper, D., & Johnson, P. (1989). Task analysis for knowledge descriptions: Theory and application in training. In J. Long & A. Whitefield (Eds.), *Cognitive ergonomics and human-computer interaction* (pp. 191–224). Cambridge, England: Cambridge University Press.

Duncan, K. D. (1972). Strategies for the analysis of the task. In J. Hartley (Ed.), *Strategies for programmed instruction* (pp. 19–81). London: Butterworths.

Fewins, A., Mitchell, K., & Williams, J. C. (1992). Balancing automation and human action through task analysis. In B. Kirwan & L. K. Ainsworth (Eds.), *A Guide to task analysis* (pp. 241–251). London: Taylor & Francis.

Fitts, P. M. (1954). The information capacity of the human motor system in controlling the amplitude of movement. *Journal of Experimental Psychology, 47,* 381–391.

Gilbreth, F. B. (1911). *Motion study.* Princeton, NJ: Van Nostrand.

Hick, W. E. (1952). On the rate of gain of information. *Quarterly Journal of Experimental Psychology, 4,* 11–26.

Johnson, P., & Johnson, H. (1991). Knowledge analysis of tasks: Task analysis and specification for human-computer systems. In A. Downton (Ed.), *Engineering the human-computer interface* (pp. 119–144). London: McGraw-Hill.

Kirwan, B., & Ainsworth, L. K. (Eds.). (1992). *A guide to task analysis.* London: Taylor & Francis.

Lim, K. Y., & Long, J. (1994). *The MUSE method for usability engineering.* Cambridge, England: Cambridge University Press.

McRuer, D. T., & Krendel, E. S. (1959). The human operator as a servo system element. *Journal of the Franklin Institute, 267,* 381–403, 511–536.

Miller, G. A., Galanter, E., & Pribram, K. (1960). *Plans and the structure of behavior.* New York: Holt.

Miller, R. B. (1953). *A method for man-machine task analysis* (Tech. Rep. No. 53–137). Dayton, OH: Wright Air Force Development Center.

Miller, R. B. (1962). Task description and task analysis. In R. M. Gagné (Ed.), *Psychological principles in system development* (pp. 187–228). New York: Holt, Reinhart and Winston.

Ormerod, T. C. (2000). Using task analysis as a primary design method: The SGT approach. In J.-M. Schraagen, S. F. Chipman, & V. L. Shalin (Eds.), *Cognitive task analysis* (pp. 181–200). Mahwah, NJ: Lawrence Erlbaum Associates.

Ormerod, T. C., Richardson, J., & Shepherd, A. (1998). Enhancing the usability of a task analysis method: A notation and environment for requirements specification. *Ergonomics, 41*(11), 1642–1663.

Patrick, J., Gregov, A., & Halliday, P. (2000). Analysing and training task analysis. *Instructional Science, 28*(4), 51–79.

Penington, J. (1992). A preliminary communications systems assessment. In B. Kirwan & L. K. Ainsworth (Eds.), *A guide to task analysis* (pp. 253–265). London: Taylor & Francis.

Penington, J., Joy, M., & Kirwan, B. (1992). A staffing assessment for a local control room. In B. Kirwan & L. K. Ainsworth (Eds.), *A guide to task analysis* (pp. 289–299). London: Taylor & Francis.

Rasmussen, J. (1983). Skills, rules, knowledge: Signals signs and symbols and other distinctions in human performance models. *IEEE Transactions: Systems, Man & Cybernetics, 13,* 257–267.

Reason, J. (1990). *Human error.* Cambridge, England: Cambridge University Press.

Reed, J. (1992). A plant local panel review. In B. Kirwan & L. K. Ainsworth (Eds.), *A guide to task analysis* (pp. 267–288). London: Taylor & Francis.

Schraagen, J.-M., Chipman, S., & Shalin, V. (2000). *Cognitive task analysis.* Mahwah, NJ: Lawrence Erlbaum, Associates.

Shepherd, A. (1998). HTA as a framework for task analysis. *Ergonomics, 41*(11), 1537–1552.

Shepherd, A. (2001). *Hierarchical task analysis.* London: Taylor & Francis.

Stanton, N., & Stevenage, S. (1998). Learning to predict human error: Issues of reliability, validity and acceptability. *Ergonomics, 41*(11), 1737–1756.

Stanton, N., & Young, M. (1998). Is utility in the eye of the beholder? A study of Ergonomics methods. *Applied Ergonomics 29*(1), 41–54.

3

Discovering How Distributed Cognitive Systems Work

David D. Woods
The Ohio State University, USA

Modern biology has demonstrated the enormous power of functional analysis coupled with naturalistic observation.

—Neisser (1991, p. 35)

Abstract

In cognitive work there is a mutual adaptation of agents' strategies, affordances of artefacts, and demands of the work setting. Regularities in cognitive work concern dynamic interactions across these sets of factors. Because cognitive work systems are not decomposable into independent basic elements, different empirical tactics are necessary, though core values of observation, discovery, and establishment of warrant remain fundamental guides.

Functional analysis is a mode of research that can cope with the unique difficulties of studying and designing cognitive work in context. Functional analysis is a process that coordinates multiple techniques in order to unpack complex wholes to find the structure and function of the parts within the whole. Central to the empirical techniques orchestrated is observation. Three families of techniques emerge that vary in how they shape the conditions of observation: natural history or in situ observation, staged world or observation of performance in simulated situations as models of what is important in situ, and spartan lab settings where observation occurs in experimenter-created artificial tasks. The chapter discusses how to coordinate these techniques in a discovery process that reveals the mutual adaptation of strategies, affordances, and demands and predicts how these dynamic processes will play out in response to change.

STUDYING COGNITIVE WORK IN CONTEXT

The study of cognitive systems in context is a process of discovering how the behavior and strategies of practitioners are adapted to the various purposes and constraints of the field of activity.
—Woods (1997, p. 7)

This single sentence both describes the enterprise of understanding and designing cognitive work and points to the difficulties and challenges embedded in that enterprise. The target phenomena of interest for distributed cognition at work are reflected in the basic founding slogans of cognitive systems engineering—adaptations directed at coping with complexity, how things make us smart or dumb, and how to make automated and intelligent systems team players. Patterns on these themes in cognitive work exist only at the intersections of people, technology, and work.

In studying cognitive work in context, we are out to learn how the more or less visible activities of practitioners are parts of larger processes of collaboration and coordination, how they are shaped by the artefacts and in turn shape how those artefacts function in the workplace, and how their activities are adapted to the multiple goals and constraints of the organizational context and the work domain (Rasmussen, Pejtersen, & Goodman, 1994). These factors of complex artefacts, dynamic worlds, cognitive work, coordinated activity, and organizational dynamics do not come to us pristine, isolated, one level at a time. Rather they come in association with each other, embodied in the particular, cloaked by some observer's vantage point. In other words, cognitive work in context comes in a "wrapped package" as a complex conglomerate of interdependent variables (Woods, Tinapple, Roesler, & Feil, 2002).

Agent–Environment Mutuality

Understanding the deeper patterns in this type of complex conglomerate is difficult because the processes of cognitive work are dynamic and mutually adapted—a case of agent–environment mutuality (Flach, Hancock, Caird, & Vicente, 1995; Woods, 1988). Agent–environment mutuality, after von Uexkull's Umwelt (1934/1957) and Gibson (1979), means that agents' activities are understandable only in relation to the properties of the environment within which they function, and a work environment is understood in terms of what it demands and affords to potential actors in that world. Agents and environment are mutually adapted, thereby establishing a dynamic equilibrium when the environment is stable (i.e., when sources of variability are not changing). When the environment is changing (i.e., when sources of variability are changing), adaptation ensues until a new equilibrium is reached.

In the case of cognitive work in context, this mutual adaptation is a three-way interplay of the strategies of agents (human and machine), the affordances of artefacts (including how they represent the processes at work), and the demands of the field of practice (both those inherent in the processes and those imposed by organizational context) as captured in the cognitive systems triad (Woods, 1988; Woods & Feil, 2002).

Partial decomposability in this triad means that methods can be used to build up a model of how any one is adapted to the conjunction of the other two. In other words, the methodological challenges build from the fact that two of the three factors will be confounded in any technique to tease open the interplay. For example, because the processes in the triad are mutually adapted, developing skill at cognitive work hides or obscures the underlying adaptive web. As the fluency law of cognitive work states (Woods, 2002), "well adapted cognitive work occurs with a facility that belies the difficulty of the demands resolved and the dilemmas balanced."

However, spartan lab experiments would simplify such conglomerates into more manageable units for experimental manipulation. Simplifying such dynamic processes into a series of static snapshots or treating this highly interconnected set of factors as separable represents a retreat from complexity that values the means of experimental tractability over the end of adding to our understanding of the phenomena of interest (Hoffman & Deffenbacher, 1993). Such efforts fall prey to oversimplification biases, which eliminate the very phenomena of interest in the process of simplification (Feltovich, Spiro, & Coulson, 1997). Furthermore, deciding what to leave out or put in, whether in an experimenter-created microworld or in a scaled world simulation, is in itself a potent model of what matters in situ.

How then can we advance our understanding of only partially decomposable, mutually adapted complexes at work? Understanding what an environment affords to agents (given their goals) and how agents' behavior is adapted to the characteristics of the environment and how this linkage changes is called *functional analysis*. Hence, the study of cognitive work in context develops a functional model that captures how the behavior and strategies of practitioners are adapted to the various purposes and constraints of the field of activity. This chapter grounds the functional analysis of cognitive work as a process of discovery built on an empirical foundation of abstracting patterns from multiple sources of observation and directed toward sparking innovative new ways to use technological possibilities to enhance cognitive work (Potter, Roth, Woods, & Elm, 2000; Woods & Christoffersen, 2002; Woods et al., 2002).

THE FUNCTIONALIST PERSPECTIVE

The factors of complex artefacts, dynamic worlds, cognitive work, coordinated activity, and organizational dynamics come to us as an adaptive web consisting of a complex conglomerate of interdependent and dynamic processes. Patterns in cognitive work do not exist separately from particular intersections of specific people, technologies, and work. Hence, field settings are more than unique domains or exotic fields of practice; rather, they function in parallel as natural laboratories where these generic patterns emerge. What approaches would help us discover the essential patterns underlying the diversity of people, technology, and work? What techniques would help us begin to unpack or partially decompose complex wholes such as these into meaningful parts and their interactions? How do we use these patterns to guide change and innovate new possibilities in concert with other stakeholders?

For studying cognitive work in natural laboratories, perhaps astronomy or evolutionary biology, not laboratory physics, provides models for progress. Independent of controls to isolate individual cause–effect relationships, in astronomy and evolutionary biology investigators are able to select and shape conditions of observation and use models to form expectations about what one might see in order to discover and understand the interplay of processes at work in the phenomena of interest. These are the areas that led Neisser (1991) to note "the enormous power of functional analysis coupled with naturalistic observation" (p. 35). The model for doing a functional analysis of cognitive work emerged at the beginning of cognitive systems engineering in several parallel lines of work (see Hutchins, 1995a; Rasmussen, 1986; and Vicente, 1998 for later descriptions).

FUNCTIONAL ANALYSIS AS A MODE OF RESEARCH

Functional analysis is a process that coordinates multiple techniques in order to unpack complex wholes to find the structure and function of the parts within the whole. Functional analysis builds a model that explains the structure in relation to the purposes, where a function specifies how structures are adapted to purposes (more formally, a functional analysis describes an object and its function as how behaviors $B_{1,...,p}$ of item $i_{1,...,m}$ function in system S in environment E relative to purposes $P_{1,...,n}$ to do C, according to Theory T and its supporting data). Functional design is the process of shaping structure to function.

Functional analysis is a mode for research that is critical to understanding systems that are only partially decomposable, dynamic, and adaptive. It is both empirical—based on a variety of observations collected in different ways—and theoretical—a model of how something functions to achieve a goal.

1. A functional model is context bound, but generic.
2. A functional model is tentative; it can be overturned or the patterns can be reinterpreted.
3. A functional model is goal oriented; purposes serve as a frame of reference or point of view; the functions of pieces change with purpose (i.e., purposes provide varying perspectives).
4. Functional analyses are inherently multilevel; phenomena at one level of analysis are situated in a higher level context and are present because of what they contribute in that context.
5. Functional analyses involve the mutual interaction of object or agent and environment (ecological).
6. Functional models do not have the status of being correct, ideal, or even fixed explanations of findings. Rather they are candidate explanations, which attempt to meet one kind of sufficiency criteria—ability to carry out the functions in question.
7. The coupling principle states the more intertwined the relationships between structure and function, the more complex the system is operationally (and the less the system is decomposable into almost independent parts).
8. The essence of good representation design is making function apparent (linkage of apparent structure and underlying function). This means functional modeling is linked to design in that functional analysis results in models of how the system in question will adapt to change and in hypotheses about what changes would prove useful.

Functional analysis generates models based on patterns abstracted from observation and that can be subjected to a critical examination through argument based on reinterpretation of what would account for patterns and on abstracting additional patterns. Functional models motivate and guide empirical confrontations with the field of practice. The functional model serves to abstract the particulars of a setting in a way that supports bounded generalities. Functions serve as a tentative condensation of what has been learned (subject to critical examination and revision) based on a set of empirical observations and investigations.

How does one take up functional analysis as a research program? The starting point is orchestrating a variety of different ways to observe cognition and collaboration at work.

SHAPING THE CONDITIONS OF OBSERVATION

Observation

The hallmark of any science is empirical confrontation—one observes to note patterns; one generates ideas to account for the variability on the surface; and one subjects these ideas to empirical jeopardy. Different methodological traditions are adapted to constraints on these processes that are related to the phenomena of interest and to be explained. All approaches have the potential to generate insight, and all are subject to fundamental sources of uncertainty that define basic trade-offs in research. Different traditions in method balance these trade-offs differently, but the ultimate test is one of productivity—do the methods help generate patterns and ideas about what the essential variables and processes are that account for the breadth and diversity on the surface?

When we see that field settings can function in parallel as natural laboratories, one relevant question becomes how can one shape the conditions of observation in naturally occurring laboratories. When one considers research as orchestrating different ways to *shape* the conditions of observation, a variety of issues arise about coordinating more field-oriented and more lab-oriented techniques.

Different field studies can become part of a larger series of converging observations and operations that overcomes or mitigates the limits of individual studies. This is based on using different techniques to examine a single field of practice and examining the same theme across different natural laboratories.

Three Families of Methods That Vary in Shaping Conditions of Observation

How can one shape the conditions of observation in naturally occurring laboratories? This question serves as a means to distinguish different classes of research methods, each based on core values, but pursuing those basic values in different ways adapted to the role of each in a full program of discovery and verification.

When one sees research methods as variations in shaping the conditions of observation, three classes of research methods appear along this base dimension (Fig. 3.1):

- Natural history methods—in situ,
- Experiments in the field or field experiments—staged or scaled worlds, and
- Spartan lab experiments—experimenter-created artificial tasks.

These three classes are distinguishable by determining the "laboratory" used. Natural history techniques are based on a diverse collection of observations in situ (Hutchins, 1995a). Experiments in the field begin with scaled world simulations that capture or "stage" what is

FIG. 3.1.   Three families of variants on shaping the conditions of observation.

believed to be the critical, deeper aspects of the situations of interest from the field (e.g., De Keyser & Samurcay, 1998). Spartan lab techniques focus on a few variables of interest and their interactions in experimenter-created (and therefore artificial) situations.

These three families differ in many ways, but all three begin with observation. The most basic role of the scientist is as trained observer—one who sees with a fresh view, who wonders, childlike—why is it this way, how does it work—making fresh connections and generating new insights. Similarly, all three classes pursue basic experimental values (Salomon, 1991), though adapted to the uncertainties and purposes of each family of methods in different ways (Hoffman & Woods, 2000):

- Means to establish warrant—How do you know? Why should one accept your findings, observations, conclusions, and interpretations?
- Standards of quality—How do you know what a competent versus an incompetent investigation is, given the uncertainties of research and modeling?
- Means to facilitate generalizability—What is this a case of? What other cases or situations does this apply to?

Also basic to all three families is *target–test mapping,* which is the mapping between the target situation to be understood and the test situation where observations are made. In natural history techniques, the target is examined relatively directly, though it is subject to a variety of choices about who observes what, when, and from where.

What is particularly interesting is that all of the constraints and issues that apply to natural history methods apply to staged world methods, plus more. In addition, all of the constraints and issues that apply to staged world methods flow to spartan lab studies, plus more. Ironically, moving from in situ to scaled worlds to artificial worlds brings sources of additional leverage, but at the cost of new dependencies related to authenticity—target–test mapping.

NATURAL HISTORY TECHNIQUES

No search has been made to collect a store of particular observations sufficient either in number, or in kind, or in certainty, to inform the understanding. . . .

—Bacon (1620, Novum Organum)

Natural history techniques are based on a diverse collection of observations, "which agree in the same nature, though in substances the most unlike," as Bacon (1620) put it at the dawn of science.

Natural history begins with the analysis of the structure or process of each case observed and then the results are compared and contrasted across other analyzed cases (requiring observations across various conditions). This is well suited to situations in which variety across multiple factors is endemic in the phenomenon to be understood.

There are no 'subjects' when one works with people who have a stake in the behavior in which they engage. Similarly, investigators become participant observers; as a result, the question of how the process of observation changes the processes observed is intensified (De Keyser, 1992).

In natural history methods, a critical constraint is that observers have to wait for what happens; however, what happens is authentic by definition, though only a sample. Shaping the conditions of observation is directed at increasing the chances or frequency of encounter of the themes of interest. For example, one might decide to observe a crossroads setting (e.g., the intensive care unit in medicine), where many diverse activities, tempos, and agents interact at different times, and then use patterns noted to choose a tighter focus for future rounds of observation.

In shaping the conditions of observation in this family, one considers the following:

- How should one be prepared to observe (doing one's homework about how the domain works)?
- What concepts and patterns help one prepare to be surprised?
- Who observes?
- What is the role of guides, informants, and others who mediate access to the field of activity?
- Where should one observe from (vantage point)? For example, points of naturally occurring disruption and change may be useful places to observe.

How does past knowledge (patterns, concepts, or models) help one form expectations about what one might see in order to discover and understand more about the processes at work? Or, *how does one prepare to be surprised?* This is the critical driving question for designing discovery-oriented research programs. Observation in situ begins a process of discovering what patterns and themes in distributed cognitive work play out in this setting. What kind of natural laboratory is this or could this be? This presupposes knowledge of patterns and themes in distributed cognitive work, which could be recognized as tentatively relevant for structuring these settings (Woods & Christoffersen, 2002).

How should one describe a culture from the insider's perspective without getting lost in the natives' views? How is one authentically connected to practice yet able to step out of the flow to reflect on practice (a reflective, but practitioner-centered account), especially when the nature of practice is under vectors of change? For example, one might observe at those points and times practitioners gather in reflective forums or one might reshape a natural event to become a reflective forum. Coming to think like a native to some extent, to understand the distinctions they make, is required. However, the observer confronts the problem of referential transparency, which refers to the difficulty in seeing what one sees with, both for the practitioners in their context and for observers themselves, given their own pressures and goals. After all and fundamentally, people may not have access to the critical factors that influence behavior, but people never fail to have some model of how they work (and of the people, devices, and processes with which they interact).

It is particularly important to keep in mind that, in cognitive work, the apparent variables are in fact complex processes in their own right, rather than mere primitive elements. Discovering or identifying the processes that drive performance and adaptation is always a critical part of research that becomes explicit in functional analysis. This relates to the problem of discovering what may be the effective stimuli that control behavior out of the infinite possible ways one could represent the stimulus world. Natural history methods are needed to identify initial places to focus on in this quest (Bartlett, 1932).

STAGED WORLD STUDIES

What has been generally labeled *field experiments* or *experiments in the field* lies in between natural history techniques and spartan lab techniques (Woods, 1993). This family is often viewed as a weak or limited version of real experiments, degraded and limited by the context and (by implication) the skill of the investigator. Rather, this area on the dimension of shaping conditions of observation is a unique, substantive family of methods that is different from lab experiments and natural history, though overlapping with each in several ways (see, e.g.,

Dominguez, Flach, McDermott, McKellar, & Dunn, in press; Johnson, Grazioloi, Jamal, & Berryman, 2001; Layton, Smith, & McCoy, 1994; Nyssen & De Keyser, 1998; Roth, Bennett, & Woods, 1987).

The key in this family of methods is that the investigators stage situations of interest through simulations of some type. The concrete situation staged (the test situation where observation occurs) is an instantiation of the generic situations of interest (the target situation to be understood). The mapping between the target situation to be understood and the test situation where observations are made must be explicit, as this represents a hypothesis about what is important to the phenomena of interest. Classically this is called the *problem of the effective stimulus*. Thus, the fundamental attribute to staged world studies is the investigators' ability to design the scenarios that the participants face. Problem sampling is the focus of this family of techniques.

Staging situations introduces a source of control relative to natural history methods, as it allows for repeated observations. Investigators can seize on this power to design scenarios and contrast performance across interrelated sets of scenarios—if something is known about what makes situations difficult and about the problem space of that domain. Though it must be remembered that although one may try to stage the situation in the same way each time, each problem-solving episode is partially unique, as some variability in how each episode is handled always occurs.

Second, in staging situations, investigators can introduce disturbances—or *probes*—and observe the reaction of the system tracing the process as the system moves toward a resolution or new equilibrium. Probes can be fashioned as complicating factors in scenarios, points of change, and the introduction and absorption of new artefacts. Probes can be built into the problems posed for practitioners, or they can be changes in the artefacts or resources one provides or allows access to, or changes in the distributed system (e.g., adding a new team member; linking a collaborator at a distance through a technology portal).

When investigators probe the nature of practice by changing the kinds of artefacts practitioners use to carry out their activities, the intent is not to evaluate the artefacts as partially refined final products, but rather to use them as tools for discovery, that is, as wedges to break into the cognitive system triad. Artefact-based methods have great potential because introducing new technology into fields of ongoing activity always serves as a kind of natural experimental intervention (Carroll & Campbell, 1988; Flores, Graves, Hartfield, & Winograd, 1988). Prototypes of new systems are not simply partially refined final products; they also can function as a kind of experimental probe to better understand the nature of practice and to help discover what would be useful (Woods, 1998). Artefact-based methods depend on a functional model of the artefacts as affordance for cognitive work or how the change in artefacts changes the cognitive work necessary to meet demands (see, e.g., Watts-Perotti & Woods, 1999). In this way functional models are a means for representing the research base in a way that is useful to frame or focus new studies and to guide provisional action.

Critically, in the final analysis, staged world studies are still fundamentally observational and discovery oriented—letting the world tell us how it works rather than playing 20 questions with nature in an artificial lab. Building on top of the fundamental observational nature of this family, investigators can engineer contrasts by designing sets of scenarios, changes in artefacts, and changes in collaborative links, again if they can draw on knowledge about the relevant problem space, affordances, and functions of distributed cognitive work. However, it is important not to be confused into now thinking these contrasts convert a staged world observational study into a spartan experiment. Each of the things contrasted (problem, affordance, and collaborative function) is still a complex conglomerate in its own right, with many degrees of freedom. Avoid the error of mistaking complex factors as primitive elements, which require no further analysis.

Process Tracing in Staged World Observation

Because the investigator designs the scenarios and probes that the participants confront, the investigator now has the ability to trace the process by which practitioners handle these situations. Process-tracing methods follow the steps specified by Hollnagel, Pedersen, and Rasmussen (1981) as sketched in Fig. 3.2 and summarized in Table 3.1 (see Roth et al., 1987 for a sample analysis in full). First, the starting point is the analysis of the process for handling the situation in its specific context. Second, shift to an analysis of the process abstracted by using concepts about the aspects of cognitive work instantiated in the concrete case (e.g., Johnson et al., 2001; Roth et al., 1987). Third, use various levels of cross-protocol contrasts (across probes, scenarios, and domains) to abstract patterns and build possible functional accounts (e.g., Woods & Patterson, 2000).

The ability to probe practice through staged world methods is a powerful tool to observe how change in demands, pressures, complexities, and capabilities triggers vectors of adaptation and resettling into new equilibria. Staged world techniques build on natural history methods, which can be used to begin to identify what the important aspects are to preserve in the scaled world (De Keyser & Samurcay, 1998; or see the linkage across Sarter & Woods, 1995, 1997, 2000).

Staging situations of interest also occurs in spartan lab methods, though these attributes may be overshadowed by new sources of control. Simulations vary from replica worlds, to scaled worlds, to the microworlds experimenters create for artificial labs (sometimes these artificial labs are given cover stories as a pretense to authenticity). In any case, what is staged, whether scaled or artificial, is a potent model of the target phenomena based on insight and evidence, tentative and subject to contention. Whereas the target is examined relatively directly in natural history techniques, target–test mapping is still subject to a variety of choices about who observes what when and from where (an extreme form of a "replica").

To summarize, staged world studies are fundamentally observational. Their power comes from the design of the situation to focus the conditions of observation on places where the patterns of interest are more likely to play out. This means not just any "interesting" situation, but working through the target–test mapping explicitly and in detail. For example, one typically does this by considering what makes the problem difficult (demands) for any system of cognitive agents. In designing the situation, probes, and the conditions of observation, one's goal is to uncover what has been encoded in practice. The general problem relates to any adapted system,

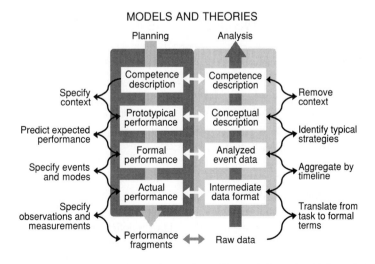

FIG. 3.2.   The steps in process-tracing analyses of cognitive work (based on Hollnagel et al., 1981).

TABLE 3.1
Checklist for Designing Staged World Field Studies

Steps	Problematic Items
1. What are the cognitive system issues in question?	Pitfalls: superficial labels
2. Explicitly map target–test situation relationship.	Problem space, scenario design, stimulus sampling Pitfalls: psychologists fallacy, cover stories
3. Shape the test situation to address the general cognitive system issues.	Scenario design Artefacts as tool for discovery Procedure to externalize cognitive activities
4. Map canonical behavior; how does one prepare to be surprised?	
5. Collect the data (update 4 as needed).	
6. Collate multiple raw data sources to construct base protocol.	Analyze each participant's or team's process Pitfalls: get lost in the raw data
7. Use concepts from 1 to build formal protocol.	Pitfalls: get lost in the language of the domain; excessive microencodings; hindsight bias Distinguishing interpretations from observations
8. Cross-protocol contrasts.	Aggregate across teams and problems Role of cognitive simulation, neutral practitioners
9. Integrate with other converging studies to build a functional model.	

Note. See Hollnagel et al. (1981) and Woods (1993).

that is, the fluency law: first, focusing observation by designing and staging situations based on the abstract patterns of interest; second, introducing disruptions and observing how the distributed cognitive system responds (probe techniques); and third, contrasting conditions relative to the artefacts, relative to the levels of experience of practitioners, relative to the difficulty of the demands of the situation, or on other dimensions.

CONVERGING OPERATIONS IN DISCOVERY

Doing a cognitive task analysis is like trying to learn the secret of a magic trick: once you know the trick and know what to look for, you see things that you did not notice when you did not know exactly what to look for.

—Woods (1997, p. 7) paraphrase of Orville Wright's comment about discovering the secret of flight

Fluency law: well adapted cognitive work occurs with a facility that belies the difficulty of the demands resolved and the dilemmas balanced.

—Woods & Feil (2002)

Different studies within and across these families become part of a larger series of converging operations that overcome or mitigate the limits of individual studies. This is based both on using different techniques to examine a single field of practice and from examining the same theme across different natural laboratories. As a result, we speak of studying cognitive work in context as a bootstrap within a field of practice and set of converging operations across natural laboratories as one hones in on a cognitive system theme (Potter et al., 2000).

Note the use of the words *weaker* and *stronger* to describe the vertical dimension in Fig. 3.1. Weak shaping of the conditions of observation is both a strength and a weakness; strong shaping of the conditions of observation is both a weakness and a strength, though each in complementary aspects—hence the bootstrap or orchestration of varieties of observation that are diverse in how they shape (and therefore distort) the processes we wish to observe and understand.

As one orchestrates a series of investigations that vary in how they shape the conditions of observation, one must evade a variety of pitfalls.

The Psychologist's Fallacy in Research on Cognitive Work

Because technical work . . . is so poorly understood, policy makers routinely fall back on stereotypes or images of work . . . in order to make sense of technical work. The potential cost of misunderstanding technical work is the risk of setting policies whose actual effects are not only unintended but sometimes so skewed that they exacerbate the problems they seek to resolve. Efforts to reduce "error" misfire when they are predicated on a fundamental misunderstanding of the primary sources of failures in the field of practice [systemic vulnerabilities] and on misconceptions of what practitioners actually do.

—Barley and Orr (1997, p. 18; emphasis added)

It is easy to commit what William James (1890) called over 100 years ago the "psychologist's fallacy." Updated to today, this fallacy occurs when well-intentioned observers think that their distant view of the workplace captures the actual experience of those who perform technical work in context. Distant views can miss important aspects of the actual work situation and thus can miss critical factors that determine human performance in that field of practice. Integrating natural history techniques is a strong guard against the psychologist's fallacy, as a functional analysis, like an ethnography, pursues access to "authenticity" in understanding practice.

There is a corollary to the psychologist's fallacy that is another pitfall in pursuit of authenticity (Law & Callon, 1995):

"Avoid assuming that those we study are less rational or have a weaker grasp on reality than we ourselves." This rule of method asks us to take seriously the beliefs, projects, and resources of those whom we wish to understand. (p. 281)

Managers, designers, researchers, and regulators often feel they are immune from the processes they study and manage, especially when the trigger for investing in studies is human "error," "limits," and "biases." This fallacy blocks effective functional analysis because it blinds the observer from seeing how behavior is adapted in a universe of multiple pressures, uncertainty, and finite resources. Learning about cognitive work begins with adopting the point of view of practitioners in the situation before the outcome is known, or what anthropologists call an *emic perspective* (Fetterman, 1989).

The Danger of Whig Histories

A "Whig" History—produces a story which is the ratification if not the glorification of the present.
—Butterfield (1931/1965, p. v)

Historians refer to the same dangers that worried James as *Whig histories,* which divide the world into the friends and enemies of progress or at least the teller's model of progress (which today is a uncritical anticipation of benefits to come from investment in the next new technology).

They look at the past to tell a story that justifies and celebrates an aspect of the present state of affairs as the best. A Whig history does not examine how participants in those events viewed the world, but picks out those parts of the past that can be seen now as anticipating or retarding development of the storyteller's model of the present. Interestingly, technology advocates often develop Whig accounts of cognitive work as a means to recruit resources and to pursue status in our culture (Butterfield, 1931/1965):

> Instead of seeing the modern world emerge as the victory of the children of light over the children of darkness in any generation, it is at least better to see it emerge as the result of a clash of wills, a result which often neither party wanted or even dreamed of, a result which indeed in some cases both parties would equally have hated, but a result for the achievement of which the existence of both and the clash of both were necessary. (p. 28)

Following Butterfield's advice places a premium on authenticity in the orchestration of methods (Woods & Christoffersen, 2002). The search for authenticity reveals a trade-off: the stronger the shaping of the conditions of observation, the greater the disruption of the thing observed and the greater the risk of a psychologist's fallacy and its analogs. Natural history methods then are not "weak" as in poor substitutes for the real methods of science or necessary evils that result from practical limitations. Rather, because they are weak in shaping the conditions of observation, natural history methods are a fundamental contributor to the goal of transcending limits to authenticity to capture how the strategies and behavior of people are adapted to the constraints and demands of fields of practice. Because natural history methods are weak in shaping the conditions of observation, they have to be coordinated with investigations that more strongly shape the conditions of observation based on tentative, partial knowledge of the essential factors in cognitive work (e.g., attempts such as those by Woods, 2002). However, staged world and spartan lab techniques are weak in the sense that they depend on strong commitments to models of how cognition adapts to artefacts and demands of work.

Discovery and Functional Analysis

Ultimately the role and value of natural history and staged world techniques in a full program of research has been downplayed because the community of science has downplayed explicit consideration of the processes of discovery in science. Reexamining these families of techniques reawakens the challenge of describing and facilitating processes of insight (remember that the corollary to the psychologist's fallacy applies to scientists who are not immune from the processes they study—limits on scientists' ability to reflect on the practice of discovery they engage in).

To focus on the role of insight, the best method is to examine sets of functional analyses of cognitive work in context in detail. For example, consider a trio of natural history studies drawn from three different domains (Cook, 1998 in medicine; Hutchins, 1995b in aviation; Patterson, Watts-Perotti, & Woods, 1999 in space mission control), and consider how each discovered (or present possible accounts of) how the behavior and strategies of practitioners are adapted to the various purposes and constraints of the field of activity (see Woods et al., 2002).

Each uses natural history methods to explore the interplay of agents, artefacts, and demands of the field of activity, but each enters the complex interplay of practice from different points. One is Richard Cook's "Being 'Bumpable'" (1998) study of the intensive care unit, which explores the significance of a phrase used by practitioners. In this study, by modeling the significance of insiders' use of a highly coded argot of practice does one begin to appreciate the adaptive power of the intensive care unit. Another is Ed Hutchins' "How a Cockpit Remembers Its Speed" (1995b) study of cockpits, which begins with the speed management tasks during a descent. The cockpit is understood as a distributed system by understanding how certain apparently insignificant artefacts turn out to be highly adapted to the pressures of critical tasks and how these artefacts can be used in ways that make cognitive work more robust. A third begins with an artefact in space mission control center—voice loops—and explores the role of this artefact in supporting successful collaborative activities when disruptions to plans occur (Patterson et al., 1999). As you examine the trio, reflect on how the eight points of functional modeling and analysis are illustrated in these specific functional analyses.

Warrant in Functional Analysis

I chose these three because of how they illustrate discovery processes at work in functional analysis and modeling. I chose three natural history studies as examples to eliminate any confusion with methods of verification.

First, note the striking absence of discussion of methods. To the degree that method is discussed at all, it is in terms of intimacy and extent of observation. In other words, claims to authenticity for natural history studies begin with intimate connection to the field of practice, that is, to be able to observe from inside. The first kind of warrant for these functional accounts is one kind of claim to authenticity derived from the close contact for observation.

Second, each study provides observations and derives patterns to be accounted for. They note deeper aspects of the field of practice that are in need of explanation in a functional account. The warrant for these is related to the sets of observations and patterns noted in previous studies of this setting and the cognitive system themes invoked (is it expanding or modifying the patterns to be explained?). The latter presupposes a means to share observations and patterns from various field studies across a community of researchers interested in those cognitive system themes in settings such as the one in question.

Third, each study also makes a claim for warrant because the study provides a coherent account of the field of practice that goes behind the veil of surface characteristics to propose deeper dynamics. Providing an account or candidate functional explanation makes a sufficiency claim relative to the sets of observations and relative to other functional accounts previously offered—how well do they cover patterns noted in observation? (Note how this links to the role of computer simulations in functional modeling; see, e.g., Johnson et al., 2001.)

Ultimately, what is most striking in this trio of functional accounts is insight. From a discovery point of view, insight has no need of justification by pure method. It is a rare and precious commodity in need of facilitation; as such, insight is its own justification. Insight does not demand repeatability or indifference to the investigator or investigative team (what some criticize erroneously as "subjectivity" in field research). Insight generates patterns from observations significant in their demand for explanation, and it generates new or modified candidate explanations significant in the claim they stake to cover the relevant patterns derived from observation.

Insight is valuable; indeed, it is primary, in itself, for science, though it is almost universally ignored as ineffable in accounts of science and in teaching of research methods. All programs of research in science embed processes of insight and methods to facilitate such insights; the methods and difficulties are generally just ignored in the descriptions of research process that are socially approved for their community.

The absence of, or even taboos on, discussion of insight and discovery do not facilitate these essential processes. As goes for any and all aspects of cognitive work, although there are individual differences in acquiring and demonstrating skill, discovery is trainable and aidable.

One can view using and orchestrating field research techniques (the top two families in Fig. 3.1) as a pragmatic program for facilitating insight.

Insight generates ideas that stand on their own and are open to argument and revision through new observations and alternative covering accounts. Insight as a discovery process is measured by models of argument (such as Toulmin's structure of argument (Toulmin, Rieke, & Janik, 1984) or accounts of abductive explanation as a model of justification in argument), not by the syntax of inductive or deductive method. In other words, insight provides the linkage between field research techniques as discovery processes and laboratory research techniques as verification processes—hence, Bartlett's dictum (1932) to begin experimental studies with observations of ongoing activities in situ.

PREDICTION AND INNOVATION IN COGNITIVE WORK

Studies of cognitive systems as dynamic, adaptive systems have revealed technology change as a process of transformation creating new roles, changing what is canonical and what is exceptional, and changing the kinds of errors and paths to failure, where practitioners and stakeholders adapt to achieve their goals and avoid failure (Woods & Dekker, 2000; Woods & Tinapple, 1999; Woods et al., 2002). This process of transformation and adaptation creates a challenge of prediction more daunting than the challenge of understanding the current adaptive balance in cognitive systems:

- How does one envision or predict the relation of technology, cognition, and collaboration in a domain that doesn't yet exist or is in a process of becoming?
- How will envisioned technological change shape cognition and collaboration?
- What are the new roles for people in the system?
- How will practitioners adapt artefacts to meet their own goals, given new capabilities and mismatches to the actual demands and pressures they experience?
- How can we predict the changing nature of expertise and new forms of failure as the workplace changes?

Asking questions such as these reveals that studies of cognitive work are going on around us constantly as developers and technology advocates make claims about the impact of objects-to-be-realized and that the adaptive response of practitioners and organizations to these changes provides data about the accuracy of those predictions (e.g., Woods & Sarter, 2000). The artefacts under development are not simply objects in the process of realization, but also hypotheses about how technology shapes cognition, collaboration, and performance (Woods, 1998) subject to empirical jeopardy. In this light, functional analysis is a program to harness and direct the ongoing natural experiments and to accelerate progress in understanding the dynamics of distributed cognitive systems.

However, studies of cognitive work are not justified solely on the basis of abstracting patterns and building explanations of the adaptive interplay of people, technology, and work. Ultimately, their purpose is to inform the design of systems for cognitive work (Woods et al., 2002). These studies occur as a part of process of organizational and technological change spurred by the promise of new capabilities, the dread of paths to failure, and continuing pressure for higher levels of performance and greater productivity (systems under "faster, better, cheaper" pressure). This means investigators, through observing and modeling practice, are participants in processes of change in meaningful fields of practice. To become integral participants who represent the goal of supporting effectiveness in cognitive work given limited resources, the potential for surprise, and irreducible uncertainty, the families of methods have to be extended to meet the challenges of design (Woods et al., 2002):

- The leverage problem—how do studies of cognitive work help decide where to spend limited resources to have the greatest impact (because all development processes are resource limited)?
- The innovation problem—how do studies of cognitive work support the innovation process (it is probably necessary but not sufficient as a spark for innovation)?
- The envisioned world problem—how do the results that characterize cognitive and cooperative activities in the current field of practice inform or apply to the design process *because the introduction of new technology will transform the nature of practice* (a kind of moving-target difficulty)?
- The adaptation through use problem—how does one predict and shape the process of transformation and adaptation that follows technological change?
- The problem of error in design—designers' hypotheses, as expressed in artefacts, often fall prey to William James' psychologist's fallacy—the fallacy of substituting the designer's vision of what the impact of the new technology on cognition and collaboration might be, for empirically based but generalizable findings about the actual effects from the point of view of people working in fields of practice.

SUMMARY

Functional analysis is a discovery program of research that can support four basic values that define practice-centered design of what will be useful in cognitive work (Woods & Christoffersen, 2002). The first value is transcending limits to authenticity to capture how the strategies and behavior of people are adapted to the constraints and demands of fields of practice. The second is meeting the challenge of abstraction to find and explain patterns behind the surface variability. The third is sparking inventiveness to discover new ways to use technological possibilities to enhance human performance, to identify leverage points, and to minimize unanticipated side effects. The fourth is creating future possibilities as participants with other stakeholders and problem holders in that field of practice. Functional analysis is the means to achieve these delicate balances—generic but relevant; finding in the particular the existence and expression of universal patterns; linking understanding to usefulness as fields of practice change in the face of new pressures and new capabilities.

ACKNOWLEDGMENTS

This work was carried out in part through participation in the Advanced Decision Architectures Collaborative Technology Alliance sponsored by the U.S. Army Research Laboratory under Cooperative Agreement DAA19-01-2-0009.

REFERENCES

Barley, S., & Orr, J., Jr. (Eds.). (1997). *Between craft and science: Technical work in US settings*. Ithaca, NY: IRL Press.

Bartlett, F. C. (1932). *Remembering: A study in experimental and social psychology*. Cambridge, England: Cambridge University Press.

Butterfield, H. (1965). *The Whig interpretation of history*. New York: W. W. Norton. (Original work published 1931)

Carroll, J. M., & Campbell, R. L. (1988). *Artifacts as psychological theories: The case of human-computer interaction* (IBM Research Rep. RC 13454). Yorktown Heights, NY: Watson Research Center.

Cook, R. I. (1998, May). *Being Bumpable*. Fourth Conference on Naturalistic Decision Making, Warrenton VA. Available at url: http://csel.eng.ohio-state.edu/woodscta, section 1.3

De Keyser, V. (1992). Why field studies? In M. G. Helander & N. Nagamachi (Eds.), *Design for manufacturability: A systems approach to concurrent engineering and ergonomics*. London: Taylor & Francis.

De Keyser, V., & Samurcay, R. (1998). Activity theory, situated action and simulators. *Le Travail Humain, 61*(4), 305–312.

Dominguez, C., Flach, J., McDermott, P., McKellar, D., & Dunn, M. (in press). The conversion decision in laparoscopic surgery: Knowing your limits and limiting your risks. In J. Shanteau, K. Smith, & P. Johnson (Eds.), *Psychological explorations of competent decision making*. New York: Cambridge University Press.

Feltovich, P. J., Spiro, R. J., & Coulson, R. L. (1997). Issues of expert flexibility in contexts characterized by complexity and change. In P. J. Feltovich, K. M., Ford, & R. R. Hoffman (Eds.), *Expertise in context*. Cambridge, MA: MIT Press.

Fetterman, D. M. (1989). *Ethnography: Step by step*. Beverly Hills, CA: Sage.

Flach, J., Hancock, P., Caird, J., & Vicente, K. (Eds.). (1995). *An ecological approach to human machine systems I: A global perspective,* Hillsdale, NJ: Lawrence Erlbaum Associates.

Flores, F., Graves, M., Hartfield, B., & Winograd, T. (1988). Computer systems and the design of organizational interaction. *ACM Transactions on Office Information Systems, 6,* 153–172.

Gibson, J. J. (1979). *An ecological approach to perception*. Boston: Houghton Mifflin.

Hoffman, R. R., & Deffenbacher, K. A. (1993). An analysis of the relations of basic and applied science. *Ecological Psychology, 5,* 315–352.

Hoffman, R., & Woods, D. D. (2000). Studying cognitive systems in context. *Human Factors, 42*(1), 1–7.

Hollnagel, E., Pedersen, O., & Rasmussen, J. (1981). *Notes on human performance analysis* (Tech. Rep. No. Risø-M-2285). Roskilde, Denmark: Risø National Laboratory, Electronics Department.

Hutchins, E. (1995a). *Cognition in the wild*. Cambridge, MA: MIT Press.

Hutchins, E. (1995b). How a cockpit remembers its speed. *Cognitive Science, 19,* 265–288.

James, W. (1890). *Principles of psychology*. New York: H. Holt and Company.

Johnson, P. E., Grazioloi, S., Jamal, K., & Berryman, R. G. (2001). Detecting deception: Adversarial problem solving in a low base-rate world. *Cognitive Science, 25,* 355–392.

Layton, C., Smith, P. J., & McCoy, C. E. (1994). Design of a cooperative problem-solving system for en-route flight planning: An empirical evaluation. *Human Factors, 36,* 94–119.

Law, J., & Callon, M. (1995). Engineering and Sociology in a Military Aircraft Project: A Network Analysis of Technological Change. In Susan Leigh Star (Ed.), *Ecologies of knowledge: Work and politics in science and technology* (pp. 281–301). Albany: State University of New York Press.

Neisser, U. (1991). A case of misplaced nostalgia. *American Psychologist, 46*(1), 34–36.

Nyssen, A. S., & De Keyser, V. (1998). Improving training in problem solving skills: Analysis of anesthetists' performance in simulated problem situations. *Le Travail Humain, 61*(4), 387–402.

Patterson, E. S., Watts-Perotti, J. C., & Woods, D. D. (1999). Voice loops as coordination aids in Space Shuttle Mission Control. *Computer Supported Cooperative Work, 8,* 353–371.

Potter, S. S., Roth, E. M., Woods, D. D., & Elm, W. (2000). Bootstrapping multiple converging cognitive task analysis techniques for system design. In J. M. C. Schraagen, S. F. Chipman, & V. L. Shalin (Eds.), *Cognitive task analysis*. Mahwah, NJ: Lawrence Erlbaum Associates.

Rasmussen, J. (1986). *Information processing and human-machine interaction: An approach to cognitive engineering*. New York: North-Holland.

Rasmussen, J., Pejtersen, A. M., & Goodman, L. P. (1994). *Cognitive systems engineering*. New York: Wiley.

Roth, E. M., Bennett, K., & Woods. D. D. (1987). Human interaction with an "intelligent" machine. *International Journal of Man-Machine Studies, 27,* 479–525.

Salomon, G. (1991). Transcending the qualitative-quantitative debate: The analytic and systemic approaches to educational research. *Educational Researcher, 20,* 10–18.

Sarter, N. B., & Woods, D. D. (1995). How in the world did we get into that mode? Mode error and awareness in supervisory control. *Human Factors, 37*(1), 5–19.

Sarter, N. B., & Woods, D. D. (1997). Team play with a powerful and independent agent: A corpus of operational experiences and automation surprises on the Airbus A-320. *Human Factors, 39,* 553–569.

Sarter, N. B., & Woods, D. D. (2000). Team play with a powerful and independent agent: A full mission simulation. *Human Factors, 42,* 390–402.

Toulmin, S., Rieke, R., & Janik, A. (1984). *An introduction to reasoning*. New York: Macmillan.

Vicente, K. (1998). *Cognitive work analysis: Towards safe productive and healthy computer based work*. Mahwah, NJ: Lawrence Erlbaum Associates.

von Uexkull, J. (1957). A stroll through the worlds of animals and men. In C. Schiller (Ed.), *Instinctive behavior* (pp. 5–7). New York: International Universities Press. (Original work published 1934)

Watts-Perotti, J., & Woods, D. D. (1999). How experienced users avoid getting lost in large display networks. *International Journal of Human-Computer Interaction, 11*(4), 269–299.

Woods, D. D. (1988). Coping with complexity: The psychology of human behavior in complex systems. In L. P. Goodstein, H. B. Andersen, & S. E. Olsen (Eds.), *Mental models, tasks and errors* (pp. 128–148). London: Taylor & Francis.

Woods, D. D. (1993). Process-tracing methods for the study of cognition outside of the experimental psychology laboratory. In G. Klein, J. Orasanu, R. Calderwood, & C. E. Zsambok (Eds.), *Decision making in action: Models and methods* (pp. 228–251). Norwood, NJ: Ablex.

Woods, D. D. (1997) Invited Talk. NATO/Office of Naval Research Workshop on Cognitive Task Analysis, Washington D.C., October 30, 1997 (see Woods et al., 2002, section 2.1).

Woods, D. D. (1998). Designs are hypotheses about how artifacts shape cognition and collaboration. *Ergonomics, 41,* 168–173.

Woods, D. D. (2002). Steering the reverberations of technology change on fields of practice: Laws that govern cognitive work. In *Proceedings of the 24th Annual Meeting of the Cognitive Science Society* (pp. 14–17). Mahwah, NJ: Lawrence Erlbaum Associates.

Woods, D. D., & Christoffersen, K. (2002). Balancing practice-centered research and design. In M. McNeese & M. A. Vidulich (Eds.), *Cognitive systems engineering in military aviation domains.* Wright-Patterson AFB, OH: Human Systems Information Analysis Center.

Woods, D. D., & Dekker, S. W. A. (2000). Anticipating the effects of technological change: A new era of dynamics for human factors. *Theoretical Issues in Ergonomic Science, 1*(3), 272–282.

Woods, D. D. and Feil, M. (2002). Laws that Govern Cognitive Work: e-pamphlet. Columbus OH: Cognitive Systems Engineering Laboratory, Institute for Ergonomics. Available: url: http://csel.eng.ohio-state.edu/laws/laws_flash1.html

Woods D. D., & Patterson, E. S. (2000). How unexpected events produce an escalation of cognitive and coordinative demands. In P. A. Hancock and P. Desmond (Eds.), *Stress, workload and fatigue* (pp. 290–304). Mahwah, NJ: Lawrence Erlbaum Associates.

Woods, D. D., & Sarter, N. (2000). Learning from automation surprises and going sour accidents. In N. Sarter and R. Amalberti (Eds.), *Cognitive engineering in the aviation domain* (pp. 327–353). Mahwah, NJ: Lawrence Erlbaum Associates.

Woods, D. D., & Tinapple, D. (1999, September). W³: Watching human factors watch people at work. Presidential Address of the 43rd Annual Meeting of the Human Factors and Ergonomics Society, Houston, TX. Available: Multimedia Production at url: http://csel.eng.ohio-state.edu/hf99/

Woods, D. D., Tinapple, D. Roesler, A., & Feil, M. (2002). *Studying cognitive work in context: Facilitating insight at the intersection of people, technology and Work.* Columbus, OH: Cognitive Systems Engineering Laboratory, Institute for Ergonomics. Available: url: http://csel.eng.ohio-state.edu/woodscta

4

Course-of-Action Analysis and Course-of-Action-Centered Design

Jacques Theureau
Centre National de la Recherche Scientifique/UTC
Université de Technologie de Compiègue, France

Abstract

This chapter presents a theoretical and methodological framework that integrates (a) an approach to cognitive task analysis, coined the *course-of-action analysis,* which considers cognitive tasks as embodied, situated, indissolubly individual and collective, cultured, and experienced; and (b) an approach to cognitive task design, the *course-of-action-centered design,* which concerns the situation (spatial, informational, technical, and organizational) as a whole, the training, and, more generally, the culture of the operators, thanks to (c) a paradigm of human cognition stemming from theoretic biology, the *enaction paradigm,* and various philosophical and scientific contributions that go well beyond analytical philosophy, cognitive psychology, and computer science.

INTRODUCTION

This theoretical and methodological framework of course-of-action analysis and course-of-action-centered design (the terms that were inspired by the user-centered system design proposed by Norman & Draper, 1986, while introducing a significant difference) developed essentially in connection with the creation of computerized and automated work situations (Pinsky, 1992; Theureau, 1992; Theureau & Jeffroy, 1994). It has also proved fruitful concerning non-computerized work situations (e.g., vine growing), some practical situations other than work situations (e.g., car driving, in particular including or meant to include different kinds of assistance systems), domestic situations (e.g., including domestic systems of energy control accessible by various media), teaching situations (in particular integrating software and human tutoring, see Leblanc, Saury, Theureau and Durand, 2001), and, more recently, different situations of high-level sport performance and sport training (see, e.g., Saury, Durand, & Theureau, 1997). These various situations have been tackled in academic and public research, through undergraduate studies and doctoral theses in ergonomics, systems control, and sporting sciences and techniques. Likewise they have been tackled in ergonomics departments in companies and consulting ergonomics groups.

The initial inspiration for the elaboration of this theoretical and methodological framework first and foremost came from the French language ergonomics tradition and dates back to Ombredane and Faverge (1955). Today, this inspiration can be summarized in one directing idea, that of the necessity of an analysis of the actual operators' activities in real work situations for the design of new work situations. However, this elaboration really started in 1979 from a second impulse provided by Newell and Simon (1972), which contributed to the foundation of both laboratory cognitive psychology and artificial intelligence. This second impulse, in contrast to the first, is shared with the other existing approaches of cognitive task analysis and cognitive task design. Indeed, in taking an essential interest in everyday cognition, cognitive task analysis can be considered as the response to the strategic criticism that was made to Simon's approach by Noam Chomsky (Piatelli-Palmarini, 1979), that of being centered on exceptional phenomena, symbolic problem solving, instead of considering the most common phenomena of human cognition. In parallel, cognitive task design can be considered as a contribution to design that goes well beyond artificial intelligence systems design. However, because this second impulse has been superimposed on the first one, it has been followed up both literally and in a critical manner in such a way that it is difficult to recognize it in its present form. We refer to this in the introductory paragraphs of the following sections.

A Literal Interpretation

At first, let us consider in Newell and Simon (1972) what is obeyed to the letter by course-of-action analysis and course-of-action-centered design. Firstly, of course, it is the proposition to study the "human system" according to "three dimensions of variation," that is, "tasks," "individual differences" (cultural), and time scale in "behavioral acts" ("performance/learning/development") (p. 2). Next, it is "to try to represent in some detail a particular man at work on a particular task" (p. 5). Furthermore, "as a scientific bet, emphasis is put on performance" and "learning is considered as a second-order effect" (p. 7). Finally, the researched theory is considered a process theory, dynamically oriented, empirical, not experimental, and nonstatistical (pp. 9–13). From this stems a mode of theory and model validation that stresses a systematic description of the verbal protocols gathered in parallel to the progression of the activity in abstract terms that express hypothetical structural invariants and gives a secondary status to the classical experiments and to the statistical treatments. This is why Newell and Simon devote nearly 200 pages of their book to discussing the difficulties in describing some verbal problem-solving protocols concerning the cryptarithmetic puzzle DONALD + GERALD = ROBERT. Another characteristic can be added to these first ones: that the limits of the protocols, that is, the variable "coverage" of the activity by the verbalizations, lead to the development of a web of inferences of which the degree of conviction depends on both the density of the coverage and the theoretical construction that is developed (p. 184).

Course-of-action analysis and course-of-action-centered design respect to the letter these characteristics concerning the study of everyday activities and the design of situations in which they are accomplished, but they radically reject the other characteristics, which are mentioned gradually in this chapter. From this stems what globally might be called a *critical obedience*.

Chapter Contents

We progressively present the various characteristics of course-of-action analysis and course-of-action-centered design that stem from this critical obedience to the impulse of Newell and Simon concerning cognitive task analysis and cognitive task design. First, we present the enaction paradigm and the consequences for cognitive task analysis that have been drawn from it: its execution in terms of course of experience and course of action. Then, we present the principles of the observatory of course of action and how they are made concrete in particular studies. The two next sections are dedicated to a semiological theoretical framework for the analysis of the course of experience. In the following section, we consider in particular the collective interlinking of the courses of action. In the penultimate section, before considering the integration of course-of-action analysis in course-of-action-centered design, we approach several related questions: that of the so-called principle of the primacy of the intrinsic in the analysis; that of knowing which kind of analytic and synthetic empirical models the course-of-action analysis aims at producing; that of the amphibology of cognitive task design and that of the object of the design considered by course-of-action-centered design; and that of the distinction between the synthetic empirical models and the synthetic models for design. In the last section, we show that the dynamic complexity constituted by the human activity imposes, for empirical knowledge as well as for design, an iterative process for task analysis and contribution to task design. We conclude by considering the process of practice of course-of-action analysis itself as a collective interlinking of courses of action.

THE ENACTION PARADIGM AND ITS CONSEQUENCES FOR TASK ANALYSIS AND DESIGN

For Newell and Simon (1972), as for nearly everybody since the criticism of behaviorism, intelligent behavior presupposes the faculty of representing the world in a certain manner. However, these authors add a hypothesis concerning this representation that can be formulated as follows: cognition consists of acting on the basis of representations of a predetermined exterior world that have a physical reality in the form of a symbolic code in a brain or a machine. It is this hypothesis that is the basis for the computer paradigm of cognition. According to Francisco Varela—who, with Humberto Maturana, put forward the so-called enaction paradigm of cognition—this is the weak point: "The most important faculty of all living cognition consists, in a large measure, in asking relevant questions which arise at any moment of our life. These are not predefined but enacted, we make them emerge on a background and the criteria of relevance are dictated by our common sense in a manner which is always contextual" (Varela, 1989, p. 90–91). Besides, Simon himself had specified that "in real life, there is no well defined, unique, and static problem but rather one which is constantly changing, whose definition is modified on the basis of information that the agents extract from their memory or they obtain through responses from their environment, to the actions that they have executed" (Simon, 1977, p. 239). On the basis of such evidence in empirical studies, the course-of-action analysis has therefore detached itself from the computer paradigm of cognition in favor of the enaction paradigm of cognition.

In course-of-action-centered design, the centrality of the analysis of real operators' courses of action in real work situations/obeys profound theoretical and epistemological reasons concerning the nature of human activity and its possibilities for scientific knowledge. The theoretical hypotheses at stake state that human activity is as follows:

- Autonomous; that is, it consists of asymmetrical interactions between the agent and the agent's environment, in the sense that his or her interactions concern not the environment as an observer from the outside could apprehend it, but his or her proper domain, that is, what, in this environment, is relevant for the internal structure of this agent at the instant t;
- Cognitive; that is, it manifests and continually develops knowledge;

- Embodied; that is, it consists of a continuum between cognition, action, communication, and emotion, to temporarily keep common sense notions;
- Dynamically situated; that is, it always appeals to resources, individual as well as collectively shared to various degrees, which stem from constantly changing material, social, and cultural circumstances;
- Indissolubly individual and collective; this means that even individual events are interwoven with collective events;
- Cultured; that is, it is inseparable from a cultural situation that is either collectively shared or individual to various degrees;
- and finally, experienced; that is, more precisely, it causes experience for the agent at the instant *t*, however partial and fleeting it might be.

These general hypotheses are to be taken in their strictest sense. For instance, in contrast to various attempts made since the public *rise of situated action* (Suchman, 1987), the dynamically situated character of the activity cannot be reduced to the much earlier methodological idea of making scientific studies of human activity in nonexperimental situations (see, e.g., anthropological field studies and Ombredane & Faverge, 1955). It is more the idea that, first, the experimental situations are doomed to miss essential phenomena of human activity—at least if it develops unconnected from the scientific research in nonexperimental situations; second, the theories and methods for studying human activity in experimental situations should also consider its situated character, even if it is only to justify the reductions operated. For instance, recognizing the cognitive character of the activity does not just mean stating the trivial fact that man thinks; it is to affirm, contrary to various scientific approaches in the social sciences, that it is essential to have notions concerning knowledge, its manifestation, and development so as to describe, understand, and explain this activity.

These theoretical hypotheses manifest the enaction paradigm in the work analysis and more generally in the analysis of everyday human practices. They have important theoretical and epistemological consequences. They imply distinguishing two phenomenal or descriptive domains of the agent's activity: the *domain of structure,* tackled by neurosciences; and the *cognitive domain* (or domain of structural coupling), susceptible to a symbolic description. They respond to the following formula: domain of structure = that of the processes that lead to the cognitive domain, with feedback at any moment from the cognitive domain to the domain of structure. The social sciences, of which some psychological aspects are parts, can legitimately concern the cognitive domain uniquely. However, to ensure that their descriptions of structural coupling have a explicative value and not only a practical advantage, they must first take into account the autonomous character of the agent that we defined earlier, and second be considered within the neurophysiological knowledge of the moment, as summarizing the processes that constitute the domain of structure. This, to evoke Varela's formula, is what makes these descriptions *admissible*. From this stems an epistemological problem that would be insurmountable now—and in the likely future for a reasonable period in the context of neuroscience—if there were no other phenomenal domain, which is the object of the second idea, and linked to the latter characteristic of human activity stated herein, which is to be experienced, to give rise to an experience at every moment for the agent.

The course-of-action analysis does in fact add the consideration of this third phenomenal domain: the *domain of experience,* that is, that of the agent's course of experience, of the constructing process of this experience at any moment, and it takes an interest in the articulation between the cognitive domain and the latter. On one hand, the knowledge of this course of experience for the agent is interesting in itself. Here we join the current thinking on naturalization of phenomenology (Petitot, Varela, Pachoud, & Roy, 1999). It could also be said, though, that the appeal made in Newell and Simon (1972) for thinking aloud all along the problem-solving process does in fact inaugurate a systematic description of this domain of experience. But in our opinion, the authors consider erroneously that "the processes posited by the theory presumably exist in the central nervous system, are internal to the organism" (p. 9), when they concern the asymmetrical interactions between the organism and its environment. On the other hand, we make the hypothesis that the description of the course of experience, if it is correct, constitutes a description of the structural coupling that is partial but admissible.

From these considerations stem the following formulas: First, cognitive domain = that of the processes that lead to the domain of experience, hence enabling to contribute to the explanation of the latter, with feedback at any moment from the domain of experience to the cognitive domain.

Second, description of the domain of experience = the key, considering the actual limits of the neuroscience, to an admissible description of the structural coupling, by means of an epistemological principle, that of the primacy of the intrinsic description of the course of experience (domain of experience) on that of the structural coupling (cognitive domain) as a whole, or more briefly, "primacy of the intrinsic."

These different formulas define levels that concern the agent–environment system and not just the agent and that are foreign to any separation between "mind" and "body." From these formulas stems the theoretical object that we have called course of action, concerning the relationship between the domain of experience and the cognitive domain. We define this as follows: it is what, in the **observable activity** of an **agent in a defined state,** actively engaged in a **physically and socially defined environment** and belonging to a **defined culture,** is **prereflexive** or again **significant to this agent,** that is, what is **presentable, accountable,** and **commentable** by him or her **at any time during its happening to an observer or interlocutor in favorable conditions.** In this definition, the essential elements (observable activity, agent in a defined state, defined physical and social environment, etc.) have been presented in boldface. The course of action is the agent's course of experience (also said to be intrinsic organization of the course of the action) and the relationships it has with the relevant characteristics (said to be extrinsic) of his or her observable activity, of his or her state, of his or her situation (including other agents and partly shared by these other agents), and of his or her culture (partly shared with other agents); these characteristics are released by an interpretation of data about them according to the principle of the primacy of the intrinsic, which is presented more thoroughly in the section on analysis, modeling, and falsification.

Hence we have the following schema of the description of the course of action: Description of the course of experience + observational data of the activity, of the agent's state, situation, and culture → admissible description of the relations between the dynamics of the constraints in the agent's state, situation, and culture, that of the structural coupling as a whole and that of the effects on the agent's state, situation, and culture.

Such a description of the course of action can be documented in natural work situations or, more generally, in practical everyday life. It is explanatory and leads, as we shall see when the course-of-action-centered design is presented, to ergonomic recommendations concerning the design of situations, taking into consideration the characteristics of the agents' states (permanent and instant, physiological and psychological) and cultures.

For example, in a series of studies of car driving aimed at designing advanced assistance systems (Villame & Theureau, 2001), taking into account the construction of the driving action in the situation, and considering action and perception as inseparable in this construction, our approach gave priority to the study of drivers' activity in a natural driving situation as a basic condition for understanding the complex and dynamic character of the driving activity and its eminently contextual dimension. We assumed in fact that driving is largely created as a function of circumstances that are never possible to fully anticipate and that are constantly changing. In addition, driving is multisensory and the driver is also almost permanently interacting with other drivers.

To take into account all of these characteristics and the construction of driving in connection with a given situation, we felt that it was essential to put drivers in real driving situations and to consider their point of view on how they carried out the activity, in order to collect explanatory data on it. These studies were mostly field studies on the open road during which a combination of quantitative and qualitative data were collected in connection with these general characteristics of the driving activity. For example, we were very systematic in collecting data on the dynamic of the vehicle and of certain other vehicles with which the driver was interacting (speed, acceleration, use of the brake, deceleration modes, combinations of speeds used, etc.), on the behavior of the driver (maneuvers, positioning in traffic lanes, action carried out on the vehicle or particular equipment, etc.), and on the context encountered by the driver (traffic, infrastructure, maneuvers of other drivers, etc.). Second, we also collected data in connection with the characteristics specific to the particular dimension of the activity for which we wanted to provide assistance. It was thus possible to collect data on lateral veering or the immediate repositioning of the driver in his or her lane of traffic in the context of a study conducted for the design of a "lane keeping" type of system. Relative speed and relative distance data were collected more particularly in the context of studies on management of speeds and distances. Similarly, data on distance in relation to an obstacle or another vehicle were collected more specifically for studies looking at how maneuvers are carried out. In all cases, in the studies that we conducted, important emphasis was given to the point of view of the driver himself or herself on his or her activity, as an access to his or her involvement in the driving situation. This emphasis took the form of collecting verbal data while the activity was actually being carried out or in a self-confrontation situation (the driver watches a film of his or her journey, the latter being systematically recorded, and comments on it to clarify his or her actions and events). As is clear from the kind of data collected, the car driver's course of action was thus considered in all the dimensions that are present in the definition just described.

If, contrary to these car driver's course-of-action studies, we only consider the part of the agent's observable activity that is prereflexive, without taking any interest in other aspects of the observable activity, we will obtain a less developed description—but one that is still interesting where empirical knowledge is concerned, and often sufficient for the design— of the structural coupling of this agent and his or her situation. It is often on this description, which we could qualify as minimal, that the course-of-action-centered design recommendations have been founded until now. Such a minimal description can explain extremely detailed phenomena. For example, in a study of the course of action of employees in an insurance company carrying out a complex procedure for the processing of sickness files, it has been possible to show that each change in gaze direction toward the computer screen and the document (and of course toward the keyboard), and also each change in the rhythm of typing, could be precisely presented, accounted for, and commented on by these employees (Pinsky and Theureau, 1987).

It is worthwhile to note that it is regarding these relevant characteristics of observable activities, agent's state, situation, and culture that the interdisciplinary nature of ergonomics in which these studies have developed reveals its necessity. What the description of the course of experience gives is, on one hand, a partial but admissible diachronic and synchronic description of the structural coupling, and, on the other hand, an orientation toward the relevant characteristics of the observable agent's activity, state, situation, and culture. This is a lot, but not enough. New hypotheses have to be added, from the most general to the most particular, not discarding any possible contribution from other research in other scientific or technological disciplines.

THE OBSERVATORY OF COURSE OF ACTION AND ITS "RUDIMENTARY THEORY"

The data in Newell and Simon (1972) are simultaneous verbalizations, qualified as "thinking aloud." As a response to several critical articles, a fundamental idea was introduced: "we must extend our analyses of the tasks that our subjects are performing to incorporate the processes they are using to produce their verbal responses. . . . Such a theory must be developed and tested simultaneously with our theories of task performance" (Ericsson & Simon, 1980, p. 216). The "rudimentary theory of how subjects produce such verbal responses" that these authors put forward is based on a theory of memory storing, which today is very much called into question to the benefit of theories of reconstructing memory and the role played by interactional and social contexts in this reconstruction (see, e.g., Edelman, 1992; Rosenfield, 1988). If, today, we can take for granted most of the arguments Ericsson and Simon established to reject "the notion that verbal reports provide, perhaps, interesting but only informal information, to be verified by other data" set forward by different authors (Ericsson & Simon, 1984, p. 3), we ought to consider these verbalizations and the methods to obtain them in other terms than just memory storing and thinking aloud without consideration of interactional and social contexts. Furthermore, the monopoly of verbal data, induced by the idea of symbolic representation in the brain that is linked to the computer paradigm of cognition, has to be challenged. To summarize, we need a new *observatory,* meaning a relationship among a system of data-collecting methods, its rudimentary theory, and a system of principles of falsification of hypotheses by such data (Milner, 1989).

This is why the course-of-action analysis is based on an observatory, which is not just reduced to verbal data, and of which the rudimentary theory is evidently different. This rudimentary theory comes from cultural and cognitive anthropology (concerning the mastery of the interaction between analyst and agent), from clinical and experimental psychology, from neuropsychology (regarding recall and becoming aware), from the study of private thinking developed by Vermersch (1994; in particular concerning verbalizations in self-confrontation), and of course from methodological experience constituted in the tradition of the study of the course of action. This rudimentary theory is made with supplementary hypotheses that cannot be validated (or falsified) by data produced in this way. It allows us to specify the material conditions of situated recall (time, place, and material elements of the situation), the follow-up and the guiding of presentations, accounts, and commentaries by the agents, and the cultural, ethical, political, and contractual conditions that are favorable to observation, interlocution, and creation of a consensus between the agent and the observer or interlocutor. Because this observatory is thus more complex, its diverse aspects are also susceptible to evolving in an unequal manner with the progress of research, through studies other than those solely focused on the courses of action themselves. Of course, among these other pieces of research there can be research on the course of action of verbalization of courses of action.

An Articulated Set of Data-Collecting Methods

A methodology has been developed to collect data on the courses of action that interferes as little as possible—or at least in a well-ordered way—in the development of the course of the activity at hand and that establishes necessary favorable conditions for observation and interlocution. It connects in a precise way, depending on the characteristics of the activities and situations to be studied, continuous observations and recordings of the agents' behavior, the provoked verbalizations of these agents in activity (from the thinking aloud for the observer or interlocutor to the interruptive verbalizations at privileged moments judiciously chosen), and the agents' verbalizations in self-confrontation with recordings of their behavior. To these data can be added the agents' verbalizations in private-thinking interviews (Vermersch, 1994), in which the agents are put back into context by strictly appealing to a guidance of their sensory recall. These kinds of provoked verbalization aim directly or indirectly at making the prereflexive phenomena of activity appear. Other kinds of verbalization, made by agents during activity analysis (called second-degree self-confrontation verbalizations to stress the fact that they are situated in the continuity of self-confrontation proper) are also implemented. Here the agents are in the position of observers and analysts, and their verbalizations constitute, not data, but agents' contributions to the analysis of their activity.

In addition to these different types of data, there is the "objective" data (i.e., from the observer's point of view): static or dynamic data on the agents' states, and on the characteristics of diverse components of the situations (e.g., the prescribed tasks and organizational roles, the existing interfaces, the workspaces, and the operating devices, but also the organization of training, the management modes, etc.) and of the cultures (general culture, work culture, local culture, or family and personal cultures).

The implementation of these different methods in a particular work situation necessitates a mutual familiarization among agents and the observers or interlocutors, analogous on a number of points to the classical ethnographical inquiry, which constitutes the central point in the preliminary study. However, the latter also has to specify the aims and methods of the study and, more generally, a collaboration contract with the agents. Despite the richness of these data, the study of the courses of action and their collective interlinking should appeal, as in historical studies, to the *rétrodiction* (French term), that is, the filling in through inferences of the holes caused by the limits of the data (Veyne, 1971).

This observatory has borrowed and continues to borrow from other different approaches, but this borrowing is generally profoundly transformed in connection with the whole epistemological schema presented herein. For instance, the verbalization methods in terms of thinking aloud of Newell and Simon (1972) have been completely reviewed; the self-confrontation method borrowed from Von Cranach, Kalbermatten, Indermuhle, and Gugler (1982) has been completely reviewed in its implementation as well as in its aim; the methods of field cultural anthropology have been assigned the role of contributing to the realization and the interpretation of observational data and of simultaneous, interruptive, and self-confrontation verbalization data; and finally the observation methods of behavior, of the agents' state, of their situation, and of their culture contribute to the modeling, not directly, but by the intermediary of the modeling of the course of experience. The method of private-thinking interview, which, contrary to the previous methods, is linked to a theoretical construction fairly coherent with that of the course of action (Vermersch, 1994), was assigned a limited role complementary to self-confrontation. If the use of video in self-confrontation does favor the situated recall of the details of action, as well as perceptions and interpretations that have accompanied it at a given moment, in periods that can be long, it is unfavorable indeed to the expression of what has been constructed through sensory modalities other than vision and audition and the expression of emotions. The private-thinking interview lacks the advantage of the video prosthesis but goes beyond these limitations. It is, however, worth noticing that, in research on sport performance in table tennis at an international level, the self-confrontation verbalizations of these exceptional individuals make some of these other sensory modalities become visible in a significant way, in particular touch and proprioception, as well as feelings (i.e., emotions significant for the agent).

Duration and Articulation of Data-Collecting Periods

Systematic data collecting by the methods just described takes time, in addition to which should be considered the time for retranscription. In particular, it takes up expert time, as it is difficult for an expert in a course-of-action study to have these data collected and transcribed in totality by others less expert than himself or herself. Many methodological choices require a personal familiarization with the situation and should be done during data collection and transcription.

The reflection on the duration and the articulation of the data collecting periods can be concentrated on the observed and recorded data. On one hand, the data of provoked, simultaneous, or interruptive verbalization are collected at the same time as the observed and recorded data. On the other hand, the time for self-confrontation and extrinsic description data collection depends on the duration of the observed and recorded data collection. The overall duration for the observed and recorded data-collecting periods depends on the time each continuous period takes, the spread over time, the articulation of these periods, and the number of these periods or articulated series of periods. This duration of the data collecting (and hence the duration of their transcription) depends on several factors: (a) the particular characteristics of complexity and variety of the courses of action, the agents, and situations taken into consideration; (b) the particular temporal characteristics of the courses of action taken into consideration; (c) the theoretical or practical aims of the study; (d) the time constraints of the study, imposed, for example, by an outstanding design process; and (e) the competencies (general and in connection with particular characteristics of the courses of action taken into consideration) of the course-of-action study expert. These factors have to be considered in each particular study, design project, and design process. However, it is possible to specify the minimal articulated series of data-collecting periods to implement for different categories of tasks.

Let us first consider activities related to discrete tasks. Most of the studies aimed essentially at software design as a central element in a support situation. In this case the solution is simple: the data collection can be organized on the basis of the treatment of a completed inquiry questionnaire, sickness file, information retrieval request, and client's phone calls, within a larger context. The case of computer-assisted telephone encounters, studied within a project for improving situations that already existed, is essentially similar to the case of activities related to discrete tasks. Many questions concerning the software and the general support situation design can be approached by taking a client's phone call as a unit of data collection. However, some clients' calls and those calls initiated by the agent to the different services in the company are organized in stories that can last for several hours or more. If, from this point of view, we want to redesign the software, and the company internal communication media, larger units must also be considered (e.g., a working day).

In activities of traffic control, for example air traffic control, the significant unit of course of action, which is vital to know for the design of the interface and the training, is the follow-up of a configuration of airplanes in a control district, that is, of a series of airplanes maintaining different interdependent and conflictual relations (often reduced to a conflict between two airplanes, or just to one out of the entire current traffic). These follow-ups of a configuration of airplanes can last for 10 to 20 minutes. Therefore, the minimal period of observed and recorded data collection to be analyzed can last for half an hour or an hour, surrounded by two half-hours that enable one to know under which general conditions the traffic took place during the half or full hour considered.

In collective recovery from breakdowns or chains of incidents (e.g., in air, metropolitan, or railway traffic control), if we aim at the knowledge of the course of action of diagnosis and repair of breakdowns or chains of incidents in order to design a support situation for this diagnosis and repair, of course the data must be collected during the whole diagnosis and repair, alongside larger data that enable us to specify the larger context. As these breakdowns and incidents happen unexpectedly (except in the case of experiments in natural situations concerning breakdowns or events benign enough to be provoked), the observer is obliged to spend quite a lot of time waiting for the incident or breakdown, which requires some courage on his or her part, and a serious involvement on the part of the operators. In the control of an accident in a nuclear reactor, as in the case of air traffic control already mentioned, it is and has been possible—and even in the first case necessary—to study simulated situations on full scope simulators (Theureau, 2000a).

With activities with wide spatiotemporal horizons, the problem changes critically. For example, the actions and communications of a wine grower have very variable horizons, ranging from the realization of a discrete task to the cultural year (or several cultural years for certain research-development actions), passing by the seasonal contexts. If the aim is a general improvement of the mastery of technical novelties by the wine grower (e.g., by improvement of the documentation and the role of agricultural technicians), the periods of data collecting must join together these different horizons. In this study, therefore, data have been collected over the period of a year, including (a) course-of-action observations (combined with interruptive verbalizations), (b) the wine growers' filling in time estimates and commenting on them everyday by phone for the observer during several 3-week periods, and, more generally, (c) the use of different methods of cultural anthropology.

In addition, let us consider training: For us to know the transformations in the course of action during an on-the-job training process and thus dispose of a basis to improve the conditions of this job training, the data collecting should be done during periods spread over the whole training period, avoiding the provoked verbalization methods that could modify the course of the training. If, in contrast, we are interested only in what happens during one session of training, the unit is the session and self-confrontation interviews can be used (Leblanc et al., 2001). Finally, as in a study of how travelers find their train in a big railway station, it is possible to develop experiments in natural situations, where we concentrate the natural course of training—usually developed through separate journeys taking place over a large span of time—on several successive artificial journeys of the same traveler.

Generality and Moving From Analysis to Design

Nevertheless, if we stay with these guidelines concerning the duration and the minimal articulation of data-collecting periods that they enable us to define, we could not take the variety of agents and situations into consideration. To ensure a sufficient degree of analysis generality, we must collect more data, focusing on those agents and situations that are representative of the variety of the possible agents and situations. The minimum number of cases that can be considered is two, to eliminate the risk of inevitably taking a particular case as a general one, and to be able to formulate some hypotheses about the invariants and factors of variation. But, of course, this is not sufficient. We consequently have to add a complementary principle to the first guidelines—a general *principle of the generality of the analysis*. Its implementation depends on preliminary hypotheses about the variety of the agents and situations, and it depends much more than the implementation of the guidelines stated earlier on the particular purpose of the study and the temporal constraints involved. Finally, we add a general *principle for stopping the study,* considering the purpose of the study and the theoretical and methodological tools at our disposal, when the marginal gain of new empirical discoveries that are efficient in terms of knowledge or design made on the basis of new data tends toward zero. The study will certainly start again later on, but on the basis of the new situation designed.

THE COURSE OF EXPERIENCE AS AN ACTIVITY-SIGN

A course-of-action analysis, thanks to this enaction paradigm of cognition and this observatory, is led to focus on and solve several crucial description problems in Newell and Simon (1972): the separation between perception–action and the cognition description problem; the multimodal and nonsymbolic perception description problem; the separation between anticipation and the perception–action–cognition description problem; and the separation between emotion and anticipation–perception–action–cognition description problems. These four description problems are linked with a more embedded theoretical problem, which is the mind–body problem. The course-of-action analysis considers all the phenomena of interpretation, reasoning, perception, action, anticipation, and emotion as both physical and cognitive and describes them at a particular level that we will specify in the paragraphs that follow.

The description of the data protocols of course-of-experience collected uses—but also leads us to modify more or less significantly in the case of a failure—a generic model of human experience, labeled *semiological framework* or *activity-sign* (inspired by the philosopher, mathematician, and scientist, C. S. Peirce, who spoke of "thought-signs"). The central notion of the description of the course of experience that we have proposed is indeed a notion of sign, qualified as hexadic because it involves six essential components. This notion links, in precise structural relations, components that are supposed to summarize the concatenated processes at work in a unit of course of experience, that is, in a unit of activity that is significant for the agent considered, whatever its duration might be. It is in rupture, like the semiotic Peircean theory from which it was inspired given notable transformations, with the signifier–signified dyadic Saussurian conception of sign, which presides over both classical cognitive psychology and structuralist semiotics.

The fact that a notion of sign presides over the analysis of human activity, even limited to the course of experience, is not surprising if we recall some facts of the history of psychology and semiology: Peirce's notion of triadic sign was already of this type, in connection with semiosis, that is, the dynamics of signs as human activity; Saussure's notion of dyadic sign had already been interpreted with profit as concatenation of psychological processes (processes of production of the signifier–process of production of the signified or concept); Lev Vygotsky had sketched out an attempt of treatment of human activity in terms of signs that has remained famous, even though it was not very fruitful empirically. However, the construction of the notion of hexadic sign has not been limited to the contribution of these authors. It has been made though the conceptual and epistemological contributions coming from diverse and varied disciplines: theoretical biology, cultural and cognitive anthropology, ethnomethodology and conversational analysis, linguistic pragmatics, psychology, psychophenomenology, theoretical semiotics, and theoretical semantics, semiotics of texts, natural logic of argumentation, philosophical phenomenology, and so on. In addition, these diverse contributions have been integrated into a coherent whole and did not emerge unscathed from this integration. Hence a conceptual vocabulary exists that testifies to numerous loans, which are more or less faithful, and also to neologisms in order to (on the one hand) embody the sign and (on the other hand) not engender confusion.

The Hexadic Sign and its Components

The notion of hexadic sign emerged in 1997 when it replaced that of "tetradic sign," inaugurated in 1986 and which underwent several improvements over the course of empirical research and theoretical reflection. The last version of tetradic sign appears retrospectively as a simplification of hexadic sign that can be useful for the application. The notion of hexadic sign describes the process of construction of a unit of course of experience or *local construction* of the course of experience. This unit can be fairly large, provided that it is significant for the agent, but the more that it is elementary, the more its description in these terms is heuristically fruitful, at least if it is based on sufficient data. The hexadic sign links together six essential components— which also correspond to processes—and builds them, thanks to metamathematical notions of relationships that we take here as established (Theureau, 2000b): monadic relation, dyadic relation, triadic decomposable and nondecomposable relations, relation of thought, and real

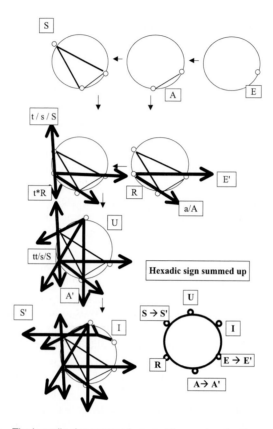

FIG. 4.1.   The hexadic sign summed up, and its construction in seven steps.

relation, which gives way, in contrast to the relation of thought, to the retroaction of the new element on preceding linked elements.

A methodological characteristic of the hexadic sign and its components is that they can be represented graphically, which allows for the construction of graphs of concatenation of hexadic signs. Figure 4.1 presents such a graphical representation of the rigorous construction of the hexadic sign in seven steps using these different kinds of relations. In this figure, open dots represent components; lines represent relations between components (thin lines for relations of thought and thick lines for real relations); arrows represent retroactions (specifications or transformations) of a new component upon its predecessors when real relations take place; the triadic real relation is represented by three convergent lines; and small arrows (horizontal and vertical) between circles represent the direction of the construction from the first step to the seventh one. The result of this construction, the hexadic sign summed up, is used in the graphs of concatenation of hexadic signs. Each sign starts from a state of preparation (*E*, *A*, and *S*) produced by the preceding sign and leads to the state of preparation (*E'*, *A'*, and *S'*) of the following sign, and so on. Such graphs correspond, given a radical change of paradigm and a corresponding development of the complexity, to the problem-solving graphs built in more simple terms of states of information and of information-processing operators by Newell and Simon (1972). The components of the hexadic sign are as follows.

1. *E*—Involvement in the situation: this is the principle of overall equilibration of the agent's interactions with his or her situation at a given moment, which is the overall closure of possibilities for the agent at this moment, coming from his or her past course of action.

2. A—Potential actuality: this is the varied expectations of the agent relative to his or her dynamic situation at a given moment, which, taking into account E, is expected (in a more or less determined way, passive or active) by the agent in his or her dynamic situation at a given moment, following his or her past course of action.

3. S—Referential: these are the types, the relationships between types, and principles of interpretation belonging to the culture of the agent that he or she can mobilize, taking in account E and A at a given moment (S is in a triadic decomposable relation of thought with E and A).

4. R—Representamen: this, at a given moment, is an effective sign for the agent (external, perceptive, or internal, proprioceptive, or mnemonic). R specifies and transforms A in a/A, that is, in a on a background of A and transforms E in E'.

5. $t*R$—R assimilated: this is an intermediary element (that we can count or not as a seventh element) constituted by the assimilation $t*R$ of R by a type t belonging to S. It is a real dyadic relationship between R and S; hence the specification of S in s/S and transformation of R into $t*R$.

6. U—Unit of course of experience: this is a fraction of prereflexive activity. It operates a transformation of A in A' and a specification of s/S into $tt/s/S$, that is, in a relationship between types tt on a background of s on a background of S. Between R assimilated ($t*R$), U and ($E' - a/A - s/S$), there is a triadic relationship decomposable into two real dyadic relationships in the sense that U, on the one hand, depends on $t*R$ and both develops it and absorbs it, and, on the other hand, depends on ($E' - a/A - s/S$) and transforms them.

7. I—Interpretant: this is construction, extension of the area or of the generality of the types and relationships between types through the production of U, and the completion of the transformation of E, A, and S into E', A', and S', which expresses the idea according to which human activity is always accompanied by some situated learning (or discovery).

These notions express the hypotheses about human activity outlined in the first section, but add a few others. Some of them stem from the definitions of the components just set. Some of them are relative to the relations between components, for example the hypothesis along which a unit of course of action changes expectations (and more precisely the agent's state of preparation, E, A, and S) and not representations; the hypothesis along which R, that is, what interests the agent at a given moment (and therefore is a sign for him or her) depends, not only on what is presented in the situation, but also on the agent's involvement (E), openings, and expectations (A) produced along the period of activity preceding that moment. A fundamental theoretical characteristic of these notions is also that they are built one on top of the other. For example, the notion U supposes all the notions preceding it in the list just given, as well as the transformations carried out along this construction. Finally, an epistemological characteristic of these notions is that they realize what is called a *literalization* (i.e., the generation of hypothetical empirical consequences from the manipulation of symbols to which we attribute a content; see Milner, 1989) of the course of experience. This characteristic of literalization of notions and hypotheses is reinforced by the dependence of the construction of the hexadic sign relative to a more general category construction.

The components of the hexadic sign emerge indeed respectively from six (seven) corresponding general categories. These general categories are inspired, given a renewal of their construction and their interpretation, from the six general categories proposed by C. S. Peirce in one of his last essays concerning categories ("A Guess at the Riddle"); they constitute the heart of what he called "phaneroscopy" (i.e., examination of the phenomenon). The interest for the analysis of the course of experience of these categories is due also to the fact that, with supplementary hypotheses, certain of which have already proved fruitful, the different components of the hexadic sign can be categorized in their turn according to the same construction. For example, a unit of course of action U can be a retaking, an introduction or transformation of an opening (or theme of action) of the agents in the situation, a multiplicity of feelings (significant emotions for the agent), an emerging interpretation, a determination (of an element object of attention), an inference, an action (which can be an ordinary action, an action on oneself, an action of search for information, or an action of communication), or an argument (or symbolic construction of new types or new relationships between types). Just as each component of the hexadic sign supposes and integrates in its construction components, which precede it in the list, each category of each of these components supposes and integrates implicitly in its construction the categories that precede it in the list.

An Example of Course-of-Action Analysis Into Hexadic Signs

As an illustration of this notion of sign, here is an example of determination of the components of the hexadic sign using data, which comes from the analysis of the course of experience of railway traffic controllers, borrowed from a recent study that was carried out by René Dufresne within a design process.

The situation is presented in which the controller of post 2 (CCF P2) asks the controller at post 5 (CCF P5) how to transfer train 147 between their respective control posts. CCF P5 informs his colleague that a foreman should occupy the northern line and, next, the southern line, but that he doesn't know when the northern line will be available. The two CCFs discuss the situation and put off the decision concerning the transfer of train 147. In addition, CCF P5 is put out by a signalman that he cannot get in contact with, and whom he needs to be able to direct train 310, which is coming into his territory.

	CCF P5:	*Yes Antoine?*
	CCF P2:	*XXX*
(*Unit 1*)	**CCF P5:**	*Yes. In a moment I'm gonna lose . . . well there, behind 21, they are going to get out the machines on the North at Saint-Hilaire.*
	CCF P2:	*OK.*
	CCF P5:	*They gonna travel up to Beloeil East with that, at the worst they gonna get down on the southern line, then they gonna change a long rail on the southern rail today.*
	CCF-P2:	*OK means we're losing err . . . X a bit XXX*
(*Unit 2*)	**CCF-P5:**	*You're going to lose. . . . well the north is going to come back just when we're going to lose the south. Means that we're going to . . . all use the northern line today. But when, I don't know!*
	CCF-P2:	*OK. Well, you'll tell me.*
(*Unit 3*)	**CCF-P5:**	*OK. For the moment, wait before aligning 147, for example.*
	CCF-P2:	*Yes, OK. No rush.*
(*Unit 4*)	**CCF-P5:**	*Err . . . well [laugh] no, there's no rush yet, but err . . .*
	CCF-P2:	*No, there's no rush yet, it's not there, then, err . . . even if your guys . . . you see, they travel . . . they go up to Beloeil, or at the worst, they can drag themselves back to Beloeil. We're going to wait for them there.*
(*Unit 5*)	**CCF-P5:**	*Err . . . it's because that would perhaps be the southern line . . . err, the line . . . err . . . yeh, well yeh, yeh, yeh.*

Let us consider unit 3 of the course of experience: "OK. For the moment, wait before aligning 147, for example." In contrast to the logical order according to which the components of the hexadic sign have been presented in the preceding section, here we follow the order according to which it is possible to document them, which starts with what is actual for the agent.

U corresponds to the verbal communication produced by CCF P5 when he replies to the question of his colleague and the perception of the good reception of this communication. We see that, in this study, we have chosen not to consider the smallest possible unit of course of experience. R includes both the question of his colleague in connection with the transfer of train 147 and the recall of the unpredictable character of the place where the foreman will be when the train arrives on his territory. It is far from being a simple stimulus. To document R, we have used his verbalization in a situation of self-confrontation. Indeed, the CCF explains that "our train was coming nearer. Then, the place where we could send our train was dependent hugely on what he was there to do; because I want him to work, that guy there."

If the determination of E is trivial here (it is enough to say that E belongs to the category "creative practical involvement," neither "search for more or less situated laws" nor "simple practical involvement"), that of A, and in particular the determination of the openings (which constitute the basis of the expectations of the controller) is essential for design. The openings documented are the approach of train 310 that must take a switching liaison; the signalman who does not reply to the radiotelephone calls of the CCF; and the occupation of the line by the foreman Després. Their documentation comes from the analysis of the course of action that precedes this particular moment, and from the verbalization of the CCF in a situation of self-confrontation. We know, because of the analysis of the past course of action, that train 310 is approaching his territory and that he has to make a liaison switching by manual commands. This is why the CCF is trying to contact the signalman. In addition, we know that the foreman Després has informed the CCF that he will occupy the line in the zone of the station of Beloeil. We also know that he will go first onto the northern line in order to fetch his equipment at Saint-Hilaire and that next he will go onto the southern line in order to change a rail. In addition, in a situation of self-confrontation, the CCF speaks to us (among other things) about the difficulty that he is experiencing in contacting the signalman; he admits that this makes him nervous: "See if he [*signalman*] makes me nervous due to that."

Let us now look at the expectations that emerge from these openings of the CCF. If we cannot pretend to have access to the set of active and passive expectations of the agent, we can at least distinguish the following: trains 147 and 310 will arrive very shortly on his territory; the foreman Després will go onto the northern line and then will work on the southern one, to give a circulation permit to train 310 if the signalman doesn't show up. These expectations are documented in two ways: by the verbalizations of the CCF and by a confrontation of these verbalizations with the expectations and the openings linked to the units of course of action preceding the unit considered. As an example, let us take the situation of self-confrontation where the CCF says, "I'll have to give his [*train 310*] permissions, and then warnings on the switchings so that he changes line." In this case, we know that the CCF is preparing himself to give an authorization to make train 310 carry on if he doesn't succeed in contacting the signalman. In a second stage, the expectations inferred with the aid of the verbalizations will be confronted with the ones set out earlier.

We did not attempt to describe S at each moment, though we sought to describe s, that is, the types and the relationships between types mobilized by the agent in relation to R. We can, at least roughly, deduce S from this, given a summation. When the CCF produces interaction unit 3, we may infer from the self-confrontation verbalization that the CCF knows that he must not direct a train onto a line to transfer it, before being certain that the line in question will well and truly be available: "I tell him to wait before aligning 147, because there, Després . . . was not transferred there/his equipment was coming out of here [between Bruno JCT and Douville]. . . . Then at some point, he needs to go 3 or 4 miles over there, to take the liaison, to go onto this line here. But all that, it's not done yet. But our train was coming close. So, where we were for sending our train depended hugely on what he was [the foreman Després] doing." In this way we documented the elements of s by inference, using the verbalization of the CCF and our previous knowledge of the activity. The types and the relationships between types named in this way were then validated in part by the CCFs.

I here is trivial. It is the simple reinforcement of the practical previous knowledge of the controller without any new contribution. Cases where I is not trivial are too complex to be presented here.

In relation to design, the direct value of this analysis in terms of hexadic signs is to specify the transversal aspects of the course of experience considered to be taken into account in the definition of interfaces, of spaces, of the communication tools, and of the training of the various agents. As one can easily imagine, such an analysis is fully developed only concerning those parts of the data collected. The rest of the data is analyzed only into significant structures, as we see in the next section.

THE COURSE OF EXPERIENCE AS A LATTICE OF SIGNIFICANT STRUCTURES RELATED TO THE ANTICIPATION STRUCTURE AT t

The notion of significant structure developed from research in semiotics and the grammar of texts, and from a difficulty identified in Newell and Simon (1972): a systematic description of the protocols, second by second, (local description) to which a notion of "episode of problem solving" was added (global description), presented by the authors as atheoretical, as purely methodological, but that nonetheless participated in a significant way in the description and the explanation of the data.

The semiological framework as a whole can be summarized in the following formula: concatenation of hexadic signs = a process leading to a set of significant structures, with retroaction at each moment between this set and the processes that lead to it. These significant structures express continuities of creation, transformation, and closure of openings o_i, that is, what constitutes the basis of the Potential actuality A (i.e., the anticipation structure at each instant). Reciprocally, the documentation of these significant structures informs us about A.

To specify this notion of opening, let us consider control activities in a full scope simulator of a nuclear reactor control room. A property of the course of action of each agent is its opening to a more or less indeterminate future. For example, an elementary action can be fully completed: thus, looking at the simplest example, the agent makes a phone call, gets hold of the right person, and gives his or her message: "You're wanted in the control room." In this case, once the agent has hung up, the operation has been carried out and completed. In contrast, if the agent makes a call and cannot speak to the right person, he or she leaves a message asking to be called back. In this case, when the agent hangs up, he or she creates an opening or, in other words, an action that has not been completed, which remains open to a future end. The same can apply in the first case too if there are other contingencies accompanying that of the arrival of the person called, such as briefing the person on the situation. In fact, this notion of opening is a very general one. Its relevance extends well beyond cases like this. As soon as a simulator test begins, an opening is created for each operator and is experienced by this operator: operation under normal circumstances that will be turned into an emergency operation, in one way or another. As soon as any operator gets involved in an emergency procedure, an opening is created—the situated follow-up and interpretation of the instructions, until they have been successfully accomplished or until the evolution of the process leads the operator to change procedure.

These openings (denoted by o), and therefore the significant structures, which express their continuity of creation, transformation, and closure, maintain diverse relationships between themselves. First, for a given hexadic sign, the Representamen R leads both to the selection of an o and the subordination to o of an o_R (the notation o_R/o specifies what we coin as a in

Fig. 4.1), concerning the extinction of the perturbation R, and U transforms A in A'. An aspect of this transformation of A is the transformation of o in o' and possibly o_R in o'_R if the opening o_R is not closed by U. Insofar as o is selected at instant t, is or is not identical to the opening o' resulting from the preceding sign at instant $t - 1$, there is temporal continuity or discontinuity.

Next, between the openings o_i and o_j selected by the Representamens of two different signs, there can be, from the point of view of the agent, three relationships:

- Dyadic diachronic or serial relationship—the openings o_i and o_j $(t_i > t_j)$ are, from the point of view of the agent, at the moment t_j considered, the same except for the determinations contributed by the course of experience between instants t_i and t_j;
- Dyadic synchronic subordination relationship (valid for a given interval of time)—o_i is subordinated to o_j if, from the point of view of the agent in this time interval, the extinction of o_i contributes to that of o_j;
- Triadic synchronic contextual relationship relative to a given opening (valid for a given time interval)—for the agent, the openings o_i and o_j are independent but both subordinated to an opening o_k. Finally, all the openings at a given moment are in a triadic synchronic relationship relative to the involvement in situation E.

The other kinds of relationships, which are not specified here, are refinements and specifications of these three sorts of relationships. All these relationships can crisscross and therefore do not necessarily produce trees, but eventually rhizomes, or trees that "rhizomatize," or rhizomes that branch. Let us remember that, by definition, in a maximal rhizome, all nodes can be in relationship with all the others. These relationships build different sorts of significant fundamental structures that cannot be specified here. The analysis in terms of significant structures of a particular course of experience, such as the analysis in hexadic signs, gives rise to a representation in the form of a graph. Such graphs express the sequentiality, the parallelism, and the hierarchical subordination of significant structures. By construction, the descriptions and the resulting graphs executed in terms of hexadic signs and in terms of significant structures are dual.

The value for design of the determination of these significant structures is that their comparison allows us to identify archetypal structures that provide corresponding scenarios for design. In this way, the analysis of railway traffic controllers' courses of experience, already quoted, enabled us to identify the following archetypal series: organize and coordinate the circulation of trains, locomotives, and maintenance vehicles; cooperate in the transfer of trains between territories; and respond to demands of occupation of the line. For each of these archetypal series, it also enabled us to identify different complementary or alternative archetypal sequences and archetypal macrosequences that may compose them.

COLLECTIVE INTERLINKING OF THE COURSES OF ACTION

Another theoretical and description problem concerns the relationship between individual and collective cognition. Indeed, human problem solving, as a result of its theoretical postulates, aims at studying the human system of information processing—hence a scientific approach that separates the individual from the individual's fellow men or women. Therefore it has been possible to qualify this scientific approach and its extension in the cognitive task analysis as *methodological individualism*.

As the work activity—and more generally the practice—has by nature a collective aspect, several solutions have been attempted. The first solution, the most prevalent, has been to consider that the collective activity constituted a superior level to that of individual activity, having the same structure. The second solution has been to consider that individual activity could only be described and explained by way of a description and explanation of collective activity. This is the interactionist approach inspired by ethnomethodology (e.g., Heath & Luff, 1998, to consider, among the numerous pieces of research of this sort, one of those that contributed significantly to our own studies concerning traffic control) and also the socially distributed cognition approach (Hutchins, 1994), that we could qualify as *methodological collectivism*. The course-of-action approach constitutes a middle ground between methodological individualism and methodological collectivism.

The characteristic of autonomy does indeed concern more than just the agent. It also concerns different more or less large collectives of agents with their interfaces. To consider this characteristic of the autonomy of a collective of agents with their interfaces, that is, to study it as such and also to draw from this study consequences for the design of collective distributed situations, we consider another theoretical object (partly inspired by the aforementioned approaches), the collective interlinking of courses of action, according to the following formula: intertwining structural individual couplings (identified according to the principle of the primacy of the intrinsic) = processes leading to the collective interlinking of the courses of action, that is, to the structural coupling between a collective with its interfaces and its material and social environment, with retroaction at any moment of this collective interlinking to the processes that lead to it. If the course of action is individual–social and enables us to consider the collective from the point of view of a given individual, the collective interlinking of the courses of actions is social–individual, and it enables us to consider the collective as such, though we cannot forget that it is the product of courses of action. Let us specify that a collective is not a given fact and that one and the same agent can participate in parallel, in diverse collectives that are more or less wide and persistent.

However, the enaction paradigm does not exclude direct study (i.e., without going through study of individual–social activity) of the collective construction of activity. A study of the collective construction of the activity can give rise to more parsimonious theoretical objects and observatories than the study of courses of action that sacrifice phenomena of the individual–social construction of the activity to acquire easier access to its collective construction.

Were we to leave things there, the interactionist studies and studies of distributed social cognition would simply appear to be more parsimonious and therefore faster but more limited than such studies of collective interlinking of courses of action, but which are sufficient in certain cases and for certain aspects of the activities. In fact, these interactionist studies and studies of distributed social cognition also consider relatively subtle phenomena of spoken and gestural interactions, which are not appreciated by the collective interlinking of the courses of action, at least when limited, as concerns the observable activity, to the part of it belonging to the courses of experience of the different agents.

In all, I believe that in the current scientific context of abandonment of the computer image of mind (or paradigm of man as an information processing system), the course-of-action analysis, the interactionist approach, and the distributed social cognition approach, in conjunction with other approaches I cannot list here, are building the various facets of what could be called cognitive anthropology or empirical praxeology. At the same time, they build the methodology for the corresponding task design. A part of the results and methods of the methodologically collectivist approaches can certainly be integrated in a fully developed course-of-action analysis. However, as such integration is not effective for the moment, the cooperation of a plurality of these approaches seems to be the right way forward.

It is not possible here to give sufficiently developed examples of studies of collective interlinking of courses of action that have been carried out. Let us just say that, in certain cases, collective activity can be considered as being that of a collective agent. This was the case in the study of activities of diagnosis and repair of computer software breakdowns in an insurance company because these activities involved agents whose competence was similar when faced with the same computer screen. Such a reduction was, on the contrary, not relevant concerning the metropolitan traffic control, which involved a dozen controllers and signalmen, with diverse competencies and roles and equipped with slightly differing computer interfaces. In such a setting, the analysis of collective interlinking of courses of action has essentially consisted of analyzing, in parallel, the individual–social courses of action of two controllers, and one controller and one of the signalmen who were associated with him, and to identify the relationships between the diverse fragments of these (Theureau & Filippi, 2000). Concerning accidental nuclear reactor control on a full scope simulator, we needed to proceed in stages, both because of the difficulty of the analysis and because of the complexity of the data to collect and analyze—and therefore also the difficulty to convince the management of the value of these. We went from a first analysis of fragments of the courses of action of reactor operators, to a first analysis of collective interlinking of the courses of action of the supervisor and the reactor operator carried out on the model of the study of metropolitan traffic control, and then to a second more developed analysis, concerning the same agents with the same data (Theureau, Filippi, Saliou, & Vermersch, 2001), and an ongoing study of a wider study concerning the supervisor, the reactor operator, and the water-steam operator. These various analyses also give way to graphs that link the elements of the situation and the agents' courses of experience, in terms principally of significant structures but also secondarily in terms of hexadic signs (Theureau, 2000c).

ANALYSIS, MODELING, AND FALSIFICATION

The systematic description of verbal protocols in abstract terms expressing hypothetical structural invariants is carried out by Newell and Simon in the form of a "problem behavior graph." Such a graph constitutes "a behavior representation of subjects solving a problem in the laboratory" that "will retain the full information about the dynamics of search, including repetitions" (Newell & Simon, 1972, p. 173).

The Primacy of the Intrinsic in the Analysis

A requirement of this kind of analysis is to start with a "description of the task-environment by the experimenter," enabling the latter, in adding diverse considerations stemming from experimental psychology, "to construct a hypothetical problem space that is objective only in the sense that all of the representations that human subjects in fact use for handling the problem can be imbedded as specialisation of this larger space" (Newell & Simon, 1972, p. 64). The authors themselves point out a difficulty in this approach: "We are, in fact, somewhat handicapped in studying the behaviour of masters and grandmasters in chess, since we cannot attain a better understanding of the task environment than such subjects" (p. 64). Moreover, from the point of view of course-of-action analysis, a work situation—or more generally a practical situation—poses the same problem: the operator has an advantage to the analyst who only passes through, even if it might be for several months: the advantage of time and practical know-how. Concerning work, such an analytical approach could only be reasonable if it might be considered a priori that the information obtained in a limited lapse of time with the engineers, managers, and operators includes all those that are used by the operator during the work period being studied. It is far from always being the case. From this claim stems the development of an alternative method of analysis that has its starting point in the operator's activity and not in an a priori "task structure."

   The principle of the primacy of the intrinsic is a principle of analysis. It makes no hypothesis whatsoever on a hierarchy of "causes." Rather, it sets up a dialectics between the description of the course of experience and the search for causes in the agent's state, his or her situation, and culture. It only concerns the analysis and not the entire methodology. In particular, it does not say that it is essential to know the course of experience before collecting the data concerning the agent's state, his or her situation, and culture. We have already seen that the ethnographical enquiry, which first concerns the culture shared by the agents, is the center of the preliminary study. To this might be added a preliminary knowledge of different characteristics of the agents and different technical characteristics of the situations, in particular the prescribed tasks. However, it is highly recommended, as a spiritual asceticism to persons used to an approach similar to that of Newell and Simon, to develop their knowledge on technical characteristics of situations, in particular the tasks, after having acquired further knowledge of the course of experience in itself.

Analytic Models

As mentioned earlier, the notions represented in Newell and Simon's problem behavior graph are those of "information processing operator" and "state of information." We must add here that the *information processing operators* are next analyzed in "productions," that is, in expressions of the form "condition → action," composing a "production system." The whole analysis can then be summarized in the search for a translation of the whole of the protocol in terms of these productions. This translation may fail: "The failure to find a production (or sequence) that fits a segment of protocol has various causes, ranging from lack of data to inability to construct an appropriate mechanism" (Newell & Simon, 1972, p. 203). It is therefore a case of evaluating these causes and, for the failures, which do not seem to stem from the lack of data, to look for "a modification of the production system" that is successful (p. 209). The epistemological problem of this approach is that the falsification by the empirical data is therefore susceptible to put into question only the concrete use of the theory and not the theory itself. The course-of-action analysis also systematically searches the falsification of its descriptive notions by data, but, in contrast, systematically uses this falsification in order to develop the theory.

This requirement for a scientific approach, added to the particularity of the theoretical objects studied (ontology), leads us to put a lot of care into specifying the epistemology of the study of courses of action and their collective interlinking, in the matter of the observatory as we saw in the description of the course of experience as activity-sign, and in the matter of the modeling as we are going to see in this section. In particular, let us make the following distinctions: distinction between the analytical empirical model (of the course of experience, of the course of action, or of the collective interlinking of courses of action) and the synthetic empirical model (of the course of action or the collective interlinking of courses of action), between the synthetic empirical model and the synthetic practical model, and between the synthetic model of the diagram type and the synthetic model of the simulation type. Modeling makes it possible, on the one hand, to benefit from the gains in connection to precision, fruitfulness, and validation or falsification of hypotheses that allow the literalization of the latter, and, on the other hand and complementarily, to contribute to technical transformation. It is taken here in its strictest sense, in which the model is inseparable from theoretical objects, empirical data, and theories. The notions of hexadic sign and significant structure lead to analytic empirical models of course of experience, but also to analytic empirical models of courses of action, depending on the relevant characteristics of observable activity and of the agent's state, situation, and culture.

Empirical Synthesis

According to Newell and Simon (1972), "The present theory is oriented strongly to content. This is dramatised in the peculiarity that the theory performs the tasks it explains" (p. 10). We saw earlier that the course-of-action analysis has the same ideal of "a theory (which) performs the tasks it explains." However, the requirement of computer simulation of the psychological process claimed by Newell and Simon has been abandoned, as well as the methodology and the model of analysis. This double abandonment is the outcome of the enaction paradigm: in order to be admissible, the description of the course-of-action protocols must respect the agent's autonomy and cannot be based on the task as seen by the observer; a computer simulation cannot express this autonomy. It is destined to serve practical interests—but, as we will see immediately, not only those—rather than scientific interests.

This double abandonment is costly from a scientific point of view. The great strength of Newell and Simon (1972), which enabled its fruitfulness for the cognitive task analysis, lay in the fact that its analytic model (states of information and information processing operators) corresponded to its synthetic model. In fact, the abandonment of the computer models is not absolute, but relative. These computer models, if they have no explanation value for the reasons given, can indeed have a predictive value and therefore contribute as such to the precision, to the fruitfulness, and to the validation or falsification of empirical hypotheses, the limited domain of application of which is defined through course-of-action analysis. The research concerning courses of action and their collective interlinking, though it has led to the design of various computer systems, has produced only a few of these kinds of computer models, because it has concentrated more on other questions. However, the emphasis put by other researchers (see, e.g., Pavard, 1994) on computer modeling of cooperative activities, so considered, without any illusion of scientific explanation, in letting it play at the same time a predictive role and thus a role in the validation or falsification of the hypotheses (and a role in the design of computer-controlled situations)—by showing the consequences for the activity of some set of design decisions—can be considered as complementary.

Parallel to this double abandonment, the research on the courses of action and their collective interlinking has looked for a new way to develop synthetic empirical models, in terms of "dynamic systems determined by their initial state," which is a return to the differential equations denigrated in Newell and Simon (1972) with a more powerful mathematical theory (concerning this way of cognitive modeling, see, e.g., Port & Van Gelder, 1995). However, the synthetic mathematical models, susceptible of being built in such a way, have to be considered more as "humility injectors" (as expressed in Cowan, Pines, & Meltzer, 1995) than as models allowing a full mastery of the phenomena.

The Ambiguity of Cognitive Task Design and the Object of Course-of-Action-Centered Design

The cognitive task analysis performed in Newell and Simon (1972) leads up to the design of artificial intelligence systems that are able to replace human intelligence in problem solving. The course-of-action-centered design, like most of the approaches of cognitive task design, aim at something more complex: the design of *joint cognitive systems,* which combine human operators and computer systems to replace the operators for certain tasks. This is what we could call the ambiguity of cognitive task design.

We already mentioned that the course-of-action analysis has the same ideal as Newell and Simon (1972) of a theory (that) performs the tasks it explains. With time, it thus opens up to systems susceptible of replacing the human operator for specific tasks or parts of tasks. However, the emphasis is put not on the replacement of the human operator for specific tasks, which today require less and less cognitive task analysis (see the design principles of the celebrated chess program "Deep Blue"), but on the design of situations for this human operator. Extending the distinction made by Woods between support and cognitive prosthesis (Woods & Roth, 1990), it considers the support situations as its objects of design; these are situations that, with given operators' states and cultures, give them the possibility of an activity that, at the same time, produces pleasurable involvement, satisfactory workload and security, adapted performance, and learning.

The Paradox of the Course-of-Action-Centered Design of Computerized and Automated Situations

As the aim is not to totally or partially replace the human operator by computer and automatic systems, but to design support situations, it can be considered that the course-of-action-centered design should not take any computer model into consideration, except, as seen in the previous section, with regard to their limited predictive and heuristic virtues. As a matter of fact, this is not the case. The course-of-action-centered design is indeed confronted with a paradox, called the *paradox of the course-of-action-centered design of computerized and automated situations*. On one hand, the designers of computerized and automated situations require design scenarios and models of the man–machine–environment system that are computerised, whether they construct them themselves or not. On the other hand, such computerized scenarios and models are unable to account for the autonomy of the agents in such a system, present or future, as already stated. The solution of this paradox is that the underlying regularities of the courses of action drawn from the analysis of the courses of experience, and the relation of these with the constraints and extrinsic effects together with the rest of the observable activity, can be translated more or less roughly into computer models whose validity is limited to determined phases of the courses of action. This translation is, to use the classical formula, a betrayal, but one that can be useful to design if it is used with care and with full knowledge of the facts. These computer models, in addition to their predictive value already underlined, play an important role in the design of support computer systems.

COMPLEXITY AND EMPIRICAL ITERATION ALONG THE DESIGN PROCESS

The Paradox of the Analysis for the Course-of-Action-Centered Design

The dynamical complexity of situated activity leads to a second paradox, the *paradox of the analysis for the course-of-action-centered design,* which concerns both the observatory and the analysis. This paradox is as follows. To elaborate design proposals for a future work situation based on the knowledge of the extrinsic constraints of the course of action, one should have no doubt about the course of action in this future situation. Within a design process, the course of action in this future situation will only be fully known when this future situation is completely designed and implemented. But then, the scope of the ergonomic contribution will be greatly reduced as the design will be totally completed. It will only be applicable for the next design process. This paradox is a result of the complexity, the variety, and the continuous transformation of the course of action and of its constraints: variety of the population of the users; great dispersion and complex organization of the relevant human characteristics; variety and complex organization of the relevant characteristics of the situation; and variety and complexity of the operators' experience. This is the reason why, if we state this paradox here in connection with the design process, we could also have stated it in connection with the empirical knowledge process in the preceding section.

The solution of this paradox is the iteration of the study of the course of action in situations approaching more and more the future situation inas much as they have been selected or constructed progressively during the design process. The closer these situations are to the future situation, the more the design proposals based on these studies of courses of action gain in validity and in precision, and the more their impact on the design becomes marginal. This enhances the importance of the first stages compared with the following ones. At each stage of the design process, the design proposals made at the previous stage find themselves more or less validated or invalidated correspondingly. It is not necessary to wait for the final product. At each stage of the design process, the contributions to the design are based on the analysis of the data obtained at this particular stage, but also on the analysis of the data obtained at previous stages. This iteration can introduce itself naturally into the design process, at least if two problems are solved (on the one hand, the problem of the construction of the analyzed situations, and, on the other hand, the problem of the supplying of these situations by the designers at each stage).

The Construction of the Situations Analyzed at Each Stage

Let us begin with the examination of the construction of the analyzed situations. They are first of all natural situations, among which it is necessary to distinguish referential situations and springboard situations. The referential situations enable the analysis of the course of action of given operators, having a given culture, in situations considered by the design project. The latter can be noncomputerized or contain an unsatisfactory computer support. They also allow the analysis of the course of action of assistance brought to an operator by other more competent operators in these situations. The springboard situations enable the analysis of the courses of action in other situations containing a globally or partially more satisfactory computer support than the one present in the considered situation. These natural situations are constructed, but not in the same sense as the ecological experimentation situations that are considered next: they are only chosen. It is a choice to consider a particular given situation, either referential or springboard. It is a choice whether to analyze a given course of action of a given agent in this situation. These choices are based, on one hand, on the knowledge obtained prior to the design project and process, and, on the other hand, on the results of the familiarization stage. These natural situations essentially enable us to define the functions of the technical system and of the essential characteristics of the environment, the population, the training, and the organization.

A further step forward in the construction of the analyzed situations is the ecological experiments in a natural situation, either referential or springboard. These ecological experiments in a natural situation are concentrated on certain aspects that are particularly important for design. They enable us to refine the results of the analysis in a natural situation and to improve their validation. With the exception of the refinement and the validation, the ecological experiments in a natural situation play the same role for the design as the natural situations. This is not the case for the ecological simulations and experiments on mock-ups or prototypes, partially or completely representing the future situation, taking place outside the natural situations. Because of the advanced stage of the design, it is then no longer possible to question certain of its aspects. These ecological experiments and simulations on mock-ups or prototypes enable us to define the specifications of the functions of the computer system and the essential characteristics of the environment, the population, the training, and the organization, identified earlier.

The same is not applicable to the situations based on prototypes in pilot sites. When such a prototype is put into a pilot site, the study of the course of action, whether in a natural situation or in an ecological experiment in a natural situation, does essentially enable us to correct certain superficial aspects of the prototypes. However, it can also enable us to refine the recommendations concerning implantation, workspace design, organization, and documentation and training of the future operators, identified earlier.

The study of the course of action during the implantation phase of a new device in natural situations enables us to validate or correct the recommendations concerning the organization, the documentation, and the training, identified earlier. Finally, the study of the course of action during the life span of the new established situation enables us to take the changes that have occurred in the situation into consideration, to formulate new recommendations with regard to organization, documentation, and training, to suggest superficial or local adaptations of the computer systems, and to prepare the design of new versions of these systems.

All these situations should include natural agents. They can be ranked according to two criteria: the distance to the natural situation (past or future) and the distance to the future situation.

The Supply of the Situations Analyzed at Each Stage

Of course, the complete development of such iteration implies that it should be possible to establish mock-ups and prototypes introducing a realistic interaction with the future operators. If this is not the case, the stages "mock-up," "prototype," and "prototype in pilot site" should be replaced by a desk study or reduced scale model involving the course-of-action-centered designers (based on the analyses of the courses-of-action) and various operators (based on their experience of the referential or springboard natural situations).

Though it is possible for a number of systems, in particular for all the office automation systems, to design mock-ups and prototypes allowing such a realistic interaction with the future operators, it is, at present, not a matter of course for most designers. The present trend is still to design mock-ups and prototypes that can be tested only from a technical point of view, which are not sufficiently developed to give a realistic interaction with the future operators. Such mock-ups and prototypes can at the very most lead to subjective reactions, proposals, and remarks from future operators. Only if the designers are convinced of the interest of the iteration of course-of-action analysis can the required complementary effort be integrated into the design process concerning mock-ups and prototypes sufficient for ergonomic experiments. This has been the case in numerous course-of-action studies in connection with the design process of office automation systems (Theureau & Jeffroy, 1994). With regard to the design of systems for nuclear process control, air traffic control, and aircraft piloting systems, it is possible to rank, in the same way, studies in natural situations, studies on full scope simulators, and studies on part task simulators (see Theureau, 2000a, for a synthesis and Theureau et al., 2001, for an example of course-of-action studies on full scope simulators).

A Refinement of Scenarios and Models for Design all Along the Design Process

The course-of-action analytical models are developed with the aim of a contribution to design, but their construction is strictly determined by considerations about their empirical evidence and, more generally, about their coherence with the existing scientific knowledge concerning the different aspects of human activity. Inversely, the scenarios and models for design concern future situations and integrate the design constraints. A model for design is a representation of the courses of action in the future situation likely to guide the design of this future situation. A scenario for design is the expression used rather for a partial model for design (see Carroll, 1991). These scenarios and models for design can be computerized or not. They are refined all along the design process, in relation with the different sorts of situations where courses of action are analyzed. Let us add that at each stage of input to design, a certain participation of the operators and the designers is required. With regard to the content, the methods, the cost, and the inputs of this participation, past experience does not enable us to extrapolate general rules. They depend on the situations to be transformed, on the given design project and process, and on their social context.

CONCLUSION

Let us consider now the design process as a whole. This design process can be considered as the interlinking of the individual–social courses of action of numerous agents. These different courses of action take into consideration the predefined procedures but do not always comply with them. These agents are, overall, the operators, the technical designers, the course-of-action analysts, and the course-of-action-centered designers. The extrinsic constraints are the actual agents' states, the (dynamic) situation of the design, and the culture, partly distributed, partly shared, of the different agents. Therefore, the problem of the course-of-action-centered design appears to contribute to a collective interlinking of the courses of action of the different design agents that has the following effects: a pleasurable and stimulating involvement in the design situation for these different agents; an efficiency of the design process from the viewpoint of the support to future production and maintenance operators; and a development of the knowledge of the different agents that will enable them to tackle in an even better way the future design processes.

From this stems a global ideal. At first, this ideal is that all the design agents take as one of the objects of design the support situation and only relate to the technical division of work as a means. Next, it is the situation of the different design agents that enables them to participate in an optimal way in the design. For example, this is the case if (a) the course-of-action analysts and course-of-action-centered designers have access to data in natural situations and to the necessary ecological means of experimentation and if the designers receive the recommendations at the right moment; and (b) the operators have access to the design information, can participate in the data collection and analysis, and dispose of time for this. It is also the case if (c) there exists a sufficient mutual training of the different agents with regard to their different languages, objects, methods, and common sense or scientific theories. Finally, it is the case if (d) the limits of competence of each agent are sufficiently determined as well as the corresponding cooperation and coordination means (meetings, means of communication). To these ideal characteristics of the design process, it is necessary to add prerequisite conditions, of which the principal is that the relevant scientific research in course-of-action analysis is being conducted and its communication and its part in the renewal of the training of the experts are ensured.

The design processes and their conditions rarely correspond to this global ideal. The individual–social course-of-action of the cognitive task designer working through the course-of-action-centered design approach should aim to approach this ideal. Independent of the degree of realization of this global ideal, he or she is therefore, at one and the same time, directed toward the overall design situation, directed toward his or her major object (articulation between design object and analysis object), and directed toward the other design agents. His or her horizons are manifold: to contribute to the ongoing design process; to develop locally or globally the course-of-action-centered design; to improve his or her own skills as a course-of-action analyst and course-of-action-centered designer; and to develop communication with the other design agents. Obviously, the full development of this individual–social action requires the active participation of the course-of-action analyst in the design, and not just the supplying of recommendations now and then. It goes without saying that success depends on the course-of-action analyst, but also on the other design agents and a great number of economical, social, cultural, and political factors. For the course-of-action analyst, as for any of us, the most difficult factor finally is to distinguish, as recommended a long time ago by the slave-philosopher Epictetus, between what is up to him and what is not.

REFERENCES

Carroll, J. M. (Ed.). (1991). *Designing interaction: Psychology at the human-computer interface.* Cambridge, England: Cambridge University Press.

Cowan, G. A., Pines, D., & Meltzer, D. (1995). *Complexity: Metaphors, models & reality,* Santa Fe Institute studies in the sciences of complexity. Reading, MA: Addison-Wesley.

Edelman, G. M. (1992). *Bright air, brillant fire: On the matter of mind.* New York: Basic Books.

Ericcson, K. A., & Simon, H. (1980). Verbal reports as data. *Psychological Review, 87*(3), 215–253.

Ericcson, K. A., & Simon, H. (1984). *Protocol analysis. Verbal reports as data.* Cambridge, MA: MIT Press.

Heath, C., & Luff, P. (1998). Convergent activities: Line control and passenger information on the London Underground. In Y. Engeström & D. Middleton (Eds.), *Cognition and communication at work* (pp. 96–129). Cambridge, England: Cambridge University Press.

Hutchins, E. (1994). *Cognition in the wild.* Boston: MIT Press.

Leblanc, S., Saury, J., Theureau, J., & Durand, M. (2001). An analysis of a user's exploration and learning of a multimedia instruction system. *Computers & Education, 36,* 59–82.

Milner, J. C. (1989). *Introduction à une science du langage.* Paris: Seuil.

Newell, A., & Simon, H. (1972). *Human problem solving.* Englewood Clifs, NJ: Prentice-Hall.

Norman, D. A., & Draper, W. D. (1986). *User centred design.* Hillsdale, NJ: Lawrence Erlbaum Associates.

Ombredane, A., & Faverge, J. M. (1955). L'analyse du travail. Paris: PUF.

Pavard, B. (Ed.). (1994). *Systèmes coopératifs: De la modélisation à la conception.* Toulouse: Octares.

Petitot, J., Varela, F. J., Pachoud, B., & Roy, J. M. (1999). *Naturalizing phenomenology.* Stanford, CA: Stanford University Press.

Piatelli-Palmarini, M. (1979). *Théories du langage, théories de l'apprentissage: Le débat entre Jean Piaget et Noam Chomsky.* Paris: Seuil.

Pinsky, L. (1992). *Concevoir pour l'action et la communication: Essais d'ergonomie cognitive.* Berne: Peter Lang.

Pinsky, L., & Theureau, J. (1987). Description of visual "action" in natural situations. In J. K. O' Regan & A. Levy-Schoen (Eds.), *Eye movements: From physiology to cognition* (selected/edited proceedings of the 3rd European Conference on Eye Movements, pp. 593–602). Amsterdam: Elsevier.

Port, R. F., & Van Gelder, T. (1995). *Mind as motion: Explorations in the dynamics of cognition.* Boston: MIT Press.

Rosenfield, I. (1988). *The invention of memory, a new view of the brain.* New York: Basic Books.

Saury, J., Durand, M., & Theureau, J. (1997). L'action d'un entraîneur expert en voile en situation de compétition: Étude de cas—contribution à une analyse ergonomique de l'entraînement. *Science et Motricité, 31,* 21–35.

Simon, H. A. (1977). *Models of discovery.* Dordrecht, The Netherlands: Reidel.

Suchman, L. (1987). *Plans and situated action.* Cambridge, England: Cambridge University Press.

Theureau, J. (1992). *Le cours d'action: Analyse sémio-logique: Essai d'une anthropologie cognitive située.* Berne: Peter Lang.

Theureau, J. (2000a). Nuclear reactor control room simulators: Human factors research & development. *Cognition, Technology & Work, 2,* 97–105.

Theureau, J. (2000b). Anthropologie cognitive & analyse des compétences. In J. M. Barbier, Y. Clot, F. Dubet, O. Galatonu, M. Legrand, J. Leplat, M. Maillebouis, J. L. Petit, L. Quéré, J. Theureau, L. Thévenot, P. Vermersch (Eds.), *L'analyse de la singularité de l'action* (pp. 171–211). Paris: PUF.

Theureau J. (2000c) L'analyse sémio-logique de couns d'action et de leur articulation collective en situation de travail. In A. Wall-Fossina & H. Benchekrown (Eds.), *Le travail collectif—Perspectives actuelle en ergonomie,* Toulouse: Octares, 97–118.

Theureau J., & Jeffroy, F. (Eds.). (1994). *Ergonomie des situations informatisées: La conception centrée sur le cours d'action des utilisateurs.* Toulouse: Octares.

Theureau, J., & Filippi, G. (2000). Analysing cooperative work in an urban traffic control room for the design of a coordination support system. In P. Luff, J. Hindmaroh, & C. Heath (Eds.), *Workplace studies* (pp. 68–91). Cambridge, England: Cambridge University Press.

Theureau, J., Filippi, G., Saliou, G., & Vermersch, P. (2001). Development of a methodology for analysing the dynamic collective organisation of the reactor operator's and supervisor's courses of experience while controling a nuclear reactor in accidental situations in full scope simulated control rooms. In R. Onken (Ed.), *CSAPC'01: 8th Conference on Cognitive Science Approaches to Process Control* (pp. 307–313). Bonn: Deutsche Gesellschaft für Luft-und Raumfahrt-Lilienthal-Oberth.

Varela, F. J. (1989). *Connaître. Les sciences cognitives, tendancy et perspectives.* Paris: Seuil.

Vermersch, P. (1994). *L'entretien d'explicitation*. Paris: ESF.

Veyne, P. (1971). *Comment on écrit l'histoire*. Paris: Seuil.

Villame, T., & Theureau, J. (2001). Contribution of a "comprehensive analysis" of human cognitive activity to the advanced driving assistance devices design. In R. Onken (Ed.), *CSAPC '01: 8th Conference on Cognitive Science Approaches to Process Control* (pp. 93–102). Bonn: Deutsche Gesellshaft für Luft-und Raumfahrt-Lilienthal-Oberth.

Von Cranach, M., Kalbermatten, U., Indermuhle, K., & Gugler, B. (1982). *Goal directed action*. London: Academic Press.

Woods, D. D., & Roth, E. (1990). Models and theories of human computer interaction. In M. Helander (Ed.), *Handbook of human computer interaction* (pp. 3–43). Amsterdam: Elsevier.

5

Cognitive Work Investigation and Design in Practice: The Influence of Social Context and Social Work Artefacts

John R. Wilson, Sarah Jackson, and Sarah Nichols
University of Nottingham, United Kingdom

Abstract

This chapter is concerned with cognitive work investigations, which are necessary as a foundation for cognitive task design. It draws on two programs of research, one concerning the work of planners and schedulers in various manufacturing industries and one concerning the work of signallers and controllers in the railway network. In particular, the chapter stresses the fundamental influence of context on the performance and subsequent understanding of cognitive work, and it stresses the need for focus on social interactions and social work artefacts, not just as part of this context but also as vital work elements in their own right. As a consequence, the case is made for a high-quality field study approach, and the need to develop new methods of measurement and assessment, for use when investigating and redesigning complex interacting and distributed sociotechnical systems.

INTRODUCTION

We have two central and interconnected arguments in this chapter. The first is that social aspects of performance are as important as the cognitive aspects in studying and understanding what is known as cognitive work and subsequently in its design. For instance, in designing the work of skilled control room operators, cognitive ergonomists have long sought to understand how best to present information on a set of integrated displays, to identify the mental model that the operators may form of the underlying state of the system, or establish the way they prioritise factors in making decisions. However, it is just as important for cognitive work design that we understand about the relationships, collaborations, and communications of the operators with each other, their supervisors, and with remote agents such as maintenance engineers, train drivers, meter readers, or site workers. For example, we may need to study their knowledge of who to telephone from their informal network about working around a problem, and how their decision making is interrupted by telephone calls or visitors, and how they deal with these.

This brings us to our second argument: performance of individuals within such social networks, use of social work artefacts and the contextually influenced decision making that follows can best be investigated, and lessons drawn for cognitive work design, in field study.

The approach taken and questions posed in this chapter lie somewhere between the broad–philosophical and the specific–analytical. We are concerned with the practical efforts of observing, interpreting, understanding, and improving people's cognitive performance at work, and in particular their interactions with the physical and social artefacts they use and with the other people in their (physical and virtual) environments. Our perspective is similar to that of Hollnagel and Cacciabue (1999) in their introduction to the first issue of *Cognition, Technology and Work*. They identify increasing support for studies of cognition that account for the situation of the people being studied, and for a proper balance between theory generation on the one hand and empirical data and field studies on the other, the latter reflecting work carried out in large sociotechnical systems. We go further and suggest that even much of what has been published to date apparently as cognitive work studies in the field are in fact carried out in highly constrained and controlled work environments, such as nuclear power plants, which have many of the attributes of a laboratory simulation.

COGNITIVE ERGONOMICS AND COGNITIVE WORK DESIGN

A number of difficult questions arise whenever the topic of cognitive ergonomics is discussed in any depth. Do we even need a *cognitive* ergonomics? Should we qualify ergonomics with such an adjective? It is difficult to conceive of an ergonomics that is of any value or interest that does not address the cognitive, just as it is equally hard to accept a purely physical or a purely social ergonomics—see Wilson (2000)—and this doubt is addressed in the discussion of cognitive task analysis that follows. If we do need a cognitive ergonomics, then how should it be distinguished from cognitive science, cognitive engineering, cognitive psychology, and engineering psychology? Cognitive ergonomics also presents us with difficult questions of detail. Can we ever truly represent elicited knowledge? How do concepts such as scripts, frames, schemata, and mental models all interrelate? Can we ever make real sense of that old ergonomics standby, the mental model? Are we any closer to that elusive quarry, mental workload? How can we account for social context in making recommendations for cognitive work design?

What much of this suggests is the need for greater care in using the label *cognitive*, and not to apply it to any entity such as task or tool in the hope of marking it out as new, rather than a perceived traditional, ergonomics. We now have industrial and software engineers, for instance, talking glibly about "cognitive X," without any apparent understanding of what they are really describing. (Similar points have been made in his introduction by the handbook editor in chapter 1).

Cognitivists first used the label to distinguish themselves from behaviourists, arguing that we *can* learn about and understand the mental processes that take place between inputs and outputs to and from the human being (but for a contrary view, see Skinner, 1985). Neuroscientists believe in making direct measurements of the brain's activity to understand performance, and experimental psychologists believe in gradually piecing together a picture of how we think in practice by means of the falsification associated with careful and clever experimental, but often artificial, procedures. By working in both laboratory and field, cognitive psychology and cognitive ergonomics address three key problems—understanding and explaining human thinking and knowledge, finding the relationship between this knowledge and human performance, and supporting the design of tasks, interfaces, and training required for reliable, safe, and effective performance.

There is no one widely accepted definition of cognitive ergonomics. Indeed, in the attempt by Wickens, Gordon, and Liu (1998) to illustrate the relationship between the different components of ergonomics and human factors, cognitive ergonomics is not mentioned as a distinct subject area. Cognitive ergonomics as a description of activity has gradually been used by those wishing to distinguish that ergonomics which is to do with people processing information and making decisions from that ergonomics which is to do with people sensing, acting, and fitting within the workplace (or knobs and dials ergonomics; see Shackel, 1996; Sheridan, 1985).

When we consider the idea of cognitive task design, we face similar questions. We could assume that by "cognitive task design" we mean design of tasks that are purely "cognitive." However, no work involves only thinking, and a more enlightened view of cognition is of embodied action, in which physical and cognitive aspects of work are considered together (Clark, 1997). Alternatively, cognitive task design could mean that we are designing tasks—people's work and the artefacts they employ—by taking a cognitive perspective, regardless of the extent of cognitive elements in the task. This would include using or developing cognitive models and cognitive task analyses (CTAs).

For those of us brought up on Singleton's (1974) explanation of total activity analysis, with its emphasis on decisions as well as actions, and who believe all work has cognitive and physical elements, albeit in different proportions, the need for a task analysis that is nominated as cognitive has always been a mystery. The idea of thought without action or action without thought is incomprehensible. Activity that may appear at a cursory glance to be purely mental, for example, monitoring control room displays for rare deviations, in fact contains substantial physical components such as maintenance of a seated posture, which may well be very influential in operator comfort, capacity, attitudes, and performance. (Only in jobs in which pure decision making is the major component of someone's job—investment banking perhaps—will there be few or no physical elements that are relevant to understanding performance.) Even more so, there are no jobs that are purely physical; walking needs perception about where to step, albeit processing information at a subconscious level; shovelling requires decisions about how much to load the spade. When we look at someone's work over a whole shift, rather than isolate a few tasks or functions, it is even more apparent that any analysis of work must embrace cognitive and physical components as an integrated whole.

There have been a large number of very different efforts in producing and using CTA tools, maybe because of the rapid increase in the amount of research and researchers in human–computer interaction (HCI), not all of whom have done their basic reading in ergonomics. In an effort to distinguish their contribution from that of software engineers, systems analysts, and programmers, who paid little if any attention to user requirements, a CTA was born. This was to help understand people's interactions with a system through a software interface, how people learned, and then how they used the system. Whether anything was really required, beyond the tools available from task analysis generally, is doubtful.

Is CTA purely the analysis of cognitive tasks, or a specially and peculiarly cognitive form of analysis of tasks? Somewhat belatedly, attempts are being made to sort out just what is meant by CTA. Hoffman and Woods (2000) regard CTA as a label for studies of cognition and collaboration in the workplace, and in fact extend it into CTA–CFR (cognitive task analysis–cognitive field research).

A realistic view of cognitive task design is that it seeks to understand and improve how people perform at work, and to understand and improve their well-being at the same time; it does this by studying how people think, how they register, comprehend, sort, and interpret information, and how they apply memory, allocate mental resources, make decisions, and learn from feedback, in relation to their physical and social settings, conditions, and influences. This latter set of contextual conditions forms one of the main themes of this chapter, which is the study of cognitive work and cognitive tasks as they happen, in actual practice. Such practical application of cognitive ergonomics should include social and organisational influences. Key social considerations in the workplace for ergonomics are the way in which we come to understand other people, and the way that others, and our relationships with them, may affect how we act—in other words, the impact of interindividual and intraindividual processes and interactions on workplace performance.

EMPIRICAL BACKGROUND

The issues and ideas raised in this chapter have emerged from a range of our cognitive ergonomics endeavours, but we draw largely on two particular sets of studies. In the first we carried out an extensive investigation of planners and schedulers in industry. How do they handle information and make decisions? How do they fit in with—and indeed catalyse—social networks in the workplace? How is their performance assessed? What makes a good scheduler and schedule?

The motivation for this research stemmed from the gap between the theory and reality of production scheduling. Scheduling is a fundamental process that is undertaken across businesses and the service sector. The general aim of any scheduling activity is the goal-directed allocation of resources over time to perform a collection of tasks. There are numerous forms of scheduling, such as personnel rotas and transport timetables; our programme focused on the domain of production scheduling. This had been intensively studied since the 1950s, mainly using "classical" scheduling theory and production control frameworks. This approach utilises mathematical modelling frameworks to address the allocation of machines to perform a set of jobs within a defined period of time; that is, it is a static and deterministic approach. However, against this established theoretical background, the manufacturing industry has been less than successful in the development and implementation of planning and scheduling systems within businesses; more significantly, it has not been successful in truly understanding the role of the human scheduler within these systems.

The human contribution to planning and scheduling is of major significance, yet in order to respond to the increasing complexity and uncertainty of the manufacturing environment, businesses have tended to concentrate on technical solutions such as decision support systems and computer-aided production and planning systems. One of the difficulties with attempts to address scheduling problems by using computer-generated solutions is that scheduling is viewed as a static mathematical problem within simple and well-defined scenarios. In contrast, the business reality is of people carrying out manufacturing scheduling as a dynamic process. This gap between the framework used by scheduling and production control theorists and actual industrial scheduling practices was the motivation for the detailed study of planners and schedulers in aerospace, steel, textiles, and electronics work, in order to develop more effective scheduling support systems.

Background to this research can be found in MacCarthy, Wilson, and Crawford (2001); a framework was presented that focuses on understanding the scheduling environment, the process of scheduling, and related performance issues. The book by MacCarthy and Wilson (2001) provided a human-centered perspective on how people in businesses manage planning and scheduling processes. Three chapters are of particular interest: Crawford and Wiers (2001) provided a review of the existing knowledge on the human factors of planning and scheduling; and Crawford (2001) and Vernon (2001) presented two detailed field studies of planners, schedulers, and their industrial practices. The theories and models developed from the authors' research can be found in Jackson, Wilson, and MacCarthy (2002), and the methodology that supported this development was presented in Crawford, MacCarthy, Vernon, and Wilson (1999).

The second programme of research comprises investigations into the work of the different functions and roles involved in railway network control. Planned changes in the technical and organisational systems used in railway signalling and zone control in the UK (and across much of Europe), as well as increased concerns for network safety following recent accidents, have prompted a full review of relevant human factors. Until recently, railway ergonomics research was mainly concerned with train drivers and driving, chiefly their ability to perceive and react to signals and the causes of SPADs (signals passed at danger). The work of train drivers is akin to that in other domains in which the tasks and environment are not greatly different from those in a simulator, and in which social influences and artefacts are limited. More recently, rail ergonomics attention has turned to the more contextually rich and open work systems of planning, signalling, and maintenance. Of relevance to this chapter are assessments we have made of situation awareness, workload and crew size, and investigations into decision-making authority and communication links, opportunities for human error and organisational failure, and safety critical communications.

For further information, Wilson et al. (2001) presented an overview of the Railway Ergonomics Control Assessment Package (RECAP) that contains a number of tools and techniques to assess aspects of railway control work in the field, including an audit instrument covering attitudes, stress, job design, work environment, interfaces, and safety culture. There are also stand-alone measures of workload and situation awareness and qualitative studies of safety culture in maintenance and of train driver route knowledge. All the work has the consistent theme of being carried out in the field, and these studies show the difficulties of maintaining the same rigour as would be achieved in an experimental context, while also obtaining data in a relatively short period of time from real workplaces. Cordiner, Nichols, and Wilson (2002) described the process undertaken in identifying the key factors that should be evaluated within RECAP; this process consisted of a number of stages, including a large amount of time spent reviewing the workplace before any actual methods were applied. Nichols, Bristol, and Wilson (2001) described the measures applied to measure workload in railway signalling, Pickup, Nichols, Wilson, and Clarke (2003) report the theory lying behind field study of workload, and Murphy (2001) described the analysis of communications during track possessions.

APPROACHES TO STUDYING COGNITIVE WORK

Laboratory studies in cognitive ergonomics have generally been far more prevalent than the field study of real human performance in complex interacting systems (or the performance of joint cognitive systems—Hollnagel, 2003). This is certainly the case to a great extent in North America (because of the sources of research funding and academic tenure systems) and to a degree in Europe, although the Francophone ergonomics tradition stands out against this, as reading *Le Travail Humain* would show (and see de Montmollin & Bainbridge, 1985). There is a need for, but general lack of, a cognitive ergonomics that can understand jobs where coping with social networks is as important as coping with computer-generated displays and computer-controlled processes. As work increasingly becomes carried out in complex interacting systems in which distributed teams of people and computers are the cooperating actors, the social interfaces become as important as the technical, but are much more difficult to replicate in the laboratory.

Cognitive ergonomics to date has been distinguished largely by searches for user models (and models of users), the use of task analytic methods on subsets of activity, artificial and molecular (e.g., studying the effectiveness of two layers of menus versus four) laboratory experiments, and laboratory simulations. There are few naturalistic descriptions that report comprehensively on real work carried out by real people in real—complex, confusing, noisy, and dynamic—environments. Where field studies are reported in cognitive ergonomics, they are often in domains where work is highly constrained and lacks the random and noisy influences that have an impact on much of real work; such domains—the military and nuclear power plants are good examples—are almost as clean as simulations.

To illustrate this concentration of cognitive ergonomics on the "artificial," consider three relevant special issues—*Ergonomics,* 32(11), 1989, *Ergonomics,* 36(11), 1993, and *Human Factors,* 42(1), 2000. These are taken as examples because they are explicitly collections of work on cognitive ergonomics in major ergonomics journals. In the earlier *Ergonomics* (1989) there were no real studies of mental work in a field setting; the papers were mainly concerned with theory or models (three), methods (four), simulations (one), laboratory studies (one), design descriptions or surveys (four), and reviews (1). In the later *Ergonomics* (1993) there was a greater proportion of empirical studies, but these were almost all laboratory simulations and trials (six—albeit some were very close to reality, in nuclear power plants for instance). Other papers were on models or theory (six) and methods (two), and only one contained some field observation reports, but only as a small part of a contribution that was set mainly in the laboratory.

Despite a brave introduction by the editors (Hoffman & Woods, 2000), the more recent special issue of *Human Factors* showed a similar distribution. As an exception to this trend, we are in sympathy with the aims of Mumaw, Roth, Vicente, and Burns (2000)—to uncover strategies that people use to cope with demands at work rather than to specify detailed mental representations of mental processes; we investigated the latter previously, and we are not convinced we had much success (Rutherford & Wilson, 1992; Wilson & Rutherford, 1989)! Mumaw et al. denoted the study of real work strategies as a distributed cognition tradition (see Hutchins, 1995; Engeström & Middleton, 1996) as against an analytical cognitivist one. However, this approach would have more persuasive impact if there were more explicit, transparent descriptions in the literature from the ethnography of distributed cognition. Exactly how do the rich accounts presented lead to the interpretations, and then how do these interpretations lead to design changes or other improvements?

Of the other contributions in the *Human Factors* special issue, Prietala, Feltovich, and Marchak (2000) reported a simulation, Schaafstal, Schraagen, and van Berlo (2000) presented an "actual system available for training," that is, not really in actual use, and Xiao and Vicente (2000) provided a framework, although they did draw in part on some fieldwork. Ehret, Gray, and Kirschenbaum (2000) reported a major military simulation (called a "scaled world") and Fowlkes, Salas, and Baker (2000) discussed methods—including event-based assessments (similar to scenario analysis).

In recent years the cognitive ergonomics field has seen the launch of journals specifically devoted to the area. Is the balance of effort between field and laboratory work any different in these new publications? Take the first six issues of the journal *Cognition, Technology and Work* (during 1999–2000). Despite the aims and focus of this journal, explicitly promoting field study, of 31 papers, 7 were reviews, 5 reported laboratory work or simulations, 10 presented models, theories, or frameworks for study, 5 reported methods, and only 4 were field studies, and some of these were somewhat constrained. This reflects the discipline, not the journal; editors can only publish those (quality) papers that are submitted.

Not all published work in cognitive ergonomics lacks field study and investigation of social as well as cognitive behaviour. As an illustration of some of the better work, two studies are mentioned here. Mackay (1999) reported two field studies of air traffic control in general, and the use of paper flight strips in particular. There is an ethnographic study of one team of air traffic controllers and then a comparison study of shorter periods spent in eight other control centres. Among the methods used in the 4-month ethnographic study were familiarisation (by means of observation and conversation), systematic videotaping, frequency counts of activities, records of situation system states at sample times, and collection of archival and physical evidence. From this study the researchers have been able to greatly inform the debate over the use of paper strips versus their potential electronic replacements, particularly by highlighting the real benefits of paper strips, which might be difficult to replicate in new computer systems. This was possible through careful investigation of the real work in context, including the cultural and social impacts on the controllers' work, and also by studying the social work factors (such as handovers and adjacent colleague support) as much as cognitive factors.

Artman and Waern (1999) studied emergency coordination, within an emergency coordination centre. They explicitly used the perspective of a sociocultural system, in which operators' actions are grounded in social practice and expectations and which "involves people as well as cultural artefacts, designed and evolved to accomplish a certain goal" (p. 238). In their field study they videorecorded work in the role of participant observers. Subsequent analysis was of various phases or types of interaction between callers and operators, involving assessment of the situation, identification of the caller, negotiation of the meaning of the call, and the coordination and distribution of tasks involving calls among different operators. From this work the researchers were able to identify central phenomena of the work in coordination centres from which to make proposals for future work and workplace design. They were also able to typify the work of the operators in terms of a number of different role and task descriptions.

We must stress that these two are not the only quality field studies in cognitive ergonomics; they are, however, examples of the few well-founded and justified pieces of work that are reported in scientific literature.

COGNITIVE WORK INVESTIGATION AND DESIGN IN PRACTICE

While carrying out our studies of cognitive work in practice, we have been struck by a number of recurring themes. When we reflect on our own research and practice, we continually ask if we are carrying out this type of research to best effect, whether we are improving knowledge and insight, and whether this knowledge is achieving better design, implementation, and maintenance of the next generation of technical and social systems.

We discuss our concerns in the remainder of this chapter, using illustrations from the scheduling and rail network control studies, and summarised under the following headings:

- Context—its influence on key aspects of cognitive work;
- Approach—the need for high-quality field study;
- Measurement and assessment—developing new methods and providing traceability of evidence; and
- Focus—on social factors and social work artefacts.

These themes are, of course, closely interdependent. The aim of discussing cognitive ergonomics research and cognitive task design by using these headings is to progress toward appropriate research approaches, and to highlight the many factors that may have been neglected by standard (laboratory-based) approaches to cognitive work analysis and cognitive system design. In this way, we hope to promote a move toward a socially situated and context-sensitive approach to cognitive task design.

Influence of Context

Within the sporadic debates about the nature of ergonomics and human factors over the past decade, there has been much discussion of the merits of field and laboratory studies, contrasting formal and informal methodological approaches and setting the advantages of control against those of veridicity (see Wilson, 2000). We work from the perspective argued by Moray (1994)—that ergonomics only makes sense in the full richness of the social settings in which people work (see also Nardi, 1993). The influence of contextual factors—such as interactions between individuals, the formation and relationships of teams, individual motivations, computer-based support systems, and organisational structures—on cognitive work in practice must be understood before we can decide on techniques for the measurement and prediction of the outcomes of people's work, or on recommendations for the design of *joint cognitive systems,* the complex networks of people and computers that do the thinking in modern organisations.

The strong influence of context was seen in our study of industrial schedulers in a number of ways. The dynamic environment has such influence on expected and actual scheduling performance and behaviour that context-rich field study is necessary before "sensible laboratory and quasi experimental research . . . [can be] carried out" (Crawford et al., 1999, p. 67). Strong contextual influences include the sheer spread of the scheduling activity over a number of physical settings; the variety of other actors with whom the schedulers had to interact, for interrogation and information provision; and the large number and variety of interruptions to their work, which are both specifically task related and more general. At an organisational level, contextual factors included training, job expertise, informal information networks, availability of up-to-date knowledge of the business situation, measurement and reward systems, and the individual's span of responsibility and authority. The general business environment was also enormously influential, for example if the business was single site or multi site, and whether there were good relationships across the business supply chain. All of these factors were shown to influence how people worked in practice.

In studies of railway control centres, and particularly the work of signallers, the influence of context was perhaps slightly less marked. The degree of influence is also to some extent a function of the different generations of signalling technology used. In newer, more automated systems, jobs are certainly somewhat less influenced by social context compared with the many lever boxes and NX (entry–exit) control panel systems still used widely; this is probably a reflection of the increased use of automated systems to support the signalling activity in newer control centres, particularly automatic route setting. However, even in the more modern systems there is significant contextual influence. This comes from the staff's previous experience, local geographic knowledge, attitudes, and perceptions—linked to the current structure of the business, its management, and public and media disquiet at safety and performance. Context is also highly influential because of the multiplicity of activities carried out in network signalling and control, such as delay attribution and management information system data collection. These activities are not primarily associated with main signaller tasks and use of interfaces, but any attempt to study such core tasks and interfaces without regard to the contextual factors will not really capture the true nature of the work. A comparison of signallers working with the newer automated systems with those using the older style lever boxes and control panel systems had to consider the signallers' work context explicitly, to be able to reliably predict signaller performance.

The influence of context will be felt when findings from empirical studies are used in cognitive task design, with aims of making work easier, more efficient, and effective, of higher quality and reliability, and more interesting and enjoyable. Business data for the schedulers were mainly handled by the computer-based information systems, but information also had to be gathered, checked, and distributed by means of a network of people, because of inadequacies in the information management systems. Therefore, although from a pure task perspective one would consider how scheduler cognitive and physical load could be reduced by the more effective design of computer-based systems, in many situations the computer-based solution was not appropriate or available. Here one needs to investigate other information management support techniques, such as meetings or e-mail support, and the standardisation of key business processes and procedures.

From a measurement and prediction perspective, we need to consider how the person's context has an impact on the outcome of his or her work. Effective measurement must be carried out to record a person's performance but also take into account the contextual noise that influences someone while they are working. Subsequent to this, effective techniques that can predict a person's work performance by accounting for the key contextual influencing factors also have to be defined. If we are aiming for transparent and structured analysis of data, there will inevitably be a limit to the number of aspects of work that can be measured—either in real-time observation or post hoc analysis. Therefore the aspects of work that are recorded must be carefully selected. For example, in signalling workload analysis, a number of visits were made prior to data collection to identify which operator activities should be recorded, ranging from phone calls received to train routes set.

Field Study Approach

The task descriptions for railway signallers and industrial schedulers demonstrate that their work comprises a complex mix of cognitive tasks and social roles; this complexity is what makes work difficult to study in practice. Our scientific orientation, the preferences of our clients or collaborators, and the recognition of the vital influence of context already discussed persuaded us of the need for much of our research to take place in the setting where people actually work, using field studies, a decision subsequently supported by the knowledge gained by carrying out early studies. "When the unit of analysis is interactions, then field research is arguably the main methodological approach for ergonomics" (de Keyser, 1992). "Methodologically explicit field research is vital for the core purpose of ergonomics, investigating and improving interactions between people and the world around them." (Wilson, 2000, p. 563).

We must also consider the possible inadequacies of cognitive work design based solely on modelling or laboratory simulations. Experimental studies of people at work have long been the dominant approach in cognitive research. They are a useful source of information and insight about isolated work variables, but arguably may not be a valid approach to understanding cognitive work in practice and certainly do not replace the need for field study. The very nature of the controlled laboratory environment means that the complexity and uncertainty of work environments are not being simulated. Even excellent—and expensive— simulations probably only capture the information display and individual control aspects of any work.

This is not an argument against laboratory research in cognitive ergonomics—it is important to acknowledge that both laboratory and field studies are necessary—but is an argument for a better balance; Vicente (2000) presents this as a continuum, with prototypical research types identified. As already noted, there are numerous published papers that present theoretical or experimental investigations of cognitive work in practice, but there is a significant lack of published field studies. Even when the net is cast wider, and perspectives such as ethnomethodology, ethnography, and naturalistic decision making are encompassed within a view of cognitive ergonomics research, this type of study is frequently carried out in very constrained or controlled work environments. In addition, from the undeniably rich and interesting accounts or histories which are produced through such approaches, it is often hard to see what they provide in terms of practical cognitive task design improvements. All too often, measurement methods and the variables measured are not made explicit in reports and, significantly, understanding cognitive work for improved cognitive task design is often not the overall objective.

It is apparent that a simulation or laboratory experimental approach would not have been a practicable research option for the scheduling study. The early results demonstrated that an experimental approach would not capture the work context variables noted earlier or enable us to examine the social factors in the work, as explained in the paragraphs that follow. We accept that certain aspects of scheduling can be examined in the laboratory, particularly the cognitive aspects of work in which the scheduler employs various scheduling support systems, and we could also perhaps study some fragments of the social interactions involved. However, the variables that determine the quality of scheduling in practice cannot easily be examined in the laboratory. Questions such as What is the scheduling function in practice?, How do schedulers perform the whole job of scheduling?, or Who are the people whose role is designated as scheduling? required that the context, and artefacts used by the people who carry out the task of scheduling, were investigated in the field.

The arguments are similar for the field-based approach we took to the work with railway control staff and signallers. A major overall goal was to move toward proposals for multi-functional and multiskilled teams in the future, comprising electrical control room operators, zone controllers, and signallers as well as various planners. Although (as with the scheduling project) some aspects of signalling work are suitable for simulation experiments (e.g., the individual operator's interactions with a variety of signalling control interfaces), the focus of the project as a whole was to examine the effect of changes in working context on the different members of these teams. To even begin to understand something about the current interactions and relationships between these functions and roles, we needed to carry out structured research in the field.

The key issue in cognitive ergonomics is the appropriateness of the approach and the setting for research. Experimental and simulation studies are an extremely important and useful approach for the investigation of cognitive work, but they cannot replace the field study of peoples' work in complex interacting systems. The nature of a controlled environment means that the complexity and uncertainty of real work environments, that should be integrated within cognitive ergonomics models of working, usually cannot be accounted for. Moreover, there is also the very real danger of a classical problem for experimental cognitive ergonomics, whereby the variables that are manipulated and found to have significant effects actually do not matter very much in practice. They may account for little of the variance in the real situation, with all the variance of interest lying within the variables that are actually controlled in the experimental setting—for instance, environmental disturbances, work flow, individual motivation, interpersonal relationships, and team relationships.

In fact, both laboratory and field research take place in a context. In the former, although it is generally difficult to replicate much of the actual work setting (physical and social), the advantage is that good experiments eliminate or at least control for any bias from the experiment, setting, or experimenter. In contrast, although field study can include all the context from the environment, supervision, motivation, and disturbances, the very presence of the investigator will create an additional context that, without care, can invalidate the study. It is not for nothing that the ethnographic community, and particularly the ethnogmethodologists, talk of the participant observer. This is not just reserved for when the observer is also a member of the group being studied; the observer also participates as an actor in the workplace by his or her very presence as a researcher, even though well planned, sensitive, and long-term studies can minimise any influences (Bernard, 1995; Denzin & Lincoln, 1998; Hammersley & Atkinson, 1995; Webb, Campbell, Schwartz, & Sechrest, 1972).

Methods for Measurement and Assessment

As part of the limited reporting of field cognitive ergonomics methods, techniques and metrics, there is particularly a low number of explicit detailed published accounts of method within the field studies. It is often not clear exactly how researchers gathered raw data and evidence, reduced and analysed data, and produced interpretations and recommendations. In other words, the interested reader often cannot find out *how* a particular finding was derived, or *how* a certain account of activity will be used to implement change.

We have spent a considerable part of the past 10 years trying to develop and justify innovative methods and measures for use in field-based cognitive ergonomics. We wanted to build methods that capture what someone is doing and why, especially in circumstances where think aloud or concurrent (and even retrospective) protocols are not appropriate. Another important consideration is that an effective analysis of such protocols, no matter what technique is used, is very labour intensive—despite developments such as exploratory sequential data analysis (Sanderson & Fisher, 1994). We also want to allow greater traceability of results, findings, and interpretations, integrating the best of ethnographic methods and ethnomethodology to the benefit of cognitive task study and design. We believe the following:

- New or adapted methods or techniques for data collection and analysis are needed for cognitive ergonomics field studies.
- These methods must allow for the measurement and assessment of *how* people work and also *why* people work in that way, that is including people's goals and the factors that influence people at work.
- Most important, the methods should allow for traceability of both the data themselves and also the interpretations to be made on the basis of them, in order to propose new understandings.

A number of new methods were developed for the study of schedulers (see Crawford et al., 1999). One method particularly relevant to this discussion is a decision probe. Early on in the study, it was evident that scheduler decision making was rarely "classical." What we mean by this is that we did not often observe the schedulers perceiving information and then assessing a finite range of possibilities prior to selecting one particular course from a number of evident alternatives, each with specific goals to be achieved. Instead, scheduler decisions tended to be mediated by a series of social interactions, and, they were rarely carried out in one location. A method to record and analyse decisions had to be developed that could capture all the factors that were involved in the scheduler's decision-making process. To validate the data captured, the method had to allow for subsequent retrospective assessment by the schedulers and researchers in discussion, and also support a process of potential amendment by the scheduler. The overall aim of the method was to be rich enough to capture all relevant decision information yet structured enough to allow us to assess the schedulers' decision making within the context of their work, so that we could make valid and useful suggestions for the design of future supportive information technology systems and of the cognitive work supported.

In the railway studies, some of the most interesting measurement and assessment issues had to do with method conversion. A number of well known methods have been developed, mainly within the U.S. military or aerospace sectors, to assess such concepts as mental workload and situation awareness, typically for use in simulation or experimental conditions, or in restricted workspaces or where the participant is in a fixed location. Typical examples of these are the NASA TLX (Hart & Staveland, 1988) and SAGAT (Endsley, 1995). We have used these methods as a starting point, to exploit the underlying understanding and reported norms that have come from their use in controlled environments, but also to produce new methods suitable for use in a civilian setting and field context. The study of actual work in context means it is rarely possible to "freeze the simulation," and the work is continuous and varying, involving a large number of different tasks and distractions, rather than being focused on one display-control system or one or two variables of interest. In addition, the length of time taken to administer even a relatively short subjective measure, such as the NASA TLX, in a real world context is not always available. Therefore, many existing well-established measures and tools, in their current form, are not directly applicable in a field context. Although the safety consequences of work in such civil settings are not as critical as in aerospace or nuclear industries, the type of work itself is in many ways more complex and therefore requires a new type of measurement and assessment technique (see Pickup, Nichols, Wilson, and Clarke, 2003).

Focus on Social Factors and Social Work Artefacts

Our main comment regarding the focus of cognitive ergonomics in practice is on the central role of social factors and performance in the majority of the cognitive work we have studied. What is not clear is how we can structure, or even completely describe, the role of social performance at present; even the terminology is difficult.

There are numerous frameworks and descriptors that we can relate to. Social cognition states that people do not just receive external information; they also process it and become the architects of their own social environment (Markus & Zajonc, 1985). Situated action has the underlying premise that work cannot be understood solely in terms of cognitive skills or institutional norms but is a product of continually changing people, technology, information, and space (Suchman, 1987). Situated cognition takes the view that there is joint development of what people perceive, how they conceive of their activity, and what they physically do (Clancey, 1997). Distributed cognition emphasises the distributed nature of cognitive phenomena across individuals, artefacts, and internal and external representations, in terms of a common language of "representational states" and "media" (Hutchins, 1995). We could also look to the fields of anthropology and ethnography, but here the concept of the social has more cultural than organisational connotations, and measurement and prediction of people's behaviour are not usually the main objectives.

We are concerned with work where skilled and successful performance depends on an individual's detailed and current understanding of his or her immediate and remote social environments. Important fragments of such understanding, or social work artefacts, may be as follows:

- Knowing who to telephone for a certain piece of information not available through formal systems;
- Using experience of previous requests and their outcomes to prioritise between two last-minute requests from colleagues;
- Knowing which colleagues can provide accurate and reliable "real" information in contrast to what is assumed to be happening according to procedures;
- Putting together temporary groupings of colleagues who would make the best temporary team to handle a problem; and
- Knowing who (and what) to ask to find out what is really happening in a supply chain.

These examples emphasise the centrality of informal networks and communications systems to the effectiveness of the day-to-day workings of organisations (Jackson et al., 2002). For many of the informal systems, knowledge of their existence and being able to get the best out of them are major parts of individuals' tacit skills and knowledge about how to work.

This focus on the conjoint social, organisational, and cognitive is, we believe, very different to domains such as office HCI and nuclear power plant control that have been traditional foci for cognitive ergonomics. In those domains, working with screen-based displays, either continually or on a monitoring and intervention basis, in order to interpret information and make decisions is the predominant task.

In scheduling, we started our research by making an assumption that we were going to study a cognitive task, and we developed early ideas for methods that would examine, in field and laboratory, how schedulers made a series of decisions constrained by various parameters including quantity, time, cost, and customer priority. However, all our early work demonstrated that we would be wrong to typify scheduling as a classic cognitive activity; the attributes that seemed to define those schedulers who were regarded by their peers and supervisors as the "experts" were mainly in the realm of social awareness and social skills. The schedulers carried out their job, not by monitoring and interrogating databases in order to calculate optimal sequences of work, but by being constantly alert for information coming from their social environment and by interrogating a wide range of other people, internal and external to the company, about what various events or pieces of information actually meant. This intelligence was then executed and facilitated by use of the schedulers' own social networks.

For the railway work, in our study of signallers we had expected the factors producing workload to be span of responsibility—numbers of trains, points systems, level crossings, and the like—and dynamic events—slow running trains, weather conditions, damaged points, and so on. However, one of the key determinants of load appeared to be the function in their job known as delay attribution. They have to find the reason for every time a train is delayed more than a certain period, and because of the multicompany organisation of the railway in Britain, with cost and penalty arrangements made between the companies, this is a function attracting high management pressure. There appears to be a marked difference between those signallers for whom delay attribution is carried out by their being continually interrupted by a series of telephone calls from delay attribution staff, based in a number of different remote sites, and those signallers for whom the delay attributor worked alongside them within the same control centre. Measures of load, stress, and general attitudes were different across the two situations, which represent different sets of social interactions and social work artefacts. In both cases the interruptions caused irritation and interference with the "main job."

The railway example shows how the social setting and social work structure adversely affect the performance and well-being of the staff. In the scheduling case the social system is an important tool of the schedulers. In both cases the social aspects of work are just as important as the cognitive.

CONCLUSIONS

Cognitive task design must be based on quality evidence about how people perform at work and how they can be supported through interfaces, training, job designs, and organisation networks. This chapter has made the case for such evidence to be obtained through field study, and for the social factors and artefacts of work to receive major emphasis.

Cognitive ergonomics must work with both the limited, if neat, abstractions of the laboratory and the logistic and interpretative nightmare of studies in the rich complexity of the field. In a "halfway house," we transfer some of the rigorous tools and approaches from the laboratory into a field context, to attempt to yield more valid data. What is needed now is a model of work in complex distributed networks, which accounts for social artefacts as well as physical artefacts and mental representations.

Elsewhere in this book, other authors have taken up the themes of structural versus functional approaches to cognition. That is, on one hand, we can view cognition from the point of view of a structural explanation such as the information processing model, and therefore concentrate on the characteristics of processes such as perception and memory. On the other hand, we can examine the function of our cognitive processes and examine such capabilities as recognition, discrimination and decision making, and their use in task performance, to help us better understand cognitive work. A third view is that "cognitive work" is a useful description of the types of jobs and performance expected of people in the complex sociotechnical networks found in business, commerce, transport, and social systems today. Used in this sense, without any cultural baggage of arguments over where cognition and cognitive work begin and end, then CTA, CFR, and cognitive task design must all embrace tools, methods, and guidance that cover the social context and social work artefacts that underpin and influence work in complex networks.

REFERENCES

Artman, H., & Waern, Y. (1999). Distributed cognition in an emergency co-ordination center. *Cognition, Technology & Work, 1,* 237–246.

Bernard, H. R. (Ed.). (1995). *Research methods in anthropology (2nd ed.). Qualitative and Quantitative Approaches.* Walnut Creek, CA: Altamira Press.

Clancey, W. J. (1997). *Situated cognition: On human knowledge and computer representations.* Cambridge, England: Cambridge University Press.

Clark, A. (1997). *Being there: Putting brain, body, and world together again.* London: MIT Press.

Cordiner, L. J., Nichols, S. C., & Wilson, J. R. (2001). Development of a railway ergonomics control assessment package (RECAP). In J. Noyes & M. Bransby (Eds.), *People in control: Human factors in control room design* (pp. 169–185). London: IEE.

Crawford, S. (2001). Making sense of scheduling: The realities of scheduling practice in an engineering firm. In B. L. MacCarthy & J. R. Wilson (Eds.), *Human performance in planning and scheduling* (83–104). London: Taylor & Francis.

Crawford, S., MacCarthy, B. L., Vernon, C., & Wilson, J. R. (1999). Investigating the work of industrial schedulers through field study. *Cognition, Technology and Work, 1,* 63–77.

Crawford, S., & Wiers V. C. S. (2001). From anecdotes to theory: Reviewing the knowledge of the human factors in planning and scheduling. In B. L. MacCarthy & J. R. Wilson (Eds.), *Human performance in planning and scheduling* (pp. 15–44). London: Taylor & Francis.

de Keyser, V. (1992). Why field studies? In M. Helander & M. Nagamachi (Eds.), *Design for manufacturability: A systems approach to concurrent engineering and ergonomics* (pp. 305–316). London: Taylor & Francis.

de Montmollin, M., & Bainbridge, L. (1985). Ergonomics and human factors? *Human Factors Society Bulletin, 28*(6), 1–3.

Denzin, N. K., & Lincoln, Y. S. (Eds.). (1998). *Strategies of qualitative inquiry.* London: Sage.

Ehret, B. D., Gray, W. D., & Kirschenbaum, S. S. (2000). Contending with complexity: Developing and using a scaled world in applied cognitive research. *Human Factors, 42*(1), 8–23.

Endsley, M. R. (1995). Direct measurement of situation awareness in simulations of dynamic systems: Validity and use of SAGAT. In D. Garland & M. Endsley (Eds.), *Proceedings of experimental analysis and measurement of situation awareness* (pp. 107–113). Daytona Beach, FL: Embry-Riddle, Aeronautical University Press.

Engeström, Y., & Middleton, D. (1996). *Cognition and communication at work.* Cambridge, England: Cambridge University Press.

Fowlkes, J. E., Salas, E., & Baker, D. P. (2000). The utility of event-based knowledge elicitation. *Human Factors, 42*(1), 24–35.

Hammersley, M., & Atkinson, P. (1995). *Ethnography: Principles in practice.* London: Routledge.

Hart, S. G., & Staveland, L. E. (1988). Development of the NASA task load index (TLX): Results of empirical and theoretical research. In P. A. Hancock & N. Meshkati (Eds.), *Human mental workload* (pp. 139–183). Amsterdam: North-Holland.

Hoffman, R. R., & Woods, D. D. (2000). Studying cognitive systems in context: Preface to the special section. *Human Factors, 42*(1), 1–7.

Hollnagel, E. (2003). Cognition as control: A pragmatic approach to the modelling of joint cognitive systems. *Special issue of IEEE Transactions on Systems, Man and Cybernetics A: Systems and Humans* (in press, and see www.ida.liu.se~eriho)

Hollnagel, E., & Cacciabue, P. C. (1999). Cognition, technology and work: An introduction. *Cognition, Technology and Work, 1*(1), 1–6.

Hutchins, E. (1995). *Cognition in the wild*. Cambridge, MA: MIT Press.

Jackson, S., Wilson, J. R., & MacCarthy, B. L. (2003). A model of scheduling based on field study. Accepted for publication in *Human Factors*.

MacCarthy, B. L., & Wilson, J. R. (Eds.). (2001). *Human factors in scheduling and planning*. London: Taylor & Francis.

MacCarthy, B. L., Wilson, J. R., & Crawford, S. (2001). Human performance in industrial scheduling: A framework for understanding. *Human Factors and Ergonomics in Manufacturing, 11*(4), 299–320.

Mackay, W. (1999). Is paper safer? The role of flight strips in air traffic control. *ACM Transactions on Computer Human Interaction, 6,* 311–340.

Markus, H., & Zajonc, R. (1985). The cognitive perspective in social psychology. In G. Lindzey and E. Aronson (Eds.), *Handbook of Social Psychology* (Vol. 1, pp. 137–230). New York: Random House.

Moray, N. (1994). "De Maximis non Curat Lex" or how context reduces science to art in the practice of human factors. *Proceedings of the Human Factors and Ergonomics Society 38th Annual Meeting* (pp. 526–530). Santa Monica, CA: Human Factors and Ergonomics Society.

Mumaw, R. J, Roth, E. M., Vicente, K. J., & Burns, C. M. (2000). There is more to monitoring a nuclear power plant than meets the eye. *Human Factors, 42*(1), 36–55.

Murphy, P. (2001). The role of communications in accidents and incidents during rail possessions. In D. Harris (Ed.), *Engineering psychology and cognitive ergonomics: Vol. 5. Aerospace and Transportations Systems* (pp. 447–454). Aldershot, UK: Ashgate.

Nardi, B. A. (1993). *A small matter of programming: Perspectives on end user computing.* Cambridge, MA: MIT Press.

Nichols, S., Bristol, N., & Wilson, J. R. (2001). Workload assessment in railway control. In D. Harris (Ed.), *Engineering psychology and cognitive ergonomics: Vol. 5. Aerospace and Transportation Systems* (pp. 463–470). Aldershot, UK: Ashgate.

Pickup, L., Nichols, S., Wilson, J. R., & Clarke, T. (2003). Foundations for measuring workload in the field. Invited paper for special issue of *Theoretical Issues in Ergonomics Science.*

Prietala, M. J., Feltovich, P. J., & Marchak, F. (2000). Factors influencing analysis of complex cognitive tasks: A framework and example from industrial process control. *Human Factors, 42*(1), 56–74.

Rutherford, A., & Wilson, J. R. (1992). Searching for the mental model in human-machine systems. In Y. Rogers, A. Rutherford, & P. Bibby (Eds.), *Models in the mind: Perspectives, theory and application* (pp. 195–223). London: Academic Press.

Sanderson, P. M., & Fisher, C. (1994). Exploratory sequential data analysis: Foundations. *Human Computer Interaction, 9*(3/4), 251.

Schaafstal, A., Schraagen, J. M., & van Berlo, M. (2000). Cognitive task anlaysis and innovation of training: The case of structured troubleshooting. *Human Factors, 42*(1), 75–86.

Shackel, B. (1996). Ergonomics: Scope, contribution and future possibilities. *The Psychologist, 9 July,* 304–308.

Sheridan, T. B. (1985). Forty-five years of man-machine systems: History and trends. *Proceedings of Analysis, Design, and Evaluation of Man-Machine Systems, 2nd IFAC/IFIP/IFORS/IEA Conference,* Varese, Italy, September 5–12.

Singleton, W. T. (1974). *Man-Machine systems.* Harmondsworth: Penguin.

Skinner, B. F. (1985). Cognitive science and behaviourism. *British Journal of Psychology, 76,* 291–301.

Suchman, L. A. (1987). *Plans and situated actions: The problem of human-machine communication.* Cambridge, England: Cambridge University Press.

Vernon, C. (2001). Lingering amongst the lingerie: An observation-based study into support for scheduling at a garment manufacturer. In B. L. MacCarthy & J. R. Wilson (Eds.), *Human factors in scheduling and planning* (pp. 135–164). London: Taylor & Francis.

Vicente, K. J. (2000). Toward Jeffersonian research programmes in ergonomics science. *Ergonomics, 1,* 93–112.

Webb, E. J., Campbell, D. T., Schwartz, R. D., & Sechrest, L. (Eds.). (1972). *Unobtrusive measures nonreactive research in the social sciences.* Chicago, IL: Rand McNally.

Wickens, C. D., Gordon, S. E., & Liu, Y. (1998). *An introduction to human factors engineering.* New York: Longman.

Wilson, J. R. (2000). Fundamentals of ergonomics. *Applied Ergonomics, 31,* 557–567.

Wilson, J. R., Cordiner, L. A., Nichols, S. C., Norton, L., Bristol, N., Clarke, T., & Roberts, S. (2001). On the right track: Systematic implementation of ergonomics in railway network control. *Cognition, Technology and Work, 3,* 238–252.

Wilson, J. R., & Rutherford, A. (1989). Mental models: Theory and application in human factors. *Human Factors, 31*(6), 617–634.

Xiao, Y., & Vicente, K. J. (2000). A framework for epistemological analysis in empirical (laboratory and field) studies. *Human Factors, 42*(1), 87–101.

6

Group Task Analysis and Design of Computer-Supported Cooperative Work

Holger Luczak, Manfred Mühlfelder,
and Ludger Schmidt
RWTH Aachen University, Germany

Abstract

Pushed by the connectivity of the Internet, the emerging power of the World Wide Web, and the prevailing nature of "virtual" organizations, i.e., locally distributed teams that collaborate by using networked computer systems (groupware), many new forms of computer-supported cooperative work (CSCW) are becoming increasingly common. The communities of science and practice dealing with the analysis and design of CSCW are truly multidisciplinary and combine knowledge from computer science, software engineering, human factors, psychology, sociology, ethnography, business administration, and training. Despite this quantitative and qualitative growth, many groupware systems still have serious usability problems. At worst, cooperation by use of groupware can be awkward and frustrating compared with face-to-face collaboration. Still, research on CSCW has only sketchy knowledge of how people *really* collaborate, and translating what people actually do into effective designs is difficult. As well, the effort involved in assessing groupware prototypes and systems is onerous. There are neither any universal evaluation techniques nor any generally approved design criteria existing. The methods of cognitive task analysis can contribute to the improvement of both *groupware design,* for example, by giving insight into demand, structure, and process of collaborative task execution, and *task design* for locally distributed teams, for example, by analyzing the effects of groupware on work structure and process, allocation of responsibility and resources, and task-bound communication. This paragraph joins the domains of task analysis and CSCW by explaining cognitive issues in collaboration and applying them in a collaborative project-planning scenario.

INTRODUCTION: THE SOCIAL BAND OF COGNITION

During the past 50 years or so, mainly after the beginning of the cognitive revolution in psychology, often associated with the studies of Miller, Galanter, and Pribram (1960), the early modeling trials of general human problem solving (Newell & Simon, 1963), and the early compendium by Neisser (1967) about the architecture of human cognition, deep insight has been gained into the structure and dynamics of intrapersonal cognition and psychic regulation processes. Among the well-known theoretical concepts with relevance for cognitive task analysis (CTA), only a few shall be named here: Rasmussen's decision step ladder (Rasmussen, 1976), action regulation theory (Hacker, 1998; Oesterreich, 1982; Volpert, 1982), goal, operator, method, selection (GOMS) rules (Card, Moran, & Newell, 1983), or, more recently, cognitive work analysis (or CWA; Vicente, 1999). Within the past 10 to 15 years, driven by evidence from social sciences and new streams of research in cognition, the focus has widened to include various social aspects. Among these innovative approaches are concepts such as "transactive memory" (Wegner, 1987), "shared mental model" (Cannon-Bowers, Salas, & Converse, 1993; Klimoski & Mohammed, 1994; Kraiger & Wenzel, 1997; Orasanu & Salas, 1993; Rouse, Cannon-Bowers, & Salas, 1992) and "collective action regulation theory" (Weber, 1997; see Thompson, Levine, & Messick, 1999, for a contemporary review of relevant literature). Newell (1990) has started to integrate a model of the social person or agent as a component or element of social systems (couples [$N = 2$], groups [$N > 2$], or organizations [$N \gg 2$]) in a draft for a "unified theory of cognition." He sets a social band on top of the basic biological, neural circuit and cognitive bands in the human cognitive architecture. The intently rational agent as modeled for example with Soar, has goal-oriented behavior and acts in multiagent situations in cooperation and conflict with other agents, developing specific abilities in order to coordinate his or her actions with those of others, performing individual tasks within the group task environment. The question is now what are the characteristics of these tasks with relevance for cognition and action regulation.

Social Organization and Cooperation Analysis

Vicente (1999) organizes the various aspects of cognitive work analysis in five sequential phases (see Table 6.1).

Within level four, social organization and cooperation, at least two aspects can be analyzed separately: *content* and *form* of cooperation. The content dimension of social organizational analysis pertains to the division and coordination of tasks. Simply put, what are the criteria governing the allocation of responsibilities and roles among actors? Rasmussen, Pejtersen, and Goodstein (1994) have identified six mapping rules for task allocation within a given work group.

The first is *actor competency*. Task demands can impose a set of heterogeneous demands on actors in terms of specific skills and knowledge, in which case it makes sense to divide work up accordingly. For example, during surgery, demands are distributed across surgeons, anaesthesiologists, and nurses, in part because each organizational role requires a different set of competencies.

The second is *access to information or action means*. In some application domains, actors do not have equal access to task-relevant information. In such cases, those actors with the most immediate access to a particular information set should be given authority and responsibility for making decisions associated with that information. Concerning the example of surgery, the surgeons themselves have the most direct access to the surgical field, that is, the part of the body being operated on, so they should have responsibility for decisions pertaining to anatomical interventions; the anaesthesiologists hold authority for any actions concerning the blood circuit parameters or metabolism.

The third is *facilitation of communication and coordination*. Dealing with task demands in shared workspaces in an efficient manner requires coordination of multiple decisions and constraints. In such contexts, it would make sense to allocate responsibility to representatives of interest groups who act on behalf of their comrades. In this way, the interdependencies that have to be coordinated can be factored in without requiring an exponentially increasing number

TABLE 6.1
Five Stages of a Complete Cognitive Work Analysis

Focus of Analysis	*Content of Analysis*
1. Work domain	Information requirements Task field and task dynamics Boundaries
2. Tasks	Goals Input–Output Constraints
3. Strategies	Work process Categories of strategic instances Information flow maps
4. Social organization and cooperation	Division and coordination of work Social organization Communication flow
5. Worker competencies	Skills, rules, knowledge Selection Training

Note. From Vincente (1999).

of communications flows between actors or group members, respectively. In sociotechnical terminology this strategy, can be called "boundary maintenance" or "boundary regulation" between work groups (Susman, 1976). For example, in software design and engineering, it is useful to create a single team that includes end users and human factors specialists, as well as hardware and software engineers and marketing professionals. Such an organization allows multiple goals and criteria to be coordinated within a group rather than across groups. This scheme facilitates the communication required for coordination compared with the case in which different interest groups get in conflict powered by in-group–out-group perceptions and motives (e.g., user representative group, human factors group, software group, hardware group, and marketing group).

The fourth is *work load sharing*. The cognitive capacity required by a given task can exceed the information processing limits of a single operator. Then, the task should be split in smaller units, if possible, and be distributed to more than a single person. In air traffic control, for example, flying objects in a crowded airspace can be separated by type (military or civil aircraft) or flight level (high, middle, or low altitude). However, this requires a clear disjunction of subtask sets.

The fifth is *safety and reliability*. In many application domains, redundancy is deliberately built into the organizational structure in order to improve safety or reliability. This redundancy can take many forms. For example, in commercial aircraft cockpits, when they are following procedures in case of emergency, the pilot and copilot are supposed to check and verify each other's actions. The rationale is that two people are more reliable than one because one person can detect the mistakes of another. Besides, there are multiple automatic safety systems, each with the full capability of bringing the total system to a stable state. By having multiple systems, a failure in one will not jeopardize safety. Thus, redundancy can involve multiple workers or multiple technical systems.

The sixth is *regulation compliance*. In some application domains, there are corporate or industry regulations and frameworks that constrain how roles can be distributed across actors.

In these situations, such regulations could shape the organizational structure. Simple examples are codified job descriptions as a result of negotiations and agreement between workers' unions and management representatives, which specify the job responsibilities of particular classes of workers.

It is evident that these criteria are not independent of each other. In any one case, decisions about task allocation can be governed by multiple criteria. For example, surgeons, anaesthesiologists, and nurses have different roles, not just because particular competencies are required for each role, but also because any one person could not alone take on the workload associated with the demands of surgery.

Criteria such as those just mentioned can help to investigate how task demands can determine the distribution of tasks and subtasks or allocation of resources across actors. By identifying the boundaries of responsibility of different actors, we must take into account the flexibility and adaptation that are required for the best human–task fit in complex sociotechnical systems. At the same time, the definition of internal and external boundaries allows us to identify communication needs between various actors on different levels concerning specific objects or task-relevant information.

Once the task demands have been allocated across actors or groups of actors, the form and social organization for carrying out the work process is still not yet decided. As Rasmussen et al. (1994) point out, various forms of cooperation can be observed. For example, one possibility is to create an *autocratic* organization, that is, one in which one worker is the head of the group and is in charge of all of his or her subordinates. Another possibility is to generate an *authority hierarchy* in which a clearly defined chain of command and report governs the execution of tasks. A third possible solution is a *heterarchic* form of social organization. There are various alternatives within this last category, ranging from (semi-) autonomous permanent task groups to virtual project teams.

The Role of Shared Cognition and Memory in Cooperation

Outside computer modeling and research laboratories, cognition is almost always cooperative by nature (Levine, Resnick, & Higgins, 1993). The human ability to act successfully depends on coordinated cognitive interactions with others, and the cognitive "products," that is, materialized artefacts, that emerge from these interactions can hardly be attributed to the contributions of single individuals. In a study of joint cognition, it is critical to examine both the process and outcomes of cognitive collaboration.

Interpersonally coordinated activities depend on *intersubjectivity* (Ickes, Stinson, Bissonnette, & Garcia, 1990; Rommetveit, 1979; Selman, 1980), that is, a shared understanding of what is being discussed or worked on. An early and classic example of intersubjectivity research is the study by Sherif (1935) on norm formation in groups. Sherif investigated how people come to share common perceptions of an ambiguous perceptual stimulus, namely the apparent movement of a stationary point of light in an otherwise dark room (the *autokinetic effect*). The judgments of individual group members converge until a shared estimate of the light's direction and distance of movement is attained, and this socially developed norm continues to influence members' judgment when they later respond alone. Subsequent work has indicated that, once established, such a norm is often maintained over several generations, during which old members gradually leave the group and new members join (Jacobs & Campbell, 1961; Weick & Gilfillan, 1971). Going beyond perceptual norms, a large body of research work indicates that a group's effort to transmit its norms and knowledge are particularly strong when newcomers are involved (Levine & Moreland, 1991, 1999). Groups are highly motivated to provide newcomers with the knowledge, ability, and motivation they need in order to play the role of full members. The freshmen, on the other side, are highly receptive to these attempts of social influence because they feel a strong need to learn what is expected from them (e.g., Louis, 1980; Van Maanen, 1977).

To the extent that socially shared cognitions are developing during social interaction, groups might be expected to perform better than individuals on various tasks, including learning, concept attainment, creativity, and problem solving. However, this is often not the case (Hill, 1982). In a review on group process and productivity, Steiner (1972) concluded that individual or group performance depends on three classes of variables: task demands, resources, and process gains, respectively process losses. Hastie (1986), for example, has demonstrated that group judgment is superior to the average individual performance on numerical estimation tasks. However, on other tasks, such as logical and mathematical brainteaser problems, group performance is normally not better than the average individual judgment, except on such problems in which the solution path is complicated. In problems of this type (in memory of Syracusian philosopher and mathematician Archimedes, often called "Eureka" problems) the average group performance is equal to that of the most competent member. Beyond the "demonstrability" and "communicability" of an unequivocal correct solution is a critical determinant of a group's ability to develop an adequate shared mental representation (Laughlin & Ellis, 1986).

Thus, information exchange is an important source for the effectiveness of joint decision making, both in the laboratory and in practice (Vinokur, Burnstein, Sechrest, & Worthman, 1985). However, research evidence by Stasser (1992) indicates that groups often do not exchange all the information available to their members. Rather than disseminating unshared information, group discussions tend to be dominated by information that members initially share and that supports initial preferences.

This overreliance on shared information points to the negative consequences of too much intersubjectivity, which can prevent groups from fully exploiting the cognitive resources of the individual group members. An extreme example of this phenomenon is "groupthink," which is defined as an extreme group behavior that produces poor and sometimes devastating outcomes. Janis (1982) argued that factors such as external threat, high group cohesiveness, and directive leadership produce symptoms of groupthink (e.g., illusions of invulnerability or pressure on dissenters), which in turn undermine members' ability to process information in an adequate manner and arrive at sound group decisions (also see McCauly, 1989). In a similar endeavor, Hutchins (1991) demonstrated by using connectionist models of group and individual thinking how the initial distribution of information within a group, in combination with applied communication patterns and decision rules for integrating distributed information, can either deteriorate or ameliorate problems that exist in individual cognitive systems, such as confirmation bias (see also Tindale, 1993).

As the research on groupthink and confirmation bias suggests, maximum intersubjectivity, with all group members possessing exactly the same knowledge and thinking pattern, fails to capitalize on the total cognitive resources of the group. To solve this problem, groups often evolve mechanisms for distributing cognitive responsibilities, thereby creating an expanded and more efficient cognitive system. Wegner (1987) has studied transactive memory, which he defines as a shared cognitive system for encoding, storing, and retrieving information. This system develops gradually when people interact over time and is composed of the memory systems of the individuals in the relationship and the communication processes that link these systems. Wegner and his colleagues have pointed out several ways in which a transactive memory improves social information processing. On the other side, they have also shown that, under certain conditions, it can have detrimental effects. For example, Wegner, Erber, and Raymond (1991) demonstrated that romantic partners exhibited lower recall than did pairs of strangers when the experiment imposed a particular memory structure (e.g., one person should remember food items, whereas the other should remember history items), whereas just the opposite pattern emerged when no memory structure was assigned. Apparently, transactive memory systems that develop in ongoing relationships can hinder recall when the partners are forced to adopt a new memory structure.

As much as general cognition, memory is social in at least two ways (Middleton & Edwards, 1990). First, the memorial content is social to the extent that it refers to past social actions and experiences (e.g., Duck, 1982; Messe, Buldain, & Watts, 1981). Second, the process of memory formation is social in such a way that it is based on symbolic communication with other people. Several studies have examined how interaction affects memory in face-to-face groups (e.g., Clark & Stephenson, 1989; Hartwick, Sheppard, & Davis, 1982; Vollrath, Sheppard, Hinzs, & Davis, 1989). Evidence indicates that three variables play an important role in collaborative recall: the degree of consensus favoring a response alternative, the correctness of the chosen alternative, and members' confidence in their responses.

In the following sections, the application of these basic research results for analysis and design of CSCW are discussed in more detail.

COMPUTER-SUPPORTED COLLABORATION

Specifics of CSCW

CSCW is a specific type of collaboration by means of information and communication technology (e.g., e-mail, shared databases, video conferencing, and real-time software application sharing). There are different scenarios for computer support of teams, such as group decision support, collaborative software engineering, computer-supported collaborative distance learning, or telecooperative product design. Each setting creates particular opportunities and restrictions on how to apply groupware to tackle the demands of the task.

Greenberg (1991) defines CSCW as "the specific discipline that motivates and validates groupware design. It is the study and theory of how people work together, and how the computer and related technologies affect group behavior" (p. 7). Hasenkamp, Kirn, and Syring (1994) identify three closely related, but separated, research fields within the domain: (a) understanding and modeling the phenomenon of cooperation and coordination, (b) developing concepts and tools for the support of cooperative work processes, and (c) using methods for the design and evaluation of CSCW.

Compared with other forms of cooperative work, CSCW has some specific features (see Table 6.2).

TABLE 6.2
Overview of Specific Differences Between CSCW and Other Forms of Interpersonal Cooperation

Criteria	CSCW	Other forms of cooperative work, e.g., (semi-) autonomous assembly or production work groups, design teams, or surgery teams
Coordination	With reference to object or computer artefact	With reference to observation of behavior of partner(s)
Communication	Computer mediated	Face to face
Context	Within network	Within organization
Contact	Restricted to media availability and compatibility	Immediate, with possibility to equalize incompatibilities (e.g., language, deafness or blindness)
Constraints	Bandwidth of network technology	Bottlenecks of human information processing capabilities

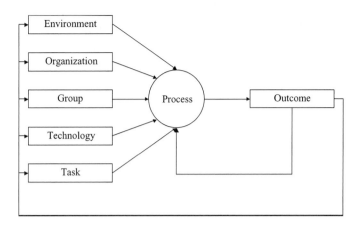

FIG. 6.1. Overview of the research space in CSCW (Olson & Olson, 1997).

In CSCW, all social interaction is more or less mediated by electronic artefacts (documents, messages, video streams, voice streams, sound, whiteboards, etc.). Thus, the computer network establishes the shared workspace, with its characteristics influencing collaboration between people. For a member of a CSCW group, the boundary is the network; that is, other people who might be working in the same organization, perhaps even in the same department, but without connection to the electronic network, are out of scope for those with access to the network. In organizational science, this process is often called "virtualization" of organizational boundaries. Physical contact is not possible in CSCW, with all other forms restricted by availability and compatibility of media. In other forms of interpersonal cooperation, immediate physical contact is possible. Even incompatibilities between coworkers, such as different languages or disabilities (e.g., deafness or blindness), can be counter balanced by the human actors, for example, by using other communication modes. Thus, the constraints for cooperation in CSCW are first and foremost of technical nature (e.g., network bandwidth); meanwhile there are mostly human cantered bottlenecks in non-computer-mediated cooperation, such as capacities of information processing.

Olson and Olson (1997) span a research framework for CSCW containing at least the following elements and causal relationships (see Fig. 6.1). In the following paragraphs we focus on the *task* element of this framework, especially the social cognitive aspects of group tasks.

Social Cognitive Issues in CSCW

Shared Understanding and Deixis. Previously, the need for any form of shared understanding in collaborative tasks was emphasized. This may take three major forms. First, there is an understanding about the task itself, that is, the goal to be achieved, the time available, the process to be followed. Second, actors need a shared understanding about the boundaries of the cooperation; that is, they need to know what is part of the cooperative task, and what exceeds the common workspace. Third, shared understanding may be a goal of the task itself. People communicate in order to discover and share some understanding in a certain domain. In the course of this process, they tend to externalize their subjective understanding, for example by producing white papers, reports, or handbook articles.

They even try to visualize their concepts by using software and hardware artefacts, such as documents, spread sheets, whiteboards, mock-ups, demonstrators, or computer simulations.

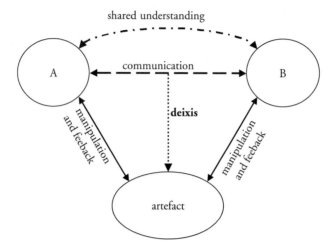

FIG. 6.2.   Deixis as reference to a common artefact during communication between two actors A and B (Dix, 1994).

These symbolic and material representations may merely be generated in order to support interpersonal communication, or they might be the primary goal of the cooperation itself.

In natural conversation these artefacts are present in the workspace; everybody is able to indicate the different objects by pointing and looking at them. The indicative movements may be accompanied by phrases such as "Put that one there!" or "Let's use this!" Dix (1994) introduced the term *deixis* for these forms of reference to particular artefacts. Deixis is shown in Fig. 6.2 within the elements of actors, artefact, communication, shared understanding, manipulation, and feedback. Deictic reference supports people in developing any form of shared understanding in the sense that they link a conversational act to a specific artefact, thus creating a pragmatic cohesion between the two. By saying "Push that!" and pointing to a round panel in the PC front, the receiver of that message learns the concept of on/off switch and how to use it in order to switch a PC on or off.

Shared Mental Models.   On the basis of the original concept of "mental model" widely used in research on human–computer interaction (HCI), many authors have refined and elaborated "shared mental models" in collaborative settings, from aircraft carrier operations (Rochlin, La Porte, & Roberts, 1987) to team decision training under experimentally induced stress (e.g., Cannon-Bowers & Salas, 1998). Rouse and Morris (1986) define mental models as "mechanisms whereby humans are able to generate descriptions of system purpose and form, explanations of system functioning and observed system states, and predictions (or expectations) of future system states" (p. 35).

Research on mental models has primarily focused on shared equipment and tasks. Consequently, the strongest empirical results have been obtained in these domains (see, e.g., Morris & Rouse, 1985). Rouse et al. (1992) introduced the third category with high relevance for collaborative tasks, that is team knowledge. Tables 6.3a–6.3c display the details in "Knowledge × Abstraction level" matrices.

Rouse et al. (1992) presented four propositions that state relationships among shared mental models and team performance:

1. Teams whose assessed mental models provide better means for forming expectations and explanations will perform better than those teams whose models provide poorer means.

TABLE 6.3a
Equipment Knowledge

Abstraction Level	What	How	Why
Low	Characteristics of equipment elements (what element is)	Function of equipment elements (how element works)	Requirements fulfilled (why element is needed)
Medium	Relationships among equipment elements (what connects to what)	Cofunction of equipment elements (how elements work together)	Objectives supported (why equipment is needed)
High	Temporal patterns of equipment response (what typically happens)	Overall mechanism of equipment response (how response is generated)	Physical principles or theories (physics, chemistry, etc.)

TABLE 6.3b
Task Knowledge

Abstraction Level	What	How	Why
Low	Situations (what might happen)	Procedures (how to deal with specific situations	Operational basis (why procedure is acceptable)
Medium	Criteria (what is important)	Strategies (how to deal with general situations)	Logic basis (why strategy is consistent)
High	Analogies (what similarities exist)	Methodologies (how to synthesize and evaluate alternatives)	Mathematical principles or theories (statistics, logic, etc.)

TABLE 6.3c
Team Knowledge

Abstraction Level	What	How	Why
Low	Roles of team members (who member is)	Functioning of team members (how member performs)	Requirements fulfilled (why member is needed)
Medium	Relationships among team members (who relates to whom)	Cofunctioning of team members (how members perform together)	Objectives supported (why team is needed)
High	Temporal patterns of team performance (what typically happens)	Overall mechanism of team performance (how performance is accomplished)	Behavioral principles or theories (psychology, management, etc.)

2. Training that explicitly enables learning, practice, and use of methods for developing appropriate expectations of likely teams behaviors will enhance team performance.
3. Training that explicitly enables learning, practice, and use of methods for determining appropriate explanations of observed team behaviors will enhance team performance.
4. With increased practice, mental models will be increasingly compiled and less accessible, which will eventually result in improvements of team performance without change of the underlying mental models.

All four propositions can be equally applied to systems and task design for collaboration. In evaluating the behavior of teams in tactical military operations, the authors of this concept have found that a missing understanding of common terminology, different ways of equipment usage, and role ambiguity increase deficient behavior of individuals and decrease smooth coordination of actions. In the following paragraph, some relevant theories for explaining the links between social cognition and human action are presented.

ACTION THEORIES WITH RELEVANCE FOR CSCW DESIGN

Activity Theory

Activity theory originated in Russia as part of the cultural–historical school of psychology founded by Lev S. Vygotsky (1896–1934), Alexej Nikolajewitsch Leontjew (1903–1979), Sergey L. Rubinstein (1889–1960), and Alexander R. Lurija (1902–1977). It is a framework for studying different forms of human activities as development processes, with both the individual and social level interlinked. Within the HCI and CSCW community, activity theory has attained increased attention, especially in Scandinavia and the United States (Bardram, 1997; Bødker, 1991; Kuuti, 1991; Nardi, 1996; Suchman, 1987).

The fundamental unit of analysis is human activity, which has three basic characteristics; first, it is directed toward a material or ideal object that distinguishes one activity from another; second, it is mediated by artefacts (e.g., tools, language, or objects); third, it is embedded in a certain culture. In this way, computer artefacts mediate human activity within a given situation or context. By acting in the world, human beings are confronted with other social agents and material objects, which are perceived and experienced through active exploration and manipulation. Thus, human knowledge is reflection about the world, mediated by activity, constituting the basis for expectations and desires about future activities. This describes a basic dialectic relationship between the human being and the world, subject and object. In this way, human activity can be described as a hierarchy with at least three levels: *activities* realized through chains of *actions*, which are carried out through *elementary operations*. It is always directed toward a material or ideal object satisfying a need. The subject's reflection of and expectation to this object characterizes the motive of the activity. Human activity is carried out through actions, realizing objective results. These actions are controlled by the subject's goals, which are anticipations of future results of the action. The activity exists only as one or more actions, but activity and action are not identical and cannot be reduced to each other. For example, for a physician, the activity of diagnosing a patient can be realized in several ways, such as by referring to the medical report of a colleague, conducting an anamnesis, or using special clinical data, such as x-rays. These are different actions, mediated by different tools, all of which realize the activity of diagnosing the patient's disease. In contrast, the same action can be part of different activities. The action of requesting an x-ray examination can be part of a diagnosing activity, but it can also be part of preparing for surgery as well, thus realizing a totally different activity. Furthermore, actions are usually controlled by more than one motive:

two or more activities can temporarily merge and motivate the same action, if the goal is part of satisfying the motives of several activities simultaneously.

Even though the goal of the action can be represented in the human mind independent of the current situation, the process of materializing the action cannot be detached from the contextual situation. Therefore, actions are realized through a series of elementary operations, each accommodated to their physical conditions. Whereas the analytical level of actions describes the intentions of an activity, that is, what results should be obtained, operations describe the behavioral level, that is, how the action is realized and adjusted to the actual material conditions of the action. For example, the way a telephone is used to order an x-ray examination depends on how the phone works (short cuts, memory function, etc.), the phone number of the radiology department, and the physical surroundings of the phone (hanging, standing, mobile, etc.). Operations can be performed without thinking consciously, but they are oriented in the world. This orienting basis is established through experiences with the object in a given situational context. It is a system of expectations and control checks about the execution of elementary operations in the course of the activity. Again, the action and the material operations realizing it are not identical and cannot be reduced to each other. An operation can be part of several actions, and the same action can be materialized by different operations.

Adaptive Structuration Theory

DeSanctis and Poole (1994) transfer concepts of sociological structuration theory (Giddens, 1984) to the usage of groupware in collaborative work. The key elements of "adaptive structuration theory" (AST) are displayed in Fig. 6.3.

Successful media usage is depending on complex feedback loops around the core concept of "social interaction." Given specific media equipment, different sources for social structures, an ideal learning process on how to use the equipment in a certain way, and decision processes that are appropriate for the cooperative task, the people will use the available tools with high intensity. The hypothetical relationships between the elements are explained in the paragraphs that follow.

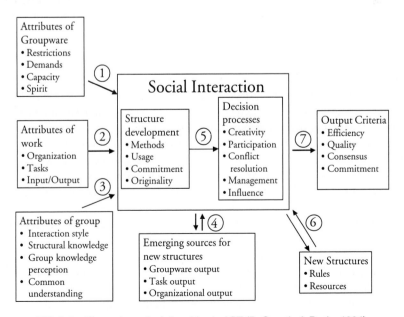

FIG. 6.3. Elements and relationships in AST (DeSanctis & Poole, 1994).

A specific groupware or tool for cooperation has immanent restrictions (e.g., synchrony–asynchrony), demands (e.g., complexity of usage), and capacity (e.g., bandwidth). However, it also incorporates a certain character or spirit. This spirit stands for the meaning of certain groupware functions and the way they are used. For example, the possibility to mark an e-mail with a notification agent when read by the receiver can be both simple feedback mechanism and supervisory control. It depends on how the function is implemented, that is, whether the receiver has any chance to influence, or at least to know whether the sender is notified.

Both the organizational context and the collaborative task itself influence social interaction in work groups. The interdependence of tasks can range from loose coupling to high interreliance, and they can be embedded either in a stable or dynamic organizational environment.

Groups differ in the way the individual members interact with each other. The group leader might be authoritarian or liberal; conflicts are covered or revealed. The individual knowledge about the group and process structures can be elaborated or superficial. Thus, the level of common understanding about task, process, and use of the media in an appropriate manner can be high or low. All these attributes have a particular, yet uncovered, impact on social interaction in CSCW.

In the course of interaction, various results are obtained. These can be either usage styles of certain groupware functions, task output (e.g., task completed), or organizational output (e.g., for other work groups). However, there is a feedback loop between these outcome variables and the social interaction process in the sense that, for example, successful behavioral patterns are reused, whereas others are abandoned.

Inside the "black box" of social interaction, at least two more detailed elements can be distinguished. People develop a social structure in an active manner by acquiring methods and principles of tool usage. They commit themselves to existent structures and decide to which extent they want to copy patterns, and where to develop new elements. These structural factors shape the way plans are generated and deployed, conflicts are managed, and so on.

During the course of the social process, new structures are co-constructed. These reconfigure the interaction and decision processes of the group by reallocating responsibilities, power, and resources.

The final outcome variables are dependent on the issue of analysis and interest. Criteria of interest are e.g., efficiency, quality, consensus about core issues, and commitment in the group.

Collective Action Regulation Theory

There are essential analogies between individual action (see e.g., Frese & Zapf, 1994) and collective action regulation in interpersonal cooperation. Both are goal oriented, related to material objects, embedded in social context, and have process character. Both are restricted by the given regulation basis, as it is predetermined by work and task organization. In action regulation theory (ART), the regulation basis of human action is the knowledge and skills necessary for carrying out the task. With increasing cognitive demand of the task, the regulation of actions becomes more complex. Weber (1997, p. 157) describes the process of interlaced individual actions in collective action structures by the following elements.

1. Generating a common plan, an estimation of the current situation, or a common problem solution:
    - by conversation, or
    - by using individual, intrapersonal regulation processes (e.g., motor scheme activations, memory, and cognition), which
    - are communicated and made public to others in the group,

- are reciprocally annotated, modified, or resigned, and
- are continually integrated into a shared goal–action program structure.

2. Interleaving of communicated and not communicated planning and decision procedures:
   - Individuals receive stimulus through the statements of other group members, which
   - generate a response in their action regulation process, which
   - effects a new action result, which
   - fits into the collective action regulation process like a missing chain link:
3. Mutual exchange and development of shared knowledge;
   - Individually bound knowledge and skills are communicated and
   - materialize in shared objectifications (common tools, archives, devices, etc.).

The last point hints to the fact that people who share common action fields, relying on the input of others for their own actions, start to transfer collective psychological action schemes into materialized artefacts. Galperin (1966, 1972) has emphasized the value of "materialized representations" of thoughts for learning and action shaping. They have an important impact during the sequential emergence of mental activities. Especially, tasks with high cognitive demand can be learned more easily when the apprentice is instructed by a materialized representation of the action process. Raeithel (1992) has elaborated how physical materializations (e.g., models, sketches, and symbols) create links between external object-oriented action and internal mental representations. Hacker, Sachse, and Schroda (1998) have examined the importance of designs and layout for communication and problem solving in engineering. They serve as collective reifications for planning, demonstration, and knowledge exchange. Thus, in collective ART, cooperation is a dialectic process, by which the partners learn to adjust their individual actions to those of others, thereby generating particular symbolic shared objectifications as part of their collective regulation basis.

Awareness Theory

An important topic of recent research on CSCW has been the investigation of awareness-oriented collaborative systems in which users coordinate their activities based on the knowledge of what the members of the collaborating group are doing at the same time, or have done before. Awareness is "an understanding of the activities of others, which provides a context for your own activity" (Dourish & Belotti, 1992, p. 107). According to Gutwin and Greenberg (1998), there are at least four distinguishable types of awareness:

1. Informal awareness of a work community is basic knowledge about who is around in general (and perhaps out of sight), who is physically in the same workspace, and where other people are located relative to one's own position.
2. Group-structural awareness involves knowledge about such things as people's roles and responsibilities, their positions on an issue, their status, and group processes.
3. Social awareness is the information that a person maintains about others in a social or conversational context: things like whether the coworker is paying attention, and his or her emotional state, skills, or level of interest.
4. Workspace awareness is the up-to-the-minute knowledge that a person requires about another group member's interaction within a shared workspace.

Because awareness information is dynamic, its maintenance entails a continuous process of gathering information from the workspace environment and integrating that information with existing knowledge. Gutwin and Greenberg (2002) have built a conceptual framework of

TABLE 6.4
Elements of Workspace Awareness in Synchronously Shared Workspaces

Category	Element	Specific Questions
Who	Presence	Is anyone in the work space?
	Identity	Who is participating? Who is that?
	Authorship	Who is doing that?
What	Action	What are they doing?
	Intention	What goal is that action part of?
	Artefact	What object are they working on?
Where	Location	Where are they working?
	Gaze	Where are they looking?
	View	Where can they see?
	Reach	Where can they reach?
When	Event history	When did that event happen?
How	Action history	How did that operation happen?
	Artefact history	How did this artefact come to be in this state?

workspace awareness that sets out its component elements, mechanisms for maintaining it, and typical ways of using it. The framework is built from a synthesis of existing knowledge about awareness (both in CSCW and in human factors) as well as from observational studies of group work (see e.g., Gutwin, 1997). The first part of the framework divides the concept of workspace awareness into components. People keep track of many things in a shared workspace, but certain elements make repeated appearances in research literature (e.g., Bannon & Bødker, 1997; Dourish & Bellotti, 1992; Sohlenkamp & Chwelos, 1994). When people work with others in a shared workspace they should know who they are working with, what they are doing, where they are working, when various events happen, and how those events occur. Table 6.4 gives an overview of the relevant elements and specific questions to be addressed in groupware design for ensuring high workspace awareness.

Implications for Groupware Design

On basis of these exemplary theoretical frameworks, the design of groupware can be instructed in different ways. First, the designers have to take into account that functionality always affects not only one isolated anonymous user but all those people connected to the shared workspace. Before and while writing the technical specification for a certain module or function, the designer should take into consideration what effects the implementation has not only on the individual user, but also on all those who can be linked to the system. For e-mail, for example, a person sending a computer virus program attached to its e-mail can hamper the work processes of millions of other e-mail users who are in no way connected by a common task or shared objective. The term *system boundaries* gains new quality in this sense. Restricted to a given group, AST suggests that the meaning and spirit of a specific function in a given context should be analyzed. The boss might use a log file function without any bad intention, but it can be doubted whether his or her subordinates share the same understanding, unless they have control about which data are logged and for which purpose. ART and collective ART suggest the importance of shared objectifications in the context of groupware design for the sake of undisturbed action coordination in collective action fields. Awareness theory underlines the necessity of orientation and notification in groups working synchronously in shared workspaces, cooperating with others and manipulating common objects. However, a construct "task awareness" is obviously missing in the awareness framework in its current status.

Summarizing on evidence from human-centered design of collaborative work, Herrmann (1994) has described several criteria for groupware ergonomics by adapting and extending the elements of ISO 9421 (part 10). He suggests the following amendments.

1. Appropriate flow of information: Often, groupware users complain about either lack or abundance of information. To provide better means of information structure and distribution, systems should provide opportunity for completion of information objects concerning, for example, history, version, or context information.

2. Structure: Not every user needs the same view on shared objects or data. A manager only needs a summary of business transactions; an operator needs detailed information concerning a specific customer, an item, or a purchase.

3. Transparency: User input not only changes a technical system state, but also the status of a shared work process, thereby being part of the control stream of another person. Both coworkers need transparency concerning system and work process state. For example, the sender of an e-mail wants to know if it has been sent or if it is still in the outbox. The same is valid for the receiver when waiting for the e-mail as input for his or her own activity.

4. Control: There are three various aspects of control in groupware ergonomics. First, control of access ensures the restriction of usage to a limited number of persons. Second, control of accessibility guarantees that individual user can block unwanted interruptions or theft of work results. Third, control of distribution ensures that only an accredited number of persons receive certain information or work objects.

5. Negotiability: Due to the possibilities of mutual influence and control, users have the necessity of negotiating the information structure, transparency, control, conflict resolution, and so on. In contrast to a single user software, for which the system is or should be configured in accordance with task and user requirements, the complexity of social interaction makes it hardly possible to think ahead of all possible situations and configurations. There are at least two different persons using the system with their own needs for personalization and adaptivity. Instead, the users should be enabled to negotiate which mechanisms and settings they deem to be relevant in a given situation. This is *task group dependent configurability*.

Beyond these software ergonomic issues, designers might be grateful for innovative concepts derived from a deeper understanding of collaborative action structures and processes. Volpert (1994), for example, suggests the following ideas for innovative CSCW tools.

1. Process pattern recognition: These could assist the experts in finding regularities and emerging structures in parallel and sequential work processes. Thus, process analysis and control in complex cooperative task settings can be improved.
2. Gap finding: An individual or group should be able to test its decision base, in order to check the suitability of decision rules and procedures.
3. Concept and design archive: Ideas and solutions for problems not relevant for the current situation, but possibly in the future, are saved and can be reused on demand.
4. Source inventories: These are knowledge bases, where group members can find links to problem- and action-relevant materials and other external sources (e.g., books, videos, and hyperlinks).
5. Computer-supported process simulation: Groups could be supported in modeling their shared workspace and test different versions until they reach the best fitting process

model for the current circumstances. In case of uncertainty, several reversible action chains could be triggered and evaluated.

Design issues for groupware have been being discussed for a long time in CSCW literature (e.g., Rosenberg & Hutchison, 1994). Unfortunately, the consequences for cognitive task design have not been granted the same attention, even though the group task characteristics have a strong impact on usability of shared systems. Properly designed groupware cannot fulfill its purpose when used in insufficient organizational contexts or for inadequate tasks. Thus, if people do not use a groupware with the expected intensity, the reason might not be the tool itself. Instead, there might be a dissatisfying task design, or a mismatch between groupware functions and task requirements.

Implications for Task Design

Dunckel (1989) applies principles of psychological work system analysis to office work and differentiates the following criteria for human-centered task design.

1. Wide scope of decision: Human beings are capable of autonomously defining goals and flexibly adjusting their activities in order to reach them, even under changing conditions. If tasks are designed with a wide scope of decision, positive effects on learning, motivation, and health have been observed (e.g., Hackman & Oldham, 1976; Karasek & Theorell, 1990; Oesterreich & Volpert, 1999).

2. Time autonomy: Task design determines the extent to which the actor can plan the schedule of task execution and completion. At least two aspects can be separated: demand for time planning, and time coupling. Tasks can vary in their demand for free scheduling the sequence and duration of actions. However, even when the actor has the chance to plan the time structure within his or her action field, the actor can be depressed by external tact cycles, *jours fixes,* or too-tight project milestones.

3. Structure: Goal-directed action requires transparency and the power to organize the work process according to current circumstances. When the structure is assessed, the internal and external relationships of a given task are analyzed. The higher the possibility to influence the organizational interfaces with the task, the better the structure.

4. Psychological strain: Hindrances in the course of task execution have detrimental effects on psychic health (Leitner et al., 1993). People have to compensate for the existence of task-immanent obstacles (e.g., missing task-relevant information) or general overtax (e.g., time pressure) by either additional effort or risky action. Both generate psychological strain when they are a permanent reaction to insufficient task design.

5. Physical activity: Human activity is centered in the physical world. All modalities of the human body serve for orientation, planning, execution, and sensual feedback. Thus, task design should provide opportunities for immediate contact with physical entities in the world, be it other humans or objects. However, it is an issue for philosophical disputation whether a perfect virtual environment can substitute for the physical world, if the actor does not realize the difference.

6. Contact with material and social task constraints: Information about a given task, such as mission, responsibilities, or available documents, enable the actor to plan and adjust his or her activities. The actor can have direct access to the material or by contact with social sources. The number and quality of information sources can be assessed separately.

7. Variability: Tasks can be quite monotonous or diversified. There are two aspects to be discerned: task variability is the degree of diversification of activities; input–output variability

describes the affiliation of a task with different work processes. The higher the variability of tasks for a certain job description is the more opportunities for learning and transfer for the jobholder exist.

8. Communication: Tasks vary in their demand for communication. For example, call center agent A has to receive simple orders from customers, simply routing the interview through a determined interview path. Another agent, B, however, could be assigned with the task of consulting customers for complex products, such as life insurance. Obviously, the level of communication in task B is much higher than in task A, with relevance for other task attributes (e.g., time consumption or mental effort).

Even though these criteria have actually been developed before the major appearance of groupware in today's offices, and the original instruments for task analysis (e.g., "KABA," Dunckel et al., 1993) must be adjusted to the new domain, they are still valid for any task design and can contribute to the assessment of group tasks with respect to their degree of performance in this scheme. The method for cognitive task analysis and design of CSCW described in the following paragraph has to be embedded in this normative framework for human-centered task design.

GROUP TASK ANALYSIS FOR CSCW

Recent groupware inspection methodologies are based on the claim that some groupware usability problems are not strongly tied to only social or organizational issues, but rather are caused by insufficient support for the basic activities of collaboration (Baker, Greenberg, & Gutwin, 2001; Pinelle & Gutwin, 2000; Steves, Morse, Gutwin, & Greenberg, 2001). According to these findings, the *mechanics of collaboration* (Gutwin & Greenberg, 2000) represent the small-scale actions and interactions that group members carry out in order to fulfill a shared task. These actions are part of the teamwork (the work of working together) rather than part of the task work (the work that carries out the task). The mechanics of collaboration can be clustered in at least seven major categories.

1. Explicit communication: Group members must be able to provide each other with information. Verbal, written, and gestural communication are cornerstones of collaboration.
2. Implicit communication: People also pick up information that is implicitly produced by others while carrying out their activities, such as information from artefacts being manipulated, or information from others' movements and actions.
3. Coordination of action: People organize their actions in a shared work space so that they do not conflict with others. Shared resources and tools require coordinated access and control.
4. Planning: Some types of planning activities are carried out in a shared workspace, such as dividing up the task, reserving areas of the workspace for future use, or plotting courses of action by simulating them in the workspace.
5. Monitoring: People generally need to keep track of others in the workspace—who is there, where they are working, and what they are doing. In addition, situations such as expert–novice collaboration require more explicit monitoring.
6. Assistance: Group members may provide mutual help. The assistance can be opportunistic and informal, when the situation makes it easy for one person to help another, or it may be explicitly requested by the one who is seeking help.

7. Protection: One danger of group work is that someone may alter another's work inappropriately. People therefore feel the need to keep an eye on their artefacts and take action to protect their work against unauthorized change.

On the basis of these elements of mechanics in collaboration, Pinelle and Gutwin (2001) suggest a new approach for analysis and design in CSCW on the microlevel of cooperation; that is, the mechanics are the vehicle in which the group work is carried out.

The addition of one or more people to a task leads to a variety of differences in task execution and performance when compared with single-user tasks. In particular, the following assumptions constrain the analysis:

*   Task descriptions should include possibilities for a variety of coupling styles—both loosely coupled and tightly coupled interaction should be considered as valid ways of carrying out particular tasks.
*   Task descriptions should allow for alternate courses of action—group task execution is more variable than individual task execution, and descriptions that are strictly linear can cause important alternate paths to be overlooked.
*   Task descriptions should be rigorously analyzed so that they include the mechanics of collaboration that are essential for carrying out the task (e.g., communication with gesturing, or monitoring another member's actions).
*   It is not practical to identify a correct sequence of actions to complete tasks in a groupware system because of the high degree of variability in group work (caused, e.g., by coupling, or interdependence of actions and goals).

Difficulties Specifying Group Tasks

It is difficult to capture the complexities of collaboration in task descriptions. Groupware must support both task work and teamwork. However, standard task analysis methodologies, as described, for example, by Luczak (1997), are oriented toward capturing task work only. It is not clear how team work should be captured and represented in those task descriptions. Many common group tasks such as planning and decision making are highly interactive and do not have a linear sequence of steps; therefore, the activities, even more their temporal order, are difficult to specify in advance. This is inconvenient for classical task modeling methods (e.g., GOMS), for which users have initial goals upon entering the system and perform a batch of "correct actions" in order to reach those goals. Some promising improvements for modeling the vagueness of collaborative task execution by using Harel's high graphs have been achieved by Killich et al. (1999) and tested in practice (e.g., Foltz, Killich, Wolf, Schmidt, & Luczak, 2001).

A user might start without clear goals and decide on his or her next actions by reacting to the actions of other users. A user's goals and actions are interdependent, and each individual uses the actions of others as a context for his or her own performance. The degree of variability in group task execution is increased by differences in levels of coupling. Coupling refers to the level of interaction among group members while carrying out a task, and can range from loosely coupled interactions, in which little group interaction takes place, to tightly coupled interactions, in which users interact intensively and frequently. Users tend to switch between these modes while carrying out group tasks. However, the mode users decide to use at any given time is hardly predictable, and exclusive task descriptions may overlook one of these modes.

Group Task Model

As a way to address the difficulties inherent in specifying group tasks, developing a task model facilitates the analysis of real-world group work scenarios. A group task model does not use hierarchical or sequential ordering as its major means of organization; instead, it comprises all individual and collaborative tasks in a given scenario, without definition of structural or temporal relationships. This voluntary vagueness accommodates the uncertainty of group task execution. A reasonable range of methods for carrying out a task can be represented in the analysis results, and then the possibilities can be examined in an evaluation. Therefore, not all identified task elements in the analysis results are necessarily needed to accomplish a task. Likewise, various tasks (and their component parts) may be repeated, carried out iteratively, or in some cases be executed in an alternate order. The major components of a group task model are scenarios, tasks, individual and collaborative subtasks, and actions. The analytical process itself is organized in a hierarchical manner, beginning from the top (the scenario definition) and ending at the action level.

Scenario.   Scenarios are commonly used as an entry point for design and evaluation. They are conceptual descriptions of the tasks that users will likely perform by using the groupware. A scenario typically contains multiple tasks and provides contextual information about the users and the circumstances under which the tasks are commonly carried out. They can contain the following information: high-level activity descriptions, user specifications, group goals, and boundary conditions. A valid scenario should specify real-world work activities that are carried out by the target users. The content can be generated by observing the users while they perform activities in a natural context, that is, without computer support. The observations should be recorded as a high-level description of the activities required to achieve a group goal. In addition to the activity description, the scenario should contain a description of the desired outcome (the group goal), a description of the group of likely users, and a description of the circumstances under which the scenario is commonly performed. It is also important to capture information about the types of users that are likely to perform the specified scenario, including their expertise and knowledge. This information can be useful to evaluators when they attempt to ascertain how different users will approach the system. Likewise, common circumstances surrounding the scenario provide important insight. For example, information about the physical location of group members in the scenario (i.e., widely distributed or colo-cated) or about timing of interactions (i.e., synchronous or asynchronous collaboration) could prove valuable to the evaluators when they assess the groupware usability in supporting the scenario. Finally, goal descriptions provide insight into the motivation of group members and also provide an important checkpoint for continuous system evaluation.

Tasks.   These are the basic elements of scenarios, and they are often explicitly stated in the scenario activity description. Tasks are high-level statements that describe what oc-curs in a scenario, but not how a task is accomplished. Each task can be analyzed at a finer granularity. It can be divided into subtasks that specify how a task is carried out. In group work, tasks can often be accomplished with different levels of coupling among group mem-bers. For this reason, tasks can be separated into individual subtasks (loosely coupled) and collaborative subtasks (tightly coupled). Some tasks are realistically accomplished with only a single level of coupling (e.g., decision-making tasks require tightly coupled interactions). However, in other tasks, both collaborative and individual subtasks can be included as practi-cable alternatives for accomplishing the task. The mechanics of collaboration provide a useful framework for specifying collaborative subtasks. Many of the teamwork aspects of real-world work are easily overlooked when one is specifying tasks and translating them into groupware specifications. For example, gesture may be an important element of completing a task, but it is more easily overlooked than verbal or textual communication. The mechanics of col-laboration provide a thorough approach for capturing these aspects in collaborative subtask specifications.

Actions.   The mechanics of collaboration do not provide a complete picture of task execution. Therefore, an additional specification should be included in the task model. Each collaborative subtask can be carried out through a set of possible actions. For example, if a collaborative subtask is the identification of an object (an explicit communication mechanic), common actions for accomplishing this may be verbal communication or pointing. Either of these actions is sufficient for the accomplishment of the identification subtask; therefore, actions are usually presented as a list of reasonable alternatives.

Group task models can be used to add task context to a variety of groupware inspection techniques. For example, one appropriate method to take advantage of the information is a "cognitive walkthrough." This is a commonly utilized usability inspection technique for software development (Polson, Lewis, Rieman, & Wharton, 1992). It enables the designers to evaluate software in the early stages of design while taking into consideration common user tasks and goals. A "groupware walkthrough" is a substantial modification of the original method when used for evaluation of multiuser systems. Changes have to be made to accommodate multiple user descriptions, uncertainty in group task performance, and group work metrics. The task descriptions from the group task model are the basis for the walkthrough. The evaluators step through the task analysis results and explore support for each task, subtask, and action within the prototype or application. A strict sequential execution of these steps is not necessary, but breadth of feature testing is important for covering a wide range of potential collaborative situations. In instances in which two levels of coupling are supported in subtasks, both should be explored as practicable alternatives. Likewise, evaluators should give consideration to alternative actions and how support (or lack thereof) will affect the usability of the system.

EXAMPLE: COMPUTER-SUPPORTED COLLABORATIVE PROJECT PLANNING AND EXECUTION

In the following example, some of the theoretical and methodological implications for human-oriented and task-centered design of CSCW are demonstrated. Whenever possible, the link from the design process to the theoretical background will be made explicit.

Scenario

Within the project management literature, the focus is largely on top-down process management (e.g., Cleland, 1998). Issues of project structure, quality maintenance, reporting procedures, and the like are of high concern for the project manager and her or his staff. This approach can be regarded from an opposite point of view, that is, bottom up from the perspective of those who actually carry out operational tasks within a project. Some daily work practices, issues, and needs can be described by the following example questions or statements.

1. Now we are here. What's next? This is process awareness. By definition, a project is rather one of a kind and temporarily restricted. However, certain process fragments and strategies can be applied independently of specific circumstances and be reused in other projects. The process owner of the succeeding or parallel product development process might be happy to find a template she or he can adapt to her or his own needs. At the same time, all persons involved in a specific process can look at their individual parts in the net of activities and become aware of the input–output relations, interdependencies, and the general status of the process.

2. Yesterday we fixed it the opposite way. This is process dynamics. Because of intensive cooperation between different partners with their particular interests and goals, the work process often appears as a negotiation between various interest groups. Within certain limits,

such as the project master plan, decisions are continuously accepted, criticized, improved, and withdrawn. This enables flexibility and quick response to changes of the product concept, but it also prevents a workflow-like a priori planning of the complete work process.

3. If you had only told me before! This is the shortage of task-oriented information with regard to upcoming decisions and actions. Because of the high interdependence of individual activities inside the project team, the dissemination of "weak" decisions, such as agreements without compulsory effect, is an extremely important function for smooth coordination of the workflow. Knowledge about such decisions can create valuable gains of time and the opportunity to adjust individual resources to upcoming tasks, even before the actual assignment.

The major group goals of a project team in such environments are to develop a shared mental model (exchange of knowledge about equipment, task, and team), to gain and maintain high awareness about the shared work space (who is doing what, where, when, and how in the project), and to find work process alternatives in case of changes to the project agenda (e.g., elimination of single activities, rescheduling, restructuring, or reassigning).

Tasks

On a more detailed level of analysis, the scenario contains at least the following tasks to be mastered by the project team members. At this point, some requirements for groupware support can be already fixed. We distinguish between the following individual and collaborative tasks.

1. Producing and altering plans in the course of work (collaborative): The experience of developing and using a plan to guide an activity under certain conditions is obtained during the activity itself. Thus, for plans to become resources for the future realization, the planning itself should be made part of this activity (*situated planning*) in a collaborative manner. Therefore, a planning tool should allow for the continuous (re)creation and modification of action plans based on experience during their execution.

2. Sharing plans within a work practice (individual): The use of shared plans can be used as a guide for action coordination in shared tasks, typical for cross-functional teamwork in projects. When all involved personnel have access to use a common plan (i.e., a shared materialized representation in the sense of collective ART), the opportunity for sharing perspectives and knowledge, and thus developing a shared mental process model for guiding collective action, is enhanced. For the design, this requires an easily understandable and unified set of symbols, which are capable of representing the activities in the shared workspace.

3. Executing plans according to the current conditions of work (individual and collaborative): Plans are never carried out as they were originally fixed, because of changes in the environment, unavailable resources, new evidence revealed by intermediary results, and so on. Thus, a planning tool should provide functionalities for splitting a master plan in atomic units (building blocks). Each block comprises a process sequence of shared activities with the same objective. When various units are combined, complex process plans can be generated bottom up. If necessary, a single block that does not fit the situation any more can be replaced by another.

4. Inspecting plans and their potential outcome (collaborative): Computer simulation of a work plan and altering the scheduling of activities can help the planning team to test various versions of their process agenda. Each process building block has its own range and focus of convenience, that is, the environmental conditions under which the plan is useful and relevant. The collaborative inspection can help to find the most appropriate solution in a given situation.

5. Monitoring the execution of plans (individual): Having an overview of the unfolding of activities is essential for a coordinated action flow. However, when the work process is initiated on the basis of a given plan, monitoring the progress of work according to the plan becomes

TABLE 6.5
Individual and Collaborative Actions of the Planning Scenario

Task	Individual Actions	Collaborative Actions
Producing plan		Inspecting constraints Discussing goals Deciding on next activities and milestones (points of decision)
Altering plan		Discussing changes Inspecting alternative plans Exploring possible consequences of alteration Deciding on alteration
Sharing	Publishing own activities Checking activities of other team members	
Executing	Performing activities Informing others about status Delivery of results (e.g., documents or information objects)	Splitting work process in individual subprocesses (personal activities) Defining process building blocks
Inspecting		Discussing process performance Reasoning about opportunities for process improvement
Monitoring	Observing process statistics (e.g., duration, or percentage of fulfillment) Alarming others in case of anomalies	

vital. This means that recognizing any deviation from the plan is particularly important and should be supported by every single team member. Mechanisms for notification of change for all team members support maintaining an up-to-date view of the current process status.

Actions

The observed behavior during task execution shows the variety of possible elementary operations in the scenario. Table 6.5 sorts the individual and collaborative actions with regard to the described tasks. Of course, these are only the most prominent actions in this scenario. In the real case, the accurate analysis reveals more subliminal actions, such as the behavior during the decision process itself. By classification of individual and collaborative tasks, some important requirements for the groupware design become obvious. Because the production of plans is always a collaborative task, any personalized planning functionalities are considered as not relevant. In contrast, sharing individual subprocesses with others is an individual task, which requires software features for notification, authorizing, and access control.

Groupware Design

In compliance with the analysis, the following design for an "agent based intelligent collaborative process construction kit (AGIP)" was specified.

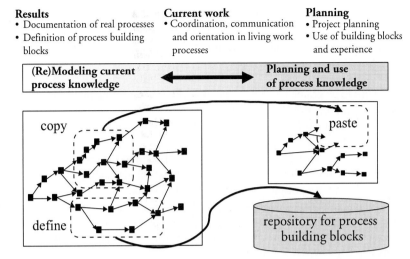

Results
- Documentation of real processes
- Definition of process building blocks

Current work
- Coordination, communication and orientation in living work processes

Planning
- Project planning
- Use of building blocks and experience

FIG. 6.4. Concept of AGIP: A combined project planning and management tool for capturing, reusing, and sharing work process knowledge.

Construction Kit. AGIP implements a modular construction kit with access to a repository of process building blocks (process knowledge base). The planning team chooses from the collection those planning units it needs at the moment and compiles a sequence of shared activities. The short-term planning horizon is determined by the stability of the work context and the complexity of the task. In case there is no suitable building block at hand, the team can generate a new one. A copy can be stored in the repository for reuse in a similar context. After various usages, a planning unit can be evaluated and adjusted to the experience of the users during the process execution. This procedure guarantees a continuous enlargement and updating of the shared work process knowledge base.

User Interface. An intuitive method for process modeling is an essential requirement for grasping the causal and temporal relationships between activities. Therefore, an easy to learn symbol language, consisting of only nine activity types ("planning," "meeting," "documenting," etc.), has been developed. The users can sketch a graphical process model with all activities and documents and their input–output relations. On demand they can call up additional information for a specific activity or document (e.g., duration, responsibility, or priority).

Coordination Functions. The modeled process is continuously reshaped and adjusted to triggering events. Some activities might be rescheduled, others are deleted, and new ones are inserted. All team members can assign an electronic agent to monitor processes, activities, or documents of interest. In case there are modifications, the agent alerts the user about the event. In order to escape notification overflow, the user can also decide who is to be informed about certain changes. For privacy reasons, team members can freely negotiate the coordination mechanisms, access rights, and modification rules in a running work process. Figure 6.4 illustrates the core of the AGIP concept.

Task Design

Applying AGIP to project management requires certain preconditions of work and task organization. First and foremost, processes must be clearly defined and assigned to identifiable people within a distinguishable team. The roles within the group have to be identified (e.g., team leader, assistant, or member), and the process boundaries must be set. In addition, the human-centered criteria of Dunckel (1989) for group task design must be fulfilled to a high degree:

- The team has to be granted a wide scope of decision concerning the coordination of their workflow within the project boundaries (e.g., project master plan or milestones).
- The team has the full time autonomy within its planning space. It can freely schedule the sequence and duration of activities.
- The allocation of activities has to be linked to the distribution of resources and the variable assignment of responsibilities to persons according to their qualifications. The planning, inspecting, and altering of plans is part of the group task.
- The high demand for task-bound communication requires an unrestricted flow of communication. Thus, any hindrances for the free exchange of messages concerning the planning, execution, or status of the shared work process cause additional effort and psychological strain for the team members, which has to be prevented.
- All members of the group have to be guaranteed unhindered access to all shared information objects (e.g., documents or spread sheets).

In the case in which the organizational context does not allow for these group task characteristics to be realized, the implementation of AGIP as a collaborative project management tool will certainly fail. However, when these task factors are guaranteed, the team can autonomously develop a common work plan, perhaps only up to the next stopover, if the situation does not allow for further planning. During the work progress, the team members are entitled to modify their work process model according to their needs and the current situation, as long as the process output is ensured to be in time with asserted quality at negotiated costs.

Inside the team process, members are free to establish their own situational coordination principles (e.g., stages of escalation, notification policy, and frequency of process review). They even can select parts of a shared process and store it in a private folder as a personal building block for new projects, perhaps in cooperation with other team members. After some time, people are able to bring their personal knowledge basis to the kick-off meeting, start to visualize their proposal on how to carry out the process for collective inspection, negotiate the best way given the general conditions of the task, and begin to combine their personal process building blocks in order to compile the complete team process in a coherent process model.

CONCLUSION AND OUTLOOK

Software design is a highly creative task; hence, there is no algorithmic procedure for designing good groupware or productive cooperative tasks. Instead, the design process is a continuous test for the original concept. It might happen that the final product has only vague similarity with the first sketch. Nonetheless, groupware design is not gambling, but rather a software engineering process with high degrees of freedom. The following points can help the design team to save time, money, and effort in the course of prototyping useful groupware applications.

Scenario-Based Design Is Not Fantasy Story Telling

Good scenarios are more substantial than mere fairy tales. They open the space for new design ideas and generate a common ground for both users and designers. A reasonable scenario contains the core tasks of a given situation in its context. It should represent a meaningful sequence of actions for accomplishing the given group tasks. The scenario should be generated by the users themselves in a free but structured way. A moderator can assist the users to build and extract the basic scenario, but should resist from corrupting these.

Socially Extended Task Analysis

Human tasks are always embedded in a social context in a double sense. First, human society assigns tasks to individuals according to qualifications, merits, experience, and the like. On a deeper level, the execution of tasks is influenced obliquely and immediately by actions of other persons. Thus, cognitive task analysis for design of CSCW must extend its focus beyond the individual user. Individual and collaborative tasks should be separated and considered with respect not only to their objectives, but also with regard to their social functions and constraints.

Integration of Task Design and Tool Design

Collaborative tasks vary in their degree of cognitive and communicative demand. Given a specific task, the need for mutual awareness, coordination of action, and structure varies over time. Thus, creating new strategies and work procedures by cooperative planning requires flexible and adaptive computer support. However, not only does the user interface have to fit the task's and user's requirements, but also complex functions, such as communication features, have to be built with respect to the nature of human communication and collaboration.

In contrast, collaborative tasks are continuously and actively reshaped by the actors themselves while creating their own workspaces and planning their task procedures. In fact, inventing and building new work processes can be an essential element of the group task itself. There is a indissoluble connection between the task characteristics (e.g., structure, transparency, and control) and groupware features. The consequence is that there is no priority of task design against tool design and vice versa. Instead, group task analysis should produce meaningful results that can guide designers during their prototyping. More methods for this purpose should be developed and tested in the future. In follow-up investigations, the impact of groupware on individual and collaborative tasks can be evaluated and compared with the situation before. New requirements for groupware design might appear then; others might vanish. This procedure could be criticized as too expensive. However, it is the most natural method to explore new ways of collaboration in risky environments, as it is a typical domain for CSCW research.

ACKNOWLEDGEMENT

The development of AGIP was supported by the German Federal Ministry of Education and Research (BMB + F) as part of the INVITE project (Research Grant 01 IL 901 B 4). We thank our research partners at the Research Institute for Applied Knowledge Processing (FAW Ulm, Germany) and at the Research Department of the BMW Group in Munich, Germany, for the procreative cooperation. For further information on INVITE, see http://www.invite.de.

REFERENCES

Baker, K., Greenberg, S., & Gutwin, C. (2001). Heuristic evaluation of groupware based on the mechanics of collaboration. In M. R. Little and L. Nigay (Eds.), Engineering for Human-Computer Interaction (8th IFIP International Conference, EHC, 2001, Toronto, Canada, May) Lecture Notes in Computer Science. Vol. 2, pp. 123–139, London: Springer.

Bannon, L., & Bødker, S. (1997). Constructing common information spaces. In *Proceedings of the 5th European Conference on Computer Supported Cooperative Work (ECSCW '97)*. (pp. 81–96). Lancaster, England: Kluwer.

Bardram, J. (1997). Plans as situated Action: An activity theory approach to workflow systems. In *Proceedings of the 5th European Conference on Computer Supported Cooperative Work (ECSCW '97)*. (pp. 17–32). Lancaster, England: Kluwer.

Bødker, S. (1991). Activity theory as a challenge to systems design. In H. E. Nissen, H. K. Klein, & R. Hirschheim (Eds.), *Information systems research arena of the 90's*. Amsterdam: Elsevier/North-Holland.

Card, S. K., Moran, T. P., & Newell, A. (1983). *The psychology of human-computer interaction*. Hillsdale, NJ: Lawrence Erlbaum Associates.

Cannon-Bowers, J. A., & Salas, E. (1998). Individual and team decision making under stress: Theoretical underpinnings. In J. A. Cannon-Bowers & E. Salas (Eds.), *Making decisions under stress: Implications for individual and team training* (pp. 17–38). Washington, DC: American Psychological Association.

Cannon-Bowers, J. A., Salas, E., & Converse, S. A. (1993). Shared mental models in expert team decision making. In N. L. Castellan, Jr. (Ed.), *Current issues in individual and group decision making* (pp. 355–377). Hillsdale, NJ: Lawrence Erlbaum Associates.

Clark, N. K., & Stephenson, G. M. (1989). Group remembering. In P. B. Paulus (Ed.), *Psychology of group influence* (2nd ed., pp. 357–392). Hillsdale, NJ: Lawrence Erlbaum Associates.

Cleland, D. (1998). *Global project management handbook*. New York: McGraw-Hill.

DeSanctis, G., & Poole, M. S. (1994). Capturing the complexity in advanced technology use: Adaptive structuration theory. *Organization Science, 5* (2), 121–148.

Dix, A. (1994). Computer supported cooperative work: A framework. In D. Rosenberg & C. Hutchison (Eds.), *Design issues in CSCW* (pp. 9–26). London: Springer.

Dourish, P., & Belotti, V. (1992). Awareness and coordination in shared work spaces. In J. Turner & R. E. Kraut (Eds.), *Proceedings of the Conference on Computer-Supported Cooperative Work (CSCW'92)*. (pp. 107–114). New York: ACM Press.

Duck, S. W. (1982). *Personal relationships 4: Dissolving personal relationships*. New York: Academic Press.

Dunckel, H. (1989). Contrastive task analysis. In K. Landau & W. Rohmert. (Eds.), *Recent developments in job analysis* (pp. 125–136). London: Taylor & Francis.

Dunckel, H., Volpert, W., Zölch, M., Kreutner, U., Pleiss, C., & Hennes, K. (1993). *Kontrastive Arbeitsanalyse im Büro. Der KABA-Leitfaden. Grundlagen und Manual*. Zürich, Switzerland: Verlag der Fachvereine.

Foltz, C., Killich, S., Wolf, M., Schmidt, L., & Luczak, H. (2001). Task and information modelling for cooperative work. In M. J. Smith & G. Salvendy (Eds.), *Proceedings of HCI International 2001. Volume 2: Systems, social and internationalization design aspects of human-computer interaction* (pp. 172–176). Mahwah, NJ: Lawrence Erlbaum Associates.

Frese, M., & Zapf, D. (1994). Action as the core of work psychology: A german approach. In H. C. Triandis, M. D. Dunette, & L. M. Hough (Eds.), *Handbook of industrial and organizational psychology* (2nd ed., Vol. 4, pp. 271–342). Palo Alto, CA: Consulting Psychologists Press.

Galperin, P. J. (1966). Die geistige Handlung als Grundlage für die Bildung von Gedanken und Vorstellungen. In J. Lompscher (Hrsg.), *Probleme der Lerntheorie* (pp. 33–49). Berlin: Volk und Wissen.

Galperin, P. J. (1972). Die Entwicklung der Untersuchungen über die Bildung geistiger Operationen. In H. Hiebsch (Hrsg.), *Ergebnisse der sowjetischen Psychologie* (pp. 367–465). Stuttgart: Klett.

Giddens, A. (1984). *The constitution of society: Outline of the theory of structuration*. Berkeley, CA: University of California Press.

Greenberg, S. (1991). *Computer supported cooperative work and groupware*. London: Academic Press.

Gutwin, C. (1997). *Work space awareness in real-time distributed groupware*. Unpublished doctoral dissertation, University of Calgary, Alberta, Canada.

Gutwin, C., & Greenberg, S. (1998). Effects of awareness support on groupware usability. In *Proceedings of ACM CHI '98* (pp. 511–518). Los Angeles, CA: ACM Press.

Gutwin, C., & Greenberg, S. (2000). *The mechanics of collaboration: Developing low cost usability evaluation methods for shared work spaces* (IEEE WET ICE 2000). New York: IEEE.

Gutwin, C., & Greenberg, S. (2002). A descriptive framework of work space awareness for real-time groupware. *Journal of Computer Supported Cooperative Work,* (in press)

Hacker, W. (1998). *Allgemeine arbeitspsychologie*. Bern, Switzerland: Huber.

Hacker, W., Sachse, P., & Schroda, F. (1998). Design thinking—possible ways to successful solutions in product development. In H. Birkhofer, P. Badke-Schaub, & E. Frankenberger (Eds.), *Designers—the key to successful product development* (pp. 205–216). London: Springer.

Hackman, J. R., & Oldham, G. R. (1976). Motivation through the design of work: Test of a theory. *Organizational Behaviour and Human Performance, 16,* 250–279.

Hasenkamp, U., Kirn, S., & Syring, M. (1994). *CSCW—computer supported cooperative work—Informationssysteme für dezentralisierte Unternehmensstrukturen.* Bonn, Germany: Addison-Wesley.

Hartwick, J., Sheppard, B. H., & Davis, J. H. (1982). Group remembering: Research and implications. In R. A. Guzzo (Ed.), *Improving group decision making in organizations* (pp. 41–72). New York: Academic Press.

Hastie, R. (1986). Review essay: Experimental evidence on group accuracy. In G. Owen & B. Grofman (Eds.), *Information pooling and group decision making* (pp. 129–157). Westport, CT: JAI.

Herrmann, Th. (1994). Grundsätze ergonomischer Gestaltung von Groupware. In A. Hartmann, Th. Herrmann, M. Rohde, & V. Wulf (Eds.), *Menschengerechte Groupware. Softwareergonomische Gestaltung und partizipative Umsetzung.* Stuttgart, Germany: Teubner.

Hill, G. W. (1982). Group versus individual performance: Are n + 1 heads better than one? *Psychological Bulletin, 91,* 517–539.

Hutchins, D. J. (1991). The social organization of distributed cognition. In L. B. Resnick, J. M. Levine, & S. D. Teasley (Eds.), *Perspectives on socially shared cognition* (pp. 283–307). Washington, DC: American Psychological Association.

Ickes, W., Stinson, L., Bissonnette, V., & Garcia, S. (1990). Naturalistic social cognition: Empathic accuracy in mixed-sex dyads. *Journal of Personality and Social Psychology, 59,* 730–742.

Jacobs, R. C., & Campbell, D. T. (1961). The perpetuation of an arbitrary tradition through several generations of a laboratory microculture. *Journal of Abnormal and Social Psychology, 62,* 649–658.

Janis, I. L. (1982). *Groupthink* (2nd ed.). Boston, MA: Houghton Mifflin.

Karasek, R., & Theorell, T. (1990). *Healthy work: Stress, productivity and the reconstruction of working life.* New York: Basic Books.

Killich, S., Luczak, H., Schlick, C., Weissenbach, M., Wiedenmaier, S., & Ziegler, J. (1999). Task modelling for cooperative work. *Behaviour and Information Technology, 18*(5), 325–338.

Klimoski, R., & Mohammed, S. (1994). Team mental model: Construct or metaphor? *Journal of Management, 20* (2), 403–437.

Kraiger, K., & Wenzel, L. C. (1997). Conceptual development and empirical evaluation of measures of shared mental models as indicators of team effectiveness. In M. T. Brannick, E. Salas, & C. Prince (Eds.), *Team performance, assessment and measurement: Theory, methods and applications* (pp. 139–162). Mahwah, NJ: Lawrence Erlbaum Associates.

Kuuti, K. (1991). The concept of activity as a basic unit of analysis for CSCW research. In L. Bannon, M. Robinson, & K. Schmidt (Eds.), *Proceedings of the Second European Conference on Computer-Supported Cooperative Work* (pp. 249–264). Amsterdam: Kluwer.

Laughlin, P. R., & Ellis, A. L. (1986). Demonstrability and social combination processes on mathematical intellective tasks, *Journal of Experimental Social Psychology, 22,* 177–189.

Leitner, K., Lüders, E., Greiner, B., Ducki, A., Niedermeier, R., & Volpert, W. (1993). *Das RHIA/VERA-Büro-Verfahren. Handbuch.* Göttingen, Germany: Hogrefe.

Levine, J. M., & Moreland, R. L. (1991). Culture and socialization in work groups. In L. B. Resnick, J. M. Levine, & S. D. Teasley (Eds.), *Perspectives on socially shared cognition* (pp. 257– 279). Washington, DC: American Psychological Association.

Levine, J. M., & Moreland, R. L. (1999). Knowledge transmission in work groups: Helping newcomers to succeed. In L. L. Thompson, J. M. Levine, & D. M. Messick (Eds.), *Shared cognitions in organizations. The management of knowledge* (pp. 267–296). Mahwah, NJ: Lawrence Erlbaum Associates.

Levine, J. M., Resnick, L. B., & Higgins, E. T. (1993). Social foundations of cognition. *Annual Review of Psychology, 44,* 585–612.

Louis, M. R. (1980). Surprise and sense making: What newcomers experience in entering unfamiliar organizational settings. *Administrative Science Quaterly, 25,* 226–251.

Luczak, H. (1997). Task analysis. In G. Salvendy (Ed.), *Handbook of human factors and ergonomics* (2nd ed. pp. 340–416). New York: Wiley.

McCauly, C. (1989). The nature of social influence in groupthink: Compliance and internalization. *Journal of Personality and Social Psychology, 57,* 250–260.

Messe, L. A., Buldain, R. W., & Watts, B. (1981). Recall of social events with the passage of time. *Personality and Social Psychology Bulletin, 7,* 33–38.

Middleton, D., & Edwards, D. (1990). *Collective remembering.* London: Sage.

Miller, G. A., Galanter, E., & Pribram, K. H. (1960). *Plans and the structure of behaviour.* New York: Freeman.

Morris, N. M., & Rouse, W. B. (1985). The effects of type of knowledge upon human problem solving in a process control task. *IEEE Transactions on Systems, Man, and Cybernetics, SMC-15,* 698–707.

Nardi, B. (1996). Context and consciousness: Activity theory and human-computer interaction. Cambridge, MA: MIT Press.

Neisser, U. (1967). *Cognitive psychology.* New York: Appleton-Century-Crofts.

Newell, A. (1990). *Unified theories of cognition*. Cambridge, MA: Harvard University Press.

Newell, A., & Simon, H. A. (1963). GPS, a program that simulates human thought. In E. A. Feigenbaum and J. Feldman (Eds.), *Computers and thought*. New York: McGraw-Hill.

Oesterreich, R. (1982). The term "efficiency-divergency" as a theoretical approach to problems of action-planning and motivation. In W. Hacker, W. Volpert, & M. Cranach (Eds.), *Cognitive and motivational aspects of action* (pp. 99–110). Berlin: Deutscher Verlag der Wissenschaften.

Oesterreich, R., & Volpert, W. (1999). *Psychologie gesundheitsgerechter Arbeitsbedingungen*. Bern, Switzerland: Huber.

Olson, G. M., & Olson, J. S. (1997). Research on computer supported cooperative work. In M. Helander, T. K. Landauer, & P. Prabhu (Eds.), *Handbook of human-computer interaction* (Rev. 2nd ed., pp. 1433–1456). Amsterdam: Elsevier.

Orasanu J., & Salas, E. (1993). Team decision making in complex environments. In G. A. Klein, J. Orasanu, R. Calderwood, & R. E. Zsambok (Eds.), *Decision making in action: Models and methods* (pp. 327–345). Norwood, NJ: Ablex.

Pinelle, D., & Gutwin, C. (2000). *A review of groupware evaluations* (IEEE WET ICE 2000). New York: IEEE.

Pinelle, D., & Gutwin, C. (2001). *Group task analysis for groupware usability evaluations* (IEEE WET ICE 2001). New York: IEEE.

Polson, P., Lewis, C., Rieman, J., & Wharton, C. (1992). Cognitive walkthroughs: A method for theory-based evaluation of user interfaces. *International Journal of Man-Machine Studies, 36,* 741–773.

Raethel, A. (1992). Activity theory as a foundation for design. In C. Floyd, H. Züllighoven, R. Budde, & R. Keil-Slawik (Eds.), *Software development and reality construction* (pp. 391–415). Berlin: Springer.

Rasmussen, J. (1976). Outlines of a hybrid model of the process plant operator. In T. B. Sheridan & G. Johannsen (Eds.), *Monitoring behaviour and supervisory control* (pp. 371–383). New York: Plenum.

Rasmussen, J., Pejtersen, A. M., & Goodstein, L. P. (1994). *Cognitive systems engineering*. New York: Wiley.

Rochlin, G. I., La Porte, T. R., & Roberts, K. H. (1987). The self-designing high-reliability organization: Aircraft carrier flight operations at sea. *Naval War College Review, 40*(4), 76–90.

Rommetveit, R. (1979). On the architecture of inter-subjectivity. In R. Rommetveit & R. M. Blakar (Eds.), *Studies of language, thought, and verbal communication* (pp. 93–107). New York: Academic.

Rosenberg, D., & Hutchison, C. (1994). *Design issues in CSCW (computer supported cooperative work series)*. London: Springer.

Rouse, W. B., Cannon-Bowers, J. A., & Salas, E. (1992). The role of mental models in team performance in complex systems. *IEEE Transactions on Systems, Man and Cybernetics, 22*(6), 1296–1308.

Rouse, W. B., & Morris, N. M. (1986). On looking into the black box: Prospects and limits in the search for mental models. *Psychological Bulletin, 100,* 349–363.

Selman, R. L. (1980). *The growth of interpersonal understanding: Developmental and clinical analyses*. New York: Academic.

Sherif, M. (1935). A study of some social factors in perception. *Archives of Psychology 187*.

Sohlenkamp, M., & Chwelos, G. (1994). Integrating communication, cooperation, and awareness: The DIVA virtual office environment, In *Proceedings of the ACM Conference on Computer Supported Cooperative Work* (pp. 331–343). New York: ACM Press.

Stasser, G. (1992). Pooling of unshared information during group discussion. In S. Worchel, W. Wood, & J. A. Simpson (Eds.), *Group process and productivity* (pp. 48–67). Newbury Park, CA: Sage.

Steiner, I. D. (1972). *Group process and productivity*. New York: Academic Press.

Steves, M., Morse, E., Gutwin, C., & Greenberg, S. (2001). A comparison of usage evaluation and inspection methods for assessing groupware usability. In *Proceedings of the ACM International Conference on Supporting Group Work* (pp. 125–134). New York: Academic.

Suchman, L. A. (1987). *Plans and Situated Actions*. Cambridge, MA: Cambridge University Press.

Susman, G. I. (1976). *Autonomy at work*. New York: Praeger.

Thompson, J. L., Levine, J. M., & Messick, D. M. (1999). *Shared cognition in organizations. The management of knowledge*. Mahwah, NJ: Lawrence Erlbaum Associates.

Tindale, R. S. (1993). Decision errors made by individuals and groups. In N. J. Castellan (Ed.), *Current issues in individual and group decision making*. Hillsdale, NJ: Lawrence Erlbaum Associates.

Van Maanen, J. (1977). Experiencing organization. Notes on the meaning of careers and socialization. In J. V. Maanen (Ed.), *Organizational careers: Some new perspectives* (pp. 15–45). New York: Wiley.

Vicente, K. J. (1999). *Cognitive work analysis*. Mahwah, NJ: Lawrence Erlbaum Associates.

Vinokur, A., Burnstein, E., Sechrest, L., & Worthman, P. M. (1985). Group decision making by experts: Field study of panels evaluating medical technologies. *Journal of Personality and Social Psychology, 49,* 70–84.

Volpert, W. (1982). The model of hierachic-sequential organization of action. In W. Hacker, W. Volpert, & M. v. Cranach (Eds.), *Cognitive and motivational aspects of action* (pp. 35–51). Berlin: Hüthig Verlagsgemeinschaft GmbH.

Volpert, W. (1994). Die Spielräume der Menschen erhalten und ihre Fähigkeiten fördern—gedanken zu einer sanften KI-Forschung. In G. Cyranek & W. Coy (Hrsg.), *Die maschinelle Kunst des Denkens—Perspektiven und Grenzen der künstlichen Intelligenz* (pp. 193–213). Braunschweig, Germany: Vieweg.

Vollrath, D. A., Sheppard, B. H., Hinzs, V. B., & Davis, J. H. (1989). Memory performance by decision making groups and individuals. *Organizational Behaviour and Human Decision Processes, 43,* 289–300.

Weber, W. G. (1997). *Analyse von Gruppenarbeit—Kollektive Handlungsregulation in sozio-technischen Systemen.* Berne, Switzerland: Huber.

Wegner, D. M. (1987). Transactive memory: A contemporary analysis of the group mind. In B. Mullen & R. G. Goethals (Eds.), *Theories of group behaviour* (pp. 185–208). New York: Springer.

Wegner, D. M., Erber, R., & Raymond, P. (1991). Transactive memory in close relationships. *Journal of Personality and Social Psychology, 61,* 923–929.

Weick, K. E., & Gilfillan, D. P. (1971). Fate of arbitrary traditions in a laboratory microculture. *Journal of Personality and Social Psychology, 17,* 179–191.

7

Cognitive Workload and Adaptive Systems

James L. Alty
Loughborough University, United Kingdom

Abstract

The development of adaptive systems is suggested as a possible way forward for decreasing cognitive workload, particularly in the control of large dynamic systems. The nature of adaptation is discussed and two types of adaptation are identified—short term and long term. In the former case the system returns to equilibrium after a disturbance. In the latter case the disturbance becomes the new equilibrium point. Early attempts at developing adaptive systems are reviewed and some construction principles are identified, including the identification of adaptive triggers and adaptive functions. Software agent technology is identified as an appropriate technology for developing adaptive systems, and a research project using these techniques—Autoadaptive Multimedia Environment Based on Intelligent Collaborating Agents, known as AMEBICA—is described in some detail. The AMEBICA project has developed an adaptive interface for a large dynamic system. The implementation is briefly discussed, and the experiences gained in using it in an electrical network exemplar are presented together with the lessons learned.

ADAPTABILITY AND COGNITIVE WORKLOAD

All work (apart from pure mental activity) involves interplay between cognitive activity and physical activity, and different types of tasks involve different mixes of the two. Some tasks are almost exclusively physical—digging the garden, for example—whereas other tasks, such as playing chess, are almost exclusively cognitive. Many tasks involve both cognitive and perceptual-motor skills—for example, carpentry or plumbing. Rasmussen's well-known Ladder Model (Rasmussen, 1986) provides a useful description of the interplay between cognitive and perceptual-motor skills, and it proposes three levels of activity—skill based, rule based, and knowledge based. At different stages, tasks may require moves between these levels. In many tasks the users may operate for much of the time, routinely, almost at the skill level;

then, when there is a serious disturbance from the norm, they may have to use rule-based or even knowledge-based techniques to solve a problem.

In this chapter, *cognitive workload* is regarded as that subset of mental workload that requires conscious effort. Although there are many tasks that require a large amount of mental activity, not all of them require conscious involvement (e.g., recognising familiar objects or driving a car) and such processes can often operate in parallel with conscious activity. The actual cognitive workload generated for a particular task will also depend on experience. One person's heavy cognitive load will be another's routine skill.

As society moves toward a more knowledge-based workplace, both the number of tasks and the work content within tasks are increasingly requiring a cognitive approach. In particular, the introduction of computer technology has been a major factor in introducing more cognitively based activity into the workplace. This means that the designer can no longer consider each system separately, but must consider the man–machine system as a whole cognitive system (Hollnagel & Woods, 1983).

When human beings execute tasks, they often adapt their approach to the task in some way to carry out the task more efficiently. Indeed, one might argue that an important aspect of human expertise is the ability to adapt to different situations. Some types of adaptation take place over a period of time. Such adjustments might be the result of the application of work-study techniques or through the introduction of affordances (Norman & Draper, 1986) to simplify the task. Hollnagel has described these as "adaptation through design" (Hollnagel, 1995). Other adaptation takes place "on the fly" to deal with unexpected or unusual situations (what Hollnagel has called "adaptation during performance"). Until recently, the responsibility for such adaptation during performance in the workplace rested solely with the human beings doing the work. This was because the workplace itself had no capability for reflecting on the work activity being done, nor did it have any memory capability, which is a prerequisite for learning. Human beings did, however, still experience work situations in which other components of the system could adapt to the current work situation. For example, any task that involves another human being already has this property, so dealing with an adaptive partner in a task is therefore not new.

With the introduction of computer technology into the workplace, the possibility that the objects in the workplace can also exhibit adaptation has become a reality, and this has important implications for cognitive task analysis. For example, if a system can adapt the way in which it presents information to users (say in order to make the information more salient) or adjust its style of interaction to fit in with particular user characteristics, the cognitive workload on the user or operator could be considerably reduced.

In most current computer applications, interface designers still choose a specific mapping between task parameters and interface modalities and media representations at design time, which results in a lack of flexibility at run time. The idea of flexible mapping, which can be adapted at run time such that optimal representations that minimise user workload can always be presented, provides a possible solution to this dilemma. We therefore have a new opportunity for minimising cognitive workload—by introducing an adaptive capability into the nonhuman aspects of the work situation.

One possible area of application for adaptive systems is the control of large dynamic processes. These are complex systems involving the presentation and understanding of a process state described by hundreds or even thousands of variables. They involve the control of a physical system that is largely hidden from the operators, who must infer its behaviour from limited information. Therefore, for any particular process situation, there will be many possible representational choices for presenting the salient variables to the operators and thus many opportunities for improving the efficiency of the total human–machine system. Currently, the mappings chosen are what designers consider to be the most effective or efficient mappings possible, given the known constraints under which the system will operate. The rejected mappings (design alternatives) are rejected because they are presumably suboptimal in relation to one or more design criteria. However, one mapping cannot be optimal for all possible system states, so there will always be an element of compromise in the final choice made. The rejected mappings, for example, could well have been more efficient for a specific set of conditions or for a particular situation. The concept of an adaptive human–computer interface for dynamic systems is potentially, therefore, very attractive indeed.

Most dynamic systems are now controlled by computer-based supervisory and control systems, which in turn are controlled by teams of operators. Thus, such a workplace offers considerable potential for the study of adaptive systems and their effect on cognitive workload. This chapter examines the nature of this adaptation opportunity with particular reference to the process control workplace.

COULD ADAPTIVE SYSTEMS ACTUALLY INCREASE COGNITIVE WORKLOAD?

There are a number of potential dangers that must be considered when adaptive systems are introduced. If the interface can modify its appearance as a result of task or operator responses, issues of interface consistency immediately arise. How will the operators or users react to an interface that changes its appearance, presenting similar information in different ways at different times? Such an adaptive interface, for example, would violate an important principle of human–computer interaction design—that of consistency (Shneiderman, 1983). Consistency is an important design principle that is often used to minimise cognitive workload. Other difficulties might arise if the interface changes its appearance suddenly while users are in the process of reasoning through a problem. Such unwanted shifts of attention could seriously affect the efficiency of the reasoning process. The third set of problems can arise because human intent is often difficult to ascertain from a set of external actions. This can result in most unhelpful switches of emphasis or context by the adaptive system, which is likely to increasing the cognitive workload rather than decrease it (e.g., by switching unconscious processing to a conscious stream).

Because human beings are adept at adaptation, the introduction of adaptive systems carries another risk, which is that of instability. In current systems only one component of the human–machine system is adapting. When both components each have the capability of adapting to the actions of the other, there is a risk of falling out of step. System A begins adapting to a particular state in system B. In the meantime, system B changes its state to adapt to a situation in system A. This can result in "hunting," with each system trying to adapt to the other, and may produce deadlocks or cycles.

The proponents of adaptive interfaces suggest three possible responses to these concerns. First, they point out that human beings are quite used to a changing world, a world that presents itself in different ways at different times. Such behaviour is quite common when one human being is dealing with other human beings. Second, adaptability might be acceptable if adaptability *itself* were implemented in a consistent way—the system adapting in the same way under the same conditions (Alty, 1984a)—thereby making its adaptive behaviour consistent. Third, although understanding human behaviour from a set of external actions is admittedly difficult, the existence of good user models may eventually simplify this problem.

People have also questioned whether the goal of adaptation will ever be feasible, either from technical, operational, or economical points of view. Thimbleby (1990), for example, argued that only complex systems could benefit from an adaptive capability, but that this very complexity means that it is not possible to provide such a capability because the patterns of usage are harder to determine. He asserted that there was simply not enough bandwidth in the user interface to accommodate the required functionality for adaptation. Similar arguments were made with respect to the cost of building an adaptive capability into an application, or the resulting technical problems such as processing speed, knowledge representation, and so on. In addition, the capability of systems to incorporate enough suitable knowledge about an individual user in order to make sensible adaptive responses, and the basis on which such user characteristics can be inferred, is also problematic.

However, the truth of these arguments has still not been resolved. For example, a very simple adaptive mechanism might be highly effective in some circumstances, or the cost associated with the implementation of adaptive systems might still be justified if it significantly improves usability and the quality of interaction even if the cost is high (in process control this is particularly true). Furthermore, the inclusion of an adaptive capability may not be such a large overhead if it arises as a natural consequence of improved approaches and metrics being applied to interactive system design.

WHAT IS THE NATURE OF ADAPTATION?

To understand the nature of adaptation better, we must examine the phenomenon of adaptation in general. According to Alty, Khalil, and Vianno (2001), the main defining characteristic of an adaptive organism is "its ability to maintain a stable state or an equilibrium in spite of disturbances and influences from the outside world. In extreme cases when the disturbance is prolonged, an adaptive system will modify its internal state so that the disturbance is now part of normal conditions" (p. 53).

This definition supports two different views of adaptation. In the first case, the organism reacts in a temporary way, reverting to its original state when the disturbance is removed. In the second case, when the disturbance persists for a long time, the internal state of the organism changes to match the nature of the disturbance so that the organism is eventually in a normal state in the presence of the disturbance (i.e., the disturbance is no longer a disturbance). In this situation the reestablishment of the original conditions will now be seen as a disturbance.

It is therefore necessary to distinguish between *short-term adaptation* (or what we might call *intelligent reaction*) and *long-term adaptation.* The term intelligent reaction is used here to emphasize the fact that true adaptation must be more than a simple reaction to a stimulus (e.g., a pop-up window appearing when a button is pressed). Although it is easy to give extreme examples that fall into either class, the dividing line between a simple reaction and an intelligent reaction is not clear. Whichever definition (short-term adaptation or long-term adaptation) is appropriate will depend on the application area. In a supervisory and control system, three obvious states are *start-up, equilibrium,* and *shutdown*. In this domain the maintenance of equilibrium is usually the most common goal and will usually always be returned to. However in, say, an intelligent game that learns from its opponent's moves, long-term adaptation will be very important and will probably involve eventually forgetting some of the experience learned in early games played.

In computational terms, adaptation is a mechanism for enabling the complete system (which involves both the users and the computer application) to still achieve its goals in the most efficient manner, even when disturbances and situations occur that could interfere with those goals. There are many ways in which the system might be disturbed. For example, the user–application system might work well under a particular configuration when the users are experts, but it may not work well when the users are novices. In such a case, the computational aspects of the system will have to adapt their procedures and interface characteristics so that the novices can still effectively interact with the system. If, later on, it detects that experts are again interacting with the system, then adaptation back to the original configuration will be required. It is clear that such a system will require both task models and user models to be effective.

In the process control situation, the main goal of the combined operator–system is normally to maintain the system's ability to perform according to its specifications, that is, to maintain its equilibrium state. This is true regardless of whether the adaptive system is a technological system, a human operator or a team, or a joint human–machine system. The joint system of human team and machine must be able to maintain performance at an acceptable level, despite potentially disrupting events. This is achieved by activating an appropriate response to the disturbance. During this response the goals of the system, and hence the nature of its performance, may change, for instance from keeping the system running in a stable state to establishing a safe shutdown state. In either case the performance must remain within acceptable levels. Adaptation happens, of course, in conventional systems, but in these cases it is the operators who do all the adapting. An adaptive approach should provide a more equitable system when the interface system itself contributes to the adaptation process as well.

In all cases described herein, we can create a more usable definition of an adaptive computer system: The primary role of an adaptive computer application is to enable the operators or users to attain their goals with minimum cognitive workload.

So, in the case of traditional computer applications, adaptation enables the system to minimise cognitive workload by adjusting to the characteristics and experience of the user and the context of the task. In the case of process control, adaptation can enable the system to present the most salient information at the right time and in the right context so as to minimise operator cognitive workload.

EARLY ATTEMPT IN DESIGNING ADAPTIVE SYSTEMS

Early attempts at creating adaptive interfaces were based on the concept of the user interface management system, or UIMS (Edmonds, 1982), which eventually developed into the Seeheim Model (Pfaff, 1985). In this model, the application code is separated from the code concerned with presenting the interface.

The idea is shown in Fig. 7.1. The interface is further separated into a dialogue component, a presentation component, and a user model. The dialogue component controls the sequence of interactions and the presentation component actually renders the interface to the operators. Early examples of dialogue systems were SYNICS (Guest, 1982), RAPID (Wasserman & Shaw, 1983), and CONNECT (Alty & Brookes, 1985). The model provides an architectural approach for implementing adaptation. With the model, different renderings can be presented for the same dialogue sequence. So, for example, in a traditional computer application such as

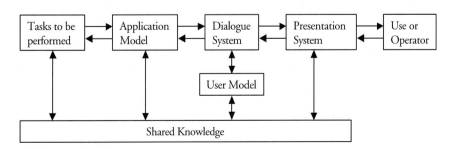

FIG. 7.1.   A UIMS.

a word processor or spreadsheet, different interfaces can be presented for experts and novice users.

One of the earliest adaptive systems that used the UIMS idea was the CONNECT system (Alty & Brookes, 1985). In this system, the dialogue component was a transition network, with each node generating renderings in the presentation system. The dialogue network actually consisted of three parallel layers for communicating with novice, intermediate, and expert users (Alty, 1984a). An expert system observed the user transitions on the network and opened or closed transition arcs between the three layers. In this way, a fairly rudimentary adaptive system was created and was tested on users. One novel feature of the system was that it used path algebra techniques to analyse the dialogue networks (Alty, 1984b).

The AID project (Browne, Totterdell, & Norman, 1990) provides another example of an attempt to create an adaptive system. The project was initiated to "research the techniques appropriate to the development of user interfaces that adapt to a particular user's ability and knowledge of a given system" (p. 2). The term *adaptive* referred to both user-tailoring and self-adaptation. The project built an adaptive front end to an electronic mail system. Later a document support system was built that provided adaptation in a number of ways. A task organiser adapted to linguistic context, the system adapted to the help preferences of a community of users, a reference information provider adapted to relevance, and an adaptive menu structure adapted to frequency of use.

One of the difficulties of discussing early adaptive systems is that similar ideas were concurrently emerging from different disciplines, which used their own terminology. This makes comparisons and generalisations difficult. Systems that were described as "intelligent" took many forms, and they were built for many different reasons and to achieve many different goals (Elkerton, 1987; Mason & Edwards, 1988). Examples of relevant application areas include intelligent support systems and intelligent help. However, claims of success were often exaggerated, and implemented systems were usually only successful in well defined, manageable areas or where they dealt with limited, more tractable issues.

The early attempts to create adaptive systems were certainly hampered by a lack of processing and input–output power, and it was only in the mid-1990s, when powerful computers became available, that consideration of how to build such systems became of interest again. The development of multiagent systems (Laurel, 1990; Maes, 1991) also focused attention on adaptive systems.

PRINCIPLES FOR CONSTRUCTING ADAPTIVE SYSTEMS

A Model of an Adaptive System

Benyon and Murray (1993) have described a generalized architecture for adaptive systems. This architecture contains a model of the user, the system, and the user–system interaction. The adaptive system uses its knowledge of these domains to determine when adaptation is necessary and what form it should take. The adaptive system takes cues for when to adapt from the user or system by matching them with rules contained within the user–system models (Fig. 7.2). The changes can be determined from utilising its user–system interaction model as a means of capturing salient interactions between the two domains and characterising them within the individual models.

The user model captures what the system believes to be the knowledge and preferences of the user(s). It can be used to generate adaptability proactively when its knowledge of the user's state indicates that adaptation may be required, or by cooperating interactively with the user to deduce when adaptation is required. For an accurate model of the user to be maintained, it

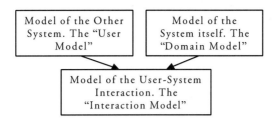

FIG. 7.2.   Overview of an adaptive system.

is essential that the user model monitors and samples user behaviour and keeps both a history of past actions and an up-to-date model of the current user state. It uses these data to attempt to deduce the user's current goal and then alters the system in some way so as to facilitate the achievement of that goal.

The knowledge represented in the user model may be acquired *implicitly* from inferences made about the user, or it may be *explicitly* elicited from the user. Explicit acquisition may be achieved through some cooperative behaviour such as asking relevant questions in specific contexts.

The domain model is required in order to define the aspects of the application that can be adapted or that define the context in which adaptation should take place. Other similar terms that have been used to describe this concept include *application model, system model, device model,* and *task model.* The domain model serves a number of purposes. It forms, with the user model, the basis for all the inferences and predictions that can be made from the user–system interaction. Therefore, it is important that the model is defined at an appropriate level of abstraction to allow the required inferences to be made.

The domain model consists of one or more abstractions of the system. These abstractions allow the adaptive system to reason about the target application, to facilitate adaptations by other agents, and to evaluate its effectiveness. For example, if the system is to be capable of adapting the screen displays, then there must a submodel describing the screen state in the domain model. If it is to adapt the functionality of the system, then the domain model must contain representations of alternative functional capabilities and the relationship between these functions. Similarly, if the system is required to alter the description of concepts, then these too must be modelled.

The benefits to be gained from having an explicit and well-defined domain model are considerable and have long been recognised in artificial intelligence research (Winston, 1992). A separate domain model provides improved domain independence that allows easy refinement of the domain model. Thus the domain model is a description of the application containing facts about the domain, that is, the objects, their attributes, the relationships between objects, and the processed involved. It is the designer's definition of all aspects of the application relevant to the needs of the adaptive system. A central question, however, in constructing a domain model is deciding what level of description should be represented.

The interaction model is a representation of the actual and designed interactions between user and application. An interaction is an exchange between a user and the system at a level that can be monitored. Data gathered from monitoring interaction can be used to make inferences about the user's beliefs, plans, or goals, long-term characteristics such as cognitive traits, or profile data such as previous experience. The system may tailor its behaviour to the needs of a particular interaction, or, given suitably "reflective" mechanisms, the system may evaluate its inferences and adaptations and adjust aspects of its own organization or behaviour. This interaction model is much closer to the notion of a discourse model or dialogue model than

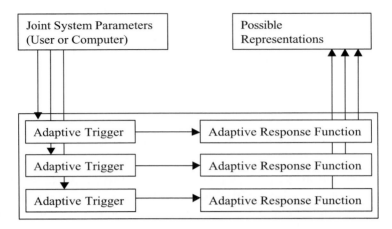

FIG. 7.3.   Adaptive triggers and adaptive response functions.

the interaction model proposed by Hutchins (Hutchins, 1989), which was a theoretical representation of human–computer interaction in general. In some representations (Alty & McKell,
1986) the interaction model was seen as a part of the domain model. However, it can also be
modelled as an entity in its own right or be seen as a function of the user model.

Triggers and Adaptive Functions

Because the purpose of adaptation is to adjust system operation so as to minimise cognitive
workload during a disturbance, the system must be able to measure, in some way, deviations
in performance so as to trigger adaptation when required. Such deviations might be observed
changes in operator or user behaviour or deviations in system performance. These deviations
(or *adaptive triggers*) define the conditions for the initiation and cessation of the adaptive
functions.

If a set of initiating conditions (or adaptive triggers) can be defined and determined, the
second main design requirement is the definition of the compensating (or adaptive) response
functions (see Fig. 7.3). These adaptive response functions must serve to further the overall
goal of the joint system. This requires a definition of more specific goals that could be derived
from consideration of a cognitive task hierarchy. For instance, if in the process control situation
the operators seem to have lost control and if the rate of system output information is very
high, then an appropriate response might be to reduce the rate of information by removing
low-priority information through filtering or providing a higher level of abstraction. As another
example, if the operators are slow in responding even though the state of the process is normal,
a reasonable adaptive response function might amplify or strengthen the annunciation of the
plant conditions that require a response.

Adapting Through Different Interface Media

One important factor that can contribute to cognitive workload is the set of representations
used in the interface between the user (or operator) and the computer system. It is generally
acknowledged that an appropriate choice of interface representation can make an easy problem
difficult and vice versa, so this is an important application area for adaptive systems. If run-time
decisions can be made, in context, that automatically switch representations to assist cognitive
problem solving, then the cognitive workload may be considerably reduced. How and when
to switch media is, of course, largely an unsolved problem.

In addition, we do know that different media in different interface situations can dramatically affect cognitive workload. The three-dimensional shape of a complex molecule is a key factor in solving many pharmaceutical problems. Presenting this information as a list of (x, y, z) coordinates of electron density values, while containing the necessary information to solve the problem, would place an enormous cognitive strain on the chemist, compared with the same information presented as a three-dimensional view of the molecule. In other situations, information presented as a movie or diagram may be understood much more easily than blocks of text. An inappropriate choice of medium for presenting information could therefore be a key contributor to cognitive workload. The problem is determining not only what is the most appropriate medium in a given context, but also the appropriate level of abstraction for presenting the information. In a dynamic system, a process and instrumentation diagram might be ideal for some problems, but higher levels of abstraction such as mass or energy flow (Lind, Larsen & Osman, 1992), or even representations, which do not have a one-to-one correspondence with process variables (Elzer & Beuthel, 1999), might be more appropriate.

The problem is not a simple one. Experiments have shown that different media affect learning, comprehension, and task efficiency (Alty, Bergan, Craufurd, & Dolphin, 1993), and that the interaction is a complex one involving the user, the task, and the medium. For a particular user, different media have different levels of expressiveness (Alty, 1999) for representing a particular task. Too little expressiveness and the medium cannot properly represent the problem set (e.g., trying to communicate the play *Hamlet* by using morse code). Too much expressiveness and the richness of the medium can contribute too much information noise (e.g., describing a complex graphical relationship by using text). Using adaptation to select the right level of expressiveness could be an important application of adaptive systems.

In cognitive workload terms, the right level of expressiveness means not only presenting to the user or operator the required information but doing so in a manner in which the information can be processed with the minimum cognitive effort. For example, there is some evidence that additional redundant information can impede learning. Moreno and Mayer (1999) added extraneous musical material to a learning situation, and this tended to adversely affect student understanding of the material. In contrast, adding relevant and coordinated auditory material—in the form of environmental sounds—did not adversely affect students' understanding. However, adding auditory material that did not contribute to making the lesson more intelligible could create auditory overload.

There is also a body of research work (called cognitive load theory) related to working memory limitations and their effects on learning that may be relevant (Mousavi, Low, & Sweller, 1995; Sweller, 1988) Moreno and Mayer (1999) showed that students learned better when the instructional material did not require them to split their attention between multiple sources of mutually referring information. In one experiment, students viewed animation with either concurrent narration describing eight major steps in lightning formation (group AN) or concurrent on-screen text involving the same words and presentation timing (group AT). Group AN generated significantly ($p < .001$) more correct solutions than group AT on the transfer test, a result consistent with the predictions of the cognitive theory of multimedia learning. Moreno and Mayer called this the split-attention principle.

Software Agents and Adaptive Systems

Over the past few years, agent-based systems (Bradshaw, 1997) have emerged from research in artificial intelligence. A multiagent system offers a good technical approach for implementing an adaptive, flexible, interface for the following reasons.

1. Modularity: Agents are proactive objects, and they share the benefits of modularity that have led to the widespread adoption of object technology. A good candidate for an agent-based approach has a well-defined set of state variables that are distinct from those of its environment, and its interfaces with that environment can be clearly identified.
2. Decentralized: Agents do not need to be invoked externally. They autonomously monitor their own environment and take action when they deem appropriate. This makes them particularly suited for applications that can be decomposed into stand-alone processes, each capable of doing useful things without continuous direction by some other process.
3. Changeable: Agents are well suited to decentralised problems because they are proactive objects. Modularity permits the system to be modified one stage at a time. Decentralisation minimises the impact that changes in one module have on the behaviour of other modules.
4. Ill structured: In process control, not all of the necessary structural information is available when the system is designed. Such a situation is a natural one for the application of agents.
5. Complex: One measure of the complexity of a system is the number of different behaviours it must exhibit. Typically, the number of different interactions among a set of elements increases much faster than does the number of elements in the set. Agent architectures replace explicit coding of this large set of interactions with run-time generation.

Agent technology therefore offers a useful implementation pathway for adaptive systems. However, one problem with the application of agent-based systems has been the lack of good implementation tools, but this situation is rapidly improving and some useful development systems are now entering the marketplace.

EXPERIENCES WITH AN ADAPTIVE SYSTEM—THE AMEBICA PROJECT

In 1998, a proposal for developing an adaptive interface for process control applications was accepted by the European Commission (Project 21972). Its name was AMEBICA. The objective of the project was to construct an adaptive interface to a process control application by using agent technology to investigate both the technical issues in construction and whether the resulting system could reduce the overall cognitive load on the operators. The project consisted of a number of academic and industrial partners: Alcatel CIT (Project Manager), Softeco Sismat SpA (an Italian software house), Elsag SpA (the Network application, Italy), Ecole des Mines d'Ales (Nimes, France), the IMPACT group at Loughborough University (UK), Iberdrola S. A. (Spain, the Thermal Power Plant), Labein Research Centre

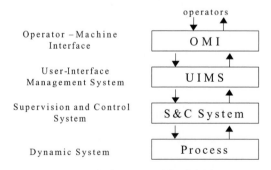

FIG. 7.4.   A conventional process control system.

(Spain), and the Institut For Energiteknikk (Norway). The project submitted its final report in March 2001.

In Hollnagel's classification of adaptive approaches (Hollnagel, 1995), the AMEBICA system *adapts through performance* rather than *through design* or *through management*. Two application domains were chosen—the control of an electrical supply network in Milan and a conventional thermal power plant in Spain.

A conventional process control system has four basic constituent parts, as shown in Fig. 7.4. The basic components in the figure are the process, the supervision and control system (S & C system), the UIMS, and the operator–machine interface (OMI).

The OMI is where actual information presentation takes place, and where the operators carry out the various control actions—both actions to control the system and actions to get additional information. The flow of information through the whole system is bidirectional. The objective of the system is to minimise system instability. A key feature of AMEBICA is an attempt to minimise operator cognitive workload through an adaptive OMI.

In conventional systems, designers choose a specific mapping of process parameters onto OMI modalities and media at design time. There are many possible mappings, from low level to high level (Lind et al., 1989), and one mapping cannot be optimal for all possible system states so there will always be an element of compromise in the choice made. The idea of a flexible mapping that can be adapted at run time is a possible solution to this dilemma.

There are two possible approaches to the provision of a flexible mapping at run time: first, choose from a set of designer predefined mappings, or, second, generate each mapping on the fly at run time. The decision of which approach to use must be based on a careful trade-off between the complexity of the solution, the requirements to the application (e.g., response time), and the level of adaptation it offers.

In AMEBICA the latter approach has mainly been adopted, though some elements of the former have also been incorporated as well. As an example of the former case, the interface designer predefines a number of possible representations for a particular process state at design time and a selection is made between them at run time. An example of the latter is when the adaptive reasoning is not related to the process state (e.g., improving the spatial layout or centering an important element on the screen). In this case, generative reasoning techniques are used. Agent technology has been adopted in both approaches.

The AMEBICA system is therefore an agent-based presentation system that adapts the content of the operator interface according to the current state of the operator (or operators) in the control room and the current process state. The basic underlying hypothesis is that different display configurations are more appropriate for particular operator–process tasks. The AMEBICA system, therefore, adapts the interface by moving to what it considers to be a more optimum presentation representation when it detects operator or system state problems.

The AMEBICA agent architecture consists of a set of cooperating agents. The way adaptation is triggered is shown in Fig. 7.5. The normal situation in AMEBICA is characterised by

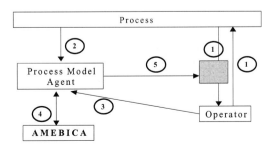

FIG. 7.5. Triggering adaptation in AMEBICA.

information flows coming directly to the operators from the process and control actions from the operator to the system (shown as 1 in Fig. 7.5). Without adaptation, default presentation scenarios are used to communicate with the operator. This situation continues until the process model agent detects some disturbance in the operator (3) or system (2) state. It decides that an adaptation trigger exists and asks AMEBICA (4) to suggest interface modifications. These then occur (5).

AMEBICA makes adaptation decisions on the basis of its combined view of what is going on in both the process and the operator responses.

The AMEBICA system itself (within the AMEBICA box in Fig. 7.5) consists of a number of cooperating agents. Each data stream from the process is controlled by a media agent, which can alter the presentation of its stream and thereby change the representation. It does this by choosing from a set of predefined rendering possibilities after receiving advice from two agents—the media allocator agent and the rendering resolution agent. The former is concerned with screen layout (advised by a presentation agent) and the effects of changes on other displayed information (through other media agents). The latter is concerned with choices between the possible representations offered by the media agent. A human factors database (a collection of human factors rules) is also consulted. The presentation agent monitors screen real estate usage and advises the other agents on placement of screen material. For a more detailed treatment of how AMEBICA reasons, see Alty et al. (2001).

Therefore, as an adaptive system, AMEBICA has a domain model in the form of the process model agent, an operator model in the form of the operator agent, and a primitive interaction model. It also has a generic operator model stored in the human factors database. The media agents contain designer options available for the adaptation of screen objects, and adaptation reasoning knowledge is stored in the rendering resolution agent.

COGNITIVE WORKLOAD ISSUES

In the design of the system, a number of steps were undertaken to minimise effects on operator cognitive workload. For example, any delay between the occurrence of a process action and the rendering of its representation on the screen to the operators could have a serious effect (like trying to drive a car when the view of the driver is a few seconds behind reality). It was therefore decided to design AMEBICA such that the adaptive approach must not interfere *initially* in any way with the basic flow of data between process and operator. In other words, the process data must always be updated immediately at the graphical interface when changes occur in a variable. Only then may the system decide to adapt the display in some way if the process–operator triggering conditions are met. This adaptation will take place a short time after the initial rendering of the data, because a considerable amount of negotiation and reasoning between agents may be necessary. Adaptation will therefore always be a few seconds behind changes in the rendering system, but this was not expected to cause problems.

Another important issue in the design of a multiagent adaptive system is the choice of intra-agent communication policies (Faratin, Sierra, & Jennings, 1997), because these influence the ultimate efficiency and responsiveness of the adaptive aspects of the system. Agent negotiation strategies were therefore deliberately kept to a minimum (compared with multiagent systems from other domains, such as distributed Internet agent systems). Fortunately, the multiagent system in AMEBICA is a closed system. Because of this, agents need not query other agents for the existence of a source of information, and therefore there are no intermediary broker-ing agents. Furthermore, the agents within the closed system are not competing for limited resources or against each other; rather, they all have the same goal (the system goal). Thus, even though the various media agents may appear to be competing for screen real estate, their internal negotiation policy will obey any overall decision made by the allocator agent in the belief that it is for overall systemwide good.

THE ADAPTABILITY MATRIX—A TOOL FOR DEVELOPING AN ADAPTABLE INTERFACE

One of the main problems we encountered when trying to implement the AMEBICA system in our two application areas was convincing designers when and how to use the adaptive approach. Although designers could understand the general need for adaptation in certain circumstances, they lacked a framework within which to make those decisions. We have therefore developed an approach that has proved a useful tool for designers and has clarified for us the nature of the adaptive process.

Adaptation will not be activated all the time, but only when the defined conditions are reached or certain thresholds have been passed (Hollnagel, 1998). Then, one or more appropriate adaptive functions are needed to provide adaptation, which will cease when another set of conditions has been fulfilled.

In AMEBICA, the process model is the source of adaptation information (combinations of operator state and process state), and the adaptive response is a set of recommended modifications to the rendering system. The process model agent observes both the operator activity and process state and triggers adaptation on defined conditions. The general form of the adaptability matrix is shown in Fig. 7.6.

The columns of the matrix are possible process-triggering conditions, and the rows are the possible operator-triggering conditions. At each intersection there will usually be an adaptable function $(\alpha, \beta, \gamma, \ldots)$. In an initial design, the triggering conditions will be quite high level (e.g., the operator is providing erratic responses, or the process is in a disturbed state with high information flows), but these can be progressively made more detailed in a stepwise fashion as the design becomes more detailed. The adaptable functions will similarly initially be rather general (e.g., accentuate presentation).

	Process Condition P	Process Condition Q	Process Condition R	→
Operator Condition A	Adaptable Function α	Adaptable Function ε	Adaptable Function φ	
Operator Condition B	Adaptable Function β	Adaptable Function γ	Adaptable Function κ	
Operator Condition C	Adaptable Function χ	Adaptable Function η	Adaptable Function λ	
Operator Condition D	Adaptable Function δ	Adaptable Function ι	Adaptable Function ν	
↓				

FIG. 7.6. The form of the adaptability matrix.

TABLE 7.1
An Actual Adaptability Matrix

Operator Response	Process Status Normal	Process State Information Rate	
		Disturbed, High	Disturbed, Low
Normal.	OK, no action.	OK, no action.	OK, no action.
Delayed (relative to expected response).	(1) Inattentive. Accentuate presentation.	(4) Overloaded. Filter information; simplify presentation.	(7) "Frozen." Repeat recent information. Try alternate representation.
Erratic (occasionally wrong display or commands).	(2) Inattentive. Accentuate presentation (specific).	(5) Overloaded. Simplify displays; remove information.	(8) Partial loss of comprehension. Switch modality.
Disorganised (constantly wrong display or commands).	(3) Confused, loss of control. Go to overview presentation.	(6) Severe loss of control. External help.	(9) Complete loss of comprehension. Go one level up; summarise info.

The matrix used in the AMEBICA exemplars has 4 operator states and 3 process states, leading to a matrix with 12 cells as shown in Table 7.1. In a real system, this matrix could be further developed and additional operator or process states added to give a finer granularity.

There was considerable debate over the nature of the adaptive functions within the project. For example, not all participants agreed that there should be no action when the process state was disturbed but the operator responses were normal. They argued that the operators should be warned. In contrast, others felt that this simply meant that the operators were in full control and that there was therefore no need for any adaptation.

When the process is in a normal state and operator behaviour is abnormal, we argue that such behaviour is probably caused by inattention or activity on another task and we therefore respond with an adaptive response that draws the operators' attention to salient features of the problem.

The adaptive responses in the High Information Rate column (this occurs usually at the commencement of a disturbance) show a filtering adaptive response. Here we argue that the operators are overloaded with information and that the response should be to filter information, simplify displays, and remove unnecessary information. Such actions should reduce the current state of cognitive overload.

In the Low Information Rate column, the adaptive functions are responding to a different situation. In this case the operator cognitive overload has resulted in loss of control, or has resulted in inactivity caused by confusion, and the system is beginning to initiate safety procedures that may eventually result in shutdown. In such cases we argue that the adaptive response should assist the operators in breaking out of their current "tunnel" vision view of the system. The response is therefore to change representations, abstractions, modalities, and viewpoints in order to encourage the operators into different lines of reasoning and consider different causes and effects.

The matrix will, of course, take different forms in different applications. However, we believe that the broad structure represents responses to different forms of cognitive overload. For example, the response in the Process Normal column for progressively disorganized operator behaviour is to accentuate presentation.

DID THE ADAPTABLE NATURE OF THE AMEBICA INTERFACE AFFECT OPERATOR COGNITIVE LOAD?

AMEBICA prototypes have been developed in two contrasting application domains—an electrical network management (ENM) in Italy, and a thermal power plant in Spain. In both cases the architecture has been implemented for a subset of the application. Because of space limitations, this section concentrates on results from the ENM application.

In ENM applications, the network (high voltage or medium voltage lines, substations, and users or loads) is continuously supervised and controlled to achieve optimum and safe operations. The graphical interface is the most important point of interaction with the system. The cognitive loads placed on the operators determine the efficiency and effectiveness of this interface. The system must provide the right information about the status of the network and allow the operator to react in a timely manner to any abnormal situation, in order to restore normal operation. Furthermore, a well-designed interface that maintains the cognitive loads at acceptable levels will reduce the stress on the operator, theoretically resulting in safer, and more economic, decision making. Typical tasks of the operator in the ENM application include network supervision, network configuration analysis, security assessment of the operations, optimising operations, network switching, scheduling recovery from emergency conditions, and planning network development and maintenance.

It is obviously not possible to implement an exemplar in an operational network control centre, so the AMEBICA framework was coupled to a simulated environment but used example inputs from real process data. An exhaustive "living" mock-up of the interface was prepared, running on a PC (using the agent technology) to provide the feeling of the real functionality under the stimulus of an event simulator. The situation prototyped was what is usually referred to as the "moving storm." This is a situation in which a moving electrical disturbance in the atmosphere develops in the network into a "storm" of alarm and error situations as effects are propagated through the network. Live data from the application were collected in a data file and the prototype and was driven by a "discrete" simulator that generated events in such a way to simulate the moving storm.

A number of adaptive functions were implemented:

- Spatial adaptation—the network diagram is positioned such that the alarm cause is at the centre of the screen. The display is reorganised (perhaps by zoom and translation operations) to ensure all relevant alarms are presented on the screen.
- Accentuation—the fault area is accentuated and a new window opened with alarm detail. The area affected by the previous alarm is highlighted.
- Change of media—to attract the operators' attention, alarms can be switch from visual to audio.

The AMEBICA software of the prototype reacted adaptively, illustrating that the *approach was technically feasible.* The prototype demonstrated acceptable response times and adaptive multimedia capabilities. However, the key aspect was whether the operators and designers viewed the approach as a potential advance or not.

Evaluation of AMEBICA

To determine the effectiveness of the approach, two usability evaluation workshops were attended by the software designers, chief operators, engineers, managers, and operators from the ENEL Distribuzione Company.

The prototype was presented to the group, illustrating how AMEBICA handled the moving storm scenario. This was followed by interview-discussions and questionnaires. The discussion was very open and helped in achieving considerable progress in the improvement of the interface design.

A questionnaire was developed that probed different aspects of cognitive workload and the likely affect of AMEBICA. For example, factors that made the operators' tasks easier were identified as follows:

- Information presented only when needed, in context and in a logical manner.
- Natural relationships made clear between data, particularly between different adaptations of the same data.
- Feedback at the appropriate time.
- Appropriate media usage.
- Spatial organisation after adaptation.

Factors that increased cognitive workload were identified as follows:

- Inappropriate, conflicting, or slow adaptation functions.
- The obscuring of important information by new information.
- Inappropriate zooming and translation.
- Misunderstanding of operator intentions, resulting in unnecessary intrusion.
- Inconsistencies caused by adaptation.

Overall, AMEBICA made a favourable impression. Operators reported that adaptation allowed them to focus on the solution to a problem without worrying about looking for specific information. They liked the optimal spatial reorganisation of the display and the use of multimedia techniques to display information. However, they still felt that the adaptive functions were often too intrusive and that the system frequently adapted when there was no need.

What was most interesting was that a general initial feeling of conservatism on the side of the final users was quickly replaced by a much more collaborative attitude, finally resolving into enthusiastic and sincere offers of collaboration for possible future developments of the technology.

It is expected that if ever the full integration of the AMEBICA in an ENS system were to be achieved, it would assist in a more efficient and cost effective supervision of electrical systems. In particular, this should allow more prompt actions in order to avoid possible failure of the system, caused, for example, by undetected degradation. Consequently, the approach may result in financial benefits by avoiding lost revenue during outages and the possibility of penalties for having failed to satisfy mandatory standards of supply.

CONCLUSIONS

We have traced the development of adaptive systems from theoretical ideas to an actual prototype. Most experiments probing adaptive approaches fail because the application areas are usually too simple and are unlikely to scale up. The AMEBICA prototype had the virtue of being close to a real (and large) system, but inevitably it was still a subset and a simulation. However, it did not really attempt to determine the operator state and sometimes got it wrong, even on the simple axes that we had defined. This is a hard problem, and it will continue to limit the usefulness of the adaptability approach.

The adaptability matrix approach, after some initial doubts, was favourably received and provided a tool for relating adaptation and cognitive workload. We believe that the adaptability matrix approach is probably one of the most useful techniques to come out of the AMEBICA experiment. We also think that that it has wider applications than process control. For example, in the learning situation, the process state might relate to stages of learning, and the operator state to the current state of user understanding.

We were also able to construct a fairly large adaptive system by using agents that had an acceptable performance, though much hand-coding had to be done to achieve this. This situation should improve as better tools arrive in the marketplace.

The system necessarily had limited evaluation results, but the comments and evaluations for operators and designers did indicate that some aspects of the system performed well. In particular, we did not experience serious problems over adaptive changes in representation. We certainly did not get it right, and there were times when operators were irritated by sudden adaptations, or being wrongly classified as "inattentive" when they were actually attending to something else. However, their support for the system did slowly grow from extreme scepticism to enthusiastic support when they began to recognise what such a system might provide.

Taking Hollnagel's adaptability criteria (Hollnagel, 1995), with respect to adapting through performance, we were able to *adapt within seconds* without destroying the immediacy of the data by using the streaming concept. In terms of *generality,* our solutions were focused at performance types and were therefore not general, but we did develop an approach that provided a measure of generality in our adaptive system. The process model provides the conversion for a particular domain-specific situation into more general activity in the AMEBICA adaptive system. The rendering system and the media agents provide the reverse path from AMEBICA generality to domain-specific displays. Certainly our approach was *resource intensive*. Hollnagel notes that for adaptation during performance, evaluation is possible but only to the extent that solutions are specific. This makes it difficult to generalise any evaluation results.

We are only just beginning to understand something about adaptation and how to build adaptive systems, so the difficulties should not be underestimated. Although techniques for assessing the process state look promising, artificial intelligence techniques still have a long way to go before we can confidently use them to assess the state of the operator. Recent advances in agent design have improved our technical capability; however, really understanding the nature of cognitive workload and how to influence it is still some way off for complex systems.

ACKNOWLEDGEMENTS

The author thanks all those work contributed to the AMEBICA project—Veronique Dauresan, Chris Khalil, Erik Hollnagel, Chris Khalil, Gianni Vianno, Daniel Biglino, Michel Crampes, Inaki Angulo, Lauro Mantoani, Paulo Frugione, and their colleagues, without whose contribution this paper could not have been written. All those involved in the AMEBICA project express their thanks to the European Commission, who provided funding under the Basic Research Initiative of Framework IV.

REFERENCES

Alty, J. L. (1984a). The use of path algebras in an interactive dialogue design system. In B. Shackel (Ed.), *Proceedings of INTERACT'84* (pp. 351–354). New York: North-Holland, 321–324.

Alty, J. L. (1984b). The application of path algebras to interactive dialogue design. *Journal of Behaviour and Information Technology, 3*(2), 119–132.

Alty, J. L. (1999). Multimedia and process control interfaces: Signals or noise? *Transactions of the Institute of Measurement and Control, 21*(4/5), 181–190.

Alty, J. L., Bergan, M., Craufurd, P., & Dolphin, C. (1993). Experiments using multimedia in process control: Some initial results. *Computers and Graphics, 17*(3), 205–209.

Alty, J. L., & Brookes, A. (1985). Microtechnology and user friendly systems: The CONNECT dialogue executor. *Journal of Microcomputer Applications, 8,* 333–346.

Alty, J. L., Khalil, C. J., & Vianno, G. (2001). Adaptable interfaces in process control: Experience gained in the AMEBICA project. In M. Lind (Ed.), *Proceedings of the XX Annual Conference On Human Decision Making and Manual Control* (pp. 47–62). Copenhagen: Technical University of Denmark.

Alty, J. L., & McKell, P. (1986). Application modelling in a user interface management system. In M. Harrison & A. Monk (Eds.), *People and computers: Designing for usability* (pp. 319–335). Cambridge, England: Cambridge University Press.

Benyon, D. R., & Murray, D. M. (1993). Experience with adaptive interfaces. *The Computer Journal, 33*(5), 453–461.

Bradshaw, J. M. (1997). *Software agents.* Menlo Park, CA: AAAI Press/MIT Press.

Browne, D., Totterdell, P., & Norman, M. (1990). *Adaptive user interfaces* (p. 226). London: Academic Press.

Edmonds, E. A. (1982). The man-computer interface: A note on concepts and design. *International Journal of Man-Machine Studies, 16,* 231–236.

Elkerton, J. (1987). A framework for designing intelligent human-computer interaction. In G. Salvendy (Ed.), *Cognitive engineering in the design of human-computer interaction and expert systems* (pp. 567–574). Amsterdam: Elsevier.

Elzer, P., & Beuthel, C. (1999). Three-dimensional representation of process values. In *Proceedings of HCI International, Munchen* (pp. 1296–1300). Mahwah, NJ: Lawrence Erlbaum Associates.

Faratin, P., Sierra, C., & Jennings, N. R. (1997). Negotiation decision functions for autonomous agents, *International Journal of Robotics and Autonomous Systems, 24*(3–4), 159–182.

Guest, S. P. (1982). The use of software tools for dialogue design. *International Journal of Man-Machine Studies, 16,* 263–285.

Hollnagel, E. (1995). The art of man-machine interaction: Improving the coupling between man and machine. In J.–M. Hoc, P. C. Cacciabue, and E. Hollnagel (Eds.), *Expertise and technology: Cognition & human computer interaction* (pp. 229–241). Hillsdale, NJ: Lawrence Erlbaum Associates.

Hollnagel, E. (1998). *Cognitive reliability and error analysis method (CREAM).* Oxford, England: Elsevier.

Hollnagel, E., & Woods, D. D. (1983). Cognitive systems engineering: New wine in new bottles. *International Journal of Man-Machine Studies, 18,* 583–600.

Hutchins, E. (1989). Metaphors for interface design. In M. M. Taylor, F. Neel and D. G. Bouwhuis (Eds.), The Structure of Multimodel Dialogue. Amsterdam: Elsevier, Amsterdam, 11–28.

Laurel, B. (1990). Interface agents. In B. Laurel (Ed.), *The art of human-computer interface design* (pp. 355–366). Wokingham, England: Addison-Wesley.

Lind, M., Larsen, M. N., & Osman, A. (1992). Applications of multilevel flow modeling. In *Proceedings of the International Conference on Design and Safety of Advanced Nuclear Power Plants (ANP'92),* Tokyo, Japan Publications, *4,* 42.3/1–10.

Maes, P. (1991). Agents that reduce work and information overload. *Communications of the ACM, 37*(7), 31–40.

Mason, J., & Edwards, J. L. (1988). Surveying projects on intelligent dialogue. *International Journal of Human-Computer Studies, 28,* 259–307.

Moreno, R., & Mayer, R. E. (1999). Cognitive principles of multimedia learning: The role of modality and contiguity. *Journal of Educational Psychology, 91,* 358–368.

Mousavi, S. Y., Low, R., & Sweller, J. (1995). Reducing cognitive load by mixing auditory and visual presentation modes. *Journal of Educational Psychology, 87,* 319–334.

Norman, D. A., & Draper, S. W. (1986). *User centered interface design: New perspectives on human computer interaction.* Hillsdale, NJ: Lawrence Erlbaum Associates.

Pfaff, G. E. (Ed.). (1985). *User interface management systems (Seeheim Workshop).* Berlin: Springer.

Rasmussen, J. (1986). *Information processing and human-machine interaction: An approach to cognitive engineering.* New York: North-Holland.

Shneiderman, B. (1983). *Designing the user interface: Strategies for effective HCI* (3rd ed.). Reading, MA: Addison-Wesley.

Sweller, J. (1988). Cognitive load during problem solving: Effects on learning. *Cognitive Science, 12,* 257–285.

Thimbleby, H. (1990). *User interface design.* Wokingham, England: Addison-Wesley.

Wasserman, A. I., & Shaw, D. T. (1983). A RAPID/USE Tutorial, Medical Information Science, Univ. of California (Son Diego).

Winston, P. (1992). *Artificial intelligence.* Reading, MA: Addison-Wesley.

8

Adaptive Automation: Sharing and Trading of Control

Toshiyuki Inagaki
University of Tsukuba, Japan

Abstract

Function allocation is the design decision to determine which functions are to be performed by humans and which are to be performed by machines to achieve the required system goals, and it is closely related to the issue of automation. Some of the traditional strategies of function allocation include (a) assigning each function to the most capable agent (either human or machine), (b) allocating to machine every function that can be automated, and (c) finding an allocation scheme that ensures economical efficiency. However, such "who does what" decisions are not always appropriate from human factors viewpoints. This chapter clarifies why "who does what and when" considerations are necessary, and it explains the concept of *adaptive automation* in which the control of functions shifts between human and machines dynamically, depending on environmental factors, operator workload, and performance. Who decides when the control of function must be shifted? That is one of the most crucial issues in adaptive automation. Letting the computer be in authority may conflict with the principle of *human-centered automation,* which claims that the human must be maintained as the final authority over the automation. Qualitative discussions cannot solve the authority problem. This chapter proves the need for quantitative investigations with mathematical models, simulations, and experiments for a better understanding of the authority issue.

Starting with the concept of function allocation, this chapter describes how the concept of adaptive automation was invented. The concept of *levels of automation* is used to explain interactions between humans and machines. *Sharing* and *trading* are distinguished to clarify the types of human–automation collaboration. Algorithms for implementing adaptive automation are categorized into three groups, and comparisons are made among them. Benefits and costs of adaptive automation, in relation to decision authority, trust-related issues, and human–interface design, are discussed with some examples.

FUNCTION ALLOCATION

Suppose we are to design a system with specific missions or goals. We first have to identify functions that are needed to accomplish the goals. We then come to the stage of function allocation. Function allocation refers to the design decisions that determine which functions are to be performed by humans and which are to be performed by machines. Various strategies for function allocation have already been proposed.

Traditional Strategies for Function Allocation

Rouse (1991) classified traditional function allocation strategies into three types. The first category is termed *comparison allocation*. The strategies of this type compare relative capabilities of humans versus machines for each function, and they allocate the function to the most capable agent (either human or machine). The most famous MABA–MABA (what "men are better at" and what "machines are better at") list may be the one edited by Fitts (1951); see Table 8.1.

The second type is called *leftover allocation*. The strategies of this type allocate to machines every function that can be automated. Human operators are assigned the leftover functions to which no automation technologies are available.

The third type is named *economic allocation*. The strategies of this type try to find an allocation that ensures economical efficiency. Even when some technology is available to automate a function, if the costs of automating the function are higher than that of hiring a human operator, then the function is assigned to the operator.

Note here that the traditional strategies just described consider "who does what." Such design decisions yield function allocations that are *static*: Once a function is allocated to an agent, the agent is responsible for the function at all times.

Traditional Strategies Are Not Always Appropriate

Suppose design decisions are made by using either the leftover or the economic allocation strategies. The strategies do not reflect any human characteristics or viewpoints, and the resulting

TABLE 8.1
The Fitts List

Humans appear to surpass present-day machines with respect to the following:
- Ability to detect small amounts of visual or acoustic energy.
- Ability to perceive patterns of light or sound.
- Ability to improvise and use flexible procedures.
- Ability to store very large amounts of information for long periods and to recall relevant facts at the appropriate time.
- Ability to reason inductively.
- Ability to exercise judgment.

Present-day (in 1950s) machines appear to surpass humans with respect to the following:
- Ability to respond quickly to control signals and to apply great forces smoothly and precisely.
- Ability to perform repetitive, routine tasks.
- Ability to store information briefly and then to erase it completely.
- Ability to reason deductively, including computational ability.
- Ability to handle highly complex operations, i.e., to do many different things at once.

Note. After Fitts (1951), Hancock and Scallen (1998), and Price (1985).

function allocation may be elusive for operators. Some operator may ask, "Am I supposed to be responsible for this function, or is the automation?" Also, there is no guarantee that the allocations provide the operators with job satisfaction.

The comparison allocation may be nicer for the operators than either the economic or leftover allocations. However, the comparison allocation cannot be free from criticisms. Price (1985) and Sharit (1997) claimed that the list by Fitts is overly generalized and nonquantitative. Sheridan (2002) pointed out that, "in order to make use of the Fitts MABA-MABA list, one needs data that are context dependent, but these data are mostly unavailable" (p. 59). He argued, referring to the ideas of Jordan (1963), "the idea of comparing the human with the machine should be thrown out but the facts about what people do best and what machine do best should be retained," and "the main point of retaining the Fitts list is that people and machine are complementary" (p. 59). A complementary strategy can be seen in KOMPASS (Grote, Ryser, Wafler, Windischer, & Weik, 2000).

Even though the operators are allocated only functions in which people surpass machines, the superiority may not hold at all times and on every occasion. Operators may get tired after long hours of operations, or they may find it difficult to perform the functions under time pressure. This implies that "who does what" decisions are not sufficient, but "who does what and when" considerations are needed for the success of function allocation.

"Who Does What and When" Decisions: An Example

Aircraft in recent years have become equipped with various automations that can perform important functions to make flights safe, smooth, and efficient. Lateral navigation (LNAV) and vertical navigation (VNAV) are such essential functions. Pilots usually take responsibilities for both LNAV and VNAV during takeoff. In the climb phase, pilots may handle LNAV and let various automations deal with VNAV. During cruise, pilots often hand both LNAV and VNAV over to automation. In descending or landing, pilots may seize back control of either LNAV or VNAV. The two functions are allocated in different ways, depending on the situation.

What happens if aircraft must be designed in a "who does what" manner? No shift of control is allowed, and there can be only four design decisions: first, no automation is needed because the pilots are to handle both LNAV and VNAV at all times during flights; second, design the automation that performs LNAV for any circumstances; third, design the automation that performs VNAV at all times and for every occasion; or fourth, design the automation that performs LNAV and VNAV all the time during flights. It is easy to see that none of these are practical.

NOTIONS USEFUL FOR "WHO DOES WHAT AND WHEN" DECISIONS

Sharing of Control

Sharing of control means that the human and the computer work together simultaneously to achieve a single function (Sheridan, 1992). Three types of sharing are distinguishable. The first type is *extension*, in which the computer may help the human so that his or her capability may be extended (e.g., the power steering or the power braking of an automobile), or in which the human extends the computer's capability (e.g., "supervisory override" for some types of aircraft, in which the pilot may add control force when the maneuver by the autopilot was not perceived to be satisfactory).

The second type is *relief*, in which the computer helps the human so that his or her burden may be reduced. A lane-keeping support system for an automobile is a good example. The

TABLE 8.2
Scale of Levels of Automation

---

1. The computer offers no assistance; humans must do it all.
2. The computer offers a complete set of action alternatives, and
3.   narrows the selection down to a few, or
4.   suggests one, and
5.   executes that suggestion if humans approve, or
6.   allows humans a restricted time to veto before automatic execution, or
7.   executes automatically, then necessarily informs humans, or
8.   informs them after execution only if they ask, or
9.   informs them after execution if it, the computer, decides to.
10. The computer decides everything and acts autonomously, ignoring humans.

---

system detects white lane markers on the road, and it generates torque to assist the driver's steering action for keeping the host vehicle approximately on the center of the lane (Kawazoe, Murakami, Sadano, Suda, & Ono, 2001).

The third type is *partitioning,* in which a required function is divided into portions so that the human and the computer may deal with mutually complementary parts. A car driver may want to be responsible only for steering by letting the computer control the velocity. Partitioning is a complementary function allocation.

Trading of Control

Trading of control means that either the human or the computer is responsible for a function, and an active agent changes alternately from time to time (Sheridan, 1992). We have already seen an example on the flight deck in which the pilots and the automation trade controls for LNAV and VNAV functions occasionally during flights.

For trading of control to be implemented, it is necessary to decide when the control must be handed over and to which agent. It is also important who makes the decision. These issues are discussed later.

Levels of Automation

Human–computer interactions can be described in terms of the *level of automation* (LOA), originated by Sheridan and Verplank (1978). Table 8.2 gives a simplified version (Sheridan, 1992).

FUNCTIONS THAT MAY BE AUTOMATED

Parasuraman, Sheridan, and Wickens (2000) described human–computer interactions by distinguishing the following four classes of functions:

1. Information acquisition,
2. Information analysis,
3. Decision selection, and
4. Action implementation.

There can be various design alternatives regarding to what extent each of the four functions may be automated. In other words, an appropriate LOA must be chosen for each function. The automated forms of functions 1–4 are called, respectively, acquisition automation, analysis automation, decision automation, and action automation.

Acquisition Automation

When the LOA is set at the lowest, humans must themselves collect every piece of information at all instances. An example of the automated system information acquisition may be radar for automobiles or aircraft, or sonar for ships. Sometimes these systems simply collect information and display it on the screen. When the computer becomes more involved, certain types of acquired information may be highlighted to attract a human's attention. Filtering is another important capability for acquisition automation. It is well recognized that transmitting every piece of information to the operator may lead to undesirable events. For instance, as a result of lessons learned following the accident at the Three Mile Island nuclear power plant, some alarms may better be suppressed in certain circumstances. This is also the case in a commercial aircraft. Suppose an engine catches on fire during takeoff. Even if the sensors detected the fire successfully, the acquired information may be filtered. The fire bell will not ring until the aircraft climbs to a certain altitude (e.g., 400 feet), or until some amount of time (e.g., 25 seconds) elapses after V1 (the takeoff decision speed). Until then, master warning lights are inhibited.

Analysis Automation

If the LOA is set at some moderate level, the computer may be able to give humans some information by processing available raw data. One example of such information is the *prediction* of the future state of a system. There are various examples of this kind. For instance, in central control rooms of recent nuclear reactors, large screens display trend graphs for various parameters with their predicted values. When using a notebook computer, we can see the predicted remaining life of the battery if we put a mouse cursor on the battery icon. In the cockpit of an aircraft at a level flight, a navigation display may indicate an arc that shows at which point the aircraft is to start a descent (or an ascent). If the pilot changes the flight plan, the computer replaces the old arc with a new one.

Another example of the analysis automation is a system for integrating multidimensional information into an easily understandable form. The enhanced ground proximity warning system (EGPWS) for aircraft is such an example. The EGPWS was designed to complement the conventional GPWS functionality with the addition of look-ahead terrain alerting capabilities. The EGPWS has worldwide airport and terrain databases, and they are used in conjunction with aircraft position, barometric altitude, and flight path information to determine potential terrain conflict. The terrain may be shown on the navigation display in patterns of red, amber, and green, where the colors indicate the height of the terrain relative to the current aircraft altitude (Bresley & Egilsrud, 1997).

Decision Automation

As researchers in naturalistic decision making say, it is often useful to distinguish *situation-diagnostic decisions* and *action-selection decisions* (Klein, Orasanu, Calderwood, & Zsambok, 1993; Zsambok & Klein, 1997). For a situation-diagnostic decision the operator needs to identify "what is going on," or to select the most appropriate hypothesis among a set of diagnostic hypotheses. Action selection means deciding the most appropriate action among a set of action alternatives. Some expert systems are equipped with capabilities to automate situation-diagnostic decisions. When an inference has to be made with imprecise information, the expert systems may give humans a set of plausible diagnostic hypotheses with degree of belief information. The LOA of the expert systems is positioned at levels 2 or 3. If, in contrast, the expert systems show humans only a single diagnostic hypothesis with the largest degree of belief among all, the LOA is set at level 4.

The traffic alert and collision avoidance system (TCAS) is an example that can make action-selection decisions automatically. When a midair collision is anticipated and no resolution maneuver is taken, the TCAS gives pilots resolution advisory, such as "climb, climb, climb." Pilots are supposed to initiate the suggested maneuver within 5 seconds. It is known, however, that the TCAS can produce unnecessary resolution advisory, though such cases do not happen frequently. Pilots may disregard resolution advisory when they are definitely sure that the advice is wrong. In this sense, the LOA of the TCAS resolution advisory is positioned at level 4.

Action Automation

There are many examples of automation for action implementation. A photocopy machine, described in Parasuraman et al. (2000), is a good example for illustrating that various LOAs can be chosen in a single machine. Suppose someone needs to quickly make copies of 10 pages for five people. He or she must decide which mode to use: automatic sorting without automatic stapling, automatic sorting with automatic stapling, or manual mode to make five copies of each sheet. In the last case, he or she must sort and staple sheets manually. The time required for giving necessary directives to the machine through a touch sensitive panel differs, as does the time needed to finish the task. Once a mode has been chosen, operation starts at one of three different levels of automation.

In aviation, the LOA of action is not set high. From the viewpoint of action automation, the LOA of the TCAS is positioned at level 4, because the TCAS itself has no mechanical subordinate to initiate a collision avoidance maneuver. The GPWS does not have capability for such a maneuver, either. It may be worth considering whether a high LOA should never be allowed for automation to implement an action. Take as an example the crash of a Boeing 757 aircraft that occurred near Cali, Colombia, in 1995. The pilots performed a terrain avoidance maneuver immediately upon a GPWS alert. However, they failed to stow the speed brake that they had extended some time before under their previous intention to descend (Dornheim, 1996). The crash could have been avoided if there had been an automatic mechanism to retract the speed brake if it had not been stowed when the pilot applied the maximum thrust. It is hard to imagine a situation where one would apply the speed brake and the maximum thrust simultaneously. When automation detects such a contradiction, it may be reasonable to allow the automation to adjust the configuration automatically (i.e., to stow the speed brake) so that the new configuration may fit well to the pilot's latest intention.

ADAPTIVE AUTOMATION

Suppose a human and a computer are requested to perform assigned functions for some period of time. The operating environment may change as time passes by, or performance of the human may degrade gradually as a result of psychological or physiological reasons. If the total performance or safety of the system is to be strictly maintained, it may be wise to reallocate functions between the human and the computer because the situation has deviated from the original one. A scheme that modifies function allocation dynamically and flexibly depending on situations is called an *adaptive function allocation*. The adaptive function allocation assumes criteria to determine whether functions have to be reallocated, how, and when. The criteria reflect various factors, such as changes in the operating environment, loads or demands to operators, and performance of operators. The adaptive function allocation is dynamic in nature. The automation that operates under an adaptive function allocation is called *adaptive automation*. The term *adaptive aiding* is used in some literature, such as that by Rouse (1988). In this chapter we treat the terms adaptive aiding and adaptive automation as synonyms.

Are Adaptive Function Allocation and Dynamic Function Allocation Equivalent?

It depends on the definition. To give contrast to static function allocation, *dynamic function allocation* may be defined as a scheme that may alter function allocation occasionally in time during system operation. Then, the dynamic function allocation is not always the same as the adaptive function allocation. Consider the following case: The pilots were flying northward to the destination airport. They had already finished supplying the computer with necessary data to make an automatic landing on Runway 01. Suddenly, the air traffic controller called the pilots to tell them that, because of extremely strong south winds, they had to use Runway 19. The pilots were not very pleased with the instruction for the following reasons: (a) they have to pass over the airport to make an approach from the north, which causes at least 10 minutes delay in arrival time; (b) Runway 19 is not equipped with the navigation facility for an automatic landing; and (c) the geography north of the airport is mountainous. It is highly stressful to make a manual approach over mountainous lands under time pressure. However, because there was no alternative, the pilots had to disengage the autopilot to fly manually, and the flight path management function shifted from the automation to the human pilots. The function allocation in this case is dynamic, but not adaptive.

The point is that the adaptive function allocation uses criteria to decide whether function must be reallocated for better or safer performance or adjustment of human workload. It would be almost useless to implement dynamic function allocation that is not adaptive. In fact, researchers use dynamic function allocation to mean almost the same as adaptive function allocation, by assuming some automation invocation criteria implicitly or explicitly.

History of Adaptive Automation

The notion of adaptive allocation can be traced to 1970s: see Rouse (1976, 1977). Later, Rouse (1988) stated, "The concept of adaptive aiding ... emerged in 1974 in the course of an Air Force Systems Command-sponsored project at the University of Illinois that was concerned with applications of artificial intelligence (AI) to cockpit automation" (p. 432). The investigators of the project were initially concerned with "getting the technology to work, rather than with how pilots were going to interact with this system" (Rouse, 1994, p. 28). During the research project, the investigators found situations in which the pilot and computer chose reasonable but mutually conflicting courses of action. "The desire to avoid conflicting intelligence and create cooperative intelligence quickly lead to questions of function allocation as well as human-computer interaction" (Rouse, 1994, p. 28). At that stage, they found inadequacies, as we did in the previous section, in making design decisions on function allocation based on Fitts' list. Rouse (1994) says, "Frustration with the MABA-MABA approach led to a very simple insight. Why should function, tasks, etc. be strictly allocated to only one performer? Aren't there many situations whether either human or computer could perform a task acceptably? ... This insight led to identification of the distinction between static and dynamic allocation of functions and tasks. Once it became apparent that dynamic invocation of automation might have advantages, it was a small step to the realization that the nature of computer assistance could also be varied with the situation" (Rouse, 1994, p. 29).

The adaptive aiding concept was further investigated in the Pilot's Associate program, a joint effort of the Defense Advanced Research Project Agency and the U.S. Air Force, managed by the U.S. Air Force's Wright Laboratory. The Pilot's Associate consists of cooperative knowledge-based subsystems with capabilities to (a) determine the state of the outside world and the aircraft systems, (b) react to the dynamic environment by responding to immediate threats and their effects on the preassigned mission plan, and (c) provide the information the pilot wants, when it is needed (Banks & Lizza, 1991; Hammer & Small, 1995). The program uncovered many gaps in technology and showed "the design of adaptive aids has to be based on a thorough understanding of the other activities and functions in the cockpit" (Rouse, 1994, p. 30).

AUTOMATION INVOCATION STRATEGIES

In adaptive automation, functions can be shared or traded between humans and machines in response to changes in situations or human performance. How can such sharing or trading capability be implemented? There are three classes of automation invocation strategies: first, critical-event strategies; second, measurement-based strategies; and third, modeling-based strategies.

Critical-Event Strategies

Automation invocation strategies of this class change function allocations when specific events (called critical events) occur in the human–machine system. It is assumed that human workload may become unacceptably high when the critical events occur. Allocation of functions would not be altered if the critical events did not occur during the system operation. In this sense, function allocation with a critical-event strategy is adaptive.

Three types of logic are distinguished (Barnes & Grossman, 1985; Parasuraman, Bhari, Deaton, Morrison, & Barnes, 1992). The first is emergency logic, in which automation would be invoked without human initiation or intervention. The second is executive logic, in which the subprocesses leading up to the decision to activate the process are automatically invoked, with the final decision requiring the human's approval. The third is automated display logic, in which all noncritical display findings are automated to prepare for a particular event, so that the human can concentrate on the most important tasks.

LOAs differ among the three types of logic. The LOA for the emergency logic is positioned at level 7 or higher, which implies that humans may not be maintained as the final authority. In the executive logic, the LOA is positioned at level 5. The automated display logic assumes sharing (or partitioning) of tasks. The computer distinguishes "noncritical" portions of the tasks from the "the most important" ones, and it allocates the former to machines so that the workload of operators may be reduced or maintained within reasonable levels. The LOA of the automated display logic is set at level 7 or higher, because it is the computer that judges whether a task is noncritical or the most important, and operators are not usually involved in the judgment. Applying a high level of automation, such as level 7 or above, can be beneficial for reducing workload or for buying time. However, it may bring some costs, such as degradation of situation awareness or automation-induced surprises. The issue is discussed in a later section.

Adaptive automation with critical-event strategies may have several operational modes with different LOAs. For instance, the Aegis system has a small rule base that determines how the Aegis system will operate in a combat environment. The following modes are available (Parasuraman et al., 1992): (a) manual, in which the system is fully controlled by the operator; (b) automatic special, in which a weapon-delivery process is automatic, but the fire button has to be pressed by the operator; and (c) fully automated, in which the ship's defensive actions are automatically implemented without operator intervention, because of the need for a short reaction time within which the operator may not complete the required actions.

Measurement-Based Strategies

Automation invocation logic of this class emerged at an early stage in the adaptive automation research. Rouse (1977) proposed the dynamic allocation of functions between operators and machines so that the moment-to-moment workload of the operators could be regulated around some optimal level. Workload levels of operators in complex systems fluctuate from moment to moment and at different mission phases. The operators may be able to achieve very high performance levels but only at the cost of high mental workload by neglecting "less critical" functions. If the situation that requires a high level of workload lasts too long, performance degradation may result. Performance may also deteriorate when other minor tasks are added. These observations give a rationale to adjust function allocation dynamically by evaluating moment-to-moment workload.

However, that does not mean that a single algorithm can be effective for all individuals. In fact, different operators will use different strategies to cope with the demands of multiple tasks under time pressure. It is thus necessary to develop *custom-tailored* adaptive automation algorithms if the system is to be compatible with and complement the strengths and weaknesses of individual operators (Parasuraman et al., 1992). Moreover, individual differences in human operator capabilities will influence the response to multiple task demands: Some operators may have sufficient resources left to cope with other tasks, whereas other operators may be operating at peak workload. This means that an algorithm developed for an "average" operator will not be suitable to either class of operators. For an adaptive system to be effective in maintaining mental workload at an appropriate level in dynamic real-time environments, it must be equipped with a workload measurement technology that is capable of (a) detecting changes in workload levels and (b) identifying what component of mental workload is under stressed or overstressed.

Adaptive automation works as follows under a measurement-based measurement strategy (Hancock & Chignell, 1988): First the task is defined and structured and subtasks are allocated to either an automated subsystem or to the operator. Next, the operator's effort is compared with the task difficulty so as to assign a criterion for adaptivity. The criterion can be expressed as a measure of mental workload, a measure of primary task performance, or a combination of both. Once the criterion is defined, the adaptive system trades task components in order to improve future measurement of the criterion. In the workload-based measurement method, adaptivity can be achieved through three main procedures: by adjusting allocation of subtasks between operators and automation; by adjusting the structure of the task; and by refining the task.

Psychophysiological measures, such as pupillary dilatation and heart rate, may be used for adjusting function allocation. The psychophysiological measures may be recorded continuously and thus be useful, unlike most behavioral measures, in measuring mental activities of human operators placed in supervisory roles that require few overt responses. In addition, the psychophysiological measures may provide more information when coupled with behavioral measures than behavioral measures would alone. For example, changes in reaction time may reflect contributions of both central processing and response-related processing to workload. Refer to Scerbo et al. (2001) for the most up-to-date guide for using psychophysiological measures in implementing adaptive automation.

Modeling-Based Strategies

Operator performance models can be used to estimate current and predicted operator states and to infer whether workload is excessive or not. The models are often categorized into three groups: intent inferencing models, optimal (or mathematical) models, and resource models.

Intent inferencing models work as follows (Rouse, Geddes, & Curry, 1987–1988): Operator actions are decoded and compared with the set of scripts. If at least one script matches, the actions are resolved. If no match is found, the unresolved actions are analyzed to identify plans. If one or more plans are found that are consistent with known goals, the actions are resolved and the scripts associated with these plans (if any) are activated. If no match is found, the unresolved actions are put into the error monitor.

Optimal models include those based on queuing theory (Walden & Rouse, 1978), pattern recognition (Revesman & Greenstein, 1986), and regression (Morris, Rouse, & Ward, 1988). For example, Walden and Rouse (1978) investigated multitask performance of a pilot, where time sharing is required between monitoring, control, and other tasks. They modeled the monitoring task as a queuing system with a "single server" and subsystem events called "customers" and with the control task incorporated as a special queue. Once the customers arrive at the control task queue they can control the service of a subsystem event. From what proceeds, a customer in this case can be defined as a "significant amount of display error." Therefore, when a major error is displayed, the subsystem service is preempted and a control action is taken to eliminate the error.

Resource models, especially the multiple resource theory (Wickens, 1984), try to describe how performance interference occurs in information processing. Suppose an operator is trying to perform two different tasks. If the two tasks require different resources, say verbal and spatial codes, then the operator will have no difficulty performing them efficiently. However, if the two tasks require the same resources, then some conflict can occur and performance of the tasks may suffer significantly. The multiple resource theory may be used to evaluate efficacy of function allocation, or to assess the impact of possible competition that may be caused by tasks requiring the same resource. The multiple resource theory is an important tool for human–machine system design, and it has already been incorporated in discrete-event simulation software, WinCrew, to evaluate function allocation strategies by quantifying the moment-to-moment workload values (Archer & Lockett, 1997).

Advantages and Limitations of Automation Invocation Strategies

Among three types of strategies, the critical-event strategies may be the most straightforward to implement, if critical events are defined properly. No investigations are needed regarding how human cognition or behavior could be modeled, what parameters must be measured to infer the human state, and how. All we have to do is develop techniques to detect occurrence of the critical event by using information available in the human–machine system. A possible limitation of the critical-event strategies is that they may reflect human workload only partially or implicitly. Subjective workload under the critical events may differ significantly among operators.

From a viewpoint of adapting to an individual who is facing a dynamically changing environment, measurement-based logic may be the most appropriate. It can change function allocation by explicitly reflecting the mental status of an operator at a specific circumstance. There is no need to predict in advance how the mental status of the operator may change. However, there are a few limitations. First, not all operators may welcome situations in which they are monitored by sensing devices all the time. Second, sensing devices are sometimes expensive or too sensitive to local fluctuations in operator workload or physiological states. Third, performance measurement occurs "after the fact" (Scerbo et al., 2001), that is, after a point in time when adaptation may be needed.

If good performance models are available, it may be possible to extract "leading indicators." Leading indicators refer to precursors that, when observed, imply the occurrence of some subsequent event. For instance, Kaber and Riley (1999) demonstrated the benefits of adaptive aiding on a primary task (a dynamic cognitive monitoring and control task) by taking, as a leading indicator, degradation of the secondary task (an automation-monitored task). It is not always easy, however, to develop a good performance model that represents the reality perfectly.

DECISION AUTHORITY

Who is supposed to make decisions concerning when and how function allocation must be altered? The human operator, or machine intelligence? Let us note here that the automation invocation strategies can be expressed in terms of *production rules*. For instance, a particular critical-event strategy may be represented as follows: "If critical-event E is detected, then function F must be handed over to the automation, if the function was dealt with by the human at that time point." In case of a measurement-based strategy, we have the following: "If human workload is estimated to be lower than a specified value, then function F must be traded from the automation to the human." Once the production rules are given, it is basically possible for machine intelligence (the computer) to implement adaptive function allocation without any help from the human operator. However, for some reasons, the reality is not so simple.

One apparent reason is reliability. It is unrealistic to assume that the computer never fails. The failure may be caused by hardware malfunction, software errors, or inappropriate data. If the computer is nevertheless given the authority to make an automation invocation decision, human operators may have to monitor the computer carefully all the time, which produces burden on the operators in addition to their original tasks.

A second reason is related to the principle of human-centered automation, which claims that a human operator must be maintained as the final authority and that only he or she may exercise decisions concerning how function allocation must be changed and when (Billings, 1997; Woods, 1989). The principle is reasonable. However, is it always the best for the human operator to bear the final decision authority at all times and for every occasion? Rouse (1988) says, "when an aid is most needed, it is likely that humans will have few resources to devote to interacting with the aid. Put simply, if a person had the resources to instruct and monitor an aid, he/she would probably be able to perform the task with little or no aiding" (p. 431). There may be cases in which it is rational that "variations in levels of aiding and modes of interaction will have to be initiated by the aid rather than the human whose excess task demands have created a situation requiring aiding" (Rouse, 1988, p. 432).

Appropriate LOAs

The decision authority issue is related to the selection of appropriate LOAs. When the LOA is positioned at level 5 or lower, the human operator is maintained as the final authority (see Table 8.2). The human-centered automation principle is violated when the LOA may be positioned at level 6 or higher. A committee of the U.S. National Research Council discussed appropriate LOAs for new civil air traffic control systems (Parasuraman et al., 2000; Wickens, Gordon, & Liu, 1998). Sheridan (2002) said, "After much debate, the committee decided that

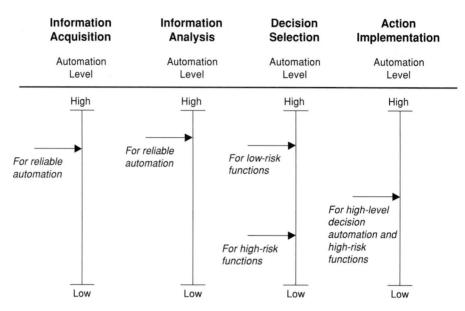

FIG. 8.1.   Recommended LOAs for future air traffic control systems (from Parasuraman et al., 2000. © 2000 IEEE, reprinted with permission).

acquisition and analysis could and should be highly automated—in fact, they already are (radar, weather, schedule information, etc.)—but that decision-making, except for certain decision aids now under development, should be done by human air traffic controllers. Implementation is in the hands of the pilots, which in turn is largely turned over to autopilots and the other parts of the Flight Management System" (p. 63). The recommended LOA for each function is depicted in Fig. 8.1.

There are some systems in which the computer may initiate invocation of automation. One such example is the automatic ground collision avoidance system for combat aircraft (Scott, 1999). When a collision against the terrain is anticipated, the computer gives a "pull up" warning. If the pilot takes a collision avoidance maneuver aggressively, then the computer does not step in any further. If the pilot does not respond to the warning, the computer takes control back from the human pilot and executes an automatic collision avoidance action. The LOA is positioned at level 6.

A clear-cut answer is hard to get for the decision authority issue. It is clear that the issue cannot be solved by qualitative discussions alone. Quantitative methods are needed for more precise understanding. The following approaches are described with some examples: (a) experiments, (b) computer simulations, and (c) mathematical modeling.

Laboratory Experiments

Laboratory experiments are usually designed under settings with multiple tasks, such as resource management, system monitoring, and compensatory manual tracking. Comparisons are made among various strategies for function allocation and automation invocation. Some results suggest the efficacy of the human-initiated invocation strategies, and others suggest the need for computer-initiated invocation.

Harris, Hancock, Arthur, and Caird (1995) compared the following conditions of a multitask environment. First, the subjects must perform all tasks manually. Second, the tracking task is performed by the automation. Third, the subjects can decide whether to invoke automation

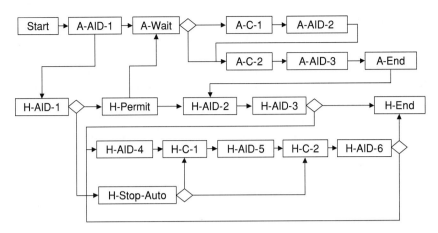

FIG. 8.2. An example of the cognitive task network.

for the tracking task. The investigators found that subjects' performances in resource management task were more efficient when automation invocation was initiated by humans. Hilburn, Molloy, Wong, and Parasuraman (1993) compared executive and emergency logic in the critical-event strategies, and they found a slight automation cost under the computer-initiated invocation.

Harris, Goernert, Hancock, and Arthur (1994) compared human-initiated and computer-initiated invocation strategies, and they found that human-initiated invocation of automation might be less beneficial than computer-initiated invocation when changes in the workload could be abrupt or unexpected for the human operator. Harris, Hancock, and Arthur (1993) also found that, when subjects became fatigued under a multiple-task environment, they were less likely to engage automation even when it was supposed to be used, which means that the benefits of automation may not be fully appreciated if human-initiated invocation of automation is adopted.

Computer Simulations

Furukawa, Niwa, and Inagaki (2001) investigated an optimal LOA for an emergency operating procedure for steam generator tube rupture (SGTR) failures in a pressurized water reactor. The SGTR failures are typical events to be assessed for safety in nuclear power plants. The emergency operating procedure for the SGTR failures has seven steps, each of which consists of several substeps. Among them there is the substep to shut the coolant flow at a particular pipe, which requires a sequence of tasks, such as "send a command to close a valve," "manipulate the valve," "examine the state of the valve," and so on. The tasks are categorized into four classes, that is, information acquisition, information integration, decision selection, and control implementation, in a similar manner as shown by Parasuraman et al. (2000). Some strategies of human–automation collaboration may be possible to perform the sequence of tasks. Furukawa et al. (2001) analyzed LOAs positioned at levels 5, 6, and 7. One of the cognitive task networks developed is shown in Fig. 8.2, in which each rectangular node represents a task that should be performed either by the automation (denoted as "A-task name") or the human operator ("H-task name"), and in which the LOA is positioned at level 5. Each diamond-shaped node is called a decision diamond, which can distinguish types of decision-making logic, such as probabilistic, tactical, and multiple decisions.

The cognitive task network was implemented with WinCrew, which is software for discrete-event simulation. WinCrew quantifies the moment-to-moment workload on the basis of the multiple resource theory (Wickens, 1984). Monte Carlo simulations were performed under various conditions of reliability for human operators and automations. It was found that the LOA positioned at level 6 was best in the sense that it could minimize the expected time for completion of the task sequence. However, the LOA positioned at level 6 was not optimal from the viewpoint of cognitive workload. The cognitive workload might be quantified in various ways, such as by the peak value, the average value, and the time-integral value over the whole period of system operation, and the time length during which the workload exceeded a specified threshold value. It was level 7 that gave the best result from the viewpoint of cognitive workload.

Another interesting computer simulation can be found in Archer, Lewis, and Lockett (1996), in which WinCrew was used to model and evaluate function allocation strategies of the bridge activities on a Navy Guided Missile Destroyer. Some function allocation strategies were defined for a crew of three and the automation, and workload was evaluated for each strategy under several scenarios to investigate the feasibility of reduction of crew size from nine (conventional size) to three.

Mathematical Modeling

Inagaki (1997, 1999, 2000) made a mathematical analysis on the decision authority issue in a rejected takeoff problem of an aircraft. Suppose an engine fails while an aircraft is making its takeoff roll. The pilot must decide whether to continue the climb-out (Go) or to abort the takeoff (No Go). The standard decision rule upon an engine failure is stated as follows: (a) Reject the takeoff, if the aircraft speed is below V1, and (b) continue the takeoff if V1 has already been achieved, and then make an emergency landing. The critical speed V1 is called the "takeoff decision speed" at which the pilot must apply the first retarding means in case of No Go. The V1 speed is a variable depending on weight, aircraft lift capability under various environmental conditions, length of runway used in achieving velocity, position on the runway at the velocity point, whether the aircraft can safely lift off with one engine failed, and so on.

The aforementioned rule for a Go–No Go decision is simple. However, that does not mean at all that the Go–No Go decision is easy to make, because only a 1 or 2 seconds are available for the pilots to complete the following tasks: (a) recognizing an engine failure alert, (b) analyzing the situation, (c) deciding if the circumstance warrants rejection of the takeoff, and (d) initiating the first stopping action (i.e., retarding the throttles of all engines) to abort the takeoff. A probabilistic model was given on the basis of the following assumptions:

- An alert is given to the human pilot when a sensor detects an engine failure. However, the sensor can give a false alert.
- The pilot's understanding of a given situation may not be correct. In reality, an engine failure is not a single factor for rejecting the takeoff. Various events such as an engine fire with or without loss of thrust, a blown tire, or a bird strike may happen during takeoff roles. Some of these events may give symptoms similar to those of an engine failure. It is not easy to identify the situation correctly when available information and time are limited. There may be some cases in which the pilot hesitates to say either that the alert is correct or that the alert is false.
- Incorrect or late decisions cause costs that vary depending on the situation.

By evaluating the conditional expected loss, provided an engine failure alert has been set off, Inagaki (1997, 1999, 2000) proved that the Go–No Go decision should be neither fully automated nor left always to a human; that is, the decision authority of automation invocation must be traded dynamically between human and computer. More concretely, (a) the human pilot must be in authority when the aircraft speed is far below V1; (b) the computer must be in authority if the aircraft is almost at V1 and if there is a possibility that the human pilot may hesitate to make decisions when he or she fails to decide whether the engine is faulty or not; and (c) when the aircraft speed is between (a) and (b), the selection of an agent in charge depends on the situation.

It was also proven (Inagaki, 1997, 1999, 2000) that, for a human pilot to be given decision authority at all times, human–interface design must be changed so that the human can be supported directly in Go–No Go decision making. The suggested human–interface design was to give either a "Go" or an "Abort" message on the display, while an "Engine failure" message appears on the display in the glass-cockpit aircraft. When receiving the "Engine failure" alert, pilots have to interpret the alert: "Engine failure" means "No Go" if it was before V1, but implies "Go" after V1. It is in the interpretation task that human pilots may sometimes make errors. Inagaki, Takae, and Moray (1999) found, by experiments, that improvements in interface design alone were insufficient to attain decision-making accuracy at levels that could be obtained with dynamic trading of decision authority between humans and automation.

BENEFITS AND COSTS OF ADAPTIVE AUTOMATION

Dynamic Alteration of LOAs

One major motivation for introducing adaptive automation is to regulate operator workload, where an operator "can control a process during periods of moderate workload, and hand off control of particular tasks when workload either rises above, or falls below, some optimal level" (Hilburn et al., 1993, p. 161). Another major benefit of adaptive automation lies in its ability to keep the operator in the control loop, which is done by altering the LOA. These characteristics contrast with static function allocation. When the LOA for a function is always positioned at high levels, the operator is likely to suffer from the *out of the control loop* phenomena, which lead to degradation of manual skill, vigilance decrements, and loss of situation awareness for the function (e.g., Endsley & Kaber, 1997; Endsley & Kiris, 1995; Gluckman, Carmody, Morrison, Hitchcock, & Warm, 1993; Kaber, Omal, & Endsley, 1999; Parasuraman et al., 1992; Wiener, 1988). When the automation or the system is perceived as being "highly reliable," automation-induced "complacency" may arise (Parasuraman, Molloy, & Singh, 1993), where complacency refers to the self-satisfaction that may result in nonvigilance based on an unjustified assumption of satisfactory system state. Occasional alteration of the LOA may be useful to avoid the out of control loop phenomena.

What happens if the LOA is altered frequently? If the algorithm were highly sensitive, the LOA would be changed by even a small perturbation in the input value to the algorithm. In extreme cases in which only manual control and full automatic control are available, frequent cycling between automated and manual control may occur, which can lead to performance degradation. The short cycling is a possible by-product of adaptivity in function allocation. Some researchers investigated the effects of short cycling on task performance by use of laboratory experiments under multitask environments. Parasuraman, Bhari, Molloy, and Singh (1991) demonstrated both benefits and costs of short-cycle automation on the manual performance of tasks and on the monitoring for automation failure. Glenn et al. (1994) investigated the effects on flight management task performance to show no automation deficits, and found automation benefits for reaction time in the tactical assessment task. Scallen, Hancock, and Duley (1995) analyzed the situations in which tracking task cycled between manual and automated control at fixed intervals of 15, 30, or 60 seconds. The investigators found that excessively short cycling of automation was disruptive to performance.

Sharing of Intentions

As in the case of conventional automation, possible failure to share intentions between the human and the computer is one of the major concerns in adaptive automation. Such failures can be classified into two types. The first type refers to the case in which the human and the computer have "similar" but different intentions. An example can be seen in the crash of an Airbus 320 aircraft in 1991 near Strasbourg, France. The pilots had an intention to make an approach using a flight path angle mode of −3.3 degrees. However the computer, which was the active agent for flying at that time moment, was given a command to create an intention to make an approach by using a vertical speed mode of −3,300 feet per minute (Billings, 1997; Sparaco, 1994). If the pilots had carefully interpreted various clues given in their primary flight displays, they could have noticed that, although the aircraft was descending, the vertical flight path was quite different from the one that they had planned.

The second type of failure in sharing intentions refers to the case in which the human and the computer have completely conflicting intentions. An example can be seen in the crash of an Airbus 300-600R aircraft at Nagoya, Japan, in 1994 (Billings, 1997). At some point during the final approach, the pilot gave a Go-Around directive to the computer *unintentionally*. The computer started its maneuver for going around. However, the pilot decided to descend for landing. The pilot knew that the autopilot was in the Go-Around mode, but he did not follow an appropriate procedure to cancel the mode. Thus the intentions of the pilot and the computer were quite opposite. The computer was ordered by the human to go around, and it tried to achieve the go-around at any cost. To the computer, the human's input force to descend was a *disturbance* that must be cancelled out by applying a stronger control input to ascend. From the viewpoint of the pilot, the aircraft did not descend smoothly and thus he applied a stronger control input. Thus the aircraft was subject to completely contradictory controls by two agents with opposite intentions.

Trust and Automation Surprises

Lee and Moray (1992) distinguished between four dimensions of trust: (a) *foundation,* which represents the "fundamental assumption of natural and social order that makes the other levels of trust possible," (b) *performance,* which rests on the "expectation of consistent, stable, and desirable performance or behavior," (c) *process,* which depends on "an understanding of the underlying qualities or characteristics that govern behavior," and (d) *purpose,* which rests on the "underlying motives or intents" (p. 1246).

For most technological artefacts (gadgets, devices, and machines' complex processes), the first dimension shall not cause serious problems. The technological systems usually satisfy requirements for the fourth dimension of trust. For instance, it is easy to answer the question, "For what purpose did the EGPWS have to be designed?" It would also be the case for adaptive automation. That is, human operators would accept the designer's motives or intents to design a technological system that can help users by regulating operator workload at some optimal level.

Respecting the second and the third dimensions of trust may not be straightforward. Because adaptive automation is designed to behave in a dynamic and context-dependent manner, its behavior may be obscure. Human understanding of the automation invocation algorithms may be imperfect if the algorithm is "sophisticated" or complicated. Suppose there are two conditions, A and A*, that differ only slightly. What happens if the human operator thought that condition A had been met, whereas it was condition A* that had become true and the automation invocation algorithm detected it? The operator would be surprised when he or she saw that the automation did not behave as he or she expected. The phenomena in which operators are surprised by the behavior of the automation are called *automation surprises* (Hughes, 1995b; Sarter, Woods, & Billings, 1997; Wickens, 1994). The surprised operators often ask questions such as what the automation is doing now, why it did that, or what is it going to do next (Wiener, 1989). When the human could not be sure of fulfillment of the second and third dimensions of trust, he or she would fail to establish proper trust in the adaptive automation. Human's distrust or mistrust in automation may cause inappropriate use of automation, as has been pointed out by Parasuraman and Riley (1997).

AVOIDING AUTOMATION SURPRISES

How should we design a system so that it may not cause automation surprises? That is the key to success of adaptive automation. One possible way to avoid automation surprises may be to design a mechanism through which operators and the automation (the computer) may communicate. Two types of communication are distinguished (Greenstein & Revesman, 1986; Revesman & Greenstein, 1986). The first is dialogue-based communication, in which the operator provides the computer with information regarding his or her action plans. The second is model-based communication, in which the computer predicts actions of the operator based on his or her performance model, and selects its own actions so as to minimize some measure of overall system cost. Dialogue-based communication was found to be advantageous in attaining precise understanding and a high level of situation awareness. However, dialogue may increase human workload (Revesman & Greenstein, 1986). Model-based communication may not increase human workload. However, it may be a new cause of automation surprises, if the assumed human performance model did not match the operator perfectly, or if it failed to cope with time-dependent characteristics of operator performance. Designing good human interface is the most basic requirement for avoiding automation surprises.

Human Interface Design

Human interface must be designed carefully to let operators know what the automation is doing now, why it did that, or what it is going to do next. Many incidents and accidents in highly automated aircraft show that even well-trained operators may fail to understand the intention or behavior of the automation (Dornheim, 1995; Hughes, 1995a; Sarter & Woods, 1995). Automation technology is spreading rapidly to wider areas of application in which designers may not assume that every operator is substantially trained. An automobile is such an example.

Example: Adaptive Cruise Control System.   Suppose an automobile company has a plan to develop an adaptive cruise control system with headway control capability, as shown in Fig. 8.3(a). The system is supposed to decelerate its own vehicle (O) when it approaches too close to the vehicle ahead (A). Designers have to design their system by enumerating various situational changes that may occur while driving. For instance, the third vehicle (C) may be cutting in between cars O and A, as shown in Fig 8.3(b). It is important for car O's driver to know whether the automation sensed car C or not. Suppose the driver thought that the

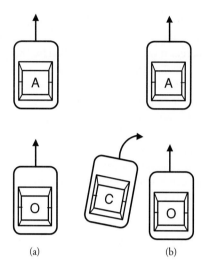

FIG. 8.3.   Adaptive cruise control: (a) headway control; (b) cutting in.

automation had sensed car C and had already changed its control law to follow car C, instead of car A. Then the driver may not hit the brakes, instead expecting the automation to do so. A collision with car C may happen if the automation had not sensed the car C and was still controlling car O longitudinally so that the distance to car A might be maintained as specified. This suggests the need for a human interface that can show the automation's recognition of the situation.

When designing a human interface for the adaptive cruise control system just described, designers first need to enumerate all the possible state vectors, (X, Y), where X denotes the car that the automation tries to follow, and Y denotes the car that the driver thinks the automation is following. Among all the possible state vectors, some of them represent cases in which the driver and the automation do not share the situational recognition: (A, C), (C, A), (−, A), and (−, C) are some such cases, where the minus sign (−) implies that the automation has lost the target to follow, which may occur at a sharp bend in the road, or when the vehicle ahead changes its lane. Each state vector may yield a collision. For instance, in case of (A, C), collision against car C may happen if the driver of car O was overly reliant on the automation by hoping that the automation would decelerate in a moment. Even in case of (C, C), car O may collide with car C. The collision shall occur if the car C was cutting in too abruptly and too closely to car O.

In general, *fault tree analysis* techniques are useful for identifying causes of possible colli- sion for each state vector. Though the methods were developed originally for assessing safety or risk of large-scale systems, they are now recognized as useful tools for cognitive task analy- ses (Kirwan & Ainsworth, 1992). Taking a state vector, say (A, C), as the *top event,* a fault tree must be constructed. The first step is to find combinations of events that may cause the top event to occur, where Boolean gates (such as AND gates and OR gates) are used in describing how events are combined. The process is repeated until no further expansion of an event is necessary. The events at the bottom of the fault tree are called *basic events,* which may represent various human factors (e.g., human errors, loss of awareness, complacency, or distrust), violations of laws, hardware–software failures, or hostile environmental conditions. The designers have to consider what means are available to prevent the basic events from occurring. The identified countermeasures must then be examined with theoretical analysis, computer simulations, or experiments. The fault trees obtained in this process are useful for systematic extraction of scenarios under which the efficacy of the design decisions must be examined.

TABLE 8.3
Design Aspects of Adaptive Automation

---

1. Sharing control or trading control
   (1) Sharing control
       (i) Extension
       (ii) Relief
       (iii) Partition
   (2) Trading
2. Automation invocation
   (1) Critical-event strategies
   (2) Measurement-based strategies
   (3) Model-based strategies
3. Decision authority
   (1) The human is maintained as the final authority at all times and for every occasion
   (2) Either the human or the automation may have the final authority, depending on the situation

---

CONCLUSION: TOWARD SENSIBLE ADAPTIVE AUTOMATION

Design aspects for adaptive automation are summarized in Table 8.3. The first aspect (sharing and trading) may be regarded as physical collaboration between human operators and the automation. Required automation differs, depending on the type of collaboration. For instance, in case of trading, the automation must be designed so that it can replace the human completely. In case of partition, the automation is to be designed so that it can cope with every possible partition of functions. In case of extension or relief, we need a mechanism to add control force on the other agent's force.

The second aspect (automation invocation) requires analyses the following: (a) availability of a clear definition on the event in which it is hard for any operators to perform assigned functions properly, (b) availability of methods to measure indices precisely without placing a heavy burden on the operators, and (c) availability and precision of performance models that may represent a wide variety of operators or that may tune the model parameters dynamically.

The third aspect (decision authority) deals with mental collaboration between operators and the automation. The aspect is closely related to the principle of human-centered automation. However, it is not wise to assume, without careful analyses, that human operators must be maintained as the final authority at all times and for every occasion.

Theoretically speaking, adaptive function allocation offers more flexible design decisions than static function allocation. This very flexibility, however, may bring operators various inconveniences or undesired results when the adaptive automation is put into use. Before implementing design ideas, adaptive automation designers have to analyze possible consequences of design decisions:

- How do humans feel or respond when they are overridden by the automation for safety reasons?
- Do humans trust in and use the automation that works well but in a slightly different manner than anticipated?
- Do not operators become so reliant on adaptive automation that they may be reluctant to take actions themselves?
- Is appropriate human interface provided with the operators so that any automation surprises can be avoided?

These are some of questions that designers have to ask in the design stage. Because adaptive automation is more sophisticated, complex, and possibly obscure than conventional automation, the cognitive task design must be done seriously. Especially, the decision authority issue can never be solved with dogmatic or qualitative discussions. The issue has to be investigated in a rigorous manner by applying quantitative methods, such as mathematical modeling, computer simulations, and experiments.

Research on adaptive automation had its origin in aviation, and various studies have been conducted in the context of aviation applications, including military applications (Bonner, Taylor, & Fletcher, 2000; Morgan, Cook, & Carbridge, 1999). Rouse (1988), Parasuraman et al. (1992), Scerbo (1996), Wickens and Hollands (2000), and Scallen and Hancock (2001) give good surveys of those efforts. The adaptive automation concept can be applied to other areas, such as process control (Moray, Inagaki, & Itoh, 2000) and automobiles. It must be stressed here again that the design considerations for automobiles must be quite different from those for aircraft, large process plants, and military applications. Car drivers are not always well trained. They may not have perfect knowledge of the automation (and its logic) in their cars. The cognitive task analysis and design must take this point into account.

This chapter took the terms *function* and *task* to mean essentially the same thing. It is sometimes difficult to discriminate between the two terms without any ambiguity (Wickens et al., 1998), although there is a sort of hierarchical relation between function and task. Those readers who are interested in the hierarchical structure should refer to Sharit (1997).

ACKNOWLEDGMENTS

The author expresses his thanks to Raja Parasuraman, Tom Sheridan, and Neville Moray for their useful discussions. Although they do not appear as coauthors, readers may find reflections of their ideas at many places in this chapter. The author also extends his sincere gratitude to anonymous referees. Without their critical but constructive comments and suggestions, this chapter could not be revised appropriately.

REFERENCES

Archer, R. D., & Lockett, J. F. (1997). WinCrew—a tool for analyzing performance, mental workload and function allocation among operators. In E. Fallon, L. Bannon, & J. McCarthy (Eds.), Proceedings of the First International Conference on Allocation of Functions (ALLFN'97). Volume II (pp. 157–165). Louisville: IEA Press.

Archer, R. D., Lewis, G. W., & Lockett, J. F. (1996). Human performance modeling of reduced manning concepts for Navy ships. In *Proceedings of the 40th Annual Meeting of Human Factors and Ergonomics Society* (pp. 987–991). Santa Monica, CA: Human Factors and Ergonomics Society.

Banks, S. B., & Lizza, C. S. (1991). Pilot's associate. *IEEE Expert, 6*(3), 18–29.

Barnes, M., & Grossman, J. (1985). *The intelligent assistant concept for electronic warfare systems.* China Lake, CA: Naval Warfare Center. (NWC TP 5585).

Billings, C. E. (1997). *Aviation automation—the search for a human-Centered Approach.* Mahwah, NJ: Lawrence Erlbaum Associates.

Bonner, M., Taylor, R., & Fletcher, K. (2000). Adaptive automation and decision aiding in the military fast jet domain. In *Proceedings of Human Performance, Situation Awareness, and Automation* (pp. 154–159). Savannah, Georgia: SA Technologies, Inc.

Bresley, B., & Egilsrud, J. (1997). Enhanced ground proximity warning system. Boeing Airliner, July–September 1997, 1–13.

Dornheim, M. (1995, January 30). Dramatic incidents highlight mode problems in cockpits. *Aviation Week and Space Technology, 142*(5), 57–59.

Dornheim, M. (1996, September 9). Recovered FMC memory puts new spin on Cali accident. *Aviation Week and Space Technology, 145*(11), 58–61.

Endsley, M. R., & Kaber, D. B. (1997). The use of level of automation as a means of alleviating out-of-the-loop performance problems: A taxonomy and empirical analysis. In *Proceedings of the 13th Triennial Congress of the International Ergonomics Association* (pp. 168–170). Helsinki: Finnish Institute of Occupational Health.

Endsley, M. R., & Kiris, E. O. (1995). The out-of-the-loop performance problem and the level of control in automation. *Human Factors, 37*(2), 3181–3194.

Fitts, P. M. (Ed.). (1951). *Human engineering for an effective air-navigation and traffic-control system*. Columbus, OH: The Ohio State University Research Foundation.

Furukawa, H., Niwa, Y., & Inagaki, T. (2001). Levels of automation in emergency operating procedures for a large-complex system. In M. J. Smith, G. Salvendy, D. Harris, & R. J. Koubele (Eds.). *Usability evaluation and interface design* (Vol. 1, pp. 1513–1517). Mahwah, NJ: Lawrence Erlbaum Associates.

Glenn, F., Barba, C., Wherry, R. J., Morrison, J., Hitchcock, E., & Gluckman, J. P. (1994). Adaptive automation effects on flight management task performance. In M. Mouloua & R. Parasuraman (Eds.), *Human performance in automated systems* (pp. 33–39). Hillsdale, NJ: Lawrence Erlbaum Associates.

Gluckman, J. P., Carmody, M. A., Morrison, J. G., Hitchcock, E. M., & Warm, J. S. (1993). Effects of allocation and partitioning strategies of adaptive automation on task performance and perceived workload in aviation relevant tasks. In *Proceedings of the International Symposium on Aviation Psychology* (pp. 150–155). Columbus, OH: The Ohio State University.

Greenstein, J. S., & Revesman, M. E. (1986). Development and validity of a mathematical model of human decision-making for human-computer communication. *IEEE Transactions on Systems, Man, & Cybernetics, 16,* 148–154.

Grote, G., Ryser, C., Wafler, T., Windischer, A., & Weik, S. (2000). KOMPASS: A method for complementary function allocation in automated work systems. *International Journal of Human-Computer Studies, 52,* 267–287.

Hammer, J. M., & Small, R. L. (1995). An intelligent interface in an associate system. *Human/Technology Interaction in Complex Systems, 7,* 1–44.

Hancock, P. A., & Chignell, M. H. (1988). Mental workload dynamics in adaptive interface design. *IEEE Transactions on Systems, Man, & Cybernetics, 18,* 647–658.

Hancock, P. A., & Scallen, S. F. (1998). Allocating functions in human-machine systems. In R. R. Hoffman, et al. (Eds.), *Viewing psychology as a whole* (pp. 509–539). Washington, DC: American Psychologic Association.

Harris, W. C., Hancock, P. A., Arthur, E., & Caird, J. K. (1995). Performance, workload, and fatigue changes associated with automation. *International Journal of Aviation Psychology, 5,* 169–185.

Harris, W. C., Hancock, P. A., & Arthur, E. J. (1993). The effect of taskload projection on automation use, performance, and workload. In *Proceedings of the 7th International Symposium on Aviation Psychology* (pp. 178–184). Columbus, OH: The Ohio State University.

Harris, W. C., Goernert, P. N., Hancock, P. A., & Arthur, E. J. (1994). The comparative effectiveness of adaptive automation and operator initiated automation during anticipated and unanticipated taskload increases. In M. Mouloua & R. Parasuraman (Eds.), *Human performance in automated systems* (pp. 40–44). Hillsdale, NJ: Lawrence Erlbaum Associates.

Hilburn, B., Molloy, R., Wong, D., & Parasuraman, R. (1993). Operator versus computer control of adaptive automation. In *Proceedings of the 7th International Symposium on Aviation Psychology* (pp. 161–166). Columbus, OH: The Ohio State University.

Hughes, D. (1995a, January 30). Incidents reveal mode confusion. *Aviation Week and Space Technology, 142*(5), 56.

Hughes, D. (1995b, February 6). Studies highlight automation "surprises." *Aviation Week and Space Technology, 142*(6), 48–49.

Inagaki, T. (1997). To go no not to go: Decision under time-criticality and situation-adaptive autonomy for takeoff safety. In *Proceedings of the IASTED International Conference on Applied Modelling and Simulation* (pp. 144–147). Calgary: IASTED ACTA Press.

Inagaki, T. (1999). Situation-adaptive autonomy: Trading control of authority in human-machine systems. In M. W. Scerbo & M. Mouloua (Eds.), *Automation technology and human performance: Current research and trends* (pp. 154–159). Mahwah, NJ: Lawrence Erlbaum Associates.

Inagaki, T. (2000). Situation-adaptive autonomy for time-critical takeoff decisions. *International Journal of Modelling and Simulation, 20*(2), 175–180.

Inagaki, T., Takae, Y., & Moray, N. (1999). Automation and human interface for takeoff safety. In *Proceedings of the 10th International Symposium on Aviation Psychology* (pp. 402–407). Columbus, OH: The Ohio State University.

Jordan, N. (1963). Allocation of functions between man and machines in automated systems. *Journal of Applied Psychology, 47,* 161–165.

Kaber, D. B., & Riley, J. M. (1999). Adaptive automation of a dynamic control task based on workload assessment through a secondary monitoring task. In M. W. Scerbo & M. Mouloua (Eds.), *Automation technology and human performance: Current research and trends* (pp. 129–133). Mahwah, NJ: Lawrence Erlbaum Associates.

Kaber, D. B., Omal, E., & Endsley, M. R. (1999). Level of automation effects on telerobot performance and human operator situation awareness and subjective workload. In M. W. Scerbo and M. Mouloua (Eds.), *Automation technology and human performance: Current research and trends* (pp. 165–169). Mahwah, NJ: Lawrence Erlbaum Associates.

Kawazoe, H., Murakami, T., Sadano, O., Suda, K., & Ono, H. (2001). Development of a lane-keeping support system. In *Proceedings of Intelligent Vehicle Initiative (IVI) Technology and Navigation Systems* (pp. 29–35). Warrendale, PA: Society of Automotive Engineers.

Kirwan, B., & Ainsworth, L. K. (1992). *A guide to task analysis*. London: Taylor & Francis.

Klein, G., Orasanu, J., Calderwood, R., & Zsambok, C. E. (Eds.). (1993). *Decision making in action: Models and methods*. Norwood, NJ: Ablex.

Lee, J. D., & Moray, N. (1992). Trust, control strategies and allocation of function in human machine systems. *Ergonomics, 35*(10), 1243–1270.

Morgan, C., Cook, C. C., & Corbridge, C. (1999). Dynamic function allocation for naval command and control. In M. W. Scerbo & M. Mouloua (Eds.), *Automation technology and human performance: Current research and trends* (pp. 134–138). Mahwah, NJ: Lawrence Erlbaum Associates.

Moray, N., Inagaki, T., & Itoh, M. (2000). Adaptive automation, trust, and self-confidence in fault management of time-critical tasks. *Journal of Experimental Psychology: Applied, 6*(1), 44–58.

Morris, N. M., Rouse, W. B., & Ward, S. L. (1988). Studies of dynamic task allocation in an aerial search environment. *IEEE Transactions on Systems, Man, & Cybernetics, 18*(3), 376–389.

Parasuraman, R., Bhari, T., Molloy, R., & Singh, I. (1991). Effects of shifts in the level of automation on operator performance. In *Proceedings of the 6th International Symposium on Aviation Psychology* (pp. 102–107). Columbus, OH: The Ohio State University.

Parasuraman, R., Bhari, T., Deaton, J. E., Morrison, J. G., & Barnes, M. (1992). *Theory and design of adaptive automation in aviation systems* (Progress Rep. No. NAWCADWAR-92033-60). Warminster, PA: Naval Air Development Center Aircraft Division.

Parasuraman, R., Molloy, R., & Singh, I. L. (1993). Performance consequences of automation-induced "complacency." *International Journal of Aviation Psychology, 3*(1), 1–23.

Parasuraman, R., & Riley, V. (1997). Humans and automation: Use, misuse, disuse, abuse. *Human Factors, 39*(2), 230–253.

Parasuraman, R., Sheridan, T. B., & Wickens, C. D. (2000). A model for types and levels of human interaction with automation. *IEEE Transactions on Systems, Man, and Cybernetics, 30*(3), 286–297.

Price, H. E. (1985). The allocation of function in systems. *Human Factors, 27*(1), 33–45.

Revesman, M. E., & Greenstein, J. S. (1986). Application of a mathematical model of human decision making for human-computer communication. *IEEE Transactions on Systems, Man, and Cybernetics, 16*(1), 142–147.

Rouse, W. B. (1976). Adaptive allocation of decision making responsibility between supervisor and computer. In T. B. Sheridan & G. Johannsen (Eds.), *Monitoring behavior and supervisory control* (pp. 295–306). New York: Plenum.

Rouse, W. B. (1977). Human-computer interaction in multitask situations. *IEEE Transactions on Systems, Man, and Cybernetics, 7*, 384–392.

Rouse, W. B. (1988). Adaptive aiding for human/computer control. *Human Factors, 30*(4), 431–443.

Rouse, W. B. (1991). *Design for success: A human centred approach to designing successful products and systems*. New York: Wiley.

Rouse, W. B. (1994). Twenty years of adaptive aiding: Origins of the concept and lessons learned. In M. Mouloua & R. Parasuraman (Eds.), *Human Performance in Automated Systems: Current Research and Trends* (pp. 28–32). Hillsdale, NJ: Lawrence Erlbaum Associates.

Rouse, W. B., Geddes, N. D., & Curry, R. E. (1987–1988). An architecture for intelligent interfaces: Outline of an approach to supporting operators of complex systems. *Human-Computer Interaction, 3*, 87–122.

Sarter, N. B., & Woods, D. D. (1995). How in the world did we ever get into that mode? Mode error and awareness in supervisory control. *Human Factors, 37*(1), 5–19.

Sarter, N. B., Woods, D. D., & Billings, C. E. (1997). Automation surprises. In G. Salvendy (Ed.), *Handbook of human factors and ergonomics* (2nd ed., pp. 1926–1943). New York: Wiley.

Scallen, S. F., Hancock, P. A., & Duley, J. A. (1995). Pilot performance and preference for short cycles of automation in adaptive function allocation. *Applied Ergonomics, 26*(6), 397–403.

Scallen, S. F., & Hancock, P. A. (2001). Implementing adaptive function allocation. *International Journal of Aviation Psychology, 11*(2), 197–221.

Scerbo, M. W. (1996). Theoretical perspectives on adaptive automation. In R. Parasuraman & M. Mouloua (Eds.), *Automation and human performance* (pp. 37–63). Mahwah, NJ: Lawrence Erlbaum Associates.

Scerbo, M. W., Freeman, F. G., Mikulka, P. J., Parasuraman, R., Di Nocero, F., & Prinzel III, L. J. (2001). *The efficacy of psychophysiological measures for implementing adaptive technology* (NASA TP-2001-211018). Hampton, VA: NASA-Langley Research Center.

Scott, W. B. (1999, February 1). Automatic GCAS: "You can't fly any lower." *Aviation Week and Space Technology, 150*(5), 76–79.

Sharit, J. (1997). Allocation of functions. In G. Salvendy (Ed.), *Handbook of human factors and ergonomics* (2nd ed., pp. 301–339). New York: Wiley.

Sheridan, T. B. (1992). *Telerobotics, automation, and human supervisory control.* Cambridge, MA: MIT Press.

Sheridan, T. B. (2002). *Humans and automation: System design and research issues.* Santa Monica, CA: Human Factors and Ergonomics Society & Wiley.

Sheridan, T. B., & Verplank, W. L. (1978). *Human and computer control of undersea teleoperations* (Man-Machine Systems Laboratory, Tech. Rep. x). Cambridge, MA: MIT.

Sparaco, P. (1994, January 3). Human factors cited in French A320 crash. *Aviation Week and Space Technology, 140*(1), 30–31.

Walden, R. S., & Rouse, W. B. (1978). A queueing model of pilot decisionmaking in a multitask flight management situation. *IEEE Transactions on Systems, Man, & Cybernetics, 8*(12), 867–875.

Wickens, C. D. (1984). Processing resources in attention. In R. Parasuraman & D. R. Davies (Eds.), *Varieties of attention* (pp. 63–101). Orland, FL: Academic Press.

Wickens, C. D. (1994). Designing for situation awareness and trust in automation. In *Proceedings of IFAC Integrated Systems Engineering* (pp. 77–82). London, Pergamon Press.

Wickens, C. D., Gordon, S. E., & Liu, Y. (1998). *An introduction to human factors engineering.* Reading, MA: Addison-Wesley.

Wickens, C. D., & Hollands, J. G. (2000). *Engineering psychology and human performance* (3rd ed.). Englewood Cliffs, NJ: Prentice-Hall.

Wickens, C. D., Mavor, A., Parasuraman, R., & McGee, J. (1998). *The future of air traffic control: Human operators and automation.* Washington, D.C., National Academy Press.

Wiener, E. L. (1988). Cockpit automation. In E. L. Wiener & D. C. Nagel (Eds.), *Human factors in aviation* (pp. 433–461). New York: Academic Press.

Wiener, E. L. (1989). *Human factors of advanced technology (glass cockpit) transport aircraft.* (NASA Contractor Rep. No. 177528). Moffett Field, CA: NASA—Ames Research Center.

Woods, D. (1989). The effects of automation on human's role: Experience from non-aviation industries. In S. Norman & H. Orlady (Eds.), *Flight deck automation: Promises and realities* (NASA CR-10036, pp. 61–85). Moffett Field, CA: NASA—Ames Research Center.

Zsambok, C. E., & Klein, G. (1997). *Naturalistic decision making.* Mahwah, NJ: Lawrence Erlbaum Associates.

9

A Search for Meaning: A Case Study of the Approach-to-Landing

John M. Flach,
Paul F. Jacques, and
Darby L. Patrick
*Wright State University,
USA*

Matthijs Amelink,
M. M. (Rene) Van Paassen,
and Max Mulder
*Delft University of Technology,
The Netherlands*

Abstract

The first section of the chapter contrasts a cognitive systems engineering (CSE) approach to task analysis with classical approaches. The critical difference is that a CSE approach is designed to provide a "survey" of the meaning landscape—what are the constraints that might guide adaptation. Classical approaches tend to focus the task analysis on "routes" or specific behaviour trajectories (the one best way) within the domain. The second section provides a brief account of an ongoing task analysis to understand the task of flying a precision instrument approach. This section provides samples of what we have learned from studying the history of aviation, reading authorities on flying, developing a synthetic task environment for landing, learning and teaching people to fly the simulation, observing and talking with experienced pilots, and talking with aeronautical engineers.

INTRODUCTION: WHERE ARE WE COMING FROM AND WHERE ARE WE GOING?

The idea of "task analysis" can be traced to the early "time and motion" studies of the "scientific management" approach to work (Gilbreth & Gilbreth, 1917; Taylor, 1911). The general approach was first to describe a manual process in terms of fundamental elements (e.g., therbligs). The second step was to improve on the process by designing methods, tools, and jigs that eliminated wasted motions and that allowed less efficient motions to be replaced with more efficient motions. The focus was on activity, and the goal was to use experimental methods to discover the "one best way." Over the years, the notion of "activity" has been expanded to include not only physical motions, but also cognitive operations (e.g., interpolates, verifies, remembers, calculates, plans, decides). The underlying theme has remained much the same, however: "the dissection of human work into 'tasks,' and the further analysis thereof" (McCormick, 1976, cited in Drury, Paramore, Van Cott, Grey, & Corlett, 1987).

Thus, work has been viewed as a sequence of activities or "tasks." The job of work designers has been to discover the optimal sequence (the one best way); and to provide the instructions, tools, and incentives required so that operators would follow the optimal "route" through the work domain. As systems have become more complex, there has been a tendency to split the responsibilities for work design between the *domain engineers* and the *human factors engineers*. The domain engineers would focus on the work process, and they would design the tools (the aircraft or nuclear power plant) and the operating procedures—that is, they would choose the *routes* that operators should take through the work domain based on an understanding of the process constraints and the safety requirements. These routes would then be communicated to the human engineers in terms of *procedures* and *regulations*. The job of the human engineers was to design the interfaces, warnings, instructions, training, and incentives necessary to guide the operators along the chosen routes (i.e., to get humans to follow the *correct* procedures).

Thus, a fundamental assumption of traditional task analysis has been the idea of "one best way." This assumption tended to work well for domains that Perrow (1984) would describe as linear ("a system with a well-established technology, standardized raw materials, and linear production system," p. 333). In these domains, the processes are well understood and predictable. Thus, it is possible to constrain behaviors to conform to predetermined routes formulated by the domain engineers. Perrow (1984) observed the following:

> When failures occur, as they inevitably will, they will not interact in unexpected and incomprehensible ways, but in expected and visible ways. The system *programs responses* for these infrequent but expected failures; the responses are dete*rmined at the top or in design,* and employees at all levels are expected to carry them out *without question.* . . . Repeated drills will insure fast and appropriate reactions, but the reactio*ns are prescribed ahead of time, by the central authority.* (p. 333, emphasis added)

In domains that Perrow (1984) characterized as complex (e.g., nuclear power plants and universities), however, the possibility of multiple failures and unexpected interactions makes it impossible for a designer or centralized authority to prescribe solutions in advance. Perrow (1984) observed that to respond appropriately to disturbances in these complex systems,

> Those at the point of the disturbance must be free to interpret the situation and take corrective action. . . . To do this they must be able to "move about," and peek and poke into the system, try out parts, reflect on past curious events, ask questions and check with others. In doing this diagnostic work ("Is something amiss? What might happen next if it is?"), personnel must have the discretion to stop their ordinary work and to cross departmental lines and make changes that would normally need authorization. (pp. 332–333)

Another dimension that Perrow (1984) used to characterize systems was the nature of the coupling or dependence among components. He noted that for complex and loosely coupled systems such as universities where individuals have a high degree of independence, the "peeking and poking around" needed to discover solutions to unexpected faults (or unexpected opportunities) is typically not a problem. This is because the "loose coupling gives time, resources, and alternative paths to cope with the disturbance and limit its impact" (p. 332). Society has been reluctant, however, to allow "peeking and poking around" in complex, tightly coupled systems such as nuclear power plants and air transportation systems with the potential for accidents of catastrophic impact. In fact, the interfaces to these systems, which have been designed to support procedures, do not facilitate exploration. The result is that we have typically tried to manage these complex systems using the "scientific management" philosophy in which a centralized authority prescribes the "one best way." I (the senior author) can remember an instance early in my career in which I was invited to visit a nuclear power plant in the southeastern United States to discuss the role of human factors. A manager argued that the solution to human factors problems was simple: "The first time an operator makes an error, slap his hand. The second time he makes the error cut off his finger." The result of this attitude is that accidents remain a "normal" feature of these systems (to paraphrase Perrow). The particular nuclear power plant that I was visiting has been notorious for its poor safety record.

The inadequacy of "procedures" for managing complex systems is graphically illustrated in Vicente's (1999) anecdote about "malicious procedural compliance." Control room operators found themselves caught in a double bind when they were being evaluated in a test simulator. When they deviated from standard procedures (in ways they knew were necessary to meet the operational goals) they were given low scores for not following procedures. When they rebelled and followed the procedures exactly (even though they knew the procedures were inadequate) they crashed the simulator and were penalized for "malicious procedural compliance."

The emergence of cognitive systems engineering (Flach & Rasmussen, 2000; Hollnagel & Woods, 1983; Rasmussen, 1986; Rasmussen, Pejtersen, & Goodstein, 1994; Vicente, 1999) reflects a recognition that classical approaches to work design will not be effective in complex systems. In these systems the "procedural" aspects of work are typically automated. The human operator is included in the system to deal creatively with the unanticipated contingencies that were not (and probably could not have been) anticipated in the design of the automated systems (Reason, 1990). The human operators' ability to diagnose faults and to respond adaptively so that these faults do not become catastrophic accidents is a critical resource for safe and effective systems. With cognitive systems engineering, the work design problem shifts from

how to constrain operators to follow predetermined routes through the work domain

to

how to provide operators with the survey knowledge of the work domain that they need to adaptively respond to unexpected events.

With cognitive systems engineering, the focus of analysis shifts from "activities" or "procedures" to "situation awareness" or "meaning." That is, the goal of work design is not to simplify or structure the operators' behaviors but to provide the operators with the information required for a deep understanding of the processes being controlled so that the operators can assemble the appropriate actions as required by the situations encountered. To do this, the cognitive engineer has to bridge between the domain engineers (who generally have knowledge about the deep structure of the processes being controlled; who know what is meaningful) and the human engineers (who are responsible for the design of interfaces, instruction manuals, and training). The goal is to structure the interfaces, instructions, and training so that they reflect the "deep structure" (provide survey knowledge) of the work domain as opposed to merely communicating procedures (route knowledge). For those not familiar with the distinction between survey and route knowledge, route knowledge specifies the path from one location to another in terms of a specific set of actions (e.g., turn right at the first stop sign, then go five blocks, etc.). Survey knowledge specifies a more general knowledge about the relative layout of the environment (e.g., a map showing the relative locations of start and destination along with alternative paths between them). In essence the goal is to "loosen" the tight coupling (in Perrow's sense) by enriching the information coupling between operator and process. By enriching the information coupling between operator and the controlled process, it becomes more feasible for the operator to "peek and poke" into the process to discover solutions to unanticipated events before they result in catastrophic consequences.

The difference between survey knowledge (needed to discover new solutions) and route knowledge (sufficient for procedure following) can be subtle, and this distinction has been difficult for many classically trained human factors engineers to appreciate. To them, the procedures are "what's meaningful." From their perspective, knowing where you are in a sequence and knowing what action comes next is all an operator should need to know. In fact, arguments have been made that it is best to restrict operators' knowledge to the prescribed procedures to protect against undesirable variability from the human element in the system. But the challenge is this: how does an operator know what to do next when the path prescribed by the procedures is blocked? What does the operator need to know when the standard procedures don't meet the functional objectives for the process? In this situation, operators who only know procedures will be lost. The challenge of cognitive systems engineering is to consider what kinds of information would be useful for operators, so that they can go beyond the procedures when required (or alternatively when desirable due to some opportunity for improvement).

Operators with survey knowledge should be able to "see" the procedures in the context of the work domain constraints. Such operators should be able to see a procedure in relation to the other opportunities for action. They should be able to discriminate when the standard procedures are the best way and when they are not. Thus, providing "survey knowledge" means making the work domain constraints "visible" to the operators through instructions, training, and interface design. It means trusting that operators with this enriched view will choose the best paths. Rasmussen and Vicente (1989) have articulated this idea using the label "ecological interface design" (EID). The general theme of EID is to use the power of representations to reveal the deep structure of processes to the operators responsible for controlling those processes (see also Rasmussen, 1999; and Vicente, in press).

For the past 4 years, we have been attempting to apply the EID approach to the development of alternative displays for the "glass" cockpit. The remainder of this chapter focuses on the cognitive task analysis process that has informed our display development. We do not describe the graphical displays here. Descriptions of the display are available in other sources (Flach, 1999, 2000b; Jacques, Flach, Patrick, Green, & Langenderfer, 2001). In this chapter, we attempt to provide a few snapshots of our efforts to discover the "deep structure" associated with flight control and share what we have learned about flying (particularly the precision approach). The hope is that our search might serve as a "case study" to illustrate a work analysis process designed to discover what is meaningful.

A SEARCH FOR MEANING

The aviation domain is interesting from the perspective of Perrow's (1984) framework in that air traffic control is classified as a linear system, but piloting is classified as a complex system. Thus, for air traffic control, safety is currently managed through a centralized authority, and the paths through the sky are tightly constrained by procedures and regulations prescribed by this centralized authority. This management style is successful because there has been plenty of buffer room within the air space to linearize the air traffic control problem effectively. With increasing air traffic density, however, the probability for unexpected interactions is increasing (the system is becoming more complex), leading to a demand for a more distributed management style, in which pilots will have greater discretion and authority (i.e., free flight; Billings, 1997; Rochlin, 1997; Woods & Sarter, 2000).

For the task of piloting, unexpected disturbances and interactions are normal. Although it is possible to automate flight from takeoff to landing, it is impossible to reduce piloting to a fixed set of actions or procedures; in piloting (as with any control task), there is no single set of actions that will produce the same desired output in every context. In other words, the actions (e.g., control manipulations) must be tuned to the particulars of each situation (no takeoff or landing will ever be identical to another). In fact, this is the essence of a control (closed-loop) problem. Powers (1998) noted the following:

> The interesting thing about controlling something is that *you can't plan the actions needed for control before hand....* if actions were not variable, if they did not vary exactly as they do vary, in detail, the same consequences couldn't possibly repeat. The reason is that the world itself shows a lot of variation, and for every variation, our actions have to vary in the opposite way if any particular consequence of acting is to occur again and again. *Controlling means producing repeatable consequences by variable actions.* (pp. 3–5)

Although piloting involves much procedure following and although automated systems are increasingly capable of closing inner control loops, this does not mean that the human is no longer part of the control system. In fact, just the opposite is true. Humans remain in complex systems to respond to disturbances that could not be anticipated in the design of automated systems and to tune the procedures to the inevitable situational variability (Reason, 1990). That is, the operators' role is to respond adaptively to unexpected events. Thus, interface design has to take a "control theoretic perspective" to try to define the information requirements of the control problems that a human supervisor may be asked to solve (Rasmussen, Pejtersen, & Goodstein, 1994, p. 8). This requires that the operators have information about the "state" of the process, about the dynamic constraints that govern the transformations from one state to another, and about the associated costs and benefits relative to the functional goals for the system.

Classical single-sensor–single-indicator interfaces (e.g., the "steam gauge" cockpit) generally were designed to represent most of the "state variables." There appears to have been little consideration to the dynamical constraints among these variables, however, or to the relation of these variables to functional goals of the system. The result is plenty of *data,* but little help for the processes of information or *meaning* extraction (Flach, 1999). For instance, airspeed would be an example of a critical state variable for flight—the airspeed indicator was one of the first instruments introduced to the cockpit. Airspeed data is displayed and "speed bugs" (see Hutchins, 1995a) can be used to identify goal speeds and to a certain extent the acceptable margins of error. The dynamic relations between airspeed and other states of the aircraft are not represented in the interface, however. So, although deviations from the desired airspeed are well represented, solutions (the ways to recover a target airspeed) are not well specified.

What should a pilot do when airspeed is lower or higher than the target speed? That is, what actions should the pilot take so that the aircraft will speed up or slow down to the target speed? Is there anything in the form of the representations used in standard cockpits that helps a pilot to "see" what to do? Of course, experienced pilots generally "know" what to do, but is this because of or despite the interface design? An important question for a cognitive task analysis is what do experienced pilots *know* (either explicitly or implicitly)? Or better still, what *should* experienced pilots *know*? The challenge of EID is, can this *knowledge* be externalized as part of the interface, so that less experienced pilots will be able to *see* what more experienced pilots *know,* or so that any pilot can *see* what they need to *know* to keep the system within the field of safe travel?

A HISTORICAL PERSPECTIVE

A survey of the cognitive task analysis literature shows that much of the recent focus has been on knowledge elicitation (Cooke, 1992) and naturalistic observation (Klein, Orasanu, Calderwood, & Zsambok, 1993). There is currently a stampede of human factors engineers rushing into the field to talk with and watch experts in their natural habitat. Although we agree that this is an essential component of any search for meaning, it is important to complement fieldwork with less glamorous library research, that is, *tabletop analysis*. We are tempted to say that a thorough tabletop analysis is a prerequisite to effective fieldwork, but recognizing the iterative and opportunistic nature of discovery, we make the more modest claim that early investments in tabletop analysis can greatly facilitate later interactions with experts in the field.

Classically, tabletop analysis has focused on instructions, procedure manuals, and training manuals to identify the desirable flow of activities. In searching for meaning, however, it is generally necessary to dig deeper to discover the rationale that guided the choice of procedures. We have found that a good place to begin this search for meaning is to study the historical development of the work domain. This is particularly well illustrated by Hutchins' (1995b) analysis of the domain of ship navigation. This analysis included a study of the historical evolution of navigation in Western culture, as well as comparisons with navigation in other cultures. This type of analysis helps to make the meaning and assumptions behind the current tools and procedures explicit. In biology, it is noted that "ontogeny recapitulates phylogeny." It is tempting to hope that the evolution of a work domain (phylogeny) may provide insights to the ontogenesis of expertise within that domain. This hope has been well substantiated in our study of the evolution of the flight domain.

Let's begin by considering the early development of flight. Anderson (1978) wrote the following:

> During most of the nineteenth century, powered flight was looked upon in a brute-force manner: build an engine strong enough to drive an airplane, slap it on an airframe strong enough to withstand the forces and generate the lift, and presumably you could get into the air. What would happen *after* you get in the air would be just a simple matter of steering the airplane around the sky like a carriage or automobile on the ground—at least this was the general feeling. Gibbs-Smith calls the people taking this approach the "chauffeurs." In contrast are the "airmen"—Lilienthal was the first—who recognized the need to get up in the air, fly around in gliders, and obtain the "feel" of an airplane *before* an engine is used in powered flight. The chauffeurs were mainly interested in thrust and lift, whereas the airmen were firstly concerned with flight control in the air. The airmen's philosophy ultimately led to successful powered flight; the chauffeurs were singularly unsuccessful. (pp. 17–18)

The chauffeur's philosophy and its futility were illustrated by Langley's failed attempts at powered flight in 1903. Twice he attempted to catapult his assistant Manly into the air. As Anderson noted, however, he "made no experiments with gliders with passengers to get the feel of the air. He ignored completely the important aspects of flight control. He attempted to launch Manly into the air on a powered machine without Manly having one second of flight experience" (p. 22). It is quite lucky that Manly was not seriously injured in either of the failed launches. Nine days after Langley's second failed attempt, on December 17, 1903, the Wright brothers made their first successful flight at Kitty Hawk.

Biographers of the Wright brothers contrast their approach to the problem of flight with that of others who had failed. For example, Freedman (1991) wrote the following:

> The Wrights were surprised that the problem of balance and control had received so little attention. Lilienthal had attempted to balance his gliders by resorting to acrobatic body movements, swinging his torso and thrashing his legs. Langley's model aerodromes were capable of simple straight-line flights but could not be steered or manoeuvred. His goal was to get a man into the air first and work out a control system later.
>
> Wilbur and Orville had other ideas. It seemed to them that an effective means of controlling an aircraft was the key to successful flight. What was needed was a control system that an airborne pilot could operate, a system that would keep a flying machine balanced and on course as it climbed

and descended, or as it turned and circled in the air. Like bicycling, flying required balance in motion. (p. 29)

In contrasting the approaches of Lilienthal and Langley with the approach of the Wrights, it is obvious that the Wrights' approach was more consistent with what today would be called a "human-centered" approach to design. Lilienthal and Langley designed wings to be passively stable, and they left it up to the pilot to solve the problems of balance and control that remained. Langley tended to be more of a "chauffeur," focusing on the propulsion problem and ignoring the control problem altogether. Lilienthal was more of an "airman," but his designs provided little support to help the human manage the control problem. "In all Lilienthal machines, the pilot hung upright between the wings, controlling the craft by shifting his own weight to alter the center of gravity" (Crouch, 1989, p. 143). The fact that Lilienthal was able to make "controlled" flights in his "hang gliders" was a testament to his athletic skill; however, it is extremely risky to engineer a system around the skills of exceptional humans.

The Wrights engineered the system around the control problem. They focused their efforts on designing "a control system that an airborne pilot could operate." They succeeded because they engineered a system that allowed manipulation of the flight surfaces. Perhaps the most significant contribution was the invention of "wing-warping" to aid in balance and control. Freedman (1991) noted that the insight for wing-warping began with observations of "buzzards gliding and soaring in the skies over Dayton, they noticed that the birds kept adjusting the positions of their outstretched wings. First one wing was high, then the other" (p. 29). Freedman (1991) continued as follows:

> It occurred to the brothers that a bird can balance itself and control its flight by changing the angle at which each wing meets the oncoming air. When the bird wants to begin a turn, for example, it tilts the leading edge of one wing up to generate more lift, making that wing rise. At the same time, it tilts the leading edge of the opposite wing down, causing that wing to drop. Its body revolves toward the lowered wing. When it has revolved as far as it wishes, it reverses the process and begins to roll the other way. (p. 29)

Wing-warping involves twisting the semirigid wing so that when one end of the wing is up, the other is down. Wilbur Wright reportedly got the idea for wing-warping while holding a cardboard box that had contained an inner tube. He noticed that by twisting the ends of the box in opposite directions, he could accomplish the control function that had been observed in the buzzards. The first practical aircraft, the *Wright Flyer III,* had three controls: One hand controlled an elevator, the other hand controlled a rudder, and a saddle device at the pilot's hip allowed him to warp the wings by shifting his hips. As Freedman (1991) noted, the Wrights' solution to the control problem remains the foundation for modern flight.

> Except for the engine, their 1902 glider flew just as a Boeing 747 airliner or a jet fighter flies. A modern plane "warps" its wings in order to turn or level off by moving the ailerons on the rear edges of the wings. It makes smooth banking turns with the aid of a movable vertical rudder. And it noses up or down by means of an elevator (usually located at the rear of the plane). (p. 64)

The Wright brothers discovered the first workable solution to the "manipulation" aspect of the inner loop control problem. Pioneers in aviation soon realized, however, that the control problem involved perception as well as manipulation. It soon became clear that the unaided human senses did not provide all the information necessary for stable control. Edwards (1988) reported the following

> Experience indicated . . . that certain basic aids were essential in order to achieve an acceptable level of control. Early amongst these was the famous piece of string tied either to the trailing edge

of an elevator or to a lateral frame member so that the pilot could avoid skid or slip during a turn by keeping the fluttering string parallel to the fore-and aft axis of the aircraft. Without this aid, turns could easily lead to a spin from which there might be no recovery. (p. 6)

Gradually, other instruments were introduced to supplement the information that was directly available to the human senses. These included airspeed indicators to reduce stalling accidents, compasses, and barometric altimeters. For the most part, however, early aviators "tended to relegate instruments to a subsidiary role, and little or no mention was made of their use in the programs of flight training" (Edwards, 1988, p. 6). This is reflected in this historical note from Ocker and Crane's (1932) book on the theory of "blind flight":

The equipment of the airplane prior to 1914 was characterized by an almost complete lack of instruments, both engine instruments and flight instruments. This was due, no doubt, to a number of causes, principal among these being that there always has been a studied dislike of instruments by the pilot. The pilot likes to feel that he is an artist at flight control, and in plying his art feels that his natural instincts are a better guide to the performance of his craft than any number of instruments. The natural result of this psychology was an airplane with few instruments. (p. 9)

It is interesting to note that there was great initial scepticism about early instruments. This scepticism about instruments was illustrated by Ocker and Crane's (1932) description of a situation in which pilots' from one "airway" company returned several turn instruments because they were defective. The instruments appeared to give correct readings in clear weather but did not seem to work in clouds. The instruments were tested and shown to work correctly. The problem, as we now know, was that the pilots were putting their faith in unreliable vestibular feedback, rather than in the reliable instruments (e.g., Leibowitz, 1988). Thus, a significant aspect of instrument training is to help pilots to transfer their trust from the unreliable "seat of the pants feelings" to the reliable instruments.

We could continue this story, but we hope that this brief review of some of the issues of early flight illustrates the value of taking a historical perspective. The "chauffeur's" approach to flight is often replicated in novice pilots who attempt to generalize their experiences controlling ground and sea vehicles to the task of flying. This is reflected in both their intuitions about control and their choice of a coordinate system for spatial orientation. Novices tend to orient to their intrinsic coordinate system (e.g., upright as signalled by their eyes under visual conditions and their vestibular system in instrument conditions), and it takes extensive training to get them to trust the world-centered (i.e., horizon-centered) coordinate system presented by their instruments. A goal for CSE and EID is to facilitate the transformation from "chauffeur" to "aviator."

ONE PILOT'S OPINION

Another significant stimulus to our early thinking about the flying problem was Langewiesche's (1994) book *Stick and Rudder: An Explanation of the Art of Flying.* This book, originally published in 1944, is still valued by some general aviation flight instructors. It was originally brought to our attention because of his insights about the visual information used in an approach to landing, which was an important inspiration for Gibson's (1966) conception of the optical flow field. Thus, his insights have not only influenced pilots but also have inspired an important area of scientific research associated with visual control of locomotion.

The following quote describes Langewiesche's goals for his book, which resonate with the motives behind CSE and EID:

It may be that, if we could only understand the wing clearly enough, see its working vividly enough, it would no longer seem to behave contrary to common sense; we should then expect it to behave as it does behave. We could then simply follow our impulses and "instincts." Flying is

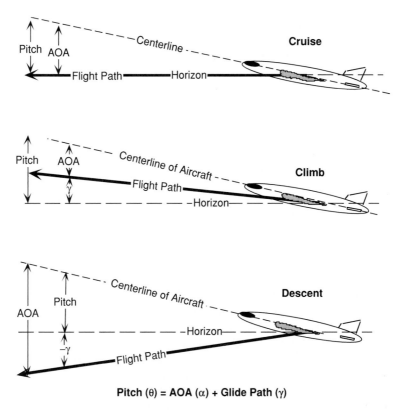

Pitch (θ) = AOA (α) + Glide Path (γ)

FIG. 9.1.   Unlike most ground vehicles, an aircraft does not always go in the direction in which it is pointed. Here are three situations in which the aircraft is pointing in the same direction relative to the horizon (pitch) but is moving in three directions (three flight paths: cruise, climb, and descent), none of which correspond to the direction it is pointing. The differences between pitch and flight path are reflected in three Angles of Attack (AOA) reflecting three airspeeds.

done largely with one's imagination! If one's images of the airplane are correct, one's behaviour in the airplane will quite naturally and effortlessly also be correct. (p. 4)

For Langewiesche, the key to understanding a wing was the angle of attack (AOA; α). He labeled the angle of attack as the "single most important fact in the art of piloting" (p. 5). The AOA is the angle at which a wing's chord line meets the relative wind. Typically, the relative wind will be parallel and in the opposite direction to the motion of the aircraft. AOA is related to but not the same as the pitch angle or the glide path angle of the aircraft. Figure 9.1 illustrates the relations among these three angles. Aircraft are generally designed to stabilize around a specific AOA dependent on the current configuration (e.g., trim). Thus, AOA is a key to understanding how the aircraft naturally behaves so that the pilot can work with the aircraft rather than against it. Also, Langewiesche felt that AOA was critical for developing intuitions about the function of the various controls. Figure 9.1 clearly illustrates that even with longitudinal control an airplane does not typically move in the direction it is pointed as do most land vehicles. The AOA is an important factor relating the orientation (pitch; θ) and the path angle (direction of motion; γ) of the aircraft. Thus, for Langewiesche, an understanding of AOA can help a pilot to see how the throttle can function as an "altitude" control and the elevator as an "airspeed" control.

One of the difficulties in grasping the AOA, as Langewiesche observed, is that it is invisible to pilots. That is, because the pilot can't see the "wind," the AOA can't be seen without instruments. Although it can be measured, most aircraft did not have an instrument to measure or to display

AOA. Because AOA is an attribute of motion, static diagrams can also be misunderstood. This misunderstanding is frequently because the motion of the wing, the direction of relative air, or both are typically depicted as abstract vectors and are less salient than the orientation of the wing within the diagram.

Langewiesche saw an understanding of AOA as the foundation for piloting skill and the failure to understand AOA as a contributing factor in aviation accidents: "the stall is the direct and invariable result of trying to fly at too large an Angle of Attack" (p. 20). A stall refers to the situation in which a wing no longer functions to support the aircraft. A wing will produce more lift at larger AOA up to a limit. Beyond this limit the wing will cease functioning. At this point, there will be insufficient lift for controlled flight and the aircraft begins to fall out of the air. In Langewiesche's day, stall was a cause of many accidents. Wilson (1979) reported that "two-thirds of all (general aviation) fatal accidents in the 1934–38 period involved stalls, spins, or both; in 1939–40, over half the fatal accidents originated in the pilots' inability to recognize or recover from stall" (p. 36). Since Langewiesche's day, however, recovering from a stall has become easier because of improvements in the design of some aircraft and stall warning systems have been introduced to help alert pilots to the problem when it occurs. For example, by 1946 (2 years after Langewiesche's book), the Civil Aeronautics Authority's (CAA) Technical Development and Evaluation Center had developed three stall-warning models, each reacting to changes in airflow over the wing by activating a horn or flashing light in the cockpit. In 1951, the CAA sent a specially equipped plane around the country to demonstrate stall recovery techniques. These steps dramatically reduced the number of stall-spin accidents, from 113 to 39 accidents, for the last quarter of 1951. Although the alarms and improved handling qualities have made aviation safer, they have not necessarily made pilots smarter; they do not help pilots to understand the development of conditions that lead to stall. From a CSE perspective, the challenge is how can we help pilots to understand and anticipate the conditions of a stall, so that the warnings rarely are needed.

From the perspective of CSE and EID, AOA attracted our interest. It seemed clear that AOA was Langewiesche's answer to the question of "what should a skilled pilot know?" Thus, we began with an initial hypothesis that it might be an important element of the "deep structure" of flight control that is not well represented in most cockpits. Early in our analysis, we began to wonder if making AOA explicit could not only help prevent stalls but also enhance general piloting abilities.

PARTICIPANT OBSERVATIONS IN A SYNTHETIC TASK ENVIRONMENT

In this phase of our search for meaning, we made observations in a flight simulation that we developed. Key to this phase was our experiences of making the simulation work to the satisfaction of the pilots (so that they could fly precision approaches consistently using their normal experiences, without additional training in the simulation). Also, the process whereby one of the experienced pilots in our group trained two of the other authors, who were not pilots, was important to our understanding of the precision approach problem.

For those in the group who were naïve participant observers trying to learn the precision approach task in the simulator, the most striking problem was what we perceived as a cross-coupling across the controls. When we struggled to align the course of the aircraft with the runway, we would lose track of our glide slope (and suddenly the ground would be looming on the screen or the runway would be passing far below us). If we focused on our glide slope, we would soon find our course helplessly out of alignment. Attention to one aspect of the approach invariably resulted in unintended deviations with respect to another dimension of the task.

It was interesting to us that this coupling problem does not get much discussion in the literature on aviation psychology. In a chapter on pilot control, Baron (1988) described the degrees of freedom for flight:

> The aircraft moves in three-dimensional space and has, from the standpoint of its basic, dynamic response, six degrees of freedom; it can translate linearly in three ways (forward-backward, left-right, up-down) and it can rotate angularly about three separate axes. The motions of the airplane are actually determined by the action of a variety of aerodynamic, gravitational, and thrust forces and moments, some of which can be controlled by the pilot using the control manipulators. For conventional, fixed-wing aircraft, there are four controls (elevator, aileron, rudder, throttle). With these, and changes in configuration, the pilot can introduce control moments about three axes (pitch, roll, and yaw) and control forces along the aircraft's forward-aft body axis (through thrust-drag modulation). Clearly for complete control of all six degrees of freedom, there must be interaction or coupling among motions. The most familiar example of this are motions in which the pilot rotates the aircraft to orient its lift force so as to produce appropriate sideways and normal accelerations (e.g., bank-to-turn). (pp. 349–350)

After introducing the degrees of freedom problem, Baron quickly dismissed it, noting that "although there is some inevitable coupling among all six degrees of motion of the aircraft, some interactions are quite strong while others may be virtually negligible" (p. 350). From this point, Baron's discussion of control tended to focus on the longitudinal (glide path) and lateral (course) control dimensions as essentially independent control tasks. A similar approach is seen in Roscoe (1980). In terms of the aerodynamics, the interactions between lateral and longitudinal control axes should generally be small, especially when the bank angles are small, as would be expected for precision approaches. Our experiences learning to fly the precision approach, however, were not consistent with this analysis. The interactions between control dimensions seemed to be a significant factor relative to the difficulty of flying a precision approach. This suggests that the major source of the interactions was our ineffective control strategy. The interactions we experienced most likely reflected poor time sharing; error was building up because of inattention to one axis while focusing intently on the other.

Figure 9.2 was our early attempt to illustrate the interactions among the control dimensions. The boxes are empty because our goal was not to specify precise quantitative relations but simply to visualize the couplings among the variables. There are two significant features of this diagram relative to other representations in the aviation psychology literature (i.e., Baron, 1988; Roscoe, 1980). First, Fig. 9.2 makes the interactions across flight dimensions explicit, whereas others illustrate the longitudinal and lateral control tasks independently. Second, the diagram is structured so that the outputs reflect the dimensions that would typically be used to "score" an approach as safe or unsafe. In other illustrations of the control tasks, the "outputs" are typically altitude and lateral displacement (other dimensions are included as internal state variables). Note that there are six output variables, these six variables span the control space (six degrees of freedom). However, note that the dimensions of the "control space" are different from the dimensions that are sometimes used to characterize aircraft state [i.e., three positions (x, y, & z) and three rotations (pitch, roll, and yaw)]. The rotational dimensions are the same, but the x (e.g., distance to runway) and z (altitude) dimensions of standard representations are replaced by airspeed and glide slope in Fig. 9.2. Also, the glide slope and course (y) variables are referenced to the touchdown target on the runway, not the aircraft. This change reflects the goals for a precision approach. Also, note the central position of AOA within the diagram, consistent with Langewiesche's emphasis on this factor. Finally, the coupling between the lateral and longitudinal control axes is "dashed" to reflect the fact that the magnitude of these interactions may depend heavily on the skill of the pilot.

As we thought more about the degrees of freedom problem, we began to think about priorities among the different dimensions of a precision approach. Figure 9.3 shows our attempt to

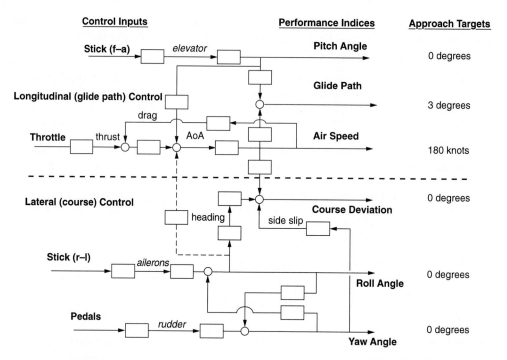

FIG. 9.2.   A schematic representation of the piloting problem designed to emphasize the couplings between control axes and the performance indices that would be used to evaluate the success of an approach to landing. f−a = fore−aft; r−l = right−left.

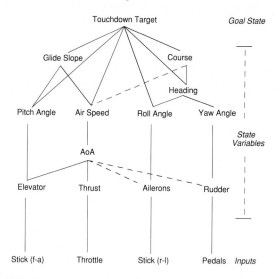

FIG. 9.3.   A hierarchical representation designed to emphasize means−ends relations associated with an approach to landing. f−a = fore−aft.

illustrate these priorities in a hierarchical means−ends graph. Again, this diagram shows the cross couplings across the input or control dimensions and the central role of AOA in mediating those interactions.

The frustrations in learning the landing task helped us to appreciate the difficulties of the flight control problem. There were clearly strong temptations to "chauffeur" the aircraft to the runway based on our experiences with land vehicles. And the adjustments and adaptations needed to become "aviators" were not readily apparent. This had all the characteristics of a "workload" problem; however, the "load" was not due to "too much data" or to a processing limit. Rather, it was due to poor attunement to problem structure (a meaning processing problem). It was difficult for us to take advantage of the natural constraints to "parse" the many dimensions of the control problem into a manageable number of meaningful "chunks."

PROCESS TRACING IN SYNTHETIC TASK ENVIRONMENTS

> When experienced pilots fly an airplane, when they put it through climbs and glides, turns, stalls and spins, takeoffs and landings, they are asking themselves all the time, "Where, at this moment, is the air coming from? And at what angle are my wings meeting it?" Most pilots would, of course, deny that they ever think such a thing, and many excellent pilots don't even know the meaning of the words *angle of attack* and *relative wind* (Langewiesche, 1944, pp. 23–14).

Our interactions with pilots confirmed Langewiesche's observations described here. Many pilots did not explicitly understand AOA, and most saw little value to a display of AOA in the cockpit. Was this a sign that we were on the right or the wrong track?

Rather than AOA, the variable that seemed to be of most concern for our experienced pilots was airspeed. Note that AOA and airspeed are tightly coupled in flight. The first thing that most experienced pilots wanted to know before attempting to land the simulation was the "final approach speed." Furthermore, tracking the final approach speed (i.e., correcting deviations from the target speed) was a top priority. Also, surprising to those of us who were still thinking like chauffeurs, the primary control action for correcting deviations from the target airspeed was fore–aft deflections of the stick, not the throttle. Pilots tended to use a "pitch-to-speed" mode of control for the large aircraft on which our simulation was based, as illustrated in the top half of Fig. 9.4. However, we have been told that there are situations in which alternative

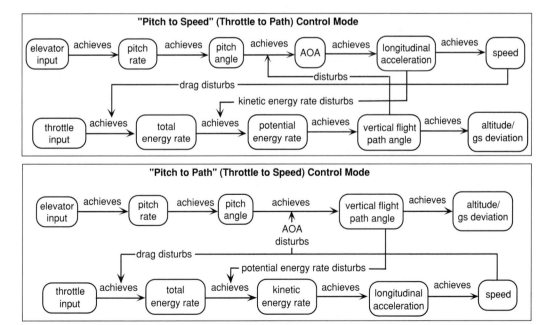

FIG. 9.4.   Two modes of controlling speed. The top panel shows the pitch-to-speed mode that pilots most commonly use. In this mode, the throttle is typically set at a fixed level, and speed is regulated using primarily the elevator (pitch). The bottom panel shows the pitch-to-path mode used in some automatic landing systems. In this mode, the path is regulated using the elevator, and the speed is tracked with continuous modulations of thrust. AOA = angle of attack.

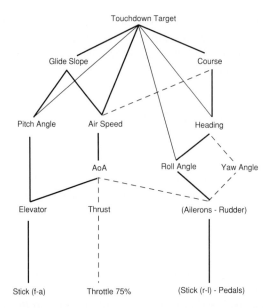

FIG. 9.5.   A hierarchical representation that illustrates constraints that skilled pilots introduce to help manage the degree of freedom problem. Dotted lines indicate paths that are constrained, and dark solid lines indicate primary functional control paths. The partitioning helps the pilot to function as a smart mechanism with a minimum number of degrees of freedom that are particularly well suited to the natural task constraints. f–a = fore–aft; r–l = right–left.

modes of control are more common (e.g., carrier landings). Also, the automated systems in some aircraft use a mode of control more similar to the "pitch-to-path" mode illustrated in the bottom half of Fig. 9.4. As Holder and Hutchins (2001) described, this can sometimes result in an "automation surprise" for pilots:

> Speed can be controlled by thrust while thrust is at idle *only* if the vertical path has been constructed such that idle thrust *is* the amount of thrust required to produce the target speeds. This computation cannot be done by any human pilot. But, it is exactly what a managed descent in an Airbus or a VNAV path descent in a Boeing does. Many pilots see this as magic because they literally cannot imagine how it is done. (p. 3)

Figure 9.5 attempts to illustrate what we learned from talking with the experienced pilots and from watching them land the simulation. The pilots tended to lock out degrees of freedom to simplify the control problem. First, the pilots set the throttle so that the aircraft would operate in the "ballpark" of the target approach speed (about 75% of full throttle for the aircraft and configuration used in our simulation). Then the pilot used fore–aft deflections of the stick to track the target airspeed and the glide slope. With the throttle fixed at the appropriate point, the airspeed and glide slope would be functionally coupled, that is, higher-than-target speed typically indicates that the approach is steeper than desired and lower-than-target speed typically indicates that the approach is shallower than desired. The pilots tended to use right and left deflections of the stick synergistically with the fore–aft stick (elevator) to control course deviations. For many of the pilots, the rudder became the primary control for correcting small deviations in the region of final touchdown. For small corrections, rudder controls give yaw angle and slip angle and thereby course changes. This is faster than aileron control, which needs an extra integration (from roll, to bank angle, and then to course change), and it tends to minimize unintended disturbances to other control targets (glide path and roll attitude).

Thus, consistent with the decompositions of the control task described by Baron (1988) and Roscoe (1980), pilots tended to manage the control problem so that interactions between the glide slope (longitudinal axis) and the course (lateral axis) were minimized. This probably reflects a more effective distribution of attention between the two axes. In this sense, skill development involves discovering a "smart" way to "chunk" the control problem to use natural constraints and minimize the dimensionality of the control problem (e.g., Runeson, 1977).

Several factors contribute to accomplishing an appropriate decomposition of the control task. First, the coupling is not symmetric. Course control using ailerons is more likely to affect the glide slope than vice versa. Course control generally requires coordinated motion of the stick right–left and fore–aft (sometimes with rudder control), whereas, glide slope corrections can be accomplished using only the fore–aft dimension of the stick. Although the interactions are small, course corrections will typically affect airspeed and glide slope; however, corrections to glide slope can be accomplished with no change to course.

A second factor affecting the decomposition of the control task is that the spillover between axes is highly dependent on the ability to control the individual dimensions. Correcting a course deviation error requires a more complex control input than correcting a deviation from glide slope. Remember that discovering the solution to "turning" the aircraft was a major obstacle to the design of the first aircraft. Course control is a fourth-order problem: There are four integrations between the control manipulation and the control variable. Glide slope control is only third order. A fourth-order control problem is extremely difficult (essentially impossible) for a single-axis tracking task for which the output is the only feedback. To manage this control task, the pilot must be able to anticipate (predict) the effects of his control actions before they show up as changes in the final control variable. From our observations of pilots and our experiences learning to fly the simulation, we discovered that attending to the heading indicator (heading and rates of change of heading) can be helpful for anticipating effects of actions to correct course deviations. The heading indicator, rather than the course deviation needle of the flight director, tended to be the primary reference for course control. By referencing intermediate states (roll and heading) the pilot can effectively reduce the fourth-order problem to multiple nested, lower order control problems (see Baron, 1988; figure 11.4, p. 355).

It is our understanding that some flight director displays actually combine heading and change of heading (or some other index of the derivative and second derivative or course deviations) to drive the horizontal needle of the flight director. In the language of manual control, the course deviation indicator is "quickened" to reduce the effective order of control from fourth to first order. When we inquired with manufacturers of these instruments, however, we were surprised to learn that the algorithms driving the flight director are "proprietary." We were also surprised that pilots show little concern or curiosity about the internal "logic" of these displays. This doesn't necessarily indicate a blind trust on the part of the pilots but rather a confidence that has evolved based on an ability to validate the flight director against other redundant sources of information (including the optical flow field and other instruments).

Our observations and discussions with pilots indicated that, for many pilots, correcting course deviations tended to be more difficult. We found from our own experience learning the task that as our skill at controlling course deviations improved, the interactions across the control dimensions were much easier to manage. This does not necessarily indicate that one type of error is more or less important than another. Rather, the point is simply that the overall control problem tends to be easier if the course errors are managed well.

As a result of our observations in the synthetic task environment, our picture of the deep structure of the precision approach task changed. We began to see AOA not as a separate critical variable, but as the nexus of the degree of freedom problem. We began to believe that adding AOA to the interface would not help to solve the meaning-processing problem. What novices needed was a better organization or configuration (i.e., chunking) of information so that the interactions among the control axes (largely mediated by AOA) were more apparent. It seemed to us that the problem was not lack of critical information, but poor organization of the information. We needed to think about better ways to "configure" the interface (Bennett & Flach, 1995) so that the meaningful interactions among the components of the task were better specified.

BACK TO FIRST PRINCIPLES

The effort to understand how an airplane flies is sometimes called "The Theory of Flight." Under that name, it has a bad reputation with pilots. Most pilots think that theory is useless, that practice is what does it. . . . What is wrong with "Theory of Flight," from the pilot's point of view is not that it is theory. What's wrong is that it is the theory of the wrong thing—it usually becomes a theory of building the airplane rather than of flying it. It goes deeply—much too deeply for a pilot's needs—into problems of aerodynamics; it even gives the pilot a formula by which to calculate his lift! But it neglects those phases of flight that interest the pilot most. It often fails to show the pilot the most important fact in the art of piloting—the Angle of Attack, and how it changes in flight. And it usually fails to give him a clear understanding of the various flight conditions in which an airplane can proceed, from fast flight to mush and stall. (Langewiesche, 1944, pp. 4–5)

The search for meaning is often complicated by the fact that each of the disciplines associated with a domain (pilots versus aeronautical engineers) has their own perspective (flying versus building aircraft) and their own specialized jargon reflecting that perspective. These differences were striking as we began to explore our understanding of the landing task and our display concepts with aeronautical engineers. This is reflected in our earlier discussion of the dimensions of the problem space: Pilots and engineers tend to use different coordinates to index the same space. Figure 9.6 illustrates what we believe is an important difference in how aeronautical engineers and pilots approach the problem of flight. Aeronautical engineers tend to think about the "step response" of an aircraft. That is, if a step input is presented through one of the controls, what will the aircraft do? In evaluating this question, they have found it valuable to distinguish between the "short-period" and "long-period" response. Thus, when we asked aeronautical engineers what factors would determine whether "pitch-to-speed" or "throttle-to-speed" approach strategies would be more effective, their first response was that it depended whether we were interested in the "short" or "long-period" response. At this point,

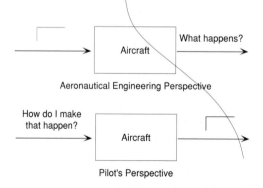

FIG. 9.6.   Two perspectives on the flying problem: Although the interests of aeronautical engineers and pilots overlap, they tend to think about the problem within different frameworks. The engineers choose a framework that emphasizes the "causal" relations associated with opposing forces. The pilots choose a framework that emphasizes the "intentional" relations associated with safe accomplishment of operational objectives.

TABLE 9.1
Symmetric Flight: Energy Management

Airspeed Altitude	Too Slow	Too Fast
Too high	Energy distribution problem, push stick (elevator) forward to trade excess PE for needed KE	Too much energy—throttle back
Too low	Too little energy—throttle forward	Energy distribution problem, pull stick (elevator) back to trade excess KE for needed PE

Note. PE = potential energy; KE = Kinetic energy.

communication was difficult. Pilots never talked about "short" or "long-period" responses. Even after the engineers explained the differences, it wasn't obvious which should be more interesting to us, although we are coming to believe that the short-period response is most critical to the pilot's control problem. Baron (1988) includes a discussion that relates this decomposition to inner-(short-period) and outer-loop (long-period) aspects of the control task. One way to think about the difference between short- and long-period responses is that the short-period response is the instantaneous reaction of the aircraft to a command input. The long-period response is the aircraft's natural motion as the aircraft settles into a steady state as its natural stability properties come into play (e.g., damping out the initial control effects). Note that in Fig. 9.2, the links emphasize the short-period interactions. To represent the long-period effects, we would need to include feedback loops to illustrate the natural stability properties of the aircraft.

In contrast to aeronautical engineers, pilots tend to think about how to produce "step outputs." That is, they want to know what control input will move an aircraft from one state (where they are) to another (where they want to be). Both groups are asking questions about the aircraft dynamics, and both perspectives are legitimate ways to ask about the dynamics; however, the translation from one perspective to the other is not trivial, and failure to make this translation can lead to great difficulties in communication between aeronautical engineers and pilots (not to mention the difficulty of communicating with the psychologists).

We found that aeronautical engineers also found it valuable to decompose the control task along the longitudinal (they used the term *symmetric flight*) and lateral (*asymmetric flight*) axes. Some further decomposed the control problem into concepts of energy. For symmetric flight, total energy (TE) could be decomposed into kinetic (KE) and potential energy (PE). Kinetic energy is a function of the speed of the aircraft ($1/2 * \text{mass} * \text{velocity}^2$), and potential energy is a function of the altitude (mass * gravitational acceleration * altitude). In this context, the throttle could be thought of as controlling the total energy, and the fore–aft stick (elevator) could be thought of as governing the distribution of energy between (KE ≈ airspeed) and (PE ≈ altitude).

This provides an alternative way for thinking about the pilots' approach strategy. The effect of fixing the throttle is to create a situation in which energy out (due to drag) is greater than energy in. The throttle setting should be set so that the rate of energy loss (drag–thrust) is at a small constant value, which corresponds to the ideal glide slope for a given aircraft. The stick then is used to make sure that the energy loss is in terms of PE (altitude) rather than KE by making sure that airspeed remains roughly constant. Table 9.1 shows how deviations

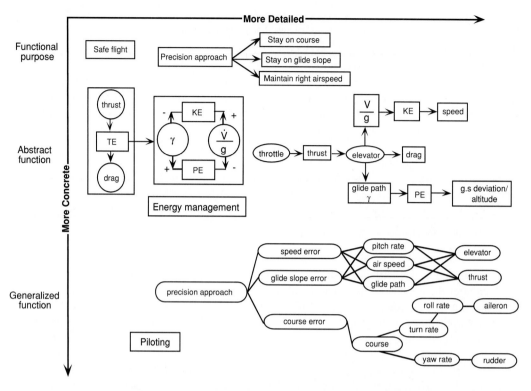

FIG. 9.7.   The abstraction–decomposition space is used to illustrate several perspectives on the precision approach problem. The perspectives span different levels of abstraction and different amounts of detail. At the abstract function level, the problem is described using energy concepts suggested by aeronautical engineers. At the generalized function level, the problem is described to reflect the pilots' view of the task. KE = Kinetic energy; PE = potential energy; TE = total energy. (γ = flight path angle: $\dot{V}/g$ = longitudinal acceleration).

from glide slope and airspeed map into the energy distribution and the implications for the appropriate response. Note that just prior to touchdown (PE near zero) a flare manuever is executed to minimize vertical velocity at touchdown. This further reduces total energy at the point of contact with the earth; the residual KE is then dissipated after the aircraft is on the runway.

At this point, we are beginning to wonder whether it might be useful to use the concept of energy and its partitioning (balance) between kinetic and potential energy as an organizational constraint in a graphical configural display. Energy is beginning to replace AOA as the lynch pin in our conceptual image of the flying problem. We are in the early stages of our discussions with aeronautical engineers. After some false starts, we are getting past the language differences so that we can think about the pilots' control problems relative to dimensions that are meaningful to the aeronautical engineers. At this point, we are shifting focus from AOA to energy concepts as a way of thinking about how to parse the control problem and how to organize a graphical representation into meaningful chunks. Our current belief is not that we need to "add" energy displays. Rather, we believe that understanding the "energy balances" may be fundamental to developing configural representations that will organize the data into patterns that better reflect meaningful constraints on the control problem.

Figure 9.7 shows our current evolving image of the precision approach problem using the top three levels of Rasmussen's (1986; Rasmussen et al., 1994; Also, see Bisantz & Vicente, 1994) abstraction hierarchy. Note that this is an "evolving" image, that is, we are not satisfied that we have captured all the meaningful constraints. At the functional purpose level, the goal of a safe flight is reduced to following a precision path, as is typically specified by an ILS beacon. From the pilot's perspective, this typically involves tracking the target landing speed and the ILS needles that provide glide path and course errors.

At the abstract function level, the precision approach can be described as an energy management problem. Flying the approach can be described in terms of a constant decrease in potential energy while maintaining a fixed level of kinetic energy. This energy management problem can be decomposed to reflect some of the state variables and controls. The abstract function level reflects the aeronautical engineer's perspective on the precision approach problem.

The pilot's perspective on the control problem can be seen at the generalized function level of description. Here the control problem is described in terms of the targets (speed, glide slope, and course), important intermediate state variables, and the control surfaces. These top levels of the abstraction hierarchy should be fairly independent of the type of aircraft or landing situation. Lower levels of the abstraction hierarchy can be used to represent the local constraints associated with specific types of aircraft, instruments, levels of automation, or situational variables.

THE STORY CONTINUES

"Forty-two!" yelled Loonquawl. "Is that all you've got to show for seven and a half million years' work?"

"I checked it very thoroughly," said the computer, "and that quite definitely is the answer. I think the problem, to be quite honest with you, is that you've never actually known what the question is." "But it was the Great Question of Life, the Universe and Everything," howled Loonquawl.

"Yes," said Deep Thought with the air of one who suffers fools gladly, "but what actually is it?"

A slow stupefied silence crept over the men as they stared at the computer and then at each other.

"Well, you know, it's just Everything . . . everything . . ." offered Pouchg weakly.

"Exactly!" said Deep Thought. "So, once you do know what the question actually is, you'll know what the answer means." (Adams, 1980; p. 183)

The reader is probably expecting "the answer" to justify our long search for meaning. Unfortunately, we have no answer. We have generated several display concepts. Some are currently being evaluated in our synthetic task environment, but all are under revision because we are not satisfied with any of our current displays. And quite frankly, although we hope to ultimately produce some improvements to the representations in the cockpit, we don't ever expect to get "the answer." The search for meaning, like skill development, tends to be an asymptotic process—a small investment early yields large improvements—but continued progress comes at the expense of increasingly larger investments, and no matter how much is invested there is always more to learn. Thus, we think it is a mistake to think of cognitive task analysis as an activity to be "completed" prior to design. Decisions about when to begin design or when a solution is good enough must take into account the practical, social, political, and economic constraints associated with the design project. The search for meaning does not end when a system is implemented, however. An important challenge for cognitive task analysis is to integrate the cognitive task analysis into the work domain so that the system continues to evolve, to learn from mistakes, and to take advantage of new opportunities. Cognitive systems are dynamic, if they are not continuously learning and adapting, they risk extinction. Meaning is a moving target.

People who are searching for simple, easy-to-understand answers will be sorely disappointed by the cognitive systems engineering approach. The search for meaning is not easy. It involves an iterative process of redefining the questions with no guarantee that answers will emerge. As far as discovering a great new interface for aviation is concerned, we are less confident now than we were at the start of our project. We hope this is because we are beginning to appreciate the depth of the problem. We think we have a much deeper understanding of the domain, although we are far from consensus and there are still vigorous debates among the authors about how particular facts fit into the big picture. One measure that gives us confidence that we are moving in a positive direction is the flavor of the interactions with the domain experts (i.e., pilots and aeronautical engineers). It is easier and (more enjoyable) to talk with either group, and we find ourselves in a position to mediate between the two groups—for example, helping pilots to appreciate "aeronautical theory" and helping aeronautical engineers to take a "pilot-centered" view of the control problem.

In describing our search for meaning in a book chapter, we are constrained to a linear narrative structure. Nevertheless, the search for meaning is not linear. All lines of investigation (reading about the history of aviation, talking to and observing pilots, talking to aeronautical engineers, flying and tuning the simulation) are all going on in parallel. It is difficult to overstate the magnitude of the interactions across the various lines of investigation. For example, we can hear the same thing over and over again from pilots and not appreciate its meaning until we experience the problem flying the simulation or grasp a construct from aeronautical theory. In responding to design deadlines, some approaches can be more efficient than other; talking with experts is typically the quickest way to learn about a domain, and published sources and manuals are probably the most accessible source of information. Every source has its limitations, however, so it is important to seek input from multiple perspectives (Flach, 2000a).

At times the search for meaning can be extremely frustrating as our first naïve answers to complex problems are crushed under the weight of new experiences. Although we don't know what the answer is, we are better able to critique our ideas and to anticipate the limitations of any particular solution. So the search for meaning will not lead to "one best way," but for now we are encouraged that we are taking small iterative steps toward better, deeper ways to ask the question. As Will Rogers said, "It isn't what we don't know that gives us trouble, it's what we know that ain't so." At this point, we are happy that less of what we know about aviation ain't so, but we are still humbled by how naïve some of us were at the start of this search and by how much still remains to be learned.

ACKNOWLEDGMENTS

Our search for meaning has been supported by funding from several sources. The Japan Atomic Energy Research Institute provided a 3-year grant to support evaluations of early display concepts. Dr. Tanabe of JAERI provided wise council that helped guide our display concepts. Funding from the Air Force Office of Scientific Research (Grant F49620-97-1-03) and the State of Ohio supported the development of the synthetic flight environment (i.e., the CAVE) that has been a critical resource for testing many of our ideas. Our approach to task analysis has been inspired by the work and ideas of Jens Rasmussen, who continues to contribute as a patient and gentle advisor to all the work we do.

REFERENCES

Adams, D. (1980). *The hitchhikers guide to the galaxy*. New York: Ballantine Books.

Anderson, J. D. (1978). *Introduction to flight engineering and history*. New York: McGraw-Hill.

Baron, S. (1988). Pilot control. In E. L. Wiener & D. C. Nagel (Eds.), *Human factors in aviation* (pp. 347–385). San Diego, CA: Academic Press.

Bennett, K. B., & Flach, J. M. (1992). Graphical displays: Implications for divided attention, focused attention, and problem solving. *Human Factors, 34,* 513–533.

Billings, C. (1997). *Aviation automation: The search for a human-centered approach.* Mahwah, NJ: Erlbaum.

Bisantz, A. M., & Vicente, K. J. (1994). Making the abstraction hierarchy concrete. *International Journal of Human-Computer Studies, 40,* 83–117.

Cooke, N. J. (1992). Modeling human expertise in expert systems. In R. R. Hoffman (Ed.), *The psychology of expertise.* (pp. 26–60). New York: Springer-Verlag.

Crouch, T. (1989). *The bishops boys.* New York: Norton.

Drury, C. G., Paramore, B., Van Cott, H. P., Grey, S. M., & Corlett, E. N. (1987). Task analysis. In G. Salvendy (Ed.), *Handbook of human factors* (pp. 310–401). New York: Wiley.

Edwards, E. (1988). Introductory overview. In E. L. Wiener & D. C. Nagel (Eds.), *Human factors in aviation.* (pp. 3–26). San Diego, CA: Academic Press.

Flach, J. M. (1999). Ready, fire, aim: Toward a theory of meaning processing systems. In D. Gopher & A. Koriat (Eds.), *Attention & performance XVII.* (pp. 187–221). Cambridge, MA: MIT Press.

Flach, J. M. (2000a). Discovering situated meaning: An ecological approach to task analysis. In J. M. Schraagen, S. F. Chipman, & V. L. Shalin (Eds.), *Cognitive task analysis* (pp. 87–100). Mahwah, NJ: Erlbaum.

Flach J. M., (2000b). *Research on information form in human machine interfaces: A meaning processing approach.* Final Report to Japan Atomic Energy Research Institute. Dayton, OH: Wright State University.

Flach, J. M., & Rasmussen, J. (2000). Cognitive engineering: Designing for situation awareness. In N. B. Sarter & R. Amalberti (Eds.), *Cognitive engineering in the aviation domain* (pp. 153–179). Mahwah, NJ: Erlbaum.

Freedman, R. (1991). *The Wright brothers: How they invented the airplane.* New York: Holiday House.

Gibson, J. J. (1966). *The senses considered as perceptual systems.* Boston, MA: Houghton Mifflin.

Gilbreth, F. B., & Gilbreth, L. M. (1917). *Applied motion study.* New York: Sturgis and Walton.

Holder, B., & Hutchins, E. (2001). What pilots learn about autoflight while flying on the line. *Proceedings of the 11th International Symposium on Aviation Psychology.* Columbus, OH: The Ohio State University.

Hollnagel, E., & Woods, D. D. (1983). Cognitive systems engineering. New wine in new bottles. *International Journal of Man-Machine Studies, 18,* 583–600.

Hutchins, E. (1995a). How a cockpit remembers its speeds. *Cognitive Science, 19,* 265–288.

Hutchins, E. (1995b). *Cognition in the wild.* Cambridge, MA: MIT Press.

Jacques, P. F., Flach, J. M., Patrick, D. L., Green, R., & Langenderfer, J. (2001). The Wright configural attitude display (WrightCAD) and the ecology of virtual instrument approaches. *Proceedings of the 11th International Symposium on Aviation Psychology.* Columbus, OH: The Ohio State University.

Klein, G. A., Orasanu, J., Calderwood, R., & Zsambock, C. E. (1993). *Decision making in action: Models and methods.* Norwood, NJ: Ablex.

Langewiesche, W. (1944). *Stick and rudder: An explanation of the art of flying.* New York: McGraw-Hill.

Leibowitz, H. W. (1988). The human senses in flight. In E. L. Wiener & D. C. Nagel (Eds.), *Human factors in aviation.* San Diego, CA: Academic Press. pp. 83–110.

McCormick, E. J. (1976). Job and task analysis. In M. D. Dunnett (Ed.), *Handbook of organizational and industrial psychology* (pp. 651–696). Chicago: Rand McNally.

Ocker, W. C., & Crane, C. J. (1932). *Blind flight in theory and practice.* San Antonio, TX: Naylor.

Perrow, C. (1984). *Normal accidents.* New York: Basic Books.

Powers, W. (1998). *Making sense of behaviour: The meaning of control.* New Canaan, CT: Benchmark.

Rasmussen, J. (1986). *Information processing and human-machine interaction: An approach to cognitive engineering.* New York: North Holland.

Rasmussen, J. (1999). Ecological interface design for reliable human-machine systems. *The International Journal of Aviation Psychology, 9,* 202–223.

Rasmussen, J., Pejtersen, A. M., & Goodstein, L. P. (1994). *Cognitive systems engineering.* New York: Wiley.

Rasmussen, J., & Vicente, K. J. (1989). Coping with human errors through system design: Implications for ecological interface design. *International Journal of Man-Machine Studies, 31,* 517–534.

Reason, J. (1990). *Human Error.* Cambridge, England: Cambridge University Press.

Rochlin, G. (1997). *Trapped in the net.* Princeton, NJ: Princeton University Press.

Roscoe, S. N. (1980). *Aviation psychology.* Ames, IA: Iowa State University Press.

Runeson, S. (1977). On the possibility of smart mechanisms. *Scandinavian Journal of Psychology, 18,* 172–179.

Taylor, F. W. (1911). *The principles of scientific management.* New York: Harper.

Vicente, K. J. (1999). *Cognitive work analysis.* Mahwah, NJ: Erlbaum.

Vicente, K. J. (2002). Ecological interface design: Progress and challenges. *Human Factors.* 44, 62–78.

Wilson, J. R. M. (1979). *Turbulence aloft: The Civil Aeronautics Administration amid war and rumors of war 1938–1953.* Washington, D.C.: US Government Printing Office.

Woods, D. D., & Sarter, N. B. (2000). Learning from automation surprises and "going sour" accidents. In N. B. Sarter & R. Amalberti (Eds.), *Cognitive engineering in the aviation domain* (pp. 327–355). Mahwah, NJ: Erlbaum.

10

Requisite Imagination: The Fine Art of Anticipating What Might Go Wrong

Anthony J. Adamski and Ron Westrum
Eastern Michigan University, USA

Abstract

Good design for cognitive tasks is critical in today's technological workplace. Yet the engineering record shows that often designers fail to plan realistically for what might go wrong. Either the design fails to meet the designer's purpose, or it holds hidden traps for the users. We call the ability to foresee these potential traps "requisite imagination." Although many designers learn to avoid such pitfalls through experience or intuition, we propose a systematic way to consider them. We argue that by the use of a sound conceptual model, designers can foresee side effects and avoid unintended consequences. This chapter presents a systematic way to practice requisite imagination.

THE FINE ART OF ANTICIPATING WHAT MIGHT GO WRONG

Beauty is that reasoned congruity of all the parts within a body, so that nothing may be added, taken away, or altered, but for the worse.

—Leon Battista Alberti 1404–1472

The need to incorporate cognitive elements into task design stems from the demanding nature of human participation in the new technological workplace. There is little doubt that we have increased the cognitive demands on the human operator in human–machine systems (see, Bellenkes, Yacavone, & Alkov, 1993; Brickman, Hettinge, Haas, & Dennis, 1998; Hoc, 2000; Hollnagel & Woods, 1983; Howell & Cooke, 1989; Militello & Hutton, 1998; Riley, 2001). Today the human operator must effectively use diagnostic skills, judgment, and decision making to perform the tasks required to operate the systems we have. These cognitive demands create new requirements for the designer of man–machine systems. No longer will a systematic training design approach alone suffice for task design. With the emergence of cognitive task design as a critical component of system design, our need to anticipate human failure has increased.

As part of the conceptual phase of design, the task designer needs to anticipate what might go wrong. When the design process fails to provide adequate anticipation, the outcome can be fateful. As the chapter explains, factors leading toward failure are many, including individual limitations such as lack of designer experience, organizational constraints such as inadequate funding or political jockeying, and environmental factors such as time pressure or workplace constraints. Although many of these factors can be identified easily, others are insidious.

Anticipating what might go wrong is critical to the design of cognitive tasks for complex man–machine systems. It is important not only to recognize potential design failure paths but also to acknowledge them. Past failures have shown that even when design flaws were identified or suspected, acknowledgment was still critical to avoiding breakdown. Such was the case of the Quebec bridge disaster (Tarkov, 1995). Structural deflection of the bridge's weight-bearing chords that grew from .75 inch to 2.25 inches in a week's time was not acknowledged as significant. A few days later, the bridge structure collapsed. Similarly, when human factors are at issue, appropriate acknowledgment means taking the steps necessary to correct the situation rather than ignoring or failing to accept the data.

The *Challenger* disaster is another such example. Acknowledgment of the *Challenger*'s O-ring problem did not occur because of political pressure, organizational mind-set, conflicting safety goals, and the effects of fatigue on decision making (see Lucas, 1997; Sagan, 1993). As discussed in the literature, during the 1980s the U.S. National Aeronautics and Space Administration (NASA) faced severe political pressure to meet its launch schedules and make the shuttle cost-effective. The failure to acknowledge the data about the O-ring's operating parameters led to the *Challenger*'s destruction. It is not difficult to imagine why NASA managers failed to acknowledge the severity of the O-ring erosion when one considers the individual, organizational, and environmental factors influencing the decision to launch (see Heimann, 1997; Vaughan, 1996).

In some cases design flaws have been ignored for economic reasons. On March 3, 1974, a DC-10 airplane operated by Turkish Airlines crashed shortly after takeoff from Orly Airport, Paris, France; there were 346 fatalities. The accident investigation revealed that the aircraft lost the rear cargo door as it was climbing through 11,000 feet. The resulting decompression caused structural damage and loss of aircraft control, and the door opened because it was not properly latched. The serviceman who had closed the door that fateful day was Algerian and could not read the door instruction placard. As a result, he failed to check that the latches were closed—as the printed instructions had advised he should do. A glance through the door latch-viewing window would have shown that the latches were not fully stowed.

A conversation (personal communication, August 13, 2001) with a past human-factor specialist for McDonnell Douglas Airplane Company indicated that in this case, the instructional placards were not printed in the language of all the users. The instructions were printed in English and French. The McDonnell Douglas Science Research Department had identified this practice as a potential cause of operator error as the aircraft was being marketed in the international arena, but management and engineering did not respond to their concerns. Apparently, it was decided that there were too many components installed on the DC-10 that used instructional placards, and any design changes would disrupt production.

There had been a "dress rehearsal" for the Turkish accident. An American Airlines' DC-10 had experienced a similar door failure 2 years earlier. The National Transportation Safety Board (NTSB) recommended that several changes to the door system be implemented and the Federal Aviation Administration (FAA) had begun to prepare an Airworthiness Directive to address the repair. This action would have grounded all DC-10s. The president of

Douglas division of McDonnell Douglas, however, approached the FAA administrator and convinced him to downgrade the urgency of the changes. McDonnell Douglas management was afraid that an Airworthiness Directive would seriously affect the marketing of the aircraft. Consequently, the FAA dropped the Airworthiness Directive initiative, and McDonnell Douglas issued three service bulletins. The bulletins, however, had not been complied with when the aircraft was turned over to Turkish Airlines.

When we reflect on how seemingly minor flaws can have such disastrous outcomes, the importance of anticipating what might go wrong becomes apparent. The fine art of anticipating what might go wrong means taking sufficient time to reflect on the design to identify and acknowledge potential problems. We call this fine art *requisite imagination*.

Requisite Imagination

Requisite imagination is the ability to imagine key aspects of the future we are planning (Westrum, 1991). Most important, it involves anticipating what might go wrong and how to test for problems when the design is developed. Requisite imagination often indicates the direction from which trouble is likely to arrive (Westrum & Adamski, 1999). It provides a means for the designer to explore those factors that can affect design outcomes in future contexts.

A fine example of the successful use of foresight and anticipation was the steam engine designs of Benjamin Franklin Isherwood, who was the engineer-in-chief of the U.S. Navy during the American Civil War. Isherwood anticipated that during the war a large number of poorly trained engineers would serve in the navy because it would be impossible to gather enough skilled ones. Consequently, he designed the navy's steam engines to compensate for this lack of skill. He designed machines that could withstand the clumsiest of mechanics. Fuel economy and power took second place to mechanical dependability (Sloan, 1967).

The need to bring requisite imagination into cognitive task design stems from the increasing numbers of man–machine systems that are emerging in today's workplace. Hollnagel and Woods (1983) argued that the design of properly functioning human–machine systems requires a different kind of knowledge from that of mechanical design—one that explores and describes the cognitive functions required for the human and the machine to interface. Today's human–machine systems have increased the use of automation. This change has shifted the demands placed on the human from emphasis on perceptual-motor skills to emphasis on analytical skills. These demands for addressing cognitive aspects place new responsibilities on the designer. Today's task designer must be able to anticipate, to a high degree, what things might go wrong with the cognitive tasks specified in the design.

The failure to use requisite imagination opens the door to the threat of unanticipated outcomes. These outcomes can be incidents, accidents, or major catastrophes. Thus, successful cognitive task design involves the use of requisite imagination to avert the unwanted outcomes that seem to hide beneath the surface of all design. We further suggest that the best way to integrate requisite imagination into the design process is by means of a conceptual design model.

Conceptual Design

The idea of conceptual design as applied to cognitive task design is similar to that of conceptual design as used in engineering. Engineering design is generally divided into four phases: (a) clarification of the task, which is the determination of design specifications of the product; (b) conceptual design, which the formulation of a concept of the product; (c) embodiment design, which is the determination of layout and forms; and (d) detail design, which is the determination of final dimensions (Al-Salka, Cartmell, & Hardy, 1998). The conceptual design phase is the key to anticipation.

Conceptual design accommodates and facilitates the way in which a user thinks about a product or system (Rubinstein & Hersh, 1984). It is based on the basic premise that users construct internal mental models of the design's outcome (Norman, 1986). This helps the designer take into account user expectations and seek alignment between the designer's world and the user's world. Much of design uses a conceptual model that rests largely in the nonverbal thought and nonverbal reasoning of the designer (Ferguson, 1977).

Cognitive task design also involves four phases. The initial step in the design solution is to formulate a conceptual model that simplifies the task and organizes data. In this initial phase, the conceptual model's main purpose is to foster requisite imagination. This is accomplished by the use of systematic foresight. The second phase is the initial design of the task and testing. The third is design revision, retesting, and fine-tuning. The final phase is implementation, which involves diffusion and acceptance. Cognitive task design should result in the creation of a product that meets both the objectives of the designer and the expectations of the user.

The Conceptual Design Model

The formulation of the conceptual model is the hardest part of design (Norman, 1999). A conceptual model is an abbreviated description of reality. It provides the designer an abstract vision of what the design entails. Such a model both explains and defines the relevant components of the design (Richey, 1986). Whether the model is intuitive or formal, a sound conceptual model is at the root of all good task design.

Donald Norman, a professor of computer science at Northwestern University, emphasizes the importance of the conceptual model in his book, *The Design of Everyday Things,* which we believe is relevant to the design of cognitive tasks:

> A good conceptual model allows us to predict the effects of our actions. Without a good model we operate by rote, blindly; we do operations as we were told to do them; we can't fully appreciate why, what effects to expect, or what to do if things go wrong. As long as things work properly, we can manage. When things go wrong, however, or when we come about a novel situation, then we need a deeper understanding, a good model. (Norman, 1989, pp. 13–14)

Eder (1998) stated that design encompasses flair, ability, intuition, creativity, and spontaneity. Schon (1983) maintains that design also requires judgment and reflection. Although the goal is clear, the road to completion consists of many paths beset with visible and invisible perils. More perilous paths can be identified and avoided when the designer incorporates foresight into the design process. The potential for human failure can occur at all stages of the design process, but as Petroski (1994) emphasized, "fundamental errors made at the conceptual design stage can be among the most elusive" (p. 15). Unhappily, many such errors manifest themselves only after the design is implemented when they produce unexpected outcomes—including catastrophic failure.

A typical cognitive task design has a clear purpose, and so often the designer's attention is narrowly focused on the primary effects while neglecting "side effects"—unintended consequences that often create the greatest problems. Side effects often result from implicit or unverified assumptions about what the users "would do." Consequently, it is the designer's task to make assumptions explicit and thus to eliminate as many side effects as possible. Unfortunately, foresight varies with the designer's skills (see Fig. 10.1). Hence, we need some kind of systematic way to assist the designer "to look forward and consider longer-range consequences" (Westrum, 1991, p. 324). This is the role of the conceptual model.

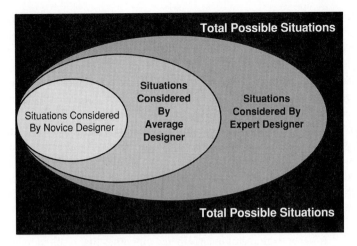

FIG. 10.1. Possible foreseeable situations.

Building a good model is key. Eugene Ferguson, a professor of history emeritus at the University of Delaware, argued that good engineering [design] is as much a matter of intuition and nonverbal thinking as of equations and computations (Ferguson, 1997). He described how American engineering students before the 1960s were expected by their teachers to envision outcomes in "the mind's eye." His concept of the mind's eye concerned more than just a quick overview of a design; rather, it involved taking the time and energy to look deeply into the design. It meant developing an intuitive feel for the way the design would interact with the material world and why the design would or would not meet expectations.

We suggest that today a similar approach is necessary for the design of the cognitive tasks used in complex human–machine systems. This approach can best be achieved with the use of an appropriate conceptual design model. This model for the process of design allows making use of one's requisite imagination. The elements that make up our conceptual design model are drawn from past design failures and successes. A review of the literature and discussions with people involved in design processes point to certain elements critical to the design process when foresight is used.

Unanticipated Outcomes

Today's designer faces many concerns, and perhaps the most insidious of these is the unanticipated event. A humorous example of the unanticipated occurred not too long ago when one of us (Westrum) received a device known as a "Pocket Terminator" as a gift. This device was a small toy about the size of a thick candy bar. It had three buttons. When one of the buttons was pushed, the toy would emit sounds associated with air combat (a bomb exploding, a machine gun firing, or the zapping of a death ray). One day, the device was shown to a 10-year-old boy. He immediately picked it up and pushed two of the buttons simultaneously. This action produced entirely different and more impressive sounds. Westrum, a well-trained adult, had followed the designer's instructions and never pushed more than one button at a time. He felt stupid at being shown new possibilities by a 10-year-old boy. This toy example suggests the difference between designer expectations and the realities of deployment.

The Sidewinder Missile.   But the same issues occur with much larger and more serious systems. One example involves the Sidewinder missile. In this case users perceived a design feature of the missile in a different way than the designers intended. A major problem of the air-to-air Sidewinder missile was stability. To counter the missile's tendency to roll about its axis, a technician named Sidney Crockett invented a fin device called a rolleron (Westrum & Wilcox, 1995). The rolleron was a small gyroscopic wheel that was rotated by the air stream flowing past the missile in flight. A rolleron was attached to a hinged tab on each of the missile's rear fins. If the missile started to roll, the wheel's precession would produce a torque on the tab, which would flip out into the air stream and stop the roll. The rollerons required precision machining, because they were small gyroscopes.

Soon after the Sidewinder missile went into operational use, the Naval Ordinance Testing Station located at China Lake, California, received a letter from the sailors who were responsible for loading and unloading the missiles on aircraft aboard aircraft carriers. The sailors wanted to thank the designers who added the "little wheels" to the missiles so that they could be pulled along the carrier deck with ease. Needless to say, this greatly upset the people at China Lake. Lack of anticipation once again led to unintended consequences.

Therac-25.   Another example of unanticipated outcomes that resulted in human fatalities and serious injuries was the Therac-25 accidents (Leveson & Turner, 1993). The Therac-25 was a dual-mode linear accelerator machine designed to combat cancer by accelerating electrons to create high-energy beams that destroyed tumors with minimal damage to surrounding healthy tissue. The machine had two settings, a low-energy 200-rem electron beam that could be aimed directly at the patient and a high-energy X-ray mode that aimed a more powerful electron beam at the patient through a thick tungsten shield. Computer programming allowed the operator to select which mode was used.

Although there had been linear accelerators with designs that did not incorporate computer technology, the developers of the Therac-25 designed it to take advantage of computer control from the outset. Older designs such as the Therac-20 had independent protective circuits for monitoring electron-beam scanning and mechanical interlocks to ensure safe operations. The Atomic Energy Commission Limited (AECL), developers of the Therac-25, is a crown corporation of the Canadian government, and it decided not to duplicate the existing hardware safety mechanisms and interlocks used on the older machines. For the Therac-25, it made a key assumption that software protection was better than hardware. The AECL apparently abandoned convention without fully testing the assumptions (Leveson & Turner, 1993).

The completely computerized version of the Therac-25 became available in late 1982, but it wasn't until March 1983 that the AECL performed a safety analysis of its design. The analysis was in the form of a fault tree, and research indicated that the software was excluded from the analysis. After the fact, it was concluded that the AECL made several assumptions that eroded the chances of detecting design flaws (Leveson & Turner, 1993). These assumptions included the following:

- Programming errors would be reduced by testing on a hardware simulator and under field conditions on teletherapy units.
- The program software would not degrade due to wear, fatigue, or reproduction processes.
- Computer execution errors were caused by faulty hardware components and by random error induced by electromagnetic noise.

Eleven Therac-25 machines were installed, and these caused six known accidents involving radiation overdoses to patients between June 1985 and January 1987. Three accidents ultimately resulted in fatalities and three in serious injury. In February 1987, the U.S. Department of Health and Human Services Food and Drug Administration (FDA) and the Health Protection Branch of Canada declared the Therac-25 defective and had the machines recalled.

The Therac-25 resulted in the worst series of radiation accidents in the 35-year history of medical accelerators. The accidents resulted from a combination of human error, the compounding of electronic and mechanical technologies, and a software coding error. Each of the events apparently involved a combination of the misinterpretation of computer display information together with a software flaw. The machine's design exhibited an overreliance on electronic technology without mechanical lockouts to prevent excessive radiation doses to the patients. Unknown to the designers, the software permitted a lethal mode that resulted from the high-energy beam being engaged when the unit's tungsten target was retracted. This lethal mode was discovered only through the accidents. The events surrounding the Therac-25 accidents point to the potential for catastrophe when all conceivable outcomes are not anticipated and considered during design and testing phases. To get the Therac-25 reapproved by the regulatory authorities, the AECL altered the software and added hardware features that prevented similar accidents from happening again. With these new features the Therac-25 was put back in service without any further problems, but three people were dead.

The U.S. Navy 1200 PSI Steam Propulsion Plant.   An example of how the lack of anticipation can lead to continuous and growing problems is the U.S. Navy's concerns with 1200 PSI (pounds per inch) shipboard steam plants in the early 1980s (Williams, Pope, Pulu, & Sass, 1982). The navy was experiencing numerous problems with the operation of these plants and undertook a study to determine why. The investigation showed that the root of the problem was poor equipment design. The skill levels of operating personnel were not sufficient to overcome the horrendous design deficiencies. The deficiencies included complexity of equipment, poor accessibility and layout of valves and gauges, unclear labeling, and inadequate operating procedures. These deficiencies increased human workload, the probability of operating errors, and difficulties in training. They resulted in 37 boiler explosions. The investigation also showed that there was a high turnover rate of shipboard engineers, which further degraded the skill levels of steam plant operators. The U.S. Navy's propulsion plant problems can be compared to the problems that Benjamin Franklin Isherwood had circumvented during the Civil War by anticipating what might go wrong in his steam plant designs.

Anticipated Outcomes

Design flaws can be easy to recognize when design involves situations previously encountered. Unfortunately, with novel systems or complex designs, it is more difficult to explore completely all failure scenarios. Although engineers are encouraged to consider failure modes systematically, we know that in novel situations this may be hard to do. We offer the following examples that demonstrate how anticipation can pay off. The first example is John Roebling, the designer of the Brooklyn Bridge, whose many talents included an ability to anticipate failure. The second example is the corrective action taken on the Citicorp building in New York City, and the last is the safety net designed for the construction of the Golden Gate Bridge.

John Roebling and the Brooklyn Bridge.   Considered by many as the "Eighth Wonder of the World," the Brooklyn Bridge is a magnificent structure spanning 1,595 feet of graceful throughway held in suspension by a tapestry of wires that links lower Manhattan and Brooklyn. Designed in 1865 by John A. Roebling, the Brooklyn Bridge was completed in the spring of 1883 under the supervision of Roebling's son Washington and officially opened on May 24, 1883. It is still in use today.

Roebling became well known as an engineer and a bridge designer primarily because of his innovation of wrapping wire to form wire rope that eventually became the cables used to suspend a bridge. Suspension bridges based on Roebling's design became a standard in bridge design. One of the reasons for Roebling's success was his skill in anticipating what might go wrong. He made a practice of exploring past engineering failures as a guide to potential failures within his designs. His practice of anticipating failures has been well documented (see Petroski, 1992, 1994, 1996).

Roebling made a particular study of the nature and causes of failure modes of suspension bridges. Rather than assume that past failures provided evidence that suspension bridges were ill suited for certain uses (such as railroad bridges), he looked at past engineering mistakes as "fortuitous experiments that revealed what most needed attention in designing against failure" (Petroski, 1994, p. 128). Roebling recognized that to build a successful bridge, he had to anticipate how it might fail. He used anticipation as much as humanly possible to identify what could go wrong and thus incorporated features in his design that averted failure. Some may believe that one must look to successes to achieve effective design, but as Petroski (1994) stated, "Any argument for the safety of a design that relies solely on pointing to what has worked successfully in the past is logically flawed" (p. 137).

John Roebling began design of the Brooklyn Bridge in 1865. The U.S. Congress and President Ulysses S. Grant approved the design for construction in June 1869. While Roebling was locating a site for one of the bridge's towers, a carelessly piloted ferryboat crashed into the pier on which he was standing and crushed his foot. He developed tetanus poisoning from the injury and died in July 1869, before even the bridge towers had been erected.

Washington Roebling took over as chief engineer and oversaw the bridge's construction. Unfortunately, the project was also to cost Washington his health. Not content with directing the work from his office, he constantly went down in the submersible bells used in the sinking of the caissons in the building of the tower foundations. In the spring of 1872, Washington Roebling became ill from compression sickness, leaving him partially paralyzed and without the use of his voice.

Rather than turn over the supervision of the bridge's construction to anyone else, he took a room on the Brooklyn bank of the East River, from which the bridge construction was clearly visible. From there, he supervised with the aid of a powerful telescope every step of the bridge's construction. He relayed his instructions to the construction foremen through his wife, Emily.

The success of the Brooklyn Bridge is rooted in John Roebling's practice of exploring past failures. Roebling's diligence in looking at these failures from the perspective of fortuitous experiments revealed what most needed attention. His successes emphasize the merits of requisite imagination.

The Citicorp Tower.   The events surrounding the discovery of structural design errors to the Citicorp Tower provide an even stronger example of how anticipating what might go wrong can avert failure. The Citicorp case demonstrates the importance of not only recognizing design problems, but of acknowledging them and taking action. Additionally, it provides an excellent example of how individual, organizational, and environmental factors can affect outcomes.

One bright spot in the particularly dismal economic year of 1977 in New York City was the opening of the Citicorp Tower. This 59-story structure was considered an engineering masterpiece that was intended to help revitalize the city's languishing public image. The building was extremely light for its size and incorporated a number of engineering innovations. When completed, the tower's height of 915 feet placed it 10th among the world's tallest buildings.

William J. LeMessurier, the project's structural engineering consultant, and Hugh Stubbins, Jr., the project's architect, had to address a unique design challenge. Standing at the corner of 54th Street and Lexington Avenue in Midtown Manhattan since 1862 was St. Peter's Lutheran Church. It controlled nearly 30% of the square block that was the chosen site for the new Citicorp

Tower (Morgenstern, 1995). The church's congregation was able to negotiate a unique deal with Citicorp. They agreed to sell the property provided that a new church would be built in place of the old with "nothing but free sky overhead" and that a plaza would be erected under the tower to continue the church's tradition of hospitality. To meet these demands, Stubbins and LeMessurier designed the building's corners to extend 72 feet out over the new church on the northwest part of the Citicorp center and over the required plaza on the southwest. This was accomplished by a daring design that called for setting the tower on four massive nine-story-high columns that were positioned in the center of each side of the building rather than at the building's corners. The design created a high visual effect through the illusion of this massive structure hovering above the street.

The building's design also incorporated a LeMessurier innovation that used large diagonal girders throughout the building, which acted as wind braces. This innovation transferred the tower's great weight to the four columns that anchored the building to the ground. LeMessurier's diagonal-brace design provided a vast weight savings, but it also created a tendency for the building to sway in the wind.

To reduce swaying, LeMessurier devised a revolutionary system called a tuned mass damper to be placed on top of the building. It consisted of a 400-ton concrete slab that floated on pressurized oil bearings and was connected by springs to the building's structure. The tuned mass damper counteracted swaying much like a shock absorber. The Citicorp Tower was the first structure to incorporate a mechanical device to counteract swaying (Petroski, 1997).

Our prime example of anticipation, however, really begins after the Citicorp Tower was completed. In June 1978, LeMessurier received a phone call from an engineering student who called to gather information about the design of the Citicorp Tower for a research paper. Apparently the student's professor told the class that the four columns that held the building up were placed in the wrong position. LeMessurier explained to the student how the placement of the four columns came about and that the placement put the columns in the strongest position to resist quartering winds—those that come from a diagonal and increase the forces on two sides of a building.

A short time later, LeMessurier decided that the information he had provided the student would also be of interest to his own students in a structural engineering class he was teaching at Harvard. Consequently, he prepared a classroom lecture that presented the Citicorp Tower's defences against swaying in severe wind conditions. He also wanted to present how strong the defences were in a quartering wind. At the time the building was designed, New York City only required a perpendicular wind calculation to meet building code. Under the construction specifications stated in the original design, the wind braces would have absorbed the quartering-wind load with no problem; however, he had recently learned that his original specifications had been modified. Instead of welded joints for the diagonal wind braces, the steel company had substituted bolted joints, because they were cheaper and faster to complete than welded joints.

Up to this point, everything still remained within accepted tolerances. From an engineering perspective, this was a legitimate substitution, and there was no requirement to notify LeMessurier. This news did not cause LeMessurier any immediate concern. It wasn't until a month after his discovery that he began to have misgivings. The individual factor surfaced here as LeMessurier's requisite imagination began to simmer. His experience and expertise fuelled his thoughts. He reflected on his design and began to question whether the bolts could withstand the loads exerted by a quartering wind. He was not sure that the issue of quartering winds had been taken into account by the people who were responsible for the substitution. LeMessurier's investigation indicated that the diagonal braces were defined as trusses and not as columns by his New York team and would not bear the projected wind loads. He than realized how precarious the situation really was.

LeMessurier had a scale model of the current structure tested and found that the Citicorp Tower could fail if sufficient wind velocity was present. Further analysis showed that the statistical probability of such a wind occurred every 16 years, which is referred to by meteorologists as a 16-year storm. The tuned mass damper system reduced the probability to a 55 year storm, but the system required electricity to function, and LeMessurier anticipated that power could easily fail in a storm.

The next hurdle LeMessurier faced was organizational in nature. The attorneys for the architectural firm and the insurance firm that insured his company advised LeMessurier not to discuss the matter with anyone until they could review his data. The review entailed pressing questions, including a determination of LeMessurier's sanity! Additional experts were brought in, and finally all agreed that they had a potential disaster on their hands. Essential time now had been lost. This added to the overall concerns because liability issues now needed to be addressed.

Repair of the Citicorp Tower meant coping with environmental factors such as wind velocity, potential storms, and the time of day. The experts recommended that an evacuation plan be devised in the event winds reached a significant level. Although LeMessurier did not think such a plan was necessary because of the tuned mass damper, others acknowledged the possibility of electrical failure and argued for the plan. Weather experts from academia, the government's Brookhaven Laboratory, and two independent weather forecasters were used to monitor the wind speed until the structure was repaired. Wind predictions and weather reports were generated four times a day. Additionally, the company that had manufactured the tuned mass damper was asked to provide an around-the-clock technician to keep the system in perfect working order. The concern was not so much for major weather situations such as hurricanes, which gave hours and days to anticipate, but for unpredictable events such as small tornadoes.

The repair involved welding 2-inch-thick steel plates over the more than 200 bolted joints. Citicorp pushed for around-the-clock repairs, but the project manager for the repair anticipated that clouds of acrid smoke might cause panic among the tenants and set off countless smoke detectors. So repairs were made after business hours. Plywood enclosures that encompassed the repair areas were built every night before the welders began their work. Before business hours the next day, the areas were cleaned and smoke detectors reactivated. The repairs were made without incident, and the structure remains today as one of the safest tall structures built by humans. Morgenstern (1995) reported that the Citicorp Tower can now withstand a 700-year storm.

LeMessurier went on to document the design errors that led to the Citicorp Tower fiasco in a 30-page document called "Project SERENE," which stood for *Special Engineering Review of Events Nobody Envisioned* (Morgenstern, 1995). The document reveals that a series of miscalculations were compounded by a failure to anticipate possible dangerous situations, such as a quartering wind.

The Citicorp Tower example shows the value of putting requisite imagination to work in the design process. Individual, organizational, and environmental factors all affect the design's outcome, and the task designer must reflect on these things to identify design flaws, acknowledge their potential, and select a path for appropriate action.

The Golden Gate Bridge.   There are often situations in which the designer does anticipate potential problems and takes action that appears adequate. Yet failure can still occur because potential outcomes were not fully anticipated. Such was the case of the safety precautions taken by Joseph B. Strauss, the chief engineer of the Golden Gate Bridge. At the time of construction, the accepted odds of worker fatalities were one death for every million dollars invested in the bridge under construction (Coel, 1987). The Golden Gate Bridge was financed with $35 million in bonds; thus, the "acceptable" number of fatalities would be 35. Strauss would not accept such a toll. Instead he took a number of precautions to avert the dangers he anticipated, based on his experience and the past accident and incident history of bridge construction. Even Strauss, however, did not anticipate all the perils that were to face the project.

When the Golden Gate Bridge opened on May 28, 1937, it was the world's highest and longest bridge. The bridge crossed one of the most treacherous channels of water on earth. Winds of up to 60 miles per hour were common, and strong ocean currents swept through the area just below the water's surface. Most bridge fatalities occur during the construction of a bridge's deck. The newspapers referred to this task as riveters doing a "dance of danger" on the steel beams (Coel, 1987, p. 16). To prevent serious injuries or fatalities, Strauss had a net of heavy hemp strung from shore to shore under the structure. It was the first safety net used in bridge construction, and during the first 2 months of construction the net had saved the lives of nineteen men. These men became known as the "Half-Way-To-Hell Club." After four years of construction, there was only one fatality among the 684 workers. On February 17, 1937, however, an event took place that essentially ruined this remarkable safety record.

Strauss's net was not designed to take more than the weight of a few men. On February 17, a section of a travelling platform carrying thirteen men plunged into the net because of the failure of a metal grip. The 10-ton platform pulled the net from its moorings and dropped into the treacherous waters. Two men jumped free and were rescued by a fishing boat, one man dangled on a girder for 7 minutes until rescued, and ten men lost their lives. No one, including Strauss, believed that such a failure was possible. Organizational safety measures were in place. Inspections were constantly conducted. But in the end, it was not enough. In fact, safety inspectors had just condemned a similar platform and were about to inspect the platform when it failed (Coel, 1987).

The "dance of danger" was not the only peril Strauss anticipated. He also required workers to wear protective headgear with tinted goggles to guard against fog blindness. This was the first time that such equipment had been used, and it became the prototype of today's construction hardhat. Strauss had a special hand and face cream made available to the workers that protected against the wind, and he had special diets developed that helped fight dizziness. Additionally, Strauss instituted some strict policies to help avert disasters. Steelworkers were not allowed to horseplay on the high beams. Fooling around meant dismissal. Workers who came to work with a hangover were required to drink sauerkraut juice to ward off any lightheadedness. Yet some factors were still not anticipated.

The first injuries the project experienced were lead poisonings. During construction of the bridge's towers, workers bolted together tower cell sections with hot steel rods. The cells had been protected with a lead-based paint to prevent corrosion. The paint melted during the process and released toxic fumes. Whole shifts of workers were rushed to the hospital, and doctors diagnosed the symptoms as appendicitis. Finally, the correct diagnosis was made, and new procedures were instituted to remove the paint before driving the hot steel rods.

Although Strauss anticipated many of the problems with the construction of the Golden Gate Bridge, he was not able to discern them all. Was this a fault with Strauss? We think not. The ability to anticipate every possibility is not likely. We, as task designers, all face the same dilemmas. First, how do we anticipate, and second, how do we know when we have devoted enough time and energy to using our requisite imagination? We recommend a conceptual model because it provides the designer a way to think systematically about things that affect his or her design. The conceptual model is an enhancement of, not a substitution for, the designer's intuition. It involves thinking not only about the tasks that the operators will perform and the physical environment in which the task will take place, but also of the social, organizational, and cultural context within the system that incorporates the design (Redmill, 1997).

The Need for Abstraction

The conceptual design model should be an accurate, consistent, and a complete representation of the task's functionality and should match the users' mental model of the task. If the initial conceptual design contains an error that goes undetected, then the error tends to be all the more difficult to identify as the design progresses through its evolution of details and modifications. Petroski (1994) referred to this deficiency as the "paradigm of error" in conceptual design. Petroski argued that the creative act of conceptual design is as old as civilization itself, and at its most fundamental stage, conceptual design involves no particularly modern theoretical or analytical component. Consequently, there is little reason to believe that there is any fundamental difference in the mistakes of our ancestors and ourselves in our conceptions and thus how we pursue flawed designs to fruition or failure.

A special problem that the cognitive task designer faces in meeting the requirements of today's technologies is to reduce the complexity and dependency of relationships that encompass the objects making up the particular task. The more complex the relationships, the more complex the design problems become. The conceptual model provides a method for the designer to reduce the number of objects and to simplify relationships by creating abstractions. This allows the designer to deal with complex task design problems by identifying design concepts and their relationships. Thus, relevant information can be abstracted from the problem situation to form a conceptual model (Kao & Archer, 1997).

In a 1989 study, researchers found that expert designers used relatively abstract representations in solving problems, whereas novices concentrated on the more concrete features of problems (Newsome & Spillers, 1989). This means that novice designers often have difficulty in dealing with abstractions because they lack design schemas based on prior design knowledge. Kao and Archer (1997) argued that the designer's use of abstraction may be expanded if prior domain and prior design knowledge are improved to enhance the design process.

COMPONENTS OF THE CONCEPTUAL MODEL

So note: Not all designers are experts or virtuosos. When the design process is forced by such factors as time pressure, political jockeying, or lack of experience or expertise, it is helpful for the designer to have at hand a basic conceptual design model to assist his or her requisite imagination. A number of elements could make up such a conceptual design model, one version of which we present in Fig. 10.2.

We have attempted to identify those factors that have aided or hindered past designs to set the stage for requisite imagination. The model is generative in that one may begin at any point to enter the process. This is not yet a systematic step-by-step process, but it represents a reflective approach to considering those things that can lead to unintended outcomes. Although the model is generative, we begin our discussion of the model with the suggestion to "clarify the task."

Clarify the Task

As in the case of the toy "Pocket Terminator," the question of "intent" is primary. The designer must answer the question of how the design is to work. To do this the designer must analyse the task by describing accurately the human actions and processes that lead to the desired outcomes. This is often accomplished by means of cognitive task analysis, which is a set of methods used to identify the mental demands needed to perform a task proficiently. Cognitive task analysis focuses on such mental demands as decision making, mental models, learning

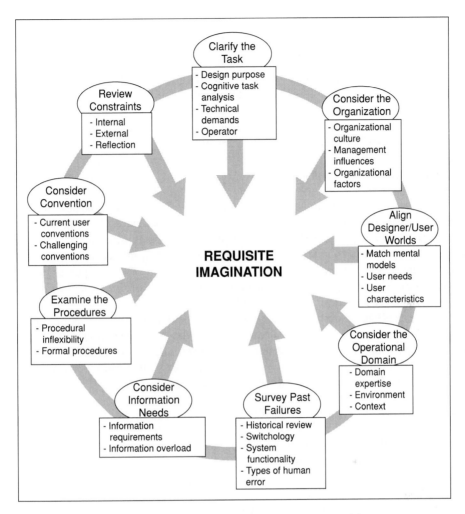

FIG. 10.2.   The conceptual cognitive task design model.

and skill development, job concepts and task components, and user characteristics (Redding & Seamster, 1994).

There are a number of cognitive task analysis methods available to the designer that include semistructured or structured interviews, protocol analysis, psychological scaling, cognitive or performance modeling, and error analysis. These methods focus on describing and representing the cognitive elements that underlie goal generation, decision making, and judgments. They often begin with high-level descriptions of the task based on observations or interviews (Militello & Hutton, 1998).

Others describe cognitive task analysis as a means to compare the cognitive structures and processes of experts with those having less proficiency (Redding & Seamster, 1995). This comparison is used to determine how a job is learned, how to expedite learning, and to describe the optimal mental models, knowledge organization, and skills for job performance. Finally, cognitive task analysis is used to fashion those requirements into a form that matches both the technical demands of the application and the operator's functional characteristics in the cognitive domain (Hollnagel & Woods, 1993). The matching of these elements provides the foundation for matching the designer's and the user's mental models.

Consider the Organization

Organization can affect a design project in numerous ways. Organization includes not only the management that oversees the design project, but also the organization's culture that defines "how things are done around here." It was an organizational factor that led to the substitution of bolts for welds, which led to the Citicorp Tower situation discussed earlier.

Good designs come about because those involved with managing their creation think intelligently about the design process. The managers of design projects need to cultivate an atmosphere of thought and creativity in their organizations. An example of how the organizational factor can negatively influence a design is the found in the development of the M-16 rifle (Westrum, 1997).

The M-16 rifle combat design was seriously compromised as a result of internal politics, and its failures are believed to have caused a number of servicemen's deaths. A congressional investigation of the M-16 program found that management of the program operated in a unsystematic way. The investigation revealed that the program's command structure was inadequate, responsibility was shifted when mistakes were made, and accountability was often absent. There was no "technological maestro" to force high standards (Squires, 1986). The M-16 scenario is an object lesson in how organizational politics can shape the outcome of design projects (see also Stevens & Ezell, 1994).

Many projects have been negatively influenced, delayed, or canceled because of management influences. Anticipating the influence of management and identifying those personnel who have a decision-making role in the project can avert many such problems. The task designer or design team must address the organizational factors that can influence the design project. Important organizational issues include the following:

- Who are members of the design team?
- Who is the customer for the project? Who is the user? Are they the same?
- What are required and available resources?
- Who will have access to project information?
- How will evaluations be conducted?
- Who has and does not have authority to alter the design?
- Who is the ultimate project decision maker?

Align Designer and User Worlds

One source of problems that affect the success of a design task is a lack of contact with the realities faced by the user. Differences normally exist between the designer's and the users' perception of tasks. Designers and users rarely share the same mental models that encompass their individual worlds (Landauer, 1995). Often the designer seems to be living in an ivory tower, and the context of the design process is far removed from the context of use (Perrow, 1983). The users are typically remote in space and time and are not psychologically present to the designers (Westrum, 1997). This leads to "designing in the dark," which results in problems such as those discussed in the U.S. Navy 1200 PSI shipboard steam plant example.

During some field studies preparatory to design of a new grinder, Ingersoll-Rand engineers were surprised to discover how defective their previous design had been. Dealers made comments such as, "We doubted our input would mean anything. We thought once again Ingersoll-Rand was going to design something in a dark room and ram it down our throats" (Kleinfeld, 1990, p. 6). Similar doubts were voiced about designers and test pilots of the Airbus 320 by French airline pilots, who felt that their own issues had not adequately been taken in account. A whopping 78% of French pilots in a 1989 survey disagreed with the statement that designers had taken their needs into account (Gras, Moricot, Poirot-Delpech, & Scardigli, 1994).

Designing in the dark often leads to meeting the needs of the designer but not the needs of the user. It is most important that the designer recognizes that these differences exist and avoids making easy assumptions about the users' knowledge, skills, and cognitive abilities. Successful cognitive task design requires the understanding of the task, the user, and their mutual interaction.

The problems associated with these differences can often be anticipated by aligning the designer's and users' worlds. Alignment can be accomplished by a variety of means including naturalistic inquiry, user feedback, and cognitive task analysis. A fruitful dialogue with users is most desirable. Regardless of the choice, the important thing is that the designers integrate the users' worlds into their conceptual design models. Designers often use a single model of the user and make little attempt to identify the actual characteristics of the intended user group (Newell & Alistair, 1993). To do so the designer must identify the following factors:

- What are the typical users' perceptions of the task purpose and outcome?
- What knowledge and skills are required to complete the task?
- What knowledge and skill differences exist among novice, intermediate, and expert users?
- What are users' perceptions of constraints or limitations of the task?
- What information or feedback do users' need?
- How does this information align with the designer's perceptions?

The experienced designer will naturally consider a broader population of users then the novice (see Fig. 10.1).

Consider the Operational Domain

In a variety of domains, advancement in automation has led to the development and refinement of sophisticated human–machine systems that have created unexpected problems. These problems are often centered around the coordination of actions required and coping with information flow. In many complex domains, the trained human is the most valuable resource for linking information and action. Consequently, it is the designer's task to use the capabilities of the human operator fully to solve the problems posed by complex work environments. To do this, it is necessary to define the facets that make up the cognitive task design problem. Identification and prioritization of these facets require domain knowledge; if the designer is a non–domain expert, we suggest getting a subject matter expert to assist the designer.

In their research, Kao and Archer (1997) found that non–domain expert designers tended to identify fewer problem facets or could identify many facets but were unable to recognize which ones were more important. The non–domain experts often identified many detailed ideas or activities but failed to group them in a meaningful way. Domain-expert designers emphasized concepts and principles and developed highly abstract concepts such as goals, objectives, and principles. The non–domain experts demonstrated difficulty in retaining a global view of their designs. Thus, an integrated and systematic overview is often the most important contribution of the domain expert.

The two major components of domain are the environment and context in which the task will take place. Environment is the physical setting that includes such facets as temperature, available light, workspace accommodations, and the technologies used. Context encompasses the conditions under which the task is to be performed (e.g., normal operations or emergency operations) including the human–machine interfaces that enhance or constrain the coordination between the human and the machine (e.g., labeling, procedures, training, etc.). Every domain (e.g., air traffic control center, hospital operating room, and university research center) has its own unique environment and context that can pose unique problems for the designer. One area that creates many problems for the cognitive task designer is the interface between the human and the machine.

Many human–machine problems are related to "communication with machines" rather than "operation of machines" (Card, Moran, & Newell, 1983). Studies have shown that computer systems can create new burdens and complexities for those responsible for operating high-risk systems (Cook & Woods, 1996). The designs of such systems often result in what is called "clumsy automation" (Wiener, 1989). Clumsy automation is a result of designs with too many features. Such technology may decrease the operator's physical workload, but it often increases the cognitive workload. Feature-rich designs may create hard-to-use products because users only understand a fraction of the available functions (Riley, 2001).

The human–machine cooperation problem is amplified in dynamic systems. Dynamic situations are uncertain in nature because they are only partially controlled by the human operator and unexpected factors can modify the dynamics of the process. Furthermore, dynamic situations are also risky, and the cost of errors can be high. Inversely, static situations are those situations in which the machine is fully controlled by the human operator and the situation is not susceptible to autonomous change (Hoc, 2000).

Good designs come about because those involved in their creation research carefully and think intelligently about the context of their use (Westrum, 1997). Most of us faced with the challenge of cognitive task design are highly trained in particular areas of expertise. But few of us are trained to think about or have easy access to the social context into which the task design will fit. Optimizing a design for a technological system does not mean that the task can easily be performed by a human being—especially the kind of human being doing that particular kind of work. Sound design requires imagination, perception, and consideration for the user and the domain in which the task is to be performed. In many cases, this means making the field trips necessary to gain a sound, intuitive basis for the design.

Context mapping encompasses exploring the conditions under which the user, the technology, and the task will mutually interact. This is often accomplished during task clarification, but it is important to revisit context to ensure that the design addresses any constraints not previously identified. Exploring the human constraints that limit actions and examining perceptions of information are critical to cognitive task design. The success of any design depends on the coordination of interactions between the constraints of actions required and information needed to perform a task (Flach & Dominguez, 1995).

Domains also possess unique stressors. Within the domain of aviation, for example, Prince, Bowers, and Salas (1994) identified such stressors for aircraft pilots as time pressure, group pressure, noxious stimuli, perceived threat, disrupted physical functions, isolation and confinement, temperature, sleep loss, turbulence, and fatigue. Stressors can affect individuals in such ways as physiological arousal, loss of motivation, increased attention demands on the self, and chronic stress reactions (Prince et al., 1994). Driskell and Salas (1991; as cited in Prince et al., 1994) stated that stress can either be ambient, and thus part of the work background (such as a noisy, hot environment) in which performance does not relieve the stress, or stress can be performance contingent, i.e., stress that is removed if the task is performed well.

In summary, it is important for designers to possess or have available domain expertise to optimize their requisite imagination. They are then able to anticipate what might go wrong when the task design is placed into its operating environment and implemented under all likely conditions.

Survey Past Failures

The literature on technological failure reveals that the systems often fail when the design process does not provide a proper environment for thought, (Westrum, 1997). We believe that this also holds true for cognitive task design. In fact, we maintain that the design of the hardware is often easier than the design of the software used for those cognitive tasks included in control of the technology. Exploring past failures can fuel one's requisite imagination and provide the framework for avoiding current mistakes and building into the design protections against previous failures. Historical review, such as Petroski's examination of Roebling's design methods (see Petroski, 1992, 1994, 1996), can provide lessons with respect to failures involving equipment, procedures, human factors, organization, and training.

An example of a past failure is a severe fire that occurred aboard the aircraft carrier, USS *Forrestal* in 1967. An air-to-ground rocket prematurely fired from a plane on the ship's deck (Wilcox, 1990). The rocket hit the fuel tank of another aircraft and started a fire that claimed the lives of 134 crewmembers and destroyed 43 aircraft. The technical cause of the accident was found to be a short circuit in the rocket; however, the task of arming the rocket was improperly performed. The postaccident investigation revealed that the missile was armed before the aircraft was readied on the catapult with its armament facing safely out to sea. The investigation further showed that the sailors who had handled the missile were not properly trained. It was assumed that all sailors knew that one never connected the umbilical cord of live rockets until the plane was in the ready position on the catapult.

Minimize Switchology.   One cause of task performance errors that seems to surface repeatedly is the inappropriate use of controls (including switches, buttons, and levers) because of deficient design or inadequate training. We refer to this design problem as "switchology." Switchology problems have many causes. Norman (1995) illustrated the depth of such problems in his explanation of the Macintosh power switch. Even though the power switch's location was considered a trivial matter, its location was constrained by numerous issues so that the design decision became complex and nearly impossible to complete. The Macintosh design problem was a result of poor organization and a confusion over who was in charge regarding design decisions. This resulted in no standardization for the power switch location among various models and created a series of false assumptions on the part of the users.

An example of switchology that takes a more serious twist was problems with the radar version of the Sidewinder air-to-air missile, the AIM-9C (Westrum, 1999). In contrast to the infrared version of the missile, the radar version required the operator of the attacking plane to keep bathing the target in radar beams that would be reflected into the missile's seeker. Pilots always had a choice to use the radar version or the infrared version, but switching from infrared to radar meant that one had to turn the seeker radar on as well as the missile itself. The result was that the pilot was more likely to keep the switch on infrared, which was simpler. Switching to radar also required a switch in mind-set, because radar did not have the fire-and-forget feature that endeared infrared missiles to pilots.

Bellenkes et al. (1993) described the catastrophic results of aircraft cockpit controls that are unusable under certain conditions. This jeopardizes the operators of fighter aircraft engaged in life-threatening situations. Numerous design problems impeded the pilot's coping with cognitive tasks. Problems included controls that were too complex because they incorporated too many switches, operator distractions caused by difficulties in locating proper switches, and the accurate identification of controls that were of similar design, color, or proximity. Santilli (1982) referred to similar problems resulting from "negative transfer." Negative transfer is the use of a procedure or performance of a task that is correct for one situation (e.g., an older model of an aircraft) but incorrect for another (e.g., a newer model of the aircraft). The result is the improper operation of controls or misinterpretation of displays.

The effects of switchology occur not only in high-risk environments or life-threatening situations, but also in tasks designed for everyday use. Riley (2001) described design problems associated with the common telephone keypad. He pointed out that a product's functions and functional logic are often defined from an engineering perspective rather than from a user's perspective. The everyday telephone-pad designs he reviewed did not provide for an intuitive procedure for the user, and there was little about the interface that made the logic of use visible. Riley maintained that designers frequently fail to develop a conceptual framework that is based on the user's existing knowledge and that encompasses the full range of required system functionality. The design problems identified by Riley point to the importance of exploring user characteristics that include skill and knowledge, during the conceptual design phase.

In its simplest form, switchology also refers to the unintended use of a switch. Simpson (1996) reported that during a meeting held at a large computer firm, he was faced with a screen that was fully retracted in a presentation room. Near the screen was a brightly backlit push-button switch that appeared to control the screen. Its actual purpose was to turn off the main power to a computer room that was located next to the presentation room. The switch's location was the result of a recent renovation. Personnel from the computer room placed a sign, hand-written in English and Spanish, near the switch that read, "Don't touch this switch." This did not deter the presenter, however; the button was pushed, and power was cut off from the computer room. According to Simpson (1996), the first warning of the loss of power was when the ventilators shut down. Even such clear written instructions do not always provide a guarantee of safety.

Switchology is a design problem rooted in poor thinking about displays of information. Controls that don't work the way they were intended or result in unintended consequences virtually always stem from inattention to user needs. Switchology is compounded when the task becomes more complex under certain conditions (e.g., nighttime operations or when the working environment becomes life-threatening or extremely stressful). Whether it is an air combat pilot attempting to optimize the selection of weapons for a given condition or a person attempting to program a new auto-dial telephone, controls and displays should support the cognitive process, not impede it.

Account for Human Error.   There is a multitude of literature on human error. Suffice it to say that human error can defeat a sound cognitive task design. It would be wise for cognitive task designers to heed the counsel of experts in the field of human error who remind us that there is no assurance the same error will not be committed again by a different person, and performers with good intentions may be forced to commit error by the way their environment influences their behavior (Van Cott, 1994). The seriousness of human error is emphasized in the words of James Reason (1990), "Now, the nature and the scale of potentially hazardous technologies, especially nuclear power plants, means that human errors can have adverse effects upon whole continents over several generations" (p. 1).

Reflecting on the potential for accidents based on the primary types of error can provide our requisite imagination with the ingredients to account for many of these user errors in the design. The primary types of error identified by Reason (1987) are as follows:

- *False sensations* are a failure to recognize the discrepancies between our subjective experience of the world and the objective reality.
- *Attention failures* result from distractions; processing simultaneous inputs; focusing on one of two concurrent messages; dividing attention between concurrent tasks; and failure to attend to monitoring, custodial, and verification tasks.
- *Memory lapses* involve forgetting list items, forgetting intentions, and losing track of previous actions.
- *Unintended words and actions* comprise all the absent-minded deviations of words, signs, and actions from their intended paths. Reason called them quintessential slips that arise from failure of execution rather than from inadequate plans.

- *Recognition failures* are misperceptions in which an erroneous cognitive interpretation is placed on sensory data. This often occurs when the sensory evidence is incomplete and when there is a strong schema-based expectation to perceive the presence or the absence of a particular stimulus configuration.
- *Inaccurate and blocked recall* involves all the errors that share the common process of recollection.
- *Errors of judgment* result from what is being judged and the nature of the task. The major subdivisions include psychophysical misjudgments, misconceptions of chance, misjudgments of risk, misdiagnoses, and erroneous social assessments.
- *Reasoning errors* result from slips, lapses, and mistakes in the cognitive reasoning process. They include errors in deductive reasoning, errors in propositional reasoning, reasoning with positive and negative instances, reasoning with abstract and concrete instances, error in concept formation, and errors in hypothesis testing.

Reason (1987) went on to distinguish three basic levels that are essential to the anticipation of error: the behavioral, contextual, and conceptual levels. These factors form the basic questions the designer needs to answer. Such questions revolve around the "What?", "How?", and "Why?" that fuel requisite imagination.

Consider Information Needs

Think about information needs. Determine what information should be made available to users and identify the factors that prevent the needed information from reaching them. Inefficient display design and multiple sources of information have impaired more than one operator of complex technology. During the Vietnam conflict, an F-4 fighter pilot was shot down because he did not hear the "bogey-on-your-tail" warning being broadcast by his wingman. His after-action report indicated that he had reached his perceptual and cognitive saturation point because of the vast amount of information he was attempting to process at the time (Bellenkes et al., 1993). The task designer must be aware of overload and a potential saturation point of the user when designing information feedback loops.

Today's more complex technologies put a greater demand on task designers. Computers generate a lot of information, and as this output increases, there is a corresponding increase in the amount of information made available to the human operator. For example, computer inputs often lead to a problem of information overload. The operator may be overwhelmed with aural and visual cues, warnings, instructions, displays, procedures, and system parameters. Information overload can be a result of limitations inherent in the tools the designer has available (Brickman et al., 1998), but it also reflects poor design (Riley, 2001). Conversely, limitations in technologies or design have also led to situations in which information underload occurs. Information underload is a situation in which necessary information is not readily available and thus impedes the operator's problem-solving or decision-making processes.

Problems also exist in the interpretation and availability of information in today's automated systems. Recent research has found such problems as an overreliance on the technology, confusion over or inefficient use of the technology, misrepresentation of information due to design based on an inappropriate model of expertise, inadequate training or lack of familiarity, and misunderstanding of the design's intent (Mosier & Skitka, 1996). Research into the role of automation in complex technologies has shown that innovations that change the extent and type of automation transform systems in unforeseen ways and have unanticipated consequences (Cook & Woods, 1996).

An example was a field study conducted by Mosier, Palmer, and Degani (1992) that explored the factors affecting the use of airline cockpit checklists. The study examined the factors affecting the successful and unsuccessful use of both a paper checklist and an advanced electronic checklist. The researchers found that the electronic checklist introduced new errors by virtue of its automaticity to the operators. Additionally, they found that crews came to rely on the automated aid as an indicator of system state rather than merely as a procedural aid. Mosier attributed this to a lack of transparency in the information most readily available to the flight crew. Transparency refers to how much an information system aids the operator in developing a mental model of the system's state.

Riley (2001) argued that one of the challenges for designers of electronic products is to simplify the number of functions made possible with today's computer technology. Riley pointed out that in the aviation industry, the evolution of the Flight Management System (FMS) typifies the functionality problem. The FMS was originally intended to optimize fuel efficiency. Since its original introduction to aviation, the FMS has become the control center for flying the airplane, and it is now the hub of operating information. Its functions have been greatly increased to include not only fuel management, but navigational planning and implementation, departure and arrival profiles, aircraft systems monitoring and operation, emergency procedures, air traffic control requirements, and much more. Today, flying the FMS is as complex as flying the aircraft. Riley (2001) reported that it takes about 1,200 hours of operational experience to learn the most rudimentary operations of a FMS.

Today's proliferation of information in the modern jet-fighter airplane cockpit has frequently led to confusion and increased workload for pilots who are required to integrate information across a wide array of displays to determine their tactical situation (Brickman et al., 1998). We can expect current problems associated with cognitive task design and information overload to become worse with the emergence of new technologies. But the designer's task will be the same—the designer must identify the information required to perform the task and develop methods to convey quickly and accurately this information, which most likely comes from multiple sources. Cognitive task design must decide how much information is necessary for the task performer to select a course of action, how the information should be presented, and how much transparency is necessary.

Examine the Procedures

One critical part of cognitive task design is the design of the procedures that will be used to guide the task. Procedures typically provide approved methods of task performance, a basis for training, and hazard avoidance. The need for effective and well-developed procedures is well accepted throughout the domains of human factors, safety, and accident prevention. Consequently, we believe that it is necessary for the cognitive task designer to be good at constructing effective procedures. The designer will most likely be faced with developing operating procedures that will accompany the task design. An *operating procedure* is a rule or guideline developed to ensure that tasks required to perform a specific function are performed in the manner intended by the designer, engineer, or manufacturer and that the activities remain within the safe operating parameters of that technology (e.g., the turbulence penetration airspeed for a specific model of airplane).

Many cognitive tasks require a set of well-defined procedures for them to be performed as intended. Complex human–machine systems encompass numerous cognitive tasks. Such tasks are usually accompanied by *standard operating procedures,* which usually require training to be properly implemented. Degani and Wiener explained the necessity of procedures used in such complex systems as follows:

> A complex human–machine system is more than merely one or more operators and a collection
> of hardware components. To operate a complex system successfully, the human–machine system

must be supported by an organizational infrastructure of operating concepts, rules, guidelines and documents. . . . In high-risk endeavors such as aircraft operations, space flight, nuclear power, chemical production and military operations, it is essential that such support be flawless, as the price of deviations can be high. When operating rules are not adhered to, or the rules are inadequate for the task at hand, not only will the system's goals be thwarted, but there may be tragic human and material consequences. Even a cursory examination of accident and incident reports from any domain of operations will confirm this. (Degani & Wiener, 1994, p. 44)

There are a number of concerns the designer should take into account when anticipating what might go wrong with procedures. These concerns are procedural inflexibility, legitimate procedures, bureaucratic procedures, and stagnant procedures.

Procedural inflexibility exists when desired performance is constrained by unnecessary or ineffective operating procedures. It is a weakness that causes the user to doubt that the procedures as presented will do the job effectively. Inflexibility encourages improvisation and work-arounds. There are certain cases in which work-arounds have actually improved operation, but they still are a sign that the designer has not properly done the job required. Improvisation and work-arounds can lead to serious consequences because the users often lack in-depth system knowledge.

Operating procedures are also categorized as formal or informal. The formal procedure is a guideline that is based on explicit rules. It is published in an official format, such as a flight operations manual and is available to all operators whose adherence is required (e.g., the minimum decent altitude without runway-in-sight allowable during an instrument approach for a low-time pilot-in-command).

The informal procedure is an emergent type of procedure that is not published or formally mandated by the organization. It typically develops (emerges) throughout day-to-day operations and becomes a "way of doing things." Informal procedures often emerge to simplify operational requirements (e.g., shortcut methods developed by pilots to hasten the cockpit preflight inspection). They may also reflect total experience not captured in the written documentation (Orr, 1996). Informal procedures are not based on explicit rules but result from contingent or situated behaviors. In what follows, we focus on the formal procedure because it is the category of procedure that is most relevant in cognitive task design.

Formal procedures may exist in one of three states. The first state we term the *legitimate procedure*. This state of procedure is an operating rule based on legitimate reasoning, facts, and the design parameters of the technology. It is a procedure that accurately describes the tasks and sequences necessary to operate a technology safely and effectively. The second state is the *bureaucratic procedure*. This state is an operating rule that has been created and implemented because of organizational politics, inadequate design, overbearing management, or inappropriate reactions to operating errors. The third state is the *stagnant procedure*. It is a legitimate or bureaucratic procedure that has lost its ability to maintain adherence. An example of a stagnant procedure is the automobile speed limit posted for U.S. highways. It has lost its bite. It is considered dangerous by many drivers to drive at the posted speed limit in many metropolitan areas because one will be dangerously tailgated or nearly overridden by the many drivers greatly exceeding the posted limit. The more bureaucratic and stagnant procedures that exist, the more the user loses faith in legitimate procedures.

The development of procedures that accompany cognitive tasks can be an arduous task for the designer. Degani and Wiener (1994) warned the designer that "Procedures do not fall from the sky, nor are they inherent in the equipment. Procedures must be based on a broad concept of the user's operation. These operating concepts blend themselves into a set of work policies and procedures that specify how to operate the equipment efficiently" (p. 47).

The design of effective procedures means meeting task requirements. This goes beyond regulatory agency or organizational requirements. Procedures should not come solely from the equipment supplier or manufacturer, nor should an individual supervisor or manager simply write them. Well-defined and effective procedures should define what the task is, when the task is to be performed, by whom, how the task is to be completed, the sequence of actions, and what feedback is provided (Degani & Wiener, 1994).

Even well-defined legitimate procedures can cause procedural inflexibility when there are too many restrictive procedures. One pitfall the task designer must take into account is the inclination for operators to perceive some procedures as illegitimate, and thus, unnecessary. This often occurs when insufficient background or foundation is provided to the operator during training to explain the basis of the procedure. An example of such a situation was the USS *Forrestal* fire previously discussed.

Today's complex technologies, such as electronic products, have placed a greater demand on designers, engineers, and manufacturers. Technological advancements and systems complexity are not the only reasons that the operations of many technologies are becoming more and more difficult to use. Procedural inflexibility is often a result of overproceduralizing operations. This occurs when those responsible for controlling a given technology respond to incidents or operational errors with the creation and implementation of more restrictive operating rules. An example of procedural inflexibility is found in the operations of two separate British Harrier airplane squadrons during the Falkland War (Westrum & Adamski, 1999).

Each squadron operated Harrier jets and used identical navigation and weapons systems. The British had over time created a multitude of restrictive operating procedures for these systems. One squadron followed the procedures to the highest degree. The other squadron adapted a philosophy of flexibility in regard to the operating rules. The result was that the squadron that closely followed every rule experienced little success in combat; in contrast, the squadron that exercised flexibility had great success. The Harrier example demonstrates that it is possible to overproceduralize tasks to the point of interfering with competent performance (Ward, 1993).

Technology when introduced into the field is typically accompanied by a basic set of operating rules. As the technology matures and is revised due to uncovered design flaws, incidents, operation errors, and so on, the controllers of the technology many times respond by creating and implementing more restrictive operating procedures. Although often desirable, this can also put the user in a situation in which the desired performance cannot be achieved. As a technology matures, allowable operating space is restricted. The user becomes more and more restricted in performance. As in the case of the Harrier squadron that had little success, the procedural inflexibility took from the pilots the space necessary to perform within the rules, and it became necessary to break the rules to succeed. The danger here becomes a question of what flexibility can be afforded to what rules? Can the necessary margin of safety be maintained with flexibility? This is a design issue that can best be addressed by requisite imagination.

Consider Conventions

Cognitive task designs often require changing users' ways of doing things to include reshaping their perceptions of the task and revising how things should be done. Habitual ways of doing a task or task-related devices that have become comfortable to use easily become conventions. Many conventions "go with the territory." They become conventions because they have been the first to be successfully marketed such as the Remington typewriter that used the QWERTY keyboard designed by Charles Latham Sholes in the 1870s (Norman, 1989). It is called QWERTY because the top row of letters spells QWERTY in the American version. Even with its arbitrary arrangement of letters, the QWERTY keyboard is still in use today on computer keyboards (Liebowitz & Margolis, 1996).

Another recurring problem is the common belief of many designers that they understand users and can act as users' surrogates (Rouse, 1991). At times, designers are former users and their experience has value, but it is limited in that the designers cannot truly capture current users' abilities and attitudes that formulate conventions. Just consider cognitive problems associated with automobile designs that involve the convention of driving on the right side or the left side of the road.

The designer should look carefully at what conventions exist within the domain of the design and have a good reason to challenge them if change is projected. There are numerous examples of designers failing to address user conventions. One example that emphasizes the need to consider user conventions occurred during World War II was related by Steven Casey in his examination of design technology and human error (Casey, 1993). During an air raid against a U.S. Army airfield by Japanese air forces, a U.S. Army Corps pilot rushed out to man one of the P-47 Thunderbolts parked on the ramp. During the confusion, he was unable to board the airplane that he had normally been flying because another pilot boarded his airplane. He quickly found another unmanned P-47, and when he sat in the cockpit, he found few of the instruments familiar. In fact, he barely got the engine started when the Japanese began strafing and dropping bombs. The U.S. Army Corps pilot was never able to get the aircraft airborne because he could not completely figure out how the controls worked; he was, however, able to taxi back and forth across the field and prevent a direct hit by the Japanese. After the war, the pilot confessed his story to aviation researchers "but he never did figure out why someone would redesign a fighter plane's instrument panel in the middle of a war" (Casey, 1993, p. 39).

Even today, a lack of convention exists within many modern airplanes. There has been a tendency for the displays in modern cockpits to become a jungle of clutter, of ill-considered symbols and text, or a dazzling presentation of color (Wiener, 1988). Much of this clutter is also found in the design of keyboards used to input commands to the various components of the automated cockpit.

Examples of errors resulting from the failure to address conventions are also found in the medical domain. For years there was no common standard for the controls or displays of anaesthesia machines (Wyant, 1978). This lack of convention led to a number of fatalities and serious injuries. In the mid-1980s, "syringe swap" was another major cause of human error among anaesthesiologists (Cooper, Newbower, & Kitz, 1984). Research indicated that these errors centered around the lack of convention. Standardized anaesthesia procedures and equipment controls barely existed. Such things as inadequate labeling, the nonstandard arrangement of drugs, and the nonstandard design of gas flow knobs were found to be contributing factors to syringe swap and serious errors. Today, conventions for labeling, controls, and displays of anaesthesia equipment are mandated by national standards.

The examples demonstrate the influences of conventions on design. It is important for the designer to identify those conventions that are in place and think twice before challenging those in existence. Furthermore, if convention is lacking, then it may become the designer's task to establish standards.

Review Constraints

The purpose of this element in the conceptual model is to serve as a reminder to the designer. As stated earlier, the conceptual model is a generative model that may be entered at any point. The various elements that make up the model overlap in meaning, but their purpose is consistent: to provide fuel for one's requisite imagination. Anticipating what might go wrong requires one to reflect on the elements of the conceptual model to identify constraints that will affect the design's outcome or constraints that will affect the design process. Such constraints come in the form of internal or external limitations.

Internal constraints are those features that negatively affect performance of the cognitive task. They create limitations for the specific cognitive task under consideration. Some of these problems may be beyond the designer's power to change. Examples would be poor physical equipment design, bad labels, or hard-to-read displays.

Sometimes constraints are hidden and may need to be discovered before being addressed. Reason (1997) in his book on managing organizational accidents reminded readers that some designs provide hidden paths to the commission of human errors and violations. Such paths are often subtle. They create constraints because for instance they provide poor or wrong feedback that leads to uncertainty in the user's mind, or the user becomes locked into a mode of action from which he or she cannot escape.

External constraints are limitations that negatively affect the design process itself. They are not an integral part of the specific task; rather, they apply to any design within that organizational environment. Organizational factors that originate upstream of the design project often act as external constraints to it. Such factors include the organization's culture, strategic goals, production pressures, resources available, personnel issues, and risk-management policies. Externally constraint-driven designs are typically dominated by budgets, credibility (e.g., having something to show the client now), inertia in the sense of past solutions and existing vested interests, and schedules (Rouse, 1991).

Identifying constraints involves what Norman (1993) called experiential and reflective cognition. We can think of experience based cognition as the designer's use of task-specific knowledge in which the design is validated through domain-specific knowledge. In the reflective mode, the designer makes use of more general principles, such as relevant memories, making comparisons between past failures and successes, and incorporating foresight about what might go wrong by creating new hypotheses. Consequently, the designer's requisite imagination looks for possible failures through identifying and acknowledging constraints.

There are no easy solutions to designing many of today's complex cognitive tasks. No matter how technology is revised or cognitive tasks re-designed, new challenges will continue to emerge. Constraints arising from human limitations associated with operating, managing, and monitoring new technologies will require creativity. Emerging problems in cognitive task design cannot be solved by reducing design to a simple routine. Designers may be forced to develop a new conceptual mind-set that modifies their concept of the task to anticipate what might go wrong.

SUMMARY

Requisite imagination is fueled both by experience and by insight. The cognitive task designer must do more than hope for insight will come. To achieve the desired outcome, a conceptual model should guide the designer. A good model allows the designer to take advantage of both experiential and reflective knowledge, and fosters the process of insight. A good conceptual model allows the designer to make the task fit the user instead of the other way around.

In view of the preceding concepts and examples, it should be clear that the designer needs to think carefully and systematically about how things might go wrong. The question becomes, what concrete design tools can the designer bring to bear to assist in sparking his or her requisite imagination? To achieve the desired outcome the designer needs to use a dependable checklist based on a sound conceptual model. An example of such a checklist is presented at the end of this chapter (Appendix). A sound conceptual model forces the designer to think about issues beyond those that immediately come to mind; it fosters the process of insight. The key is to adopt a systematic approach so that nothing critical gets missed.

Exactly how this should be done will vary from case to case. There is good motive here to use a "lessons learned" book in the process. Interviews with senior designers in a specific area could very well add to the list of lessons or provide more specification for model's components. Although the conceptual model we have developed here is a good start, a more custom-built model is needed for specific tasks.

Nor should the designer be content once the process is finished and the cognitive task design is released. The designer needs to be open to feedback from the field about how the design is working, and the designer should cultivate the channels that lead to such feedback.

Above all, the designer should have a restless mind. What situations have not been foreseen? What has been forgotten? What users might the design encounter? What situations will undo it? The Citicorp building is an example that should give the designer some restless moments. What if LeMessurier had not been so diligent? Consider that the action took place after the building had been built and occupied. The design was apparently done, but execution had created flaws. It might be useful to think of 95% of the design being accomplished at the point of release, with another 5% to be done when feedback comes in. Further experience may suggest weaknesses unseen at the point of original creation. Not everything can be foreseen. We must maintain a "questioning attitude." The price of failure may be too big to pay in today's complex technologies.

Although our checklist in the Appendix reflects the components of our conceptual model, your own personal conceptual model may incorporate different components. Such a tool can ensure that all your design activities have been addressed and assist you in successful cognitive task design.

APPENDIX
Cognitive Task Design: Conceptual Model Checklist

Clarify the Task	Define the intent of the design. Conduct cognitive task analysis. Determine technical standards and users' functional characteristics? *Have you thoroughly defined the task?*
Consider the Organization	Assess organizational culture. Address organizational factors that may affect design project. Define design team members, customers, and users. Determine resources needed and available. Establish evaluation methods and criteria. Define decision makers. *What organizational constraints have you identified?*
Align Designer and User Worlds	Determine users' perceptions of the tasks. Avoid designing in the dark. Beware of assumptions. Identify the actual characteristics of perspective users. Consider range of users from novice to expert. *Have you matched the designer's world with the users'?*

(Continued)

APPENDIX
(Continued)

Consider the Operational Domain	Determine domain expertise for the task. Identify environmental factors of domain. Review conditions under which the task will be performed. Consider unique stressors. Avoid "clumsy automation." *Did you consider all facets of the domain?*
Survey Past Failures	Conduct a historical review of past failures. Minimize "switchology." Assess users' perceptions of controls and displays. *Reflect on potential for human error.*
Consider Information Needs	Determine what information is required by user to perform task. Consider filtering factors. Beware of information overload and underload. *Do you have the right information conveyed to the right person?*
Examine Procedures	Develop well-defined procedures. Assess allowable procedural flexibility. Ensure that all procedures are legitimate. Determine necessary level of user training. *Will the procedures meet foreseeable circumstances?*
Consider Conventions	Review existing conventions. Challenge conventions with reason. Establish standards when necessary. *Did you adequately address existing conventions?*
Review Constraints	Anticipate what might go wrong. Reflect on each component of the conceptual model. Address internal constraints and external constraints. *Did you practice requisite imagination?*

REFERENCES

Al-Salka, M. A., Cartmell, M. P., & Hardy, S. J. (1998). A framework for a generalized computer-based support environment for conceptual engineering design. *Journal of Engineering Design, 9,* 57–88.

Bellenkes, A. H., Yacavone, D. W., & Alkov, R. A. (1993, April). Wrought from wreckage. *Ergonomics in Design,* 26–31.

Brickman, B. J., Hettinge, L. J., Haas, M. W., & Dennis, L. B. (1998). Designing the super cockpit. *Ergonomics in Design, 6*(2), 15–20.

Card, S. K., Moran, T. P., & Newell, A. (1983). *The psychology of human-computer interaction.* Hillsdale, NJ: Erlbaum.

Casey, S. (1993). *Set phasers on stun.* Santa Barbara, CA: Aegean.

Coel, M. (1987). A bridge that speaks for itself. *American Heritage of Invention and Technology, 3,* 8–17.

Cook, R. I., & Woods, D. W. (1996). Adapting to new technology in the operating room. *Human Factors, 38,* 593–613.

Cooper, J., Newbower, R. S., & Kitz, R. J. (1984). An analysis of major error and equipment failures in anaesthesia management: Considerations for prevention and detection. *Anesthesiology, 60,* 34–42.

Degani, A., & Wiener, E. L. (1994). Philosophy, policies, procedures and practices: The four P's of flight deck operations. In N. Johnston, N. McDonald, & R. Fuller (Eds.), *Aviation psychology in practice* (pp. 44–67). Aldershot, England: Avebury Technical.

Driskell, J. E., & Salas, E. (1991). Overcoming the effects of stress on military performance: Human factors, training, and selection strategies. In R. Gal, A. D. Mangels Dorff, & D. L. Login (Eds.), *Handbook of military psychology* (pp. 183–193). New York: Wiley.

Eder, W. E. (1998). Design-modelling: A design science approach (and why does industry not use it?). *Journal of Engineering Design, 9,* 355–371.

Ferguson, E. S. (1977). The mind's eye: Nonverbal thought in technology. *Science, 197,* 827–836.

Ferguson, E. S. (1997). *Engineering and the mind's eye*. Cambridge, MA: MIT Press.

Flach, J. M., & Dominguez, C. O. (1995, July). Use-centered design: Integrating the user, instrument, and goal. *Ergonomics in Design,* 19–24.

Gras, A., Moricot, C., Poirot-Delpech, S., & Scardigli, V. (1994). *Faced with automation: The pilot, the controller, and the engineer*. Paris: Sorbonne.

Heimann, C. F. L. (1997). *Acceptable risks: Politics, policy, and risky technologies*. Ann Arbor, MI: University of Michigan Press.

Hoc, J. M. (2000). From human-machine interaction to human-machine cooperation. *Ergonomics, 43,* 833–843.

Hollnagel, E., & Woods, D. D. (1983). Cognitive system engineering: New wine in new bottles. *International Journal of Man-Machine Studies, 18,* 583–600.

Howell, W. C., & Cooke, N. J. (1989). Training the human information processor: A look at cognitive models. In I. L. Goldstein (Ed.), *Training and development in work organizations: Frontiers of industrial and organizational psychology* (pp. 121–182). San Francisco: Jossey-Bass.

Kao, D., & Archer, N. P. (1997). Abstraction in conceptual model design. *International Journal of Human Computer Studies, 46,* 125–150.

Kleinfeld, N. R. (1990, March 25). How strykeforce beat the clock. *New York Times (business section,)* pp. 1, 6.

Landauer, T. K. (1995). *The trouble with computers: Usefulness, usability, and productivity*. Cambridge, MA: MIT Press.

Leveson, N., & Turner, C. L. (1993). An investigation of the Therac-25 accidents. *IEEE Computer, 26*(7), 18–41.

Liebowitz, S., & Margolis, S. E. (1996, June). *Typing errors*. Reason online. [Website] Retrieved Oct. 21, 2001 from http://reason.com/9606/fe.qwerty.html

Lucas, D. (1997). The causes of human error. In F. Redmill & J. Rajan (Eds.), *Human factors in safety-critical systems* (pp. 37–65). Oxford, England: Butterworth-Heinemann.

Militello, L. G., & Hutton, R. J. (1998). Applied cognitive task analysis (ACTA): A practitioner's toolkit for understanding cognitive task demands. *Ergonomics, 41,* 1618–1641.

Morgenstern, J. (1995, May 29). The fifty-nine story crisis. *New Yorker, 71,* 45–53.

Mosier, K. L., Palmer, E. A., & Degani, A. (1992, October 12–16). *Electronic checklists: Implications for decision-making*. Paper presented at the 36th Annual Meeting of the Human Factor Society, Atlanta, GA.

Mosier, K. L., & Skitka, L. J. (1996). Humans and automation: Made for each other. In R. Parasuraman & M. Mouloua (Eds.), *Automation and human performance: Theory and applications* (pp. 37–63). Mahwah, NJ: Erlbaum.

Newell, A. F., & Alistair, Y. C. (1993, October). Designing for extraordinary users. *Ergonomics in Design,* 10–16.

Newsome, S. L., & Spillers, W. R. (1989). Tools for expert designers: Supporting Conceptual design. In S. L. Newsome, W. R. Spillers, & S. Finger, (Eds.), *Design Theory '88* (pp. 49–55). New York: Springer-Verlag.

Norman, D. A. (1986). Cognitive engineering. In D. A. Norman & S. W. Draper (Eds.), *User-centered system design: New perspectives in human-computer interaction* (pp. 31–61). Hillsdale, NJ: Erlbaum.

Norman, D. A. (1989). *The design of everyday things*. New York: Doubleday.

Norman, D. A. (1993). *Things that make us smart*. Reading, MA: Addison-Wesley.

Norman, D. A. (1995). Designing the future. *Scientific American, 273,* 194–197.

Norman, D. A. (1999). Affordance, conventions and design. *Interactions, 6*(3), 38–43.

Orr, J. E. (1996). *Talking about machines: An ethnography of a modern job*. Ithaca, NY: ILR Press.

Perrow, C. (1983). The organizational context of human factors engineering. *Administrative Science Quarterly, 28,* 521–541.

Petroski, H. (1992). *To engineer is human: The role of failure in successful design*. New York: Vintage Books.

Petroski, H. (1994). *Design paradigms: Case histories of error and judgment in engineering*. Cambridge, England: Cambridge University Press.

Petroski, H. (1996). *Invention by design*. Cambridge, MA: Harvard University Press.

Petroski, H. (1997). *Remaking the World: Adventures in engineering*. New York: Alfred A. Knoff.

Prince, C., Bowers, C. A., & Salas, E. (1994). Stress and crew performance: Challenges for aeronautical decision making training. In N. Johnston, N. McDonald, & R. Fuller (Eds.), *Aviation psychology in practice* (pp. 286–305). Aldershot, England: Avebury.

Reason, J. (1987). A framework for classifying errors. In J. Rasmussen, K. Duncan, & J. Leplat (Eds.), *New technology and human error* (pp. 5–14). New York: John Wiley & Sons.

Reason, J. (1990). *Human error*. Cambridge, England: Cambridge University Press.

Reason, J. (1997). *Managing the risks of organizational accidents*. Aldershot, England: Ashgate.

Redding, R. E., & Seamster, T. L. (1994). Cognitive task analysis in air traffic controller and aviation crew training. In N. Johnson, N. McDonald, & R. Fuller (Eds.), *Aviation psychology in practice* (pp. 190–222). Aldershot, England: Avebury.

Redding, R. E., & Seamster, T. L. (1995). Cognitive task analysis for human resources management in aviation personnel selection, training and evaluation. In N. Johnston, R. Fuller, & N. McDonald (Eds.), *Aviation psychology: Training and selection* (Vol. 2, pp. 170–175). Aldershot, England: Avebury.

Redmill, F. (1997). Introducing safety-critical systems. In F. Redmill & J. Rajan (Eds.), *Human factors in safety-critical systems* (pp. 3–36). Oxford, England: Butterworth-Heinemann.

Richey, R. (1986). *The theoretical and conceptual basis of instructional design.* London: Kogan Page.

Riley, V. (2001). A new language for pilot interfaces. *Ergonomics in Design, 9*(2), 21–26.

Rouse, W. B. (1991). *Design for success: A human-centered approach to designing successful products and systems.* New York: John Wiley & Sons.

Rubinstein, R., & Hersh, H. (1984). *The human factor: Designing computer systems for people.* Maynard, MA: Digital Press.

Sagan, S. C. (1993). *The limits of safety.* Princeton, NJ: Princeton University Press.

Santilli, S. (1982, December). *The human factor in mishaps: Psychological anomalies of attention.* Paper presented at the Survival and Flight Equipment 18th Annual Symposium, Las Vegas, NV.

Schon, D. A. (1983). *The reflective practitioner: How professionals think in action.* New York: Basic Books.

Simpson, R. (1996). *Don't touch this switch.* Retrieved October 22, 2001, from http://catless.ncl.ac.uk/Risks/18.65.html

Sloan III, E. W. (1967). *Benjamin Franklin Isherwood, naval engineer.* Annapolis, MD: United States Naval Institute.

Squires, A. (1986). *The tender ship: Governmental management of technological change.* Boston, MA: Birkhaeuser.

Stevens, R. B., & Ezell, E. C. (1994). *The black rifle: A M-16 retrospective.* Cobourg, Canada: Collector Grade Publications.

Tarkov, J. (1995). A disaster in the making. In F. Allen (Ed.), *Inventing America* (pp. 86–93). New York: American Heritage.

Van Cott, H. (1994). Human errors: Their causes and reduction. In M. S. Bogner (Ed.), *Human error in medicine* (pp. 53–65). Mahwah, NJ: Erlbaum.

Vaughan, D. (1996). *The Challenger launch decision: Risky technology, culture, and deviance at NASA.* Chicago, IL: University of Chicago Press.

Ward, S. (1993). *Harrier over the Falklands.* London: Orion Books.

Westrum, R. (1991). *Technologies and society: The shaping of people and things.* Belmont, CA: Wadsworth.

Westrum, R. (1997). Social factors in safety-critical systems. In F. Redmill & J. Rajan (Eds.), *Human factors in safety-critical systems* (pp. 233–256). Oxford, England: Butterworth-Heinemann.

Westrum, R. (1999). *Sidewinder: Creative missile development at China Lake.* Annapolis, MD: Naval Institute Press.

Westrum, R., & Adamski, A. J. (1999). Organizational factors associated with safety and mission success in aviation environments. In D. J. Garland, J. A. Wise, & V. D. Hopkin (Eds.), *Handbook of aviation human factors* (pp. 67–104). Mahwah, NJ: Erlbaum.

Westrum, R., & Wilcox, H. A. (1995). The making of a heat-seeking missile. *Inventing America* (pp. 136–143). New York: American Heritage.

Wiener, E. L. (1988). Cockpit automation. In E. L. Wiener & D. C. Nagel (Eds.), *Human factors in aviation* (pp. 433–462). San Diego, CA: Academic Press.

Wiener, E. L. (1989). *Human factors of advanced technology (glass cockpit) transport aircraft* (NASA 117528). Springfield, VA: National Aeronautics and Space Administration.

Wilcox, R. K. (1990). *Scream of eagles: The creation of top gun and the U.S. air victory in Vietnam.* New York: John Wiley & Sons.

Williams, H. L., Pope, L. T., Pulu, P. S., & Sass, D. H. (1982). *Problems in operating the 1200 psi steam propulsion plant: An investigation* (NPRDC SR 82–25). San Diego, CA: Navy Personnel Research and Development Center.

Wyant, G. M. (1978). *Medical misadventures in anaesthesia.* Toronto, ONT: University of Toronto Press.

PART II
Methods

PART III

Methods

11

Cognitive Factors in the Analysis, Design, and Assessment of Command and Control Systems

Jonathan Smalley

QinetiQ, Malvern Technology Centre, United Kingdom

Abstract

This chapter looks at how to analyse, design, and assess the interaction between people and computers in a command and control system with respect to their cognitive tasks, using analytical models, demonstrators, or simulators to represent the design as it is progressed. The focus is on situation awareness for process operators to improve their cognitive contribution or decision making. The philosophies and approaches outlined in this chapter should provide a basis for defining the experimental context and conditions necessary to produce assessments that may be used to verify and clarify system requirements. Once the system requirements have been finalised, they may be translated into system design specifications for implementation.

INTRODUCTION

Command and control, or the management of an organisation, is about the cognitive tasks of decision making, whether this be the navigation of a ship, piloting an aircraft, determining the control settings for a nuclear power plant, or simply selecting the route home after a long day at the office. Wherever there are decisions to be made, there are cognitive tasks for which to design.

There is a common theme through all of these tasks, illustrated in Fig. 11.1, which portrays the development of an original concept of Birmingham and Taylor (1954). This summarises the essential functional elements of a cognitive task in terms of how it affects the system that may be designed to support it. This figure depicts a generic information flow for any cognitive task, from the interpretation of the information displays to the execution of the decisions through the controls for effective system performance.

This chapter is derived from practical experience, using simulators as a development tool and focusing on the cognitive component of the command and control system. The ideas and

FIG. 11.1.   The use of situation awareness.

syntheses have emanated from or been used in studies and investigations ranging from bridges of trawlers, control rooms for steel works, warships, nuclear power plants, air traffic control systems, mission systems for military aircraft, and other air defence applications.

Task analysis in general is well covered by Kirwan & Ainsworth (1992), and a good exposition of equipment system design issues is given by, for example, Ivergård (1989) for control rooms or by Shneiderman (1998) for the user interface. Pheasant (1986) touched on the question of psychological space in the first edition of his definitive *Bodyspace* which addresses biomechanical and anthropometrical questions of equipment design. Singleton (1974) provided an introduction to human factors in system integration and opens up the cognitive domain. This developed the question of system reliability assessment (for example, Kirwan, 1994) and paved the way for a total system approach to take account of the sociotechnical context of the cognitive task. The purpose now is to focus on the cognitive factors relating to such systems and the use of prototypes and simulators to support the process of cognitive task design.

Definition of a System

There is some confusion in the use of the word *system*; a good place to start is with some working definitions. In the abstract, a system relates to the collection of operations. In reality, it refers to the composite of equipment and software that supports those operations. At this point, it is helpful to distinguish between supersystems and subsystems with respect to the design of an interface system. The super-system contains everything that is subject to design, including the interface, thus defining the context that directly drives the operational requirement for the interface. Subsystems comprise the sets of co-ordinated components for performing the tasks. This may be expanded by the concept of a hypersystem, the system in its most extended form which provides the fixed context for the whole system implementation and provides a fixed frame of reference for the cognitive task. For cognitive tasks, this includes the cultural context in which decisions are made and the background of training that may be brought to

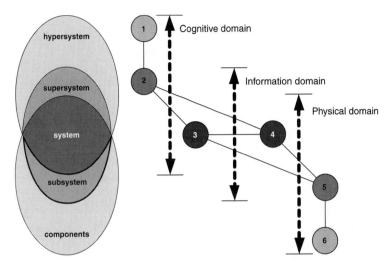

FIG. 11.2. A five-layered cognitive task structure.

bear on the task. This provides a label for the abstract components of the cognitive task, such as memories, beliefs and principles, or even jottings in private note books that do not translate directly into system implementations but are highly relevant to the cognitive processes of the human operator. At the other extreme are the component elements. This generates a five-layered model centred on the system operations, with the layers in the context of an organisation or human–machine system as follows:

- Hypersystem: for example, industrial or military context—steel industry
- Super-system: for example, organisation or service—steel ingot production
- System: for example, operators or service personnel—steelmaker
- Subsystem: for example, workplaces or consoles—control cabin
- Component elements: for example, controlled processes—electric arc furnace

Figure 11.2 shows these five layers as a generic structure to identify different modes of interaction between human and machine in terms of the intersection of the cognitive (hyper-system, super-system, and system levels), information (super-system, system, and subsystem levels), and physical (system, subsystem, and component levels) domains of concern.

If we look at these intersections as interactions, we can identify six such types of interaction between the domains, depicted by the circles to the right of the diagram in Fig. 11.2. These identify six types of cognitive processes to consider. These connect to each other through the domains that separate them, as shown by the seven connecting lines.

Take the example of driving a car. The system is the interface of direct physical interaction between driver and vehicle. This interaction must be interfaced in some way to the underlying mechanical process that is being controlled, for example, the chain of mechanical linkages between road wheels and steering wheel through which the driver's intentions for the direction of travel are expressed and translated into the direction of movement that the vehicle will take. Similarly, on the driver's side of the interface, there are cognitive processes to link the physical interaction at the system level (with the steering wheel) with the longer term destinations toward which the immediate intentions are directed.

The driver of a motor vehicle has the task of juggling between the control of the vehicle (5) and the operation of its components (6) in the physical domain and navigating the route to take (1 and 2 in the cognitive domain). That leaves the driver in the information domain to extract the immediately relevant navigational information (3) and maintain awareness of where the vehicle is heading (4). The driver reconciles these two concerns to control the vehicle (3–5 and 4–5) or update the navigation and route plan. Hence, for this example, we can attribute concerns to these circles as follows:

1. The destination
2. Route planning
3. Reading ahead from the route map while driving.
4. Reading the road ahead from the features actually encountered.
5. Control of the vehicle
6. Vehicle operation (for example, operating foot pedals or light switches)

The physical domain includes the vehicle; the abstract domain includes the knowledge of where the journey is to end and at what time (1). From this would be generated the route plan with its list of roads to follow, clues about features to look out for that will confirm which turns to take at the junctions encountered and which exits when roundabouts are negotiated (2). Some of these may be familiar to the driver (3), sufficiently so to allow mental rehearsal of particular segments of the journey (2–3) or the blind execution of well-practised manoeuvres without reference to the actual road conditions (3–5).

The information domain is something of a hybrid between the cognitive and the physical. It provides the means of describing the different ways in which they interface.

In the physical domain of the vehicle control (6), there are the skills of simply operating the equipment so that the wipers don't wipe when the trafficators are to be activated or so that gear selection is cleanly co-ordinated with clutch control (5–6). Then all these activities must be co-ordinated, for example, indicating before turning (5), and performed according to the demands of the road conditions (4) or navigational directions (3).

For this example, the cognitive domain covers abstract road knowledge, the information domain covers the visual picture of the road, and the physical domain covers the nonvisual kinaesthetic and aural cues used for driving. In a rally car, the driver's task is divided between driver and navigator. We could use the model to look at the cognitive task interactions between team members and their functions as follows:

- 1–2: Navigator determines the primary direction for where to go and why.
- 2–3: Navigator reads ahead in the abstract, studies the maps, and generates the route plan.
- 2–4: Navigator provides directions beyond what the driver can immediately perceive, ahead of the horizon.
- 3–4: Correlation of the most detailed picture available to the navigator and the broadest picture available to the driver: if this were all, the driver would be driving blind, without regard to the immediate conditions.
- 3–5: Navigator provides information for dynamic interpretation—keep driving at seventy for 30 minutes, or follow the signs for London: driver in motorway mode—hoping there are no potholes or slower vehicles in the way.
- 4–5: Navigator provides positional information—turn left after 10 metres (car-park mode driving, i.e., driver is in parking mode, only concerned with the immediate position of the vehicle with respect to the surrounding situation, for example, as in a parking lot).
- 5–6: Driver operates the vehicle "by second nature."

Similar types of interaction would exist in an orchestra for conductor, musician, and instrument. Cognitive task design is relevant to all these types of interaction; it relates to how the thinking is done and how that thinking should be supported.

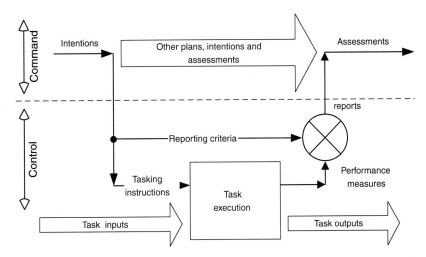

FIG. 11.3. Command and control schematic.

Provision of Capability. Whether by specific system design or left for interpretation by the operator, the aim is to ensure that, in effect, subsystem components should perform within the constraints imposed by the super-system environment. Relevant criteria include availability for parallel needs, avoidance of lockout situations (inability to be in two places at once, Catch 22), reliability in sequential operation, performance in terms of speed, and accuracy.

Simulators provide a highly effective means to investigate cognitive tasks. Figure 11.3 provides a summary of how they are used. A simulation run needs tasking instructions and reporting criteria for tasks to be executed using prototypes and then to filter the resulting performance measures and focus them against reporting criteria (for example, prompt lists) to generate assessments to feed the further intentions for system development. Figure 11.3 also illustrates some important elements of what a command and control system is or of what an organisation does, for example, industrial process control tasks, running of simulation exercises, or military operations.

Command and Control Centre Functionality

A generic scheme for the cognitive tasks of a command and control system is shown in Fig. 11.4. This diagram was originally devised at a time when second-generation computers were emerging and such automated systems were only just becoming a reality for use on ships and other marine craft. The aim here is to depict the cognitive tasks and concerns of a skipper of a stern trawler, from the most remote to the most immediate concerns. This follows the systems thinking approach advocated by Singleton (1967) and generates the skeleton for the functional analysis required by his methodology. Thus, at the highest level, there are predetermined questions of *policy,* such as the type of fish required to satisfy the market demand and political constraints on where they may be harvested. Once at sea, there are *operational decisions* about which fishing grounds to trawl and when. In making such decisions, the skipper will take past *experience* into account, together with the *current situation* with respect to other vessels in the area or the state of the weather; also the more immediate concerns about how to work the ground (*method of execution*). The current situation includes information about the state of the *mission environment and controlled processes.* This may bear on decisions concerning the method of execution to adopt (for example, whether to continue towing or haul in the nets); it may provide *warnings* of dangers to be taken into account when *supervising* the execution

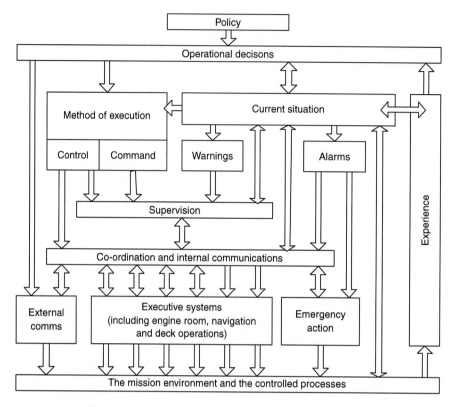

FIG. 11.4.   Information flows at a command and control node.

of the plan of action (for example, low fuel levels) or *alarms* to be acted on immediately (for example, fire alarms) when *co-ordinating* the various personnel and activities to be concerted toward the whole fishing operation. This may require the use of *internal communications*. There may be a separate radio officer responsible for the *external communications equipment* to maintain the umbilical contact with base and the outside world. Other personnel will be responsible for the machinery in the engine room, navigation equipment, watchkeeping, or deck operations.

In short, this depicts a spectrum of cognitive tasks from policy and operational decisions at the top to controlled processes at the bottom, which might require cognitive task design. This general scheme has proved to be useful in many investigations of system integration required to support the human–system interaction (HSI).

THE DESIGNER'S TASK

A command and control system exists within an environment which is defined by the influence it can have: some aspects are of known effect, some of unknown effect, and others are targets to be effected or protected. This environment provides direction (wanted aspects) and impediments to be counteracted (unwanted aspects). If an influence cannot be traced through to its effect on the system (whether long or short term), there is nothing that can be predictably done for it, so it might as well be ignored because there is no way to know whether a design will improve or degrade it. That leaves only the known and predictable influences to be considered when designing how the system will achieve its goals.

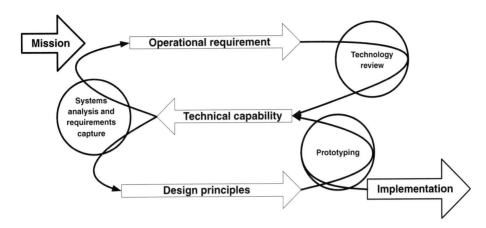

FIG. 11.5. From mission to implementation.

It is an abstract world with which the cognitive task designer has to contend. One way to understand what the designer has to juggle is to recognise the triangle of two-way-street relationships between mission (for example, steel production), operator (for example, steelmaker), and system functionality (for example, electric arc furnace) that affect the required command and control system. With respect to the impact of the operator on the system performance, there are four interactions to take into account:

- The impact of human performance on the mission
- Impact of the mission on the operator
- Impact of the operator on system functionality
- Impact of functionality on operator performance

Typically there are two parties involved: those who make the equipment and those who use the equipment. The role of the designer is to act as a go-between. The equipment makers ask to be told what is wanted because they don't need to know every facet of use for their equipment; likewise, the users won't know what they want without some awareness of the constraints of what is possible. Hence, the designer operates in this world that links the two—moving from the abstract towards implementable solutions. This can be translated into the following broad questions about any proposed system:

- Does each operator role have all the functional support that is required?
- Is that support as good as, if not better than, what already exists?
- Are all the requirements of the job supported?

Figure 11.5 illustrates this context of analysis to generate an operational requirement against which to assess technological capability and then evaluate it on prototype implementations and models to generate design principles that may be applied to generate effective implementations.

Elements of Analysis

Demonstrators allow this approach to be carried into a role-play exercise, in which people can represent people, with all the subtleties of human nature that cannot be subjected to mathematical modelling. This usually requires the specification of scenarios to draw out what affects whether the personnel can fulfil their roles and execute their tasks and then to identify the performance influences, key assessment parameters, and performance measures. These

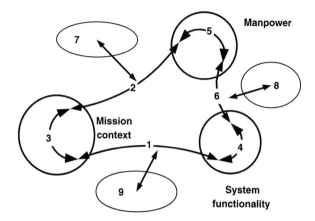

FIG. 11.6.   Analysis components. See text for an explanation of the nine steps. 1 = mission context; 2 = personnel contribution; 3 = scenarios; 4 = personnel involvement; 5 = operating constraints; 6 = performance influences; 7 = performance measures; 8 = the analysis; 9 = implications.

requirements are then translated into trials instructions (for everybody who may be involved in the simulation exercise), together with the criteria against which comments are sought and possibly a user's guide to describe what the system is supposed to do.

The central issue here is the support required for the cognitive task, whether it is the selection of a particular control setting in a nuclear power station or the decision concerning what weapon to select. Figure 11.6 depicts the three main components of this analysis: the mission, the equipment (system functionality), and the human interests that affect the cognitive task design. These components interact, so it is helpful to have some systematic approach to avoid causing problems as fast as others are solved. The following nine steps, keyed to Fig. 11.6, have proved useful for such purposes, whether using operational analysis or simulation runs to investigate the cognitive tasks.

1. *Mission context*: The first step is to consider what mission purpose is to be served by the functions that, when implemented, will support the execution of the cognitive tasks to that end.

2. *Personnel contribution*: Consider the mission and process in terms of what the personnel contribute to the organisation. This should identify the requirement for cognitive task design with respect to training, safety, or performance issues.

3. *Scenarios*: These represent the military or commercial operational context for the cognitive tasks. The analysis here should focus the requirements for a demonstrator, for scenarios to drive it, and for the assessment and information-gathering procedures to be conducted. This analysis will identify what questions to ask to assess the quality of the support provided for the cognitive task; from this, in turn, comes the requirement and specification for what the scenario brief and the synthetic environment should provide so that the trials operators will be enabled to focus on the cognitive task and how the system may be developed.

4. *Personnel involvement*: This step requires job descriptions and cognitive task analyses that define how people and technology should interact, that is, what is their cognitive tasking?

5. *Operating constraints*: This covers the particular procedures that must be followed— for example, standard operating procedures or legal constraints. The converse of this is the personnel specification and the target audience description required to characterise personnel performance and to identify what training might be needed to help them fulfil the roles defined by the personnel contribution (2) and maintain the level of involvement (4) demanded by performance measures (7). This step must also take into account usability measures (8) and implications for further development resulting from the impact of personnel executing

cognitive tasks according to the requirements of the mission using the system available to them (9).

6. *Performance influences*: What is required here is to generate the functional description for how the system design contributes to the mission purposes defined in Point 1 and by the personnel involvement at defined in Point 4. This covers all implementation-driven performance factors as well as the psychological factors that may predispose toward good or bad performance that are open to design.

7. *Performance measures*: This analysis reviews the cognitive task for the consequences of potential errors or other shortfalls with respect to defined performance criteria.

8. *The analysis*: The analysis covers the particular performance calculations and measures made, calculated, or estimated. This may employ modelling or synthetic environments and simulations.

9. *Implications*: The final step is to bring the results of the first eight steps together and determine whether the mission purpose identified at the outset can be achieved with the system design in question and what impediments should be redressed. This step includes the process of rating the prototype for how well the cognitive task is supported and collecting comments to support further evolution of the design.

Cognitive Task Definition

There is a fair degree of semantic confusion in the definition of a number of useful terms relating to human–system integration: mission, function, job, duty, task, subtask, task element, workspace. A working definition of these terms emerges in the following discussion.

Whether allocated to human performance or to be implemented by the system, a task is the requirement to achieve a function in a specific context in accordance with a specific procedure. The *objective,* which represents what is to be achieved, and the *context,* which defines the events to be faced to achieve the objective, are part of the mission context. The *procedure* is that part of a task that defines the means used to attain the objective.

During system analysis and design, the human involvement is resolved in two main steps. First, tasks requiring human participation are identified for each function, and the information exchange with the operator is defined. Second, the possible operator involvement is identified for each function, reserving judgement when either human or machine implementation could perform the operation.

"System" concerns may be separated to identify the independent concerns of the technical and mission context (see Fig. 11.7) leaving an overlap of remaining system concerns, the

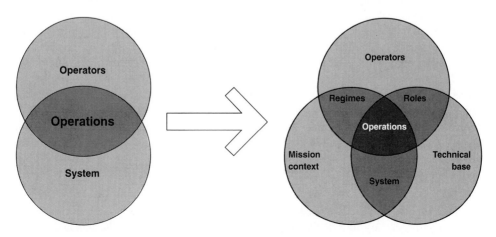

FIG. 11.7. Splitting the system into mission and technical components.

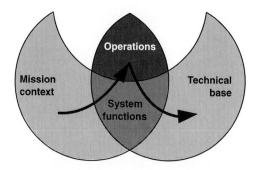

FIG. 11.8.   Normal top-down system development process.

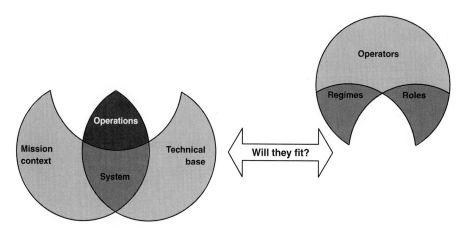

FIG. 11.9.   How well will they fit?

operators associated with objectives (the regimes defined by the mission and the roles occupied by the personnel), and the technical context with data and procedures.

The problem at the outset is one of the degrees of automation, and this boils down to how technology should be introduced into the organisation. The usual top-down process then runs from the mission context to define the system then the operations, then back through the system functions to determine a specification of the technical base components, which can then be implemented (Fig. 11.8). This leaves out an important part of the interaction—and an important question to answer: Will the solution work when the people are brought in (see Fig. 11.9)?

There are six ways to look at the central core of operations that are common to all three aspects (mission, technical, or human), depending on which aspect provides the components, which provides the hypersystem context, and which provides the implied frame of reference. From these six possibilities, there are only four realistic options because the mission context is a hypersystem almost by definition. The four possibilities represent what might be termed an *organisation* view with reference to the technology (Fig. 11.10), a "cage" (human components or slaves) or an "HCI" view (human hypersystem or direction) with reference to the require-ments of the mission (Fig. 11.11), or an "HSI" view of the system of operations with respect to the cognitive task of the operators (Fig. 11.12).

Bottom-Up Evolution and Top-Down Specification.   It is tempting to start with the technical base that is to be introduced and progressively evolve into the systems and operations of the organisation, testing out the regimes of the mission context to be handled (i.e., to

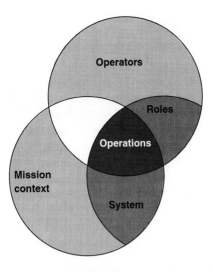

FIG. 11.10.   Impact of the technical base (organization).

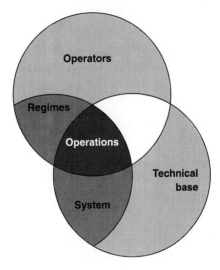

FIG. 11.11.   Impact of the mission context (HCI/cage).

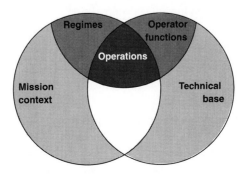

FIG. 11.12.   Impact of the operator (HSI).

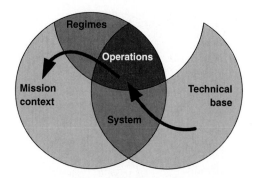

FIG. 11.13.   Bottom-up evolution.

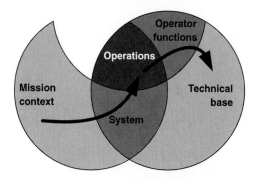

FIG. 11.14.   Top-down specification.

demonstrate the technical options to potential users in a reasonable plausible synthetic environment; Fig. 11.13). The results of this approach would need to be checked for compatibility with the overlap of human contribution and the technical component. A more complete approach is to filter the operations through the capability of the operators to use the technology, thus running from the mission context through the system functions, the operations, and the operator functions before coming up with a specification of the technical base components (Fig. 11.14).

The point here is that there is no specific analysis of the human contribution; people are used to represent people, and the system should be taken care of after the required operations have been defined with respect to the human contribution that will be required.

What is required is a task analysis of the operator functions and task sequences. This provides a reference framework against which to assess the prototypes, to select "trials operators," and to ensure relevance and validity of the performance assessments to the progress of the system design.

THE COGNITIVE TASKING COMPONENT

There is one other split to make, and that is to divide the operator component into the organisation of cognitive tasks or duties defined by operator roles and the people who will fulfil those roles (Fig. 11.15).

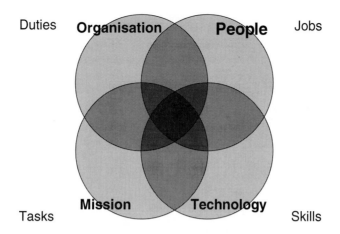

FIG. 11.15.   The organisational component.

The organisation provides the co-ordination, direction, and control that will enable people to use the technology to achieve the mission purposes.

Situation Awareness

The issue for cognitive task analysis is to identify any impediments and assess the requirements to remove them and produce the ideal interface between people and system. This tends towards a scheme that maps onto the following four levels of concern for the command and control processes:

- The context super system (between jobs and duties)
- The coordinating command system (between duties and tasks)
- The controlling subsystem (between tasks and functions)
- The controlled components (between functions and technology)

Ultimately the question "why?" will identify the mission context and, insofar as this affects the operation of the system, this is represented by people. Ultimately the question "how?" will identify the technology context and the equipment that is to be used in the system. Between these are the ambivalent functions that could be fulfilled by either people or machines. This middle area represents their tasks for the equipment.

One way for cognitive task design to addresses each of these components (as identified in Fig. 11.15) is through the following four phases, as illustrated in Fig. 11.16:

- Requirements capture, with respect to the cognitive tasks of the organisation (duties)
- Identify options, with respect to operator roles (jobs)
- Prototyping, with respect to the technology (skills)
- Assessment, with respect to the mission requirements (tasks)

Usually, for a cognitive task design project, there will be a history of previous studies and aspirations to be taken into account. This will bear on factors that may have some influence on the performance of the cognitive tasks of the system in question.

Requirements capture is based on previous systems and aspirations for improvements. The methodology described in this chapter supports the review of legacy problems and subsequent development of the case for the improvements, in terms of control room layouts, workspace designs, and HCI definition. The options for further investigation are derived from these, and

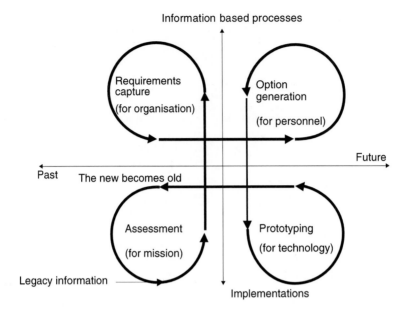

FIG. 11.16.   Role analysis or the simulation project cycle.

prototype implementations may be demonstrated and assessed. This operational assessment then contributes to the assurance that all possibilities are covered, in order to re-orientate the requirements around the future possibilities rather than simply correcting the defects of previous implementations.

Requirements Capture for the Cognitive Task

Cognitive tasks are required for any process that is not entirely automated, and, for any cognitive task design exercise, it should be a sound aim to maintain performance or training of the operators. The following six steps, shown by the arrows in Fig. 11.17, provide a way to analyse such tasks for the options for implementing the necessary support:

- A: The first step is observation for familiarisation. This should identify cognitive tasks and supporting functions from a functional analysis.
- B: The second step is to identify criteria for where to locate such tasks in the organisation with respect to issues such as the mission requirements, manpower and support equipment, command structures, or operational effectiveness. This step requires a task synthesis to derive what is required to be done by the system and how it must be done. Synthesising the task does get closer to why the task needs to happen. It enables prediction for novel situations to be made and consequently challenges the constraints. The next logical step is to justify them and the reasons for the predicted outcomes or perhaps to trace back from the derived outcomes to those areas where redress is possible. The converse of this is to determine what effect any particular change or combination of changes will produce, both in terms of the effect on task performance capability and on the further redress required. In other words, there is a trade-off to determine the most effective mix of the means of achieving the required output.
- C: The third step is to look at task loading resulting from alternative implementations in terms of acceptable workload and then to consider how to redress what is unacceptable and reiterate the assessment with the revised implementation. This step derives preferred locations for the cognitive tasks.

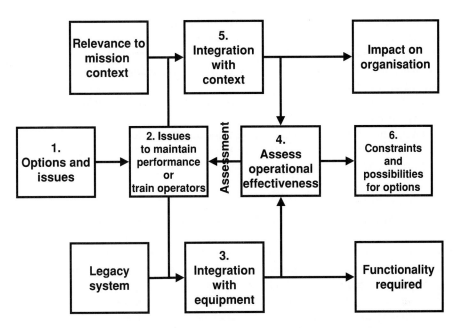

FIG. 11.17.   System assessment and decision analysis algorithm.

- D: The fourth step is to identify the necessary supporting functions, or to redress overload.
- E: This fifth step is to check the viability of the options, amend the task synthesis, and reevaluate task-loading implications.
- F: The sixth step is to review the viable options.

The general approach is for the technical integration issues with respect to the mission requirements to come first, followed by the human factors issues of integrating the proposed solution with the context. These factors combine to allow the operational effectiveness of the proposed options to be assessed against the mission context. This cycle is repeated to address the performance and training issues arising from the need to integrate both technical and human components. The outcome will be at the mission level for operational impact, at the operator level for constraints and possibilities for the interface between personnel and system, and at the technical level for functional requirement and definition.

The usual way to make progress in this process of system evolution includes the collection of comments from subject matter experts, focusing on deriving the following information:

- Required capability—what degree of benefit is required?
- User and system requirements—these are the top level and bottom level requirements between which lie the midlevel objectives that require attention for people to interact with the system.
- Interface requirements—these are the particular issues to cater for the man in the loop in order to support the functionality identified.

Questions to ask. The preceding requirements capture process centres on six key questions:

- What is the issue or option? This should define the issue or option and define the status quo.

- Why is it relevant? How does it contribute to the system? This should specify what the requirements are and what are the cost and performance trade-off issues.
- What are the issues and deficiencies in the present system? An overview should be expanded to describe the specific issues of who is affected, where they are affected, and when they are affected.
- What does the system need to achieve in these areas in terms of performance, maintainability, operability, and trainability?
- What functionality should the system provide to support the cognitive tasks?
- What are the constraints on providing a solution, and what is available? This should identify conflicting requirements and mandated requirements.

PROTOTYPING, DEMONSTRATORS, AND SIMULATORS

During the requirements capture and analysis phases, the question for the cognitive task designer is how to harness the background and experience of subject matter experts. One way is for them to act as "trials operators" who, being familiar with the realities of the application, can determine or offer an opinion on whether the concept under investigation will be adequate for the projected conditions of use. The primary concern may be an interactive display suite, for which a simulator can provide enough functionality to illustrate possibilities and to allow essential processes to be executed in a setting that allows pertinent discussion and comment. Such a simulator is referred to as a demonstrator.

In this respect, the frame of reference for the cognitive tasks that links the people, equipment, and mission is the organisation. Hence, the assessment focuses on how well the organisation or equipment allow the people to contribute to their execution of the mission requirements. Usually, individual assessments are collected concerning the prototype equipment with comments to clarify such deficiencies as are found and to indicate what improvements could be made.

Cognitive Task Analysis

A task analysis is, in the first place, simply a statement of the functional requirement for people in a system. This can then be amplified to derive the functions to support them in those purposes.

The manifold reasons for (cognitive) task analysis range from the analyst's need to gain familiarisation, through providing a baseline before organisational restructuring (reallocating authority and responsibility), for resolving questions of the allocation of function (with respect to the level of automation to be used), to generating models that will allow operational performance to be predicted. Performance prediction is required for the following four purposes with respect to the organisation, people, technology and mission (see Figs. 11.15, 11.16 and 11.17).

- Task analysis is vital to functional requirements capture for system design with respect to information and equipment usage, and supports scrutability for design decisions. The focus is on the capture of requirements in relation to the duties of the organisation and its relevance to the mission context.
- Task analysis is vital for identifying the possibilities for including personnel in a system design, particularly when employing techniques such as human reliability assessment to support safety case arguments. The focus is on the generation of options in relation to the jobs of the personnel and their impact on the organisation.

- Task analysis is vital in determining both the requirements of the organisation for teamwork and the demands of the system for technical skills in order to analyse the training needs for the personnel that the system will employ. The focus is on the trial by prototyping the technology in relation to the skills employed and the functionality required for their support.
- Operability assessments, performance analyses, and the derivation of system maintenance requirements also require task analysis. The focus is on the assessment of the system performance of mission tasks given the context of the unchanged context of legacy systems.

Issues that may be relevant include decision making under stress (for example, Flin, Salas, Strub, & Martin, 1997), teamwork and culture (for example, Helmreich and Merritt, 1998), or the modelling and assessment of performance capability and role interactions, which is the focus of this chapter.

System Design as a Decision Process

The design process is naturally iterative, if only because at the outset the user does not know what is technically feasible, nor does the technologist know what users might require if they knew what could be provided. The aim is to end up with the users' being in command of such technology as can be put at their disposal, so that it will function as required and inform as necessary about what it is doing. Although workload is the driving factor in the end, this is caused by taskload, which has direct correspondence with system design and operation parameters. For cognitive tasks, the issue to consider is which decisions can be grouped and given to a single individual, and which might demand separate manning or even separate locations: do the allocations of decision-making duties generate part-tasking or overload conditions? The "route map" for achieving this, shown in Fig. 11.18, basically shows how the design process iterates between the user requirements and specifying subsystem components in three passes. This involves five basic phases, which we can describe as follows:

- capture of primary functional requirements
- derivation of constraints and secondary requirements from models of the possibilities
- review of feasible options
- review options against criteria of acceptability
- derive functional definition for the system and specify the design.

Primary Requirements Capture.   This is concerned with answering the question *"why?"* The draft system requirement and criteria is determined at Point 1 in Fig. 11.18. The functions required of the components of the system are determined at Point 2.

Derivation of Constraints and Secondary Requirements for the System.   This includes model implications derived at Point 3 by whatever models or simulations are appropriate to enable discrimination between alternative options for final implementation.

Feasibility Checking for the System.   The phase involves determining whether the system could work (Point 4) in terms of whether the components will be compatible. The check on system implications at (Point 5) concerns whether their output as a subsystem is compatible with what the super-system requires.

Criteria for Acceptability.   Basically, these criteria concern the implications of the system for the host organisation. Having been satisfied that the super-system requirements can be met

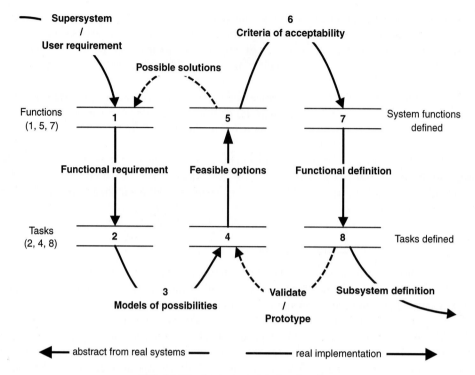

FIG. 11.18.   Analysis and design iteration.

by the proposed system design, this is confirmed at Point 6 so that the super-system can be reconfigured to receive the new system and its subsystems. The system functions are defined at Point 7.

Specify Design.   This really bottoms out *how* the system will be implemented, iterating from Point 3 with model implications, determining feasibility, and so forth to refine the possibilities for components. When these are reduced to one option at Point 8, this defines the requirement for the design at the next level of detail.

FUNCTIONAL ANALYSIS OF AN ORGANISATION

An organisation functions through role interactions. There are system interactions that relate to the functional descriptions, and there are interactions between roles which may be summarised as a role interaction matrix for an organisation.

With respect to how well a system is integrated or an organisation is supported, two key questions to ask of any such interaction are the following: Why is it included? and How does it contribute? The answers inevitably fall within a limited number of "principal system functions," of which there are seven required for the organisation to operate at any time between determining its purpose and its final disposal (Steinhausen, Orton, & Smalley, 1978):

- Command and communications (for example, radio)
- Prime task integration (for example, guns, missile launchers)
- Maneuvering and transportation (for example, tractors and trailers)
- Environmental defence (for example, weather protection)

- Common support (for example, power supply)
- Life support and habitability (for example, clean air)
- System monitoring, maintenance, and repair (for example, food, sleep)

Any command and control (C2) system must integrate with these principal system functions in terms of both sources of command and items for control. They are what are known as "imbedding dimensions"—they run right through any command and control system from top to bottom and apply at any level. They are both mission related and system related because the system that is implemented (and the HCIs that provide for operator interaction) must be justified by its mission purpose—there must be a reason it is there. This provides the basis for assessing the significance of the decision making and, consequently, of the functions of the interface that support them. Thus, these "system functions" provide a basis on which to characterise or analyse the context, or "total system" requirements, for the decisions, both in terms of their operational environment and of the systems they control.

These principal system functions provide the top-level headings for all possibilities and constraints for implementation and also for identifying all the functional purposes of an integrated system. This means that we can analyse the principal system functions according to how people, mission drivers, and technology enable an organisation to achieve its purposes and what they have to do in terms of jobs, tasks, and functions.

By continually asking "how?" designers arrive at answers that define the support that these jobs, tasks, and functions will require and, for cognitive tasks, define the requirement against which to assess the capability of the design. Consequently, the first step in cognitive task design is to review the requirements for change—asking the question "how?" to determine the implementation that will enable its execution or asking "why?" to determine the greater purpose behind it and eventually define the roles in terms of jobs to be complemented.

The question "how" focuses on the implementation required to execute the mission tasks, which usually generate a hierarchical structure that maps to the nature of equipment assemblies, for example, desk, office, building. They also map the nature of the environmental influences around a workspace, and they define a context within which cognitive capabilities must function.

This then focuses the issue on the physical domain of the mission (why) and the equipment (how) that is tractable to measurement and design. Then one can ask "how well?"

Measurement or Synthesis

One way to derive a functional description of a task is to observe and record what actually happens; another way is to synthesise what should or could happen under different circumstances. Observation and recording of what actually happens leaves no doubt about the reality of the events. What does happen, what should happen, and what could happen are different, however. Even when this is possible for a cognitive task, simply observing and recording what actually happens still provokes the questions of why things happen that way, what constraints there might be, and what could be done better or differently.

This is the usual requirement, for the new system to be as good as, if not better than, the existing or some other reference system. Ultimately, the aim is to be assured that equipment is operable, that training is adequate, and that the system is safe. How should this assurance be derived? Whether by absolute measurement on actual systems or a synthetic assessment approach, the essential points of difference are as follows:

- The absolute measurement approach looks for benchmark tests on commonly used interactions with the system. The synthetic assessment approach looks for baseline assessments against an existing system and concentrates on critical interactions.

- In terms of criteria for the interactions, the dichotomy is between absolute measurement (of speed and accuracy of interaction) and the analytical assessments of the synthetic approach in the first instance.
- In terms of criteria for performance with respect to scenarios, the contrast is between measuring the number of errors made and identifying the possibilities for errors, between measuring mission success and identifying potential impediments to success.

Either measurement or synthesis may use subjective rating. However, both also require some form of task analysis, whether implied or explicit, to structure and focus the assessment.

Operator Roles

To integrate the three essential components—technology, people, and mission—characterised by the cognitive tasks, operator roles, and scenario contexts on which the analysis is based, one can start with the functional analysis of the organisation. This should bring out the need to support human intervention, focused by the concept of operator roles. This allows the issue to turn on the need to support the operator roles, which can be more objectively and explicitly defined than personnel can be. To do this one needs a clear definition of a role.

The notion of a role, although intuitively obvious, requires careful definition to make a robust concept for assessment alongside those of functions and tasks. The dictionary definition of a role would say it is an actor's part; one's function, what one is appointed or expected or has undertaken to do. Thus, a role defines purpose and objectives outside the system, what the individual has to be or represent to do the job. It is an implied entity tied up with image and responsibility. It is concerned with the end rather than the means. There are various ways to look at this.

- The mission view of a role defines it in terms of the responsibility for achieving a high level function; it has the connotation of tasking.
- The organisation view uses a role to define purposes and objectives outside the system and is concerned with the end rather than the means; it has the connotation of duties.
- The human-centred view is that a role reflects what the individual has to do, be, or represent to do the job.
- The technical view is that each role identifies a set of activities that require skilled operation.

The Functional Analysis of a Role.   So what does a role do? A function is an output. Roles are supported by various types of functions: jobs, tasks, duties, or skills—all functions of different types of system components: people, missions, organisations, technology. So one can have a function of a person, of a machine, of an organisation or of a mission. When considering how these functions are achieved, one can say the following:

- Jobs are a consequence of human functions. They are shaped by organisation, mission, and technical factors. People occupy jobs that are defined by job specifications.
- Skills are supported or met by mechanical functions. They are shaped by mission, organisation, and people factors. The technical components of the whole are there to fulfil tasks; where human skill is involved, this requirement is covered by a person specification.
- Tasks are a consequence of mission functions, with performance shaped by organisational, human, and technical factors. The duties of missions require human contributions defined by job descriptions.

- Duties are a consequence of the organisation, with performance-shaping factors derived from mission, technical, and human factors. Operator roles describe the requirements for people in an organisation; their capabilities to fulfil those roles and consequent requirements are defined by target audience descriptions, discussed in the next section.

Job Analysis.   When decomposed, a job takes the form of a hierarchy subdivided first into a number of duties, each of which is made up of a number of tasks. These tasks comprise a number of skills, which in turn depend on a number of items of knowledge. The skills can be broken down further into subskills and skill elements.

Strictly speaking, Job analysis is the resolution of a job into its component duties, whereas task analysis is the resolution of a task or series of tasks into component skills and supporting items of knowledge. Hence, in broad terms, a job analysis is concerned with purpose, the task analysis is concerned with the means of execution.

A task analysis identifies just how the operators will perform these jobs and duties and identifies their constituent tasks. The tasks should map directly to implementable functions that may then be analysed for changes in demand, conflict, and safety criticality.

THE POSSIBILITY SPACE FOR AN ORGANISATION

Recognising that predicting actual human outputs is not realistic, if only because every individual is different and even then does not always produce consistent performance under apparently similar circumstances, the issue should be whether a particular performance standard can be achieved within the constraints of the task. The analysis should then focus on the time and space within which successful performance must be achieved and the probability of success that can be expected within that space—in fact, the probability that the organisation will achieve success.

This possibility space, within which an organisation may operate, is defined by the purpose and constraints of its people, technology, and mission context with three levels of interest characterised by the technologists and the hardware, the operator and the console or workstation, and the user and the operational context of the control room. Figure 11.19 should put this in perspective. How all three interests can be accommodated to best advantage is reflected in the corresponding types of system analyses that may be required:

- Mission analysis—to review changes from past practice
- Function analysis—to review changes from past implementations
- Manpower and task analysis—to review for any changes in roles, tasks, and task allocations, and to identify critical tasks

A mission is defined as what the system is supposed to accomplish in response to the stated operational requirement. Mission definitions provide the baseline against which the adequacy of the system may be judged. This governs operating limits and occasions. The mission analysis in Fig. 11.19 is concerned with how the system or organisation will be used and the doctrine for conducting the mission, which shapes the purpose of the system and the geographical, functional, and departmental constraints imposed on the cognitive tasks and decision making.

There is inevitably a legacy system component that contributes to defining the "space" available for the proposed system. This, combined with the input from the operational context, provides the basis for a task analysis, which is subsequently focused by the cognitive tasking requirements. It is then further refined, through prototyping, by the realities of what can be implemented, using a suitable mix of stimulation (real equipment), emulation (for real

Baseline   Assessments   Where outputs apply

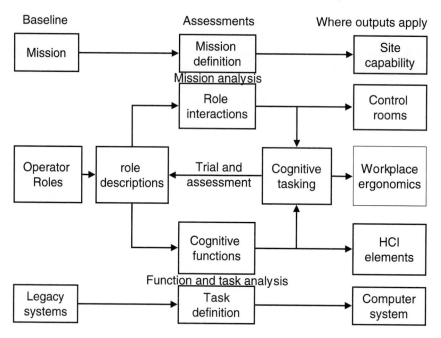

FIG. 11.19.   Task analysis in perspective.

software), or simulation (for real operators) to represent the equipment, software, or people who might be employed in the target application.

Operator Characterisation

The question of how to characterise the operator is essentially how to convert a task definition into a performance prediction through either a target audience description (TAD) or a personnel specification. A TAD is a description of the characteristics of the personnel for whom the system design is intended (i.e., the resource pool from whence the operational staff will be drawn). A personnel specification is a statement of the abilities and personal characteristics that a person should possess to enable them to perform the job or fulfil a role. This suggests a notional equation such as the following:

$$Personnel\ Specification = TAD + Training - Job\ Aids$$

This is broad brush, and requiring development to determine quite how the personnel specification may be compared to a role definition and how the TAD, training, and job aid components actually combine. In this respect, the baseline TAD for U.K. military personnel is embodied in Def-Stan 00-25; for civilians it is embodied in ISO 9241. These are not prescriptive standards but provide a characterisation of human contributions to system functioning so that the impact may be assessed. They allow the questions why and how to be answered in terms of how the personnel might affect the issues. They also provide some conditional prescriptive requirements for system designs.

Target Audience Description

System designers require information about roles, skills, and capabilities because this affects the system design and development. A TAD defines, with respect to the operator roles required

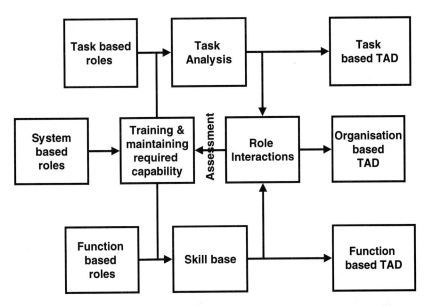

FIG. 11.20. Evolution of a target audience description (TAD).

for the mission purpose, the discriminant qualities of the personnel, both in terms of the demands of the roles and in terms of their impact on system design. It meets several purposes, including validation of initial system concepts, recruitment and career planning, or training needs analysis.

The TAD is also a basis for discussion to clarify operator roles and characteristics that could affect system implementations. It should be obvious that the organisational structure behind the cognitive tasks devised to execute the functions of the system must not conflict with the rank structure of those employed in it. Within this structure, however, each role must be adequately supported in all respects by providing suitably qualified personnel with adequate facilities and training. The TAD may be used to identify possible implementations and should provide a useful input to their assessment.

Figure 11.20, an adaptation of Fig. 11.17, illustrates the components that contribute to a TAD and how they interact and evolve.

Cognitive Tasks and the TAD. The decision-making issues of cognitive tasks should be reflected in the TAD. This should define user skills, knowledge, and abilities and identify the possible users in the system. This provides a baseline for performance modelling to support, for example, safety case arguments; alternatively, to identify the reasons for any performance deficits to identify the necessary support or training needs.

In the first instance, during the conceptual stages of a project, a TAD has a dual purpose. The first is to provide an historical perspective of current roles, skills, tasks, and capabilities of legacy system users and operators. This tells system designers what range of users could be involved in the future system and what their capabilities might be. Some of the extant roles, skills, and so forth may be still valid, particularly with respect to the integration of such legacy systems. These will be augmented by the new roles, skills, and capabilities required as the system evolves. The second initial purpose is to support the preliminary training needs analysis to assist designers in their analysis and definition of the alternative functions, tasks, and roles for which training facilities might be required.

In short, the TAD is a focus to assess the capability and capacity of an operator to carry out a potential task efficiently; as such, it should provide a crucial input to considering whether a task or process should be automated or not.

At the outset, the baseline TAD may be refined by information based on the career paths open to the operator role in question and defining the recruitment, selection, and training criteria applied, hence indicating the capabilities that personnel may be expected to bring to the performance of their tasks. As the system design progresses, the TAD should be updated to support career planning and recruitment of new operators and thus becomes a source of information for subsequent project phases. Three such strands of evolution for a TAD, each with its own particular purpose, are as follows (see also Fig. 11.20): task-based TAD (mission), organisation-based TAD (cognitive requirements), and function-based TAD (equipment based).

Task-Based TAD.   This reflects the command and control structure and the communications infrastructure that affects the tasks to be performed and hence the roles required.

Organisation-Based TAD.   This focuses on the organisational structure, pursuing, as far as the established task analysis and skill base will allow, the extant pool of personnel (by system-based roles) from whom the future system may draw. This would describe the operator positions in terms of their functions, roles, and tasking within the organisation. Subsequently, the relevant role interactions may be assessed, taking skills and tasking into account to determine the needs for training and support to maintain desired performance.

Function-Based TAD.   This may be termed a generic user TAD. This is based on the functions that any implementation might require of its operator roles and may be developed as far as the established functionality and required skill base might allow. It is system driven.

COMMAND AND CONTROL MODES

There are several ways in which judgement is exercised in command and control. For system design purposes, two complementary aspects to the human involvement in the C2 processes are the exercise of judgement and the implementation of process flows. These focus the following useful structure for the analysis of the system functions performed by human intervention.

There is a parallel with the principal system functions in terms of the principal reasons for including human operation within a system design. These include command and planning, communications (information exchange and status reporting), "navigating" and "piloting," situation awareness (SA) and tactical decision making, system operation, system monitoring, and operational co-ordination.

All of these items are relevant to the trawler skipper mentioned at the outset, as he decides where and how to operate his vessel. They all require the exercise of judgement, which requires its decisions to be supported in execution. These different thinking methods or modes of interpreting and using the information are defined as command and control functional support modes or just "C2 modes." Basically, these are the arrows in Fig. 11.21 between the clouds of so-called operational support functions. There are 10 of these modes to consider:

1. *Primary situation awareness* is about answering *why* is this system here and doing what it is doing. It is about collecting all the information, which focuses on this answer.

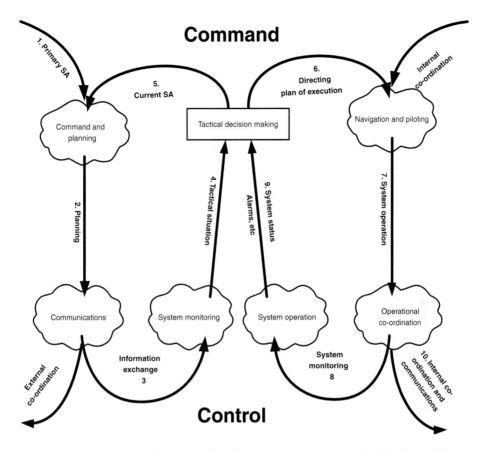

FIG. 11.21.   The command and control functional support modes for tactical decision making.

2. *Planning* is about taking in the current state of the mission field and determining what aims to strive for. This combines the "why" of primary situation awareness with the "where" of current situation awareness as the basis for deciding what the targets should be and how to get there.

3. *Information exchange* is about *who* the other players are, what they might be doing, and what their intentions might be.

4. *Tactical situation status reporting* is about the status of the mission context and players.

5. *Current situation awareness* is about maintaining awareness of the immediate state of the mission field, locating where specifically items are. This information merges with the primary situation awareness to drive the planning process of Mode 2, which drives the information exchanges with other players and outputs the status plotting and reporting of Mode 4.

6. *Directing plan of execution* is about *when* events are to happen, cued by the status of the mission field and conditioned by the state of the available technology.

7. *System operation* is about the hands-on equipment operation in response to directions from Mode 6. This is where specific protocols, sequences of operations, and the dynamics of control or perception become relevant.

8. *System monitoring* is about maintaining general awareness of system performance and capability, what reserves are left, and approaches to decision points, danger areas, and so forth. This concerns what is being used for the programme of events.

9. *System status (alarms, alerts, and warnings)* are system-status information not normally in the span of attention that will affect the process of directing the plan of execution.

10. *Internal co-ordination and communications* is about liasing with other systems under control for reconfiguring the system, reloading new software, stage changes, and so forth— basically configuring how achievement is achieved.

Figure 11.21 should clarify what the C2 modes are. The left half of the diagram is concerned with the driving influences of the outside world; the right half is concerned with the system under control. The flow lines represent the thinking processes or modes of using the information for command and control purposes, running from primary situation awareness through to internal system coordination. The upper half is concerned with command and direction; the lower half is concerned with the situation as it is found.

Functional Support Requirements

These command and control modes are generic. They provide a basis of duties to determine what functions are required by a role that may then be translated into human–system and human–computer interaction specifications. For each such function, the focus is then on the following questions:

1. Is an operator needed (and why), or can the process be automated?
2. Where in the system can, should, or must the operator interaction take place and why?
3. What data are needed before the operator can carry out the task? where do these data come from and in what format will, could, or should they be presented?
4. How does the operator complete an action?
5. Does an operator have to enter data or an action into the system? If so, what sort of data, and how should it be done?

Here Points 1 and 2 relate to manpower support, Point 3 to situation awareness, and Points 4 and 5 to system issues.

Assessment of Capability

The questions to consider now are how significant are the constraints that the design might place on the execution of the cognitive task, and what is at risk with respect to completing the immediate task, and what are the further consequences for refining the design.

For design by exception, focusing on the differences from a known task or system, the analysis of a role may focus on changes to the arrival and departure route whereby personnel enter and leave their roles (for example, changes in recruitment, selection, training, and career progression) and changes to the task or system from a known baseline task or system that may affect their performance while occupying their roles.

Command and control systems may be assessed from two complementary points of view: health and safety of crewmembers and operational performance. This split between "performance with respect to health and safety" compared with "health and safety with respect to performance" is important. It underlies the difference between the cost in terms of Health and Safety for the performance, which the system will deliver, compared with the cost of integrating these factors with the design intent to achieve the required system performance.

This is where a rating scheme helps with assessing the goodness of fit, both in terms of "operability" of the equipment with respect to the role of the operator and "utility" for supporting the regimes of operator involvement in the mission. It provides a means of prioritising and negotiating design requirements through the system development process.

ASSESSMENT CRITERIA

Hierarchies are a natural consequence of the necessity to avoid crossed authority and responsibility—particularly in a military structure where such ambiguities are intolerable. Hence, hierarchies have as much to do with preventing conflict as they do with ensuring concerted effort toward the higher purpose for which any organisation exists, whether commercial or military.

The following factors are obvious in a physical context but have equivalent consequences for cognitive tasks, which should not be ignored:

- The environment should be suitable to the task in hand.
- The operator should be able to perceive the relevant context of the situation (to enable situation awareness).
- The operators should have the means to perform their required tasks and to do so harmoniously. For example, operational components used in concert should be accessible together.
- Avoid interference between tasks (except to provide a deliberate interlock), which may require partitioning of the system.
- Provide flexibility in the system for it to be reconfigured to perform alternative tasks. This may require replication to provide the various facilities where needed.

The first question is to assess the significance or magnitude of the issue. The means adopted may range from a purely subjective assessment through to objective measurements. The rating scheme described here aims to generate structured subject matter expert judgements.

The second question is to evaluate the nature of the issue or breach of constraint—whether legal, performance, or humanitarian—and hence the justification for providing a solution that may be purely technical, may involve both human and machine, or may be a purely human or social. The converse is that these constraints govern a decision and hence define how to judge their operational effectiveness when executed. For a military system, the cultural context and basis on which the decisions are to be made is defined by their concept of operations and doctrine.

The third question is to consider the alternative human and machine solutions. Purely technical solutions range from selecting appropriate machinery, through modification of existing machinery, to better utilisation of existing machinery. "Human and machine" solutions range from redesign and selection of appropriate equipment, through filtering to isolate either machinery or individual, to selection and training. The purely organisational solutions range from work design through personnel training to personnel selection. All are options to be considered and assessed.

Performance Measures

In terms of cognitive tasks, the performance criteria revolve around four factors: comprehension, discrimination, time (speed of execution), and accuracy. In this respect, measures of performance might include the level of situation awareness achieved, the level of confidence in tactical information, the impact of including a human in the loop, the time taken to correct operator errors, the impact on operator workload, the impact on operator reaction times, the functionality coverage, the co-ordination of roles, the number of operators, or the flexibility of manning.

Although most of these concerns can be readily defined, there is still scope for developing metrics for situation awareness and the level of confidence in tactical information; these factors

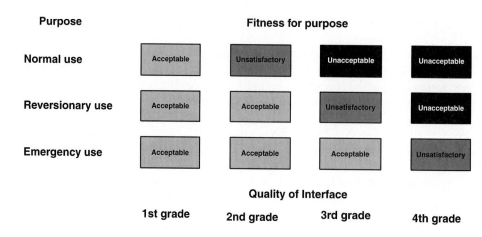

FIG. 11.22.   Rating the Interface.

are identified here as a prompt list of items to consider when designing simulation trials to support the development of a system.

Such measures link system goals to the degree to which the implemented system matches up to these requirements. The most basic assessments are against simple binary requirements that are either met or not. Such simple ratings may suffice at the outset but should be consistent with any refinement, required to provide more precisely quantified performance scores. The key, questions, however, usually are how far from adequate is the extant state of the design?, or how much still needs to be done, especially when the concept is transposed into a real working environment.

When we look at the reasons for errors of interaction with a computer system—whether caused by inappropriate task load, conventions and expectations, interface complexity, displays or by inadequate feedback, error recovery capability, or co-ordination between operators—these reasons focus what is required to mitigate their consequences and hence redress design failures.

Figure 11.22 illustrates a rating system for assessing the operational significance of required functions and the operability of proposed solutions. This is a simple adaptation of the Cooper-Harper rating scale (see Cooper & Harper, 1969). Having made the rating, it is not difficult to pursue the nature of the deficiencies in terms of their failure modes.

Cognitive Task Failure Modes.   In essence, failure mode analysis is about providing a useful set of prompts for possible shortfalls and, when found, to determine how to make good those shortfalls. For mechanical systems, any component can be assessed for its potential failure according to a small number of "failure modes": seizure, fracture, contamination, blockage, and leakage. The consequences of such failures can then be analysed for their impact on the system. The implications for system designers are then to determine what can be done to defend against such failures.

A similar philosophy can be applied to the interfaces associated with cognitive tasks. When analysed for what can be done about them, the issues come down to a small number of so-called human factors integration domains: personnel, human engineering, environmental ergonomics, training, accommodation and habitability, health hazards, and system safety.

Commissioning a Simulation

To conclude this chapter, the following prompt list covers items to consider before a simulation exercise.

- Define the context of the system concept that is being simulated. This provides the focus for why the simulation is being run at all.
- Pick up the particular questions to be resolved by the simulation run.
- Define the alternative ways in which the eventual operator may be confronted with or handle these issues.
- Focus on what the simulator should provide. Cater to those who write the software and provide the hardware.
- Provide the run-time instructions for scenario selection, simulation control, and automated data collection requirements.
- Further briefing material is required for the trial operators, as are aids and materials for collecting comments and assessments.

The actual questions and methods of collecting comments and assessments will depend on circumstances that the material presented in this chapter should help to identify and to develop a strategy for tailoring assessments to the particular requirements of the occasion. The key is to focus the operators' desires through the requirements of the task and to diagnose deficiencies before seizing on to solutions that may be of only local relevance.

CONCLUSION

For those who wish to integrate the design of the interface to a cognitive task, this chapter provides a way to approach the matter by exception and by using people to represent people. This obviates the need for complex mathematical models of the human operator and focuses on the system design as the "object of research," to be defined and specified to the degree necessary for implementation.

At the outset, some concept of what is required is necessary to function as the default design to be modified according to the specific requirements. Ideally, this default should provide the minimal skeleton to be fleshed out rather than a complete solution. The first section of this chapter aimed to provide such a skeleton.

The next question is how to flesh out the skeleton for each particular case. This is *the designer's task* and must take place in the context of the system evolution, integrating and balancing the contending mission and system and human interests. This chapter provided definitions for systems thinking and recognition of several points of view when discussing the design for the cognitive task.

Given such different views and interests, it is clearly necessary to have a clear conception of *the cognitive tasking component* in the system being designed. Some further definitions of the functional components of tasks, jobs, duties, and skills were discussed. I then recognised that, for equipment system designers, the cognitive task is tied up with situation awareness: The means of capturing the requirements for situation awareness are described.

It is one thing to catalog the discontents with existing systems and benefits that alternatives might provide; however, the proof of the pudding is, as they say, in the eating. There is nothing quite so convincing as to the viability and acceptability of a proposed system design as to be able to try it for real. This is *where prototypes and simulations* can provide extremely cost effective *demonstrations,* allowing a design to be tested and brought to maturity before its final implementation. The issues to be addressed, identified by task analysis, can only be progressed through the windows of opportunity offered by the overall system development and procurement process as this chapter has outlined.

The first sections of this chapter dwelt on the context of system development within which the cognitive task designer must work to integrate the cognitive factors into effective system designs. The next sections focused on cognitive task designs with respect to their organisation.

Any *organisation functions* through the interaction between its roles. In this respect, the two questions "why?" and "how?" lead to a limited number of so-called principal system functions under which headings the context and means of execution of the cognitive task may be defined. These provide a skeleton that may be fleshed out according to the operator roles that define why people are required to contribute to the system and organisation and what that contribution should be.

A key point is to recognise that it is not realistic to expect to model and predict exactly how people will behave. However, it is realistic and necessary to determine the boundaries within which they are to perform (for example, widths of roads) and their likelihood of success. This aligns with the well-established engineering practice of designing to tolerances. This concept is developed as *the possibility space for an organisation*. A design may then be developed according to the capability of its components to function within that space. A target audience description describes what may reasonably be expected of the human operator and with what likelihood. In effect, standards such as Def-Stan 00-25 or ISO 9241, provide such information, and this may be used to ensure that all prospective users are accommodated, regardless of the (cognitive) task.

How the people will contribute to the case in question is defined by the "Command and Control modes," which were discussed in the next section. These are generic tasks that must be fulfilled for the organisation and may be expanded according how they will be executed. These become the functional requirements that the (cognitive) design must fulfil, and it is against these requirements that the design may be assessed. Here one may consider whether it is the requirements of the TAD for a particular task or the performance that may be achieved given those requirements that is under inspection. This is where some method of system assessment comes into play, to fine-tune the design on the prototypes and simulators.

With respect to the *assessment criteria* to use, the well-established Cooper–Harper methodology has proved to be an effective tool for identifying design deficiencies, prioritising their significance, and giving a measure of how far a proposed design is from being an adequate solution. It forces the designer to focus on the purpose of the design; it allows the trial operators to express how well the design achieves that purpose for them and what should be done to move it toward acceptable perfection.

ACKNOWLEDGEMENT

Any views expressed are those of the author and do not necessarily represent those of QinetiQ ltd. QinetiQ is the former Defence Evaluation Research Agency of the U.K. MOD. The Airspace Management Systems Department at Malvern provides systems engineering research, project support, and consultancy on all aspects of civil air traffic control and military airspace management, ranging from research through requirements capture, operational assessment, demonstration, and system assessment and including the production of guidelines for the design of human–computer interfaces and their validation through simulation and prototyping. The ideas contained in this chapter have been brought to maturity in this environment of integrating cutting-edge technology with the needs of the user.

REFERENCES

Birmingham, H. P., & Taylor, F. V. (1954). *A design philosophy for man-machine control systems. Proceedings of the Institute of Radio Engineers. 42;* 1748–1758.

Cooper, G. E., & Harper, R. P. (1969). *The use of pilot rating in the evaluation of aircraft handling qualities.* NASA-TN-D-5153. NASA: Ames Research Center, Moffett Field, CA.

Flin, R., Salas, E., Strub, M., & Martin, L. (Eds.). (1997). *Decision making under stress—emerging themes and applications.* Aldershot, England: Ashgate.

Helmreich, R. L., & C. M. Merritt (1998). *Culture at work in aviation and medicine—national, organizational and professional influences.* Aldershot, England: Ashgate.

International Standards Organisation. (1992). *Ergonomic requirements for office work with visual display terminals (VDTs).* ISO 9241. Geneva: Author.

Ivergård, T. (1989). *Handbook of control room design and ergonomics.* London: Taylor and Francis.

Kirwan, B., & Ainsworth, L. (Eds.). (1992). *A guide to task analysis.* London: Taylor and Francis.

Kirwan, B. (1994). *A guide to practical human reliability assessment.* London: Taylor and Francis.

Ministry of Defence. (1984–97). *Human factors for designers of equipment.* Def-Stan 00–25 (parts 1–14). Kentigern House. Glasgow.

Pheasant, S. T. (1986). *Bodyspace* (1st ed.). London: Taylor and Francis.

Shneiderman, B. (1998). *Designing the user interface.* Harlow, England: Addison Wesley Longman.

Singleton, W. T. (1974). *Man machine systems.* Middlesex, England: Penguin Education.

Singleton, W. T. (1967). *The systems prototype and his design problems.* In W. T. Singleton, R. S. Easterby, & D. Whitfield (Eds.), *The Human Operator in Complex Systems* (pp. 20–24). London: Taylor and Francis.

Steinhausen, J. L. P., Orton, J. N., & Smalley, J. P. A. *A structured approach to man/machine interface design for command and control of ships machinery.* Proceedings of the 5th Ship Control Symposium, Annapolis, MD, October–November 1978. Vol 4. Session 01, 1-1 to 1-16.

12

Basic Cognitive Principles Applied to the Design of Advanced Supervisory Systems for Process Control

Bernard Riera
University of Reims, France

Serge Debernard
University of Valenciennes, France

Abstract

Supervision of highly automated processes is an activity in which human beings, despite the increasing automation of the recent years, are still present, and thus the design of efficient supervisory systems requires a different approach to automation integrating cognitive task design, one that is more systemic than analytic. With this approach, cognitive task design can lead to original automation solutions based on the concept of human–machine cooperation. In this chapter, we explain how to support cooperation between a human operator and a technical system in the domain of supervisory control. Cognitive models and three generic forms of cooperation (augmentative, integrative, and debative) are the foundations to develop a common frame of reference supporting human–machine cooperation. We propose two design applications of advanced human-adapted supervisory systems. The first application involves air traffic control, a process with a low level of automation. The second application involves a nuclear fuel reprocessing system, a process with a high level of automation. In both cases, cognitive task design has led to original automation solutions and has been evaluated with experiments performed by experienced operators.

INTRODUCTION

The supervision of highly automated processes is an interdisciplinary research area. Knowledge in the fields of automation, process knowledge, machine engineering, workplace ergonomics, cognitive ergonomics, working psychology, sociology, and so on are necessary to design efficient supervisory systems. This is why supervision is an activity in which human beings, despite the increasing automation of recent years, are still present.

In this chapter, we describe how cognitive task design (CTD) in the domain of human–machine cooperation (HMC) can be used to design advanced human adapted supervisory systems (AHASS).

The first part of the chapter shows that a design approach that takes into account the whole system (human, machines, and interactions) is necessary in the context of supervision. In our approach, CTD is an explicit part of the design activity, rather than something done in an unsystematic manner; it can be seen as a systemic (instead of the usual analytic) approach of automation.

The second part of the chapter explains how to support cooperation between a human operator and a technical system. It starts with definition of cooperation from cognitive psychology and expands on it by simulating the agents' activities as they interact. These primitives are grouped in two main classes: managing interference between the agents' goal and facilitating the agents' goals. To be exhaustive, it has been assumed that the three generic forms defined by Schmidt (1991) augmentative, debative, and integrative, can be combined to describe any cooperative situation. The idea of a common frame of reference has been used and developed to support HMC and particularly debative cooperation, which is often forgotten by designers.

In the last part of the chapter, two versions of these theoretical works about HMC are applied to the design of AHASS. The first application concerns air traffic control, a process with a low level of automation (LOA). The second application deals with a process with a high LOA: a nuclear fuel reprocessing system. In both cases, experiments have been performed with experienced operators. It is shown in this chapter that debative cooperation is more efficient than the integrative cooperation that designers of supervisory systems usually consider.

COGNITIVE TASK DESIGN: A SYSTEMIC APPROACH TO AUTOMATION

Why and when to automate? The technical centered approach assumes that automation has to be performed when it is possible. The humanist approach is to automate when the task is boring, physically risky, or otherwise unpleasant and undesirable for a human being to perform. The economist approach is to automate when it is cheaper than human labor. The cognitive task design approach views automation not as simply a means to withdraw the human operator from the control–command loop. This approach assumes that automation can vary across different levels (manual to fully automatic) and that in most case, automated systems have to work conjointly with one or several human operators. To take the human being into account and to determine how best to automate, it is necessary to take a CTD approach to automation, which is much more systemic than analytic.

The Systemic Approach

A new "tool" affects not only how work is done but also how it is conceived of and organized. This has consequences for other aspects of work and may lead to unforeseen changes with either manifest or latent effects. This is particularly true in the context of supervisory control of complex systems and means that an original design is required. A systemic approach to design can lead to human-adapted solutions. Table 12.1 proposes a comparison made by de Rosnay (1975) of analytic and systemic approaches.

As the table indicates, the systemic approach is the application of the system concept to solve problems. This approach supplies a strategy of decision making of which the principal aspects are as follows:

- Insistence on identification, definition of goals and objectives of the system and a listing of criteria enabling performance evaluation
- Analysis of the different aspects of the system

TABLE 12.1
Comparison of Analytic and Systemic Approaches to Automation

Analytic Approach	*Systemic Approach*
Concentrate on the elements	Concentrate on interactions between elements
Consider the nature of interactions	Consider the effects of interactions
Consider the precision of details	Consider the global perception
Precise models but difficult to use	Models not sufficiently rigorous but easy to use
Deals with an education by specific subject	Deals with a multidisciplinary education
Deals with a planned action	Deals with an action by objectives
Knowledge of details; goals not well defined	Knowledge of goals; details not well defined

- Identification of best possible functional and structural alternatives for performing the system objectives
- Analysis of the full system as subsystems by locating inputs, outputs, variables of transformation, and the interfaces between subsystems

We think that by adopting principles of the systemic approach, CTD leads to original solutions in matters of automation.

CTD, Supervisory Control, and the Systemic Approach

CTD for automation can be performed keeping the following points in mind.

- The main objective is the improvement of the overall performance of the human–machine system (not only the technical system).
- Consequently, induced effects of artifacts must be taken into account. The human–machine system is studied, considering human characteristics (requiring human decisional models) as well as technical aspects.
- Alternatives to technically centered approaches for automation must be considered. HMC and dynamic tasks allocation (DTA) are examples of original solutions proposed and evaluated by human-centerd engineers.
- An evaluation stage is necessary but difficult to perform because the human operator's cognitive behavior is not directly observable.

In summary, the CTD approach focuses on automation that integrates a human operator and studies interactions between humans and the technical system. Consequently, design of supervisory tools and particularly human–machine interfaces requires process analysis, cognitive models, and a choice of function allocation between humans and machines. Of course, solutions are linked to the kind of process for which the design will be applied, and it is for this reason that we detail two applications in this chapter. However, supervisory control of dynamic processes implies that a human operator is far from the process, and works with human machine interfaces. Cognitive models can supply a framework that is useful for the designer.

Cognitive Models

Cognitive science and some results from cognitive psychology might thus provide us with a cognitive model relevant to the human operator's cognitive activity. A model can be seen as any kind of representation. Models thus take different shapes not only according to the nature of the phenomenon to be modeled but also according to the goal of the modeling activity. To clarify the review of the various cognitive models, we differentiate between frameworks, theories, architectures, and models (Amat-Negri, 1999):

- Cognitive frameworks (or framework models) gather a set of concepts, a collection of the major statements (or axioms) on human thinking, and define the boundaries of a subject.
- Cognitive theories are less general than frameworks and more focused on the description of one object or a phenomenon based on a hypothesis. A framework can give birth to many theories according to the set of hypotheses chosen. A paradigm is a theoretical orientation.
- Cognitive architectures provide structures and comprehensive overviews of the components of and rules for reconstructing human cognition.
- Cognitive simulation, or preimplementable models, are either applications of cognitive architecture to a particular task (or domain) or ad hoc models that are not based on an explicit architecture but that are implementable.

This classification can also be seen as ordering models from more to less abstract, from the more conceptual to the more applied. The same concept can be explained by many theories: behind architecture there is a theory, and architecture can generate several models. Consequently, a model could, in principle, be described at various levels of abstraction. In practice, the number of theories is much greater than the number of architectures.

In this chapter, we only deal with cognitive frameworks and theories. Several theories on memory, knowledge representation, attention, and decision making have been developed and provide the basis for modeling cognitive activity. But only Rasmussen's (1983) framework (Fig. 12.1), together with Hoc's (1996) extension of it, are considered to represent well the human decisional process. Rasmussen defined three kinds of human behavior in a static model: skill-based behavior (SBB), rules-based behavior (RBB), and knowledge-based behavior (KBB). Hoc added feedback loops to represent delays and hypothesis tests. In the sections that follow, we describe how Rasmussen's model can be used to characterize HMC.

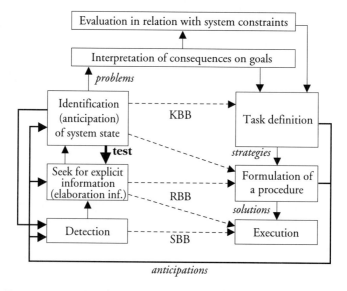

FIG. 12.1.   Rasmussen's model, as completed by Hoc (1996). Thin lines represent aspects present in Rasmussen's original model; thick lines represent Hoc's adaptations to it.

In addition, models of cognitive control must include higher level analytical problem solving at one extreme and the control of action at the other. This can be illustrated with the concept of LOA, which has an impact on the operator's information requirements. When a function is performed by an automatic system, operators do not need the same information as they would if they had to carry it out without the help of technology. Bye, Hollnagel, and Brendeford (1999) indicated that a human operators require less information if they are monitoring a task than if they were performing it. Consequently, automation may reduce information requirements, which over time can lead to negative consequences on the operators' process knowledge, and thereby making them less able to respond to unexpected events. The notion of feedback is fundamental here. Cognitive operator models such as that proposed by Neisser (1976) explicitly show the connection between past and future events. Indeed, feedback information modifies current understanding of a situation, which controls the goals to be performed through action, which in turn supplies information (Neisser, 1976). The important concept that an engineer must keep in mind is that an automated artifact is going to have an impact on human "situation awareness" and "decision making." Endsley (1988) defined situation awareness as "the perception of the elements in the environment, within a volume of time and space, the comprehension of their meaning and the projection of their status in the near future." Decision making can be defined as the "choice of actions having to be performed." As has been shown in various models, boundaries between situation awareness and decision making are not completely defined. Human operators build a picture of a process by means of the information available to them. Their situation awareness directs their decision making to a considerable degree. Today, supervisory systems are not adapted to human monitoring and diagnosis tasks because they do not facilitate situation awareness and decision making. From a theoretical point of view, HMC is a way to improve, facilitate, or support human situation awareness and decision making.

SUPPORTING COOPERATION BETWEEN HUMAN OPERATOR AND TECHNICAL SYSTEMS

Human–Machine Cooperation

For a complex system, the choice of the appropriate level of automation is a difficult one, for which human requirements must be considered precisely. Indeed, if excessive automation can affect human abilities, insufficient automation may not meet human's support needs or maintain system reliability and productivity (Garland, 1991). Automation can be achieved along two main lines. The first is the static function allocation between the human operator(s) and the control system (Hollnagel, 2000). The second, which can be complementary to the first, focuses on HMC.

Defining exactly what HMC is remains a difficult task. Hoc offered the following perspective:

Two agents are in a cooperative situation if they meet two minimal conditions:

- Each one strives towards goals and can interfere with the others on goals, resources, procedures, etc.
- Each one tries to manage the interference to facilitate the individual activities and/or the common task when it exists.

The symmetric nature of this definition can be only partly satisfied. (Hoc, 2001)

In this definition, the notion of *goal* does not refer to the global goal to reach when supervising or controlling a process, but to the goal for achieving a particular task. The word *interference* refers to the normal interaction between the activities of several agents but also to conflicts between the agents, on the results of their activities, or on the means for achieving their tasks.

Each agent can be characterized by his know-how but also by his know-how to cooperate (Millot & Lemoine, 1998). The latter comprises two main functions: The first allows agents to manage interference between their goals, resources, and so on, and the second allows agents to perform their own activities, taking into account the activities of the other agent to facilitate those activities.

In HMC, it is necessary to improve interactions between agents to allow for the correct fulfillment of their tasks and to minimize conflicts. Nevertheless, this definition shows the limits of HMC, especially if the "machine" or computer must facilitate the human activities, because it will be necessary to prevent or detect conflicts with the human operator.

Royer (1994) noted that a system is more cooperative when cooperation integrates several factors: perception, analysis, decision, and action. Cooperation is not only a coordination of actions between several agents; it also depends on the merging of perceptions, on the confrontation of situation analyses, and on the convergence of decisions.

To model these interactions, it is possible to use the three forms of cooperation defined by Schmidt (1991):

- Augmentative cooperation: Cooperation is augmentative when agents have similar know-how, but there must be multiple agents to perform a task too demanding for only one. The task is then divided into similar subtasks.
- Integrative cooperation: Cooperation is integrative when agents have different and complementary know-how, and it is necessary to integrate their contributions to achieve a task.
- Debative cooperation: Cooperation is debative when agents have similar know-how, are faced with a unique task, and compare their results to obtain the best solution.

For Grislin and Millot (1991), cooperation may be described as a combination of these three forms. But the authors noted that tasks (or subtasks) are performed entirely by individual agents, even if there are some interactions between them before or after the tasks, in accordance to the three forms of cooperation. We argue that any situation of cooperation may be described as a combination of the three forms of cooperation, if we consider the human activities (or the artificial functions performed by an assistance tool) instead of the task (Debernard & Hoc, 2001).

Characterization of Human–Machine Cooperation

To characterize HMC, it is possible to start from the know-how of an agent. Rasmussen's model can be used because it enables us to characterize the internal symbolic information "produced" by a human operator (e.g., problems, strategy, solutions). As mentioned previously, this model comprises four main activities (Rasmussen, 1985), the information elaboration, the identification, the intervention decision, and the action. We decompose these classes of activity into subclasses describing the evolution of the state of information.

- *Information elaboration:* This is usually an acquisition of data coming from sensors, and displayed on human–machine interfaces. This acquisition can be spontaneous during the monitoring of the process—which allows the detection of an abnormal situation—directed by an objective during an activity of situation identification, or decision making, or may finally be useless if the operator remembered this information before.

- *Identification:* This is identification, inference, or testing of the process state. The identification is the interpretation of information into a category. Inference is also an interpretation, but it is uncertain. Testing is the test of an inference with new information.
- *Decision making:* A decision is schematic or precise and will be evaluated. A schematic decision specifies a goal to be reached. It reflects the schematic feature of intervention decision planning. A precise decision is used when the decision is fully specified. A decision evaluation corresponds to the control or evaluation of one's own or another agent's action.
- *Action:* This activity corresponds to the implementation of one or several commands in accordance with the precise decision made in the decision-making activity.

This list of subclasses is certainly not exhaustive, and the number of activity states is dependent on the type of process and the control the agent is able to have. Human agents plan their activities to manage their internal resources according to the time pressure and stress they are under (Hoc, 1996). For managing situations, human agents build a frame of reference, which contains several attributes (Pacaux-Lemoine & Debernard, 2001): *information* (stemming from activities of information elaboration), *problems* (stemming from activities of identification), *strategies* (stemming from activities of schematic decision making), *solutions* (stemming from activities of precise decision making), and *commands* (stemming from activities of implementation of solutions).

When several human operators supervise and control the same process, they elaborate and maintain a common frame of reference (Decortis & Pavard, 1994) that is a common representation of the environment and the resource's team. Common frame of reference concerns factors such as common goals, common plans, role allocation, and representations of the process. To cooperate, human agents exchange information, problems, strategies, and so forth to share their individual frame of reference, building a virtual common frame of reference (Fig. 12.2).

The characterization of cooperation can be done by using this common frame of reference and Schmidt's (1991) three forms. Nevertheless, to optimize verbal communication, one agent can have an interpretation of the frame of reference of the other agent, and this interpretation can be false. One way to avoid this is to support cooperation by implementing the common frame of reference. We call this implementation common work space (CWS). Moreover, in the case of HMC, CWS can support cooperation between human and artificial agents, but it

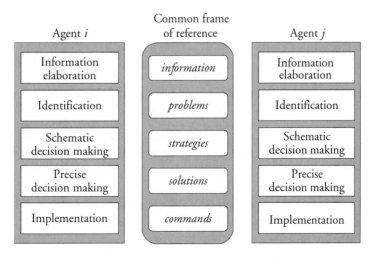

FIG. 12.2. Example of a common frame of reference.

is necessary to define which activities and functions will be performed by which agents and which form of cooperation (augmentative, integrative, and debative) must be implemented (Lemoine, 1998).

Cooperative Support

Human–computer interactions are still a problem for designing AHASS, especially when HMC must be integrated in the overall system. For example, Hoc (2000) described some failures in human–computer interactions: loss of expertise, complacency, trust and self-confidence, loss of adaptability. For a new AHASS, the functions allocated to the support tools and the human–machine interfaces must be studied with attention.

When a human operator must interact with tools, it is necessary first to implement the CWS on a machine and then to define the human–computer interactions. Many disciplines have already worked about the CWS idea (Bentley, Rodden, Sawyer, & Sommerville, 1992; Decortis & Pavard, 1994; Jones & Jasek, 1997; Royer, 1994). For example, in the case of air traffic control, Bentley et al. presented a shared work space that provides an adapted presentation of air traffic to different users, on different machines, to facilitate the use of shared entities (Bentley et al., 1992). Decortis and Pavard (1994) defined the shared cognitive environment as a set of facts and hypotheses that are a subset of each agent's cognitive universe.

The design of a cooperative human–machine system requires defining the effective content of the CWS, which depends on the domain and the tasks to be achieved. Moreover, the use of a CWS generates new activities for the human agent (but also some functions for the assistance tool), which correspond to the know-how to cooperate. These activities (or functions) must allow the following:

- Update of the common frame of reference: Activities and functions allow updating the attributes of common frame of reference.
- Control of the common frame of reference: The activities and functions aim to compare the common frame of reference and their individual frame of reference. These activities correspond to the mutual control and allow detection of interference on one or more attributes (information elaboration activities, but on the CWS data, not on the process information).
- Management of the interference: First, the activity and function corresponds to a diagnosis of the differences between the CWS and an agent's frame of reference. Second, a solution for the interference must be devised.

The agent may use three forms of solution (Pacaux-Lemoine, Debernard, 2001): negotiation, acceptance, and imposition. These forms imply human agent cognitive and communication costs. Negotiation aims at reducing the differences between the CWS and the agent's frame of reference by modifying one of them on the basis of explanations between the agents. Acceptance is the update of the frame of reference from the CWS. Acceptance is selected when the cost of a negotiation is too high or when an agent wants to facilitate the activities of the other. Imposition corresponds to the opposite of acceptance.

The implementation of a common work space between an artificial agent and one or several human agents brings with it some constraints, in particular, constraints for negotiation. Two human agents, when they negotiate, may use symbolic explanations, which are efficient. An artificial agent needs an explicit explanation on the base of operational information; at present, it is difficult to implement in an artificial agent capacities for negotiation with human agents. Another problem will appear when human agents have to accept a artificial agent's solution without being able to modify it.

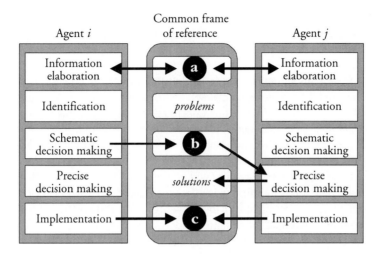

FIG. 12.3. Interaction between agents through common work space.

To model all these cooperative activities and functions and their allocation for performing the process tasks, it is possible to use the three forms of Schmidt for defining all the interaction between the agents (Fig. 12.3).

In the debative form (see case a in Fig. 12.3), all agents supply the CWS with new data (for one task and for one activity or function), and when some interference appears, they can negotiate. In the integrative form (see case b in Fig. 12.3), only one agent supplies data to the CWS, and the other takes this data into account for performing the next activity or function. In the augmentative form (see case c in Fig. 12.3), the agents perform the same activity or function and update the CWS, but for different tasks in accordance to their allocation.

To illustrate these concepts, two applications with different LOA processes (low and high) are presented. As noted earlier, in air traffic control, the LOA is low. To manage the increase in the traffic, several solutions can be conceived. The technical approach consists in a full automation of human activities. In contrast solution tested by human engineers deals with a dynamic allocation of tasks between human controller and automated tools. If we noting that LOA is not fixed, designers can use this degree of freedom to propose original control–command devices that take human factors into account. In this case, HMC is augmentative. In the case of supervisory control of automated processes, because it is not possible to withdraw the human operator from the automated system, design of supervisory tools requires integrative and debative structures of cooperation. These studies are detailed in the following sections.

APPLICATION TO AIR TRAFFIC CONTROL (LOW LOA)

This section deals with dynamic task allocation in air traffic control. It presents the results of previous studies in this area. These results have led us to design a new assistive device for air traffic controllers based on the principle of task delegation and the implementation of a CWS for improving human–machine cooperative activities.

Dynamic Task Allocation

DTA is initially an augmentative form of cooperation. The purpose of DTA is to assist the human operator first with the integration of an automated system that is able to perform some tasks and second by allocating the tasks between each agent in a dynamic way. The automated system must integrate all the functions necessary for performing a task in its entirly, from information elaboration to solution implementation (Debernard, Vanderhaegen, Debbache, & Millot, 1990).

An optimal DTA aims at devising task sharing that optimizes process performance and that takes into account the two decision makers' abilities and capacities. DTA allows dynamic modification of the LOA to keep the human operator "in the loop"; this is why DTA is also called adaptive automation. The difficulty arises in defining the set of the shareable tasks in real time that corresponds to the intersection between the set of tasks belonging to human abilities and those tasks within the realm of assistance tool abilities.

In DTA, the allocation of the shareable tasks is performed by a module called an allocator, of which the main function is to inform each decision maker about who performs which task and how. This allocator is controlled by a dispatcher (Crévits, Debernard, & Denecker, 2002). This function can be performed by an automated system or by the human operator (Millot, 1988) as follows:

- In explicit DTA, the human operator achieves the dispatcher function with a specialized interface. The human operator allocates the shareable tasks according to his or her own criteria.
- In implicit DTA, an automated system achieves the dispatcher function, which performs the allocation in accordance with an algorithm and with criteria defined by the designer. These criteria can require measurement or estimation of human–machine parameters such as human workload, the system performance, and so forth.

Implicit DTA has the advantage of discharging the human operator from allocation management. Nevertheless, this mode is more complex to implement than the explicit mode, because the choice of algorithm and criteria requires serious study to integrate correctly the human operator in the human–machine system. To avoid this difficulty, it is possible to adopt an intermediate mode called assisted explicit DTA (Crévits, 1996). In this mode, the initial allocation is performed by the automated system (as in the implicit mode), but the human operator can modify it if he or she disagrees with the proposal for any reason (as with the explicit mode, but this factor takes priority in the implicit mode).

We have implemented a software platform that integrates a realistic air traffic simulator and evaluated with professional controllers different DTA modes in the air traffic control domain.

Air Traffic Control

The French air traffic control is a public service in charge of flight security, the regulation of air traffic, and the flight economy for each aircraft that crosses into French airspace. Security is, of course, the main objective and consists of detecting potential collision between aircraft and preventing it. These situations are called conflicts, and controllers must avoid them.

Our studies concern the "en route" control more particularly, in which the airspace is divided into several sectors. Each is supervised by two controllers who have the same qualifications but not the same tasks. An analysis of this domain must allow definition of the main tasks and their goals (Crévits, 1996; cf. Fig. 12.4).

The first controller has a tactical role and is called the radar controller (RC). He or she must supervise the traffic to detect conflicts between aircraft and then resolve them by modifying the initial trajectory of one or several aircraft by sending a verbal instruction (heading, flight level, etc.) to pilots. The second controller has a strategic role and is called the planning controller (PC). This role consists of avoiding an overload at the tactical level (i.e., an overload of the

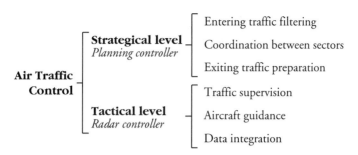

FIG. 12.4.   Main tasks of air traffic controllers.

RC tasks). The PC must predict the future traffic that the RC will manage and negotiate the entry and exit flight levels of aircraft with the adjacent sector PC so that the RC will not have too much conflict to resolve. The PC performs "traffic filtering," which has some limitations because it is impossible to increase the number of aircraft without careful consideration, and changing aircraft flight level is not an economical solution (Crévits, 1996). To face increase of air traffic, one solution consists of giving active assistance to the RC to increase the tactical level capacity.

Initial Studies: SPECTRA Projects

In an initial study called SPECTRA V1 (Debernard, 1993), a DTA was implemented at the tactical level between the RC and an assistance tool called SAINTEX (Angerand & Le Jeannic, 1992). SAINTEX is able to resolve simple conflicts between only two aircraft (called binary conflict or duel). These conflicts are called shareable conflicts because the two decision makers are able to resolve them in an autonomous way. SAINTEX has its own strategy and performs the resolution task completely, from the detection to the implementation of instructions, including rerouting. Two modes have been evaluated: an explicit and an implicit mode.

The explicit mode concerns a preemptive, explicit task allocation. In this mode, human air traffic controllers manage the task allocator through a dialogue interface (radar image and electronic stripping interface). They estimate their own performance and workload and then allocate tasks either to themselves or to SAINTEX. Shareable conflicts are indicated with a specific color on the operator dialogue screen and radar image. The deadline for the conflict to become unsolvable by SAINTEX as well as the deviation order it proposes are displayed on the strip of the aircraft.

Initially, all the aircraft are allocated to the controller. If the controller feels overloaded, he or she can select a conflict and transfer it to SAINTEX. In preemptive allocation, the controller can modify the allocation of a shareable conflict. But at that time, if the order deadline given by SAINTEX is over, this conflict can no longer be solved by SAINTEX.

The later mode is a non-preemptive, implicit task allocation controlled by an automatic management system implemented on a calculator. This allocation depends on the intrinsic abilities of the two decision makers. For SAINTEX, those abilities are functional ones: it can only solve the easiest conflicts. For the human RC, these abilities are related to his or her workload. For the moment, only the task demands are assessed in real time. This estimator takes into account all tasks to be performed by controllers (assuming an aircraft, detecting and resolving a conflict, monitoring aircraft). A value is given to each task, and the estimator sums these values. This estimator has been validated with an online subjective evaluation of the controller's workload (Vanderhaegen, Crévits, Debernard, & Millot, 1994). When those demands are too high, the shareable tasks are allocated to SAINTEX. In this case, the conflict with the nearest order deadline is chosen and transferred to SAINTEX, which displays its solving strategy on the strips. Otherwise, the conflict is allocated to the human controller.

The scenarios have been created with heavy air traffic to test the usefulness of dynamic allocation. They generated about 20 conflicts per hour of different nature. These experiments have been performed with nine qualified air traffic controllers. The task load index (TLX) method (Hart & Staveland, 1988) has been used to calculate the global workload for each controller and for each experiment.

The main result of these experiments (Debernard, Vanderhaegen, & Millot, 1992) is that a dynamic task allocation mode (either implicit or explicit) seems to improve the air traffic control task and to reduce the overall workload. According to two criteria related to the performance, the economic and security criterion, the implicit mode seems to be the most efficient. Despite these interesting results, SPECTRA V1 exemplifies the problem of gaining controllers' acceptance of the assistance. Because they have limited trust in SAINTEX, the RCs prefer the explicit mode, which allows them to take back conflicts allocated to SAINTEX. With this mode, solutions are not imposed on the controller so he or she is not simply a supervisor. But the problem of the explicit mode is the controller's workload, which could be increase because RC has to oversee the additional task of allocation. Thus, the RC takes a strategic role by detecting and allocating conflicts. These conclusions have led our research to examine the air traffic control organization to see how we could introduce the strategic level into the human–machine system.

In the second study, SPECTRA V2 (Crévits, 1996), a DTA was again implemented at the tactical level between the RC and SAINTEX. The goal of this project was to keep the advantages of implicit allocation (avoiding excessive workload due to the allocation management) but adding the opportunity for controllers to take back a shareable conflict. Because the allocation management is a strategic task, this function has been given to the PC.

At the strategic level, a new assistance tool called PLAF (a French acronym for "planning of allocation") has been implemented. PLAF helps the PC to anticipate the influence of aircraft entering the space of existing traffic and to estimate the future workload of the RC. When an overload of the tactical level is detected, PLAF can propose allocation of one or several shareable conflicts to SAINTEX. To justify these propositions of allocation, a dedicated screen displays the tasks detected by PLAF, the conflicts detected by SAINTEX, and the period of overload. All of this information is presented according to two axes: the time (about 30 minutes of prediction) and the flight levels.

Two modes have been evaluated: an explicit and an assisted explicit mode. In the explicit condition, allocation can be controlled by either the PC or the RC. Shareable conflicts are allocated, by default, to the RC. They can be allocated to SAINTEX before a deadline defined by temporal constraints. In addition to SAINTEX, traffic and conflict prediction is performed by PLAF. In the assisted explicit condition, PLAF automatically proposes conflict allocations on the basis of workload prediction. When it allocates a conflict, the PC can change the allocation, but the RC cannot directly control the allocation, except by a request to the PC.

The results of an activity analysis are in favor of the assisted explicit condition (Hoc & Lemoine, 1998; Lemoine, Debernard, Crévits, & Millot, 1996). The performance is better, and the cooperation between the agents seems to be more operational. The assistance tools appear to make the controllers' activities more proactive and less reactive to events. Moreover, the cooperation is concerned more with goals and less with communication. The PC participates in the decision making through suggestions to the RC. The effects of the allocation mode, which prescribes a precise allocation of tasks, and the involvement of an agent (the PC) facilitates cooperation between one artificial agent (SAINTEX) and one human agent (the RC). All these results may be explained by the capacity of assistance tools to reduce cognitive workload.

Another main point of these results is the information carried by graphical displays, which constitute a reduced common work space, in which SAINTEX and PLAF give their results regarding traffic analysis and in which the RC and PC could display symbols to mark important information, especially conflict detection. Because the controllers are used to work together, and because they display a subset of their understanding of the situation on the reduced common work space, a controller does not need to communicate constantly with the other to make a decision. But this reduced CWS is insufficient because it doesn't integrate all the "results" of the controllers' activities, such as information elaboration, diagnosis, and decision making.

Nevertheless, the SAINTEX strategy can hinder controllers' reasoning because they do not share the same representation of a conflict—SAINTEX has its own strategy, and the RC can take another aircraft in a conflict into account to resolve it in accordance with his or her abilities and workload.

Furthermore, in the assisted explicit mode, a complacency phenomenon (Smith, McCoy, & Layton, 1997) led the RCs to relax their monitoring of the overall traffic because they were less involved in the allocation than they were in the explicit mode.

The following section presents the project AMANDA (Automation and MAN–machine Delegation of Action) in which a complete CWS has been defined and implemented.

AMANDA project

In this project, we studied two primary principles studied (Debernard, Crévits, Carlier, Denecker, & Hoc, 1998): Controllers' task delegation to a system called STAR (French acronym for tactical system for assistance to resolution) and a CWS that allows controllers and STAR first to communicate with each other and second to share the same representation of the problems (i.e., the conflicts).

As was the case with the SPECTRA studies, we tried to help controllers by allocating some tasks to an assistance tool (i.e., augmentative cooperation), but the system will not have its own strategy for resolving a conflict. STAR would take the controllers' strategy into account (i.e., integrative cooperation) and then try to transform this strategy into a new trajectory that allows conflict resolution. If the controller agrees with the proposed solution, he or she could allocate the task to STAR; if the strategy given by controllers does not allow STAR to calculate a "good" trajectory because another (interfering) aircraft is present, STAR integrates this aircraft into the problem and informs controllers in accordance with a debative cooperation. Controllers can negotiate with STAR by introducing another strategy or constraint. This kind of cooperation is called task delegation—a task (a goal to achieve under constraints) that is explicitly allocated to the assistance tool, but this allocation integrates some constraints from controllers under a strategic form. The assistance tool does not perform the task of conflict resolution entirely on its own as it did with SAINTEX.

The cooperation activities have been built around a CWS (see Cooperative Support), which is common for all the agents (the two controllers and STAR). Each decision maker supplies this space in accordance with his or her own competencies and with the initial function allocation. Then, each agent can peruse information that is presented on the CWS, to check them (mutual control activity) or to negotiate if their representations are incompatible (Fig. 12.5; Pacaux-Lemoine & Debernard, 2001).

The content of the CWS was defined from a cognitive task analysis on an initial human–human cooperation experiment in which two RCs manage the traffic together (Carlier & Hoc, 1999; Debernard, Crévits, Carlier, Denecker & Hoc, 1998). This experiment allowed studying the cooperative activities and the symbolic information exchanged. For controllers, a problem is a set of conflicting aircraft, that is, at least two aircraft competing for the same airspace (binary conflict) and other aircraft that can interfere as well. An aircraft interferes with a

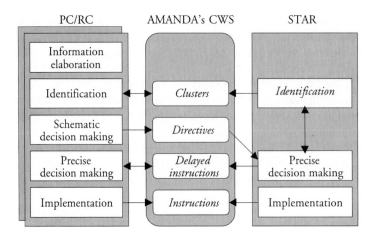

FIG. 12.5. AMANDA's common work space. PC = planning controller; RC = radar controller; CWS = common work space.

conflict if its resolution generates another conflict. The problems are called clusters. A strategy is modeled as one or several "directives" for resolving a problem. For example, a directive can be "turn AFR365 behind AAL347." A directive does not indicate a place and a value as it would in a normal instruction. A solution is a delayed instruction; for example, "turn AFR365 30° to the left at 22:35." In January 2003, these principles will be evaluated with professional controllers with the platform AMANDA (Debernard, Cathelain, Crévits, & Poulain, 2002).

In this section, some studies have been presented in the context of air traffic control. Our CTD approach consists of implementing human–machine cooperation to adapt the LOA to human needs. As described Table 12.1, this approach can be categorized as systemic because interactions between different elements of the human–machine system were studied and the effects of these interactions evaluated. In addition, the models of human activities used are not exhaustive but sufficient to design assistant tools and human machine interfaces. The next part of the chapter deals with an example of a process with a high LOA, which requires specific HMC.

APPLICATION TO A NUCLEAR FUEL REPROCESSING PLANT (HIGH LOA)

This case study (Gentil, Riera, Lambert, & Montmain, 1997) concerns a workshop composed of two head-to-foot columns and their supply system. Each column ensures the separation of combined uranium–plutonium from the fission products that are considered as wastes. The uranium, plutonium, and fission products are first dissolved in an acid solution. The first column, the extraction column, receives this loaded solution from which the uranium and the plutonium are extracted through an organic solution. The acid solution, which is denser than the organic solution, is put at the top of the column; the organic solution is placed at the bottom. Within each column, pulsation pressure increases the chemical exchanges, making a more efficient emulsion. At the exit of the extraction column, the aqueous outflow contains only residues. At the top, the overflowing organic solution contains, in addition to the uranium and the plutonium, some traces of fission products; once it is recuperated, it is sent to the second, or washing, column. An acid solution and the action of pulsation pressure create the physicochemical processes; the aqueous outflow, which contains the residual fission products, is reinjected toward the extraction column. Finally, the organic solution flowing out of the second column contains only the uranium and the plutonium. This process is a complex one to supervise because of several feedback loops with important time delays.

Human Supervisory Tasks

Only "real-time" aspects of process supervision are considered in this section. In this context, human supervisory tasks are at a level between strategic functions (e.g., planning) and tactic functions linked to the command part of the production system. A human operator receives continuous (measures) and discrete (states and alarms) information coming from the supervisory system and acts on it as necessary. Supervisory tasks require a high level of knowledge and can be divided into three classes (Rouse, 1983):

• *Command tasks:* An operator can act on the process in different ways, for example, by modifying set points (e.g., tuning) or changing a mode (e.g., manual mode). He or she acts on the process during specific phases of the process in normal (e.g., start up or the shut down) or abnormal (failures, perturbations) situations. When possible, the operator has to correct or compensate for a failure.

• *Monitoring tasks:* An operator supervises the process and filters available information to evaluate the process state and its progress. During this activity, he or she is an observer. Indeed, the operator looks at process information without acting to modify the current state. The main preliminary goal of supervisory tasks is to detect abnormal functioning. The secondary goal is to update the operator's mental model of the process to make satisfactory decisions when a problem occurs. Abnormal function (Fig. 12.6) can be detected either by means of an alarm (passive detection) or by the operator's observation of the process (active detection). Usually, the designer, by thinking only about the improvement of alarms, sees HMC only as integrative. If the supervisory tools are able to facilitate active detection, the HMC is seen as debative. In supervisory control, an operator is, in a way, out of the control–command loop because he or she does not get sufficient information at all times about what happens in the process. A human operator sometimes receives either no information or too little information to maintain a good picture of the system state (that is, the operator has insufficient situation awareness).

• *Diagnosis tasks:* After detecting a problem, the operator has to diagnose the process state. This consists of determining the primary causes but also the effects on the process. First, he or she has to check the validity of the problem. A large-scale system is supervised by means of numerous sensors. Some of them are used for feedback control; others exist only to inform the operator about the process. The operator must be able to predict whether the process will be able to reach its steady state. Indeed, automatic control can compensate for disturbances and stabilize the process. Capacity of abstraction and the fact that people may make decisions even in the case of great uncertainties explains partially why it is not possible to withdraw humans from the supervisory loop.

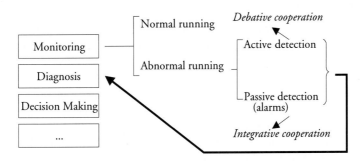

FIG. 12.6. From monitoring to diagnosis.

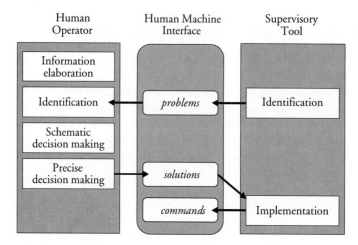

FIG. 12.7.   Interaction between human operator and actual supervisory tools.

Tools to perform these supervisory tasks have not evolved since the development of the first numerical control–command systems. One can essentially find three kinds of human–machine interface:

- Piping and instrumentation diagram (P&ID): This involves copying the physical structure of the process, including real-time measurements. This display gives an instantaneous picture of a part of the process.
- Trend graphs: These represent time-related information. This kind of displays is important because it enables human operators to understand what has happened and to predict the future state of the process.
- Alarms: These are based on the triggering a sensor or a simple comparison of a measurement with a fixed threshold. This involves well-known problems of oscillation, "Christmas trees," and relevance. In addition, line-based alarm display (as a list) does not allow the human operator to recognize the process state quickly. Alarms also characterize past events.

Figure 12.7 shows the interactions between a human operator and an actual supervisory system. The cooperation is only integrative, and the human operator has to elaborate on information and make decisions. Cognitive models show that this approach is inefficient. Active detection is not taken into account by designers. The next section deals with technical improvements.

Technical Improvements

Engineers in feedback control work to improve the supervisory system, in particular, to propose algorithms to improve the alarms quality. In this case, it is only the passive detection that will be improved. In this section, we have adopted the following definitions:

- Detection: the detection stage is an automatic decision, which consists in determining if a failure has occured. Failure is different from breakdown because the process does not cease to function but only malfunctions.
- Location: In the context of diagnosis, this word has multiple definitions. For our purposes, location consists of determining the origin of the failure, and, more particularly, it is the

first measured variable that characterizes the origin of the problem. The words *structural location* are used when the physical component or source of the problem is determined.

- Isolation: This consists in identifying and characterizing the failure, which means to determine the cause of the failure. For this purpose, a model of the failures must be available.

Algorithms involved in these three decision activities have been called *fault detection and isolation algorithms* in automatic control. Usually, people involved in supervisory control (cognitive or technical engineers) use different words with different meanings: *supervisory systems, alarms systems,* or *diagnosis systems.* In each case, all the decisional activities can be included, from detection to reconfiguration. Algorithms must enable a human operator or an automatic system to make the right decisions and take the most appropriate actions. It is interesting to note that the operator is not taken into account in the design of these algorithms. In a CTD approach, we believe it is not possible to withdraw the human completely from supervisory loop. As a consequence, even if automatic control becomes increasingly robust to disturbances, and fault detection and isolation algorithms more efficient, the human operator is still the final decision maker in the process. Hence, algorithms to design better alarms and support systems must be seen as tools for human operators.

Diagnosis methods can be classified in two groups: external and internal methods (Ghetie, 1997). External methods are applied when analytical models of the process are not available. In this case, techniques coming from artificial intelligence are often used to model and process the knowledge (Dubuisson, 1997). Internal methods are used when knowledge of the system is sufficient to conceive of an analytical model of normal and the abnormal running of the system. The principle of methods based on a normal running of the process is to perform redundancy analysis (Mazzour, Héraud, & Alfonsi, 1996). It is, in fact, a generalization of material redundancy. An analytical redundancy relation is deduced from a model, using variables of available measurements. Different techniques are possible. The first consists of estimating the observed model parameters from the systems' input and output. These methods are called *parameter estimation.* Theoretically, the differences between the theoretical and the observed models are representative of failures. The other possibility is to compare values coming from the theoretical model with those obtained by measurements. Theoretically, to avoid false alarms, the normal running model must be updated in real time. This comparison is called *residue generation.* When residue values are null, it implies that the process is running normally. When values are not null, the deviation comes from noise or failure. To eliminate noise, it is necessary to filter measurements. These values constitute a signature, or a vector identifying the failure. After the detection has been performed, it is necessary to compare the signature with a theoretical "signature library" to characterize and to isolate the failure (Frank, 1995). All these methods are attractive from a theoretical point of view; however, they require identification of reliable dynamical models that are difficult to obtain for large-scale systems. Nonetheless, these methods have been designed to improve passive detection. If alarms are well designed, the operator will react to them. If the operator rarely has to act on the process, and only when the automatic supervisory control cannot compensate for the disturbance, problems of workload and lack of vigilance can occur. One can find here all Bainbridge's (1983) "ironies of automation" and the importance of CTD in the field of supervisory control of automated processes.

Integrating Cognitive Principles into the Design Stage

The main cognitive principle a designer must keep in mind is that the current understanding of a situation is updated according to the perception of real objects, but at the same time, the current

Normal running

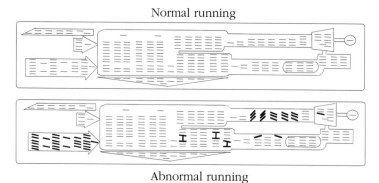

Abnormal running

FIG. 12.8.   Example of mass data display.

understanding of the situation guides attention toward specific objects. This is why it is vital to supply the right information to the operator because the available information guides the operator in his or her reasoning. In fact, a perfect supervisory tool is one that gives the human operator the necessary information at each instant to perceive the actual state of the process. It would be difficult and unrealistic to design interfaces that adapt their content according to context. First, this solution implies that the context and the process state are continuously well known. This is not true. Second, this kind of solution implies only integrative cooperation. The designer assumes a human operator is unable to determine the process state without assistance. Another solution is to supply interfaces adapted to the different tasks (monitoring and diagnosis). In this case, human–machine interfaces are seen as tools and the operator selects the one that is needed. Supervisory tasks have different objectives and require different kinds of information. Monitoring requires a global view of the process. On the other hand, diagnosis requires multiple points of view at different levels of abstraction. There is a great deal of research concerning new modes of representation (e.g., ecological interfaces; Vicente, Christoffersen, & Pereklita, 1995), functional interfaces (Prætorius & Duncan, 1991), mass data display (MDD; Beuthel, 1995). The MDD idea in particular is interesting. This concept appeared in the 1980s, and its basic concept is the presentation on a single display of all the data, allowing an operator to supervise a process. Each variable is symbolically represented by at least one of its intrinsic characteristic. Figure 12.8 provides an example of MDD (Boussoffara & Elzer, 1998) in which the variables are represented by sticks of which angle characterizes the trend. MDD has been designed to support human pattern matching.

In addition, the perception of system state requires giving the operator different points of view about the system. The description of a complex system stated so far is a structural one, which can be viewed as the organization of the system in space (Modarres, 1996). It corresponds to a picture of the system showing the whole system. This structural perspective cannot describe time-based properties of complex systems. The system organization in time can be described with functional and behavioral models. For some authors (Lind, 1994; Modarres, 1996), the functional hierarchy is the backbone of the system model describing the system's state–time behavior. Behavioral hierarchy shows the ways that functions are realized. In addition, structural hierarchy shows system parts that perform functions. Consequently, supervisory tools, in order to favor an active perception of process state, must enable the human operator to see the process according to different points of view: structural, functional, and behavioral.

With regard of the numerous variables, probability that the human operator monitors the right variable at the right time is low. Thus, if supervisory tools are not well designed, the human operator could monitor the process with alarms (i.e., discrete information) that are activated when a threshold is triggered. These thresholds are like tolerance widths, which are difficult to design. If they are too narrow, there will be a lot of false alarms; if the tolerance width is large,

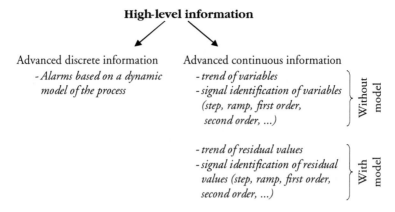

FIG. 12.9. High-level information.

detection will be delayed. It will be much more efficient for the human operator to work with a target value for each variable. Target values are easy to define for a regulated variable in a steady state; it is the set point. In addition, even in large-scale processes, behavioral models of the normal running of some parts of the process (used for fault detection and isolation algorithms) can be available. These models can supply in real time several target values of process variables. In our approach, advanced supervisory systems have to be designed to give useful information about process state to the human operator. New kinds of discrete and continuous high-level information can reach this goal (Fig. 12.9; Riera, 2001).

Advanced discrete information is, in fact, an advanced alarm. They are based on the comparison between a measure and a value coming from a dynamic model of the process. If the model is correct, advanced alarms will be triggered earlier and will be more sensitive to actual process failures. Nonetheless, the problems of passive detection are still present.

Advanced continuous information is based on real-time measurements but give to the human operator synthetic information that would have required a cognitive processing to be humanly obtained. In other words, advanced continuous information uses the good capabilities of perception of human beings. Different kinds of advanced continuous information can be designed. For instance, the first classification deals with using or not using a dynamic process model. In the first case, advanced continuous information can be based on the residue, which is the difference between the variable value and its normal value coming from a model; in the second case, advanced continuous information can be processed with the measure. The second classification depends on the signal processing performed: real time or along a temporal window. In this last case, the signal identification can, for instance, determine the shape of the signal (step, ramp, first order, second order, etc.). We argue here that advanced continuous information must improve active detection and support debative cooperation. Integration of continuous information in a supervisory system can considerably modify the human operator's work, usually based on measurements and alarms.

Figure 12.10 represents interaction between the human operator and an advanced supervisory tool. We see that CWS is designed to support human information elaboration and human identification of the process state by means of advanced continuous information and advanced discrete information.

An Example of AHASS

The AHASS developed for the nuclear fuel reprocessing plant includes all the specifications just described (Lambert, 1999; Riera, 2001): integration of advanced supervisory algorithms,

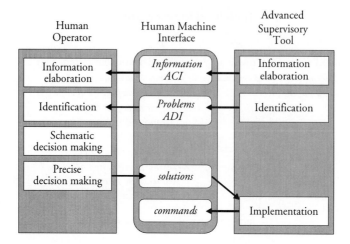

FIG. 12.10.   Interaction between a human operator and human adapted supervisory tools. ACI = advanced continuous information; ADI = advanced discrete information.

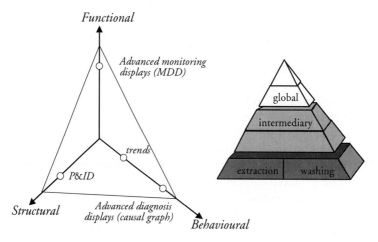

FIG. 12.11.   Points of view and levels of abstraction. MDD = mass data display; P&ID = piping and instrument diagram.

supervisory tools dedicated to each task (monitoring and diagnosis), a process representation according to different points of view (functional, structural, and behavioral) and different levels of abstraction, alarms and continuous high-level information, and explanation capabilities. The idea is to propose to the operator different interfaces representing the process following different points of view (functional, structural, and behavioral) and at different levels of abstraction: from the whole process to the different parts (Fig. 12.11).

The AHASS is composed of two parts (Lambert, 1999; Riera, 2001): a classical part (PID, trends, classical alarms) and an advanced part. The latter is composed of advanced alarms, one advanced MDD monitor that includes high-level functional information, two classical PID displays representing the two parts of the process (washing and extraction), four monitoring displays similar to the MDD on which several symbolic variables are replaced by measures, four causal graphs of the process at different levels of abstraction (including high-level information) four diagnosis displays at different levels of abstraction on which the results of location algorithm are presented, one trend display (with the normal value) available for each process

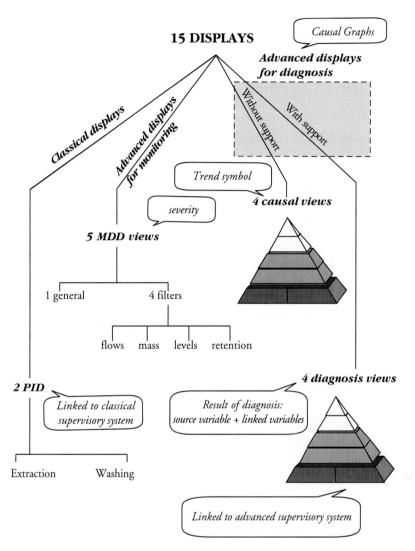

FIG. 12.12.   Architecture of the advanced human adapted supervisory systems, MDD = mass data display; PID = piping and instrument diagram.

variable, and one display for alarms report (Fig. 12.12). Advanced displays characterize the CWS, enabling positive interactions between the operators and the advanced support systems (algorithms).

High-level information supplied to operators (called *severity*) is relative to the trend of the residues. Hence, "normal" means that differences between the theoretical model and measurements are null along a temporal window. "Very serious" means that the measurement is diverging significantly from the model. Hence, severity is an advanced continuous flow of information, from 0 (*normal*) to 1 (*very serious*) characterizing the trend of the residue. Computation of severity is based on fuzzy logic (Evsukoff, Gentil, & Montmain, 2000).

In the MDD (Fig. 12.13), each variable is represented by a colored quadrilateral whose two attributes, size and color, reflecting respectively its significance within the function and its severity. This symbol has a range of colors with progressive contrasts (from green to yellow to red). Green means that the variable is in a normal state (values of the residue are null or near

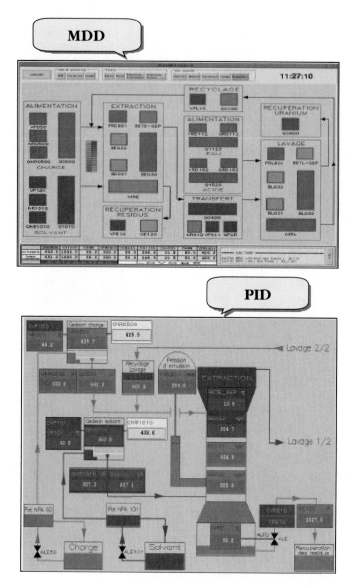

FIG. 12.13.   Mass data display (MDD) and piping and instrumentation diagram (PID).

null), whereas red indicates a very serious state. The range of 30 colors between green and red allows whole states to be translated, in a gradual manner, from normal to abnormal.

For the design of diagnosis views (Fig. 12.14), two points were considered: their general organization and their information content. Because of the informational density characterizing the causal graph of the process, a hierarchical organization following three levels was chosen. The top level supplies a general view of the process; the intermediary level, a median view; and the bottom level, two complementary views of the process, the extraction and washing views.

Each variable is represented by a node composed of two entities: an exterior crown, displaying the MDD color (severity) already seen in the monitoring views and inside the interior disk, and a trend symbol representative of the trend of the residue also called the *drift indicator.*

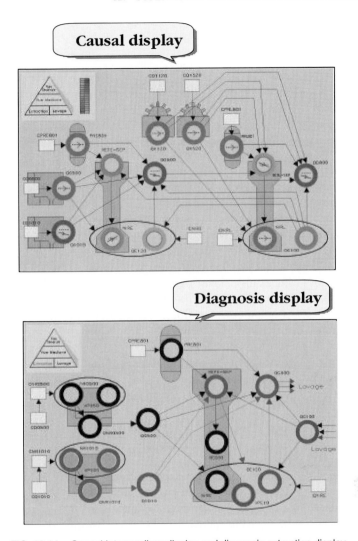

FIG. 12.14. Causal intermediary display and diagnosis extraction display.

This symbol is selected within one of nine qualitative primitive states representative of normal, suspicious, serious, and very serious (Fig. 12.15).

One can notice that the proposed advanced supervisory system has capacities of cooperation and explanation. The human operator can easily check the advanced information supplied by the system because the same model (causal graph) is used for the algorithms and to present the results. The concepts of a common frame of reference have been applied to support HMC.

In summary, advanced human–machine interfaces have been designed to facilitate debative cooperation with the operator by means of detection and location algorithms adapted to the human operator. In addition, the human operator can check all information coming from the advanced supervisory tools.

The results of these experiments are encouraging and show the potential for applying these new concepts to design supervisory systems to improve HMC. The most interesting result is with regard to detection time. Active detection has been improved with the MDDs, including provision of advanced continuous information. All subjects detected the problem of slow trend

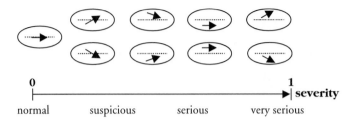

FIG. 12.15.   Nine primitive states.

in less than 5 minutes, whereas a classical alarm is triggered after 26 minutes and the advanced alarm after 9 minutes. Without the advanced displays (Context 1), the human operator detected the problem between 9 and 15 minutes. According to the operators' comments, it appears that the operators understood and appreciated the dynamic process model and its representation, which was judged very useful for drift detection. Moreover, according to the operators, the MDD interfaces facilitated process monitoring through symbolic information displayed, such as the drift indicator; giving a quick picture of the progress of the variable state allowed them to follow the evolution failures. Finally, the causal structure was judged to be well adapted to human diagnostics even if in our case, the different levels of abstraction are not particularly pertinent. The support decision system, which allows the human operator to control the action, was appreciated for its ability to check hypotheses or to search for possible sources of failures. The operators' comments show that the approach is interesting because the border between support and assistantship is never crossed. It seems that debative cooperation in a high-LOA situation has been accepted, understood, and used by the human operator.

CONCLUSION

Supervision of highly automated processes is an activity in which human beings, despite the increasing automation of recent years, is still present. Consequently, the design of efficient supervisory systems requires a different approach of automation that integrates CTD and is more systemic than analytic. With this approach, CTD can lead to original automation solutions based on the concept of human–machine cooperation. In this chapter, we have explained how to support cooperation between a human operator and a technical system in the case of supervisory control. Cognitive models and three generic forms of cooperation (augmentative, integrative, and debative) are the foundations for developing a common frame of reference supporting human–machine cooperation. We proposed two design applications for advanced human adapted supervisory system. The first application concerns air traffic control, a process with a low LOA. Dynamic task allocation concepts have been implemented in different projects to evaluate a dynamic level of automation adapted to human operators' needs. From our methodology, we have defined the contents of a CWS adapted to the air traffic control domain and to the characteristics of several assistance tools. The first, SAINTEX, is able to perform an overall task: the detection and resolution of conflicts between aircraft. The second, STAR, is able to perform the same tasks but takes into account the strategies supplied by the controllers.

The second application deals with a process with a high LOA: a nuclear fuel reprocessing system. The main idea in this case is to use debative HMC by adapting and integrating advanced detection and diagnosis algorithms. The idea to propose high-level information to facilitate situation awareness and detection of abnormal running of the process by the human operator has been validated with experiments in which experts detected faster failures with advanced human–machine interfaces. In addition, CTD enables the design of supervisory tools that never cross the boundary between the right level of support, and assistance that removes the human operator from the supervisory loop.

In both cases, CTD has led to original solutions in automation. All principles of cooperation explained in this chapter can be applied to systems in which human beings are still involved. Applications can be conceived, for example, the automotive field (low LOA) or in the management of computer networks (high LOA).

ACKNOWLEDGMENTS

The studies presented in the section on air traffic control applications were performed in collaboration with the Centre d'Etudes de la Navigation Aérienne, the French air navigation study center. Concepts presented in the section on nuclear fuel reprocessing were evaluated at the Commissariat à l'Energie Atomique of Marcoule (France) with professional supervisory operators (six experts and one novice).

REFERENCES

Amat-Negri, A.-L. (1999). *Simpatico: A cognitive model of the pilot—from concepts to specifications.* PhD dissertation. University of Valenciennes and Hainaut-Cambrésis, France.

Angerand, L., & Le Jeannic, H. (1992). Bilan du projet SAINTEX. Report R92009. Centre d'Etudes de la Navigation Aérienne (CENA), France.

Bainbridge, L. (1983). Ironies of automation. *Automatica, 19,* 775–779.

Beuthel, C. (1995). *Advantages of mass-data-display in process S & C.* IFAC analysis, design and evaluation of man-machine systems (pp. 439–444). Cambridge, MA: MIT Press.

Bentley, R., Rodden, T., Sawyer, P., & Sommerville, I. (1992, November). *An architecture for tailoring cooperative multi-user display.* Proceedings of Conference on Computer Supported Cooperative Work: Toronto, Canada.

Boussoffara, B., & Elzer, P. (1998). *About human pattern matching and time information in S&C of large technical systems.* Proceedings of the 17th European Annual Conference on Human Decision Making and Manual Control (pp. 161–166), Valenciennes, France.

Bye, A., Hollnagel, E., & Brendeford, T. S. (1999). Human-machine function allocation: A functional modelling approach. *Reliability Engineering and System Safety, 64,* 291–300.

Carlier, X., & Hoc, J. M. (1999, September). *Role of a common frame of reference in cognitive cooperation: sharing tasks in Air-Traffic-Control.* Proceedings of Conference on Cognitive Science Approaches to Process Control. Villeneuve d'Ascq, France.

Crévits, I. (1996). *Répartition dynamique de tâches dans les procédés complexes. Modélisation de la répartition anticipée: application au contrôle de trafic aérien.* PhD dissertation, University of Valenciennes and Hainaut-Cambrésis, France.

Crévits, I., Debernard, S., & Denecker, P. (2002). Model building for air-traffic controllers workload regulation. *European Journal of Operational Research, 136,* 324–332.

Crévits, I., Debernard, S., Vanderhaegen, F., & Millot, P. (1993, September). *Multi-level cooperation in air traffic control.* Proceedings of the Fourth International Conference on Human-Machine Interaction and Artificial Intelligence in AeroSpace (pp. 28–30), Toulouse, France.

de. Rosnay, J. (1975). *Le macroscope: vers une vision globale.* Paris: Seuil (Eds).

Debernard, S. (1993). *Contribution à la répartition dynamique de tâches entre opérateur humain et système automatisé: Application au contrôle de trafic aérien.* PhD dissertation, University of Valenciennes and Hainaut-Cambrésis, France.

Debernard, S., Cathelain, S., Crévits, I., & Poulain, T. (2002, June). *AMANDA Project: delegation of tasks in the air traffic control domain.* In M. Blay-Fornarino, A. M. Pinna-Dery, K. Schmidt, P. Zaraté (Eds.), *Cooperative systems design,* pp. 173–190. Proceedings of COOP'2002, Antibes, France. The Netherlands: 105 Press.

Debernard, S., Crévits, I., Carlier, X., Denecker, P., & Hoc, J.-M. (1998, September). Projet AMANDA: Rapport final de la première phase (final report). N° CENA/NR98-185. Convention de Recherche LAMIH/CENA 96/C007.

Debernard, S., & Hoc, J.-M. (2001). Designing dynamic human-machine task allocation in air traffic control: Lessons drawn from a multidisciplinary collaboration. In M. J. Smith, G. Salvendy, D. Harris, & R. Koubek (Eds.), *Usability evaluation and Interface design: Cognitive Engineering, Intelligent Agents and Virtual Reality* (Vol. 1, pp. 1440–1444). London: Erlbaum.

Debernard S., Vanderhaegen F., Debbache N., & Millot P. (1990) *Dynamic task allocation between controller and AI systems in air-traffic control.* Proceedings of the 9th European Annual Conference on Human Decision Making and Manual Control, Ispra, Italy.

Debernard, S., Vanderhaegen, F., & Millot, P. (1992, June). *An experimental investigation of dynamic task allocation between air traffic controller and AI system.* Proceedings of the 5th IFAC/IFIP/IFORS/IEA Symposium on Analysis, Design and Evaluation of Man-Machine Systems. The Hague, The Netherlands.

Decortis, F., & Pavard, B. (1994). Communication et coopération: de la théorie des actes de langage à l'approche ethnométhodologique. In B. Pavard (Ed.), *Systèmes Coopératifs de la modélisation à la conception* (pp. 29–50). Toulouse, France: Octarès.

Dubuisson, B. (1997). Détection et diagnostic de pannes sur processus. *Techniques de l'Ingénieur, Traités Mesures et Contrôles* (Vol. R.8, p. 7597). France: Editions Techniques de l'Ingénieur.

Endsley, M. (1988). *Design and evaluation for situation awareness enhancement.* Proceedings of the Human Factors Society: 32nd Annual Meeting (pp. 97–101). Santa Monica, CA: Human Factors Society.

Evsukoff, A. (1998). *Le raisonnement approché pour la surveillance de procédés.* PhD dissertation, University of Grenoble, France.

Evsukoff, A., Gentil, S., & Montmain, J. (2000). Fuzzy reasoning in co-operative supervision systems. *Control Engineering Practice, 8,* 389–407.

Frank, P. M. (1995). Fault diagnosis in dynamic systems using analytical and knowledge-based redundancy—a survey and some new results. *Automatica, 26,* 459–474.

Garland, D. J. (1991). Automated systems: The human factor. In NATO ASI series, Vol. F73 (pp. 209–215). *Automation and Systems Issues in Air Traffic Control.*

Gentil, S., Riera, B., Lambert, M., & Montmain, J. (1997). Filtrage d'alarmes en temps réel. Final report DRET. Convention DRET Scientifique et Technique N°93.34.098.00.470.75.0, Mai, France.

Ghetie, M. A. (1997). *Vers le concept de redondance algorithmique pour le diagnostic; le projet DIALOGS.* PhD dissertation, University of Henri Poincaré-Nancy 1, France.

Grislin, E., & Millot, P. (1999, September). *Specifying artificial cooperative agents through a synthesis of several models of cooperation.* Proceedings of the Seventh European Conference on Cognitive Science Approaches to Process Control (pp. 73–78). Villeneuve d'Ascq, France.

Hart, S. G., & Staveland, L. E. (1988). Development of NASA-TLX (task load index): Results of empirical and theorical research. In P. A. Hancock & N. Meshkati (Eds.), *Human mental workload* (pp. 139–183). Amsterdam: Elsevier Science.

Hoc, J. M. (1996). *Supervision et contrôle de processus: la cognition en situation dynamique.* Grenoble, France: Presse Universitaire de Grenoble, Collection Sciences & Technologie de la connaissance.

Hoc, J.-M. (2000). From human-machines interaction to human-machine cooperation. *Ergonomics, 43,* 833–843.

Hoc, J.-M. (2001). Towards a cognitive approach to human-machine cooperation in dynamic situations. *International Journal of Human-Computer Studies, 54,* 509–540.

Hoc, J. M., & Lemoine, M. P. (1998). Cognitive evaluation of human-human and human-machine cooperation modes in Air Traffic Control. *The International Journal of Aviation Psychology, 8,* 1–32.

Hollnagel, E. (2000). Principles for modeling function allocation. *International Journal of Human-Computer Studies, 52,* 253–265.

Jones, P., & Jasek, C. (1997) Intelligent support for activity management (ISAM): An architecture to support distributed supervisory control. *IEEE Systems, Man and Cybertenics, Part A: Systems and Humans, 27,* 274–288.

Lambert, M. (1999). *Conception centrée sur l'Homme d'un système de supervision avancé—Application à un procédé de retraitement de combustibles nucléaires.* PhD dissertation, University of Valenciennes and Hainaut-Cambrésis, France.

Lemoine, M.-P., Debernard, S., Crévits, I., & Millot, P. (1996). Cooperation between humans and machines: First results of an experimentation of a multi-level cooperative organisation in air traffic control. *Computer Supported Cooperative Work: The Journal of Collaborative Computing, 5,* 299–321.

Lemoine, M. P. (1998). *Coopération hommes-machines dans les procédés complexes: Modèles techniques et cognitifs pour le contrôle de trafic aérien.* PhD dissertation, University of Valenciennes and Hainaut-Cambrésis, France.

Lind, M. (1994). Modeling goals and functions of complex industrial plants. *Applied Artificial Intelligence, 8,* 259–283.

Mazzour, E. H., Héraud, N., & Alfonsi, M. (1996). Utilisation des relations de redondance analytique pour la tolérance de défaillances de capteurs. *RAIRO-APII, JESA, 30*(9), 1273–1288.

Millot, P. (1988). *Supervision des procédés automatisés et ergonomie.* Paris: Edition Hermès.

Millot, P., & Lemoine, M.-P. (1998, October). *An Attempt for generic concepts toward Human-Machine Cooperation.* Proceedings of the IEEE SMC'98, San Diego, California.

Modarres, M. (1996). *Functional modeling for integration of human-software-hardware in complex physical systems.* Proceedings of the Fourth International Workshop on Functional Modelling of Complex technical System (pp. 75–100). Athens, Greece.

Neisser, U. (1976). *Cognition and reality.* San Francisco, CA: W. H. Freeman.

Pacaux-Lemoine, M.-P., & Debernard, S. (2001). Common work space for human-machine cooperation in air traffic control. *Control Engineering Practice, 10,* 571–576.

Prætorius, N., & Duncan, K. D. (1991). Flow representation of plant processes for fault diagnosis. *Behaviour and Information Technology, 10,* 41–52.

Rasmussen, J. (1983). Skill, rules, and knowledge; signals, signs and symbols, and other distinctions in human performance models. *IEEE Transaction on Systems, Man, and Cybernetics, 15,* 257–266.

Rasmussen, J. (1985). The role of hierarchical knowledge representation in decision making and system management. *IEEE Transactions on Systems, Man, and Cybernetics, 15,* 234–243.

Riera, B. (2001). Specifications, design and evaluation of an advanced human adapted supervisory system. *Cognition, Technology and Work, 3,* 53–65.

Rouse, W. B. (1983). Models of human problem solving: Detection, diagnosis and compensation for systems failure. *Automatica, 19,* 613–625.

Royer, V. (1994). Partage de croyances: Condition nécessaire pour un système coopératif? In B. Pavard (Ed.), *Systèmes Coopératifs de la modélisation à la conception* (pp. 253–270). Toulouse, France, Octarès.

Schmidt, K. (1991). Cooperative work: A conceptual framework. In J. Rasmussen, B. Brehmer, & J. Leplat (Eds.), *Distributed decision-making: cognitive models for cooperative work* (pp. 75–110). Chichester, England: John Willey.

Smith, P. J., McCoy, E., & Layton, C. (1997). Brittleness in the design of cooperative problem solving-systems: The effects on user performance. *IEEE Transactions on Systems, Man and Cybernetics. Part A: Systems and Humans, 27,* 360–371

Vicente, K. J., Christoffersen, K., & Pereklita, A. (1995). Supporting operator problem solving through ecological interface design. *IEEE Transactions on Systems, Man, and Cybernetics, 25,* 529–545.

Vanderhaegen, F., Crevits, I., Debernard, S., & Millot, P. (1994). Human-machine cooperation: Toward and activity regulation assistance for different air-traffic control levels. *International Journal of Human-Computer Interaction, 6,* 65–104.

13

Cognitive Task Load Analysis: Allocating Tasks and Designing Support

Mark A. Neerincx
TNO Human Factors, The Netherlands

Abstract

I present a method for cognitive task analysis that guides the early stages of software development, aiming at an optimal cognitive load for operators of process control systems. The method is based on a practical theory of cognitive task load and support. In addition to the classical measure of percentage time occupied, this theory distinguishes two load factors that affect cognitive task performance and mental effort: the level of information processing and the number of task-set switches. Recent experiments have provided empirical support for the theory, showing effects of each load factor on performance and mental effort. The model can be used to establish task (re-)allocations and to design cognitive support.

This chapter provides an overview of the method's foundation and two application examples. The first example is an analysis of the cognitive task load for a future naval ship control center that identified overload risks for envisioned operator activities. Cognitive load was defined in terms of task demands in such a way that recommendations could be formulated for improving and refining the task allocation and user-interface support. The second example consists of the design and evaluation of a prototype user interface providing support functions for handling high-demand situations: an information handler, a rule provider, a diagnosis guide, and a task scheduler. Corresponding to the theory, these functions prove to be effective, particularly when cognitive task load is high. The user interface is currently being implemented on the bridge of an icebreaker.

The two examples comprise an integrated approach on task allocation enhancement and design of cognitive support. The theory and method are being further developed in an iterative cognitive engineering framework to refine the load and support model, improve the empirical foundation, and extend the examples of good practices.

INTRODUCTION

In different work domains, such as defense, public safety, process control, and transport, the need to improve the deployment of human knowledge and capacities is increasing. Safety requirements increase (due to both company and general public policies), while fewer personnel should be able to do the necessary tasks by concentrating work in one central control room and applying further automation. For example, in the future, the risk for injury and loss should be minimized, while fewer personnel will have to manage high-demand situations and supervise complex automated systems in ships built for the Royal Netherlands Navy. Tasks that were allocated to separate jobs are currently being combined into new, extended jobs (e.g., adding platform supervision to navigation tasks on a ship's bridge). In addition to selection and training, adequate (dynamic) task allocation and computer support can help to realize the required human performance level. This chapter presents a method for refining task allocations and designing support functions that extend human knowledge and capacities.

The proposed method for cognitive task load design is based on recent research in the field of human–computer interaction (HCI) and the notion that one needs both a model of the environment and the cognitive processes involved to enhance the design of human–computer work. We can make effective use of theories of cognitive processing if we have also validated theories or descriptions of the world on which cognition operates including the interactions (Green, 1998). When we have an adequate description of the environment in which cognition operates, then a human viewed as a behaving system might prove to be quite simple. Or, as Simon (1981; p. 95) stated, "The apparent complexity of human behavior over time is largely a reflection of the complexity of the environment in which a person finds himself." Nonetheless, cognitive theory is still necessary, however simple it may be, to distinguish the environmental components that affect human cognitive task performance. Such a theory should not purely focus on task performance at the microlevel and only be validated with isolated laboratory tasks as common in basic (experimental) psychological research. The validation of the models should incorporate essential interactions with the real-world environment in which the tasks are performed at the level of real-world operational requirements. Such descriptions enable statements on human task performance by accounting adequately for how context and actions are coupled and mutually dependent (cf. Hollnagel, 1998). Unfortunately, there is no single, context-independent, comprehensive theory on human cognition that can be applied for a complete and doable analysis of complex work demands, and we will not have one in the near future. For example, detailed specifications of cognitive processes such as the unified theories of Newell (1990) and the multiple-resource model of Wickens (1992) insufficiently address the dynamic task demands on human problem solving in a naval command center. The solution is not to wait until such a theory has been developed, but to develop limited or practical theories that apply to the specific domain or environmental description that is part of it (cf. Green, 1990). Such a theory should include accepted features of cognition, such as limited processing capacity; be validated in the context of a specific domain and possibly group of task performers; and provide predictions of the task performance within this domain (cf. the simple model of cognition; Hollnagel, 1998).

This chapter presents a practical theory of cognitive task load and computer support for process control tasks. The theory has been integrated in a cognitive engineering framework consisting of the specification and assessment of computer-supported work (Fig. 13.1). According to this framework, assessments guide the iterative HCI refinement process (including possible adjustments of operational requirements) and provide empirical data for improving the theory and its quantification in a specific application area (e.g., a "mental load standard" for railway traffic control; Neerincx & Griffioen, 1996). First, I summarize the theory on cognitive task load and provide an example of the corresponding method for task-allocation assessment

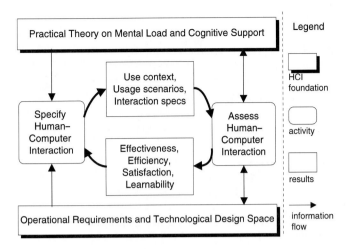

FIG. 13.1.   Addressing human-factors knowledge and environmental constraints in an iterative system development process.

in an early system development stage. Subsequently, I extend the theory with cognitive support concepts and show that the theory and method can also be applied for user-interface design. For explanatory objectives, task allocation and user-interface design are dealt with separately. At the end of the chapter, I elucidate that the two examples comprise *one* practical theory and method to guide an iterative process of human task and interface design.

COGNITIVE TASK LOAD

Task Demands

Recently, we developed and validated a practical theory for cognitive load and support in process-control tasks (Neerincx, van Doorne, & Ruijsendaal, 2000; Neerincx & van Besouw, 2001). The theory includes a model of cognitive task load that comprises the effects of task characteristics on performance and mental effort. According to this model, cognitive task load is a function of the percentage time occupied, the level of information processing, and the number of task-set switches. The first classical load factor, *percentage time occupied,* has been used to assess workload in practice for time-line assessments. Such assessments are often based on the notion that people should not be occupied more than 70 to 80% of the total time available (Beevis, 1992). To address the cognitive task demands, our load model incorporates the skill–rule–knowledge framework of Rasmussen (1986) as an indication of the *level of information processing.* At the skill-based level, information is processed automatically resulting in actions that are hardly cognitively demanding. At the rule-based level, input information triggers routine solutions (i.e., procedures with rules of the type "if <event/state> then <actions>") resulting into efficient problem solving in terms of required cognitive capacities. At the knowledge-based level, based on input information the problem is analyzed and solution(s) are planned, in particular, to deal with new situations. This type of information processing can involve a heavy load on the limited capacity of working memory.

To address the demands of attention shifts, the model distinguishes *task-set switching* as a third load factor in the performance of process control tasks. Complex task situations consist of several different tasks with different goals. These tasks appeal to different sources of human knowledge and capacities and refer to different objects in the environment. We use the term

Abstraction Level

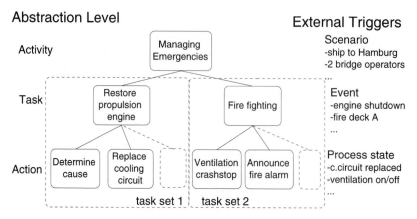

External Triggers

FIG. 13.2. The descriptive framework for multiple task performance by one operator (who must deal with an engine shutdown and fire on deck A in the example).

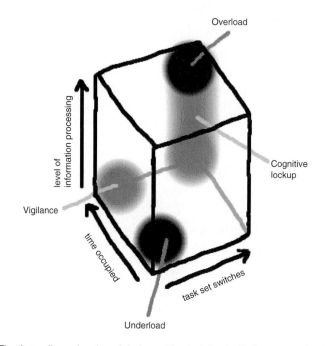

FIG. 13.3. The three-dimensional model of cognitive task load with four general problem regions.

task set to denote the human resources *and* environmental objects with the momentary states, which are involved in the task performance. Switching entails a change of applicable task knowledge on the operating and environment level. Figure 13.2 shows a model for multiple task performance that I use here to specify the goal-directed and situation-driven elements of computer-supported human behavior. Three abstraction levels of human behavior are distinguished. First, an *activity* is the combination of tasks and actions that are performed to accomplish a general goal in a definite period and for a specific scenario (e.g., damage control on a cargo ship during a nighttime storm). A scenario consists of an initial state of the ship and environment and a sequence of events that trigger tasks. Second, *tasks* are performed to accomplish a subgoal (e.g., restore the propulsion engine). Task set is used to denote the composite of goal, knowledge, and processing resources of the operator and the corresponding objects in the environment. A task is activated by an event (e.g., engine shutdown) or a predefined goal of the task executor. Third, *actions* are the elements of observable behavior that affect the state of a specific object in the environment. The process state determines which action is active or should be activated.

Three-Dimensional "Load Space"

The combination of the three load factors determines the cognitive task load: the load is high when the percentage time occupied, the level of information processing (i.e., the percentage knowledge-based actions), and the number of task-set switches are high. Figure 13.3 presents a three-dimensional "load" space in which human activities can be projected with regions indicating the cognitive demands that the activity imposes on the operator. It should be noted that these factors represent task demands that affect human operator performance and effort (i.e., it is not a definition of the operator cognitive state). In practice, operator activities will not cover all possible regions in the cube of Fig. 13.3. A higher level of information processing may cause the time occupied to increase. Also a larger amount of task-set switches may cause the time occupied to increase because the costs of these switches are so severe that the operator needs more time to execute the task. The cognitive task load analysis of this chapter aims at a cube that is "empty" for the critical regions that are described in the following section. For the remaining critical situations, it aims at empowering the operators so that they can meet the specific demands.

 It should be noted that the effects of cognitive task load depend on the concerning task duration (see Table 13.1). In general, the negative effects of under- and overload increase over time. *Underload* will only appear after a certain work period, whereas (momentary) *overload* can appear at any moment. When task load remains high for a longer period, carryover effects can appear reducing the available resources or capacities for the required human information processing (Rouse, 1988). *Vigilance* is a well-known problematic task for operators in which the problems increase over time. Performance decrease can occur after only 10 minutes when an operator has to monitor a process continuously but does not have to act (Levine, Romashko, &

TABLE 13.1
Overview of the Four Negative Effects of Cognitive Task Demands for a Certain Task Period

	Task Performance Period		
	Short (<5 min)	Medium (5–20 min)	Long (>20 min)
Time occupied low Info processing low Task switches low	No problem		Underload
Time occupied high Info processing low Task switches low		No problem	Vigilance
Time occupied high Info processing all Task switches high		Cognitive lockup	
Time occupied high Info processing high Task switches high		Overload	

Fleishman, 1973; Parasuraman, 1986). Vigilance can result in stress due to the specific task demands (i.e., the requirement to continuously pay attention to the task) and boredom that appears with highly repetitive, homogeneous stimuli. Consequently, the only viable strategy to reduce stress in vigilance, at present, appears to be having the freedom to stop when people become bored (Scerbo, 2001). Recent research on *cognitive lockup* shows that operators have fundamental problems managing their own tasks adequately. Humans are inclined to focus on one task and are reluctant to switch to another, even if the second task has a higher priority. They are stuck in their choice to perform a specific task (Boehne & Paese, 2000) and have a tendency to execute tasks sequentially (Kerstholt & Passenier, 2000).

In a sequence of experiments, we validated the three-dimensional load model. For example, a study in the high-fidelity, "one-to-one," control center simulator of a Royal Netherlands Navy frigate showed substantial performance decrease in the problem areas of Fig. 13.3 (i.e., for specific load values). In general, empirical research should provide the data to establish the exact boundaries of the critical regions for a specific task domain (e.g., by expert assessments, operator performance evaluations, or both).

Analysis Method

Neerincx et al. (2000) described an analysis method for human–computer task performance to establish the task demands in terms of the three load factors in Fig. 13.3. This method combines specifications of task-set classes with specific task-set instances (i.e., activities). Figure 13.4 shows the specification process of the goal- and situation-driven task elements with their mutual relations.

A *task decomposition* describes the task classes. It defines the breakdown structure of tasks and provides an overview of the general task objectives assigned to persons and related to detailed descriptions of human actions with the corresponding information needs.

An *event list* (hierarchically ordered) describes the event classes that trigger task classes and provides an overview of the situation-driven elements. An event refers to a change in an object that can take place in specific situations and has specific consequences.

Scenarios describe sequences of event instances with their consequences on a time line and the initial state of the objects that events act on. They show a particular combination of events and conditions that cause one line of user actions. Scenarios can be a valuable tool to

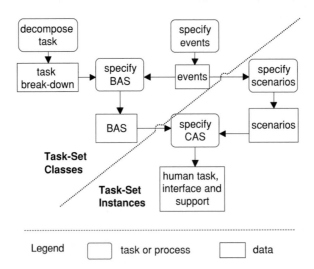

FIG. 13.4. Processes and data flows for the specification of task demands in the analysis stage of system development.

envision computer-supported work and to estimate the costs and benefits of the support for the human task performance (Carroll, 1995). In particular, scenarios can provide a "bridge" between the software engineering focus on software's functionality and human factors focus on users' goals and information needs (cf. "use cases" in object-oriented software engineering). In our method, scenarios consist of a sequence of events that occur in a specific state. The purpose of this description is not restricted to the general mission of an abstract function, such as damage control. Scenarios are formulated for a large set of action-triggering events and, therefore, they can be rather specific.

A set of *basic action sequences* (BASs) describes the general relationships between event classes and task classes as general procedures without situation specifics. Action sequence diagrams define the dynamic or control aspects of task performance (i.e., the events and conditions that trigger task execution sequences). These diagrams are a combination of specifications for time lines (e.g., Kirwan & Ainsworth, 1992), operational sequences (e.g., Kirwan & Ainsworth, 1992), and cooperation processes (e.g., Neerincx & de Greef, 1998).

A set of *compound action sequences* (CAS) describes the relationships between event instances and task instances for specific situations as activities with the corresponding interface support. Per scenario, the BASs of the events are instantiated and integrated into a CAS that do not contain feedback loops and selection mechanisms (such as if *x* then *y*), so that the time line can be established per chart (for a proposed CAS format, see Fig. 13.7).

ANALYSIS OF TASK ALLOCATIONS: AN EXAMPLE

Recently, the Royal Netherlands Navy developed the Integrated Monitoring and Control System (IMCS) for a new ship: the Air Defence and Command Frigate. The IMCS is a large and complex system that is used in diverse situations and work settings for supervision and control of the platform systems and damage control. In the first phase of the IMCS development, the high-level system requirements were provided as government-furnished information. This section gives a brief overview of the method for cognitive task analysis that was applied to assess these system requirements (for details, see Neerincx et al., 2000).

Specification

To make the task demands explicit for the IMCS operators, we followed the specification process of Fig. 13.4. First, we specified the *task-set classes*. At the top level, the task decomposition distinguished four platform functions: provide survivability, provide mobility, support hotel functions, and support weapons and sensors. In the first instance, the task breakdown stopped when all task allocations to the control-center crew and IMCS could be designated within it. After establishing the jobs as a set of tasks that have to be performed by one person, decomposition continued until the subtask can be mapped on a specific IMCS support function of the Air Defence and Command Frigate or defined as a "pure" human action, so that either the HCI or an observable human action is specified.

Subsequently, we transformed the general task-set specifications into *task-set instances* and "envisioned" about 40 scenarios. Two critical scenarios that differed from each other fundamentally were selected for further analysis. The first, "fire in the galley in harbor," consists of an "unexpected" calamity with extra complications occurring in a quiet situation and a small crew; the second consists of severe damage in wartime, comprising a hectic situation that the complete crew must be able to handle. The two selected scenarios were transformed into CASs using the BASs for handling the events of this scenario. Navy experts estimated for each action the fastest and slowest performance and the level of information processing according to the skill–rule–knowledge framework. This resulted in two CASs per scenario: one consisted of the fastest performances and the other the slowest. The actual action sequences consisted of a number of actors, the action times were presented in them, and the individual BASs were coded separately to get an overview of the number of switches between task sets in an activity. For Scenario 1, the fast CAS lasted about 18 minutes, the slow version about 56 minutes. For Scenario 2, the fast CAS lasted about 25 minutes and the slow version about 72 minutes.

Assessment

In the first step of the assessment, general patterns and extremes of task load were identified. The percentage time occupied, the percentage knowledge-based actions, and the number of task-set switches for each person of the future ship control-center crew can be directly derived from the CASs. Differences in the time a person is occupied appeared mainly between scenarios. There was a large variance, between 8% and 70% occupied. Time occupied did not appear to be a cause of overload on its own, because it remained below the critical level of 70 to 80% (Beevis, 1992). Overall, the tasks of managers and operators showed a large knowledge-based component for which system and process knowledge is required. The work is complex and cognitively demanding for the complete crew. Managers' tasks comprise primarily planning, supervision, prioritizing, and coordination, whereas the operators have to assess, diagnose, and solve specific technical problems. The future situation in the Air Defence and Command Frigate requires that the operators have knowledge of specific parts of the platform control system and tasks with which they are involved. In the cognitive task analysis method, a task set is defined as a BAS for a specific event. For example, when the same BAS appears more than once in a CAS, they are viewed as different task sets, because they apply to different events (i.e., different objects in the environment). Task-set switching proved to appear rather often. In particular, for Manager 1, the number of task-set switching was high in the fast Scenario 2: 54 switches in 1 hour (i.e., a switch every 67 seconds). The number of switches increased in Scenario 2 when action times were longer, because the operational requirements were more difficult to satisfy in this condition.

In the *second step* of the assessment, situations of momentary overload were identified. The CASs show the action times of each person and the interrelationships between the actions: the critical path. Often, more than one person is on the critical path, so that suboptimal performance of one person at a specific moment will often have a major effect on the overall ship control center crew performance. Therefore, it is of utmost importance to detect possible peak loads for all persons. Compared with the general load, for momentary peak loads the timescale of occurrences of almost continuous knowledge-based actions with a large number of switches is much shorter (between 5 and 15 minutes) and the load limit is higher. For example, the momentary load of Operator 2 in the fast condition of Scenario 2 proved to be relatively high. In a period of 5 minutes, the operator should have to switch every 20-seconds to a new task set that almost always comprises a knowledge-based action. It can be expected that the operator will not be able to fulfill these task demands and, because he or she is on the critical path, this will have an impact on the overall crew performance. Further more, in this period the operator performs 19 interactions with the IMCS. For this specific action sequence of Operator 2, it is important that the dialogue with the IMCS is as efficient as possible (i.e., the user-interface structure should map well on this sequence).

Conclusions

The assessment of the four CASs showed possible risks for overload that are mainly caused by the composite of measures on time occupied, percentage of knowledge-based actions, and the number of task-set switches. Because cognitive load was described in terms of task and interface characteristics, recommendations could be formulated for task allocation and interface design to diminish these risks. The IMCS specification for the future Air Defence and Command Frigate should be improved by describing a general coherent user-interface structure and establishing the dialogue principles for its components. In particular, the interface should enable efficient task-set switching and may even provide support to keep track of task sets that "run in the background" and to return to these task sets. It should be noted that the IMCS requirement specifications defines a number of individual support components from sensor-fusion and filtering mechanisms to damage control advice functions. Each function will probably have a positive effect on the local task performance. However, an overview on the interrelationships and the combined effects of these functions is lacking. For example, a general "high-level" user-interface structure is lacking and the management of the support functions can be a load factor in itself (i.e., the control of the envisioned information presentations). To establish support for the managers of the crew, to diminish peak loads, and to prevent human biases such as cognitive lockup, the combined effects and integration of the support in the overall task performance should be defined explicitly. Human-centerd development of interactive systems requires an iterative design process in which cognitive engineers provide the required human factors input in terms of guidelines, user-interface concepts, methods, and facilities (cf. ISO 13407). Because the IMCS development had already been started and the development process defined, it was difficult to bring these insights into this process.

HIGH-DEMAND SITUATIONS AND COGNITIVE SUPPORT

Neerincx and de Greef (1998) proposed a model-based design approach that aims at human-computer cooperative problem solving by integrating cognitive support into the user interface. Based on this approach and the load theory (Fig. 13.3), we developed a framework that distinguishes a small set of user-interface and cognitive-support functions with specific high-level design principles. In this section, I present the framework, and in the subsequent section, I provide an example design consisting of these functions.

Time Occupied

There is a trade-off between the benefits of support facilities and the interaction costs. In particular, when the time pressure is high and the user has only a small part of his or her cognitive resources available to manage and consult such facilities, the benefits should, substantially outweigh the costs. The additional time required for interacting with the support facility should be small compared with the execution time for the primary task. Four general design principles reduce the interaction load that applies to the *user interface*:

- *User adaptation.* The user interface design should account for the general characteristics of human perception, information transfer, decision-making and control, and the specific user characteristics with respect to education, knowledge, skills, and experience.
- *Goal conformance.* The functions and function structure of the user interface should map, in a one-to-one relation, on users' goals and corresponding goal sequences. Functions that users don't need should be hidden.
- *Information needs conformance.* The information provided by the user interface should map, in a one-to-one relation, on the information needs that arise from users' goals. Irrelevant information should not be presented to the users.
- *Use context.* The HCI should fit to the envisioned use context or situation (e.g., a speech interface should not be designed for noisy environments).

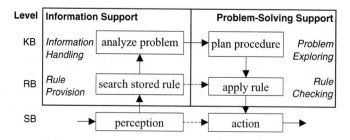

FIG. 13.5.   Four cognitive support functions at the knowledge-based (KB), rule-based (RB), and skill-based (SB) levels. The broken arrows represent "shortcuts" in the human information processing based on training, experience, and support.

Level of Information Processing

Based on the skill–rule–knowledge framework of Rasmussen (1986), I distinguish four support functions: rule provision, information handling, problem exploring, and rule checking. This section discusses these functions, following the human information processing steps of Fig. 13.5.

The *Rule Provision* function provides the normative procedure for solving (part of) the current problem. Due to training and experience, people develop and retain procedures for efficient task performance (i.e., they apply the rule-based "shortcuts" of Fig. 13.5). Performance deficiencies may arise when a task is performed rarely (so that procedures are never learned or are forgotten) or when the information does not trigger the corresponding procedure in human memory. For these situations, rule provision aims at supplementing human procedural knowledge. We distinguish four design principles:

- The support function should take the initiative to provide information at the right time. Consequently, the user does not need to know when and how to search for information and does not need to invest in these actions.
- Rule provision should consist of context-specific, procedural task knowledge. The advice is minimal; no more than necessary is provided. Each individual (sub)procedure should however describe a complete problem-solving path to accomplish the (sub)goal.
- The user interface of the rule provision facility should be a well-integrated part of the human–machine dialogue. A minimal and consistent interaction requires little knowledge and contributes to efficient task performance.
- The advice should be provided in such a way that the user remains in the loop of the overall activity by interactive involvement in the process of action executions as part of a task procedure.

Information handling support filters and combines information to improve situation awareness (i.e., knowledge of the state of the system and its environment; Endsley, 1995). Because of the increasing availability of information, situation awareness can deteriorate without support. Sensor information should therefore be combined into alarms that are structured according to their function, such as fire control, propulsion, and energy supply. Furthermore, information handling can support the operators in keeping overview by making the structure of the complete system explicit at a global level and by indicating functional relationships between system components. Information handling support should enhance information acquisition and recall in such a way that situation awareness is optimal for the current task performance. Three design principles for this type of support can be formulated:

- An information handling support function should provide an overview of state and process variables, showing the correspondence to system's components (i.e., structure) and the fluctuations in time (history).
- Alarms should be prioritized according to the current situation and provide information about how to (begin to) solve the problem. Important alarms should stand out, and irrelevant alarms should be hidden.
- The support should enable fast, easy access to the requested information with adequate orientation cues and state explanation. It should correspond to the optimal search strategy for the specific task and situation (i.e., support several accurate information acquisition processes of users).

Problem exploring comes into play when there is not a complete (executable) procedure available to deal with the current alarms and situation. First, the problem and solution space has to be analyzed. Subsequently, based on information about the environment (state, process) and information from memory, a procedure must be planned for solving the problem. Based on a mental model (i.e., an internal representation of the system), the person sets local goals, initiates actions to achieve them, observes the extent to which the actions are successful and, if needed, poses new subgoals to minimize the discrepancy between the present state and the desired state. A problem-exploring function consists of a knowledge-based component that can execute some problem-solving activities such as the generation of hypotheses and the selection of an urgent and most promising one. Another possibility is to provide predictions of future states based on the current user actions (e.g., predictive displays; Wickens, 1992). The benefits of such predictions can be great for a number of tasks if the "predicted path" is explicitly presented and well integrated into the overall presentation of state information. There are three design principles for problem exploring support:

- The user must understand what the support function is doing and why, so that, for example, the user remains in the loop of task execution.
- The problem-solving process of the support function should be compatible with user's problem-solving process and enable the involvement of specific user's capabilities.
- For providing predictions of future states, the "predicted path" should be explicitly presented and well integrated into the overall presentation of state information.

Rule checking functions recognize when the human operator has strayed from the normative problem-solving path and help him or her to reach a correct task outcome (Silverman, 1992). As task difficulty increases, however, a point will be crossed at which subject-matter experts can no longer be assisted by rule checking alone. Thus, under conditions of high task load, this kind of support seems not to be optimal. A further restriction is that the users must have some knowledge to start their task execution. If they do not know which goals to achieve, then they cannot be critiqued. In general, the first three principles that were identified for rule provision also apply to rule checking.

Task-Set Switching

Task-set switching support should comply with the following design principles:

- For the momentary activity, it should provide an up-to-date overview of the tasks with corresponding actions, including the current status of the activity and the status of each task.

- The current priority of each task should be shown. Changes in priority should be communicated to the users, so that they can keep a correct, up-to-date situation awareness.
- Humans are inclined to focus on the tasks they are working on, neglecting tasks that may have higher priority ("cognitive lockup"; see the previous section). The support functions should confirm whether users do the required abstraction from action level to the task and activity level.

DESIGN OF COGNITIVE SUPPORT: AN EXAMPLE

The previous section distinguished three load factors and a small set of corresponding support functions that can reduce the negative effects of each factor on task performance and mental effort. This section provides an example design of the user's task, support, and interaction for the ship control system manager that is being developed in the European ATOMOS and DISC projects. To exemplify the design method, I provide small (simplified) task descriptions for four support functions that reduce the negative effects of a specific load factor on an integrated ship's bridge (see Table 13.2):

- An *information handler* that supports task-set integration for keeping overview of the overall system's state. Based on the system structure and current events, this function provides an overview of alarms and a set of integrated views (i.e., system overviews).
- An *emergency scheduler* that supports task-set switching by providing an overview of prioritized alarms that have to be handled or are being handled, based on the overview of alarms.
- A *rule provider* that supports task-set switching by providing the context-specific procedures for each emergency with the current state of each procedure. Context information comprises the state of the objects in the task set, such as the position of the ship (e.g., harbor at open sea), the location of an emergency (e.g., a fire in room 12) and the maintenance data of a machine (e.g., pump X replaced last week). The combination of emergency scheduler and rule provider provides the action plan for the operator.
- A *diagnosis guide* that supports task-set integration for alarm handling. This guide consists of an overview of possible symptom-cause relations between alarms (e.g., ordered by probability). Based on context information, the alarm overview, and the system structure, relations between the alarms are proposed, and based on the settled relations, the context information is refined.

User Interface Design

In general, user interface design follows the general top-down process of software development resulting in user interface specifications at three abstraction levels (Neerincx, Ruijsendaal, &

TABLE 13.2
Four Example Support Functions That Reduce the Negative Effects of Each Load Factor

Load Factor	Support Type	Support Function
Time occupied	Information handling	Information handler
Level of information processing	Rule provision	Rule provider
	Problem exploring	Diagnosis guide
Task-set switching	Task managing	Emergency scheduler

Wolff, 2001). On the first level, based on users' goals and information needs, the system's functions and information provision are specified (i.e., the *task level* of the user interface is determined). On the second level, the control of the functions and the presentation of the information is specified (i.e., the "look-and-feel" or the *communication level* of the user interface is established). On the third level, the interface design is implemented in a specific language on a specific platform (i.e., the actual interface or the *implementation level* of the user interface is established). The distinction between these three specification levels is not intended to advertise the "old" waterfall software life cycle. It is based on the notion that software development is a top-down and an iterative process. For example, specific software components can be developed down to the implementation level after which a reanalysis phase starts for these or new components (or both). The next two subsections present the specifications of the user interface at the task and communication level respectively.

Task Level

Usability at the task level is established by mapping user's goals—and corresponding goal sequences—on system's operations, and mapping user's information needs on system's information provision. First, such an analysis must specify the functions that users need for specific goals and the corresponding goal- and situation-driven action sequences. Functions not needed should be hidden from the users (i.e., minimal interface; Carroll, 1984). Second, the mapping of users' information needs on the system's information provision is performed by means of an information analysis. Such an analysis establishes what information is required to accomplish a specific goal at a specific moment. Corresponding to the example analysis of task allocations in the Air Defence and Command Frigate, we followed the specification process of Fig. 13.4 for the design of cognitive support in the integrated ship's bridge.

Decomposition of Tasks.   We specified and subsequently enriched the task decomposition with data flow diagrams showing the information that has to be communicated between the tasks. The data flows between human tasks and machine tasks provide a high-level specification of the user interface (see Fig. 13.6).

Specification of Events.   A distinction can be made between internal events (i.e., events that arise in the ship itself such as a malfunction in the electricity supply) and external events (i.e., events that arise in the ship's environment such as the appearance of another ship). These classes can be subdivided further with increasing level of detail. Table 13.3 shows part of an event list containing events of task sets that must be accomplished with different applications, so that they can be used to show the interface-support functions of the system manager.

Specification of Basic Action Sequences.   This section offers an example action sequence in which two simple procedures (BASs) for fire control and dealing with an engine shutdown are integrated. The fire control procedure consists of alert crew, close section, plan attack route, close doors, extinguish fire, plan smoke removal, close doors and remove smoke. The shutdown procedure consists of detection, determine cause, solve problem, and restart engine.

Specification of Scenarios.   To specify the interface support of the system manager, scenarios are created with multiple events from different domains. Table 13.4 presents an example of such a scenario: its initial state, the events with time of appearance, location,

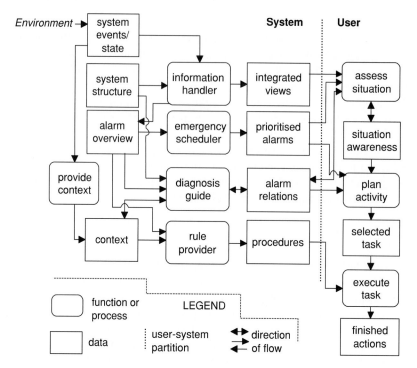

FIG. 13.6.   The processes and data flows for the support of operator's situation assessment, activity planning, and task execution.

details, consequences, and source (i.e., the detector that provides the event to the operator). This scenario contains a number of different events that require integration of (sensor) information, diagnosis, procedural task knowledge, and task-set switching. Therefore, this scenario can be used to envision the support functions of the system manager. A short circuit in a cooling pump causes a fire, and the corresponding fire alarm is presented to the operator. This short circuit also causes a malfunctioning cooling circuit, so that the temperature of the engine increases and is above the maximum level for a specific duration, resulting in the presentation of the corresponding temperature alarm. The high temperature causes an engine shutdown, and the corresponding alarm is presented to the operator. First, the scenario imposes task-set switching between dealing with the fire and the engine shutdown. Second, the scenario imposes task-set integration: the operator has to abstract from the fire and shutdown tasks to infer the relations of these separate alarms. Subsequently, the shutdown task can be continued with the correct procedure.

Specification of Compound Action Sequence.   Scenario 1 is transformed into a CAS using the procedures (BASs) for handling the events of this scenario. The CAS provides an overview of actions and processes performed during the scenario. Figure 13.7 presents the CAS for Scenario 1. The following components can be distinguished:

• *Actors.* In this CAS, there are two actors (the system and the operator). The system actor is subdivided in two columns (the applications and the system manager). The operator column is subdivided in columns representing executions of three tasks: the fire, shutdown, and assess-and-plan task.

TABLE 13.3
Example Event List

Event Category			Basic Event	Example Event	Example Consequence
External	Weather	—	—	Storm X approaching	Deviation from route
	Sea state	Current	—	—	—
		Traffic	Vessel passing	—	none
			Vessel gives way	Vessel X from port Y gives way	—
			Vessel changes to collision course	Vessel X from port Y changes to collision course	A collision occurs
Internal	Cargo		Gas leak	Release of toxic fumes	—
	System failure	Propulsion	Engine shutdown	A temperature rise caused shutdown engine X	Ship cannot maintain its current speed
			Temp. above maximum level	Engine X's temperature exceeds set point	Engine speed is limited
		Collision avoidance	—	Collision sensor X malfunction	A collision occurs
		Navigation	—	Ship off course	Ship can't arrive at estimated time
		Electric	Short circuit	Short circuit in cooling pump X	Engine's cooling circuit out of order
		Fire		Fire in the engine room	Engine speed is limited

- *Time line.* The vertical axis shows the time line. It should be noted that the time line is not linear (a block sometimes represents 6 minutes, whereas at another place it represents .5 minute), so that the diagram can zoom in on a time period where more actions are performed.
- *Events.* A column is used to present events that occur during the scenario. Figure 13.7 shows four events: short circuit, fire, maximum temperature engine exceeded, and engine shutdown. The first is not detected by the actors (i.e., no flow of actions or processes result directly from this event).
- *Actions.* At a specific moment, the operator performs actions belonging to a task (column).
- *Processes.* The system manager and applications run several processes to support the users with their task performance.
- *User–system partition.* The broken line between the two actors shows the user–system partition (i.e., the human–machine interaction at the task level). The user interacts with multiple processes at the same time. For example, during the fire task, the operator uses the rule provision process of the system manager. At the same time, the user performs operations using the fire control system application. In particular, Fig. 13.7 shows the interaction with the support functions for task-set switching and task-set integration: information handler (IH), scheduler (SC), rule provider (RP), and diagnosis guide (DG).

TABLE 13.4
Scenario 1, Fire in the Cooling System of the Engine

Initial state: Ship is en route to Hamburg; there are two operators present on the bridge

Time	Event	Additional Information
21:54	Short circuit	Location: Engine's cooling pump in engine room Details: Short circuit causes a fire in the pump, which is located in the cooling system Consequences: Cooling system will not work, and the engine temperature will increase Source: None (event is not detected by system)
22:03	Fire	Location: Engine room Details: A pump in the engine room is on fire Consequences: Unknown Source: Smoke detector of fire control system
22:06	Max. engine temperature	Location: Engine room Details: The temperature of the engine increased beyond the set point Consequences: The engine shuts down after a period of high temperature Source: Propulsion management system
22:08	Engine shutdown	Location: Engine room Details: The temperature was too high for the critical period Consequences: The vessel cannot maintain its current speed Source: Propulsion management system

Communication Level

The previous subsection specified the user interface at the task level: a high-level description of the HCI and the role of the system manager. This specification will be transformed to a low-level description of the presentation and control elements based on user requirements, HCI standards, and technical constraints. The plain interface is an interface for operators who always have sufficient knowledge and capacities available to execute their tasks (Neerincx & de Greef, 1998). The system manager integrates context-specific task support into this interface to complement possible knowledge and capacity deficiencies, resulting into the support interface of the system manager. First the plain interface of the application manager is described and then the support that is integrated into this interface by the system manager.

Plain Interface. Figure 13.8 shows the plain user interface that consists of three parts. The *status area* is shown on the top part of the screen. This area has a fixed size, is always available and is never covered by overlapping windows or dialogue boxes. The main role of the status area is to present real-time status and alarm information. An "alarm bell" is shown on the left side of the status area. The alarm bell is used to indicate the alarm status: no emergencies (grey), emergencies (red), new emergencies (a short period of blinking red and grey accompanied by a modulated sound signal), and priority raise of emergencies (blinking and sound).

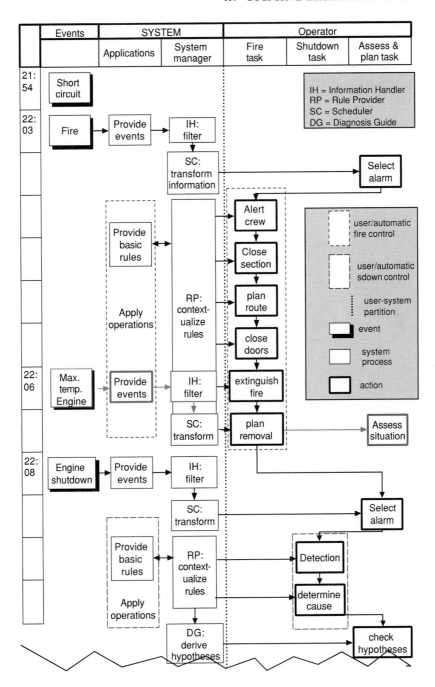

FIG. 13.7. Part of the compound action sequence for Scenario 1.

The *application presentation area* in the middle of the screen is used to present the active application (Fig. 13.8 shows the navigation application).

The *common control area* at the bottom of the screen has a fixed size, is always available, and is never covered by overlapping windows. Its main use is to switch from one application to another. The first button on the left is added to the application manager interface to switch the system manager's support on or off (as described later).

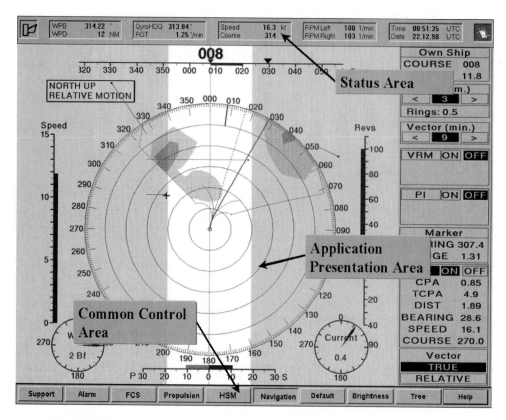

FIG. 13.8.   The "plain" interface of the application manager (with an extra support button).

Support Interface.   Figure 13.9 provides an example of the user interface in support mode (i.e. an example interface state during fire control). Two new areas can be identified in this mode. An emergency presentation area situated directly underneath the status area and a procedure presentation area on the left side of the application presentation area. The application presentation area is somewhat smaller than in the plain interface to make room for the two new areas.

A first type of information handling appears in the emergency presentation area combining low-level alarms (e.g., sensor values) and categorizing the resulting emergencies into groups (in the current example, fire, propulsion, hull stress, and navigation). In this way, the amount of alarm information on the screen is reduced compared with "classical" alarm lists. This chapter presents an example for the system manager's user interface for a workstation with only one display. When the operator can use two displays, the emergency presentation area can be presented on a separate display. Another type of information handling support is offered by integrated views in the application tree (for example, an integrated view of the propulsion, electricity, track control, and cargo to show the interdependencies). Such a view offers an overview of state and process variables, the fluctuations in time, and the relevant relations between different systems (applications).

Scheduling support is located in the emergency presentation area. It is possible that two emergencies of the same type occur simultaneously. The emergencies are ordered in the appropriate group according to time of occurrence. Each emergency is presented as a hyperlink that "loads" the corresponding procedure in the procedure presentation area. Selection of an emergency is indicated by "inverted" video. Next to each emergency, a number is given to indicate the priority of that particular emergency. Next to the group name, a priority indicator

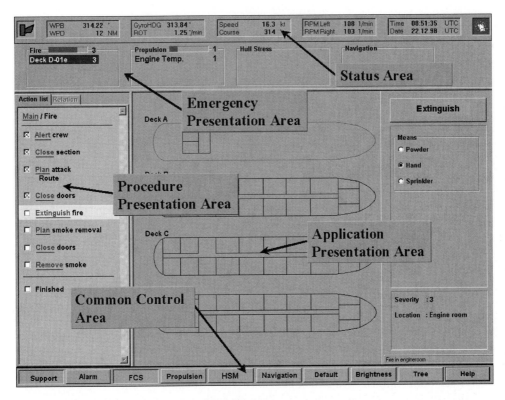

FIG. 13.9. The user interface in support mode.

(horizontal bar) and the corresponding number are given, showing the highest priority of that group (the fire group is assigned a priority of 3 in the example of Fig. 13.9).

The emergency presentation area provides the overall work plan for emergency handling. It can be viewed as a shared work plan if multiple operators are involved in such tasks, for example, one operator who executes a fire procedure and another who prevents a collision. These operators can help each other or switch tasks. Suppose that the operator who is engaged in a fire-fighting procedure needs advice from his colleague. The colleague can then obtain the same view as the other operator (but to prevent confusion he cannot operate the other operators procedure or application) and advise him. When the original "fire fighter" changes to another emergency (for example, the collision prevention), the procedure is released, and the other operator can continue at the right place in the procedure because of the check marks made by the other operator.

The procedure presentation area provides Web-browsing functionality and contains two tabs: "Action List" and "Relation." Rule Provision support is provided by the first tab, which presents a list of all actions that must be performed to deal with the selected emergency:

- For every emergency, the content of the page will be dynamically created: actions that apply to the current emergency and context are selected, the objects of these actions are instantiated, and the result is presented on the page. In this way, a context-specific concrete procedure can be formulated (Neerincx & De Greef, 1998). Each procedure step can contain hyperlinks that refer to the corresponding subprocedure or to an explanation of the content.
- Checkmarks can be placed in the appropriate checkboxes to indicate that a specific step in the procedure was completed. The background of the following step in the procedure is

highlighted while the background of the other actions is grey (i.e., the first non-check-marked step is highlighted). It is important to note that the user and the system manager can place check marks (in different colors). Each procedure ends with a "Finished" step to ensure that the emergency is really solved and to prevent an emergency from "suddenly" disappearing from the scheduler without the operator's being aware of it. States of procedures are preserved (by means of the check marks) until the emergency is over.

• The relevant application for the current (i.e., highlighted) step is activated and presented automatically in the application presentation area.

The Relation tab in the procedure presentation area contains *diagnosis guidance* (i.e., hypotheses about relations between two or more of the present emergencies; one of these emergencies must be the currently active emergency). Unlike the Action List tab, which is always available, the Relation tab is usually not available (indicated by a dim appearance). It becomes available (the name blinks a few times and become brighter) whenever one or more hypotheses about emergencies are discovered by the system manager (in the procedure presentation area of Fig. 13.9):

• Each hypothesis can contain hyperlinks that refer to a procedure to evaluate that hypothesis (when no procedure is available, more information about the hypothesis will be presented).

• A toggle switch is presented before each hypothesis to indicate whether the hypothesis is true (Y), not true (N), or still open (no button pushed). Each diagnosis of the relations ends with a "Finished" checkbox.

• After the diagnosis is finished, the system manager will indicate and explain whether the choices were consistent in a dialogue box. When the choices were not consistent, the Relation tab is shown again after this box. When the choices were consistent, the dialogue box proposes changes in the context model that corresponds to the hypotheses assessment. The operator can either approve of these changes (by clicking <OK>) or not approve it (by clicking <Not OK>). When the operator presses <OK>, the changes are made and immediately used to dynamically improve the procedures; the operator is returned to the Action List tab that contains the improved procedure. When the operator presses <Not OK>, the changes are not made and the operator is returned to the 'Relation' tab.

Evaluation

A storyboard visualizes a number of possible consecutive steps within an interaction sequence situated in typical use context. We created storyboards for the support interface to collect feedback from domain experts and possible users and to prioritize the support functions taking technological design constraints into account. Our storyboards consisted of a sequence of screen captures (such as Fig. 13.9) for a specific scenario or compound action sequence with a narrative description of the human–computer dialogue. In the first implementation stage, another project partner implemented the most cost-effective and feasible functions in a demonstrator. For example, the implementation of a diagnosis guide was postponed, because empirical foundation of its effectiveness was lacking and the required technology is rather advanced.

Corresponding with the cognitive engineering approach of the introduction in this chapter (see Fig. 13.1), we did an internal analytical assessment of the user-interface specification, up to the level of storyboards, to test if it complies with the user requirements and HCI guidelines. It was shown that the guidelines were generally well addressed. The user-interface design can be viewed as a manifestation of current human–computer interaction knowledge and corresponds with new interface proposals for naval ship control centers and manned space labs (Neerincx et al., 2001). The support is expected to diminish cognitive task load in critical situations, to facilitate dynamic allocation of tasks to multiple users, and to keep users in the loop of human–machine task execution. A small number of experiments validated central design principles and showed positive effects of the rule provider (Neerincx & de Greef, 1998; Neerincx et al., 2001). In an evaluation with 57 navy cadets of the Royal Netherlands Navy, the support functions led to a substantial increase in operator performance, especially at high task load. We did not find negative "out-of-the-loop effects" of the support, such as "automatic" acceptance of incorrect advices or decreased situation awareness (Grootjen, Neerincx, & Passenier, 2002). Part of the support system is currently being implemented on the bridge of an icebreaker in the European ATOMOSIV project.

DISCUSSION

This chapter presented an overview of a practical theory on cognitive task load and support that is being developed in a cognitive engineering framework (Fig. 13.1). Recent experiments in laboratory and real-world settings showed the large effects of the three load factors on task performance and mental effort and provided an empirical foundation of the cognitive support functions. The theory has been integrated in a method for cognitive task analysis that can be used in the early stages of system development processes. In this chapter, I chose to provide an overview of the theory and method and some example applications. The empirical foundation is provided elsewhere (e.g., Neerincx & van Besouw, 2001; Grootjen et al., 2002).

Recently, the cognitive task load model was integrated into an ERGOtool for the assessment of (future) assembly lines in industry via questionnaires and scenario observations. The ERGOtool is being applied and tested at various European companies (van Rhijn, de Looze, van Lingen, Tuinzaad, & Leskinen, 2001). In addition to this tool, we are developing simulations of the Compound Action Sequences that can be used to "refine" the task allocation and to assess the effects of support functions in an early software development stage. As another example, we applied the analysis method for cognitive task analysis to assess the task load of the operator in the future control room for a 6-kilometer motor traffic tunnel (the Westerscheldetunnel, Zealand, The Netherlands). The analysis identified a number of bottlenecks that should be resolved to guarantee adequate operator performance. In sum, current and near-future research will provide results to improve the method and its theoretical and empirical foundation. Persons involved in its development have mainly applied the method so that we have (as yet) limited experience of its "usability" (but see, for example, van Rhijn et al., 2001). The "complete" cognitive task analysis method comprises a lot of information with respect to the "load and support" theory, the specification and assessment techniques, the relevant user interface guidelines and standards, and the increasing set of examples. For the European Space Agency (ESA), we developed a customized version of the method for designers of the user interfaces in the International Space Station. To make the method easy to access, maintain, and update, it is being provided as an interactive guide (a Web-based usability handbook), providing sample applications and validations (i.e., enhancing the HCI foundation for space lab interfaces). We expect to assess its usability in the near future.

Our analysis approach focuses on (external) cognitive task factors at a "meso-level" (i.e., we do *not* deal with macrolevel work schedules in terms of days or microlevel dialogue issues in terms of milliseconds). A current project will further explicate the relations of task demands (e.g., of communication tasks) with internal cognitive and emotional states. More elaborate microlevel analyses, as supported by the information processing model of Hendy and Farrell (1997), are related to ours but are difficult to apply for complex real-world task environments (cf. Gray & Kirschenbaum, 2000; Nardi, 1996). In future, they may become more practical (e.g., by simulation environments) and help to further analyze task elements for specific details.

CONCLUSIONS

This chapter presented a practical theory on mental load and cognitive support in process-control tasks. The theory comprises a model of cognitive task load that describes load in terms of three behavioral factors: the percentage time occupied, the level of information processing, and the number of task-set switches. The higher the value of each factor, the higher the load will be. Furthermore, the practical theory distinguishes specific cognitive support functions that affect these values. For example, procedural support will diminish the *level of information processing* if it complies with a number of guidelines. A second example is task management support that will diminish the task-set switching demands if it fulfills a compound set of user requirements. The theory of mental load and cognitive support comprises the effects of task and interface characteristics on performance and mental effort to enable an iterative process of human task and interface design. For this objective, the theory is integrated into a method for cognitive task analysis that can be applied in the first phase of software development for the assessment of task allocations and high-level system specifications, and for the design of cognitive support functions.

In a first example, the method was applied to the assessment of the specifications of the Integrated Monitoring and Control System of the future Air Defence and Command Frigate. As a result of the analysis, the large number of task-set switches was identified as an overload risk for envisioned activities in the future ship control centers. Because mental load was described in terms of task and interface characteristics, recommendations could be formulated for task allocation and interface design.

In a second example, the cognitive task analysis method was applied to develop a support interface for the integrated ship's bridge. An evaluation with a prototype interface ·showed the benefits of the support (Grootjen et al., 2002). Based on this design and evaluation, the ATOMOSIV project is currently implementing a support interface on an icebreaker.

Taken together, these examples comprise an integrated approach on task allocation enhancement and design of cognitive support. Recent results of this approach are promising, providing indications that cognitive support can help to realize adequate (dynamic) task allocations.

ACKNOWLEDGMENTS

Research is teamwork. We are grateful to the Royal Netherlands Navy for their contribution to the cognitive task analysis. In particular, we would like to thank Eddie Flohr and Wim Helleman for their provision of domain knowledge and their comments on previous parts of the method. At TNO Human Factors, Jasper Lindenberg, Mark Ruijsendaal, and Sasja van Besouw provided substantial contributions to this research. The example design and evaluation of a prototype user interface is an outcome of the European DISC project.

REFERENCES

Beevis, D. (Ed.). (1992). *Analysis techniques for man-machine systems design, Vol 1 & 2*. NATO/Panel 8-RSG.14, Technical Report AC/243(Panel 8)TR/7. Brussels: North Atlantic Treaty Organization.

Boehne, D. M., & Paese, P. W. (2000). Deciding whether to complete or terminate an unfinished project: A strong test of the project completion hypothesis. *Organizational Behavior and Human Decision Processes, 2,* 178–194.

Carroll, J. M. (1984). Minimalist design for active users. In B. Shackle (Ed.), *Proceedings of Interact'94* (pp. 219–225). Amsterdam, North-Holland.

Carroll, J. M. (Ed.). (1995). Scenario-based design: Envisioning work and technology in system development. New York: John Wiley & Sons.

Endsley, M. R. (1995). Toward a theory of situation awareness in dynamic systems. *Human Factors, 37,* 32–64.

Gray, W. D., & Kirschenbaum, S. S. (2000). Analyzing a novel expertise: An unmarked road. In J. M. C. Schraagen, S. E. Chipman, & V. L. Shalin (Eds.), *Cognitive task analysis.* Mahwah, NJ: Erlbaum.

Green, T. R. G. (1990). Limited theories as a framework for human-computer interaction. In D. Ackerman & M. J. Tauber (Eds.), *Mental models and human-computer interaction (Vol. 1).* Amsterdam: Elsevier Science.

Green, T. R. G. (1998). The conception of a conception. *Ergonomics, 41,* 143–146.

Grootjen, M., Neerincx, M. A., & Passenier, P. O. (2002). *Cognitive task load and support on a ship's bridge: Design and evaluation of a prototype user interface.* Presented at the INEC 2002 Conference "The marine engineer in the electronic age." INEC 2002 Conference proceedings (pp. 198–207). London, England: Imar EST.

Hendy, K. C., & Farrell, P. S. E. (1997). *Implementing a model of human information processing in a task network simulation environment.* Report DCIEM No 97-R-71. North York, Ontario, Canada: Defence and Civil Institute of Environmental Medicine.

Hollnagel, E. (1998). *Cognitive reliability and error analysis method (CREAM).* Oxford, England: Elsevier Science.

ISO 13407 (1999). Human-centered design processes for interactive systems. Geneva, Switzerland: International Organization for Standardization.

Kersholt, J. H., & Passenier, P. O. (2000). Fault management in supervisory control: the effect of false alarms and support. *Ergonomics, 43,* 1371–1389.

Kirwan, B., & Ainsworth, L. K. (Eds.). (1992). *A guide to task analysis.* London, England: Taylor & Francis.

Levine, J. M., Romashko, T., & Fleishman, E. A. (1973). Evaluation of an abilities classification system for integrating and generalizing human performance research findings: An application to vigilance tasks. *Journal of Applied Psychology, 58,* 149–157.

Nardi, B. A. (Ed.). (1996). *Context and consciousness. Activity theory and human-computer interaction.* Cambridge, MA: MIT Press.

Neerincx, M. A., & de Greef, H. P. (1998). Cognitive support: Extending human knowledge and processing capacities. *Human-Computer Interaction, 13,* 73–106.

Neerincx, M. A., & Griffioen, E. (1996). Cognitive task analysis: Harmonising tasks to human capacities. *Ergonomics, 39,* 543–561.

Neerincx, M. A., Ruijsendaal, M., & Wolff, M. (2001). Usability engineering guide for integrated operation support in space station payloads. *International Journal of Cognitive Ergonomics, 5,* 187–198.

Neerincx, M. A., & van Besouw, N. J. P. (2001).Cognitive task load: A function of time occupied, level of information processing and task-set switches. In D. Harris (Ed.), *Engineering psychology and cognitive ergonomics, volume six: industrial ergonomics, HCI, and applied cognitive psychology* (pp. 247–254). Aldershot, England: Ashgate.

Neerincx, M. A., van Doorne, H., & Ruijsendaal, M. (2000). Attuning computer-supported work to human knowledge and processing capacities in Ship Control Centres. In J. M. C., Schraagen, S. E. Chipman, & V. L. Shalin (Eds.), *Cognitive task analysis.* Mahwah, NJ: Erlbaum.

Newell, A. (1990). *Unified theories of cognition.* Cambridge, MA: Harvard University Press.

Parasuraman, R. (1986). Vigilance, monitoring, and search. In K. R. Boff, L. Kaufman, & J. P. Thomas (Eds.), *Handbook of perception and human performance,* volume 2: cognitive processes and performance, chapter 43. New York, Wiley.

Rasmussen, J. (1986). Information processing and human-machine interaction: an approach to cognitive engineering. Amsterdam, Elsevier.

Rouse, W. B. (1988). Adaptive aiding for human computer control. *Human Factors, 30,* 431–443.

Scerbo, M. W. (2001). Stress, workload and boredom in vigilance: A problem and an answer. In P. A. Hancock & P. A. Desmond (Eds.), *Stress, workload and fatigue.* Mahwah, NJ: Erlbaum.

Silverman, B. G. (1992). Human-computer collaboration. *Human-Computer Interaction, 7,* 165–196.

Simon, H. A. (1981). *The sciences of the artificial (2nd ed.).* Cambridge, MA: MIT Press.

van Rhijn, G., de Looze, M., van Lingen, P., Tuinzaad, B., & Leskinen, T. (2001, September). International Federation of Automatic Control Symposium on Human Machine Systems (HMS 2001), Kassel, Germany.

Wickens, C. D. (1992). *Engineering psychology and human performance (2nd ed.).* New York: HarperCollins.

14

Cognitive Task Modelling for System Design and Evaluation in Nonroutine Task Domains

Nikolaos K. Tselios and Nikolaos M. Avouris
University of Patras, Greece

Abstract

This chapter concerns task-based design and evaluation of computer environments to support nonroutine tasks, such as exploratory learning and problem solving. Nonroutine tasks are characterised by exponential growth of expected task structures. Ways of tackling the complexity of this task structure is the objective of the described approach. The proposed development life cycle involves model-driven requirement specifications, design and implementation phase, followed by a field study evaluation phase. During the latter phase, evaluation of the effectiveness of the system in supporting the expected typical tasks is measured, and new requirements are derived that modify the original design. This spiral design process is supported by appropriate tools. Such tools typically allow expression of the designer expectations concerning user cognitive tasks. During field studies, the tools then support tracking of user interaction and inferring cognitive task models, which are matched against the original designer's assumption on the user cognitive models.

In this chapter, a survey of applications of cognitive modelling techniques in the design of systems involving nonroutine tasks is offered. Subsequently, an outline of a proposed design and evaluation framework and the related tools are described. Existing cognitive task modelling techniques such as hierarchical task analysis have been adapted, in the frame of the reported framework, to be used effectively for expression of both the designer model and the user task model concerning observed interaction. Examples of application of the developed techniques and tools in the design and evaluation of open problem-solving environments are also described.

INTRODUCTION

One class of computer-based environments that recently has drawn the attention of the research community is that of *open problem-solving environments*. These are typically interactive systems that support users' solving classes of problems with emphasis on the active, subjective, and constructive character of human–computer interaction.

In these environments, the user activity cannot be easily reduced to a sequence of predefined tasks and drills, as in many traditional industrial and other application environments. Specification of lower levels of abstraction in a task modelling process for such environments is necessary for good application design. However, the number of lower level tasks in this process expands exponentially and becomes impractical to address using traditional task analytic methods, as Miller and Vicente (1999) also observed.

These environments provide a context to let users actively explore certain abstract concepts. A set of tools is provided that the user can use to construct solutions to an open set of problems. Typically, such environments are described according to the following interrelated essential features: (a) a set of primitive objects, elementary operations to perform on these objects, and rules expressing the ways the operations can be performed and associated are defined; (b) a domain that relates objects and actions on the underlying objects to phenomena at the "surface of the screen." This determines the type of feedback the environment produces as a consequence of user actions and decisions, implementing direct manipulation of computational objects. By using these features, the user can study an object in a dynamic way by experiencing its properties.

The design and evaluation of these computer-based environments, which involve nonroutine tasks, such as problem solving and exploratory learning, is a nontrivial process. User interaction with these environments can take many forms and the tasks that can be performed with them are diverse and not fully anticipated. The design and evaluation involves study of possible alternative user strategies to face typical tasks and alternative ways in which the users could interact with the provided tools. The users can approach problems to be solved in different ways. They can also make mistakes during the process. Often these mistakes are a valuable part of the process; for instance, from the constructivist perspective of learning, these mistakes can be treated as opportunities to learn. Contrary to routine task environments like typical control room and other industrial human–machine systems, in our case detailed modelling of interaction and the cognitive processes involved is not possible because human–machine interaction can vary widely, depending on the characteristics of the specific problem and the user. Some observations on user interaction with these systems recognise the exploratory nature of the process, through which the user can advance the knowledge of the system and the tools concerned.

A question to be addressed in this chapter is the applicability of cognitive design approaches in this context. According to Inkpen (1997), existing human–computer interaction (HCI) design approaches primarily concern users engaged in typical workplace environments. As a result, design specifications and developed prototypes are often poor and need enrichment with observational data generated during field studies, thus necessitating an iterative, user-centred development process. In particular, one issue worth investigation is the applicability of well-established *task modelling* techniques in this context. Task and goal modelling have been used extensively in HCI research and practice to build representations of users' understanding, knowledge, goals, and task execution. The proposed models, based on cognitive science theoretical and experimental work, have already been applied both during design and requirements capturing as well as during user interface evaluation, as discussed in the overview of the state-of-the-art section of this chapter. Typically, however, these techniques attempt to model execution of routine tasks with ideal users trying to offer a mechanistic view of the user for prediction purposes instead of analysing actual users' behaviour. In this chapter, we define and demonstrate use of a design and evaluation framework for open problem-solving environments, based on adapted task modelling techniques. A number of tools that have been developed to support this process are also presented. This framework involves incremental application of cognitive task modelling techniques, starting from an expected ideal-user task model, defined during initial design, and subsequently comparing the observed user behaviour to this original model during field evaluation. Demonstration of this proposed methodology is also provided in this chapter. This is done through description of the evaluation and redesign of an open problem-solving environment.

This chapter focuses on our experience with design and evaluation of such open environments. In particular, we define a design framework based on task modelling, which involves a strong field-evaluation phase that has been the subject of long-standing research and experimentation. This is particularly suitable for nonroutine task design. Tools to support this process have also been built and described. The framework and the developed tools have led to promising results during the evaluation of an open problem-solving educational environment and are presented together with results obtained from the study.

This chapter is structured as follows: An overview of challenges in design and evaluation of open environments is provided first. Subsequently, a brief review of existing task design and evaluation techniques is outlined. A design and evaluation framework for open problem-solving systems is presented next, together with discussion on its applicability. Next, a short description of tools to support this framework is provided: the *Usability Analyser* (UAtool), which supports the field evaluation phase and the *Cognitive Modeling Tool* (CMTool), which facilitates the task modelling and evaluation phase. A case study of application of the proposed framework is presented together with examples of representative tasks. This concerns the design and evaluation of a geometrical concepts learning environment. Some conclusions on the advantages of the proposed technique and its applicability in the design of open problem-solving systems are also included in the final sections of this chapter.

DESIGN AND EVALUATION OF OPEN SYSTEMS

Design and evaluation of effective open problem-solving systems remains a challenge for the HCI field. Differences in problem-solving environments compared with workplace environments, on which the HCI field has primarily focused, are significant. In these environments, the objective for the user is not to carry out routine tasks but to use a combination of tools effectively in a nontrivial way. The interaction differs from the one mainly observed in typical environments and cannot be anticipated as a sequence of predefined actions to carry out typical tasks. Therefore the traditional metrics, which are used to evaluate the usability of a goal-oriented process such as execution time, frequency of errors, efficiency, and effectiveness should be redefined and extended to evaluate the quality of interaction with problem-solving systems. The requirement for educational use of many of these systems, like the one described in our study, further complicates the design process. Many theories of learning such as socio-constructive approaches promote the process of self-constructing knowledge during a process in which experimentation and learner errors play a key role in deeper understanding of concepts and exploring for a problem solution. Also in most design-support and problem-solving environments a trial and error approach is often used, thus errors in the solution should be accommodated and become easily visible. On the other hand, in typical routine tasks, user interaction design endeavours to minimise user errors. In most routine task working environments, such as process industry, transport systems, and so on, the designers of these systems strive to prevent errors that may bear considerable cost for the process. So there seems to be a considerable disparity between the two classes of systems. In the former case, a clear distinction should be made between errors that are the result of misconceptions on the domain knowledge required to solve the concrete problem and errors that result from inadequate design of the tools. The first kind of user error, at the semantic level of interaction, should not be constrained, whereas the second kind of error inhibits problem solving and should be not allowed. The system design should exclude severe errors in the syntactic-verbal level of interaction (i.e., those that are related to usability of the tools). A key issue in this class of systems is the transparency of available tools concerning the embodied knowledge in them. Therefore the possibility of application of various representations and multiple solutions, thus increasing the motivation and reducing the frustration from the users' perspective, is a key requirement.

Another aspect of these nonroutine problem-solving tasks relates to applicability of well-established usability heuristics. The users are expected to be supported by a large selection of tools even if these can lead to unforeseen faults in problem solving (Lewis, Brand, Cherry, & Rader, 1998). However, following the well-known heuristics relating to memory overload, one would expect that only the essential tools should be provided at any given time to the user to minimise memory load for the specific tasks to be successfully carried out. There seems to be a contradiction relating to this design issue. Availability of a multitude of tools and options can increase the creativity and the possibility of arriving to a solution by the user involved in the nonroutine task, even if this means that the short memory overload heuristic is not strictly followed.

From the examples provided in this discussion, one can deduce the requirement for a new set of design and evaluation principles that need to be defined and applied during development of open problem-solving software environments. It is exactly this kind of evaluation method that is described in the context of this chapter.

Current Status in Evaluation of Nonroutine-Task Software Environments

A variety of methodologies have been deployed for the evaluation of nonroutine-task-related software tools. In general, a series of questions challenges researchers who need to plan these evaluation studies. As already discussed, some of these concern aspects relating to software usability redefinition. Others concern domain-related matters as specific benefits and support to the perspective users. So, for instance, new theoretical frames for evaluating usability in combination with educational value have been defined for educational problem-solving software (Squires & Preece, 1999). Additional questions concern whether, during evaluation, domain or usability experts will be involved, whether it will be conducted in the laboratory or in the typical working environment, and whether typical users will be active participants in this process. The product of the evaluation studies is heavily influenced by the definition of objectives defined during such a study.

A class of these studies is inspired by scientific research methods and is conducted in the laboratory in an effort to isolate and control certain characteristics of the software under evaluation. These are frequently not driven by any specific theoretical framework, and they are mainly characterised by a quantitative approach (Bates, 1981). Even if the importance of those techniques is undeniable, their support by field studies is essential, because the environment of inquiry is somewhat artificial. Consequently, it does not provide enough information for what is expected to happen when the software is eventually going to be used by real users (Gunn, 1995).

Different methodologies involve use of questionnaires that capture the view of the users. Such an approach involves typically conducting pre- and postquestionnaires with the users to determine whether the software resulted in improvement in execution of typical problem-solving tasks. However, the questions incorporated in most typical questionnaires could not easily relate to specific issues of the system or fully identify additional factors that have an influence on the problem-solving process. Moreover, through this methodology, one cannot easily detect the individual differences of problem solvers expressed through the process (Marton & Saljo, 1976), which are of particular importance in nonroutine tasks. These approaches therefore are often accompanied by qualitative evaluation approaches. The latter provide a special emphasis on what and how the user solves a given problem (Marton, 1981, 1988). Moreover, the need of compatibility between the theoretical frame of cognitive aspects of problems and its evaluation framework emerges (see Kordaki & Avouris, 2001).

Evaluating Open Problem-Solving Environments

Open problem-solving environments present increased difficulties with regard to planning their evaluation. According to Jonnasen et al. (1993), such an environment embodies three main characteristics: it is need driven, user initiated, and conceptually and intellectually engaging. A typical example is educational environments that are theoretically based and motivated on a socioconstructive framework. Typical paradigms of such environments are the Logo language environment (Papert, 1980), the geometry experimentation tool CABRI-geometry (Laborde & Strasser, 1990), and the environment of semiquantitative models construction and manipulation ModellingSpace (www.ModellingSpace.net). Also most design tools (e.g., architectural design, graphic design, electronic circuits design tools) also share many common characteristics. The evaluation and prediction of interaction in those environments presents significant difficulties, given the fact that the problem-solving process cannot be fully anticipated from the phase of software planning and design.

Moreover, it is possible for these environments to be extended during their usage by the users, who discover novel ways to manipulate objects (Hoyles, 1993). From that point of view, evaluation through field studies is indispensable. Additionally, certain topics related to the planning of these environments as problem-solving tools exist that should be identified by usability and domain experts before the field studies take place. The so-called heuristic evaluation methods (Avouris, Tselios, & Tatakis, 2001; Nielsen, 1994) mainly deal with these aspects. These are often adapted to the specific characteristics of the specific domain, like, for instance, the proposed framework by Squires and Preece (1999). This adapts well-known usability heuristics so that evaluation of usability can be performed together with evaluation of the educational value of the software environment. These heuristic methods are conducted at the laboratory by experts, and their value and applicability in various problem-solving software evaluation experiments seems to be significant (Avouris et al., 2001).

The evaluation of problem-solving software in the field constitutes a methodological approach according to which knowledge is produced on the basis of an analytical process involving suitable organisation of collected data and not on the use of an inductive method that is based on the validation of a hypothesis (Babbie, 1989, pp. 261–290). Consequently, more complete and clearer simulation of the real situation plays a significant role in reaching a conclusion with relatively high confidence. For this reason, various types of apparatus that collect data of multiple types such as video, sound, and click streams, create valuable sources on which a more complete processing could take place to lead to more representative conclusions. The multiple forms of data are organised according to different users so that it is possible to fully study the externalisation of knowledge and interaction as the user reflects on it. Consequently, it is possible to analyse the collected data on the basis of the aggregate conclusions of the overall user population or for each user individually.

Although field-study-based evaluation is widely acknowledged, it often is not applied, mainly because it can be conducted only at the final design stages of a product. Additionally, it lacks a formal method to integrate observations in the procedure of redesigning the system. Finally, it requires thorough planning, significant time for preparation and experiment execution, and rich supportive equipment for capturing observation data and processing and analysing the results (usability laboratories). In conclusion, a strong cognitive framework can be used to guide the observation data collection and analysis. In the next section, an overview of task analysis approaches is provided with a discussion on their applicability in open problem-solving design and evaluation studies.

TASK ANALYSIS DESIGN AND EVALUATION TECHNIQUES

The main objective of task analysis from an HCI perspective is to provide a systematic approach to understanding the user's tasks and goals by explicitly describing what the users' tasks involve (Kieras, 1996). Another definition from an HCI perspective defines it as the process of gathering data concerning the tasks people perform and managing those data in a meaningful way to acquire a deeper understanding of the interactions that take place (Preece et al., 1994). In general, the objective of task modelling techniques is to formulate a hierarchical representation of users' task and goal structure. The primary task described and decomposed is considered to have been accomplished when all subgoals have been satisfied, according to plans that describe in what order and under what conditions subtasks are performed.

Together with establishment of cognitive theories such as the motor human processor model (Card, Moran, & Newell 1983), HCI community has focused during the last years on investigation of underlying supporting theories and further development of task analysis methods and techniques, based either on empirical data or psychological models. See the work of Walsh (1989), Barnard (1987), Shepherd (1989), Moran (1981), Card et al. (1983), John and Kieras (1996b), Johnson and Johnson (1991), and Diaper (1989). The outcome of application of these techniques varies according to their theoretical foundation and scope of analysis. For example, GOMS (goals, operators, methods, selection) is based on the theoretical framework of motor human processor, which tries to model human behaviour in a similar way to that of information processing artefacts. A typical GOMS analysis leads to hierarchical decomposition of goals to primitive low-level sequences of actions (the operators), together with methods describing the way this decomposition takes place. Selection rules exist to decide on the method to be used according to certain circumstances.

Extensive research in the field has produced strong evidence that task analysis approaches may substantially contribute to the successful design and evaluation of user interfaces (Lansdale & Ormerod, 1994; Richardson, Ormerod, & Shepherd, 1998). A number of successful case studies have been reported in a variety of design projects (Ainsworth & Pendlebury, 1995; Gong & Kieras, 1994; Haunold & Kuhn, 1994; John & Wayne, 1994; Olson & Olson, 1990; Paterno & Ballardin, 2000; Paterno & Mancini, 2000; Umbers & Reiersen, 1995).

In summary, the advantages of task modelling and analysis methods, relate to the fact that the design can be described more formally, analysed in terms of usability, and can be better communicated to people other than the analysts (Abowd, 1992; Johnson & Johnson, 1991; Richardson et al., 1998). Wilson, Johnson, Kelly, Cunningham, & Markopoulos (1993) proposed such models to be used as a core approach for task-centred design of interactive systems. Task analysis approaches have also received some criticism, relating to the applicability of the proposed methods, the lack of a detailed methodology for applying most specific task analysis methods into the system design cycle, and the large effort required for applying them. Some concerns are also related to the lack of tools supporting the proposed techniques. John and Kieras (1996a) argued in favour of use of cognitive modelling methods such as GOMS and NGOMSL for evaluation, based on the sound theoretical foundations of the technique and promising results of specific case studies. Additionally, Lim (1996) addressed some of the criticisms, especially its requirement for an existing system to be optimal, its focus on analysis rather than design, and its inadequate documentation of design outputs. Lim also provides a positive example of application of these techniques by describing how task analysis is incorporated explicitly throughout the design cycle in the MUSE method for usability engineering.

The techniques mentioned here often use different notations; however they share a common foundation and key assumptions, the most important of which is that they attempt to model ideal user performance. Additionally in most cases these techniques are used when the task structure of the typical system use is well defined, as happens in command line interfaces and routine tasks environments. So in cases in which the interaction is not a trivial sequence of actions, as in open problem-solving environments, the suitability of these techniques is questioned. Also there are many application areas, like learning environments in which flaws in user performance are often the rule and through them the users build new knowledge as discussed in the previous section. Designs that do not cope with such "deviations" can present considerable difficulties to users. Despite these concerns, the wide acceptability of these techniques led us investigate their applicability in the design and evaluation of nonroutine tasks support systems such as open problem-solving environments. In the next section, the main phases of the development framework and the supporting tools, based on an extension of hierarchical task analysis, are presented, followed by a case study of their use.

A FRAMEWORK FOR EVALUATION AND DESIGN

In this section a methodological framework for the design and evaluation of open problem-solving environments is described. According to this framework, design specifications and prototypes of the software environment are developed, based on assumptions on expected typical problem-solving strategies. These are subsequently revised and adapted following a field evaluation phase. The revision of the design specifications during the field evaluation study is a model-driven process. The most important aspect of this framework is the iterative development of a task model by the designer during the first phase, which is subsequently modified by taking into consideration the user task models, as observed during field evaluation. To develop this model, typical users' interaction with the problem-solving environment is closely monitored and observed. To infer models of the cognitive processes relating to problem solving of the individual user, field studies techniques are used. Alternatively, observations of interaction at keystroke level or video capture observation and review (or both) are used to enhance the applicability of the framework.

The task analysis method adopted for the process is the *hierarchical task analysis (HTA)*, proposed by Shepherd (1989) and accordingly modified as discussed here. First, it should be mentioned that our intention is to build a conceptual framework reflecting the way the user views the system and the tasks undertaken to accomplish set goals. Even though HTA bears many similarities to goal-oriented cognitive modelling techniques, such as GOMS, the main objective of our analysis is to reflect on the observable behaviour of users and identify bottlenecks in human–system dialogues rather than explicitly understand the internal cognitive processes as one performs a given task. A GOMS-like approach is more appropriate and can encourage more mechanistic views of problem solvers because it tends to look at them more as experienced users carrying out routine tasks. Instead, using a more general approach like HTA, we attempt to pay special attention to specific tasks, where true problem-solving activity takes place, such as the effort required to combine previous domain knowledge with existing tools and objects to tackle novel challenging tasks. For this reason, we include in our analysis even incorrect problem-solving approaches, caused either by limitations in the provided tools design or by conceptual misunderstandings related to domain knowledge and use of available

TABLE 14.1
Classification of Errors Identified During the Proposed Analysis Framework

Symbol	Type of Error	Description
	Severe syntactic error	Incorrect interaction, causing deadlocks, or requiring considerable effort to recover
	Minor syntactic error	Incorrect interactions, often due to incorrect user expectations on how a given task is accomplished by the system
	Conceptual error	Interaction and behaviour detected due to misinterpretation of domain entities
	Inconsistency	A task similar to another one but carried out in a different way or a task needed to be handled in a different way under certain circumstances
	Identified design principle violation	Violations of usability principles, which are not necessarily captured during the field study but can influence system usability

tools. These errors may cause unsatisfactory solutions to a given problem, but often contribute to better and deeper understanding of concepts.

Through the proposed task analysis technique, the typical or expected use of the open problem-solving environment is reduced, during design, to a sequence of tasks. As already stated, this is not a trivial process, but this task analysis can be achieved if the right level of abstraction is used. Kirwan and Ainsworth (1992) showed some methods of accomplishing this, especially in the context of studies of maintenance training and quantifying safety inspections. Boy (1999) in his book on cognitive function analysis discussed relevant issues involved in labelling tasks. Thus, any open exploratory process can be abstracted to a point where it can be represented as a sequence of predefined tasks.

One important aspect of our analysis framework is the classification of observed unexpected or incorrect user behaviour. Through the application and analysis of this technique, we have identified and classified five main categories of errors, presented in Table 14.1. These errors can be identified during observation and analysis of problem-solving activity in field studies, as discussed in more detail in the next section.

One of the significant benefits of task modelling is the process of creating the model itself. A model is relatively easy to communicate across a design team from different backgrounds and cultures (e.g., software engineers, usability engineers, domain experts). We need to pay special attention to ensure the adoption of a relatively easy-to-understand formalism, through which the model can be expressed. Description of plans tends to use simple expressions, incorporating special symbols such as (!) and (x!) to denote subtasks containing errors, and { } to indicate occurrence of unforeseen tasks according to the original designer model. ⟨ ⟩ contains tasks that should be extended to support unforeseen demands of users thus extending the overall functionality of the system. AND and OR operators are used to connect tasks that are executed together, or one can be selected instead of the other to accomplish a higher level goal. Instead of the AND keyword, the comma mark (,) could be used to describe tasks that should be all executed together but in any order. SEQ is used to describe tasks that must be executed in sequence, and PAR indicates tasks that should be executed in parallel. So, for instance, SEQ (1 (!), 2 (!), (3)*2 (!), 4 (x!)) indicates that tasks 1, 2, 3, 4 are executed as a sequence, task 3 has been repeated twice, and all tasks contained errors of various severity.

Inferring users' goals from observed behaviour and user's motivation to select specific subtasks in a given situation is not a trivial process. Mechanisms to increase the effectiveness of the process and reduce the search space need to be defined, especially because the number of tasks to be undertaken and the number of combinations of subtasks that can be executed to accomplish them can be enormous in a nonroutine task environment. To facilitate and improve the effectiveness of the method, we complement the cognitive task modelling phase with a leading heuristic usability analysis. Through this process, usability of the system is preevaluated mainly at the lexical level, and expectations on possible areas of errors can be identified. The Field studies thus can concentrate on specific tools and tasks that can relate to these areas, whereas interpretation of observed behaviour can be based on these preliminary findings. Heuristic evaluation is a well-proven and effective technique, used to measure usability aspects of a system design against 10 general usability principles (the heuristics). The value of heuristic evaluation that is performed by usability experts without participation of the users has been shown in many application areas. We do not expect to rely exclusively on this technique to evaluate usability of an open problem-solving environment. Users of such environments externalise their intended actions in often-unexpected ways, and a key assumption is their exploratory interaction with the system. The heuristic usability evaluation method serves well in assessing key usability aspects of the system, especially in certain areas such as terminology used and metaphors applied to shape existence of tools in the typical environment. Through this early analysis phase, we can concentrate on expected errors during the following field study and infer the reasons for errors or on unexpected behaviour of the users. For example, if erroneous usage of a tool is detected, it is easier to make assumptions on whether the problem exists due to the so-called gulf of evaluation (i.e., possible erroneous understanding of metaphors used or improper placement of tools) or gulf of execution (unexpected actions or input device usage due to a different user expectation), see Norman (1988).

Additional advantages relate to the method's relative low cost in terms of human effort and time resources required as well as its applicability even in the earliest stages of design. This aspect increases the overall scope of the framework because this can be applied to earlier phases of the design process and can be related both to user interface design (heuristic evaluation) and HCI design (task modelling), two interrelated aspects that are not necessarily identical.

With regard to the task modelling technique to be used, one possible solution could be the adoption of a more sophisticated task modelling technique than HTA. This technique could describe explicitly responses of the system to every task carried out by the user, such as Methode Analytique Des Taches (MAD; Sebillsote, 1988) or Task Analysis for Knowledge Descriptors (TAKD; Diaper, 1989). However, the complexity and the effort required to model systems with these techniques poses questions about their practical value in realistic situations involving large number of tasks. Together with the fact that those techniques require expertise in the HCI domain, the possibility of wider adoption and mainstreaming of the framework proposed would be tremendously reduced, especially by domain-related evaluators who do not possess HCI-related skills.

One issue worth investigating concerns the level of detail of analysis during application of the HTA technique. Based on our experience with the application of this modelling technique, we tend to conclude that there are no explicit rules on this matter. This decision is largely a matter of evaluator judgment. One particular approach, which seems to fit well with many evaluation studies is to concentrate initially on a breadth-first instead of a depth-first decomposition of tasks, or the fine details of what the user actually does at a detailed level of interaction. Plans should be considered thoroughly to present always an answer to why the user is engaged in accomplishing those tasks and later to focus on how those tasks are carried out explicitly. Two rules that could be adopted to help us decide if we need to carry out the analysis into

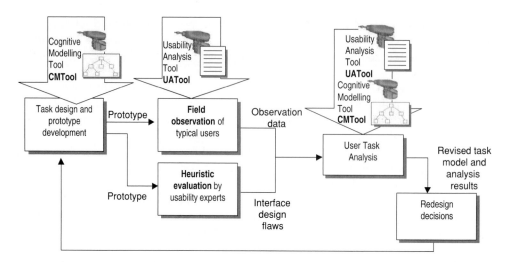

FIG. 14.1.   Outline of the evaluation framework.

a lower level are the problem solving rule and the PxC rule. With the problem-solving rule, we may decide to stop further decomposition of a task if it does not engage the user in a problem-solving process. Thus, such processes should be decomposed once, except if our analysis is focused heavily on the way users use specific tools and with what frequency. PxC rule mentions that "if the probability (P) of making a mistake in the task multiplied by the cost (C) of the mistake is below a threshold then stop expanding" (Dix, Finlay, Abowd & Beale, 1998). So, simple tasks need not be expanded, unless they are critical. In this case, alternative design solutions and task implementation approaches should be examined in order to reduce this possibility.

The main phases of the proposed framework are shown in Fig. 14.1. These involve the following:

1. The design and prototype development phase—development of a model of the anticipated task execution by users, called designer task model (DTM)
2. Heuristic usability evaluation of the prototype by usability experts
3. Field evaluation—observational data acquisition and data interpretation and analysis
4. Definition of a user task model (UTM), modification of DTM according to the developed UTMs, and generation of new design specifications

These phases are discussed in more detail in the following sections.

Definition of the Original Designer's Task Model

During this phase, the designer defines typical characteristics of human–machine interaction, which include the DTM. This activity is based on background theories and knowledge, domain-specific requirements, the expected users' characteristics, and the social environment in which the system is expected to operate. The DTM is defined by considering typical high-level tasks that are expected to be accomplished by the user with the designed system. These are decomposed in subtasks that must be accomplished to perform the main tasks. These often relate to available tools and their functionality. This model represents the way the designer expects typical users to interact with the problem-solving environment to accomplish typical tasks. It is assumed that this model is made explicit using the CMTool discussed in the following

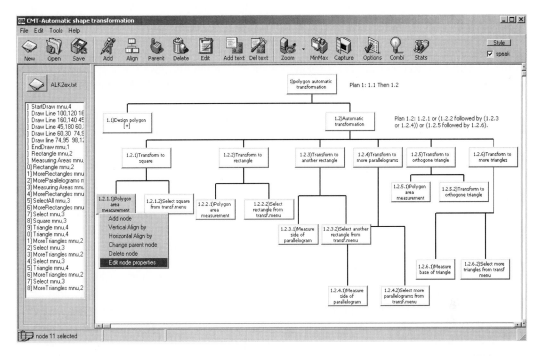

FIG. 14.2.   Cognitive modelling tool: The main task model building space. Part of the designer model of the task discussed is shown.

section. The tool is used to manipulate a graphical representation of the task model (see Fig. 14.2). This tool is used for the definition of the expected high level *goals* of the user, further decomposed into subgoals necessary to accomplish the high-level goals. Generally, in an open problem-solving environment, the individual user approaches a given problem in many ways. The most typical of these should be reflected in the designer's task model, which should therefore contain many OR-related, alternative goal hierarchies. Special effort must be made to ensure that the proposed way of interaction embodies the desirable knowledge the system should deliver; thus, this model should be used in prototype development.

Following this initial phase, a heuristic usability evaluation of the developed prototype is conducted. During the heuristic evaluation session, the evaluator runs the interface several times, inspects the various dialogue elements, and compares them with a list of recognized usability principles (the heuristics) proposed by Nielsen (1993). The result of this second phase is a list of identified usability problems in parts of the interface. These drive the specification of the field evaluation study, discussed in the following.

Field Evaluation Study

Prototype field evaluation is the phase of the proposed framework during which the designer effectively receives feedback from typical users of the software to adapt the original design to support actual users' needs. The primary hypotheses concerning the way that the users are expected to interact with the system are based often on invalid assumptions. According to Orhun (1995), effective communication between user and designer and user and devices requires mechanisms that allow adaptation of the original assumptions. Explicit representation of the DTM is essential to devise and support this adaptation process. In our case, this field evaluation process serves to validate original designer's assumptions against actual users' task execution.

This field study may take place in a usability laboratory or in a typical work environment. Participants in the experiments must be representative users with the same characteristics as actual users. The users receive a list of representative tasks, in terms of the system's proposed functionality, to accomplish. Their behaviour should be closely observed and log files of interactions produced. Desirably, if appropriate equipment exists, the field evaluation process can be videotaped for further analysis.

If the possibility of a field evaluation phase involves many experiments, in all of which evaluators cannot be involved to supervise data collection, user task modelling is often performed by inferring the cognitive model of the user from the keystroke behaviour. This is not a trivial task and requires some expertise to master. Additionally, in computer-supported open-learning environments, the problem-solving strategies applied by the users are particularly difficult to detect. This is more an issue in field studies in which user interaction with the computing environment is logged. Thus posterior analysis of these logging data is a particularly tedious process.

Automation of this process technique using Bayesian belief networks (Tselios, Maragoudakis, Avouris, Fakotakis, & Kordaki, 2001b) has been proposed in order to classify click stream data, thus facilitating the user goal-inferring process. According to this approach, problem-solving strategies are diagnosed automatically using a Bayesian network that is able to relate user activity to predefined goal sequences and problem-solving strategies to be applied even in cases in which physical presence of the evaluator is not possible. However, such approaches remain in the realm of research rather than phases of the mainstream framework.

A fair question to be asked is how many users should be involved in the study to have enough data to produce a comprehensive evaluation of the system. As Nielsen (1994) stated, in a field study evaluation five users typically unveil about to 80% of possible errors. This statement has been validated across a variety of field studies and formulated by Nielsen and Landauer (1993) into a mathematical expression:

$$ProblemsFound(i) = N(1 - (1 - l)^i),$$

where *ProblemsFound(i)* indicates the number of different usability problems found by aggregating reports from i independent evaluators, N indicates the total number of usability problems in the system, and l indicates the proportion of all usability problems found by a single evaluator. The values of l in Nielsen and Landauer's studies ranged from 0.19 to 0.51 with a mean of 0.34 and the values of N ranged from 16 to 50 with a mean of 33. According to the equation, for typical values, *percentage of errors found* $= ProblemsFound(i)/N$ equals to 87.4% for five users. For the worst case ($l = 0.19$), we obtained 65.1%.

By observing the individual user's interaction at the keystroke level (e.g., through log files) the user's model, reflecting his or her conception of the system, is constructed. Also user's views about missing functionality of the systems can be detected. For each user, $n = t * s$ models need to be built, where $t =$ number of high-level tasks undertaken and s is the number of different strategies used to solve the tasks. The *Usability Analyzer* tool (UAtool) supports the keystroke-level monitoring and analysis process.

User Interaction Modelling and DTA Refinement

The user deviations observed during the field study are monitored at the keystroke level and are mapped at the task level during this phase. If a usability-related mistake is detected, the task where the wrong interaction occurred is specially annotated with a detailed description and a grading of the severity of the problem. If an unanticipated problem-solving strategy is observed, this is added incrementally to the original designer model. Remarks on additional functionality required to support this unforeseen usage of the tools are also made. Conceptually wrong problem-solving strategies can also be included in the user's interaction model, because they are often useful for educational purposes or can be used as typical misconceptions to be tackled by the system. Having the results of this analysis in mind, the evaluator assigned to carry out the task-modelling process could reason more easily about the possible cause of errors detected during user interaction modelling.

Following this process, a combined user task model is derived, which contains all the users' solution strategies that were detected and analysed. The purpose of this process is to unveil new ways of interaction that the original designer model could not support and determine the gap between the designer and the user models. After this analysis, the original DTM is revised and augmented, which can lead to modification of the original design. This new designer's model is expected to deal more effectively and in new diverse ways with the system design.

Tools to Support the Framework

The considerable effort required for constructing a task model and for maintaining various instances of it, like the one discussed here, is one of the main reasons for discouraging the wider adoption of task modelling, according to Kieras (1996). This observation inspired us to develop appropriate tools, which could support use of the proposed design and evaluation framework. In this section, the tools developed for this purpose are presented.

The Cognitive Modelling Tool.   CMTool has been developed to facilitate the task modelling process. A more detailed description of the tool can be found in Tselios, Avouris, and Kordaki (2001a). The need for realisation of this tool emerged because little tool support to carry out the required task modelling process was identified. There are some widely available tools like the concurtasktrees builder (Paterno and Mancini, 2000) and Euterpe for Groupware Task Analysis (van Welie, van der Veer, & Eliens, 1998). Despite the fact that both these tools are powerful enough and usable, construction of task models, according to the defined framework, was found to be limited in many respects.

Task models are structured in a hierarchical way, as shown in Fig. 14.2. CMTool uses a direct manipulation approach for editing and modifying this task hierarchy. Any specific node represents a task relating to a user goal. The subtasks, which serve to accomplish this goal, are associated to the node. These subtasks can be related through a specific plan containing semi-formal descriptions involving OR, AND, IF, PAR, and SEQ operators already discussed. This plan is usually shown next to the specific task (or subtask). Additional information regarding comments on the goal accomplishment can be attached to the node, especially characterising the nature of problems identified, as shown in task 1.2.2.1 of Fig. 14.6 (properties box). Finally, each task node could be associated with specific tools used, providing additional dimensions and possibilities to postmodelling analysis. For example, problems associated with specific tools or different tool usage behaviours across groups with different user characteristics such as sex, age, or skill could be the focus of analysis.

An individual user task model can be annotated with comments concerning task execution. A special notation has been defined, extending the HTA plan notation (Shepherd, 1989), with tokens referring to user deviation from expected task execution. Token (!) marks a nondestructive syntactic deviation, whereas (x!) marks a not-completed task execution. Tasks that should be realised in a different way to meet the expectations are marked as ⟨task-id⟩. Introduction of new unforeseen tasks in a plan by a user are marked as {task-id}, as shown in Plan 1.1 of Fig. 14.5.

Both textual and graphical notations can be used for task representation. This depicts an additional advantage of the CMTool: It can support communication across a design team (often consisting of designers from different disciplines and backgrounds) because of its ability of exporting task models in various possible representations; such as tree view, sequential view, and structured report view, which are produced automatically, are consistent with each other, and thus support various phases and needs of the design process. If the analysis is focused on the time required to accomplish a task, time-related information can also be stored on each task, and appropriate calculations regarding the time required for a task to be accomplished can be carried out automatically. Display of a keystroke log file, shown at the left of Fig. 14.2, next to the corresponding task model structure, is supported. So low-level interaction details (keystrokes) and cognitive goal hierarchies are displayed simultaneously to the user of CMTool. The possibility of dragging an event of the log file to the goal structure, which results to the introduction of a new node, filled with the action description, automates the process of goal structure building.

Task models are stored in a relational database, grouping the various tasks analysed with additional identification information (designer's model or revised DTM or UTM). In addition, quantitative analysis tools are supported to extract useful metrics related to the analysed tasks. Examples of these metrics are the number of keystrokes required to achieve a specific goal or subgoal, the mean time required, and the interaction complexity of specific user model compared with the original designer's expectations or to the revised and adapted model.

A novel and timesaving functionality of the CMTool is its ability to automate synthesis of task structures already stored in the database. Using this feature, various subgoal structures can be combined or temporarily hidden away in a task model, according to the degree of detail required in the context of a particular study. Additionally, because task models are stored in a relational database schema, interesting comparisons can be made through database querying related to various aspects of modelling, such as execution times and frequency of usage patterns, both in absolute or relative values (e.g., compared with other task structures). In CMTool the evaluator or the designer can select parts of a task structure representing a specific problem-solving strategy, which can be stored for future reference or comparison with other users' problem-solving strategies. Additionally, CMTool supports storage of various users' characteristics, such as age, gender, and so on, so that further analysis can be supported focused on them. In CMTool, the evaluator or the designer can select parts of a task structure representing a specific problem-solving strategy, which can be stored for future reference or comparison with other users' strategies. The possibility to analyse system usage in five dimensions is a contribution of the tool to the evaluation of open environments both in usability and learning terms. These five dimensions are as follows:

1. General problem-solving strategies
2. Users
3. Specific activities to carry out a strategy
4. Tools
5. Usability problems detected in interaction

This analysis is carried out through a visual stored information query environment, shown in Fig. 14.3. Modelling experiments could be analysed across any possible element of the five dimensions (e.g. "Show all encountered problems related to tool X," or "Identify all usability problems of users age 15 and older related to problem-solving strategy Y," etc.). Complex and in-depth analysis can be carried out according to any of these dimensions, supporting in-depth

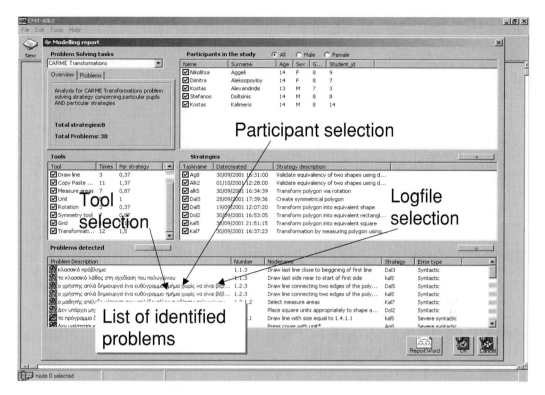

FIG. 14.3.   CMTool: Visual observation data query construction environment.

study and analysis of encountered problems. For example, a usability engineer could focus on a classification of problems detected concerning particular tools.

Together with the frequency of usage of those tools, one could decide on which errors should be paid special attention to, in a possible redesign effort. Domain experts could study and reason about the intraindividual differences presented between groups of users (i.e., male against female users, older vs. younger users, etc.). Differences are shown both in the adoption of different problem-solving strategies and in the usage of different tools. Finally, results of such an analysis could be automatically exported and shared across design members.

Usability Analyser.   An additional tool that was developed to support our framework has been the UAtool. A typical screenshot of the UAtool environment is shown in Fig. 14.4. UAtool supports the analysis of the keystroke log files. It comprises a log file parser, which converts difficult-to-understand keystroke sequences, mouse movements, and click streams captured during the field study into information readable by humans. This is achieved through an association table between the environment functionality and the relevant keystrokes. In addition, screenshots, captured during the logging process, can be shown, thus extending the traceability of observed interaction. Comments on the interaction activity, made by field observers, can also be associated to log-files and screenshots. Additionally, UAtool can execute a massive parsing against all the collected log files and store extracted information into a database, containing information such as number of actions, frequency of specific keystroke sequences, and performance data. The results can be communicated to the user of the tool (e.g., the evaluator) either in the form of graph charts or structured text reports. It has been observed that this complementary tool enhances the adoption of the proposed framework, facilitating

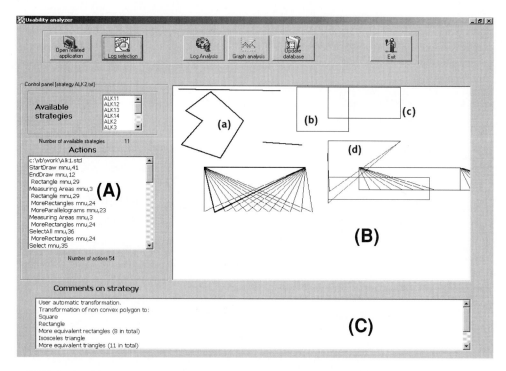

FIG. 14.4. UAtool: On the left (a) the log file of the user. In window (b), the corresponding user screenshots. At the bottom (c), evaluator's comments. The screenshot is related to the task discussed in the case study section.

integration of user keystroke logging data, evaluators' comments, and interaction screenshots, and a new version of the tool can incorporate sources of streaming data such as audio and video.

CASE STUDY: DESIGN AND EVALUATION OF A PROBLEM-SOLVING ENVIRONMENT

In this section, we present a case study using the presented framework and tools. The objective of the study was to evaluate and redesign the open problem-solving environment C.AR.ME (conservation of area and its measurement), an open problem-solving environment designed to support 12- to 14-year-old pupils in exploring the concepts of conservation of area and its measurement (Kordaki & Potari, 1998). C.AR.ME. offers the opportunity to explore a variety of geometric representations of these concepts, which can be produced in many ways. For instance, areas can be measured by selecting and iterating units to cover them. In the case of conservation of area, the pupil can manipulate geometrical shapes to split areas into parts and recompose them to produce equivalent areas, and tools are provided for dynamic transformation of a polygon to different shapes of equivalent areas. To support the users in creating these representations, the environment provides a set of tools. By studying these representations, C.AR.ME users have the opportunity to experience the conservation of an area in shapes that are difficult to produce with paper and pencil, and to discover the common properties of these shapes.

The original task model (DTM), containing many alternative approaches for solving the problem of measuring an area of a geometric shape or transforming it into an equivalent one, was developed. The evaluation process took place in a secondary school computer laboratory. Thirty students aged 13 and 14 took part in the experiments. Each one interacted individually with the software. Students were given two typical "conservation of area and its measurement" tasks. The first involved transformation of a nonconvex polygon to another shape with equal area using all possible ways; the second task involved comparison of this polygon to a square not easily comparable by "eye." One of the researchers participated as an observer without intervening in this experiment and recorded the students' solution strategies. The students' interaction with the environment was recorded in log files, and screenshots of their workspaces were also saved for future reference. Some 114 log files containing various strategies from the 30 students were obtained. This experiment has been performed both for studying domain-related educational concepts, related to learning conservation of area and area measurements, and for usability evaluation of the tools. The latter use of the experiment data is reported in this chapter.

After the experiment, evaluation of interaction and tool usability took place. A heuristic usability evaluation phase was conducted independently from the field evaluation. This study unveiled 48 usability problems. The nature of the identified problems was primarily related to the way the system delivers and receives responses during a user–system interaction dialogue to the user rather than how this dialogue actually takes place. Problems at the lexical level of interaction, such as feedback semantics, the general terminology used, the general menu design, and the quality of feedback concerning interaction errors are difficult to discover through task analysis.

In addition, the usability problems discovered during the heuristic evaluation phase were used to explain specific user responses in various instances of interaction, thus helping the task modeling process. As a result, we found heuristic usability evaluation to be a useful, complementary method to support the main model-based framework. According to the analysis, presented in the field evaluation study introduction section, we selected the problem-solving strategies of six users concerning the task that involved transformation of a nonconvex polygon to another shape of equal area. Eight strategies were obtained and modelled with the aid of CMTool. Task models were inferred from click streams, which were analysed with their corresponding screenshots and field evaluators' remarks. In some cases, the progress of inferring the low-level actions led to non-one-way task actions. A valuable support to address this problem was provided by the field observers, who used field evaluation protocols that involved significant interaction with the users during the field study (e.g., think-aloud protocols, interviews, etc.). Strategy recording and relevent field notes during the observation helped us infer intentions and build user task models.

This task analysis study unveiled both syntactic deviations, identified at the lower levels of the task hierarchy (mostly interpreted as inappropriate usage of tools due to usability issues attributed to the initial design of the interface) and semantic deviations. Semantic issues such as unpredicted goal sequences were detected either due to misconceptions or users' attempts to set up strategies that were not predicted in the original designer model. The analysis resulted in 25 interaction problems (30 in total, but 5 were detected more than once). In some problems, if a similar problem was detected in the heuristic analysis, it helped us reason about the type of and the reasons for the specific problem. Some of them were related to unforeseen ways of handling and manipulating objects, which could not be captured without involving actual usage of the environment.

Additionally, this task-modelling-based evaluation study provided the opportunity to identify patterns of frequency of tool usage and relate them to specific usability problems. Results from analysis of tool usage are shown in Table 14.2. Efficient system design should promote usage of tools, which are of key significance for carrying out tasks that embody important domain concepts. For example, as explained in the examples discussed in this section, the

TABLE 14.2
Frequency of Tool Usage in Transformation Task

Tool Name	# Used	Per Strategy
1. Cognitive tasks	61	7.62
2. Polygon	30	3.75
3. Transformations	12	1.5
4. Copy/paste tools	11	1.37
5. Units	8	1.00
6. Measure areas	7	0.87
7. Symmetry tool	7	0.87
8. Draw line	3	0.37
9. Rotation	3	0.37
10. Grid	1	0.12
Total	143	17.875

polygon design task is expected to be carried out by three task actions; however, our analysis indicated 3.75 task actions per different strategy, an indication of poor usability of the tools used to carry out this specific task. This is an indication that users may have paid unnecessarily excessive attention and put too much effort into routine tasks instead of reflecting on useful constructions and geometric object manipulations. Another result is that task actions related to specific tools seem to follow an exponential distribution, with a small number of tools being highly popular and all others relatively less so. Tools concerning measurement of area suffered from five problems. Four problems were detected in tools concerning polygon design and direct shape manipulation.

Following a consultation phase involving the evaluators and designers, individual UTMs and a combined UTM were produced using the CMTool. The combined model helped us to understand the nature of tasks and the way the users conceptualised them, thus providing substantial aid in our effort to capture requirements and support the redesign process. In the following sections, two extracts of this process are described to illustrate how the framework has been used for this purpose.

Example 1

The task under study involved the creation of a polygon and the generation of geometric shapes of equivalent area, which could be squares, rectangles, or triangles. In the C.AR.ME environment, this can be achieved using various tools, such as gridlines, measuring units, or automatic transformation of a reference shape. In this example, a user opted for the last approach. A short extract of this user's interaction with the system is shown in Table 14.3.

The designer model for this task, shown in Fig. 14.2, contains two subgoals: *(1.1) Design of a reference polygon* and *(1.2) Generation of an equivalent shape using relevant commands.* The workspace of this particular user during this process is shown in Fig. 14.4. The reference shape is polygon (a). Equivalent shapes built subsequently are shown as (b), (c), and (d) in Fig. 14.4. Analysis of task execution by this user is discussed in the following section. Three specific cases of observed deviations of the user task execution in relation to the DTM are included.

TABLE 14.3
Example Log File and Comments on the Interaction Behaviour

	Example Log File	Comments
1.	StartDraw mnu,41	Actions [1–7]: design of the reference polygon (a) of figure 14.4. Last segment
2.	Draw Line 100,120 160,140	drawn through action [6].
3.	Draw Line 160,140 45,180	
4.	Draw Line 45,180 60,30	
5.	Draw Line 60,30 74,95	
6.	Draw line 74,95 98,120	
7.	EndDraw mnu,12	
8.	Rectangle mnu,29	Actions [8–10]: Selection automatic transformation of rectangle menu
9.	Measuring Areas mnu,3	option. The system responded with a request to measure area of (a) first.
10.	Rectangle mnu,29	The user selected area measurement [9] and then proceeded with rectangle creation [10]. Successful completion of task (1.2.2).
11.	MoreRectangles mnu,24	Actions [11–17]: The user attempted unsuccessfully to build more equivalent
12.	MoreParallelograms mnu,23	shapes, i.e., Tasks (1.2.3) and (1.2.4). To pursue these goals, the user needs to achieve goals (1.2.3.1) and (1.2.4.1) first. This is not clear by system
13.	Measuring Areas mnu,3	response.
14.	MoreRectangles mnu,24	
15.	SelectAll mnu,36	
16.	MoreRectangles mnu,24	
17.	Select mnu,35	
18.	Square mnu,39	Action [18]: The user successfully completes Task (1.2.1).
19.	Triangle mnu,44	Action [19]: The user successfully completes Task (1.2.5).
20.	Triangle mnu,44	Actions [20–28]: Unsuccessful attempt to achieve Task (1.2.6). The system
21.	MoreTriangles mnu,25	requests manual measurement of the base of the triangle first. The user
22.	Select mnu,35	abandons the goal.
23.	MoreTriangles mnu,25	
24.	Select mnu,35	
25.	Triangle mnu,44	
26.	MoreTriangles mnu,25	
27.	Select mnu,35	
28.	MoreTriangles mnu,25	

The *reference polygon drawing* Task (1.2.1) was originally designed as follows: The user is expected to draw all sides of the polygon except the last one and then select "end draw" to complete it. From user actions [1–7] of the log file in Table 14.3, it is deduced that the user attempted to complete the polygon by drawing the last point of the final segment close to the starting point, subsequently selecting the appropriate command "end draw." This was marked as a low severity syntactic deviation in the corresponding user task model. The additional Task (1.1.3) was included in the task execution plan, as shown in Fig. 14.5. One user remarked that the C.AR.ME system should interpret drawing the end of a line near to the starting point as an attempt to complete the polygon, thus providing more support for direct manipulation.

The second misconception detected was related to the fact that to produce the geometric equivalent shape automatically, the user was requested, through a relevant message, to manually measure the area of the original polygon (a). Demanding this manual activity in the frame of an automatic transformation task confused the user, as seen in actions 8–10 of the log file

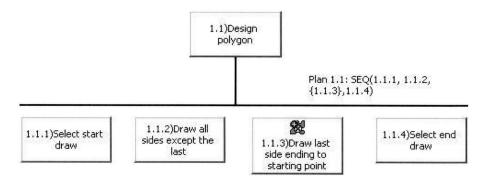

FIG. 14.5.   User task analysis of subgoal: Polygon drawing.

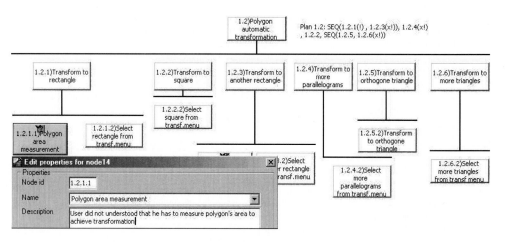

FIG. 14.6.   Extract of user task model. Subgoal: automatic polygon transformation.

(Table 14.3). Finally the user, after measuring the area of (a), produced some equivalent shapes (see Fig. 14.6, Tasks 1.2.1, 1.2.5, and Table 14.3 actions 18, 19). A comment was added to Subtask 1.2.2.1 in the UTM, shown in Fig. 14.6, describing the observed user behaviour, and the relevant annotation mark (!) was assigned to the task execution.

Subsequently, when the user attempted to create more equivalent shapes, the system prompted her to measure one side of a previously drawn shape to use it as a base, which does not clearly relate to the task objective. The result was that the user could not carry out the task and began trying different ways of interacting, resulting in a deadlock (actions 20–28). Thus, Tasks 1.2.3, 1.2.4, and 1.2.6 were marked as unaccomplished in the UTM, as shown in Fig. 14.6.

By reviewing the interaction extract, one can conclude that because of flaws in system design, only 10 of 28 listed actions (35%) were related to the expected task execution, and ultimately the given task was not fully accomplished.

Additionally, low-level interaction details are not captured in the log files. So, as identified from lower level interaction analysis, during "manual area measurement" and "segment measurement" task execution, user control problems and inappropriate system feedback were observed, thus reducing user concentration on the learning process. This finding is consistent with similar findings of the heuristic usability analysis conducted earlier.

User observation analysis both of screenshots and log files produced large quantities of feedback about the current status of usability, but the analysis procedure was tedious and

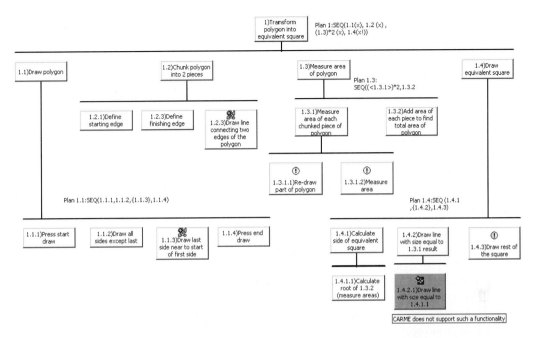

FIG. 14.7. User's task model. Task: Transformation of a polygon into an equivalent square.

time-consuming. Frequent errors were identified through the log files. An important finding of this process was that this association of low-level user actions to task and goal models, as expressed in the form of DTM, revealed flaws in the original prototype design and easily associated these flaws to specific tasks and tools. Also, the user task models of various users were compared, and the syntactic, and semantic deviations of user interaction was visualised in reference to the DTM, as shown in Fig. 14.6.

Example 2

The task discussed in this second example concerns creation of a polygon and its transformation into an equivalent square. During interaction with the C.AR.ME environment, the observed user did not try to achieve transformation through automated methods, as the student in the previous example did. First, he tried to draw a polygon (Task 1.1, model shown in Fig. 14.7). The same misconception emerged as in example (a). The user attempted to complete the polygon by drawing the last point of the final segment close to the starting point and then selected the appropriate command "end draw."

This is one of the most consistently observed syntactic errors, related to the existence of "start draw" and "end draw" statements. These statements are considered incompatible with the direct manipulation nature of the environment. This statement was marked once in all user models concerning the same subtask.

Subsequently, the user attempted to measure the polygon in a manner the designers had not predicted: He split the polygon into two parts (Task 1.2). The task of drawing a line to divide a shape into two pieces was not realised appropriately. The user has to define two points belonging to a different side of a shape and then draw a line connecting them. The problem is that the user draws a line without receiving any confirmation that the start and end points are actually recognised as parts of the shape to be divided. Indeed, the user draws a line overshooting the polygon. The syntactic deviation was marked to Task 1.2.3.

To measure the area of the polygon, the user then attempted to measure the area of each shape produced by the division of the polygon into two shapes and then to add them. This behaviour was unexpected and marked appropriately. In general, to measure the area of a shape, the shape has to be selected first and then the "measure areas" command must be issued. Both subtasks are characterised as problematic. The selection does not occur by pressing anywhere in the geometric shape; instead, the user has to redraw it. The feedback provided by the system is a form containing all previous measurements. This could lead to misconceptions, because it is unclear which value belongs to which area. No interaction errors were observed, but clearly both functions could lead to significant problems. In this example, both violations were considered as principle usability deviations and were marked appropriately (1.3.1.1, 1.3.1.2).

Finally, the user had to design a square with the same area as the polygon. So he had to design a square with a side equal to the square root of the value of measured area of the polygon. But the user discovered that the C.AR.ME environment does not support the drawing of a line with a given length, so he made several attempts to draw lines and measure their length until he achieved the desired length. We did not attempt to describe all these actions in the model because of this design limitation. Instead, we marked this as a severe syntactic mark on Task 1.4.2.1. By simulating this task, we observed that C.AR.ME does not appropriately support the drawing of a square (for example, by drawing a line and then completing the rest of it automatically). This finding, which was not actually deduced by field study observation and modelling, was registered as well (Task 1.4.3).

CONCLUSIONS

We have proposed design and evaluation framework in this chapter that captures the details of user interaction with problem-solving environments, at both the cognitive and the keystroke level. The tools presented support application of the proposed framework at both levels. Our approach attempts to manage some of the difficult issues related to the design and evaluation nonroutine task application.

We have claimed that this methodological approach successfully addresses the key issues that need to be taken into consideration to incorporate task modelling in software artifact design and evaluation, as proposed by Paterno (1999). Our framework provides flexible and expressive notations; it incorporates systematic methods to support specification, analysis, and use of task models; and it provides support for the reuse of good design solutions and tools to support the modelling framework.

As demonstrated in the case study, detailed analysis of users' problem-solving behaviour, building of the individual user task models, and aggregating these models into a single revised designer task model can lead to a bottom-up redesign phase, following the original top-down design specification phase. For instance, we have produced new interface design requirements (e.g., related to polygon drawing task) and new functionalities (e.g., drawing a line with a given length) as a result of this process.

Building individual task models and combining them into an abstract user model is a tedious process that requires considerable effort. This is particularly the case with open problem-solving environments in which individual users' problem-solving strategies differ, sometimes considerably. However, our case study reveals that the number of users necessary for discovering significant interaction problems does not need to be so large as to cover all possible task patterns. In our example, analysis of five users' interactions captured a large portion of the interaction problems, as a considerable overlap of discovered problems revealed, despite the fact that these users did not cover the full extent of possible problem-solving strategies.

Moreover, the proposed framework combines a heuristic usability evaluation approach that identifies interface design problems early in the development of the design. This way, the cognitive modelling phase that follows can be more focused on the use of specific tools and on parts of the user interface (i.e., those in which most usability problems have been discovered). This way, the issue of the large number of user problem-solving strategies that can occur in a nonroutine task environment is further reduced. The heuristic evaluation approach is also used to interpret findings related to the task modelling process, because tasks with many interaction problems are related and thus explained though the known usability problems of the tools involved.

The tools provided for automating task design, task annotation, and task analysis were proven particularly useful in the case of multiple users and multiple tasks analysis, further facilitating the process. As an indication of the manpower required for application of the described method and tools, the extensive field study experiment of environment C.AR.M.E. did not necessitate more than a few days of work, whereas the heuristic evaluation of the same tool took a few hours of evaluators' time. The analysis of the models and of the findings of the studies required more time, but this factor depends on the objectives of the study, the experience of the analysis team, and so forth.

The observations made during the task modelling phase proved to be very rich, and their capturing and classification under the syntactic and semantic redesign perspectives define a solid framework for iterative design of open problem-solving environments. Additional advantages of this framework are the possibility of deriving quantitative and qualitative results concerning the problem-solving process, focused on different dimensions at both the individual user level and at the group level. In addition, this model-based approach permits deeper understanding of the nature of tasks during their decomposition and easy control of consistency across task structures. This approach can also be integrated with a constructivist perspective of learning evaluation process because it supports the study of the development of the individual user problem-solving strategies. It also supports the study of individual user problem-solving strategies in relation to the generic user model that was built by taking into account the strategies of all users who participated in a particular study.

Some shortcomings of the proposed approach relate to the complexity of some of the phases involved. Some difficulty was observed in inferring the cognitive model of the user from the keystroke logfiles. This problem was tackled by using adequate field evaluation protocols that involved deep interaction of evaluators with the user during the field study (e.g., think-aloud protocol, interviews, etc). Also, an automated strategy classification technique using Bayesian belief networks (Tselios et al., 2001b) has been proposed to classify click-stream data, thus facilitating the user goal inferring process in cases of large-scale field studies. More research and experimentation is required in this direction.

ACKNOWLEDGEMENTS

The framework described in this chapter was developed in part through the PENED 99ED234 research project, Intelligent Distance Learning Tutoring Systems, funded by GSRT (General Secretary of Research and Technology) and the ModellingSpace (project IST-2000-25385) funded by the European Commission. Special thanks are due to M. Kordaki for providing valuable field data and continuous encouragement and support and to anonymous reviewers for constructive comments on an earlier draft.

REFERENCES

Abowd, G. D. (1992). Using formal methods for the specification of user interfaces. *Proceedings of the Second Irvine Software Symposium, 92,* 109–130.

Ainsworth, L., & Pendlebury G. (1995). Task-based contributions to the design and assessment of the man-machine interfaces for a pressurized water reactor. *Ergonomics, 38,* 462–474.

Avouris, N. M., Tselios, N. K, & Tatakis, E. C. (2001). Development and evaluation of a computer-based laboratory teaching tool. *Journal Computer Applications in Engineering Education, 9,* 8–19.

Babbie, E. (1989). *The practice of social research.* Belmont, CA: Wadsworth.

Barnard, P. J. (1987). Cognitive resources and the learning of human-computer dialogues. In J. M. Carroll (Ed.), *Interfacing thought: Cognitive aspects of human computer interaction.* Cambridge, MA: MIT Press.

Bates, T. (1981). Towards a better research framework for evaluating the effectiveness of educational media. *British Journal of Educational Technology, 12,* 215–233.

Boy, G. A. (1998). *Cognitive function analysis.* Westport, CT: Ablex.

Card S., Moran T., & Newell, A. (1983). *The psychology of human computer interaction.* Mahwah, NJ: Erlbaum.

Diaper, D. (1989). Task analysis and systems analysis for software development. *Interacting With Computers, 4,* 124–139.

Dix A., Finley J., Abowd G., & Beale A. (1998). *Human–computer interaction.* Englewood Cliffs: Prentice Hall.

Gong R., & Kieras D. (1994, April). A validation of the GOMS model methodology in the development of a specialized, commercial software application. *Proceedings of Association of Computing Machinery Computer-Human-Interaction 1994, 351–357.*

Gunn, C. (1995). Usability and beyond: Evaluating educational effectiveness of computer-based learning. In G. Gibbs (Eds.), *Improving pupil learning through assessment and evaluation* (pp. 168–190). Oxford Centre for Staff Development.

Haunold P., & Kuhn W. (1994). A keystroke level of a graphic application: Manual map digitizing. *Proceedings of Association of Computing Machinery Computer-Human-Interaction 1994* (pp. 337–343).

Hoyles, C. (1993). Microworlds/schoolworlds: The transformation of an innovation. In C. Keitel and K. Ruthven (Eds.), *Learning from computers: Mathematics educational technology* (pp. 1–17). Berlin: Springer-Verlag.

Inkpen K. (1997, June). Three important research agendas for educational multimedia: Learning, children, and Gender. *Association for the Advancement of Computing in Education World Conference on Educational Multimedia and Hypermedia 1997* (pp. 521–526).

John, B., & Kieras, D. (1996a). The GOMS family of user interface analysis techniques: Comparison and contrast. *ACM Transactions on Computer-Human Interaction, 3,* 320–351.

John, B., & Kieras, D. (1996b) Using GOMS for user interface design and evaluation: Which technique? *Proceedings of Association of Computing Machinery Transactions on Computer-Human Interaction, 3,* 287–319.

John B., & Wayne G. (1994). GOMS analysis for parallel activities (Tutorial). *Proceedings of ACM Transactions on Computer-Human Interaction 1994* (pp. 395–396).

Johnson, H., & Johnson, P. (1991). Task knowledge structures: Psychological basis and integration into system design. *Acta Psychologica, 78,* 3–26.

Jonassen, D., Mayes, T., & McAleese, R. (1993). A manifesto for a constructivist approach to uses of technology in higher education. In T. M. Duffy, J. Lowyck, & D. H. Jonassen (Eds.), *Designing environments for constructive learning,* Vol. 105 (pp. 189–212). NATO ASI Series, Series F: Computer and systems sciences. Berlin: Springer-Verlag.

Kieras, D. (1996). Task analysis and the design of functionality. *CRC Handbook of Computer Science and Engineering.* Boca Raton, FL: CRC Press.

Kirwan, B., & Ainsworth, L. K. (1992). *A guide to task analysis.* London: Taylor & Francis.

Kordaki, M., & Avouris, N. M. (2001, December). Interaction of young children with multiple representations in an open environment. *Proc. Panhellenic Conference on Human-Computer Interaction 2001,* Typorama publ, Patras (pp. 312–317).

Kordaki, M., & Potari, D. (1998). A learning environment for the conservation of area and its measurement: A computer microworld. *Computers and Education, 31,* 405–422.

Laborde, J. M., & Strasser, R. (1990). Cabri-Geometry: A microworld of geometry for guided discovery learning. *ZDM, 5,* 171–177.

Lansdale, M. W., & Ormerod, T. C. (1994). *Understanding interfaces: A handbook of human-computer interaction.* London: Academic Press.

Lewis, C., Brand, C., Cherry, G., & Rader, C. (1998). Adapting user interface design methods to the design of educational activities. *Proceedings of Association of Computing Machinery Computer-Human-Interaction 98 Conference on Human Factors in Computing Systems* (pp. 619–626).

Lim, K. Y. (1996). Structured task analysis: An instantiation of the MUSE method for usability engineering. *Interacting with Computers, 8,* 31–50.

Marton, F. (1981). Phenomenography-describing conceptions of the world around us. *Instructional Science, 10,* 177–200.

Marton, F. (1988). Phenomenography: Exploring different conceptions of reality. In D. M. Fetterman (Eds.), *Qualitative approaches to evaluation in education: The silent scientific revolution* (pp. 176–205). New York: Praeger.

Marton, F., & Saljo, R. (1976). On qualitative differences in learning: I—outcome and process. *British Journal of Educational Psychology, 46,* 4–11.

Miller, C. A., & Vicente, K. J. (1999). Task "versus" work domain analysis techniques: A comparative analysis. *Proceedings of the Human Factors and Ergonomics Society 43rd Annual Meeting* (pp. 328–332).

Moran, T. P. (1981). The command language grammar: A representation for the user interface of interactive computer system. *International Journal of Man Machine Studies, 15,* 3–50.

Nielsen, J. (1993) *Usability engineering.* London: Academic.

Nielsen, J. (1994). Usability inspection methods. In J. Nielsen & R. L. Mark (Eds.), *Usability inspection methods.* New York: John Wiley.

Nielsen, J. & Landauer, T. K. (1993, April). A mathematical model of the findings of usability problems. *Proceedings Association of Computing Machinery/International Federation of Information Processing International Computer-Human Interaction'93 Conference* (pp. 206–213).

Norman, D. (1988). *The psychology of every day things.* New York: Basic Books.

Olson J., & Olson G. (1990). The growth of cognitive modeling in human-computer interaction since GOMS. *Human Computer Interaction, 5,* 221–265.

Orhun, E. (1995). Design of computer-based cognitive tools. In A. A. diSessa, C. Hoyles, & R. Noss (Eds.), *Computers and exploratory learning* (pp. 305–320). Berlin: Springer.

Papert, S. (1980). *Mindstorms: Pupils, computers, and powerful ideas.* New York: Basic Books.

Paterno, F., & Ballardin G. (2000). RemUSINE: A bridge between empirical and model-based evaluation when evaluators and users are distant. *Interacting with Computers, 13,* 151–167.

Paterno, F., & Mancini C. (2000, winter). Model-based design of interactive applications. *ACM Intelligence,* 27–37.

Paterno, F. (2000). *Model-based design and evaluation of interactive applications.* Heidelberg-Springer-Verlag.

Preece, J., Rogers, Y., Sharp, H., Benyon, D., Holland S., & Carey, T. (1994). *Human–computer interaction.* Workingham, England: Addison-Wesley.

Richardson, J., Ormerod T., & Shepherd, A. (1998). The role of task analysis in capturing requirements for interface design. *Interacting With Computers, 9,* 367–384.

Sebillotte, S. (1988). Hierarchical planning as a method for task-analysis: The example of office task analysis. *Behaviour and Information Technology, 7,* 275–293.

Shepherd, A. (1989). Analysis and training in information technology task. In D. Diaper (Ed.), *Task analysis for human computer interaction* (pp. 15–55). Chichester: Ellis Horwood.

Squires D. & Preece, J. (1999). Predicting quality in educational software: Evaluating for learning, usability and the synergy between them. *Interacting with Computers, 11,* 467–483.

Tselios, N. K., Avouris, N. M., & Kordaki, M. (2001a, December). *A tool to model user interaction in open problem solving environments. Panhellenic Conference on Human-Computer Interaction 2001* (pp. 91–95). Typorama Publ., Patras, Greece.

Tselios, N., Maragoudakis, M., Avouris, N., Fakotakis, N., & Kordaki, M. (2001b). *Automatic diagnosis of student problem solving strategies using Bayesian networks.* Proceedings of the 5th Panhellenic Conference on Didactics of Mathematics and Informatics in Education, M. Tzekaki (ed.), pp. 494–499, Aristotle Univ. Publication, Thessaloniki, Greece.

Umbers, I. G., & Reirsen, C. S. (1995). Task analysis in support of the design and development of a nuclear power plant safety system, *Ergonomics, 38,* 443–454.

van Welie, M., van der Veer, G. C., & Eliens, A. (1998, August). Euterpe—tool support for analyzing cooperative environments. *Proceedings of the 9th European Conference on Cognitive Ergonomics.*

Walsh, P. A. (1989). Analysis for task object modeling (ATOM): Towards a method of integrating task analysis with Jackson System Development for user interface software design. In D. Diaper (Ed.), *Task analysis for human computer interaction* (pp. 186–209). Chichester: Ellis Horwood.

Wilson, S., Johnson, P., Kelly, C., Cunningham, J., & Markopoulos, P. (1993). Beyond hacking: A model based approach to user interface design. In J. Alty, D. Diaper, & S. Guest (Eds.), *Proceedings of Human-Computer Interaction '93* (pp. 217–231). Cambridge, England: Cambridge University press.

15

Design of Systems in Settings with Remote Access to Cognitive Performance

Oliver Sträter
Eurocontrol, Belgium

Heiner Bubb
University of Technology Munich, Germany

Abstract

In some fields of cognitive task design, the user, or user group, cannot be involved directly in the design process. Typically this is the case in incident investigations and assessments of human reliability, which require remote prediction of cognitive processes, and the system design relies on remote access to cognitive performance. We call these remote settings. This chapter describes a method for proceeding in such situations. We describe how to obtain information about cognitive demands, possible human behaviour, and contextual factors and how this information can be used to improve the design of technical systems. We provide examples from incident analysis in nuclear power plants, from design of alarm systems for steer-by-wire systems in automobiles, and from human reliability assessment in regulatory tasks for nuclear industry. Finally, we reflect the method presented here with respect to possible implications of cognitive modelling.

INTRODUCTION

Cognitive task design requires a close link to the persons (or groups) being investigated to obtain information about the cognitive processes that occur during task performance. Usually methods such as rapid prototyping are used, and potential user groups are confronted with the design of the new system. Their performance is then measured according to constructs such as workload or situational awareness. Cognitive task analyses are usually performed to support and facilitate the process of prototyping. Various techniques have been developed to evaluate such investigations, including video confronting, eye tracking, and various workload measures (e.g., the workload measure NASA Task Load Index).

For several cases of cognitive task design, such techniques cannot be applied because the direct access or contact to the "piece of interest" (the human being, or the group of persons, whose cognitive performance is being investigated) is missing. In such cases, none of the usual cognitive design methods can be applied with sufficient validity.

This is the case in various settings that have a high interest in explaining and predicting cognitive aspects to derive design suggestions for technical systems. Among these settings are the following:

- Assessment of humans' cognitive performance on events based on evaluation of past experience in which the access to the person(s) directly involved in the original events is lacking
- Assessment of cognitive performance in early design stages in which an experimental setting must be prepared, a user group is not specified, or a detailed specification of the design of the technical system is not available so that none of the usual techniques can be applied
- Safety assessment of cognitive performance in situations in which system failures must be assessed but no failure has ever occurred and probably never will

These settings are typical application areas for the remote prediction of cognitive processes. They have in common the requirement of performing cognitive task analysis without direct access to a certain user or user group. No direct measurements of human behaviour can be performed. Consequently, other data sources must be exploited to generate predictions about human performance and to derive design suggestions. In addition, such cases require a method that is robust enough to work in remote settings and that allows description and prediction of cognitive processes with a certain level of detail. Methods such as these are usually called human reliability assessments (HRAs).

The method presented in this chapter describes design recommendations for remote settings based on predictions of human behaviour under given circumstances by using human performance data from operational experience (i.e., events in which human behaviour led to incidents or accidents). It is based on experiences and research performed in nuclear industries, aviation, automobile design, and occupational safety. It demonstrates that technical settings can be evaluated and design recommendations can be drawn to a certain extent without any direct investigation with a particular user. This chapter also demonstrates how the method allows description and prediction of the cognitive aspects of complex group performance. Examples from incident analysis, as well as from system design in the automobile industry, are presented. Finally, the work presented here is linked to existing frameworks and models of cognitive performance.

METHOD FOR DEALING WITH COGNITIVE PERFORMANCE IN REMOTE SETTINGS

Figure 15.1 outlines the overall approach of our method for studying cognitive performance in remote settings. An important distinction is made in the figure regarding "classification concerning cognitive coupling" and "experience-based human performance model." According to the distinction between phenotype and genotype of human behaviour (Hollnagel, 1993), cognitive coupling represents the phenotype of cognitive behaviour, and the experience-based human performance model represents the genotype part of cognitive behaviour.

Describing the Coupling Between Operational Task and Cognitive Performance in Remote Settings

Any technical solution (regardless of whether it is cognitively well designed) asks the human for a specific cognitive behaviour and for specific cognitive coupling with the technical system

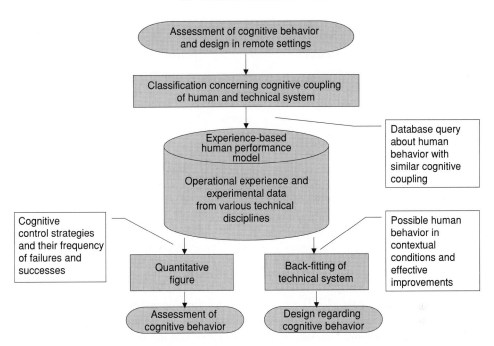

FIG. 15.1.   A method for assessing cognitive behaviour in remote settings.

when one is concerned with task-related human cognitive performance. Consequently, the approach enabling the description, assessment, and design of remote settings is based on the description of the process coupling in both the technical and the cognitive systems. It uses the basic assumption that any task-related cognitive performance is related to the characteristics of the technical task that a person or group must to perform. The examples offered in the chapter later show that this assumption is valid.

Elements for Describing the Coupling.   The properties for description, assessment, and design of the cognitive aspects of technical systems are neither fully technical nor fully cognitive. The properties have cybernetic or system ergonomic properties (Bubb, 1993). Any human role in a technical system is preceded by an exchange of information between the human and the technical system. This exchange can be initiated by the mission (i.e., the set of tasks or required functions) a system designer may have prescribed to deal with a technical system, an action of the human on a technical system, or the feedback the human receives from the system. All of these things happen in a certain physical environment and under certain situational conditions (such as time of day). The same holds for a human–human relationship, in which we have the opportunity to listen, speak, and, most of the time, gain a common understanding by establishing a relationship between our internal worlds. From the cognitive point of view, the exchange process distinguishes an internal and an external world (e.g., Neisser, 1976). Figure 15.2 shows the principal links between both worlds.

A complex working system consists of several human–human and human–machine systems. It is essential to recognize that the classification of cognitive coupling provides a classification of neither the external nor the internal world. It provides a classification of the principal ways in which they can be coupled. Based on Fig. 15.2, several couplings can be distinguished. Table 15.1 provides an overview of the coupling modes. They are described in the following sections and are related to some extent to the cognitive control modes proposed in CREAM (the cognitive reliability and error analysis method; Hollangel, 1998). Many of the distinctions mentioned here are also related to the work represented in Wickens (1992).

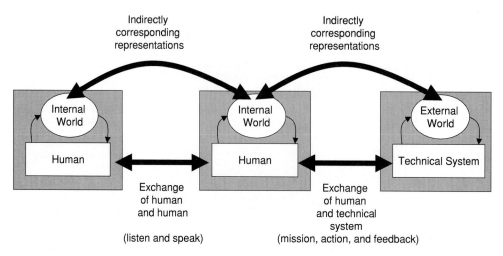

FIG. 15.2.   A human–human and human–machine system.

Isolated Versus Involved.   The first important distinction is whether a human is involved in the technical system or is performing in an isolated mode. In the involved mode, the internal and the external world are coupled. Piaget (1947) assigned the term *accommodation* to the isolated mode and distinguished this from the *assimilation* in which people are in an actively involved mode and gain new experiences from the external world. Assimilation is also strongly related to top-down or data-driven thinking, whereas accommodation is usually top-down or goal driven.

The isolated mode means that the human cognition is busy with itself. No external information enters the picture. Humans need such a "timeout"; we cannot be coupled with the external world at all times. Attempting to do so would result in serious psychological problems, as demonstrated in research on sleep deprivation (e.g., Schandry, 1981). The isolated mode does not imply that thinking is abandoned but that the different experiences of a day or during active involvement are ordered into a homogeneous, overall picture within the internal world. Sleeping, being mentally absent at a meeting, or looking out the window are all necessary cognitive behaviours to order our internal world.

During examination of operational experience, for instance, failures made during maintenance tasks were spontaneously recovered in the "after work phase." The subjects had time to accommodate their experiences within their existing internal world and to "think about the day." They realized errors they had made and informed the personnel on duty. This process was observed as typically lasting about 90 min after the active involvement in the task. After 90 min off work, considerable recoveries could be observed in nuclear power plant events (Sträter & Zander, 1998).

This decoupling of the technical system can also be observed during communication. It is a well-known fact from knowledge acquisition techniques that people are unable to express their cognitive activities if they are ordering their internal world in a problem-solving situation (Sträter, 1994).

Monitory Versus Active.   Independent from a human's currently mode, the task requires either an active or monitoring behaviour. Both require to human to be in an involved mode of information processing (assimilation).

Closed Versus Open Loop.   This describes any task requires humans to act in a certain way, namely, controlling or initiating certain functions. Whereas control tasks require working

TABLE 15.1

Summary of Aspects Describing the Different Types of Cognitive Coupling of Humans with Technical Systems or Humans with Humans

Type	Visualization	Description
Type of involvement (+)		Involved: Human is in the assimilation state (gaining new information) either by perceiving new aspects of the external world or by testing out internal aspects as hypotheses → interact
(−)		Isolated: Human is in accommodation state (housekeeping of internal world) □ → reflecting
Type of task (+)		Monitoring: The human is monitoring the process by collecting information and deciding about the process state (perceiving the feedback from a technical system) → monitor
(−)		Active: The human is actively involved in process control by collecting and combining information and process interacting; 1:1 relation between information and goal, top-down processing → perform
Type of control (+)		Closed: the task is changing over time and depends on timely demands (e.g., tracking task) → track
(−)		Open: the task is independent from timely demands → operate
Number of dimensions (+)		Multidimensional: the technical system is characterized by more than one parameter that has to be brought into relation to another; more than two goals, top-down processing → imagine
(−)		One-dimensional: the technical system is characterized by one parameter; one goal, top-down processing → expect
Necessary Operation (+)		Simultaneous: A simultaneous operation of several controls is required; parallel processing required → coordinate
(−)		Sequential: The control must be performed in a defined sequence; serial processing required → follow

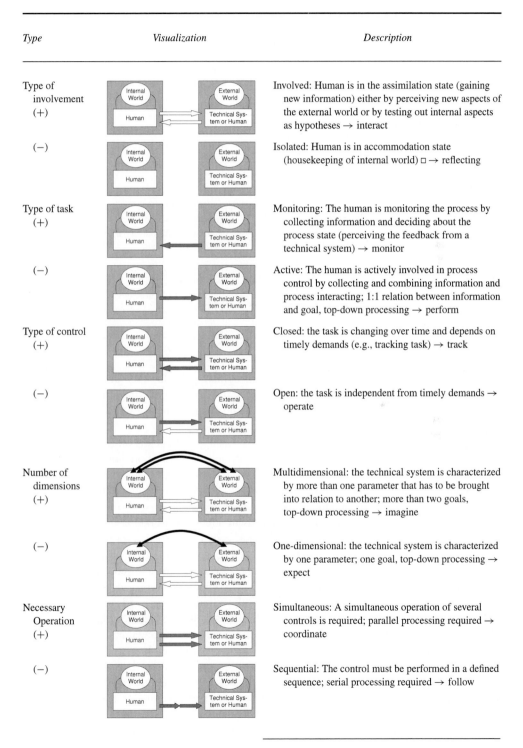

(Continued)

TABLE 15.1
(Continued)

Type	Visualization	Description
Type of Presentation (+)		Compensatory task: The system provides information for on how large the difference between task and the actual state is → identify
(−)		Pursuit task: The system provides information on the difference of task and actual state → recognize
Primary Compatibility (+)		Internal: Compatibility of the mental model of a person with external information (e.g., learned, stereotype behaviour): compatibility of goals and information → align
(−)		External: Compatibility of different external information (e.g., displays and controls); compatibility of different information → compare

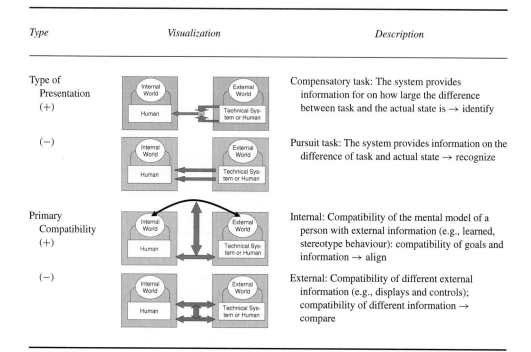

in a permanent loop with the technical system (e.g., regulating the flow of a valve or maintaining the altitude of a plane), initiation of a function only requires the completion of a certain action, which is, in principle, possible to achieve without confirming the outcome of the action performed (e.g., switching a pump on).

Whether an operation is perceived as closed or open loop depends on the time frame of the processing loop between human and machine (or, in communication, human and human). Typically, the range where humans see a closed coupling of their actions with the technical system ranges from about 200 to 2000 msec as shown in software ergonomics by Card, Maran, and Newell (1983). From several investigations of driver behaviour, it also was observed that a driver typically shifts between areas of interests in about 200 to 600 msec while scanning the scenery (Schweigert, Fukuda, & Bubb, 2001). If drivers are isolated for longer than about 2000 msec from the scenery, their actions seems be decoupled from the environment.

In Schuler (1995), investigations of communication processes showed that "good communication" requires that a closed loop between the communication partners be established, typically by short interruptions of the speaker by the receiver using words such as "Ok" and "Aha" at intervals of about 2000 msec. If such feedback is missing or slightly delayed (to about 2600 msec), misunderstandings or even bad communication outcomes were observed.

This typical time range of about 200 to 2000 msec for closed-loop performance seems to be related to the memory span. Miller (1956) found that memory is typically limited to 7 ± 2 information units (chunks). In highly dynamic environments, such as driving a car, this memory span is reduced to about 3 chunks (Sträter, 2001), and the time frame is also reduced to about 200 to 600 msec. In time frames of 200 msec or less, this span is even reduced to 1 chunk (e.g., the effect of tunnel vision in emergency situations as observed in aircraft accidents or nuclear incidents). This reduction of the memory span is also often observed in attention research paradigms like dual task investigations.

As a consequence, the memory span reflects a continuum rather than a dichotomy or a fixed number (e.g., Wickens, 1992). It can be shown that this continuum appears to be a nonlinear stepwise functional that obeys the rules of information theory (e.g., Sträter, 1994). The distinction between short-term memory (STM) and long-term memory (LTM) is therefore an artificial than a real correlate (cf. Neumann, 1992; Strube, 1990).

One to Multiple Dimensions.   One property of the external world with importance for internal representations is the dimensionality of the task with which a human must cope. Investigations of operational events showed that humans usually tend to reduce their internal world to one-dimensional concepts because there is a clear understanding about the outcome of a control action in this case.

In the case of multidimensional characteristics in a technical system, the human needs to imagine the interrelationships of the various dimensions. If operators in a nuclear power plant, for instance, must control temperature and pressure in a vessel, this is a two-dimensional task. The operators have to imagine how these two parameters are related and how they may interfere with each other during start-up of a plant.

Whether a technical system has one-dimensional or multidimensional properties, the human only can perceive what instruments the technical system provides. An interface that does not provide support to offer representation of the interrelations of several dimensions of the system complicates the process of understanding these interrelations. If, for instance, temperature and pressure are presented independently from each other (e.g., by two analogue instruments) and can only be controlled by two independent controls (e.g., two switches), this cognitively demanding task is far more complicated for the operator than a two-dimensional display or a two-dimensional control element. In an airplane, the control of position (turn and bank) and the attitude to the horizon (pitch) are controlled by a single input device, the yoke. Requiring two independent control elements for the horizontal position and slope would be a completely inappropriate design.

The complexity of a technical system is therefore related to the number of dimensions it has. Functional complexity has to be distinguished from interface complexity, that is, whether these dimensions are presented simultaneously or sequentially.

Simultaneous Versus Sequential Processing.   Simultaneous processing requires maintaining all the information necessary for of the simultaneous tasks in memory (e.g., maintaining position and speed on a road requires information on the road path as well as weather and road conditions). The simultaneous processing of all information requires coordination of the cognitive processing to perform all tasks correctly.

Sequential tasks require performance of the action in a prescribed order to perform the task successfully (e.g., while driving a car with manual transmission, the order of actions must be to declutch, change gears, and then release the clutch). In this case, the cognitive processing system has to recall certain rules or schemas stored in memory to correctly follow the procedure, predefined by the layout of the technical system.

Simultaneous and sequential actions can be at either a conscious or an unconscious level.

Compensatory Versus Pursuit.   Alarm systems usually provide information about a system that is mismatched on some predefined parameter (e.g., "pressure in vessel XY too high"). Alarms usually lead to compensatory tasks and require the human to identify the meaning of the information based on the knowledge about how high the pressure in the vessel is allowed to be. Gauge displays showing a pressure in a vessel XY require that the operator recognize the difference as a "greater than normal" deviation and are called pursuit tasks (e.g., to detect the difference between 100 and 101 psi as more that allowable).

Compatibility Versus Incompatibility.   Several elements of the external world (e.g., a switch for a pump and a switch for a generator in a power plant) can be presented to the human in a compatible or in an incompatible way. External compatibility refers to this similarity within the external world. It requires matching all pieces of information provided by the technical system against each other (e.g., displays and controls).

A certain element of the external world can be compatible or incompatible with the internal representation a human has about functioning. Internal compatibility refers to this similarity between a certain external element and its internal representation (e.g., a switch for a pump and the mental model the operator has about its functioning). Internal compatibility always implies alignment of the incoming information with one's understanding of the way the technical system works (see also McCormick & Sanders, 1983).

Compatibility is one of the golden rules in cognitive task design (e.g., in control-room upgrades, the controls in the new control room should be compatible with the ones used in the old control room). However, cognitive design does not mean striving for compatibility all the time. Information should be presented in an incompatible way, if the devices are used for different technical functions.

Table 15.1 summarises a set of aspects that describe cognitive coupling between humans and a technical system. Arrows indicate a coupling direction. Dark arrows indicate a trigger. White arrows indicate that this path is necessary but not synchronized with trigger. The (+) symbol indicates high-demand cognitive tasks; the (−) symbol indicates low-demand cognitive tasks. The next section outlines how these aspects of cognitive coupling are used for human performance modelling.

EXPERIENCE-BASED HUMAN PERFORMANCE MODEL

Based on the cognitive coupling, certain human behaviour can be predicted if an appropriate approach, for a cognitive model of human performance in remote settings. Ideally, a model for remote prediction follows the criterion of "requisite variety" (Hollnagel, 1998): The cognitive model should be built in a way that it is able to describe the variety of the aspects in its application field. An additional criteria of "computational effectiveness" is addressing its completeness: "Any serious model of cognition will have to provide a detailed computational account of how such nontrivial operations can be performed so effectively" (Shastri, 1988, p. 336). According to the latter position, a cognitive model must be conceived at a very detailed level, subsuming experimental findings, cognitive modelling approaches, and neurobiological science within a single approach.

Law of Uncertainty in Cognitive Modelling

Taking both criteria (requisite variety and computational effectiveness) into consideration leads to the dilemma that any performance model has to deal with uncertainty either with respect to the detail with which it models cognitive performance or with respect to the uncertainty in the validity of the model. Norman described this uncertainty of psychological modelling as follows: "Each new experimental finding seems to require a new theory" (Norman, 1986, p. 125).

It can be concluded in general that cognitive approaches are more untenable the more they are designed to predict certain cognitive effects. In accordance with quantum theory in physics (Heisenberg, see, e.g., Kuhn, Fricke, Schäfer, & Schwarz, 1976), this may be described as a law of uncertainty that results from the complexity of humans' cognitive behaviour: The more a certain cognitive function is assumed, the more uncertain is the stage (within a psychological model) that exhibits this function and—vice versa—the more a certain stage is assumed, the more uncertain is the function of this stage. This law of uncertainty should not lead one to conclude that cognition cannot be predicted, but the uncertainty in cognitive modelling must be considered appropriately.

Connectionism Approach for a Human Performance Model

A connectionism approach that is representing the complexity of human cognitive processes is used here to cope with the dilemma of uncertainty (i.e., that no cognitive model can be certain). It represents the relationship of cognitive processes to human performance, its interdependencies, and the relationships to contextual and situational conditions in some kind of knowledge-management system. The connectionism approach assumes that the brain is a complex net of cells (i.e., a more realistic assumption compared with classical block models like input-processing-output models).

The approach for the human performance model is called CAHR (Sträter, 1997, 2000), or *connectionism assessment of human reliability*. Connectionism is a term describing methods that represent complex interrelations of various parameters (which are known for pattern recognition, expert systems, and modelling of cognition). By using the idea of connectionism, the CAHR method recognises that human performance is affected by the interrelation of multiple conditions and factors (of an internal as well as an external nature) rather than by singular factors that may be treated and predicted in isolation. This enables the model to represent and evaluate dependencies of influencing factors and context of situational aspects on the qualitative side. It further suggests a quantitative prediction method that considers the human error probability (HEP) as driven by human abilities and the difficulty of a situation.

The connectionism approach is able to represent two important properties of artificial intelligence: polymorphism (that the same semantic information can be interpreted differently in different contexts) and encapsulation (that specific semantic information can be processed based on various experience). The approach resolves here with the problem of the dependency of cognitive reasoning on previous information (dynamic modelling).

CAHR therefore represents experience-based representation of cognitive performance. The knowledge base of this model is subsequently built up by a systematic description of human performance and human errors. In its original application, it contained information about human performance during 232 incidents and abnormal events occurred in nuclear power plants. This information included cognitive coupling, behaviour, and errors, as well as situational conditions and causes for errors.

Relation of the Cognitive Coupling and the Connectionism Approach for Human Performance

The human brain collects information and compares it with the internal representations about the world. This is sometimes a conscious process but most of the time is unconscious. Neisser (1976) called this process the *cognitive mill*. The metaphor can be used to describe the way in which cognitive coupling and connectionism fit together in our approach for modelling remote settings (Fig. 15.3). Figure 15.3 illustrates how cognitive coupling and connectionism are related to generate predictions about cognitive behaviour. As the applications of this model

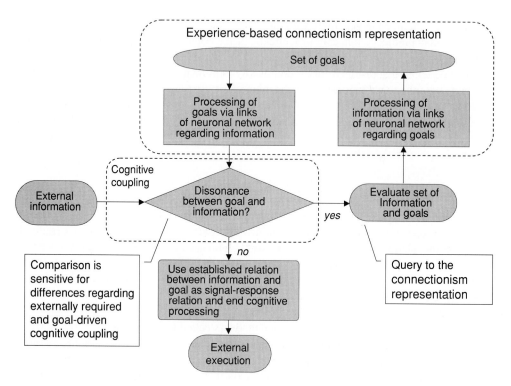

FIG. 15.3.   Coherency between coupling processes and experience-based knowledge.

shows, the connectionism representation also provides situational conditions under which behaviour is performed and the causes of possible errors.

The comparison is sensitive to differences regarding externally required and goal-driven cognitive coupling. This element of the model is essential: If a human is involved in a dynamic task, a static task cannot be performed at the same time. He or she must stand back and take time to gain a broader perspective of the information. We show in the application section that this holds not only for control tasks but also for communication within groups. This basic assumption also has a considerable link to the cognitive streaming theory (Jones & Farmer, 2001). The theory states that, assuming human is busy with a serialization task (e.g., counting), he or she is unable to serialise other information, even if the additional serialisation is required on a different information channel (visual vs. auditory) or in a different modality (spatial vs. temporal). In addition to this basic maxim, the model has four additional maxims, described as follows.

Cognition Is Perception and Goal Driven.   Cognitive performance is linked to the goals a human intends to achieve. Goals lead to attitudes one has toward given situational conditions. Depending on slight deviations of these conditions, completely different cognitive strategies may be used. This link is often ignored in investigating and predicting human error. Classical methods for error prediction, for instance, attempt a prediction by assuming a rather stable set of information-processing stages that is used by an operator with more or less success under a given situation (see Organisation for Economic Cooperation and Development, 2000).

Cognition Needs Cognitive Dissonance.   Cognition needs some kind of energy to proceed. Festinger (1957) named this energy *dissonance*. In the connectionism approach followed here, dissonance is operationalised as a mismatch between the externally required and goal-driven cognitive couplng. For instance, dissonance may occur if a task requires an operator to monitor the system, but he or she needs to be active in an emergency situation. Note the relationship to concepts of mismatched activity of orientation reactions in neuropsychological research (e.g., research on electroencephalogram, Sommer, Leuthold, & Matt, 1998).

A mismatch is a necessary but not a sufficient precondition for conscious processing. In many situations, we are able to cope with situations on a subconscious level, even though dissonance is present. This is the case if the situation can be matched to some learned experience without the need of adaptation to the situation. When the situation cannot be matched successfully, a higher level of cognitive behaviour is required. According to Rasmussen (1986) and Reason (1990) the levels of behaviour range from skill based, to rule based, to knowledge based. The approach of Fig. 15.3 assumes, however, according to the theory of cognitive dissonance, that the skill-, rule-, and knowledge-based levels and the switch between the different levels have a common mechanism, a common location, and common properties. It implies that properties such as signal-to-noise thresholds and a certain amount of sluggishness while moving form one level of behaviour to another arise from one part of the cognitive processing system.

Cognition Is Experience Based.   Every cognitive act happens in the memory of the human brain. This memory is built up and based on experience. Even purely mathematical and logical processes need experience of the elements used in the logical reasoning (e.g., numbers must be learned). Experience is not necessarily of external origin, however. We are, for instance, able to develop mathematical and logical rules without using external information but by using internal experiences that we gained by thinking. This feature enables abstract reasoning as well.

Experience-based cognition also implies that humans tend to gain new experience. The new experience is transferred to an internal representation with the well-know modes of assimilation (of new experiences) and accommodation (inserting the new experience into the existing memory structure). New information must be assimilated before it is accommodated, but the relevance of the new information in the framework of the existing internal world cannot be conscious before accommodation. Therefore, risk-taking behaviour is a result of the experience-based nature of cognition.

Experience-based cognition also explains that humans have the property to argue completely differently in different situations because experience is context specific.

Cognition Means Reduction of Cognitive Dissonance.   In cases of mismatch, human beings try to cope by reducing it. This means humans try to impose *stability* to their internal world after a mismatch has been experienced; this does not imply a stability of the external world. This aspect of cognitive processing is a decisive maxim: If there is a technical disturbance, we can operate a plant, drive a car, or fly a plane only if we achieve internal stability. Situational aspects not fitting the stability (countercues, side information) that disturb our approach to achieve stability will be ignored or even rejected (confirmation bias). This mechanism is in full accordance to the theory of epistemology stating that we assume something is true if it fits into our internal world without contradiction (Keller, 1990).

APPLICATIONS

The CAHR method was implemented as a database application for investigating nuclear power plant events in Germany. To build the experience-based knowledge of the model, 232 operational events with 439 subevents were analysed regarding human performance and related contextual conditions (Sträter, 1997, 2000; Sträter & Bubb, 1998). The knowledge base of the method is under continuous development (see International Atomic Energy Agency, 2001; Linsenmaier & Sträter, 2000).

The predictive part of the method was applied to the nuclear, aviation, automobile, and occupational health industries for predicting human reliability and for finding appropriate designs regarding cognitive aspects.

Assessment of the Cognitive Performance of Humans

When an event or incident happens, usually no observer is present to immediately investigate the cognitive behaviour of the person(s) who performed the error. When the incident leads to fatalities, as is the case in serious disasters, it is not possible to ask the person(s) afterward about the cognitive processes that led to such consequences. Instances of such a situation are Chernobyl, Tokai Mura, or the loss of the crew in an aircraft crash. In addition to these public events, there are many incidents or abnormal events recorded in several industries that can be exploited for further system improvements. Usually these are filed in written event descriptions, which may be enriched with further information, such as voice or video recordings.

In all these settings, an investigation of the cognitive processes must take place but cannot be evaluated by detailed interrogation of the persons involved. This is also the case when the people involved are available for interview. Because investigations are delayed for hours or days, the persons interrogated are less able to reproduce the details of the cognitive aspects of the event. Even rapid investigation teams cannot investigate all incidents. This means that incidents are selected for investigation depending on specific reporting criteria (event threshold). Any cognitive analysis in these settings is therefore remote. Nonetheless, past experience is a key information source for improving any type of system. New methods of HRA, for instance, put great effort into acquiring information about cognitive aspects of human error (OECD, 2000).

Causes of Human Errors. An important question on the applicability of the CAHR method is whether it is useful and valid for remote settings—either for describing cognitive processes or for predicting cognitive errors. This validity can be addressed by comparing the prediction of the causes for cognitive errors using the cognitive coupling with information about cognitive processing that is directly achievable from incident analysis.

Figure 15.4 shows this comparison. The influencing factors for human errors using the coupling approach are in the same rank order as if information about cognitive errors that directly available from the incident descriptions was used. The importance was determined by calculating the frequency of an influencing factor related to the number of events in which it occurred. Positive values reflect the relative importance of a highly demanding coupling; negative values reflect the importance of less demanding couplings.

Influencing factors may be summarised as follows (the number of observations is mentioned in parentheses, see Sträter, 2000, for details and for the discussion of the interrelations and interdependencies of the factors):

- **Mission or task**
  - Task preparation (13): lack in planning, organisation or preparation of task
  - Task precision (7): lack of precision of a task
  - Task complexity (6): task was too complex for a given situation
  - Time pressure (6): time pressure caused trouble in performing the task
  - Task simplicity (3): task deviations in a given situation were not considered
- **Communication or Procedure**
  - Completeness of procedures (24): information on procedures missing
  - Presence of procedures (7): procedures do not exist for a given situation
  - Design of procedures (4): procedures are poorly designed ergonomically
  - Precision of procedures (3): procedures lack preconditions for action
  - Content of procedures (1): procedures contain incorrect information

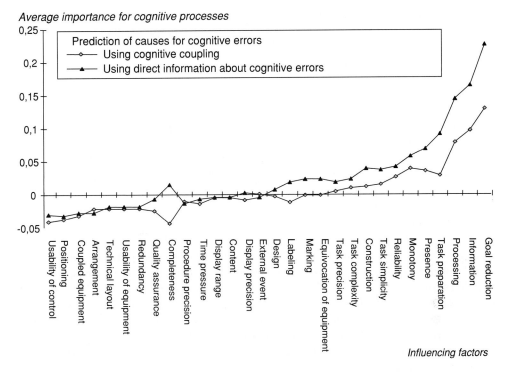

FIG. 15.4.   Profile of Influencing Factors in Relation to Cognitive Demand.

- **Operator**
  - Goal reduction (11): task and goal to be accomplished was simplified
  - Information (8): operator ignored information provided by the system
  - Processing (7): fixating on a task led to the error
- **Perception or action**
  - Usability of control (12): control systems are difficult to handle for operators
  - Positioning (7): control systems cannot be placed in a certain position
  - Quality assurance (7): quality assurance was not used
  - Equivocation of equipment (6): equipment can easily be confused
  - Usability of equipment (4): operators cannot handle equipment
  - Monotony (4): action is monotonous, typical vigilance effects occur
- **Feedback**
  - Arrangement of equipment (6): equipment is badly arranged
  - Marking (6): semantic meaning of system states are not marked on the display
  - Labelling (5): display is not labelled or is badly labelled
  - Display precision: display does not precisely show process parameter (4)
  - Reliability (2): display is unreliable and hence operators' trust in it is low
  - Display range (1): display is not capable of showing a process parameter
- **System**
  - Construction (8): a system (e.g., diesel, conduit) is badly constructed and obstructs operation
  - Coupled equipment (6): electrical equipment that is supposed to be redundant conflicted with each other or was coupled
  - Technical layout (4): layout of system is beyond required specification
  - Redundancy (4): several redundancies are effected (common causes)
  - External event (1): an unforeseen external event caused technical failure

Based on the correspondence between coupling modes and direct information, it can be concluded that the approach of systematic description of cognitive coupling between human and technical system provides a basis for prediction of the cognitive control strategies and hence for predicting cognitive behaviour.

Cognitive Tendencies to Reduce Information.   Three main cognitive reduction strategies can be distinguished on the level of the cognitive coupling. Figure 15.4 mentions these as processing, information, and goal reduction. A fourth reduction strategy can be deduced from the relationship among certain influencing factors (e.g., task complexity) and omission errors, and these represent hesitation during information or goal overload. Mosneron-Dupin et al. (1997) call these reduction strategies *cognitive tendencies*. The four reduction strategies can be described as follows:

- Fixation on established coupling (*processing* in Fig. 15.4): The desire for consistency of the internal world is the driving factor for errors in cognition (Festinger, 1957). This means that any cognitive act leads to a fixation on the decision drawn during a cognitive act.
- Information ignorance or reduction (*information* in Fig. 15.4): We cannot process all our knowledge (internal world) and all the information that the external world provides in every second of our lives. A well-known human characteristic (e.g., from political decision making) is the strategy to ignore information available from either the external world or the internal world if this information will lead to internal inconsistencies in a given situation.
- Goal reduction: (More difficult than ignoring information is ignoring goals, attitudes, or intentions. During an incident in the Davis–Besse nuclear power plant, USA, for instance, a prescribed safety procedure, which would have made the plant inoperable in the future, was ignored to achieve a recovery action, which allowed further operation of the plant (Reer, Sträter, Dang, & Hirschberg, 1999). In many other incidents, the conflict between money and safety often leads operators to ignore the safety rules instead of the economic ones. Once goals are abandoned, this usually leads to polarization and more negative statements about the former position, because this is necessary to achieve internal consistency of one's own experiences and therefore reduces dissonance.
- Information or goal overload: When high demands on cognitive coupling coincide with dissonance, the operator tends to be reluctant to act until he has found an appropriate solution for the mismatching constellation between information and goals.

Table 15.2 summarises these tendencies and relates them to the concept of dissonance and to the coupling modes. The relation presented can be observed in a relatively stable manner across human errors in other industrial settings such as aviation (cf. Linsenmaier & Sträter, 2000) and occupational health (cf. Bubb, 2001). For an explanation of the distinction between low- and high-demand coupling, see Table 15.1.

First, a person may not consider a mismatch (e.g., selective or focussed attention, fixation to be critical). In such cases, the difference between available external information and the internal representation of the world concerning the aspect of interest in not recognised. For example, when driving the car to the office by a usual route, one may fail to notice differences in the external world.

Second, a mismatch may be perceived but not all of the relevant information is collected (bounded or uncompleted search, eagerness to act, confirmation bias). In such cases, the available external information implies more opportunities than the operator(s) currently want(s) to cope with. If the operators have, for instance, already a certain hypothesis about why a technical system failed, they may complement information supporting the hypothesis and ignore information that contradicts this hypothesis. For example, when driving by car to the

TABLE 15.2
Relation of Cognitive Dissonance and Situational Complexity

Cognitive Dissonance Cognitive Coupling	No Dissonance	Dissonance
Low demand	Fixation ♦ Omission (e.g., no action) ■ Marking, labelling ● Fixation Example: Driving a car to the office as usual day by day can lead the driver to ignore surroundings	Goal reduction ♦ Omission (e.g., no action) or commission (e.g., wrong action) ■ Reliability and equivocation of equipment ● Frequency-oriented reasoning Example: Headlight failure is assumed a spent bulb, not fuse failure
High demand	Information ignorance ♦ Quantitative commissions (e.g., too much/too little) ■ Precision and design of procedures ● Eagerness to act Example: Driving by car to the office and finding the usual way is blocked; the driver takes a known alternate way	Information and goal overload ♦ Delay (e.g., too late) ■ Arrangement of equipment, reliability ● Reluctance to act Example: A car isn't running and the driver has to decide between the two alternatives, both with pros and cons: buying a new one, repairing the old one

Note. ♦-principal error types; ■-typical influencing factors; ●-cognitive behaviour.

office and the usual route is blocked, a driver often takes a known alternate way without determining whether it is better to stay in the traffic jam.

Third, a mismatch may be perceived, but the hypotheses are not elaborating sufficiently in the diagnosis process (known as frequency gambling, bounded rationality, and routine reaction). In such cases, the internal representation of the world concerning the aspect of interest is providing more opportunities then are currently relevant given the available external information. For example, headlight failure is more likely to be assumed the result of a spent bulb than a fuse failure.

Finally, a mismatch may be perceived, but the appropriate information is not collected and the hypotheses are not elaborated well enough in the diagnosis process (i.e., a mixture of Cases 2 and 3). For example, if one's car isn't running, he or she must decide between the two alternatives that have pros and cons—buying new one or repairing the old one.

The cognitive tendencies approach suggests systematically combining situational conditions and cognitive aspects by considering the coping strategies an operator may choose. This approach therefore may be considered as a way of combining the usual concept of processing stages (e.g., perception, decision, and action) and processing phases (skill, rule, and knowledge based). It has been successfully applied in various second-generation HRA methods (e.g., OECD, 2000).

Crew Performance and Cognition

Crew performance and communication is strongly related to cognition. This relationship of communication and cognition is obvious because any communication needs prior cognitive

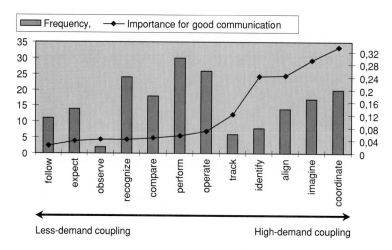

FIG. 15.5.   Impact of cognitive activities on communication quality.

processing. The link between cognitive aspects of communication behaviour has been shown in much psychological research (e.g., Eberleh, Neugebauer, & Sträter, 1989). Verbalisation of knowledge using the thinking-aloud method, for instance, shows similar effects to those observable in operational experience. Communication gets more difficult the higher the demands of the situation and the more difficult the coping for the operator.

The same operational events investigated in the study related to Fig. 15.4 indicate that not only perceptual aspects (such as precision of task or clarity of order), but also attitudes such as willingness to send or openness to receive information in the current situation are important factors for communication. The cognitive control strategies (information ignorance, goal reduction, and fixation on processing) were also observed as driving communication behaviour (Sträter, 2002).

The evaluation of the events regarding cognitive activities during communication errors is outlined in Fig. 15.5. It shows the relative impact of cognitive coupling on communication quality. The impact was defined as the number of events in which a cognitive activity was observed in relation to communication errors divided by the number of events in which the cognitive activity was mentioned. The cognitive verbs are related to Table 15.1. The figure shows that cognitive activities are a distinctive characteristic for communication problems.

The impact on communication increases in the different cognitive activities from "follow" to "coordinate." In situations in which the operators are receptive to information from outside their internal world (i.e., low-demand coupling), there are fewer communication problems compared with situations in which the operators are busy applying their own thinking to a situation (high-demand coupling). The more items a person must consider during operation, the more difficult it is to get information from outside into the cognitive loop (the cognitive mill). In cognitive states in which operators are faced with several tasks and several cognitive aspects that have to be matched (i.e., during coordination), communication problems are greater.

Cognitive Tendencies in Communication.   The following are categories of communication failures discernible from a cognitive point of view. They correspond to the cognitive tendencies outlined in Table 15.2.

- Lack of communication generally results from the receiver incorrect identification of the information (fixation). Focusing on performing the task within the technical system or on complexity and situational efforts may lead to a communication breakdown due to prioritising the technical task instead of the communicative task on the receiver side. In this case, the receiver is no longer able to identify the meaning connected to information that is sent.

- The receiver associates information with a system that differs from the one the sender intended (frequency-based reasoning leading to goal reduction). In cases of an overlap in tasks, combined with lack of experience, the receiver may misunderstand the information (which leads to wrong decisions), especially if the sender provides abstract input (assuming that the receiver may know what he or she means).

- Communication breakdown may be due to lack in coordination between receiver and sender (sender's eagerness to act combined with receiver's information ignorance). If several of the communication problems in this list appear together, poor information exchange is the probable result.

- The fourth case in Table 15.2 (reluctance leading to either goal reduction or information ignorance) is known as one of the most critical communication situations. It is characterized by reluctance of the information sender to transmit information to the appropriate receiver, even though the sender is aware that the receiver needs this information. The sender is unable to imagine the needs of a potential receiver because this would mean giving up one's own goal. Such breakdown situations were not observed in the operational experience but are quite common psychological effects. Mob situations also belong into this category.

In addition to these four cases, operational experience shows that good communication can be observed in high-workload conditions (i.e., high objective workload). The assumption of a simple relationship such as "high workload leads to bad communication" does not sufficiently explain this observation. Detailed analysis of events shows that communication problems can be explained by the concept of cognitive coupling. If the operational task is in consonance with the communication task, good communication can be observed. If the operational task is in dissonance with the communication task, bad communication occurs.

Cognitive Tendencies in Adding Information. The cognitive tendencies observed during incidents show the bounded rationality of human beings (Kahneman & Tversky, 1979). Although we are capable of performing many things in parallel, humans are one-dimensional information processors. They need stability in their internal world to perform. As the incident analysis shows, this is achieved in problem-solving situations by either fixating on an established solution by ignoring information, or by simplifying the goals. However, the literature has also identified modes in which humans add new information or imagine new goals. These modes were not observed in the incident analysis because they require deeper analysis of the information. Nevertheless, they are in full accordance to the overall approach of the cognitive coupling. The following main effects on can be distinguished:

- Information-adding modes: In interviews with incident witnesses, it can be observed that they add information to their testimonies that they did not perceive from the external world but added from their internal worlds. Witnesses have said that they saw how two cars collided, but deeper investigation shows that their recollection couldn't be right. They could not see the event and added visual information based on what they heard (e.g., Semmer, 1994).

- Goal-adding modes: The same holds for the adding or redefining goals. Dörner (1997) showed in various experimental settings that people add new goals or attitudes to their strategy in problem solving if they fit a better overall picture of the results they have achieved.

For all of these modes, sluggishness of the cognitive processes can be assumed to lead to hysteresis effects between the information considered and the goals followed. Hysteresis means that if a goal or an information is considered once in a cognitive act, it needs additional counterarguments to shift the act into a different direction than would be required if the cognitive act were still open for new information or goals.

Assessment of the Cognitive Performance in Early Design Stages

A second remote setting is the early design stage, when a product is still in a conceptual state or prototypes are unavailable. In such cases, the layout of the system may not be known. Only its functional requirements are specified so that the information necessary for usability testing and user involvement is missing. A designer's question would be, for instance, what would be the impact on the driver's cognitive processing and decision making in critical traffic situations if I introduce steer-by-wire in the next-generation vehicle?

Such a problem was addressed in Theis (2002) for a large automobile manufacturer in Germany. The issue was that drive-by-wire cars (i.e., electronic actuation of steering and braking) are in the early design, phase when the failure warning has not been designed and possible driver reactions have not been investigated because the system exists only in prototype form. The prototypes are occupied for technical developments and are not currently available for time-consuming human factor analyses. On the other hand, the human reaction on system failures has to be investigated to get information about the impact of inappropriate human reactions on the safety of the driver–vehicle system.

Prediction of Cognitive Aspects.   The CAHR method was used to assess the possible reactions of drivers if a steer-by-wire system failed. The method was used in the manner described in Fig. 15.3 to assess probable cognitive coupling and possible driver reactions. Cognitive tendencies were predicted based on the situational conditions. Quantitative ratings were performed to determine the combination of traffic situation (contextual aspect) and control strategy (cognitive tendencies). Probabilities of human errors were assessed for the most critical combinations in terms of safety (see Table 15.3). Note that the tendency to add information was included in this study.

The combination of driving manoeuvre and type of travel related to fixation was considered the combination with the most considerable impact on safety. Types of travel comprise whether the driver is forced to reach a certain destination (e.g., the goal is to get to an important appointment) or whether the driver has an unspecified mission (e.g., weekend excursion). The combination of ignoring an alarm of the steer-by-wire system due to fixation was predicted to have probability of .25 for high-demand situations and .22 for low-demand situations. The prediction is more-or-less equal for both high-demand and low-demand traffic situations. The quantitative correspondence leads to the conclusion that the human behaviour is less related to the cognitive demand than to the type of travel (i.e., the driver's goal).

The quantitative figure of .22 to .25 appears to be high compared with human error probabilities used in classical human reliability assessment methods. Nevertheless, the research in the field of errors of commission shows that human error probabilities for cognitive behaviour appear to be generally on this order of magnitude (Reer et al., 1999; OECD, 2000). The classical method of THERP (technique for human error rate prediction; Swain and Guttmann, 1983) would have assessed this situation with a human error probability of less than $P = \sim.001$. This value is a misassessment by a factor of ~220 (ranging from ~110 to ~330). Such uncertainty has a considerable impact on licensing steer-by-wire cars for public use.

The bold face in Table 15.3 indicate those predictions that were validated in the experiment using two main conditions, one condition in which drivers had to reach a certain goal and

TABLE 15.3

Application of the CAHR Method for Predicting Possible Driver Reactions to Failures of Automobile Steer-by-Wire Systems (according to Theis, 2002)

Contextual Aspect	Cognitive Tendencies/ Cognitive Coupling	Fixation	Eagerness to Act	Frequency-Oriented Reasoning	Reluctance to Undertake Unusual Actions	Information Adding Leads to Increased Confidence
Road quality and traffic situation	High demand			.13		
	Low demand				.67	
Situational and environmental Factors	High demand	.10				
	Low demand				1.00	
Driving manoeu- vre and type of travel	High demand	**.25**				
	Low demand	**.22**			.22	.22

Note. Bold face probabilities were validated with subjects under two conditions drivers who had to reach a specified goal and drivers on a weekend excursion.

another in which they were asked to make a weekend excursion without any appointment. The alarm was set up unambiguously to exclude possible errors due to not perceiving the alarm. The experiment was performed in a full-scope car simulator. Figure 15.6 shows the results (see Theis & Sträter, 2001, for additional discussion). The experiment revealed a corresponding quantitative figure as predicted by the CAHR method for this combination: for fixation error (goal orientation), $6/18 = .33$; for fixation error (no induced goal orientation), $2/18 = .11$; and for fixation error (total) $8/36 = .22$.

The quantitative result shows that 33% of drivers ignored a critical warning and preferred to reach an appointment. Even when no specific goal had to be accomplished, 11% of drivers ignored the critical system failure. Overall, a probability of .22 was observed for fixating the goal and ignoring information that would contradict or inhibit accomplishment of the goal.

The results have a significant impact on the design of vehicles because a safety assessment based on classical models would have revealed a far lower human error probability. The result would have an even greater impact on the licensing of nuclear facilities.

From Prediction to Design. The experimental setting revealed a corresponding quantitative figure as predicted by the CAHR method for the critical combination ($\sim.22$ was predicted; $\sim.22$ to $\sim.25$ was observed). The study confirmed that cognitive processes observed in nuclear power plant events are transferable to decision behaviour of drivers in a given situation for both the qualitative approach of cognitive coupling and the quantitative prediction of human behaviour. The same cognitive errors and error mechanisms were found in the nuclear events as well as in the car-driving experiment under given conditions. Overall, this work shows a successful application of the method outlined in this chapter because the qualitative and quantitative expectations fit to the experimental ones.

This fact, although likely surprising from an engineering perspective, is obvious from the psychological perspective because humans are involved in both settings and, despite differences

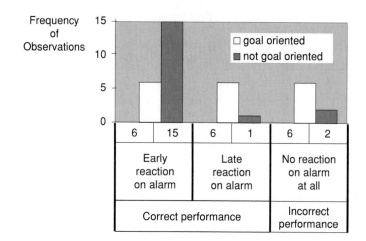

FIG. 15.6.   Failure frequency for ignoring a critical alarm in a steer-by-wire car.

among humans, cognitive coupling and tendencies appear to be similar. The results achieved during the investigation allow using the approach to predict cognitive control strategies in other industrial settings when the description is based on the coupling between human and technical system.

The CAHR method allows in a second step to arrive at suggestions on how to improve the design of the technical system. The contextual conditions and contributing factors are known for each quantitative figure the method derives. The conditions and factors can be ranked according to their importance. Fig. 15.4 is an example of such a rank profile. The ranking for ignoring a critical alarm would lead to the following contributors and design suggestions:

- Time pressure, situational pressure (i.e., to reach a specified goal), and complexity of a situation are circumstances that reinforce this behaviour.
- Improving the ergonomic design of an alarm system cannot eliminate the tendency to ignore it.
- The ignore tendency to an alarm could be reduced by improving the explanation on how steer-by-wire systems work, by preparing the driver for possible failure scenarios, and by providing an unambiguous prescription on how to behave in critical situations.

Assessment of the Cognitive Performance in Safety Assessments.   As the steer-by-wire example showed, cognitive science would be relatively easy if people would do what they expected to do from system designers' perspectives and if system designers would be able to foresee all possible problems a designed technical system may be exposed to in the future. In this case, we would have a clearly specified environment, the action would be prescribed, and the system behaviour would be determined.

This is certainly not the case. The system designer commits errors in design and constructions before a technical system is even in operation. Accompanied by additional constrains of management, these lead to human errors during operation. The *Titanic* is a classic example of design, construction, management, and operational errors.

Human Reliability in Safety Assessments.   Safety assessments are common approaches to judge the safety of complex systems. To include errors on the human side (e.g., design, construction, management, and operation), these assessments must include human reliability assessments. Regulatory bodies are typically customers of this technique because they have an interest in the overall safety of the systems they have to regulate. Insurance companies and industry also use it if information about the safety status of a certain system is needed.

Safety assessments are a classic example of remote settings because cognitive performance must be predicted. The scenario of failures in the system and the human reactions to the failures cannot be observed directly. Often such situations cannot even be simulated if they require extensive simulation facilities. Simulations of nuclear power plants, for instance, cannot simulate complex scenarios in which operators must perform in the control room and in the plant because simulators only represent the control room. Therefore, the way humans perform cognitively in such a setting cannot be simulated with sufficient validity.

The method as presented in this chapter and as outlined in Fig. 15.3 has been used in various safety assessments to overcome the limitations of current safety assessments and to reveal mitigation measures for safety-critical behaviour. The procedure to predict human behaviour was identical to that presented earlier in the chapter regarding the early design case. Among the applications are the following:

- The assessment of group performance in so-called low power and shut down states in which the safety significance of a complex maintenance task in a nuclear power plant had to be assessed and suggestions for improvements regarding collaborative cognitive performance had to be suggested (Müller-Ecker, Mayer, & Sträter, 1998; Sträter, 1998, 2001)
- The prediction of errors of commission in nuclear power plants and the analysis of causes for cognitive errors (Reer et al., 1999)

OECD (2000) described these examples in more detail. The experiences regarding system design led to a critical issue of safety assessment, the so-called errors of commission.

Omissions and Commissions.   A serious issue for any safety assessment is the omission of critical contributions to the overall risk. Because the field was dominated by assessing the risk of technical failures, the overall risk of human contributions to hazardous technologies is usually misassessed. Chernobyl is such an example. Overconfidence of the crew in its competency was never considered in risk assessments but was one of many contributing factors in the accident: The operating crew at the time of the accident was the most effective until they destroyed the reactor. Observations like these can be made for other technologies as well.

To make a realistic safety assessment, the human contributions must be included (e.g., design, construction, management, and operation). An important distinction in safety assessments in this respect is the one between omission and commission errors. An omission is when a certain task or mission is not performed, by either a human or a system. A commission error is an incorrect action taken by a human or machine. For omissions, we can observe a many-to-one relation: many elements of an entire system can fail, with a certain probability, to omit their required function. Many-to-one relations can be analysed and predicted easily because the outcome is defined. Any commission, whether committed by operator or machine, has the potential to change the course of accidents, (e.g., the intervention of the crew at Chernobyl initiated the accident; without their intervention, the accident would not have happened).

Compared with omissions, for which the problem space is precisely defined, the problem space of commissions is ill defined. The prediction and prevention of commissions is consequently much more difficult: There is a one-to-many relation between a commission and its possible outcomes. These relations are of a dynamic nature and are strongly context dependent.

The approach presented in this chapter can deal with both properties (dynamic and context representation) and can reveal design recommendations (as demonstrated by the steer-by-wire example). The method therefore can be used to avoid discussions between analysts of the technical system and of the humans involved. Most times, these discussions lead to inappropriate assessments and unsuitable construction of safety assessment models (e.g., worst-case modelling). The assessments often try to cover quantitatively the human impact on risk without having any feeling for or proof of whether the modelling or the assessment is actually covering all human influences.

DISCUSSION

The method described in this chapter provides an approach for understanding the principal cognitive aspects for the coupling of human of technical systems; it does not provide a detailed model for cognition. However, the method implies that a cognitive model requires certain properties.

Implications for Cognitive Modelling

Although human cognitive performance is a highly parallel process, a cognitive act appears to be one-dimensional (i.e., a stable relationship of the internal and external world). This desire for consistency limits cognitive human performance. This conclusion does not call for a comeback of the central processor hypothesis of cognition but for complementing models of parallel and sequential processing. It also calls for making no distinction between short- and long-term memory but to see them as different views of the same system.

The investigations presented in this chapter demonstrate that cognitive coupling is a useful tool for investigating the cognitive aspects of human–machine interaction and group communication. Humans' internal cognitive worlds (e.g., operators' goals and attitudes) have a decisive impact on our behaviour. Goals influence our actions and perceptions; habits, heuristics, and skills influence what we perceive and what we look for in a given situation. With this in mind, we need a better understanding of the cognitive processes that integrate the internal world of human beings with the external world.

A Common Database for Human Reliability

Often we fail to recognize that similar, or even the same, psychological processes that have resulted in human error can be applied to a totally different domain, perhaps helping us to avoid negative incidents in the future. The approach described in this chapter has been applied to several industries; we have shown here that predictions about the cognitive performance of nuclear power plant operators can be applied successfully to predict the cognitive performance of automobile drivers.

It can be concluded that this approach makes possible the transfer of the cognitive control strategies of human beings, independent from aspects of the technical system (nuclear power plants, automobiles, airplanes, and occupational health sites have been investigated so far with the method presented here). A comparison across industries of similarities and differences in human error mechanisms is a powerful source of knowledge, allowing us to learn from the experiences of others and to avoid the negative impact of human error in various industrial settings.

REFERENCES

Bubb, H. (1993). Systemergonomie. In H. Schmidtke, (Ed.), *Ergonomie*. Munich: Hanser.

Bubb, H. (2001). Menschliche Zuverlässigkeit und Arbeitsschutz—eine Möglichkeit für effizienten Unfallschutz. BAUA 27. *Internationaler Kongress für Arbeitsschutz und Arbeitsmedizin Innovation und Prävention (A + A)*. Düsseldorf: BAUA.

Card, S. K., Moran, T. P., & Newell, A. (1983). *The psychology of human-computer interaction*. Hillsdale, NJ: Erlbaum.

Dörner, D. (1997). *Die Logik des Mißlingens, Strategisches Denken in komplexen Situationen*. Hamburg: Rowohlt.

Eberleh, E., Neugebauer, C., & Sträter, O. (1989). *Modelle des Wissens und Methoden der Wissensakquisition: Eine Übersicht*. Arbeitsbericht Nr. 1–55. RWTH Aachen: Institut für Psychologie.

Festinger, L. (1957). *A theory of cognitive dissonance*. Stanford, CA: Stanford University Press.

Jones, D., & Farmer, E. (2001). *Applying the cognitive streaming theory to air traffic management: A preliminary study*. Brussels: Eurocontrol.

Hollnagel, E. (1993). *Human reliability analysis: Context and control*. London: Academic Press.

Hollnagel, E. (1998). *Cognitive reliability and error analysis method—CREAM*. New York, Amsterdam: Elsevier.

IAEA. (2001). *Guidelines for describing of human factors in the IRS* (Human actions and related causal factors and root causes IAEA-J4-CS-10).

Kahneman, D., & Tversky, A. (1979). Prospect theory: An analysis of decision under risk. *Ecometrica, 47,* 263–291.

Keller, A. (1990). *Allgemeine erkenntnistheorie*. Stuttgart: Kohlhammer.

Kuhn, W., Fricke, B., Schäfer, K., & Schwarz, G. (1976). *Quantenphysik*. Braunschweig: Westermann.

Linsenmaier, B., & Sträter, O. (2000). *Recording and evaluation of human factor events with a view to system awareness and ergonomic weak points within the system, at the example of commercial aeronautics*. St. Louis, MO: Mira.

McCormick, E., & Sanders, M. (1983). *Human factors in engineering and design*. Auckland, NZ: McGraw-Hill.

Miller, G. A. (1956). The magical number seven plus or minus two: Some limits on our capacity for processing information. *Psychological Review, 63.*

Mosneron-Dupin, F., Reer, B., Heslinga, G., Sträter, O., Gerdes, V., Saliou, G., & Ullwer, W. (1997). Human-centered modelling in human reliability analysis: Some trends based on case studies. *Reliability Engineering and System Safety, 58,* 249–274.

Müller-Ecker, D., Mayer, G., & Sträter, O. (1998). Probabilistic safety assessment for non-full-power states of NPPs in Germany. In: *International Conference on Probabilistic Safety Assessment and Management,* London: Springer-Verlag.

Neisser, U. (1976). *Cognition and reality*. San Francisco: W. H. Freeman.

Neumann, O. (1992). Zum gegenwärtigen theoretischen Umbruch in der Kognitionspsychologie. *Merkur—Zeitschrift für europäisches Denken, 1,* 48.

Norman, D. A. (1986). New views on information processing: Implications for intelligent decision support systems. In: E. Hollnagel, Mancini, & D. D. Woods, (Eds.), *Intelligent decision support in process environments*. Berlin: Springer.

OECD. Organisation for Economic Cooperation and Development (2000). *Errors of commission in probabilistic safety assessment* (NEA/CSNI/R, 2000)17. Paris: OECD/NEA.

Piaget, J. (1947). *Psychologie der Intelligenz*. Zürich: Rascher.

Rasmussen, J. (1986). *Information processing and human-machine interaction*. New-York: North-Holland.

Reason, J. (1990). *Human error*. Cambridge, England: Cambridge University Press.

Reer, B., Sträter, O., Dang, V., & Hirschberg, S. (1999). *A comparative evaluation of emerging methods for errors of commission based on applications to the Davis-Besse (1985) Event* (Nr. 99-11). Schweiz: PSI.

Schandry, R. (1981). *Psychophysiologie. Körperliche Indikatoren menschlichen Verhaltens*. Munich: Urban & Schwarzenberg.

Schuler, H. (Ed.). (1995). *Lehrbuch Organisationspsychologie*. Bern: Hans Huber.

Schweigert, M., Fukuda, R., & Bubb, H. (2001). Blickerfassung mit JANUS II. In *Arbeitsgestaltung, Flexibilisierung, Kompetenzentwicklung*. Dortmund: GfA Press.

Semmer, N. (1994). Der menschliche Faktor in der Arbeitssicherheit: Mechanismen, Verhütung und Korrektur von menschlichen Fehlhandlungen. SVA (Hrsg.) *Ursachenanalyse von Störfällen in Kernkraftwerken*. Ausbildungsseminar. 3.3.1994. Switzerland: Brugg-Windisch.

Shastri, L. (1988). A connectionist approach to knowledge representation and limited inference. *Cognitive Science, 12,* 331–392.

Sommer, W., Leuthold, H., & Matt, J. (1998). The expectancies that govern the P300 amplitude are mostly automatic and unconscious. *Behavioral and Brain Sciences, 21,* 149–150.

Sträter, O. (1994). An expert knowledge oriented approach for the evaluation of the man-machine interface. In: T. Ruokonen (Ed.), *Fault detection, supervision and safety for technical processes.* SAFEPROCESS '94. Finnish Society of Automation (Vol. 2, p. 673 f). Helsinki, Finland.

Sträter, O. (1997). *Beurteilung der menschlichen Zuverlässigkeit auf der Basis von Betriebserfahrung.* GRS-138. Cologne: GRS-Gesellschaft für Anlagen—und Reaktorsicherheit.

Sträter, O. (1998). Problems of cognitive error quantification and approaches for solution. *PSAM5: International Conference on Probabilistic Safety Assessment and Management.* New York: Springer-Verlag.

Sträter, O. (2000). *Evaluation of human reliability on the basis of operational experience.* GRS-170. Cologne: GRS-Gesellschaft für Anlagen—und Reaktorsicherheit.

Sträter, O. (2001). Modelling and assessment of cognitive aspects of the task of a driver. In *The driver in the 21st Century.* Düsseldorf: VDI.

Sträter, O. (2002). *Group interaction in high risk environments—communication in NPP.* GRS Report No. A-3020. Cologne: GRS-Gesellschaft für Anlagen—und Reaktorsicherheit.

Sträter, O., & Bubb, H. (1998). Assessment of human reliability based on evaluation of plant experience: Requirements and their implementation. *Reliability Engineering and System Safety, 63,* 199–219.

Sträter, O., & Zander, R. M. (1998). Approaches to more realistic risk assessment of shutdown states. In *Reliability Data Collection for Living PSA* (p. 236 ff). NEA/CSNI/R (98) 10. Organisation for Economic Cooperation and Development–Nuclear Energy Agency. Paris: OECD-NEA.

Strube, G. (1990). Neokonnektionismus: Eine neue Basis für die Theorie und Modellierung menschlicher Kognition? *Psychologische Rundschau, 41,* 129.

Swain, A. D., & Guttmann, H. E. (1983). *Handbook of human reliability analysis with emphasis on nuclear power plant applications.* Washington, DC: Sandia National Laboratories, NUREG/CR-1278.

Theis, I. (2002). *Das Steer-by-Wire System im Kraftfahrzeug—Analyse der menschlichen Zuverlässigkeit.* Munich: TU-München/LFE.

Theis, I., & Sträter, O. (2001). By-wire systems in automotive industry—reliability analysis of the driver-vehicle-interface. In E. Zio, M. Demichela, & N. Piccini, (Eds.), *Esrel 2001—Towards a Safer World.* Turin: Politecnico Di Torino.

Wickens, C. D. (1992). *Engineering psychology and human performance.* New York: HarperCollins.

16

Applied Cognitive Work Analysis: A Pragmatic Methodology for Designing Revolutionary Cognitive Affordances

William C. Elm, Scott S. Potter,
James W. Gualtieri, and James R. Easter
MaxTech Aegis Research Corporation, USA

Emilie M. Roth
Roth Cognitive Engineering, USA

Abstract

The Applied Cognitive Work Analysis (ACWA) methodology has been specifically tailored to be the cognitive task design portion of a high-quality, affordable systems engineering process. In this chapter, the methodology is presented in its entirety from its knowledge elicitation beginnings to hand-off to the software development team. Each step in the methodology is presented in "how-to" format, including design review and quality checkpoints that bring cognitive task design to the same level of maturity as the best software engineering implementation practices. A small-scale problem that has proven to adequately challenge novice practitioners is used to illustrate the key concepts while preserving the limited space of this chapter for the methodology discussion. We present the ACWA methodology as the current "best of breed" in the actual engineering use of cognitive task design. It adapts the various academic approaches and concepts into a repeatable engineering process in use by the Cognitive Systems Engineering Center as the heart of our system development process for implementing everything from small decision support tools to full-scope command and control advanced decision support systems. The extensive list of references provides the sources of the underlying research and academic approaches that formed the basis for this practical, proven methodology for cognitive task design.

PREFACE

This chapter describes the current state of a work in progress: the Applied Cognitive Work Analysis (ACWA) methodology. It has evolved through use over several years and numerous applications, proving its value in delivering revolutionary decision aids, and evolving as it did so. The current state could be described as effective, and yet calling for additional refinements and enhancements as the edge of the envelope progresses and reveals yet more to be discovered. ACWA was specifically developed to satisfy the two critical challenges to making cognitive task design (CTD) part of the mainstream systems engineering efforts: (a) CTD must become a practical, repeatable, high-quality engineering methodology that is on par with the Software Engineering Institute's efforts for software engineering processes, and (b) the results of the CTD must seamlessly integrate with the software engineering and overall system development processes used to construct decision support systems.

To satisfy those challenges, ACWA was constructed as a series of relatively small, manageable engineering transformations, each requiring the skilled application of the methodology's principles rather than requiring a design epiphany at any point in the process. This chapter presents the methodology's steps as a practical discussion with small examples. These examples were selected to reflect the points being made and for that reason do not constitute any single complete use of the methodology from start to finish. This chapter should be read to understand the techniques, characteristics, and basis for the methodology. The small-scale problem has been described as a "toy" domain, yet it fully reveals the difficulties in discovering abstract concepts from a physical domain and in fact has been used for extensive hands-on sessions that novice practitioners found extremely challenging. This toy domain is robust enough to extend to discussions of adding artificial intelligence agents, alarms systems, and so on into the CTD. For examples that better reflect a contiguous sequence of design products for real-world applications see the case studies in chapter 27.

INTRODUCTION

There is a tremendous need for powerful decision support systems to support humans in the increasingly complex work domains they face. The methodology presented in this chapter will have obvious connections to the techniques and approaches that form its roots—those of Rasmussen, Lind, Woods, Hollnagel, Vicente, and others. It is also significantly different from each of them: It has been tailored and adapted to be a pragmatic, effective process for the design and development of revolutionary decision support systems. The methodology profited greatly from the multidisciplinary staff that influenced it, both cognitive psychologists and software and system engineers. The methodology continues to be graded against a difficult standard: It has invented revolutionary new "cognitive affordances" that can be efficiently constructed and deliver dramatic leaps in the net decision-making performance of the human–machine team. The term *affordance* is normally used to describe devices that intuitively "fit" the physiological characteristics of the human user (e.g., Norman's examples of shower control knobs that increase with a clockwise rotation of the right hand). Here it is used analogously to mean decision support system features that intuitively fit the perceptual and cognitive processes of the human user.

The term *decision support system* has a wide variety of connotations, from formulae embedded within a spreadsheet to sophisticated autonomous reasoning "agents." For the purposes of this chapter, the term decision support system is used to mean a computerized system that is intended to interact with and complement a human decision maker. In this role, the ideal decision support system is one that accomplishes the following:

- It provides the information needed by the human decision maker, as opposed to raw data that must be transformed by the human into the information needed. If the data can be perfectly transformed into information by the decision support system, no human cognitive effort is required for the transformation and no cognitive work is expended on the data, allowing total focus on the domain's problem solving.
- It can be controlled effortlessly by the human. It presents the information to the human as effortlessly as a window allows a view of a physical world outside. In this sense the decision support system is "transparent" to the user. If the interaction with the decision support system

is completely effortless, no human cognitive effort is required to manage and interact with the decision support system, allowing total focus on the domain's problem solving.

- It complements the cognitive power of the human mind. In this way, the decision support system not only avoids creating a world that is ripe for human decision-making errors but can include features that complement the trends and power of human cognitive processes. Decision-making errors categorized by such terms as *fixation, garden path,* and so on are avoided by the form of the decision-making world embodied by the decision support system.

- It supports a wide variety of problem-solving strategies, from nearly instinctive reactions to events to knowledge-based reasoning on fundamental principles in a situationally independent manner.

The methodology presented in this chapter to "bridge the gap" between cognitive analysis and the design and development of such revolutionary decision support systems is based on some fundamental premises chaining back from the goal of building effective decision support systems—ones that "make the problem transparent to the user" (Simon, 1981). These fundamental premises formed the underlying basis for the techniques described in this chapter. By understanding these premises, the reader will be able to understand the motivation for each of the many tailoring decisions that were made along the way.

To achieve a decision support system that is intuitive, and in fact "transparent" to the user, we began with *Premise 1: Humans form a mental model of the domain as part of their understanding and problem solving.* These mental models form their internal understanding of the domain, on which they employ a variety of problem-solving strategies to reason and make decisions. These strategies vary both by situation and by variations between human decision makers. They are a mix of sensory-motor responses (skill-based), actions based on stored rules and experience (rule based), and knowledge based on an internal representation of underlying, fundamental behavioral characteristics of the work domain (knowledge based). Rouse and Morris (1986) defined mental models as "the mechanism whereby humans are able to generate descriptions of system purpose and form, explanations of system functioning, observe system states, and prediction of future system states." As a corollary to this premise, experts employ "better" mental models—models with richer domain knowledge, more structure, and interconnections and a better basis in the underlying principles of the work domain (Glaser & Chi, 1989; Gualtieri, Folkes, & Ricci, 1997; Jonassen, Beissner, & Yacci, 1993; Woods & Roth, 1988).

Assuming the user forms such a mental model, the desire to communicate effortlessly with the user—in effect to effortlessly communicate knowledge into the user's mental model—leads to *Premise 2: The decision support system must itself embody a "knowledge model" of the domain that closely parallels mental models representative of expert human decision-making.* By paralleling the two mental models, the communication from the decision support system will be in a form effortlessly resolvable into the user's internal mental model, eliminating the cognitive burden of transforming from one to the other. This is consistent with the notion of common frame of reference (Woods & Roth, 1988). The more closely the models parallel, the less cognitive effort it will take on the part of the human to "transform" the data of the decision support system into the information needed to mentally solve the problem (Rouse, Cannon-Bowers, & Salas, 1992). As the degree of match approaches one, the concept of "transparent" becomes a reality. In addition to providing an effortless match for the expert operator, the knowledge model within the decision support system can be viewed as an analytically derived, a priori starting point guiding the formulation of a novice user's own embryonic mental model of the domain (Glaser, Lesgold, & Gott, 1991; Klimoski & Mohammed, 1994).

To design such a knowledge model for the decision support system leads to *Premise 3: An effective decision support system knowledge model is composed of functional nodes and relationships intrinsic to the work domain.* It depicts the functional abstractions and relationships of the domain (rather than part–whole representations, temporal work task representations, etc.) as the best representation to achieve the needed richness and support to the spectrum of problem-solving strategies. For each of these functional concepts, additional domain richness is essential: explicit denotation of the cognitive tasks and the information needed to support the user in each cognitive task. This decision support system functional knowledge model therefore defines the basis for both the structure and content of the decision support system itself; it becomes the overall design specification (Woods & Roth, 1988).

To be able to analytically develop such a functional knowledge model requires *Premise 4: An adaptation of Rasmussen's abstraction hierarchy provides the needed representation of the abstract functional concepts and relationships to form the basis for the decision support system functional knowledge model.* The functional abstraction network described in this chapter is a pragmatic adaptation of that approach but nonetheless is based on the same premise: An abstract model of the functional nature of the domain provides the necessary exploration of highly abstract concepts of the domain while being sufficiently robust (independent of the physical particulars of the situation) to avoid brittleness in unexpected problem-solving situations (Vicente, 1999). With this premise, the problem has come full circle—the domain defines the fundamental skeleton of the decision support system and ultimately the form of the decision support system knowledge model communicated to the human practitioner controlling the domain.

These premises were presented earlier as a backward-chaining goal decomposition from the starting requirement of "build a decision support system to deliver improved human–machine decision-making effectiveness." They provide a sense for the rationale underlying the various tailoring decisions that created the methodology. The remainder of the chapter is presented in the order the methodology is actually practiced for development of a particular decision support system: This chapter is dedicated to answering the question: When presented with a complex, real-world domain, what is the process to design and deliver a truly effective decision support system that is much more than an incremental improvement to existing systems?

There has been growing interest in using cognitive task analysis as the principal component of CTD to understand the requirements of cognitive work and to provide a foundation for design of new decision-support systems (Schraagen, Chipman, & Shalin, 2000; Vicente, 1999). Although cognitive task analysis techniques have proved successful in illuminating the sources of cognitive complexity in a domain of practice and the basis of human expertise, the results are often only weakly coupled to the design of support systems (Potter, Roth, Woods, & Elm, 2000). A critical gap occurs at the transition from cognitive task analysis to system design, where insights gained from the cognitive task analysis must be translated into design requirements.

In this chapter, we describe an approach to bridging the gap between analysis and design. The ACWA methodology emphasizes a stepwise process to reduce the gap to a sequence of small, logical engineering steps, each readily achievable. At each intermediate point the resulting decision-centered artifacts create the spans of a design bridge that link the demands of the domain as revealed by the cognitive analysis to the elements of the decision aid.

DEVELOPING A DESIGN THREAD FROM COGNITIVE ANALYSIS TO DECISION AIDING

The ACWA approach is a structured, principled methodology to systematically transform the problem from an analysis of the demands of a domain to identifying visualizations and decision-aiding concepts that will provide effective support. The steps in this process include the following:

- using a *functional abstraction network* (FAN) model to capture the essential domain concepts and relationships that define the problem-space confronting the domain practitioners;

- overlaying *cognitive work requirements* (CWR) on the functional model as a way of identifying the cognitive demands, tasks, and decisions that arise in the domain and require support;
- identifying the *information and relationship requirements* (IRR) for successful execution of these cognitive work requirements;
- specifying the *representation design requirements* (RDR) to define the shaping and processing for how the information and relationships should be represented to practitioner(s); and
- developing *presentation design concepts* (PDC) to explore techniques to implement these representation requirements into the syntax and dynamics of presentation forms to produce the information transfer to the practitioner(s).

In the ACWA analysis and design approach, we create design artifacts that capture the results of each of these intermediate stages in the design process. These design artifacts form a continuous design thread that provides a principled, traceable link from cognitive analysis to design. The design progress occurs in the thought and work in accomplishing each step of the process, however, by the process of generating these artifacts. The artifacts serve as a post hoc mechanism to record the results of the design thinking and as stepping-stones for the subsequent step of the process. Each intermediate artifact also provides an opportunity to evaluate the completeness and quality of the analysis and design effort, enabling modifications to be made early in the process. The linkage between artifacts also ensures an integrative process; changes in one artifact cascade along the design thread, necessitating changes to all. Figure 16.1 provides a visual depiction of the sequence of methodological steps and their associated output artifacts, as well as an indication that the process is typically repeated in several expanding spirals, each resulting in an improved decision support system. The remainder of this chapter describes each step of this approach.

Representing the Way the World Works: Building a Functional Abstraction Network

ACWA begins with a function-based goal–means decomposition of the domain. This technique has its roots in the formal, analytic goal–means decomposition method pioneered by Rasmussen and his colleagues as a formalism for representing cognitive work domains as an abstraction hierarchy (e.g., Lind, 1993; Rasmussen, 1986; Rasmussen, Pejtersen, & Goodstein, 1994; Roth & Mumaw, 1995; Vicente, 1999; Vicente & Rasmussen, 1992; Woods & Hollnagel, 1987). A work domain analysis is conducted to understand and document the goals to be achieved in the domain and the functional means available for achieving them (Vicente, 1999). The objective of this functional analysis is a structured representation of the functional concepts and their relationships to serve as the context for the information system to be designed. This abstraction network is intended to approximate an "expert's mental model" of the domain. This includes knowledge of the system's characteristics and the purposes or functions of the specific entities. The result of the first phase is a FAN—a multilevel recursive means–ends representation of the structure of the work domain. This structure defines the shape of the decision support system's knowledge model and thereby the shape, nature, and ultimately the content and form of the communication between the system and the domain practitioner.

The challenging aspect of this task is the systematic discovery of the most abstract of the concepts (Rasmussen, 1986), as the ones most indicative of expert understanding of the domain, most challenged during unexpected situations, and least likely to be captured in existing documentation or decision aids. The physical aspects of the domain can be viewed, touched, and inspected; the abstract functional concepts require understanding. An effective

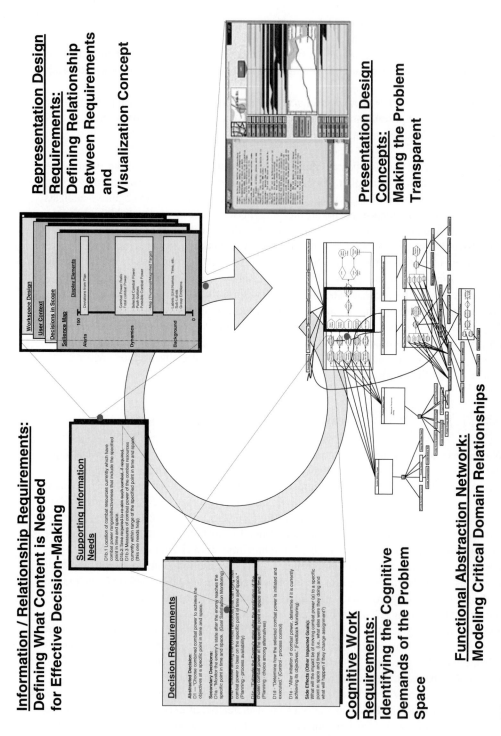

FIG. 16.1. A sequence of analysis and design steps creates a continuous design thread that starts with a representation of domain concepts and relationships through development of decision support requirements to creation of visualization and aiding concepts and rapid prototypes with which to explore the design concepts.

FAN properly discovers these essential, elusive abstract concepts and how they relate to each other.

The work domain analysis is performed based on extensive interactions with expert practitioners in the domain and includes face-to-face interviews with the experts, watching the experts work in the domain, verbal protocol techniques, and other cognitive task analysis and cognitive work analysis methods of CTD (cf. Potter et al., 2000; Vicente, 1999). In fact, many of these practices are relatively common "knowledge elicitation" practices. The distinction is in the nature of the knowledge being sought and in the nature of the analysis of the raw data collected resulting in design progress. Virtually any of the knowledge elicitation practices can form the input data; the true value comes in resolving it against the functional point of view represented in the FAN and the resulting conversion of that data into the formalisms of the FAN. In practice, building a FAN is an iterative, progressively deepening process. It starts from an initial base of knowledge (often very limited) regarding the domain and how practitioners function within it. Then, complementary techniques are used to expand and enrich the base understanding and evolve a function-based model from which ideas for improved support can be generated. This process is highly opportunistic. The FAN can be initialized from training materials, observations, interviews—whatever is the richest initial source of domain information. Once initialized it becomes a working hypothesis to be used in resolving each additional piece of domain knowledge acquired.

The phrase *bootstrapping process* has been used to describe this process and emphasize the fact that the process builds on itself (Potter et al., 2000). Each step taken expands the base of knowledge, providing opportunity to take the next step. Making progress on one line of inquiry (understanding one aspect of the field of practice) creates the room to make progress on another. For example, one might start by reading available documents that provide background on the field of practice (e.g., training manuals, procedures); the knowledge gained will raise new questions or hypotheses to pursue that can then be addressed in interviews with domain experts. It will also provide the background for interpreting what the experts say. In turn, the results of interviews or observations may point to complicating factors in the domain that need to be modeled in more detail in the FAN. A FAN can also provide the decision-centered context necessary to create scenarios to be used to observe practitioner performance under simulated conditions or to look for confirming example cases or interpret observations in naturalistic field studies.

The resulting FAN specifies the domain objectives and the functions that must be available and satisfied to achieve their goals. In turn, these functions may be abstract entities that need to have other, less abstract functions available and satisfied that they might be achieved. This creates a decomposition network of objectives or purposes that are linked together from abstract goals to specific means to achieve these goals. For example, in the case of engineered systems, such as a process plant, functional representations are developed that characterize the purposes for which the engineered system has been designed and the means structurally available for achieving those objectives. In the case of military command and control systems, the functional representations characterize the functional capabilities of individual weapon systems, maneuvers, or forces and the higher level goals related to military objectives.

The origins of ACWA's FAN uniquely include the work of Woods and Hollnagel (1987). Figure 16.2 depicts an example that illustrates these essential characteristics of a functional abstraction hierarchy as originally described by Woods and Hollnagel:

- Processes may affect more than one goal; these side effects govern the operation of a process to achieve the goal of interest.
- Each process can be modeled qualitatively to represent how it works to achieve a goal.

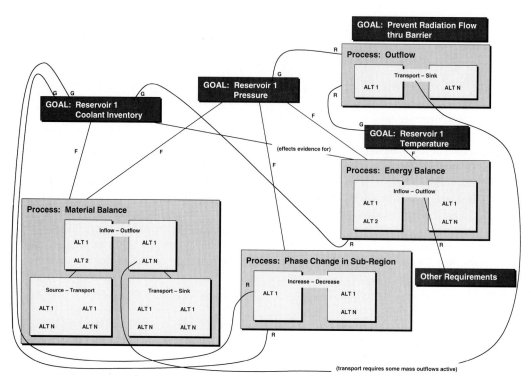

FIG. 16.2. A sample goal–means decomposition from Woods and Hollnagel (1987) for one part (primary-system thermodynamics) of a nuclear power plant.

- Relationships within the model can be recursive; processes can have requirements that are supported by a more "abstract" process.
- Moving up through the network defines supported processes and impact on goal achievement; moving down defines supporting processes and requirements for goal achievement.

The ACWA's FAN has tailored the details of this functional model to improve its repeatability and consistency, its applicability across a variety of domains, and to improve its efficiency in designing large-scale decision support systems. Figure 16.3 shows the ACWA refinements made from the original research reflected in Fig. 16.2:

- A single node is composed of a goal–process pair as opposed to considering goals and processes as separate entities. Therefore, the relationship between a goal and its process model are exactly 1:1
- Supporting functional nodes are associated with a precise component of the functional process, that is, they support the process in a precise, localized way (Rasmussen, 1986) rather than the whole of the process
- The term *hierarchy* is actually a misnomer; the structure of the model is actually a network.

Both Rasmussen (1986) and Woods and Hollnagel (1986) provided a good description of the "rules" of modeling a domain into the FAN, including the rules of "flow modeling" (see also Lind, 1991). The details of this activity will not be repeated here but are summarized as follows:

- A supporting function enables the process being supported but is not in itself part of that process. The typical example is the way a road provides a critical path for a transportation

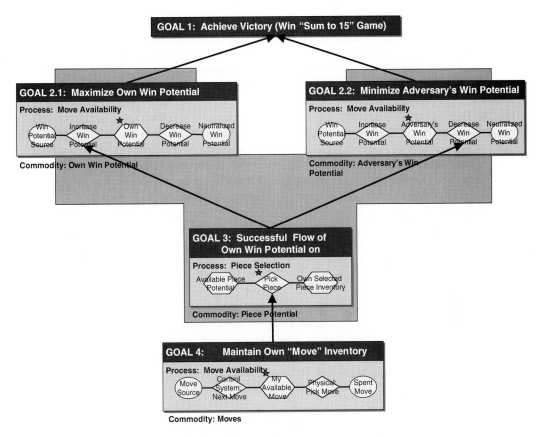

FIG. 16.3.   Refinements to the goal–means decomposition in Fig. 16.2. The shaded area indicates the coverage of the functional space allocated to this particular visualization.

process to haul its cargo from Point A to Point B. The overall goal is for the cargo to move, in fact, the truck and the road both support that functional process of moving cargo.

- The functional process model uses primitives of Source, Storage, Transport and Conversion, and Sink to represent the functional operation of both abstract and relatively physical processes. (See the process models within the nodes of Fig. 16.3.)

An Example Using ACWA for the Game of 15

The Game of 15 was selected for this chapter to provide specific instances for each step of the methodology description within the space limitations. More in-depth cases are presented in the companion case study chapter (Potter et al., chapter 27, this volume) with limited discussions of how the cases relate to the methodology. Figure 16.3 reflects the FAN produced using ACWA for the Game of 15. This simple domain will be used throughout the remainder of the chapter to provide similar examples for each step of the ACWA process. As a domain, the rules of the game are simple:

- Alternating choices, players pick one of nine game pieces, numbered 1 through 9.
- Once chosen by either player, the piece cannot be chosen by the opponent.
- The objective is to collect pieces, any three of which sum to exactly 15 (that is you can have more than three pieces in your "collection" but only count three of them into the sum to equal 15). The first player to 15 wins.

While using this domain for training in the use of ACWA, several trends occurred among the cognitive task design novices that reflect the characteristics of using the FAN modeling in ACWA.

These novice designers tended to fixate on the more obvious physical items (the process of selecting pieces) or concepts as explicitly mentioned in the rules documentation (i.e., "Select a winning triple"). Some focused on the adversarial nature, trying to define goals such as "win" and "defeat opponent." Notice that the final FAN shown in Fig. 16.3 includes such issues but located in their proper relative positions.

Novice designers also tended to question the use of the "functional process" portion of the nodes but when encouraged to use them saw the expected benefits: Explicitly modeling the functional process helped add order to identifying supporting concepts. The process depictions tended to make "mixed commodity" errors obvious, where the practitioners would try to pack large portions of the domain into a single model (often in the enthusiasm for the modeling effort.) Modeling the functional processes made it easy for the trainer to prompt further modeling with the simple support–supported relationship question: "What other thing must be achieved for any particular portion of this process to operate?"

Novice practitioners often tend to want to "see" decisions, information, and presentation concepts that became evident during the Knowledge Elicitation session and attempt to put those issues directly into the FAN. It is important not to do this, but the pressure to capture ideas opportunistically results in a typical ACWA process that tends to work all the steps discussed in this chapter in a more parallel fashion, making annotations on all of the artifacts to capture the brainstorming as it occurs yet representing them in the right artifact. This can be beneficial for opportunistically completing the ACWA effort as long as all of the brainstormed items are eventually resolved into a continuous thread through the entire process (all decisions jotted down are eventually associated with a node of the FAN, etc.) This "resolution" step is often useful in finding discrepancies or incomplete areas of the process. In some cases, particularly for the abstract concepts at the "top" of the FAN, it is often impossible to create (or agree on) any "commodity" or functional process. When this occurs, we reluctantly concede the point (only after making a concerted effort) at the risk of losing the benefits provided by the process portion of the modeling (as described earlier).

Table 16.1 provides some simple indicators that can be used throughout the development of the FAN both to guide further modeling and as a post hoc design review of the results of this step. These indicators often elicit discussions among the ACWA practitioners during the modeling process. As a practical question in any engineering process, management will always ask, "When will the FAN be complete and ready to be baselined (or frozen, or signed off, or whatever) so we can mark this task complete?" As in any engineering process the answer is always less obvious than the question. The following are some factors to keep in mind:

- It is strongly recommended that all the steps of the ACWA process be "open" in parallel, so as design thoughts for CWRs, IRRs, or PDCs occur they can be recorded and offer additional traction for the construction of the FAN.
- The FAN doesn't have to be 'perfect' to be extremely useful in the practical design of a decision support system. In many ways, this is learned through experience, but the "shape" of the FAN and initial thoughts on how much of the FAN will be covered on which display element and which display (another argument for working the steps in parallel) often indicate when a FAN is "good enough." Such indicators are a sense that all the issues are represented, even if they could be modeled in other ways. In particular, does there seem to be places for all the CWRs of the domain (often not fully known until the CWR stage of the ACWA process is "complete").

TABLE 16.1
Worksheet 1: Indicators of a Successfully Constructed FAN.

Bottom Line: Does the FAN contain all the concepts and relationships to be an effective "knowledge model" of the domain?

1. Are the goals' titles "success oriented"?
   - Good: Successfully control thermal energy of the beverage
   - Bad: Gain situational awareness
2. Do the goals relate to a functional "commodity" that can be modeled?
3. Does the functional process involve only one such commodity?
4. Is the process a functional description, not a physical map of the domain?
5. Are supporting functions actually supporting (Rasmussen, 1986)? They should not be more detail and decomposition of the supported process. Look for a different functional "commodity" in the supporting process.
6. The FAN should not contain decisions (see cognitive work requirements). You should not see goals such as "choose combat power."
7. Are circularities evident (are lower level functions supported by higher level ones in the structure)? This is typical of most real-world domains; expect it.
8. Does the FAN recursively apply Rasmussen's five categories of abstraction? (lower level, less abstract nodes supported by highly abstract concepts.) This is also typical of most real-world domains.
9. Often one physical component will affect many functions within a domain; expect to see a single physical entity in several locations of the FAN.
10. The FAN should not contain explicit physical anomalies but rather their functional impacts on the domain.
11. Remember the FAN is only the first step, decisions and information should not be in the FAN; they'll be covered in later artifacts. (If they come to mind during FAN development, start developing those artifacts in parallel with the FAN.)

- The management process must allow the baselined FAN to be revised later in the process, because work to identify the cognitive tasks, information requirements, and visualization elements will inevitably provide new insights that will necessitate modifications to the FAN.

Discussion.   The FAN shown in Fig. 16.3 reveals some interesting aspects of the domain:

- The issue of "alternating turns" is modeled as a supporting function to the "movement of a piece into your control." In essence, you must have an inventory of "moves" in your possession (storage) that are moved to "spent" (sink) to support the selection of a piece. Your "turn" is a commodity in the supporting process that is consumed in support of the process of acquiring a "piece."
- The offensive and defensive competition for the use of "select your next piece" is explicitly evident in the two support links (one from "win" and one from "prevent defeat"), setting support goals into the "flow of win potential on my piece" node. (In playing the game, it is often observed that players "target fixate" on one goal or the other, sometimes missing the opportunity to pick the winning piece or failing to pick the piece that their opponent uses to win on their next turn. Players have even failed to recognize that they have a winning collection of pieces. Having made those observations during knowledge elicitation sessions, seeing such a structure in the FAN model is reassuring when it occurs.)

Summary of FAN Step.   There are a growing number of examples of successful systems that have been developed based on a work domain analysis. Examples of functional abstraction networks and how they were used to design new visualizations and decision support systems can be found in Roth, Lin, Kerch, Kenney, and Sugibayashi (2001) and Potter et al. (2000). Examples of the application of this approach to modeling individual cognition and team collaboration and to develop new online support systems in time-pressured tasks such as situation assessment, anomaly response, supervisory control, and dynamic replanning include domains such as military intelligence analysis (Potter, McKee, & Elm, 1997), military aeromedical evacuation planning (Cook, Woods, Walters, & Christoffersen, 1996; Potter, Ball, & Elm, 1996), military command and control (Chalmers, Easter, & Potter, 2000), railroad dispatching (Roth, Malsch, Multer, Coplen, & Katz-Rhoads, 1998), and nuclear power plant emergencies (Roth et al., 1998).

Modeling Decision Making: Deriving Cognitive Work Requirements

With the FAN representation of the work domain's concepts and how they are interrelated as the underlying framework, it is possible to derive the cognitive demands for each part of that domain model. The ACWA methodology refers to these as cognitive work requirements, meaning all types of recognition, decision-making, and problem-solving reasoning activities required of the domain practitioner. Thus, the term *decision* is sometimes used as a synonym for CWR, but in a broad sense, meaning much more than a choice between multiple options.

Similar to the discussion of the process to develop the FAN above, the discovery of CWRs is also an opportunistic, bootstrapping process of mining various knowledge elicitation data captures and converting them to CWRs in the form used by ACWA. As the second step in the process, this discovery has two major advantages over the construction of the functional abstraction network: (a) the FAN itself can be analyzed to reveal CWRs of the domain and (b) many common CWRs can be found across a variety of domains, so a "template" of previously developed generic CWRs can be tested for each node of the FAN and instantiated if appropriate to this domain. (See Table 16.3 for some generic CWRs.) In executing the ACWA methodology, great efficiency is gained by performing this enumerate of the "obvious" CWRs allowing the resource intensive knowledge elicitation processes to focus on the domain specific cognitive tasks as well as validation or tailoring of the analytically derived CWRs. In performing this analysis, each node in the FAN, each goal and component of the functional process models are considered against each of the "generic" CWRs. Note that during a knowledge elicitation process, a verbal protocol analysis of a domain practitioner typically produces evidence most closely resembling a CWR of the ACWA process. By attempting to fit these CWR-like observations into the structure provided by the FAN, actual CWRs can be determined. This nonliteral use of the knowledge elicitation output is a critically valuable point in the ACWA process. The FAN provides a foundation for "understanding" the knowledge elicitation output in a way not possible by literally recording it verbatim.

In discussion-based or observational knowledge elicitation techniques, information captured about "what the practitioner was thinking" should be located as the CWR associated with the appropriate node of the FAN and could reflect a cognitive activity that should be recorded as a CWR for that node. In the ACWA methodology, the addition of the CWRs to the design repository is described as "thickening" the analysis. The nodes are those of the FAN; each CWR is then attached to a node as an enrichment of the understanding of that domain concept.

Based on the underlying premises of the ACWA methodology, these CWR center around either goals or functional process directed behavior, such as monitoring for goal satisfaction and resource availability, planning and selection among alternative means to achieve goals, and controlling functional process (initiating, tuning, and terminating) to achieve goals (Roth & Mumaw, 1995). With minor extensions to the techniques discussed here, the ACWA process can be used to identify (and then support in the resulting decision support system) specific CWRs associated with the collaboration activities in team settings (Gualtieri, Roth, & Eggleston,

TABLE 16.2
Worksheet 2: Indicators of Success for Cognitive Work Requirements (CWRs)

Bottom Line: Have you successfully identified the actual decisions?

1. Does the CWR contain a "decision verb" such as choose, adjust, monitor?
2. The CWR should *not* contain phrases such as "develop a plan" or "gain situational awareness."
3. Are the typical CWRs (see Table 16.3) evident at every node in the FAN?:
   - Are all goals being "monitored"?
   - Where the functional processes show parallel processes, are "choices" indicated?
   - Are processes being "tuned" to achieve goal constraints and setpoints?
   - Are automated systems being "chosen" to be in either manual or one of the available automatic modes?
   - Are the actions and outputs of the automatic decision-making agents being monitored for correctness?
4. Do the CWRs include comparisons to design limits, administrative limits, procedure trigger points, etc.?
5. The CWR should *not* contain words about the information it takes to make the decision (they go in the next step or artifact of the process).

2000). By organizing the specification of operator cognitive requirements around nodes in the FAN rather than organizing requirements around predefined task sequences (as in traditional approaches to task analysis), the representation helps ensure a consistent, decision-centered perspective. Often in the course of focusing on the CWRs changes needed in the FAN will be discovered based on the new insights and level of understanding achieved from the CWR perspective.

The cognitive demands that are captured at this step of the analysis and are tied directly to nodes in the FAN constitute another intermediate design artifact that captures an essential part of the design: the explicit identification of the cognitive demands placed on the domain practitioner by the domain itself. They constitute an explicit enumeration of the cognitive tasks to be supported by the decision support system. Because many of the abstract functional nodes in the FAN are "untraditional" representations, it is common also to see CWRs that are unlike any elements explicitly visible in documentation or training materials. Table 16.2 contains some indicators of when CWRs have been correctly identified that can be used during the ACWA analysis itself as well as during the design review baselining of this phase of the analysis.

The FAN forms the basis for the structure of the decision-making activities that will be reflected in the CWRs. For example, every goal node in the FAN has associated "goal monitoring" types of decisions. Likewise, processes have associated "process monitoring" decisions. Similarly, there will always be "feedback monitoring" decisions related to assessing whether actions are achieving desired results. Depending on the relationships between nodes in the FAN, there will also be decisions related to prioritization of goals, selection of alternative means to achieve a particular goal, and monitoring side effects of actions. Table 16.3 contains an example template of these generic CWRs that can be tested at the appropriate nodes of the FAN (adapted from Roth & Mumaw, 1995).

The key issue here is that this template is not meant to be a rote, "turn the crank" type of process. Rather, these questions are meant to be a guide to stimulate thinking about relevant decision making in the context of a FAN model of the target work domain. Each domain is unique in the decision-making demands imposed on the human operators. As such, each work domain will require slightly different instantiations of these questions but in general many commonalities in the cognitive requirements can be expected even across widely disparate domains. Successful elucidation of decision requirements will also depend on corroboration from multiple data sources, including case studies, interviews, observations, and so on. In

TABLE 16.3

Worksheet 3: Generic Cognitive Work Requirements to Be Tested for Instantiation at Each Node
of the FAN

Bottom Line: Learn from previous domains, test for "common" decisions
Goal Monitoring
  1. Goal satisfaction: Are the function related goals satisfied under current conditions?
  2. Margin to dissatisfaction: Are explicitly specified goal limits restrictions being approached?
Process monitoring
  3. Active processes: What processes are currently active? What is the relative contribution of each of the active processes to the functional goal achievement? Are the processes performing correctly?
  4. Process element monitoring: Are the automated support systems functioning properly? What goals are the automated support systems attempting to achieve? Are these appropriate goals?
Feedback Monitoring
  5. Procedure Adequacy: Are the current procedures achieving the desired goals?
  6. Control action feedback: Are the operator control actions achieving their desired goals?
Control
  7. Process control: How is the process controlled for process deployment, tuning for optimum performance, termination, etc.?
  8. Manual take over: If intervention is required, what action should be taken?
Abnormality Detection and Attention Focusing
  9. Limit crossing: Has any goal or process component exceeded an established administrative, procedural, or design limit?
  10. Procedure "trigger point": Has any goal or process component reached a value that a procedure uses as a triggering event?
  11. Automatic system "trigger points": Has any goal or process component reached a value that an automatic system uses an initiating value?

---

addition, guiding insights can come from research on similar work domains as well as basic research on human cognition, decision making, biases, and errors. For example, previous work on decision-making in dynamic, high-risk worlds can guide analysis and interpretation of analogous worlds in terms of potential points of complexity, typical decision-making difficulties and strategies, and critical characteristics of difficult problem-solving scenarios.

An Example: Cognitive Requirements for the Game of 15.   Table 16.4 contains an example of some of the CWRs identified for the FAN shown in Figure 16.3 for Goal 2.1: Maximize Own Win Potential. The first two CWRs relate to the goal portion of the node, the third to the process. They can be seen to be instantiations of some of the "generic" CWRs shown in the Table 16.3. Note the unique language that was carefully selected to avoid subliminally luring the designer into "thinking physical." For example, the term *piece* is zealously avoided to preserve the focus on the abstract concept ("quanta of win potential" that functionally rides on physical things called pieces). Interestingly, the importance of this becomes obvious if any of the isomorphic transformations of the game are played where the piece's "name" is not equal to its numerical value in the calculation. These isomorphs exhibit entirely different physical descriptions and languages yet are represented in the identical ACWA representation. For this simple domain, that value may not be evident, but this explicit identification of the functional decisions is essential to the effective design of truly effective decision-supporting features.

Novices find this step particularly difficult because it is dependent on having an internalized understanding of the functional thinking of the ACWA process. Novices' CWRs will typically be enumerations of the mechanical tasks being performed (somewhat like a procedure or time–motion study), which will be virtually impossible to resolve into associated nodes of

TABLE 16.4
Cognitive Work Requirements (CWRs) for a Portion of the Game of 15

CWR 2.1-1: Monitor own "win" potential (goal)."
CWR 2.1-2: Monitor "win" potential "quanta" source available for selection into own "win" potential (process).
CWR 2.1-3: Select the "win" potential "quanta" source that will maximize (be best for) own "win" potential.

---

the FAN. Novice designers also find it difficult to analyze the structure of the FAN to find CWRs associated with the relationships within the FAN. For example, Goal 3 of Figure 16.3 has explicit support links to movement both of friendly and opponent "win potential." This establishes a pair of goal setpoints that must be managed within the domain, that is, that establishes a CWR to manage the resolution of the two goal setpoints by the single functional process node.

Summary of the CWR Step.   From the analysis of the FAN as well as the resolution of knowledge elicitation data into CWRs, the explicit enumeration of the cognitive tasks to be supported by the decision support system is established. In many ways, this makes the CWR task of ACWA the point where the decision support system requirements take shape. As with any other task in ACWA, this design step makes the next analytical step across the gap toward the development of the decision support system a pragmatic, repeatable analytical step. Also as with other ACWA tasks, it may result in changes to previous design artifacts (e.g., the FAN) and opportunistically reveal thoughts for future tasks (information and relationship resources or presentation design concepts) that should be captured as they occur. Successfully crafting the CWRs in functional, decision-centered terms across the many levels of abstraction of the FAN is essential to the final resulting decision support system.

Capturing the Support Needs: Identifying Information and Relationship Requirements

The next step in the process is to identify and document the information required for each decision to be made. Information and relationship requirements (IRRs) are defined as the set of information elements necessary for successful resolution of the associated CWR. This set of information constitutes a third span in the bridge across the gap between cognitive analysis and decision support system design. The focus of this step in the methodology is on identifying the ideal and complete set of information for the associated decision making.

IRRs specify much more than specific data elements; it is data in context that becomes information (Woods, 1988, 1995). The relatively awkward name of this design output reflects both the nuances of this step and its importance. Identifying the information and how it relates to other pieces of information to establish a meaningful context is the value added output of this step in the ACWA process. The data-to-information relationship can be complex and require a significant amount of computations and transformations. For this reason ACWA is a design approach that has a much deeper impact on the entire decision support system architecture than merely the look and feel of the final graphic user interface. For example, in the case of a thermodynamic system, an IRR might be "flow coefficient with respect to appropriate limits." This requires the estimation of the parameter "flow coefficient" derived from model-based computations and sensor values and the comparison of that parameter against a limit referent. The degree of transformation required can vary from simple algebra to complex, intelligent

TABLE 16.5
Worksheet 4: Some Quality Indicators About Information Relationship Requirements

---

Bottom Line: Is it information or just a list of data elements?
1. If you're listing sensors, you're wrong.
2. Does it clearly describe what's needed (not physical data) e.g., "Present deviation from target of <commodity x> overtime." Note how different a presentation that is than "Present <commodity x> over time."
3. Does it have an intent–functional 'feel' to it (i.e., what are you *really* asking for)?
4. Expect phrases and sentences, not database tags.
5. Don't be afraid to let it force data transformations within the overall system design. That's a good thing—even if its as exotic as an embedded state estimator for things that can't be sensed.
6. Don't be afraid to ask for new sensors to collect new data. That's a good thing.

---

algorithms. Potter et al. (1996) provided an example of IRRs that could only be satisfied by an advanced planning algorithm and significant data transformations. Some indicators of correctly identified IRRs are shown in Table 16.5.

In addition, it is important to note that identifying IRRs is focused on satisfying the decision requirements and is not limited by data availability in the current system. In cases in which the required data is not directly available, ACWA provides a rationale for obtaining that data (e.g., pulling data from a variety of previously stove-piped databases, adding additional sensors, or creating "synthetic" values). This is a critical change from the typical role that human factors engineers have had in the past (designing an interface after the instrumentation has been specified). Consequently, this type of an approach is fundamentally broader in scope than other approaches to interface design that do not consider the impact of IRRs on system architecture specifications (Vicente, Christoffersen, & Hunter, 1996).

An interesting anecdote related to this occurred in an interface design effort for a thermo-dynamic system (Potter, Woods, Hill, Boyer, & Morris, 1992). At this point in the process, the information requirement of "predicted liquid level in the accumulators versus current level over time" was identified to compensate for significant lags in the system in monitoring for system integrity. One of the engineers argued "but we don't have any way to sense 'predicted level'—that's our fundamental problem!" Slowly, another engineer in the room raised his hand and offered, "My high-fidelity simulation of the system calculates that exact thing. I've just never talked about it because I didn't think it was of any value to anyone except me."

Example IRRs for the Game of 15. For this relatively simple domain, the IRRs appear intuitive. An example is shown in Table 16.6. The complexity is in the combinatorial problem implied by the words *relative value* in the example. Work artifacts constructed by game players during many knowledge elicitation sessions showed an attempt to enumerate the winning triples but then had difficulty in attempting to determine which next move was tactically the best decision in terms of contributing to successful outcomes. For example, the 5 Piece is involved in four possible winning triples (at game start when all pieces are available), whereas several pieces are involved in three, and the remainder in two. The value remaining as some of the pieces are taken is a dynamic calculation, depending on the exact situation. Note that the IRR does not dictate that a "score" or "value" metric be calculated; the best form for indicating "relative value contribution" is left to the final system design step. The IRR identifies the required information, not how to depict it.

Novices typically find this task more frustrating than difficult. The temptation to list data available is almost overpowering. The indicators in Table 16.4 can be useful to break out of that rut and determine if the IRRs have been identified correctly. Novice designers also typically

TABLE 16.6

Sample Information and Relationship Requirements (IRRs) for one Cognitive Work Requirement (CWR) of Game of 15

CWR 2.1-3: Select the "Win" Potential "Quanta" Source that will Maximize (be best for) Own "Win" Potential.
  IRR-2.1-3-1: Contribution of all remaining "quanta sources" (pieces) toward friendly winning outcome, either "no value" for those not leading to a win or "relative value" of those that have a positive contribution toward a win.

---

want "the answer" to be depicted explicitly (i.e., an integer value score). At this point it is critical to realize that the intent is to communicate this IRR to the user "effortlessly," success is defined not by rank-ordered list to be painted on a display device, but by the user having in his or her mind an understanding of the "relative value contribution."

Summary of the IRR Task.   Just as the FAN representation provided the framework for the derivation of decision requirements, the CWRs provide the essential context for the determining IRRSs that are essential to successfully making the decisions. For example, in a "choice among alternative resources"-type of CWR, the choice requires information about the availability of the alternatives (exercising the supporting relationships of the FAN), current tasking of those alternatives, the other functional affects of selecting it for the task under consideration (side effects), and specific performance capabilities of the alternatives (lower level functional properties of the alternatives). Properly defined IRRs become a critical resource in the design of the physical aiding concepts implemented in the decision support system.

Linking Decision Requirements to Aiding Concepts: Developing and Documenting a "Model of Support" Using the Representation Design Requirements

The FAN and its associated CWR and IRR overlays constitute a solid foundation for the development of the aiding concepts to form the decision support system. The design of the system occurs at two levels: at a micro level to ensure that each presentation element effortlessly communicates its information to the user and at the macro level to ensure that the overall collection of presentation design concepts (the decision support system in a holistic sense) is organized in an intuitive way that does not add its own "manage the decision support system" cognitive burdens to those of the domain.

The solution to the macro–holistic issue harkens back to the selection of a FAN as the organizing first step: It represents an expert's "knowledge model" of the domain's abstract concepts. Therefore, designing the virtual "information space" explicitly replicating the domain structure captured in the FAN results in a decision support system that is organized according to the model of the domain; the decision support system parallels the mental model. To accomplish this objective, this step develops the mapping between information on the state and behavior of the domain (i.e., CWRs and IRRs) and the syntax and dynamics of the visualization or decision aid being developed. Figure 16.4 shows how the functional space can be allocated to one presentation element (implying the remainder of the information space is allocated to other presentations).

The micro design issues center on revealing the critical IRRs and constraints of the decision task through the user interface in such a way as to capitalize on the characteristics of human perception and cognition. This approach is consistent with cognitive engineering principles that have variously been called representational aiding (Bennett & Flach, 1992; Roth, Malin, &

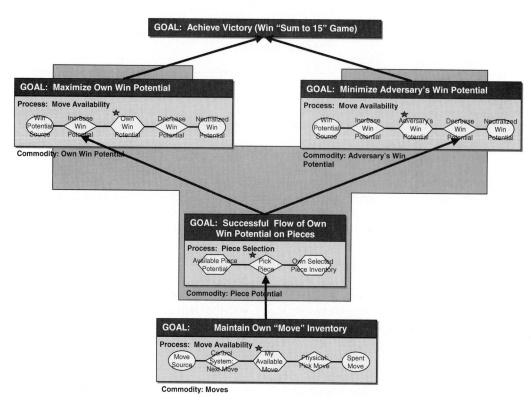

FIG. 16.4.   Allocating functional space to a presentation element. The shaded area indicates the coverage allocated to this particular presentation element—in this case, visualization.

Schreckenghost, 1997; Woods, 1995) and ecological interface design (Reising & Sanderson, 1998; Vicente & Rasmussen, 1990, 1992).

This step in the ACWA process develops the specification of the display concept and how it supports the cognitive tasks and is captured in representation design requirements for the eventual development of presentation design concepts. The RDR defines the goals and scope of the information representation in terms of the cognitive tasks it is intended to support (and thus a defined target region of the FAN). It also provides a specification of the supporting information required to support the cognitive tasks. An RDR is another span of the bridge that helps to link the decisions within the work domain to the visualization and decision support concepts intended to support those decisions. In many cases, multiple design concepts may be generated that attempt to satisfy the RDR's requirements. Typically, other supporting artifacts are generated at this step in the process as required to specify such issues as presentation real estate allocation, attention management (salience) across the information to be presented, and so on.

The RDR also represents a critical system design configuration management tool, critical for ensuring coverage of the functional decision space across all presentations and presentation design concepts. The RDR begins the shift in focus from "what" is to be displayed to "how," including annotations on relative importance that maps to relative salience on the visualization. A complete RDR is actually a set of requirement "documents," each describing the requirements for the intended representation of the IRRs. It contains descriptions of how all presentation mechanisms of the domain practitioner's workspace are to be coordinated, how available audio coding mechanisms are allocated—similarly for visual, haptic, and any other sensory channels to be employed. The RDR is not only a compilation of information

TABLE 16.7
A Small Portion of the Representation Design Requirements

Context: The Game of 15 Main Display will form the primary visualization element to both plan offensive and defensive strategies, and understand the dynamic state of the game from both those perspectives.

CWRs and IRRs	*Representation Design Requirements*
CWR 2.1-3: Select the "win" potential "quanta" source that will maximize (be best for) own "win" potential. IRR-2.1-3-1: Contribution of all remaining "quanta sources" (pieces) toward friendly winning outcome, either "no value" for those not leading to a win or "relative value" of those that have a positive contribution toward a win.	• Provide a depiction of pieces remaining available for selection. • Present an indication of currently active winning "paths" and how they relate to remaining pieces. • Present an indication of which pieces are involved in the greatest number of winning paths still viable. . . .

---

developed earlier, it has the added value of a more complete description of the behaviors and features needed to communicate the information effectively as well as an allocation of the Information/Relationship Resources across the entire set of displays within the workspace. When done correctly, it is still in the form of a "requirement" and not an implementation. This artifact becomes a key transition between the cognitive system engineer, the system developer, and the system (effectiveness) tester.

The RDR also provides one important ancillary benefit, as long as the domain remains unchanged, the RDR serves as an explicit documentation of the intent of the presentation independent of the technologies available and used to implement the decision support system. As newer technologies become available, and as their interaction with human perception become better understood, the technologies used to implement the RDR requirements can evolve. A sample of RDRs for a single IRR is shown in Table 16.7.

Summary of RDR Development.   The RDR requires designers to be more explicit about the specific cognitive activities that a given presentation design concept must support. Thus, it becomes an objective specification for the display; eliminating a substantial amount of the subjective and artistic arguments during display design. RDRs specify the user's cognitive requirements to be supported and the information (not data) that must be conveyed. Explicit links are made between particular aspects of the display concepts and specific cognitive demands they are intended to support. As such, they constitute explicit hypotheses—*a model of support*—that can be empirically evaluated. As a consequence, RDRs enable more objective, informed, and pointed testing of the effectiveness of the proposed aiding concepts separate from the aesthetic debates. One can ask, "are the support objectives of this display correct and complete?" and "does a particular embodiment of the display concept achieve the intended support objectives?" This is a key aspect of bridging the gap between analysis and design.

Developing Presentations: Instantiating the Aiding Concept as Presentation Design Concepts

From the RDR's specification of how information is to be represented within the decision support system, the next step of the ACWA process is the explicit design of presentation design concepts for the decision support system. (A similar process is used for the design of auditory, visual, or other senses' presentations of the RDR's specification.) This final step requires an understanding of human perception and its interaction with the various presentation techniques and attributes. As such, it requires considerable skill and ability beyond cognitive work analysis. The actual design of a revolutionary aiding concept is probably one of the largest "design gaps" that is needed to be bridged within the ACWA process. The ACWA design practitioner must be fluent in the various presentation dimensions: color, layout, line interactions, shape, edge detection, and so on. Essentially the designer must understand what characteristics of presentation implicitly specify about the interaction with the user's perception. The conversion of the requirements in the RDR to a sensory presentation form in a PDC requires considerable skill and background in these areas (Norman, 1993; Tufte, 1997). With the RDR as a guide, the sketches, proposals, and brainstorming concepts can all be resolved against the display's intent and requirements. The issues of how it is perceived can best be done with empirical testing of prototypes and often requires considerable tuning and adjustment to achieve the representational capabilities specified in the RDR.

Example Presentation Design Concepts

A presentation of the raw physical nature of the Game of 15 would be typical of those found in any game handbook. For actual use during a Game of 15, it might look something like that shown in Fig. 16.5. This presentation has been used during informal testing of various presentation approaches and was originally adapted from working aids composed by test subjects themselves. Subjects both marked the pieces at the top in various ways and kept working notes in the blank area at the bottom.

In fact, the presentation chosen as best depicting the RDR for the Game of 15 is based on an affordance familiar to virtually everyone, as shown in Fig. 16.6. By marking the selections with X's and O's, both the offensive and defensive potentials are explicitly depicted and yet remain intuitively evident and almost effortlessly managed. Note many critical RDR elements are not depicted by numeric values, but rather by the spatial relationships and proximity of the dynamic marks (X's and O's). It is exactly this elegant design concept to satisfy the RDR's needs that provides the value in this ACWA step. In effect, users employing this simple decision support system are playing Tic-Tac-Toe—a significant transformation from the originally stated game rules and apparent physical components. During testing, users with this decision support system make none of the errors found among users of Fig. 16.5. Users of Fig. 16.6 demonstrate such attributes as dramatically quicker decision-making performance, ability to reconstruct the conduct of the game post hoc, and the ability to discuss the game at higher levels of abstraction.

Interestingly, even the "correct" PDC must be carefully designed to correctly employ the visualization technologies. In fact, in an attempt to improve the aesthetics of the display, it is tempting to overdue the aesthetics at the expense of its ability to support the user during actual decision making. One such aesthetic makeover is shown in Fig. 16.7.

Notice that even with the same "dynamic state indicators" (the X's and O's) reused identically, the background of the physical pieces overwhelms what made the presentation effective for communicating those win potentials in Fig. 16.7. In actual use, many of the advantages discussed for the overall concept were not evident; users had difficulty "seeing" the state of the game, made more errors, and took longer to make decisions.

Discussion of Presentation Design Concepts. In practice, the presentation shown in Fig. 16.6 is remarkably effective at making the Game of 15 trivial to play and win. In fact, a user can simultaneously oppose 10 or more opponents playing from the physical display shown in Fig. 16.5. Of all the steps in ACWA, this final presentation development requires

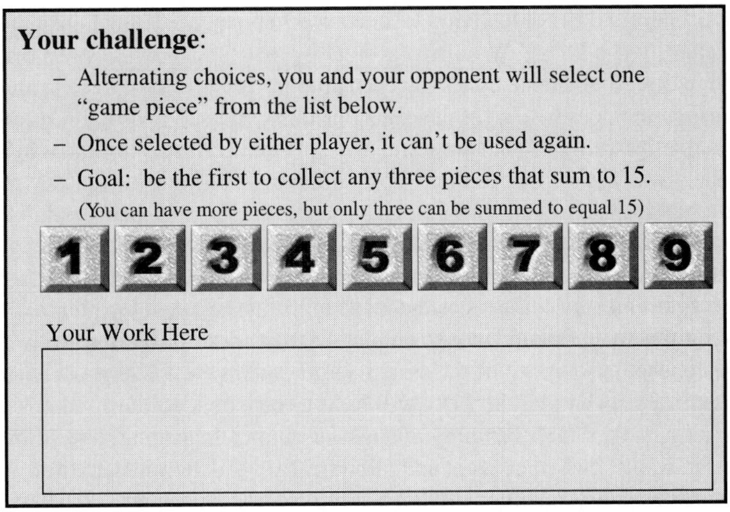
Your challenge:
– Alternating choices, you and your opponent will select one "game piece" from the list below.
– Once selected by either player, it can't be used again.
– Goal: be the first to collect any three pieces that sum to 15.
(You can have more pieces, but only three can be summed to equal 15)
1 2 3 4 5 6 7 8 9
Your Work Here

FIG. 16.5. A conventionally designed Game of 15 Display.

8
X
3
4
1
5
O
9
6
X
7
2

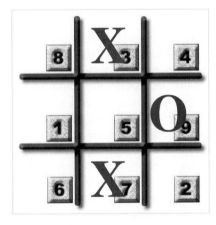

FIG. 16.6. Effective decision support system presentation design concepts.

8
X
4
1
5
O
6
X
2

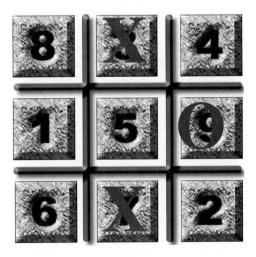

FIG. 16.7. The impact of an ill-advised aesthetic makeover.

a significant background in presentation technologies, human perceptual characteristics, and how they interact. The other ACWA artifacts, notably the RDR, provide a test basis to iterate the presentation design concepts. By testing each proposed display prototype against the single indicator question of "does it support the decisions it is supposed to as defined by the RDR?" it is possible to at least identify unsuccessful attempts and continue to design toward a more successful one. This last step across the gap is often difficult, but the ACWA methodology has made it a much smaller step, from a much more solid footing than would be the case if attempting to directly design a presentation without its RDR precursor.

The term PDC is used to convey the maturity of the design at this point. The PDC is one of the main system requirements documents handed off to the system development team. Although communicating the functionality more completely than more traditional software analysis methods, the concept lacks much of the system requirements completeness needed by a fully designed system. For example, the PDC will have to be extended during the system design to address issues such as fault handling, manual or automatic scaling, and so on. The PDC captures the innovative design concept and allows early validation testing of it in storyboard or prototyped forms, but is intentionally not a complete system design.

USING ACWA IN THE DESIGN OF ENVISIONED WORLDS

The introduction of new technology inevitably transforms the work domain and the demands placed on domain practitioners, often in unanticipated ways. New technology introduces new error forms; new representations change the cognitive activities needed to accomplish tasks and enable the development of new strategies; new technology creates new tasks and roles for people at different levels of a system. Changing systems change what it means for someone to be an expert and change the kinds of errors that will occur. Often these new technologies are introduced as part of a total redesign of business processes or as part of an activity never previously attempted. In this case, the decision support system designer cannot depend on a rich, deep pool of subject matter expertise. Methods that depend on such available expertise cannot be used effectively.

Given this transformation, developers face the "envisioned world" problem of the unforeseen impacts of the introduction of new technology (Dekker & Woods, 1997; Smith et al., 1998; Woods, 1998). A similar phenomenon has been noted in the human–computer interaction literature where it is referred to as the "task-artifact cycle" (Carroll, Kellogg, & Rosson, 1991). Concepts for new visualizations and decision-support systems (aiding concepts) represent hypotheses about what will provide effective support to domain practitioners in the envisioned world. Rapid prototypes of aiding concepts that implement the RDRs become tools for discovery (Potter et al., 2000). By exploring the impact of prototypes that embody aiding concepts, it becomes possible to evaluate the effectiveness of the envisioned support systems (i.e., test the hypotheses) as well as to identify additional support requirements and unanticipated consequences of the introduction of the new technologies.

The envisioned world problem means that system developers must face a challenge of prediction:

- What will be the cognitive demands of the envisioned world?
- How will the envisioned support concepts shape cognition and collaboration?
- How will practitioners adapt artifacts to meet their own goals given mismatches to the actual demands and the pressures they experience?
- How will the new technology impact a domain that doesn't yet exist or is evolving?

The goal of such predictions is to influence the development process so that new decision aids provide effective, robust decision support. This can be accomplished by developing prototypes of the visualization and decision support concepts as specified in the RDR. These prototypes provide a concrete instantiation of the aiding concepts specified in the RDR. They can be used to explore the viability of the aiding concept. Then, each opportunity to assess the utility of the prototype can also provide additional understanding of the requirements for effective support. Thus, these assessments can serve to enrich and refine the initial FAN and identify additional decision and IRRs that were missed in the original analytic process. Note that extending the analysis to encompass exploration of the envisioned world in a closed-loop, iterative manner contrasts with the narrow view of cognitive analysis as an initial, self-contained technique, the product of which is handed off to system designers in a waterfall model approach.

BENEFITS OF THE ACWA APPROACH

As an adaptation of the various cognitive task design approaches developed by Rasmussen, Woods, et al., ACWA maintains the power of the abstraction taxonomy as the expert-like model of the domain's concepts and relationships. This structure provides a powerful organizing framework for the knowledge elicitation process, the discovery of cognitive work requirements, as well as the design of the eventual "information space" presented to the user through a variety of sensory channels. The functional perspective of this analysis fosters innovative thought as well as a uniquely powerful perspective for supporting skill, rule, and knowledge-based behavior in a "full spectrum" decision support system.

The pragmatic adaptations made to the approaches developed by researchers makes ACWA a real-world component of mainstream system development processes, filling a critical niche in defining revolutionary new decision support concepts during the "what to build" phase of systems engineering. By reducing the process from one great design epiphany to a series of manageable, albeit nontrivial, design transformations, the ACWA process allows quality control measures similar to modern software engineering practices.

The explicit enumeration of cognitive work requirements along with the necessary information to make those decisions provides an essential separation of "what" is to be presented to the domain practitioner from the "how" the presentation is to appear. By finally separating the content from the aesthetics, design of effective presentations using any sensory channel can be evaluated objectively rather than purely subjectively. User interface design then becomes objective rather than personal preference.

Other adaptations made within ACWA make it directly compatible with advanced software engineering methodologies and practices, enabling it to be part of the critical path mainstream and not viewed as a distraction or ancillary effort.

ACWA has been used across a wide range of domains, from classically designed process control to so-called "intentional" domains such as military command and control. In each case, it has developed decision support concepts that in hindsight appear intuitively obvious (as an ideal decision support system should), yet remained undiscovered prior to the application of ACWA.

SUMMARY

In this chapter, the steps in the cognitive engineering process are presented as if they are performed in a strictly sequential order: first domain analysis, then decision requirements analysis, then supporting information analysis, culminating with design concepts. It is presented this way for expository simplicity. In practice, the process is much more parallel, opportunistic, and iterative in nature. For example, it is not unusual for an initial presentation concept to emerge before the complete decision requirements it is intended to support or the supporting information needs have been clearly articulated. The order in which the artifacts are produced is not as important as the fact that all artifacts are eventually produced to provide a functional description of the cognitive and decision tasks that the display is intended to support and the information and display elements that provide the required support. As mentioned previously, the process of generating these artifacts, not the artifacts themselves, constructs the multiple spans of the bridge. The key point is that it is the *process* of generating these artifacts that forces the cognitive analysts to think about the problem in a systematically transformed manner and capture the evolving requirements in a manageable sequence of steps, each step requiring a value-added effort by the designer, no single step requiring an "epiphany" that cannot be assured.

The generation of intermediate artifacts that model the structure of the work domain, the demands and cognitive activities to be supported, the information supporting these cognitive activities, and presentation design concepts to provide this support are needed to provide a traceable link from analysis, to design requirement, to display concept. The ACWA approach outlined in this chapter offers a means for using a model of the underlying, fundamental behavioral characteristics of the work domain in a principled manner to generate well-grounded decision support concepts for the cognitive demands facing the human–machine decision-making team. This type of approach is essential to bridge the gap between a cognitive analysis of the work domain and the development of innovative decision aids for "envisioned-world" types of problems to provide highly effective and robust decision-making performance.

ACWA has been used across a wide range of domains, from classically designed process control to so-called intentional domains such as military command and control. In each case, it has developed decision support concepts that in hindsight appear intuitively obvious (as an ideal decision support system should), yet remained undiscovered prior to the application of ACWA.

Only when decision-centered approaches are cost-effective, directly contribute to the value of the end decision support system, and have the reliability and credibility of an engineering process rather than a miraculous art will cognitive task design make an effective impact on the systems being built every day. ACWA represents the best practice in use today to satisfy those requirements and bring cognitive task design into the mainstream of systems engineering.

REFERENCES

Bennett, K. B., & Flach, J. M. (1992). Graphical displays. Implications for divided attention, focused attention, and problem solving. *Human Factors, 34,* 513–533.

Carroll, J. M., Kellogg, W. A., & Rosson, M. B. (1991). The task-artefact cycle. In J. M. Carroll (Ed.), *Designing interaction: Psychology at the human-computer interface* (pp. 74–102). Cambridge, England: Cambridge University Press.

Chalmers, B. A., Easter, J. R., & Potter, S. S. (2000, June). Decision-centered visualisations for tactical decision support on a modern frigate. In *Proceedings of 2000 Command and Control Research and Technology Symposium, Making Information Superiority Happen,* Naval Postgraduate School, Monterey, CA.

Cook, R. I., Woods, D. D., Walters, M., & Christoffersen, K. (1996, August). Coping with the complexity of aeromedical evacuation planning: Implications for the development of decision support systems. In *Proceedings of the 3rd Annual Symposium on Human Interaction with Complex Systems.* Dayton, OH.

Dekker, S., & Woods, D. D. (1997, October). The envisioned world problem in cognitive task analysis. Paper presented at the *ONR/NATO Workshop on Cognitive Task Analysis,* Washington, DC.

Glaser, R., & Chi, M. (1989). Overview. In M. Chi, R. Glaser, & M. Farr (Eds.), *The nature of expertise.* Hillsdale, NJ: Erlbaum.

Glaser, R., Lesgold, A., & Gott, S. (1991). Implications of cognitive psychology for measuring job performance. In K. Wigdor & B. Green (Eds.), *Performance assessment for the workplace* (Vol. II, pp. 1–26). Washington, DC: National Academy Press.

Gualtieri, J., Folkes, J., & Ricci, K. (1997). Measuring individual and team knowledge structures for use in training. *Training Research Journal, 2,* 117–142.

Gualtieri, J. W., Roth, E. M., & Eggleston, R. G. (2000). Utilizing the abstraction hierarchy for role allocation and team structure design. In *Proceedings of HICS 2000—5th International Conference on Human Interaction with Complex Systems* (pp. 219–223). Pascatawag, NJ: IEEE.

Jonassen, D., Beissner, K., & Yacci, M. (1993). *Structural knowledge: Techniques for representing, conveying and acquiring structural knowledge.* Hillsdale, NJ: Erlbaum.

Klimoski, R., & Mohammed, S. (1994). Team mental model: Construct or Metaphor? *Journal of Management, 20,* 403–437.

Lind, M. (1991). Representations and abstractions for interface design using multilevel flow modelling. In G. R. S. Weir & J. L. Alty (Eds.), *Human-computer interaction and complex systems,* London: Academic Press.

Lind, M. (1993, July). *Multilevel flow modelling.* Presented at AAI93 Workshop on Reasoning About Function, Washington, DC.

Logica Carnegie Group. (2000). *Command post of the future: cognitive modelling and engineering.* 1999 Year End Report prepared for U.S. Army Research Laboratory, Human Research and Engineering Directorate.

Norman, D. (1993). *Things that make us smart: Defending human attributes in the age of the machine.* Cambridge, MA: Perseus Books.

Potter, S. S., Ball, R. W., Jr., & Elm, W. C. (1996, August). Supporting aeromedical evacuation planning through information visualization. In *Proceedings of the 3rd Annual Symposium on Human Interaction with Complex Systems* (pp. 208–215). Dayton, OH: IEEE.

Potter, S. S., McKee, J. E., & Elm, W. C. (1997). *Decision cantered visualization for the military capability spectrum project.* Unpublished technical report. Pittsburgh, PA: Carnegie Group.

Potter, S. S., Roth, E. M., Woods, D. D., & Elm, W. (2000). Bootstrapping multiple converging cognitive task analysis techniques for system design. In J. M. C. Schraagen, S. F. Chipman, & V. L. Shalin (Eds.), *Cognitive task analysis* (pp. 317–340). Mahwah, NJ: Erlbaum.

Potter, S. S., Woods, D. D., Hill, T., Boyer, R. L., & Morris, W. S. (1992, October). Visualization of dynamic processes: Function-based displays for human-intelligent system interaction. In *Proceedings of the 1992 IEEE International Conference on Systems, Man, and Cybernetics* (pp. 1504–1509). Chicago, IL.

Rasmussen, J. (1986). *Information processing and human-machine interaction: An approach to cognitive engineering.* North Holland Series in System Science and Engineering. New York: Elsevier Science.

Rasmussen J., Pejtersen, A. M., & Goodstein, L. P. (1994). *Cognitive systems engineering.* New York: John Wiley & Sons.

Reising, D. V., & Sanderson, P. (1998). Designing displays under ecological interface design: Towards operationalizing semantic mapping. In *Proceedings of the Human Factors and Ergonomics Society 42nd Annual Meeting* (pp. 372–376). Santa Monica, CA: Human Factors and Ergonomics Society.

Roth, E. M., Lin, L., Kerch, S., Kenney, S. J., & Sugibayashi, N. (2001). Designing a first-of-a kind group view display for team decision making: A case study. In E. Salas & G. Klein (Eds.), *Linking expertise and naturalistic decision making* (pp. 113–135). Mahwah, NJ: Erlbaum.

Roth, E. M., Lin, L., Thomas, V. M., Kerch, S., Kenney, S. J., & Sugibayashi, N. (1998). Supporting situation awareness of individuals and teams using group view displays. In *Proceedings of the Human Factors and Ergonomics Society 42nd Annual Meeting* (pp. 244–248). Santa Monica, CA: Human Factors and Ergonomics Society.

Roth, E. M., Malin, J. T., & Schreckenghost, D. L. (1997). Paradigms for intelligent interface design. In M. Helander, T. Landauer, & P. Prabhu (Eds.), *Handbook of human-computer interaction* (2nd ed., pp. 1177–1201). Amsterdam: North-Holland.

Roth, E. M., Malsch, N., Multer, J., & Coplen, M. (1999). Understanding how train dispatchers manage and control trains: A cognitive task analysis of a distributed planning task. In *Proceedings of the Human Factors and Ergonomics Society 43rd Annual Meeting* (pp. 218–222). Santa Monica, CA: Human Factors and Ergonomics Society.

Roth, E. M., Malsch, N., Multer, J., Coplen, M., & Katz-Rhoads, N. (1998). Analyzing railroad dispatchers' strategies: A cognitive task analysis of a distributed team planning task. In *Proceedings of the 1998 IEEE International Conference on Systems, Man, and Cybernetics* (pp. 2539–2544). San Diego, CA.

Roth, E. M., & Mumaw, R. J. (1995). Using cognitive task analysis to define human interface requirements for first-of-a-kind systems. In *Proceedings of the Human Factors and Ergonomics Society 39th Annual Meeting* (pp. 520–524). Santa Monica, CA: Human Factors and Ergonomics Society.

Rouse, W., Cannon-Bowers, J., & Salas, E. (1992). The role of mental models in team performance in complex systems. *IEEE Transactions on Systems, Man, and Cybernetics, 22,* 1296–1308.

Rouse, W., & Morris, N. (1986). On looking into the black box: Prospects and limits in the search for mental models. *Psychological Bulletin, 100,* 349–363.

Schraagen, J. M. C., Chipman, S. F., & Shalin, V. L. (Eds.). (2000). *Cognitive task analysis.* Mahwah, NJ: Erlbaum.

Smith, P., Woods, D. D., McCoy, E., Billings, C., Sarter, N., Denning, R., & Dekker, S. (1998). Using forecasts of future incidents to evaluate future ATM system designs. *Air Traffic Control Quarterly, 6,* 71–85.

Simon, H. (1981). *The science of the artificial.* Cambridge, MA: MIT Press.

Tufte, E. R. (1997) *Visual explanations.* Chesire, Canada: Graphics Press.

Vicente, K. (1999). *Cognitive work analysis.* Mahwah, NJ: Erlbaum.

Vicente, K. J., Christoffersen, K., & Hunter, C. N. (1996). Response to Maddox critique. *Human Factors, 38,* 546–549.

Vicente, K. J., & Rasmussen, J. (1990). The ecology of human-machine systems II: Mediating direct perception in complex work domains. *Ecological Psychology, 2,* 207–249.

Vicente, K., & Rasmussen, J. (1992). Ecological interface design: Theoretical Foundations. *IEEE Transactions on Systems, Man, and Cybernetics, 22,* 589–606.

Woods, D. D. (1988). *The significance messages concept for intelligent adaptive data display.* Unpublished technical report. Columbus, OH: Ohio State University.

Woods, D. D. (1995). Toward a theoretical base for representation design in the computer medium: Ecological perception and aiding human cognition. In J. Flach, P. Hancock, J. Caird, & K. Vicente (Eds.), *Global perspectives on the ecology of human-machine systems* (pp. 157–188). Hillsdale, NJ: Erlbaum.

Woods, D. D. (1998). Designs are hypotheses about how artefacts shape cognition and collaboration. *Ergonomics, 41,* 168–173.

Woods, D. D., & Hollnagel, E. (1987). Mapping cognitive demands in complex problem-solving worlds. *International Journal of Man-Machine Studies, 26,* 257–275.

Woods, D. D., & Roth, E. M. (1988). Cognitive engineering: Human problem solving with tools. *Human Factors, 30,* 415–430.

17

Decision-Centered Design: Leveraging Cognitive Task Analysis in Design

Robert J. B. Hutton, Thomas E. Miller,
and Marvin L. Thordsen
Klein Associates Inc., USA

Abstract

Decision-centered design (DCD) is a design approach that focuses on understanding and supporting cognitive tasks. Research in the areas of naturalistic decision making and expertise has provided the cognitive models and analytic methods that enable this approach. DCD provides a means for communication and understanding between designers and the individuals for whom interventions are being designed. It targets the critical, often challenging cognitive tasks confronting these individuals, including individual tasks in which cognitive performance may break down and team-level tasks that require collaboration and coordination. The approach consists of task analysis, design, and evaluation.

The task analysis portion of DCD is made possible by continuing advances in analytic methods that enable subject-matter experts to articulate complex cognitive activities, analyze task performance, and describe the environments in which performance occurs. Analysis is also grounded by existing and developing models of applied cognition that guide observations and interviews (such as the Recognition-Primed Decision Model, Advanced Team Decision-Making model, and evolving models of macrocognition). The output of the analysis is an understanding of cognitive requirements illustrated by critical incidents and an understanding of the strengths and weaknesses of existing strategies and supports for these cognitive tasks.

The design aspect of the approach creates designs for decision-centered information technology, organizations, task design, and training. Design interventions are focused on reducing the cognitive challenges, improving error management, and supporting the acquisition of skilled performance, raising novice performance to the level of experts. The design approach also emphasizes the collaboration of designers, analysts, and subject-matter experts through design prototypes and mock-ups using challenging scenarios that provide the context for performance of these cognitive tasks.

The iterative design approach requires continuous evaluation, testing, and feedback. The decision-centered design approach provides the input for cognitively challenging test and evaluation scenarios as well as context-sensitive measures of performance. Subject-matter experts' feedback is elicited and focused on the cognitive aspects of task performance, and the data are fed back into the design process.

This chapter provides an overview of the decision-centered design approach, its critical phases, methodologies, and theoretical underpinnings. It then provides two context-rich case studies where the approach has been successfully applied. Decision-centered design explicitly addresses the need for task and technology design required for the increasing cognitive complexity of modern work environments.

INTRODUCTION

In a world of increasingly complex tasks and task environments, designers have not been provided with the appropriate tools or guidance to design effective supports and aids. These task environments are placing increasingly complex cognitive and perceptual demands on workers. The intent of this chapter is to outline an approach that leverages the findings of cognitive task analysis (CTA) in support of the design of systems, interfaces, training, and organizations. We call this approach decision-centered design (DCD). DCD is intended to provide the designer with techniques to work with subject-matter experts (SMEs) to help them articulate cognitive task performance. It then supports the analysis of the task in terms of its decision-making and judgment components. Finally, it uses those findings to guide design decisions. DCD relies on the findings from CTA in the form of the critical decisions and judgments, the decision strategies used to make them, and the information requirements of those decisions and judgments. The subsequent designs are thus explicitly intended to support the operator in the cognitively challenging aspects of the task.

DCD is intended to support the design of cognitive tasks and cognitive activities of human operators. These activities include subtle perceptual judgments, assessment of complex situations (characterized by fluidity, ambiguity, and uncertainty), decision making, problem solving, troubleshooting, and planning. DCD is an approach that is intended to build on traditional system design approaches that characterize the required tasks, functions, and system goals and that describe procedures and standard operating procedures (SOPs). It is intended to help the analyst or designer understand the user's cognitive challenges and to characterize and describe the complex aspects of the task.

We consider DCD to be a subset of user-centered or human-centered design approaches. It is also one approach of several that can be described under the cognitive systems engineering approach (Woods, Johannesen, Cook, & Sarter, 1994). *The key distinction that separates the DCD approach from these other approaches is the focus of the design on describing, analyzing, understanding, and supporting complex perceptual and cognitive activities.* User-centered design and human-centered design take the perspective of the human user or operator, but their focus is not on cognition. These approaches do not provide the tools, methods, or strategies to effectively understand and support the human operator (or operators) in increasingly complex and demanding cognitive task environments. Their focus is on how to design the physical work space to address human physical and behavioral strengths and limitations. They do provide best practices and support SOPs that begin to optimize the human role with respect to technology, but that is no longer enough. Human cognitive and perceptual activities, capabilities, and limitations must be explicitly recognized and accounted for in modern systems—not only for individuals but for multioperator cognitive tasks (e.g., planning, distributed decision making). DCD builds on the traditional design approaches and adds the missing focus.

THEORETICAL UNDERPINNINGS

DCD has evolved within a research approach called naturalistic decision making (NDM); (Klein, 1998; Klein, Orasanu, Calderwood, & Zsambok, 1993). The NDM approach has attempted to describe and understand decision making in "operational" environments (as opposed to laboratory environments) with experienced decision makers using field research methods and interviewing methods.

The NDM approach has focused on understanding how decisions are made by experienced decision makers facing time stress, conflicting goals, dynamic events, organizational constraints, uncertainty, and team interactions. The findings from NDM have suggested alternate models of decision making that have implications for the design of cognitive tasks and technologies to support those tasks. DCD grew out of the development of methods to help decision makers articulate the critical components of cognitive tasks and critical decisions (what the decisions are and how they are made). These methods have provided a means to articulate complex decision-making tasks to the design community such that designers can now understand the decision requirements for which they are designing and the contexts in which these decisions have to be made and supported.

A general finding from this area of research has been that to describe and understand decision making and cognitive tasks in real-world environments, a higher level of analysis and description has been required than is typically described in cognitive psychology. This level has been termed "macrocognition" (Klein, Klein, & Klein, 2000; Klein, in press). Macrocognition describes cognitive activity at the level of tasks as they pertain to real-world assessment, diagnosis, planning, problem solving, and decision making. The DCD approach is intended to explicitly acknowledge that people need support at this level of cognitive tasks and provides the means to articulate the macrocognitive activities that must be supported.

Some of the macrocognitive models and descriptions of cognitive tasks that underpin the DCD approach include recognitional decision-making models (such as the Recognition-Primed Decision (RPD) Model (Klein, 1997b; Klein, 1995; Klein, 1989a), models of uncertainty management, mental simulation, sensemaking, situation awareness, attention management, problem detection, planning and "on-the-fly" replanning, and option generation. It is beyond the scope of this chapter to describe the research and models that have evolved to explain these cognitive activities, and research is still required to understand and support them. The goal of DCD is to support the articulation, analysis, and understanding of these activities such that we can provide the appropriate choice and design of advanced technologies.

One key theoretical model on which the articulation of decision and information requirements is based is the RPD Model. The RPD Model is a model of situation assessment (assessment decisions and judgments) and course of action (COA) evaluation (action decisions).

The RPD Model has three versions: simple RPD, diagnosis RPD, and COA evaluation (mental simulation RPD). The simple version of RPD places the decision maker in an evolving situation. The decision maker assesses if the situation is typical (based on his or her experience). If it is perceived to be typical, the decision maker is able to focus attention on the most relevant cues (or patterns of cues) and information, identify plausible goals for that situation, generate expectancies about how the situation got to its current state and how the situation will develop, and, most important, what course of action will work in this situation. In the simple match case, where a situation is assessed as typical, the course of action decision "falls out" of the assessment; it is obvious to the decision maker what has to be done. The only criterion decision for the course of action is, "Will it work (in this situation)?"

If the situation is not immediately recognized as "typical," for example, if an anomaly or deviation from an expectancy is perceived, then the decision maker must reassess the situation and engage in some diagnosis through either a pattern-matching or story-building process (or both). This is the diagnosis RPD (Variation 2). Once the assessment has been resolved, the components of recognition (relevant cues, plausible goals, expectancies, and course of action) become evident. The decision of "what to do" becomes trivial once the situation has been recognized.

In Variation 3 of the RPD model (COA evaluation/mental simulation), the evaluation of the COA criterion of "will it work?" is negative. The decision maker may have some reservations or discomfort with the true nature of the situation and whether the COA will in fact accomplish the intended goal. In this case, the decision maker must engage in mental simulation to evaluate whether the COA will work. This evaluation by mental simulation is based on the decision maker's experience and refined mental model of causality in the situation. If the COA is deemed to be unworkable, then the decision maker either modifies the COA and reevaluates it or returns to reassess the situation (based on new insights about the situation gleaned from the mental simulation). This is a brief description of the RPD Model. For an expanded description of RPD and expert decision making, see Hutton and Klein (1999).

The RPD Model contains multiple aspects of cognitive activity that are part of decision making, including assessments, judgments, and COA decisions. It has also spawned several more detailed models of, for example, problem detection (Klein, 1997a) and mental simulation (Klein & Crandall, 1995), as well as improvements in the methods of cognitive task analysis. DCD has thus drawn heavily from the RPD Model and the resulting models and methods that have been generated from it.

OVERVIEW OF THE DCD APPROACH

DCD is an evolving approach that is tailored to the specific design problem being addressed (Klein, Kaempf, Wolf, Thordsen, & Miller, 1997a). The core concept, however, is the use of knowledge engineering techniques—such as cognitive task analysis, work domain analysis, and functional analysis methods—to identify the critical decision and information requirements for current and envisioned tasks, for individuals and teams. Decision requirements are defined as the critical decisions and the strategies and information used to make those decisions. These critical decisions and information requirements are then used as the key guides and drivers for the design of displays, interfaces, system modes, function allocation strategies, system integration schemes, decision support systems, and decision aids.

This approach to design contrasts with more traditional approaches to systems design, such as data-centered approaches (or technology-centered) and system-centered approaches.

DCD Versus Traditional Approaches

Data-Centered Approaches. Many approaches to system design have been driven by the information-processing power of emerging technologies. The capabilities of these high-power systems permit access to vast amounts of raw data. Any or all of these data may be made available to, and sometimes even imposed on, the user. In other words, the display and control design is "data-centered" and technology driven. This is not surprising because there is great temptation, in the face of this awesome power, to provide as much information to the operator as is technologically possible. The problem with a data-centered approach is that it does not take into account what is meaningful to users (what they want or need), when they can make best use of it, or how it should be represented. We have seen many situations in which users of complex systems begin by turning *off* various "support" and "warning" systems because they say these are distracting. Vast amounts of context-nonspecific data can be distracting, result in information overload, and force the operator to use valuable time sorting through unneeded data to find the one or two important pieces. More is not always better.

Data-centered approaches involve feeding the user every bit of information the technology can generate "just because it's there." Data-centered design has contributed to the demise of many a noble attempt at decision support systems. Without some frame within which to present available information, the support system interfaces become almost chaotic to the users, with irrelevant information masking or distracting the user from critical elements.

System-Centered Approaches.

Although system-centered design represents a vast improvement over the data-centered approach, it also is not without limitations. An example of system-centered design would be the displayed representation of the fuel or hydraulic system status onboard a vessel. This approach frames information in a much more useable fashion than does data-centered design, but it still fails to consider the overall tasks that the users are trying to accomplish and the fact that the information required for critical decisions is spread over a number of systems (e.g., fuel, hydraulics, and navigation). Information about systems is almost always needed in the context of a larger decision-making or problem-solving task. The crew does not want to know that there are 627 gallons of fuel in the tank; they want to know if they have enough fuel to reach a particular alternative port. So, for example, a display might provide something like distance available before fuel is exhausted or time before fuel exhaustion. Granted, there are times when specific system status information is what is needed (e.g., when troubleshooting a hydraulic problem), but more often, the system schematic will be but one piece of the overall information a crew must draw on to help make appropriate decisions or assessments for a particular situation. This approach often leads to the development of multiple, stove-piped systems that are system-centered. They are not integrated, and therefore the user cannot make assessments and decisions about larger issues that are critical to them but that cut across specific systems. This leads to problems of information management, integration, situation awareness, and the big picture.

The systems-centered approach does not recognize that this type of data generally plays a "supporting role" to the cognitive processes that the operators employ and require to achieve their goals. To better support the operator's cognitive processes, the data presentation needs to go beyond being data- and system-centered and be presented in the context of the operator's decision making, that is, it needs to be decision centered. To achieve this, the information must be presented in a functionally meaningful way with extraneous data "filtered" from the user. The information must be framed by the nature of the operator's cognitive requirements within a particular context or situation. These cognitive requirements must first be understood and then supported by the design of displays, interfaces, decision supports, and decision aids developed by the system designers.

The Decision-Centered Design Approach.

DCD, as the name implies, anchors design around the decisions that will confront the user or users while involved with the system's tasks. DCD itself can be viewed as a variation of cognitive systems engineering (CSE) in which decision requirements (the most critical and difficult decisions and judgments and the strategies used to make them) provide the foundation for the generation of the design principles and recommendations.

In general, both CSE and DCD focus on the application of findings from the areas of cognitive science, cognitive psychology, and NDM in the design of systems so that the cognitive strengths of the human operators are promoted and their cognitive weaknesses are supported. In the past, this perspective has provided a framework for designers to create training systems, decision support systems, and human–computer interfaces in which human perception, judgment, thought, reasoning, and action capabilities are treated explicitly and become an integral part of the final product. CSE and DCD are inherently user-centered approaches to design. To learn how decision makers handle the complexities and confusion inherent in operational environments, CSE and DCD researchers have moved their research away from highly controlled and predictable laboratory studies and into the field to study domains that are complex and challenging.

The general DCD approach is intended as a critical supplement to current systems engineering and human-systems integration processes and practices. However, there is a critical shift in focus from behavioral aspects of human performance (e.g., anthropometric recommendations, workstation layout, legibility, color contrast, luminance values, other more traditional human factors and ergonomics issues, etc.) to the critical cognitive requirements of advanced cognitive tasks. The shift is toward understanding and supporting the increasingly cognitive activities that these advanced systems require. We do have to account for the other system design considerations for human performance, but we miss the mark if we do not emphasize and design for the operator's capacity to take in, assimilate, filter, assess, judge and make meaning of, and share information. We must also understand those critical assessments and judgment skills and the skills required to problem solve, troubleshoot, plan, replan, project, mentally simulate, generate courses of action, and command and control within the working environment.

METHODS FOR SUPPORTING DECISION-CENTERED DESIGN

One critical aspect of the DCD approach is the elicitation, capture, and analysis of information from SMEs about the decision requirements, that is, the key decisions, the decision strategies, and the information requirements that are critical to effective performance in their work domain context. It is this focus on these critical decisions and judgments that extends the approach from purely CSE into DCD. It is also a critical strength of this approach that the methods allow the task SMEs to articulate the cognitively complex aspects of the task, the decision, and information requirements to the analyst-designer team. This critical strength provides the design team with an understanding of the complex contexts in which their system or systems will be used. This process mediates the relationship between the user group and the designers.

These cognitive activities are what we refer to as "decision requirements." In DCD, decision requirements are the drivers of design. To capture these decision requirements, we need to understand the cognitive activities, knowledge, and strategies of experienced personnel as well as the constraints on cognition presented by the work domain. To capture the decision requirements and decision strategies, we primarily use cognitive task analysis techniques (Cooke, 1994; Klein, Calderwood, & MacGregor, 1989; Militello & Hutton, 1998; Schraagen, Chipman, & Shalin, 2000) and work domain analysis techniques (Rasmussen, Pejterson, & Goodstein, 1994; Vicente, 1999), coupled with an understanding of the cognitive mechanisms that underlie the performance of the cognitive tasks (macrocognitive models and theories).

CTA techniques have been used to identify cognitive requirements in a variety of domains, including tactical decision making at the Aegis Combat Information Center (Kaempf, Wolf, Thordsen, & Klein, 1992; Klein, McCloskey, Pliske, & Schmitt, 1997b), command and control in the Airborne Warning and Control System (AWACS) aircraft (Klinger et al., 1993), target weaponeering (Miller & Lim, 1993), flight crews in fixed and rotary wing aircraft (Kaempf, Thordsen, & Klein, 1991; Thordsen, Klein, & Wolf, 1990), and U.S. Navy Electronic Warfare Specialists (Crandall, Klein, Militello, & Wolf, 1994).

Cognitive task analytic techniques are a set of methods and tools to uncover expert decision-making strategies, critical cues and patterns of cues, critical decisions and how they are made, what makes them difficult, what current technology supports they have, and how technology may support improved performance. The key goal for CTA is to identify the decision requirements through interviews with SMEs; through observations of real and simulated performance in the field or in a simulator, in real operations or exercises; through focused simulation exercises intended to flex the most challenging cognitive aspects of the job; and through simulation exercises that look at envisioned world (future-world) problems, challenges, and next-generation technologies. CTA techniques cover a broad range of methods (Cooke, 1994; Klein et al., 1989; Klinger et al., 2000; Militello, Hutton, & Chrenka, 1998). For a more thorough exploration of several CTA and knowledge elicitation and analysis techniques, see Schraagen et al. (2000).

Several methods support the DCD process. The primary methods come from the NDM community and the cognitive systems engineering community. These methods broadly describe CTA methods. Here, we describe the Critical Decision Method and the Knowledge Audit as two examples. Another method, recently developed, is a hybrid systems engineering–NDM method that focuses CTA resources. This method is called Cognitive Function Modeling. Finally, Work Domain Analysis further supports the understanding of the work domain constraints that impact cognitive performance. The following sections briefly describe these four methods.

Critical Decision Method

A prime example of a CTA technique is the Critical Decision Method (CDM). CDM interviews, based on Flanagan's (1954) critical incident technique, are organized around an initial, unstructured account of a specific incident. The incident account is generated by the interviewee in response to a specific open-ended question posed by the interviewers, and it provides the structure for the interview that follows. The nature and content of the opening query is determined by the research goals of the particular study but is always asked in terms of an event the interviewee has personally experienced. For example, in a study of neonatal intensive care unit (NICU) nurses' clinical judgments (Crandall & Getchell-Reiter, 1993), each nurse was asked to select an incident in which her patient assessment skills had made a difference to the patient's outcome. In several studies of fireground command decision making, participants were asked to recall an incident in which their expertise as a fireground commander was particularly challenged (Calderwood et al., 1987; Klein, Calderwood, & Clinton-Cirocco, 1986).

Once the participant identifies a relevant incident, he or she recounts the episode in its entirety, with no interruptions from the interviewer. The interviewer takes on the role of an active listener at this point. The respondent's account, solicited in this noninterfering way, provides the focus and structure of the remainder of the interview. By requesting personal accounts of a certain type of event and structuring the interview around that account, potential interviewer biases are minimized. Once the report of the incident has been completed, the CDM interviewer leads the participant back over the incident account several times, using probes designed to focus attention on particular aspects of the incident and solicit information about them. For example, in the study of NICU nurses' assessment skills, a nurse related an incident in which she described a baby as "looking bad that day." This incident was probed for information about what aspects of the baby's appearance caught the nurse's attention (e.g., "his color") and specific, detailed descriptors regarding each cue ("his color was grey-green, not cyanotic blue"). CDM probes are designed to elicit specific, detailed descriptions about an event, with particular emphasis on perceptual aspects (e.g., what was actually seen, heard, considered, remembered) instead of asking people for their general impressions or for explanations or rationalizations about why they had made a particular decision. The probes are designed to progressively deepen understanding of the interviewee's account.

Solicited information depends on the purpose of the study but might include presence or absence of salient cues and the nature of those cues, assessment of the situation and the basis of that assessment, expectations about how the situation might evolve, goals considered, and options evaluated and chosen. Information is elicited specific to a particular decision in the context of the particular incident under review.

Although memories for such events cannot be assumed to be perfectly reliable, the method has been highly successful in eliciting perceptual cues and details of judgment and decision strategies that are generally not captured with traditional reporting methods. The CDM has been demonstrated to yield information richer in variety, specificity, and quantity than is typically available in experts' unstructured verbal reports (Crandall, 1989). The information obtained through this method is concrete and specific, reflects the point of view of the decision maker, and is grounded in actual incidents. For this reason, the methods have been found to provide an excellent basis for the development of instructional materials and programs. Detailed descriptions of CDM and other work surrounding it can be found in Klein (1989b), Klein et al. (1989), and Hoffman, Crandall, & Shadbolt (1998).

Knowledge Audit

The Knowledge Audit was developed while working under a contract with the Navy Personnel Research and Development Center in San Diego, California. The objective was to develop a relatively inexpensive and simple method for applying CTA to the process of training development. The Knowledge Audit focuses on the categories of knowledge and skills that distinguish experts from others. These categories include metacognition, mental models, perceptual cues and patterns, analogs, and declarative knowledge. The Knowledge Audit provides an efficient method for surveying the various aspects of expertise. The method does not attempt to find whether each component of expertise is present for a given task. Rather, it employs a set of specific probes designed to describe the type of knowledge or skill and to elicit examples of each based on actual experiences. The specific probes must be identified for each domain. The primary strength of the Knowledge Audit is that it enables us to survey rapidly the nature of expertise in a given domain.

The Knowledge Audit draws directly from the research literature on expert–novice differences (Chi, Feltovich, & Glaser, 1981; Dreyfus, 1972; Dreyfus & Dreyfus, 1986; Hoffman, 1992; Klein & Hoffman, 1993; Shanteau, 1985) and CDM studies (Crandall & Getchell-Reiter, 1993; Klein et al., 1989; Klinger & Gomes, 1993; Militello & Lim, 1995) of expert decision making. The Knowledge Audit has been developed as a means to capture the most important aspects of expertise while streamlining the intensive data collection and analysis methods that typify studies of expertise.

The Knowledge Audit employs a set of probes designed to describe types of domain knowledge or skill and elicit appropriate examples. The goal is not simply to find out whether each component is present in the task, but to find out the nature of these skills, specific events for which they were required, strategies that have been used, and so forth. The list of probes is the starting point for conducting this interview. For example, a probe may sound like this: "Novices may only see bits and pieces. Experts are able to quickly build an understanding of the whole situation—the big picture. Can you give me a specific example of what is important about 'the big picture' for this task? What are the major elements you have to know and keep track of?" Then, the interviewer asks for specifics about the example in terms of critical cues and strategies of decision making. This is followed by a discussion of potential errors that a novice, less-experienced person might have made in this situation.

The incident examples elicited with the Knowledge Audit do not contain the extensive detail and sense of dynamics that more labor-intensive methods such as the CDM incident accounts often do. They do, however, provide enough detail to retain the appropriate context of the incident.

Cognitive Function Model (CFM)

The Cognitive Function Modeling software tool was developed for the U.S. Navy. The purpose of the tool is to help design teams break down jobs into their meaningful functions and tasks so that the design team knows which of them are likely to be cognitively challenging for a human operator. The CFM tool supports DCD by using a systems engineering breakdown of tasks and functions, a functional decomposition representation of behavioral and cognitive aspects of performance, and identifying which of these tasks or functions are the most cognitively challenging. This method provides a high-level CTA-like analysis that helps the systems analysts and designers to focus their attention on the cognitively challenging aspects of the job that should be the focus for a more thorough CTA. This is where DCD will have the most impact on the design of the system to support performance.

The CFM tool consists of two modules. The first contains a tutorial and the software required to build an Operator Function Model (OFM). The OFM is a human–system engineering representation of the human role within a system context, in terms of functions and tasks (Mitchell, 1987). The second module identifies the challenging cognitive aspects of those tasks to further focus design efforts. This module, called the Cognimeter, assists the analyst in identifying which tasks or functions within the OFM are cognitively challenging and should be explored in more depth using CTA. These tasks and functions require supporting CTA data for the design engineer to understand the human–system integration design issues.

CFM serves as the starting point and road map for subsequent CTAs. It identifies the range of goals, functions and tasks, and then highlights the cognitively challenging aspects of the system that require further CTA methods. For further detail on the Cognimeter and the CFM tool see Anastasi, Hutton, Thordsen, Klein, & Serfaty, 1997; Anastasi et al., 2000; Chrenka et al., 2001a; Chrenka et al., 2001b; Hutton et al., 1997; Thordsen, Hutton, & Anastasi, 1998.

Work Domain Analysis

Work Domain Analysis is used as a supplement to the CTA to provide a perspective on the constraints of the work environment. Work Domain Analysis comes from a cognitive work analysis perspective (Rasmussen et al., 1994; Vicente, 1999) and provides a set of methods to characterize the constraints and leverage points of the work domain, including descriptions of functionality from high-level system purpose to the physical form of current systems (an example of this technique is Rasmussen's abstraction hierarchy (Rasmussen et al., 1994)). The purpose of the Work Domain Analysis is to capture the aspects of the task domain that affect the cognitive aspects of performance, focusing on the environmental constraints and affordances presented to workers.

As examples, there may be aspects of the sensor technology capabilities that limit the display of certain kinds of information or fusion technologies that provide only certain types of fused data, or there may be aspects of the work space and team organization that currently create stove-piped data systems that constrain the sharing of situation awareness within a functional team. Other issues may relate to the "physics" of the domain that limit the possibilities for interpreting certain kinds of information by operators, such as the "chaotic" nature of weather that places inherent limits on a forecaster's ability to forecast, or the lack of "physics" that drives the intentions of an intelligent adversary and limits the ability of systems to "display intent." The purpose of Work Domain Analysis is to complement the CTA by providing some of the constraints and opportunities in the work domain that affect the operator's cognitive activities.

Examples of specific methods to accomplish Work Domain Analysis are captured in Rasmussen et al. (1994) and Vicente (1999).

DECISION-CENTERED DESIGN PROCESS

The DCD process is dependent on the specific nature of the design problem and the intended design solution. A DCD approach can be used to specify design concepts and specifications for a range of system solutions, from the design of specific display elements of a single display to the integration rationale for a suite of systems to be incorporated into a combat information center, control room, or vehicle crew station. In theory, DCD can be used for a number of design tasks from specifying the requirements of a decision support system to the rationale for task and function allocation in a large system such as a ship self-defense system.

The general DCD approach will always require an analysis of the cognitive aspects of the tasks to identify the key decision requirements that must be supported by the design. DCD provides a framework on which to base early design decisions and generate early design concepts. Many of the design specifics will be governed by the CTA as well as sound human factors and ergonomic guidelines and recommendations. The basic steps of the DCD process include the following:

- Analysis
  - Background preparation and domain familiarization
  - Observations and knowledge elicitation: Cognitive task analysis (CTA)
  - Definition of decision requirements and decision strategies
- Design
  - Transformation of decision requirements and decision strategies into design concepts
  - Development of key design recommendations
- Test and evaluation
  - Development of evaluation scenarios and contexts
  - Development of context-sensitive metrics
  - Development of real-world outcome evaluation criteria

Analysis

Background Preparation and Domain Familiarization. Every DCD project is tailored to the specific design problem. For purposes of preparation, key issues include the following: what is the intended application, who is the envisioned user, what are the envisioned tasks and functions, what will be the organizational and informational constraints, and what are the existing (or envisioned) technologies. This information frames the scope of the subsequent CTA effort by identifying key domain and task expertise and the key aspects of the task to pay attention to and what CTA techniques may be most appropriate for the analysis. For example, if the end product is a situation display, the focus for the CTA will be more on information requirements, critical cues, meaningful patterns and constellations of cues, key subtle judgments, and perceptual skills that need to be supported. If the end product is a function allocation strategy and manning plan, then the focus may be more on critical decisions and how they are made, team interactions, communication, and information flow. The background preparation leads into the CTA to derive the decision requirements that will drive the design.

Observations and Knowledge Elicitation. As described earlier, there are many approaches to CTA, and the specific focus of the CTA is guided by the design problem and other design and domain constraints. The CTA portion will consist of data gathering, data reduction, and analysis to identify the critical decisions and how they are made. Subject-matter expert interviews, observations, and decision exercises will provide the required data. The analysis is conducted by skilled analysts with backgrounds in human factors (psychology and engineering) and cognitive psychology. These analyses lead to a breakdown of the tasks and functions into critical decision requirements: the critical, challenging, and frequent decisions that are made and the information, knowledge and strategies that go into making those decisions.

Definition of Decision Requirements and Decision Strategies. The decision requirements, or cognitive demands, and supporting data are a product of the CTA. The decision requirements capture the key difficult, challenging, and critical decisions within a specific task domain. They can vary in specificity from "develop situation awareness" to "assess the intent of an unknown track" or from "revise a course of action" to "project the enemy situation in 12 hours." These descriptions are supported by a host of relevant information about why the decision or cognitive activity is challenging or difficult, what strategies are used to make the decision, what information or patterns of information support the decision making, what critical background knowledge is required to support the decision, what are common errors or difficulties in coming to a decision, and so forth. A critical aspect of the design process with respect to the decision requirements is what are the decision strategies used by the decision maker to make these judgments or decisions? Strategies may include feature matching, holistic matching, story building, mental simulation, analogical reasoning, comparative option selection, and so forth. These strategies are derived from the decision-making literature, decision-making theory, and decision models (Zsambok, Beach, & Klein, 1992). The decision requirements encompass the "whats" and "hows" of decision making. What are the critical decisions? What pieces of information are critical to making those decisions? How do decision makers make these decisions? The design must support the decision requirements. These pieces of information are the key drivers for design solutions.

Iterative Design

Transformation of Decision Requirements into Design Concepts. The decision requirements and decision strategies are the drivers of the design concept solutions. Often design concepts fall out of, or emerge from, the CTA and the decision requirements analysis, even though these are not necessarily specific objectives of the CTA itself. Sometimes the expertise and creativity of the design team is required to envision potential solutions to support these tough cognitive activities. Sometimes, new technologies are required or recommended to solve the design problem, and in other cases the findings drive research to solve problems that currently are beyond the technologies and human skills that exist.

Development of Key Design Recommendations. Design recommendations are the result of combining the decision requirements with the appropriate design concepts to address the need. The design recommendations could range from specific display elements to support situation assessment, to the complex integration of multiple systems to support information gathering, filtering, and display in a command post or operations center environment. The decision strategies will provide information about what processes must be supported, and the key decisions and information requirements will provide the input for what must be supported and what content must be provided. Again, other design factors will play into the design process, including human factor guidelines, ergonomic issues, organizational design issues, doctrinal and SOP factors, and standards and guidelines from other disciplines, but these recommendations will have grown from the goal of supporting the key decision requirements for a specific user, group of users, or work team.

These design recommendations can be represented in a number of ways including a systems requirements document, a functional specification, paper storyboards, electronic storyboards, and functional prototypes (varying from parts of the system to a whole system prototype).

Iterative Test and Evaluation

Having generated a design concept and developed a prototype or mock-up of the system, the decision requirements once more come into play. Testing of the system or design should address the decision-making environment in which the system is to perform. The decision requirements and CTA can provide valuable insight into the key challenges that must be presented to the system to flex it, and the operators using it, through the design of evaluation scenarios and contexts.

Performance of the operators and technology must be measured based on their ability to accomplish the cognitive activities that were identified in the CTA as critical to effective performance. The CTA can provide scenario contexts, context-sensitive metrics, and evaluation criteria. The CTA can provide details for measures of performance that reflect the ability of the system to support effective decision making. The system testing provides additional information on the ability of the operators to perform with their equipment in a challenging decision-making environment, and data from the test and evaluation can be fed back into the iterative design process. SME feedback should be sought on the design's ability to support the critical decision-making activities that were originally identified in the CTA. SME involvement in this process and subsequent iterations throughout the design process is critical, and the CTA methods support subsequent SME interviewing regarding the effectiveness of the system design.

VALUE ADDED

Many traditional task analysis methods only address the procedural steps that must be followed; decision requirements offer a complementary picture of the critical and difficult judgments and decisions needed to carry out the task. Decision making in this context is not limited to choices between options but includes judgments, situation awareness, and problem solving.

To design better systems, we need to understand the task decision requirements. Otherwise, system designers are left without useful guidance on how to organize a display or identify the required underlying architecture. A task listing is usually not sufficient for guidance in human–computer interface design. When designers are not given a good sense of the decision requirements of the task, they may fall back on a technology-driven strategy of adding in the newest technology that is available or a data-driven strategy of packing the most data elements into the display to ensure that nothing essential is left out. The technology-driven approach results in initial enthusiasm, often followed by disillusionment as the operators find they still must wrestle the interface. The data-driven approach is safe but creates frustration when operators cannot find important data items or detect trends and thereby are unable to make key judgments under time pressure.

Landauer (1995) described the disappointing results of technology- and data-driven approaches and argued that a user-centered design approach is needed to identify the needs of the users and to make systems conform to these needs. But the task of identifying and accounting for the users' needs and expectations is daunting. User-centered design may have trouble being put into practice without guidelines and recommendations to make the process more manageable. The challenge is now to develop methods for efficiently capturing and supporting user needs.

COST OF IGNORING COGNITIVE CHALLENGES

Not understanding the decision requirements of a task risks developing a system that does not enhance fluid performance, or worse, that obfuscates the user's tasks. An example of a decision requirement of a task is judging intent of an air contact. The commander of a U.S. Navy Aegis cruiser may have to judge the intent of an approaching aircraft that has taken off from a hostile country and ignores radio warnings to change course. At earlier stages in an incident, the commander may search for a strategy to warn the aircraft away. At some point, the commander may have to decide whether to fire a missile at the aircraft. Throughout the incident, the commander will be trying to understand the intent of the pilot flying the aircraft. Designers can use this understanding to suggest ways of configuring a display that makes it easier to infer intent.

For the decision requirement *determining intent,* commanders and tactical action officers have heightened sensitivity to trends in a track's speed, altitude, and range, as well as a need to detect sudden changes in these trends. The Aegis character read out (CRO) display for the mid-1990s, for example, only displayed four-digit alphanumerics. For operators to get trend data, they were required to track changes in the CRO mentally and perform mental arithmetic to arrive at the necessary trend. The cost of not paying attention to and understanding the key decisions, why they are challenging, and how they are made is that the decision supports, interfaces, information flow, and allocation of functions are inadequate to deal with the complexity of decision making in that environment and the organization incurs losses of time, money, or, in the worst case, lives.

EXAMPLES

Because of the context-specific nature of the application of DCD, we use two examples to illustrate the approach. The first is the design of a tool for use by air campaign planners in the development and evaluation of their plans; the second is the design of a real-time display for the Navy's landing signal officers (LSO). Both environments are cognitively complex but have different challenges. The LSO is faced with an extremely time-pressured judgment based on perception and situation awareness. The air campaign planners are not faced with the same time pressure as the LSO but are faced with large amounts of dispersed data, conflicting goals, and unclear criteria for the development and evaluation of plans. The CTA in each environment revealed the keys to critical components of appropriate decision aiding systems.

Supporting the Development and Evaluation of Robust Air Campaign Plans

Computer-based technology can rapidly generate detailed plans for complex situations. However, these plans may be rejected by planning experts who judge dimensions such as the "robustness" of the plan using operational rather than computational criteria. Our goal in this project (Miller, 2001; Miller, Copeland, Heaton, & McCloskey, 1998) was to first capture the tactical and strategic concerns of air campaign planners and then incorporate this understanding into planning technology to assist with filtering out the unacceptable options as a plan was being developed. Specifically, the project team focused on identifying characteristics of quality plans and how these characteristics were judged in operational settings. We relied on CTA knowledge elicitation techniques to identify the process of plan evaluation and the factors underlying judgments of plan robustness. Our research drew on observations and interviews in a variety of settings. The primary data sources were joint military exercises, in-depth interviews, and a simulation exercise with Pentagon planning staff. The insights from the CTA formed the foundation of a software tool, the Bed-Down Critic, which highlights potential problem areas and vulnerable assumptions and summarizes aspects of quality as the plan is being developed. Supporting the tactical and strategic concerns as plans are developed and evaluated was central to this project.

Background Preparation and Domain Familiarization.   To allow us to understand both the high-level conceptualizations and the low-level information gathering and interpretation that takes place in this domain, we used several specific knowledge elicitation tools: (a) observations at joint military exercises, (b) interviews using the Knowledge Audit and Critical Decision Method (CDM), and (c) a simulation exercise titled "Air Counter '97."

Exercise Observations.   Our research team attended three live exercises in which U.S. Air Force and Navy planning staff developed air plans that were actually flown during the exercise. The first exercise we attended was "Roving Sands," for which we had one observer stationed with the planning staff at Roswell, New Mexico. Because this was the first exercise we observed, it was used primarily to make contacts within the planning community and to become familiar with the planning process.

The second exercise we attended was a joint task force exercise in which the planning staff was stationed aboard the *USS Kittyhawk,* which was docked at San Diego for the exercise. Two research staff members observed this exercise and conducted interviews with the planning staff as opportunities arose. This second exercise allowed us to explore issues identified from the first exercise and to deepen our understanding of the existing process.

The third exercise was conducted by the *USS George Washington* carrier battle group at sea in the Atlantic Ocean. One researcher spent seven days with the planning staff aboard the *USS Mount Whitney,* which is the command and control ship for the carrier battle group. This was the highest fidelity exercise observed, because being at sea adds to the stress, fatigue, and reality of 24-hour operations. Again, this exercise was used to further explore issues identified in earlier exercise observations but also to specifically focus on the characteristics of quality plans, how the staff knows when they have a good plan, and where in the process there are opportunities to evaluate the quality of the plan. During this exercise, we also made contacts with the planning staff at Barksdale Air Force Base, who were responsible for developing the air plan. We conducted follow-up interviews with six members of the planning staff at Barksdale to explore specific plan evaluation issues.

In all of these exercises, members of the planning staff were separated into several teams, each performing one or more tasks within the overall goal to develop air plans. Such an arrangement made it difficult for individual members of the planning staff to maintain a broad understanding of the intent behind the plans. The observers of these exercises were in the fortunate position of not being assigned to any particular planning team and thus were able to make observations across the various teams and at different times in the daily cycle.

Individual Interviews.   To help us understand both the high-level conceptualizations and the low-level information gathering and interpretation that takes place in planning, we used two specific knowledge elicitation tools in the individual interviews: the Knowledge Audit and the CDM.

The team conducted interviews with three groups of experienced military personnel. The first were opportunistic interviews during the exercise observations. We call them opportunistic because these interviews were conducted with the exercise planning staff when they were not performing other duties.

The second interviews were with planning staff at Barksdale AFB. These were follow-up interviews with staff members who had put the initial plans together for one of the exercises we had observed and who modified and developed the plans as the exercise was conducted. The interviews were semistructured, CDM sessions lasting approximately 2 hours. The sessions began with a discussion of the interviewee's background in the U.S. Air Force and the path by which he or she became an 8th Air Force planner. Our goal was to understand the knowledge and experience base possessed by the typical planner. We then continued our questioning by eliciting specific incidents experienced by the interviewee either during the course of an exercise or in a real-world operation. Our goal here was to probe events in which the planning process had fallen apart to gain an understanding of the barriers to effective planning. Finally, we initiated a discussion of the characteristics of quality plans. We asked the interviewees about incidents in which they felt their plans were strong and incidents in which their plans had to be altered significantly following briefings to higher level personnel. The purpose of this line of questioning was to identify aspects of plans that make them robust and thus acceptable for use in an operation.

The third sets of interviews were conducted with high-ranking military officers, some of whom had intricate involvement in developing plans for the Gulf War in 1991. For example, we interviewed General Charles Horner, who was in a key position during the Gulf War to evaluate air campaign plans as they were developed. In his role during the war, no plans were implemented without his approval. For our purposes, we interviewed him to discuss specific incidents in which plans were not approved to gain an understanding of what criteria were not met. We were interested in what was missing or erroneous about these plans and how these plans could have been better.

Simulation Exercise. The Air Counter '97 (AC '97) exercise was the last and most extensive of our knowledge elicitation work. In knowledge elicitation work prior to AC '97, we found that there was no formal (explicit) aspect of the planning process in which evaluation takes place. To study plan evaluation, we designed AC '97 to engage experienced planners explicitly in evaluating plans. We did this by having two teams of human planners separately develop plans against the same challenging scenario. The situation in the scenario demanded rapid air force response to the massing of troops from a non-allied country along the border of an allied country. After the teams developed their respective plans, the two teams commented on and evaluated each other's plans.

During the exercise, members of the research team focused their observations on the incremental development of the plans and on any indication that the developing plans were being evaluated. The planning process was observed to document informal evaluation processes and criteria and to identify areas in which human planners had difficulties within the planning process. The exercise also helped us identify a specific problem area (i.e., initial "bed-down" of resources) for technology development to support one aspect of plan development and evaluation.

AC '97 was a three-day planning exercise hosted by a planning office in the Pentagon. The exercise was designed to simulate many attributes of actual planning events. Data produced from the planning teams in the exercise included their maps of the "Initial Preparation of the Battlefield," the bed-down plans, and briefing slides containing the plans from the two teams.

During the first part of Day 1 of the exercise, the planners were all present and briefed on the situation. Planners jointly performed the initial preparation of the battlefield and an analysis of possible enemy courses of action. After the planners had a shared understanding of the situation, they were split into two teams of three to four people each. During the second part of Day 1, the planning teams worked on the concept of operations, and a strategic approach to the scenario. Each team worked independently in developing a plan. The planning continued through the morning of Day 2.

On the afternoon of Day 2, each team briefed its plans to the other team for critique. The briefings were open to questions and discussion by both the planners and the observers to provide insights into specific issues primarily involving evaluation.

On the morning of Day 3, the research team conducted individual interviews with all of the planners who participated in the exercise. The purpose of the interviews was to (a) clarify issues raised during the observation portion of the exercise, (b) elaborate on the evaluation criteria used both when developing a plan and when a plan is briefed formally as well, and (c) elicit opinions and issues not expressed in the open forum the day before.

Definition of Decision Requirements.   The first stage of analyzing the data from the exercise was to do a preliminary sweep though the entire data set. The purpose was to identify a comprehensive set of variables and content categories for inclusion in a systematic examination and coding. We also developed thematic categories to examine higher level processes and decision events that are not captured by more discrete measures (Hoffman et al., 1998). For example, the results of this thematic analysis resulted in the following data categories that impacted the decision requirements:

- The planning process
- Enemy course of action analysis
- Use of the map
- Timing and phasing
- Resource use and availability
- Tasks
- Flexibility
- Measurability
- Marketing and briefing the plan

This section briefly summarizes our CTA findings, which are organized around three aspects of planning that helped us to anchor the design recommendations and system goals and functions: aspects of quality plans, plan evaluation, and the bed-down planning process. For detailed results, see Miller et al. (1998).

Aspects of Quality Plans

To be robust, a quality air plan must strike a balance between various characteristics of the plan. Failure to consider these aspects explicitly can yield a plan with vulnerable approaches to achieving the objectives stated. A key component to evaluating a plan during its development includes an understanding of the inherent trade-offs between certain characteristics. Specifically, our CTA data suggested that robustness of air campaign plans needs to consider at least the following plan characteristics:

- Risk to forces
- Coordination between plan components
- Flexibility in applying the plan
- Sensitivity to geopolitical issues
- Resource use

- Communicating intent of the plan
- Measurability of plan progress
- Ownership of initiative.

Considerations of risk include understanding risks to friendly forces (i.e., location and proximity, and configuration of friendly command and control structures), risks to bases (i.e., location and proximity to special forces, missile, and other air attacks), risks to aircraft and mission success (i.e., mobile air defences), and so forth. Planners are faced with the task of knowing where the sources of risk are and then balancing this risk to friendly forces with the potential consequences of not taking certain actions.

Coordination issues are difficult for any large-scale operation, but they are greatly complicated by joint or coalition military forces. Coordinating activities in the planning stages can help alleviate confusion and miscommunication that can occur during the execution of a plan.

Having adequate flexibility is the next aspect of quality plans. The ability to change with dynamic conditions and not be locked to a certain action, regardless of inopportune timing or conditions, is imperative to successful planning. One general we interviewed described this as being a prisoner to the plan. Using plan components in multiple ways is one way to build flexibility into a plan. In the air campaign planning domain, for example, one way to build in flexibility is to use aircraft that can perform multiple missions. That is, specific aircraft are tasked with certain missions, but if the need arises, these same aircraft can be redirected to another, more time-critical task such as providing combat air support to ground troops. If the need does not arise, the aircraft performs its primary mission.

Understanding geopolitical, cultural, and political motivations of the opponent (i.e., "being able to plan through the enemy's eyes") enables the planner to predict the opponent's reaction to actions taken. Without this understanding, actions may not have the intended effect.

A critical aspect of any quality plan is the efficient use of resources. Knowing which resources have limited availability, and are therefore constraints to the set of feasible plans, is critical to building sustainable plans. For example, during the Gulf War, cruise missiles were a limited resource, and planners needed to use them sparingly and efficiently.

Throughout the planning process it is also important that the planners consider how to communicate the intent behind the development of the plan. Planners may develop a series of tasks that they have clearly linked to stated objectives, but fail to document or otherwise communicate these links. Without an understanding of the intent behind a task downstream in the planning and execution process, those who implement the tasks will be unable to improvise if the situation requires adaptation of the plan.

The ability to measure the progress of implementation should be built into the plan. Tasks should be stated in such a way that realistic observations can be made that are relevant to assessing progress. Measuring plan progress is an area that is currently evolving. For example, measures of effectiveness of the plan have traditionally been tied to the reduction of enemy forces or their resources. However, current thinking links implementation of tasks with an effect on opponent functionality or capability. For example, it may not be necessary to reduce an opponent's aircraft supply by 50% if a few well-placed precision weapons can disable command and control of those aircraft. This current thinking about measures of effectiveness has increased the need for clearly specified measures that are built into the plan.

Several participants interviewed during our CTA discussed the importance of having and maintaining the initiative in a plan. Although difficult to measure, having the initiative refers to the ability to force the opponent to react to your actions and not being forced to react to your opponent's actions.

Plan Evaluation

Our research team found that plan evaluation takes place both informally and in more formal settings. The most obvious arena in which plan evaluation takes place is in formal settings such as when a plan is briefed to the commander. Briefing the plan gives superior officers the opportunity to weigh whether the given plan meets the required military, political, and national objectives. Thus, during formal evaluations it is not only imperative that the plan address the necessary objectives, but that the plan is presented such that it is understandable to those who are doing the critique. Thus, marketing the plan is as much a part of the evaluation process as is satisfying the evaluation criteria.

We found that evaluation can also be informal in the sense that it is iterative, tightly coupled to plan development, and continuous throughout the planning cycle (Klein & Miller, 1999). Evaluation occurs within the development of the plan, and we found that members of the planning staff take the responsibility for questioning the plan or pointing out discrepancies within it. Therefore, computer-based planning systems that generate plans automatically cut out this vital linkage with the human planners.

This incremental evaluation is concurrent with plan development. We observed team members noticing details in the plan that could become problems downstream in the process, such as having the wrong kind of support equipment for F-15s at a certain air base. We also observed instances of backing up in the process to reconsider aspects of the plan or reviewing the entire plan to maintain situation awareness. Another form of incremental evaluation is the use of specialized software tools. For example, there is software available to help planners "de-conflict" flight paths of aircraft in a crowded sky.

The Bed-Down Planning Process

One of the goals of this work was to use results from the CTA to identify how artificial intelligence techniques could be used to support the human planners with respect to plan evaluation. We identified a slice of the planning problem, called "bed-down." This was the basis for the prototype support tool that would assist planners in critiquing the quality of their bed-down plan. "Bed-down" refers to the initial placement of resources in theater. For purposes of this work, it referred primarily to the placement of aircraft and logistic support at friendly air bases.

AC '97 produced a wealth of information on the planning process, sources of data used to develop plans, and the evaluation of plans. This exercise was instrumental in shaping the development of a computer-based planning system. One of our observations from this exercise was that the placement of initial resources in the theater of operation occurs early in the planning process and that errors made here will propagate throughout the rest of the process until detected and corrected. If detected far downstream, these errors can significantly delay implementation.

During AC '97 it became clear that the initial bed-down of resources was important to the development and refinement of an air plan. Situational, political, and logistical factors all contributed to the unique problem of bedding-down resources in a theater of battle. The bed-down problem has become more important since the end of the Cold War as a result of the increase of rapid response missions in which the United States now participates. During the Cold War, our assets were already forward positioned. Since the Cold War, however, there has been a dramatic reduction in U.S. forces and in the number of overseas bases available to U.S. forces. These circumstances made the initial bed-down plan more difficult to develop and more critical to the success of the mission.

The bed-down plan is not simply placing assets at various bases; it involves having an understanding of and strategy for how limited assets can be used most effectively and where those assets should be located to accomplish tasks in the most economical and compatible fashion, all the time maintaining low levels of risk to the mission success (e.g., loss of life, loss of assets, loss of time, etc.). Just as the development of strategy will drive the initial bed-down, refinements or changes to it can also facilitate changes in the strategy. In this sense, the bed-down is instrumental to the transition between the strategy and the implementation of the plan.

During the development of the bed-down plans in AC '97, the planners relied on maps on the walls to help answer questions they had about the developing plans. From our observational data, and from the interview data, we found that these questions fell into the following four categories:

1. Risk to air bases and to the mission that can be flown from the bases
2. Time to complete objectives in the plan
3. Logistics constraints
4. Compatibility of aircraft and aircraft support to airbase capabilities.

We developed a prototype planning system to assist the human planners explicitly in evaluating these characteristics of the initial plan. This Bed-Down Critic is described next.

Development of Key Design Recommendations.   The prototype Bed-Down Critic (BDC) is a planning system that assists planners in allocating resources and provides high-level plan evaluation functions. The concept emerged from the Air Counter '97 simulation exercise. The BDC can either be used to modify an existing bed-down or support the development of an initial bed-down plan. The BDC system also supports a strategy-to-task approach. It is a tool that allows the human planner to evaluate the planning process as the bed-down is being developed.

The plan evaluation functions are driven by manual (interactive) inputs on a map-based interface. The BDC provides estimates of feasibility, effectiveness, and threat to guide force allocation. The tool provides high-level evaluation estimates, as well as the intermediate information on which these estimates are based. The user can see more of the process and understand the trade-off issues and other factors that were created within the development of the bed-down plan.

The BDC supports the development and evaluation of the bed-down process. It provides feedback for (a) total aggregate risk, (b) time-to-completion, (c) logistics constraints, and (d) compatibility issues. An influence diagram of the system architecture is shown in Fig. 17.1.

The user can physically move squadrons to different locations and create one-to-one squadron-to-target pairings (assign squadrons to targets) (see Fig. 17.2). The user can also label squadron, base, target, and threat names on the map for rapid identification.

The user has the ability to set and modify theater–level variables such as the likelihood of attacks from special forces, supply reserves and resupply rates, and the required level of destruction required when attacking targets. Most of the theater–level variables are qualitative judgments that should (and normally would) be provided by the human planner.

The BDC supports rapid assessment and evaluation of the bed-down via graphical representations of evaluation criteria. An overall evaluation window is shown in Fig. 17.3. Within this window, the user can see aggregated evaluation of the logistics availability, time-to-completion, and aggregate risk for the currently planned bed-down. This window displays text-based feedback in a browserlike environment that allows the user to drill down into the details of higher level, aggregated assessments.

For example, in Fig. 17.3 we see that the completion time for Visalia Soc (an enemy target) is 5 days. In the test scenario from which this screen was taken, however, the commander specified that this target should be nonoperational within 2 days. The user drills down into the

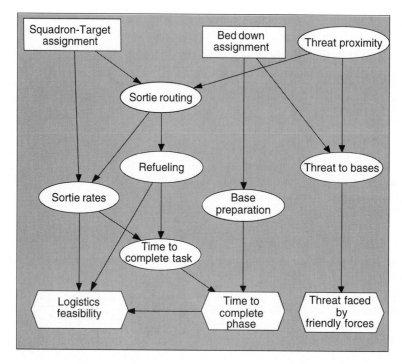

FIG. 17.1.   System architecture for the Bed-Down Critic planning and evaluation tool.

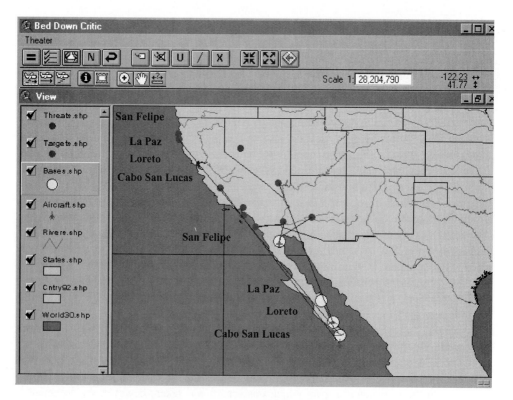

FIG. 17.2.   Map-based user interface.

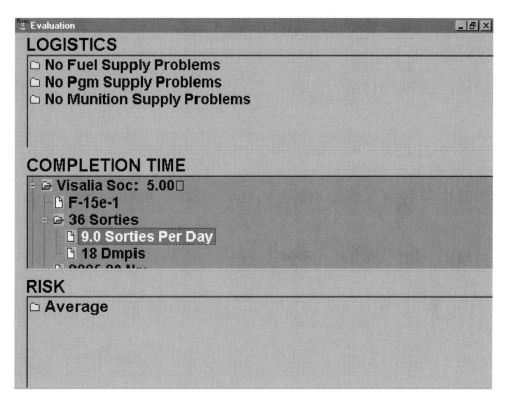

FIG. 17.3.   Evaluation window for the Bed-Down Critic planning tool.

Visalia Soc by clicking on it to reveal more detail. We can see that the Visalia Soc is targeted using F-15e-1, that it will take 36 sorties, and that there will be nine sorties per day. The human planners know that nine sorties per day is a very low rate and that this is the source of the long completion time. The planner now has several options to increase the sortie rate, including actions such as locating the fighter aircraft closer to the target or allocating more fighter aircraft to the target.

Other forms of rapid evaluation feedback are implemented in the form of tripwires and agendas. Besides the aggregated evaluation window, the system also alerts the user when specific constraints or trade-offs have been identified. These warnings are displayed when specific actions are taken (i.e., the user moves a squadron to another base) and a constraint is violated (i.e., there is not enough ramp space to accommodate a squadron).

These warnings do not require the user to correct the problem immediately. The BDC provides an "agenda" feedback. All warnings or alerts that represent a constraint violation and are yet to be resolved are recorded in the agenda. As a warning is corrected or addressed, that issue is removed from the agenda list. The user can refer to the agenda periodically to determine which issues have yet to be resolved.

The use of the agenda allows the user to keep track of factors that violate some specific constraint. The implementation of this feature does not force the user to solve each problem as it arises. The user can review the agenda when they are ready and address each item in an order they choose. The BDC system does not force the user to address each and every item (unaddressed items will remain on the agenda, but this will not prevent the user from continuing to develop the bed-down). This is especially important when the user wishes to continue with a bed-down with which the BDC detects potential problems but that may be suitable for the user's needs.

The tripwire and agenda features are implemented using intelligent agents, where each agent is responsible for specific constraints. They monitor user actions and when one of these actions violates a constraint, the agent immediately informs the user (tripwire). The agenda queries every agent simultaneously and presents these warnings to the user.

Test and Evaluation Using Decision Requirements.   The BDC has not been formally evaluated yet. Throughout the design process, planner SMEs were consulted, and feedback was received on various iterations of storyboards and prototypes. When the final prototype was completed, the BDC was demonstrated to the user community on five occasions, and initial reception of the BDC has been positive. The system has been integrated into a DARPA Technology Integration Experiment (TIE) which highlights nine planning technologies.

A future evaluation could be conducted using the planning environment from the AC '97 exercise. This evaluation would focus on performance of the tasks relating directly to the key decision requirements that were documented during the project and which drove the design process. The challenge of evaluations such as this is to generate measures of effectiveness that capture performance on the key decisions and judgments and to evaluate whether the required decision strategies are supported. This requires that the overall system effectiveness performance measures, such as evaluations of "goodness" of the plan or time to complete the plan, must be supplemented with measures that explicitly account for the planning process and whether the tool supports the critical decisions and judgments that have to be made during plan development and concurrent evaluation.

Summary.   The CTA in this project identified a need to support time-pressured replanning, and it identified factors that planners use to evaluate certain types of plans. For example, the initial bed-down plan is evaluated in terms of (a) risk to air bases and to the missions that can be flown from the bases, (b) the time to complete objectives in the plan, (c) logistics constraints, and (d) compatibility of aircraft and aircraft support to air base capabilities.

We further discovered that the most common form of plan evaluation occurs during plan generation, not after the plan is developed. This finding is problematic for automated plan generation systems because the automated systems short-circuit the most common form of evaluation that currently takes place in this domain. Automated plan generation not only leaves the human planner out of the loop as plans are being developed, but it also denies the planner opportunities to critique the plan as it is developing.

The CTA on the bed-down problem allowed us to build a prototype plan evaluation system within the context of a critical yet constrained portion of the overall planning process. The BDC is a planning system that assists planners in allocating resources and provides high-level plan evaluation functions.

Supporting the Landing Signal Officer Under Extreme Time Pressure

The landing signal officer (LSO) is part of a small team responsible for the "safe and expedient recovery of aircraft" on an aircraft carrier. They stand on the deck of the aircraft carrier and watch each aircraft as it comes in to land, assessing whether the conditions are appropriate to let the aircraft land (See Fig. 17.4).

Background Preparation and Domain Familiarization.   The CTA on the LSOs began in the initial phase of a Navy-sponsored Small Business Innovative Research project.[1] At that time, the goal was to look at the overall task of recovering aircraft aboard U.S. aircraft

---

[1] Klein Associates was a subcontractor to Stottler Henke Associates Inc. (SHAI) on Phase I and II.

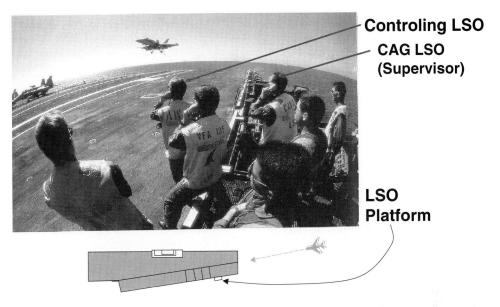

Controling LSO

CAG LSO (Supervisor)

LSO Platform

FIG. 17.4.   The landing signal officer (LSO) platform and key players.

carriers. It was a fairly broad approach that did not restrict itself to only the LSOs, but also took into consideration the roles of enlisted personnel on the LSO platform, as well as those individuals in the tower and air operations (air ops). During Phase I, we did generate some display recommendations for the controlling LSO (the individual who actually controls the aircraft through its final approach to the recovery). It was decided to focus the Phase II effort on specific design concepts to support the LSO recognition of pilot trending and oscillation. We continued the CTA development with specific attention given to the decision requirements associated with pilot trending and oscillation recognition.

Observations and Knowledge Elicitation (Cognitive Task Analysis).   The knowledge elicitation portion of this project was conducted using multiple sources of data including subject-matter experts, participation and observation, and archival records.

Our primary subject-matter experts for the CTA were the instructors at the LSO Training Center at Naval Air Station (NAS) Oceana, Virginia Beach, Virginia; a retired U.S. Navy Commander who was a former U.S. Navy aviator; and a Carrier Air Group (CAG)–qualified LSO. The commander had taken part in nearly 16,000 recoveries and was employed as a pilot for a major commercial airline. During Phases I and II, we interviewed the commander on multiple occasions totalling around 100 hours. Many of the sessions were one to two days in length.

In addition to sessions with the commander, we visited the LSO school on eight occasions, over a 30-month period, and performed knowledge elicitation on both the instructors and students. During these sessions, the instructors and students also reacted and commented on design concepts and previous data analyses we had conducted. Through the course of the Phases I and II, nine CAG-qualified LSO instructors from the school took part in the refinement of the interface recommendations. In addition, reactions were elicited from approximately 30 students who were taking courses at the LSO school. In the early stages of the project, the LSO school personnel served as additional sources of data, and in the latter stages, they provided feedback on the interface designs. During the course of these various trips and sessions, we estimate that around 60 interviews took place. These visits to the school also offered our researchers the opportunity to view multiple videotapes of recovery incidents and to interview the senior LSOs on duty for two of those. Both of these individuals were CAG-qualified LSOs. We also were able to observe six students as they worked in the LSO simulator. Finally, in addition to conducting knowledge elicitation on these individuals, one researcher had the opportunity to sit in on the Advanced LSO course.

In addition to the data elicited during the CTA, we examined several other data sets. These data provided us with critical information about data and display implementation issues (for example, given that we need to provide the LSO with a specific kind of information, where exactly is that piece of information going to come from? A real sensor? An algorithm using several pieces of sensor data? And so forth). These data sets are now briefly described.

- *SPN-46 Radar Data.* The SPN-46 radar tracks several aspects of each approaching aircraft and the ship at a rate of approximately 15 to 20 samplings per second. For each aircraft, it collects plane position information using three coordinates: distance from the ship, horizontal position with respect to the ship (lineup), and altitude. In addition, it processes some of these data to also provide the aircraft's closing speed and its sink speed. The SPN-46 radar also tracks the ship's pitch and roll in degrees. Header data includes the time-hack, pass number, the radio channel in use, and aircraft side number.
- *LSO Comments and Grades.* Every pass is graded by the controlling or backup (senior) LSO. These grades, or comments, describe the aircrafts' location and characteristics for each segment of the approach, at least from the beginning of the approach in to the wires.
- *Automated Performance and Readiness Training System (APARTS) Program, Data, and Reports.* APARTS is a program used by the LSOs to input records of every recovery pass. It works off of a Microsoft Access database and can generate a variety of reports, both for the individual pilot and for the squadrons. A key piece of data the APARTS uses is the LSOs' grade and comments for each pass. It is from these inputted APARTS data that our system extracts the information about a pilot's trends and history.
- *Accident Summary Reports.* The U.S. Navy Safety Center, located at the Norfolk Ship Yards, Norfolk, Virginia, keeps a complete set of summaries of all incidents that occur in the U.S. Navy. We requested a set of recovery incidents involving fixed-wing aircraft in which the LSO was mentioned. Several hundred incidents were retrieved. We studied these for any possible patterns and information that might prove to be useful in this project. They gave us a feel for the range of incidents that occur on the carriers, beyond the ramp strike and ditching we had learned about through other data sources.
- *Video Tapes of Recovery Incidents and Accidents.* The instructors at the U.S. Navy LSO school at NAS Oceana provided us with a video compendium of carrier recovery incidents. These were extremely informative—and sobering. The video provided us with a sense of the acute time pressure within which the LSOs work and the dangerous nature of the aircraft recovery process.

Definition of Decision Requirements.   The initial CTA revealed that there were several key judgment and decision areas required of the LSOs to perform their jobs successfully. These include judgments about the following:

- Deviations of the aircraft glide path from the glide slope, dictated by the basic angle of the day
- Deviations of the aircraft speed from the ideal, based on its attitude and the particular aircraft type

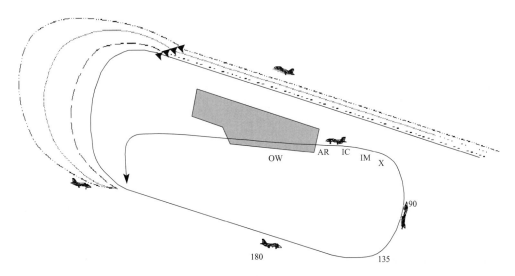

FIG. 17.5. The approach pattern for landing on an aircraft carrier. A flight of four aircraft, 45 seconds apart, come in to land, with an additional flight 2 minutes behind them. The time for making a wave-off decision varies from about .5 to 4 seconds. X = the start; IM = in the middle; IC = in close; AR = at the ramp; OW = over the wires.

- Deviations of the aircraft line up from the centerline of the landing area on the angled deck
- Abnormal aircraft configuration, based on the aircraft type
- Whether the deck is foul (obstructed or not ready to receive an aircraft)

For each of these points (glide path, attitude, lineup, and configuration), the CTA also brought out the important conditions under which the landings were occurring (Case 1, 2, or 3). Case 1 (day) and 2 (night) refer to good weather, visibility, and sea conditions. Case 3 refers to poor weather, visibility, or sea conditions.

In addition, two other criteria were identified as important influences on the decision making of the LSO: the type and model of aircraft and the experience and competence of the pilot. The type and model dictates the characteristics and responsiveness of the aircraft. This in turn influences how much the LSO can let it drift off centerline (the larger the wingspan, the less drift that can be allowed), how quickly the engine can spool up (the slower the spool up time, the further out the "power" call must occur), and so forth. The LSO classifies the experience and competence of the pilot using three categories: new guys (NG), average pilots, and top pilots. Two additional initial categories included "problem children" (those pilots who were having a lot of difficulties) and "staff/COD pilots" (those pilots who did not get to land that often on the carrier). We later determined that problem children and staff/COD pilots are treated similarly to new guys, so we classified them under the NG category.

We collected data on a number of different dimensions. In Phase I, it also became evident that the LSOs consistently visualized the overall recovery of aircraft by segments that are related to where the aircraft is in its approach pattern. Flight segment is the first dimension. For the pilot trending and oscillation recognition components for Phase II, the LSOs indicated that the critical areas begin once the aircraft is roughly 1 mile out from the ramp. We also collected data starting at about 3 miles out. The LSO terms for these segments are as follows (see also Fig. 17.5):

- The 180 (Case 1 and 2 approaches) or 3 NM (nautical miles) out (Case 3)
- The 135 (Case 1 and 2 approaches) or 2 NM out (Case 3)

TABLE 17.1
Example of a Decision Requirements Table for the Landing Signal Officer

Case	How Can Deviate?	Indicators	Specific Cues/ Indicators	Discrimination Ability	When Does Deviation Become a "Problem"?	Differences for Pilots/ Pilot Types?
Glide slope/ glide path 1,200 feet descent begins Altitude/speed Lineup Configuration of aircraft						

- The 90 (Case 1 and 2 approaches) or 1 NM (Case 3)
- The start (X), approximately .5 to .75 NM from the ramp
- In the middle (IM): approximately .25 NM from the ramp
- In close (IC): approximately NM from the ramp
- At the ramp (AR): right at the ramp (stern of the ship)
- In or over the wires (IW or OW): the area where the arresting wires cross the landing area; IW implies the aircraft has been trapped, whereas OW implies it missed the wires

The second dimension refers to the recovery case (Case 1, 2, 3). The third dimension comprises the key judgments required of the LSO as described earlier. To simplify matters, we refer to these judgments as *glide slope, attitude and speed, lineup,* and *configuration.* Within each combination of the three dimensions, we collected the following data:

- How can they deviate (e.g., go high, low, left, right, etc.)?
- What are the indicators of these deviations (angle of bank, altitude, etc.)?
- What are the specific cues or indicators to the LSO (how much right wing tip is visible, etc.)?
- At which point can the LSOs discriminate the deviation (e.g., ±30 feet off glide slope)?
- At what value or point does a deviation become a problem (e.g., 5 feet low)?
- Are there differences (when it becomes a problem) for the three pilot types?

For example, we collected data about how an aircraft can deviate from the glide slope (from Dimension 3), under Case 2 conditions (Dimension 2) within flight segment IM (Dimension 1). We refer to the tabulation of information across dimensions as a decision requirements table. Table 17.1 shows a sample decision requirements table that has not been completed.

The decision requirements tables and cognitive task analyses from Phases I and II identified the critical decisions and judgments around which all design recommendations were focused (i.e., decision-centered design).

Transformation of Decision Requirements into Design Concepts. Having identified the critical decisions and judgments that the LSO must perform, and having identified other

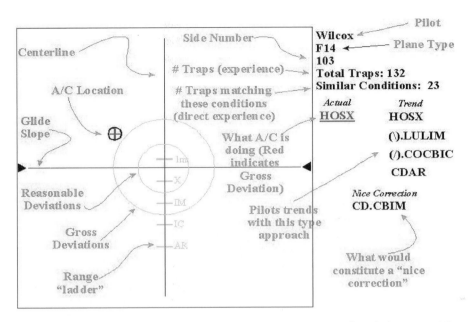

FIG. 17.6. An early storyboard representation of key display elements intended to support the decision requirements derived from cognitive task analysis.

environmental constraints and information that specify those decisions, it was time to develop the display concepts. In Phase I of the work, we had identified the controlling LSO as a key player in the LSO team. His attention, however, had to be focused on the next aircraft in the approach pattern. He could not take his eyes off the aircraft for sufficient time to look down and interpret a "head-down" display. In Phase I, we had proposed a monocle-type display that would present critical information in the controlling LSO's peripheral visual field. For Phase II, the Navy asked us to focus on the supporting tasks of the CAG LSO and the backup LSO. This meant we could provide a head-down display on the LSO platform. The design concepts were therefore generated for that purpose.

Annotated Storyboards. The following storyboards were created with the specific intent of conveying LSO display designs more easily. The concepts presented here are not final versions.

The storyboard display presented here (Fig. 17.6) was initially designed to be a separate CRT accessible for the CAG or backup LSO; it was not developed for the controlling LSO. The goal was to take advantage of the expertise of the CAG or backup LSO, to support this individual's decision making, and to then permit the individual to make the judgment as to whether some information should or should not be passed on to the controlling LSO.

Development of Key Design Recommendations. The display concepts focused on several key decision requirements. The specifics of the implementation of the display concepts came from the decision requirements tables and environmental constraints that provided the specific values that informed the display concepts (such as the glide-slope angle and acceptable deviations or specific ranges that are equivalent to the LSO's categorizations of "distance from the wire"). The key display concepts were as follows:

- Approach range ladder: indicating distance from the trap-wire (in segments specified by the LSOs)

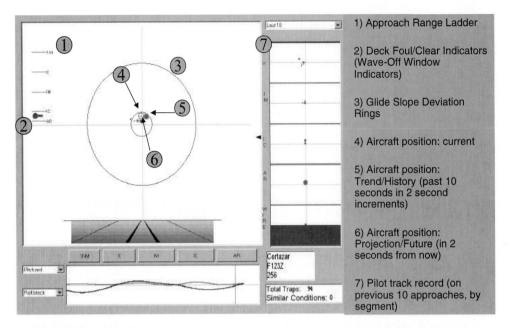

FIG. 17.7.   Landing signal officer platform display panel. The display elements derived from decision requirements are labeled 1 to 7.

- Deck foul–clear indicators: signaling to the LSOs when the deck is clear of obstacles and when the trap-wire mechanisms are set and ready
- Glide-slope deviation rings: indicating the aircraft's position relative to the glide-slope, acceptable deviation, and unacceptable deviation (dependent on aircraft type)
- Aircraft position indicators:
  - Current aircraft position: indicating the aircraft's current position relative to the glide-slope
  - Aircraft trend and history information: a "tail" of dots indicating aircraft position from 2 to 10 seconds previously (decreasing in size and contrast with latency)
  - Aircraft projection and future position information: a projection of the aircraft's position 2 seconds from the present time.
- Pilot track record by approach segment information: indicating the approaching pilot's previous 10 (or more) approaches and where he lined up with respect to the glide-slope at each segment of the approach (X, IM, IC, AR, and In or Over the Wire (IW or OW)).

Figures 17.7 and 17.8 illustrate the final prototype display elements (implemented by SHAI) that were developed to address the decision requirements. Figure 17.7 illustrates the whole display space with a panel indicating the critical display elements. Figure 17.8 is a magnification of those display elements, extracted from the full display.

The display consists of three main display areas. The first is the range ladder (on the left) that illustrates the decision window based on distance of the aircraft from the ramp as well as the state of the deck. As the aircraft gets closer to the deck, the two deck-foul indicators get closer to the critical wave-off window. If the indicators do not turn green before they reach the critical point, then the LSO must wave-off the approaching aircraft.

The second is the central display area indicating the primary aircraft position indicator and glide-slope rings. This display area shows the aircraft's past, present, and future positions relative to the ideal glide-slope for that aircraft. The glide-slope rings indicate the amount of

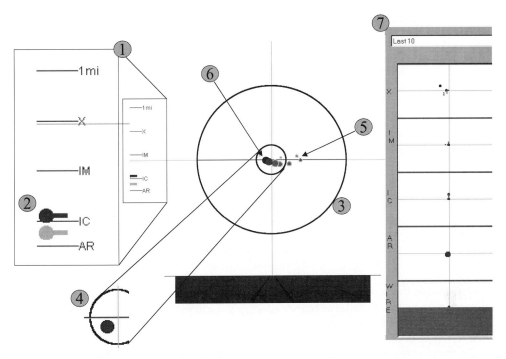

FIG. 17.8. The primary elements magnified from the actual display. 1. The range ladder (distance of the aircraft from the wire), 2. deck clear–foul indicators, 3. glide-slope deviation rings, 4. aircraft current position indicator, 5. aircraft past position indicator, 6. aircraft past position history trend dots, 7. (black circle) projected aircraft position, and 8. pilot trend history (last 10 traps).

acceptable deviation (inner circle) and deviation-requiring corrections (second circle), and the area outside the second ring indicates unacceptable and unrecoverable deviations. The track history dots indicate past trends and are indicative of future trends. The track projection dot indicates the aircraft's future position 2 seconds from the present time and provides the LSO with an indication of whether the current aircraft corrections will bring the aircraft within acceptable glide-slope parameters for a safe landing.

The third display area is the pilot's trend history panel that is segmented by approach segment (as identified by the LSO SMEs). This provides a historical record of where the pilot has lined up at each stage of the approach. The LSO can choose how many data points are provided (e.g., past 10 traps) and under what conditions those traps occurred (past 10 traps at night or in low-visibility conditions). This provides the LSO with information about what that pilot tends to do and whether the current deviations are acceptable based on what he has done in the past.

The general concepts for each display element were driven by a need to support the decision requirements of the task. The decision requirements do not necessarily have a one-to-one mapping to the display elements, however. It may be that several display elements are required to address a single decision requirement or that several decision requirements are supported by a single display element (see Fig. 17.9). It should not be assumed that a simple one-to-one mapping falls out of this process with one decision requirement leading to a single display intervention. Likewise, the specific parameters used to implement each display element require that the data be available (i.e., that a sensor exists to measure distance of the aircraft to the wire, or to measure the roll of the deck, for example) and that the display elements reflect the physical aspects of the situation (i.e., reality).

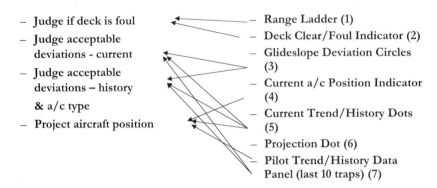

Decision Requirements

Display Elements

- Judge if deck is foul
- Judge acceptable deviations - current
- Judge acceptable deviations – history & a/c type
- Project aircraft position

- Range Ladder (1)
- Deck Clear/Foul Indicator (2)
- Glideslope Deviation Circles (3)
- Current a/c Position Indicator (4)
- Current Trend/History Dots (5)
- Projection Dot (6)
- Pilot Trend/History Data Panel (last 10 traps) (7)

FIG. 17.9.   The mapping between decision requirements and display elements is unlikely to be one to one.

Furthermore, the design of the display elements themselves still requires a more traditional human factors evaluation and adherence to display guidelines and standards. For example, the colors used, the brightness of the screen, the size of the characters, and so forth all needed to be designed to meet the required standards and follow the industry standards for display design. DCD is not intended to be a complete solution to designing interfaces, displays, or systems. It is intended to drive the key design concepts, early in the design process, and reflect a decision-centered approach that supports the key, difficult decisions and how they are made. It is intended to complement more traditional design approaches.

Test and Evaluation Using Decision Requirements.   Testing and evaluation of the interface concept occurred throughout the development process. We consulted primarily with a project consultant or SME (a retired commander) and four LSO instructors at the LSO school, NAS Oceana. We visited the LSO school on four occasions and interviewed more than 20 active LSOs about the interface design, focusing specifically on the pilot trending and oscillation recognition aids. We received feedback during each of the four trips (two separate trials), and used the data to enhance features on the interface concept. The evaluation method was an informal process.

There were two major LSO trials. Each trial was conducted using an informal method of evaluation and feedback. The process entailed interviewing highly experienced LSOs about the design concepts followed by LSOs' reactions to specific features of the design. Each LSO who took part in the trials was instructed to run through the display demonstrations and to subsequently provide feedback.

Although we were most interested in collecting feedback on how the information was presented, if it was useful, and if it provided support for their decision making, we were also interested in other aspects of the display, which included factors such as color, size, and location of features on the display. The trials were not conducted on an actual simulator; however, the LSOs were able to "put themselves in the moment" and comment on potential environmental factors that would affect the use of the interface. They identified factors such as sun glare, proper lighting at night so as not to blind the operator, water on the screen, and effective use of colors so the features would stand out during a quick glance at the screen. In addition to these display issues, the first trials conducted at the LSO school primarily concentrated on information produced by pilot trending. The second trials, conducted at the

LSO school, primarily demonstrated the results of the oscillation recognition research and interface implementation. Following each LSO trial, the software was enhanced based on the LSOs' feedback and discussion.

For a full description of the LSO design project see Stottler and Thordsen (1997) and Thordsen (1998).

Summary. The DCD process began by identifying the key decisions and judgments in the LSO domain. We employed CTA for this purpose. CTA provided us with a set of methodologies for eliciting general as well as specific domain knowledge of the critical decisions made in the LSO operational environment. The CTA allowed us to go beyond the procedural textbook knowledge and behavioral aspects of a task that are traditionally elicited and represented by a behavioral task analysis. The CTA developed our understanding of the cognitive aspects of the operator's behavior. The CTA particularly developed our understanding of the judgments and decision making and problem-solving skills that are so critical in time-pressured, uncertain, and highly dynamic operational environments such as that of the LSO.

CONCLUSIONS

DCD is a subset of human-centered or user-centered design; however, the emphasis is shifted toward understanding and supporting the cognitive aspects of the task—the judgments, assessments, decision making, problem solving, and planning. An explicit treatment of these activities is required as advanced technologies accomplish the easy tasks and leave the tough, complex tasks to the human operator. DCD relies on the use of cognitive task analysis techniques to uncover the key decision requirements and cognitive demands of the task and to provide information about why these are difficult, what information is required to accomplish the task, and in what format, where the potential errors or barriers are, and how to overcome them. The CTA also helps the analysts and design team to discover how experts accomplish these complex cognitive tasks and thus discover different ways of supporting them through systems design, display design, organizational design, or training.

Two examples illustrated how two problems were addressed using the DCD approach. The first supported the complex task of military planning and plan evaluation, the second illustrated a real-time display intended to support the extremely time-pressured and high stakes task of landing multiple aircraft on an aircraft carrier within a fixed time window. The common theme was the use of CTA to identify the key decision requirements followed by the use of those decision requirements to drive key early design concepts. The specifics of DCD are domain- and task-specific, making it difficult to provide a "cookbook recipe" for how to do it. Efforts are still underway to try to formalize the process, or at least to capture it in a way that supports its use by human factors professionals as part of the design team.

ACKNOWLEDGMENT

The authors would like to acknowledge the review and editorial comments of staff of Klein Associates, who provided input to this chapter, especially Terry Stanard, Gary Klein, Mika Uehara, and Barbara Law. Revisions and preparation of this chapter were partially supported by and improved through participation in the Advanced Decision Architectures Collaborative Technology Alliance sponsored by the U.S. Army Research Laboratory under Cooperative Agreement DAAD19-01-2-0009.

REFERENCES

Anastasi, D., Hutton, R., Thordsen, M., Klein, G., & Serfaty, D. (1997). Cognitive function modeling for capturing complexity in system design. *1997 IEEE International Conference on Systems, Man, and Cybernetics, 1,* 221–226.

Anastasi, D., Klinger, D., Chrenka, J., Hutton, R., Miller, D., & Titus, P. (2000). CFM: A software tool to support cognitive function modeling in system design. *Human Performance, Situation Awareness and Automation: User-Centered Design for the New Millennium Conference Proceedings* (CD-ROM; pp. 41–46). Savannah, GA.

Calderwood, R., Crandall, B. W., & Klein, G. A. (1987). *Expert and novice fireground command decisions* (Final Report under contract MDA903-85-C-0327 for the U.S. Army Research Institute, Alexandria, VA). Fairborn, OH: Klein Associates.

Chi, M. T. H., Feltovich, P. J., & Glaser, R. (1981). Categorization and representation of physics problems by experts and novices. *Cognitive Science, 5,* 121–152.

Chrenka, J. E., Hutton, R. J. B., Klinger, D., & Anastasi, D. (2001a). The cognimeter: Focusing cognitive task analysis in the cognitive function model. *Proceedings of the Human Factors and Ergonomics Society 45th Annual Meeting,* 1738–1742.

Chrenka, J., Hutton, R. J. B., Klinger, D. W., Miller, D., Anastasi, D., & Titus, P. (2001b). Cognitive Function Modeling: Capturing cognitive complexity for system designers. (Final Report under contract N00024-99-C-4073) Fairborn, OH: Klein Associates.

Cooke, N. J. (1994). Varieties of knowledge elicitation techniques. *International Journal of Human-Computer Studies, 41,* 801–849.

Crandall, B. (1989). A comparative study of think-aloud and critical decision knowledge elicitation methods. *ACM SIGART, 108,* 144–146.

Crandall, B., & Getchell-Reiter, K. (1993). Critical decision method: A technique for eliciting concrete assessment indicators from the "intuition" of NICU nurses. *Advances in Nursing Sciences, 16*(1), 42–51.

Crandall, B., Klein, G., Militello, L., & Wolf, S. (1994). *Tools for applied cognitive task analysis* (Technical Report prepared for Contract No. N66001-94-C-7008 for the Naval Personnel Research and Development Center, San Diego, CA). Fairborn, OH: Klein Associates.

Dreyfus, H. L. (1972). *What computers can't do: A critique of artificial reason.* New York: Harper & Row.

Dreyfus, H. L., & Dreyfus, S. E. (1986). *Mind over machine: The power of human intuitive expertise in the era of the computer.* New York: Free Press.

Flanagan, J. C. (1954). The critical incident technique. *Psychological Bulletin, 51,* 327–358.

Hoffman, R. R. (Ed.). (1992). *The psychology of expertise: Cognitive research and empirical AI.* New York: Springer-Verlag.

Hoffman, R. R., Crandall, B. W., & Shadbolt, N. R. (1998). Use of the critical decision method to elicit expert knowledge: A case study in cognitive task analysis methodology. *Human Factors, 40,* 254–276.

Hutton, R. J. B., Anastasi, D., Thordsen, M. L., Copeland, R. R., Klein, G., & Serfaty, D. (1997). *Cognitive function model: Providing design engineers with a model of skilled human decision making* (Contract No. N00178-97-C-3019 for the Naval Surface Warfare Center, Dahlgren, VA). Fairborn, OH: Klein Associates.

Hutton, R. J. B., & Klein, G. (1999). Expert decision making. *Systems Engineering, 2*(1), 32–45.

Kaempf, G. L., Thordsen, M. L., & Klein, G. A. (1991). *Application of an expertise-centered taxonomy to training decisions* (Contract MDA903-C-91-0050 for the U.S. Army Research Institute). Fairborn, OH: Klein Associates.

Kaempf, G. L., Wolf, S., Thordsen, M. L., & Klein, G. (1992). *Decision making in the AEGIS combat information center* (Contract N66001-90-C-6023 for the Naval Command, Control and Ocean Surveillance Center). Fairborn, OH: Klein Associates.

Klein, D. E., Klein, H. A., & Klein, G. (2000). Macrocognition: Linking cognitive psychology and cognitive ergonomics. *Proceedings of the 5th International Conference on Human Interactions with Complex Systems* (pp. 173–177). Urbana–Champaign: University of Illinois at Urbana–Champaign, The Beckman Institute; U.S. Army Research Laboratory, Advanced Displays & Interactive Displays, Federated Laboratory Consortium.

Klein, G. (1997a). Nonlinear aspects of problem solving. *Information and System Engineering, 2,* 194–204.

Klein, G. (1997b). The recognition-primed decision (RPD) model: Looking back, looking forward. In C. E. Zsambok & G. Klein (Eds.), *Naturalistic decision making* (pp. 285–292). Mahwah, NJ: Lawrence Erlbaum Associates.

Klein, G. (1998). *Sources of power: How people make decisions.* Cambridge, MA: MIT Press.

Klein, G., Kaempf, G., Wolf, S., Thordsen, M., & Miller, T. E. (1997a). Applying decision requirements to user-centered design. *International Journal of Human-Computer Studies, 46,* 1–15.

Klein, G., McCloskey, M. J., Pliske, R. M., & Schmitt, J. (1997b). Decision skills training. *Proceedings of the Human Factors and Ergonomics Society 41st Annual Meeting,* pp. 182–185. Santa Monica, CA: HFES.

Klein, G., & Miller, T. E. (1999). Distributed planning teams. *International Journal of Cognitive Ergonomics, 3,* 203–222.

Klein, G., Ross, K. G., Moon, B. M., Hoffman, R. R., Klein, D. E., & Hollnagel, E. (in press). Macrocognition. *IEEE Intelligent Systems.*

Klein, G. A. (1989a). Recognition-primed decisions. In W. B. Rouse (Ed.), *Advances in man-machine systems research* (Vol. 5, pp. 47–92). Greenwich, CT: JAI Press.

Klein, G. A. (1989b). *Utility of the critical decision method for eliciting knowledge from expert C debuggers* (Contract JL20-333229 for the AT&T Bell Laboratories, Middletown, NJ). Fairborn, OH: Klein Associates.

Klein, G. A. (1995). *Naturalistic decision making and wildland firefighting.* Proceedings of the U.S. Forest Service Conference. Missoula, Montana.

Klein, G. A., Calderwood, R., & Clinton-Cirocco, A. (1986). Rapid decision making on the fireground. *Proceedings of the Human Factors and Ergonomics Society 30th Annual Meeting, 1,* 576–580.

Klein, G. A., Calderwood, R., & MacGregor, D. (1989). Critical decision method for eliciting knowledge. *IEEE Transactions on Systems, Man, and Cybernetics, 19,* 462–472.

Klein, G. A., & Crandall, B. W. (1995). The role of mental simulation in naturalistic decision making. In P. Hancock, J. Flach, J. Caird, & K. Vicente (Eds.), *Local applications of the ecological approach to human-machine systems* (Vol. 2, pp. 324–358). Mahwah, NJ: Lawrence Erlbaum Associates.

Klein, G. A., & Hoffman, R. (1993). Seeing the invisible: Perceptual/cognitive aspects of expertise. In M. Rabinowitz (Ed.), *Cognitive science foundations of instruction* (pp. 203–226). Mahwah, NJ: Lawrence Erlbaum Associates.

Klein, G. A., Orasanu, J., Calderwood, R., & Zsambok, C. E. (Eds.). (1993). *Decision making in action: Models and methods.* Norwood, NJ: Ablex.

Klinger, D. W., Andriole, S. J., Militello, L. G., Adelman, L., Klein, G., & Gomes, M. E. (1993). *Designing for performance: A cognitive systems engineering approach to modifying an AWACS human-computer interface* (Technical Report AL/CF-TR-1993-0093). Wright-Patterson Air Force Base, OH: Department of the Air Force, Armstrong Laboratory, Air Force Materiel Command.

Klinger, D. W., & Gomes, M. G. (1993). A cognitive systems engineering application for interface design. *Proceedings of the Human Factors and Ergonomics Society 37th Annual Meeting,* 16–20. Santa Monica, CA: HFES.

Klinger, D. W., Phillips, J., & Thordsen, M. (2000). *Handbook for team cognitive task analysis* (Report prepared under Aptima Contract #0009). Fairborn, OH: Klein Associates.

Landauer, T. K. (1995). *The trouble with computers: Usefulness, usability, and productivity.* Cambridge, MA: MIT Press.

Militello, L. G., & Hutton, R. J. B. (1998). Applied Cognitive Task Analysis (ACTA): A practitioner's toolkit for understanding cognitive task demands. *Ergonomics, Special Issue: Task Analysis, 41,* 1618–1641.

Militello, L. G., Hutton, R. J. B., & Chrenka, J. E. (1998). You can't teach what you can't describe: One experience in developing CTA instruction. *Proceedings of the Human Factors and Ergonomics Society 42nd Annual Meeting, 1,* 390–394.

Militello, L., & Lim, L. (1995). Patient assessment skills: Assessing early cues of necrotizing enterocolitis. *Journal of Perinatal & Neonatal Nursing, 9*(2), 42–52.

Miller, T. E. (2001). A cognitive approach to developing tools to support planning. In E. Salas & G. Klein (Eds.), *Linking expertise and naturalistic decision making* (pp. 95–111). Mahwah, NJ: Lawrence Erlbaum Associates.

Miller, T. E., Copeland, R. R., Heaton, J. K., & McCloskey, M. J. (1998). *Robustness criteria and evaluation of air campaign planning* (Contract F30602-95-C-0216 for Rome Laboratory, Griffiss AFB, NY). Fairborn, OH: Klein Associates.

Miller, T. E., & Lim, L. S. (1993). *Using knowledge engineering in the development of an expert system to assist targeteers in assessing battle damage and making weapons decisions for hardened-structure targets* (Contract DACA39-92-C-0050 for the U.S. Army Engineers CEWES-CT, Vicksburg, MS). Fairborn, OH: Klein Associates.

Mitchell, C. M. (1987). GT-MSOCC: A research domain for modeling human-computer interaction and aiding decision making in supervisory control systems. *IEEE Transactions on Systems, Man, and Cybernetics, 17,* 553–570.

Rasmussen, J., Pejterson, A. M., & Goodstein, L. P. (1994). *Cognitive systems engineering.* New York: Wiley.

Schraagen, J. M. C., Chipman, S., & Shalin, V. (Eds.). (2000). *Cognitive task analysis.* Mahwah, NJ: Lawrence Erlbaum Associates.

Shanteau, J. (1985). *Psychological characteristics of expert decision makers* (Vol. 85). Manhattan: Kansas State University.

Stottler, R., & Thordsen, M. L. (1997). *Artificial intelligence techniques for the Pilot Approach Decision Aid Logic (PADAL) System* (Contract No. 68335-96-C-0205 for Naval Air Warfare Center, Aircraft Division, Lakehurst, NJ). San Mateo, CA: Stottler Henke Associates.

Thordsen, M. (1998). Display design for navy landing signal officers: Supporting decision making under extreme time pressure. In E. Hoadley & I. Benbasat (Eds.), *Proceedings of the Fourth Americas Conference on Information Systems* (pp. 255–256). Madison, WI: Omnipress.

Thordsen, M., Hutton, R., & Anastasi, D. (1998). The cognitive function model. *Proceedings of the Human Factors and Ergonomics Society 42nd Annual Meeting, 1,* 385–389.

Thordsen, M. L., Klein, G. A., & Wolf, S. (1990). *Observing team coordination within Army rotary-wing aircraft crews* (Contract MDA903-87-C-0523 for the U.S. Army Research Institute, Aviation Research and Development Activity, Ft. Rucker, AL). Fairborn, OH: Klein Associates.

Vicente, K. J. (1999). *Cognitive work analysis: Towards safe, productive, and healthy computer-based work.* Mahwah, NJ: Lawrence Erlbaum Associates.

Woods, D. D., Johannesen, L. J., Cook, R. I., & Sarter, N. (1994). *Behind human error: Cognitive systems, computers and hindsight* (CSERIAC SOAR 94-01). Wright-Patterson Air Force Base, OH: Crew Systems Ergonomics Information Analysis System.

Zsambok, C. E., Beach, L. R., & Klein, G. (1992). *A literature review of analytical and naturalistic decision making* (Contract N66001-90-C-6023 for the Naval Command, Control and Ocean Surveillance Center, San Diego, CA). Fairborn, OH: Klein Associates.

18

A Cognitive Framework for Operation of Advanced Aerospace Technologies

Kevin M. Corker
San Jose State University, USA

Abstract

The international aviation community is advocating goals for the first quarter of the new millennium that compel a radical revamping of the practice of aircraft and air traffic management. This vision finds expression in both the European and U.S. commissions and offices as well as in corporate statements of intent to invest in a requirements analysis of an international air traffic management, coordinated through satellite-based information-exchange processes and reduced constraint of aircraft movement. Other envisioned advances include increased use of airspace to support space access with single-stage-to-launch vehicles, as well as space access ports distributed throughout the United States and around the world. These visions assert new modes of operation and technological requirements. Essentially without exception, these technologies fundamentally change the processes of air traffic and airspace management. The advances include redistribution of information and control among human operators and automation systems, altering decision and execution modes and optimization processes. The changes in the work of air transportation operations require an approach to analysis sensitive to the change in the cognitive processes supporting the work in context.

INTRODUCTION

We are in the early stages of creating a new process of aerospace use; however, the initial framework of the proposed changes such as trajectory-controlled airspace and increased autonomy will establish the pattern for the subsequent sociotechnical implementations. It is critical that a cognitive task perspective be brought to these developments. Earlier work (Sarter & Woods, 1995) recognized the impact of active automation on the flight deck; advances in air traffic control automation are bringing the same message to technological development for air traffic management (Bresolle, Benhacene, Boudes, & Parise, 2000). Billings (1996) and Wickens,

417

Mavor, Parasuraman, & Mcgee (1998) pointed out the dangers in tightly coupling automation and human systems in air–ground integration without consideration of the many issues of stability, information currency, and control in multiagent management.

In addition to the development and insertion of automation for improved air transport operation, both the United States and Europe are engaged in exploration of revolutionary airspace management concepts that systematically reduce constraints in air traffic and flight path management (Busquin, 2001; National Aeronautics and Space Administration (NASA) Aerospace Technology Mission Vision, 2001; Federal Aviation Administration (FAA) National Airspace (NAS) Modernization Plan, 2000). Distributed control and responsibility for flight operations is being explored in relation to shared separation authority, shared flight path determination, collaborative decision making in establishing schedules, and cooperative management of various disturbances (weather, traffic restrictions, etc.). In support of these changes, technologies are being developed to relax system constraints and integrate flight deck and ground-based operation.

These affect human performance research in two ways. First, the decision-making process becomes distributed. This distributed decision process differs from current operations in the number of participants, in the timing of their communication, in their situational awareness, and in the span of their authority, which has direct impact on crew and team resource-management processes. Second, the dynamic concept of operations provides new coordination challenges for human operators of a given system. In current operations, the roles of the constituents in airspace operations are relatively fixed; implementation of technologies to share information and authority in air traffic management (ATM) relaxes that stability. The human operators (pilots, air traffic controllers, and airline operations personnel) must monitor and predict any change in the distribution of authority or control that might result as a function of airspace configuration, aircraft state or equipage, and other operational constraints. The operators make decisions and share information not only about the management of the airspace, but also about the operating state of that airspace and about the operating behaviors of the automation that is in place to aid them. In so doing, they are moving from direct participatory to supervisory roles and from supervisory roles to automated system monitoring and control by exception effectors. They are supported in these roles by automation and are, by design, given information that is intended as advisory or contributory to "situation awareness." In these situations the human-system monitoring of response to an air traffic command may need to consider the information state of the participants in the system, the behavior of the automation that is aiding those participants, the current seat of authority in implementing the control command, and the discretionary behaviors permitted in the current system operating conditions Klein and Anton (1999). Finally, the notion of anomaly in these interrelated but independent systems is made complex. What is an anomaly in a system of distributed discretion, in which pilot action (although perhaps different from a controller's intention or preference) is allowed under relaxed self-separation operations? Or, to take the another view, what is an actionable alert when layers of automated information system provide (to all participants) levels of alert and warning for potential future anomalies or conflicts that are subject to change through the actions of any of several agents? These are arenas in which little is known about cognitive bottlenecks, human–system interaction, and system dynamics and stability. Nevertheless, these are the operating modes of the future air traffic system. Such complexity requires a new process of definition that includes cognitive task design to ensure that the system's behavior exploits but does not stress or exceed the cognitive and information capabilities of any of the agents in the joint cognitive system.

Herein I undertake the description of an initial application of cognitive task analysis for those systems that have proceeded through prototype and suggest cognitive task design for those operational concepts that are still emerging. I also provide and identification of relevant studies performed in response to projected developments in aerospace technologies. This chapter attempts to provide the outlines of the sociotechnical system of air transportation operations and to identify the human–system performance issues brought on by the system's goals and performance requirements from the perspective of cognitive task analysis of a representative set of technologies.

Finally, I address the development of computational models to support cognitive and contextually sensitive analyses. As I have indicated, the scope of work in the air transportation system is broad, encompassing all pilotage, air traffic control, systems (ground and air) maintenance, corporate policy, and regulatory and security policy and practice. Full analyses of all these in appropriate detail is not possible in this chapter; thus, I concentrate on examples of interactions among flight crews, air traffic management service providers, and airline operations undertaken by commercial airlines and air traffic management service providers. I also suggest how and in what capacity cognitive task design may provide an alternative approach to the system function.

OVERVIEW OF AN AIR TRAFFIC MANAGEMENT WORK SYSTEM

ATM can be defined as the control of aircraft and airspace system assets to ensure safe, efficient, and flexible movement of aircraft and passengers through international airspace. The process of ATM is fundamentally a coordination of vehicle and airspace control. The process is mediated by three primary instrumentalities: (a) the air traffic service provider (air traffic control), (b) the airlines and air transportation service operations, and (c) commercial flight crews—general aviation and military pilots. In addition to these primary constituents, there are a large number of ancillary service organizations that provide maintenance, communications, weather information, scheduling, training, and infrastructure to support ATM. National organizations of regulation and certification impose and police operational requirements (in all aspects of flight, maintenance, airport and system security). In current international operations, the International Civil Aviation Organization (consisting of state delegates) assigns responsibility for provision of air traffic services within flight information regions and establishes standards and operating practices for operation in the international airspace (largely those operations taking place over oceanic environments).

It is a dynamic and distributed control system. Operation is managed in time frames that extend from seconds to months. The system operates globally in a dynamic range from surface movements at tens of miles per hour to en route operations with a range to 60,000 feet altitude and Mach 1.8 velocities. As I discuss, the process of air transportation of goods and people operates as a closely coupled and complex time-varying system. Distributed operators and organizations with varying spans of control authority must cooperate to ensure effective and efficient provision of services. The final common path of these global operations is the human operator who specializes in work in the various elements of air transportation.

HUMAN PERFORMANCE CONSIDERATIONS

The human performance issues associated with so complex a system are similarly complex and diverse. Human performance research is traditionally performed in the following broad categories:

- Selection and training of personnel
- Design and development of controls and displays for human use

- Development of automation and optimization decision support tools for the operators
- Analysis of human performance contribution to safety and human–system error propagation
- Human response to psychophysiological stressors, situation awareness, vigilance, crew and team coordination
- System dynamics and response to perturbation and stability

The research paradigms brought to bear to address the issues are similarly broad, extending from computational analyses and simulation through empirical studies in laboratory and simulation to field studies and large-scale institutional demographic and ethnographic studies. These analyses tend to be focused on the impact of local process changes or singular technology insertions.

Nonetheless these sociotechnical systems are tightly linked. Changes brought on by simple modifications of communications practice have been observed to propagate goal-based effects across the entire human-system (Corker, Fleming, & Lane, 2001). A global cognitive task view is needed both to understand the implications of the tasking required and to support cognitive task design to provide system that is sensitive to the cognitive requirements of the operators of the system. As a research community, we have sufficient experience to express the nature of changes in practice with classes of "automation aids," and cognitive task analyses and design provides a framework to understand the linked modification of the artifacts of control and the sustained change to the practice of human–system interaction. It is my assertion that linked air–ground company operations will only be stable and meet the goals of safety and efficiency if a common expression of the changes brought on by advanced automation and linked operations is undertaken. In this view, the air transport system represents a joint cognitive system in which human operators (separated in time and space) coordinate with, in, and through automation that has, in its own right, cognitive capacities, goals, and constraints.

AIR–GROUND COORDINATION: CURRENT EXAMPLES

A good way to get a sense of the effect of interdependencies, linkages, side effects, and technological constraints of a complex system (sometimes illustrated as the ripple effects of a technological pebble—or the occasional boulder—dropped into a calm pond; Hancock, 1997; Woods, 2001) is to look at situations in which expectations of performance clash among multiple practitioners in the system. As an illustration, consider what seems to be a simple request-and-response processes.

A pilot of a commercial airliner requests a change of altitude from air traffic control for reasons of efficiency or comfort of ride (in the case of light-to-moderate turbulence at the flight level the aircraft currently occupies). The air traffic control response is that the request will be deferred and cannot be granted at this time. Why?

The answer lies in understanding, at the level of cognitive task analysis and design, what the process, roles, and responsibilities are for each of the participants in this simple exchange. The FAA currently oversees the air traffic management process in the national airspace, a process designed to ensure the safe and efficient movement of aircraft, integrating air traffic and flow control. Air traffic control is concerned with maintaining aircraft separation and order in the national airspace. Traffic flow aims to optimize the flow of traffic, especially in areas of high traffic density, when demand for airspace usage as well as for landing and takeoff slots may be in excess of capacity. Air traffic control is currently implemented from 476 air traffic control towers, 194 terminal radar approach controls, and 21 en route air route traffic control centers

FIG. 18.1. An example of the physical workspace in an air traffic control center. The controllers work singly or in pairs, with a set of controllers commonly working together in crews. The workforce is stable, and crews and teams form expectations and a calibration of their own and others' competencies. These assessments form the basis of decisions on communications, interruptions, and coordination among teams and between individuals.

segmented in response to some analysis of usage; I return to this point later in the chapter. Traffic flow is managed from a single facility, the Air Traffic Control System Command Center. The relationship between the management of low and the control of traffic locally is one of the significant evolutions in the national airspace.

The air traffic control specialist controller (termed "certified professional controller" in the United States) has the responsibility of providing safe, efficient and orderly flow of traffic through the area of their responsibility. Controller teams consisting of a "radar" R-side controller and a "data" D-side controller are used in sectors where the flow rate, complexity, or simple density of traffic require two human operators for their management. Controller task loading can, for some periods of time, get so high that three controllers may be assigned to airspace to handle "rushes." The additional "set of eyes" is intended to support the primary task of separation assurance, but a three-person team imposes communication and coordination burdens. New methods for dealing with such temporary load developments are being considered and are discussed later in Limited Dynamic Resectorization. Now, the flight crew response for an altitude change may occur such that the dynamics of climb would have the aircraft in another adjacent airspace at the completion of the maneuver. Coordination between controllers is required to respond to the aircrew's request. This coordination may require simple verbal communication, electronic communication (data messaging), or telephonic coordination depending on the physical layout of the center's radar display systems (see Fig. 18.1).

FIG. 18.2.   A typical flight management computer interface. Route changes are entered into the system through a series of function select and alphanumerics. Feedback is minimal for incorrect entries.

The controller coordination may be successful or not, depending on any number of variables, such as workload, work break rotations, or traffic flow impositions and restrictions imposed by a national traffic flow command center. Thus, a seemingly simple request is, by timing and chance, embedded in local and national structures, the state of which at any time is essentially unknown to the requestor.

The flight crew that made the requests, of course, understands the complexity of the system, understands the notion of airspace boundaries, and has, from experience, developed reasonable physical and dynamic models of the airspace through which they are flying and into which they have inserted their request. The flight crew will at times, however, find the deferment of their request annoying and will on occasion infer at least a lack of cooperation in that response. Consequently, what seems simple from one perspective is complex from another.

To provide an example, an air traffic controller may request a simple route offset by an aircrew (to move the aircraft some number of miles to the right or left of current route and then resume prior heading). This seemingly simple request on the part of the controller requires significant activity on the flight deck of most current generation aircraft because the aircrew needs to reprogram their flight management computer systems. The simple request is made even more difficult if the air traffic controller also adds a time restriction at some point in space for the aircraft. Why?

Flight management computer systems were developed with concern for accuracy and efficiency in operations in a structured airspace and on established routes (either established by the air traffic management authority or by the airline operations facilities). Modification of that route structure requires, in most current-generation flight management systems, extensive interaction with the system through specialized keyboard (see Fig. 18.2). Modification of route has some potential to produce unintended side effects (Sarter & Woods, 1995)—and in certain circumstances even disastrous consequences (FAA, 1996). Controllers also have some notion of the requirements of flight management systems; however, the impact of the interaction on crew operations during certain phases of flight may not be fully appreciated, leading to a mind-set that wonders "what's the big deal" with compliance to this request.

These examples provide a glimpse of the complex system that is presently being lumbered with significant changes. The following sections describe the evolution of ATM, identify the human–system performance issues associated with current and future operations, and discuss the methodological implications of these issues.

SOCIOTECHNICAL AFFECTS ON ADVANCED AIR TRANSPORTATION MANAGEMENT

The evolution of processes of air transportation is an interaction of social demand, economic requirement, technological development, human performance, and valuation of the resultant system response. Figure 18.3 illustrates this cycle.

Research on human performance within air traffic management has tended to focus on the following factors: training and selection of air traffic controllers and pilots, definition and development of technology for presenting information (e.g., traffic, weather, systems status, ground proximity) to human operators in the system and on provision for control by the human operator, and technology for transmission of control input to appropriate parts of the system. However, the drive toward efficiency and flexibility has provided technologies to make the process of ATM distributed and decentralized. In response, the foci in human performance research must shift to include consideration of collaborative decision making in response to schedule and planning for aviation operations, automation aiding for airspace flow and facility optimization, distributed control, and coordination of roles and responsibilities in dynamic systems as the system explores the concept of exchange between air and ground of authority for separation from other aircraft. The issues raised in response to these demands are primarily focused on the cognitive components of task performance and the response of the practitioner to the new performance requirements. A summary of these topics that are integral to advanced system operation is provided here.

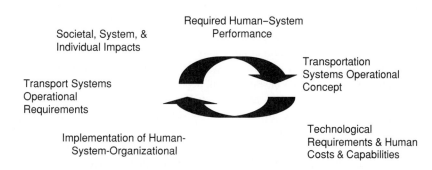

Societal, System, & Individual Impacts

Required Human–System Performance

Transport Systems Operational Requirements

Transportation Systems Operational Concept

Implementation of Human-System-Organizational

Technological Requirements & Human Costs & Capabilities

FIG. 18.3.   A cycle of requirement and implementation in the air transportation system.

Schedule and Planning

Airline operations schedules are established months in advance of flight operations. The schedules are based on market demographics and on the particular operations being scheduled (e.g., optimal cargo airline schedules vs. those of passenger and revenue operations). In addition to market demands, human performance issues associated with fatigue, vigilance, and duty cycle play a significant role in mediating operation duration and frequency and in determining crew complement. The performance impacts of circadian desynchronization are the focus of a significant body of research in aviation and air traffic management human factors (Aviation Space and Environmental Medicine, 1998). The issues of duty time and cycle are being extended from flight crew to maintenance and air traffic controller schedules. There are significant implications for human performance in scheduling for evolving NAS operations, including the possibility of ultra-long-range operations. Flexibility in operations, schedule, and routing may impose variations in the flight crew's and maintenance operator's rest and duty cycle. It is well documented that these variations in circadian rhythm lead to fatigue-based errors in performance. Although these issues in human performance are critical to the safe operation of the airspace, they are not generally within the purview of the technological mediation, and I do not deal with them extensively here.

There is significant effort in human–technical systems approaches to improving the adaptive and real-time response of the air space system to traffic requirements. A radically different concept for encouraging information sharing and cooperation among competing commercial entities, termed collaborative decision making, offers important insight into the role of increased transparency of operations and shared information among airspace users. I review this approach and its impact on airspace operators in the next section.

Air Traffic Control Flow and Facility Optimization

The national and international airspace and the airports that serve commercial and general aviation operations are a constrained resource. Human operators are called on to manage these scarce assets to provide predictable and timely performance while maintaining a margin of safety to ensure system robustness in response to environmental or other sources of operational disturbance. Within the U.S. national airspace, 50,000 flights operate daily. The issues of large-scale optimization and decision making come quickly to the fore. Airspace structure in both the assignment of aircraft to routes and the fixed structure of the sector airspace must be reconsidered. Automation and information aids are likely necessary to ensure that all factors are accounted for in the decision-making process. Automated checks for information integrity and currency are critical to the air traffic service providers and to the operators. Appropriate resolution of human performance issues associated with human–automation integration and distributed decision making are fundamental to effective operations. The argument has been made that the airspace will be better managed and provide more capacity when human operators in that airspace are supported through advanced automation aiding (Billings, 1997; Wickens et al., 1998). I propose that such joint cognitive systems require cognitive task analysis and empirical contextually embedded research to support appropriate cognitive task design and provide examples of such research to support that design process.

Specific human performance issues are motivated by the kind of service provided and the types of tools provided to the pilots, airline operations centers, and controllers to manage those services. These services are provided by the delivery of clearance and advisory information from the ground to the aircraft and by the provision of operational state information from the aircraft to the ground service providers. Human–automation integration, training for operation, interface development, procedure development, and coordination in strategic decision making and communication processes are the key areas of concern for cognitive task design in air traffic control.

Cognitive task design is a process that supports the development of appropriate definitions of the information, procedures, and contextual support for roles and responsibilities of the final common path for ATM. The process of flight management seeks to ensure safe timely and efficient operation of aircraft. The human–system interaction issues attendant to the evolution of flight control and flight management as the integration of human and machine is well documented in Billings (1997). The management of flight has evolved from simple manual control with augmentation to almost fully automated operation. The development and evolution of flight deck automation has compelled significant development in the theory and practice of joint cognitive systems. This fundamental research in human automation integration in flight deck operations serves as an excellent precursor for similar automation integration issues in ground-based air traffic control and as a platform to approach issues of distributed air ground operation.

ADVANCED ATM COGNITIVE TASK ANALYSES AND DESIGN: REQUIREMENTS AND APPLICATIONS

The complexity of current air traffic management challenges analyses of human–system interaction. The potentially revolutionary impact of operational changes in advanced air traffic management requires fundamental and extensive cognitive task analyses to ensure operational integration, management of automation, assurance of appropriate situation awareness, safety assurance, and interface optimization (FAA, 1996; Wickens, Mavor, & McGee, 1997; Wickens et al., 1998). In addition to these human–technology interactions, the context of operations for these technologies is international in scope. Language and culture will have a significant impact on the integration and operation of technologies for worldwide use. Research in the interaction of automation and culture is in its infancy in terms of theory and method but is a critical component to developing and assessing technical systems with worldwide societal impact.

What defines the specific requirement to cognitive task analysis in this vast arena of system development? The basis for our assessment is guided by the definition and analyses of systems provided by Charles Perrow (1984). The human-system interaction functions in such a way as to have emergent behavior that is not predictable from a simple analysis of the system components. Joint cognitive issues requiring cognitive task analysis in ATM systems are a function of the following components of the joint cognitive system being created or changed:

- Information required to perform the task
- Procedures required to perform the task
- Roles and responsibilities in task performance assigned to the operators or the automation
- The impact of a component's off-nominal operation on the system (loosely or tightly coupled)
- The influence of organizational factors private and public on task performance

Safety Metrics and Human–System Performance

The assessment of the impact of these created or changed relationships among the human operators and the aviation system is focused on efficiency, security, and safety of operation.

Efficiency is defined through the process of organizational and societal calibration of the requirement for air transportation and its availability, predictability, and timeliness. The achievement of efficient and effective transportation has been hampered in the United States (according to government and commercial studies, cf. RTCA, 1995) by constraints in schedule and airspace usage. Relaxation of these constraints is postulated to improve efficiency and decrease costs on the order of $1.5 billion annually. The movement to reduce constraints and increase demand for a scarce resource (the human–system intelligence to manage the NAS) may have an impact on the other two metrics commonly applied to operations.

Safety and security as metrics of airspace operation are not so clearly or cooperatively defined. Safety of operation is held ultimately to be the responsibility of the human operators in the system, and so it is not unusual to find it reported that human error accounts for 80% of commercial aircraft accidents (Boeing Commercial Airplane Group, 1995). The process of assignment of human error as a cause for an aircraft incident or accident is a process of heated debate. Unfortunately, as noted by Hollnagel and Amalberti (2001), this fundamental debate with reference to erroneous human performance is often couched in a vocabulary supporting different connotations and denotations for the same words. Therefore, the approach to human error, as a cause of an accident (or as a measure of system effectiveness in its reduction), that I use needs some elaboration before continuing with this analysis.

First, there is the issue of defining causes for error. As noted in Hollnagel and Amalberti (2001), there are several dimensions to the attribution of cause. Generally, it has been standard practice to assume that attribution is supported by the *modus ponens* logical form that B implies A. It is discovered or asserted that B is true, and it is therefore concluded that A is true. Also considered appropriate is attribution supported by the *modus tollens* form in which it is asserted that A implies B. It is asserted or discovered that B is false, and it is therefore inferred that A is false. In the analysis of accidents and incidents, the discovery or assertion that a human performed or failed to perform an action does not support the logical requirements for attribution; the human acts in the system as a motivated, active agent attempting (one assumes for trained professional operators) to control and adapt the joint cognitive system of which they are a part according to circumstance. Also, the process of analysis tends to disembody the actors and the proximal action from the full context of behavior in this specific incident and more critically from the process of behavior across the full range of performance of those agents in that system across the many multiples of times the operators have performed the action without consequence. This isolated and disconnected analysis of error is inappropriate in an approach that considers the evolving performance to be the playing out of design decisions, rules or thumb, past experience, and the interplay of natural and designed goals and circumstances. Such integrated and contextually sensitive analysis of incidents (cf. Dekker, 2002) is a much sounder basis for appropriate analysis, and the results from same provide the "lessons learned" in a form appropriate for cognitive task design, in which one does not design to the last accident but abstracts the principles of joint cognitive systems that allow design to the next (and the $n + 1$) incident.

From available data, one may discern the fact that a human performed a given action but his or her intention when that action was performed usually cannot be derived in this way. (It has been suggested by Dekker, 2001, that accident investigation would be better focused on the "why" of an action under given circumstances than in the judgment of its appropriateness or adequacy from the perspective of post hoc analysis.) In the attribution of cause, there is the further issue of the anisotropic nature of examining events in time. The retrospective view yields a serial and implied causal structure that the nature of the aviation operation makes unlikely at the time of occurrence. As action unfolds, complexity makes the system difficult to understand from a single perspective, and tight coupling spreads the problems in unpredictable ways once they begin (referred to by Tenner, 1996, as the "revenge factor" of technologies).

The issue of intention in aviation accidents is a matter of some fairly explicit legal definition because culpability and subsequent legal responsibility are significant issues in aviation operation. The relevant standard is embodied in Federal Aviation Regulation 91.13, which reads as follows: (a) Aircraft operations for the purpose of air navigation. No person may operator an aircraft in a careless or reckless manner so as to endanger the life or property of another.

In this definition, legal opinion has held that carelessness can be considered "simply the most basic form of human error or omission ... simply *ordinary negligence* a human mistake" (National Transportation Safety Board, 1986; emphasis added).

This view that humans are commonly the source of ordinary negligence has dominated reviews and analyses of human error in system operation. Dekker (2001) dubbed it "the bad apple theory." This view, then, asserts that the human operator in a system is a source of error, and as such the system design should work to exclude the opportunity for error and minimize its consequence when it is inevitably performed.

There is a view of "human error," different in its perspective, that I and others (cf. Dekker, 2002; Dekker & Woods, 1999; Hollnagel, 1993; Kirlick, 1998; Snook, 2000) consider more valid and better suited to the purposes of prudent design of joint cognitive systems. In this view, the human operator (often included in systems to account for unanticipated and unpredicted contexts in operation) acts to maintain system safety, to adapt to new requirements, to discern signal from noise, and to balance societal valuation against system performance requirements. In my opinion, the human acts to keep the system safe. An accident or incident, then (to state it strongly), is possibly a failure in design to provide sufficient resource and control to allow the human to perform that job successfully.

I continue this analysis of the joint cognitive systems from this perspective and evaluate the human operators interaction with the systems in terms of this safety role.

Cognitive Impact of Schedule and Planning Technologies

The purpose of the operation is to allow airspace users to plan their flight operations in anticipation of NAS capacity and traffic conditions and to minimize congestion or possible delays. This capability is supported by comprehensive information made available by the NAS-wide information system. The introduction of the Collaborative Decision Making in Traffic Flow Management was initiated in 1995 to make the FAA Ground Delay Program more responsive to individual airlines' needs and schedules. The notion of collaborative decision making in a broader scope for optimization of routing and schedule in operations other than ground delay program is a more recent development and is integrated as part of the Free Flight Phase 1 program (Wambsganns, 1997). The system supporting the more elaborated collaborative decision making will include up-to-date information such as capacity and aggregate demand at airports and other NAS resources, airport field conditions, traffic management initiatives in effect, and Special Use Airspace status. Collaboration is used to negotiate a revised flight trajectory in real-time.

The human performance issues associated with collaborative decision making are very much a part of the common set of issues dealing with an expanded reliance on distributed and dynamic information to link multiple human and automatic agents in coordinated control. As such, they share a common form with issues associated with self-separation processes, dynamic resectorization processes, and other negotiated or collaborated exchanges as outlined in the distributed air–ground traffic management concept elements. There is a large literature on human factors research associated with collaboration and team activity and distributed decision making. The application of these team and teamwork issues as applied to the process of air transportation, dubbed "Crew (or Cockpit) Resource Management," has also had a long and prolific research history. A review can be found in Wiener, Kanki, and Helmreich (1993). Application of these techniques to controller teams has been much less active, with some exceptions (cf. the Air Safety System Enhancement Team report [Jones 1993] or analyses by Sherman [1992], or Bailey [1997], or Baker [1997]). Research into team and collaborative decision making specific to air transport operation is even rarer (Cambell, Cooper, Greenbaum, & Wojcik, 2000; Klein & Anton, 1999), although research has been performed in the analog environment of military decision making in large-scale battle management (Pew & Mavor, 1998).

Based on sensitivity to cognitive task requirements, one can say that the joint cognitive system aspects of collaborative decision making are significant. These issues fall into categories associated with how to maintain awareness of the situation of the NAS and how to keep the information of independent and collaborative decision making in synchrony. In collaboration, there is also a requirement to keep collaborators aware of the "fairness" and gain functions to support continued collaboration. Finally, there are issues associated with the development of information displays and manipulation methods to enable traffic flow managers to formulate traffic flow management solutions.

The process of implementation of collaborative decision making is well underway; data collection and analysis of its effects have just begun. One aspect of interest is the process of performance modification under reduced constraint. This has been studied in the implementation of a program to reduce constraints on aircraft operation above certain altitudes, the national route program. At this writing, the implementation of this program has had 5 years of operational experience on which to draw in the exploration of the impact of the relaxation of constraints. In this process, it has been noted that the shift in operating mode has engendered requirement for new types of information and, interestingly, has supported the development of new methods of competition among airspace users (Smith et al., 1997).

COGNITIVE IMPACT OF AIR TRAFFIC CONTROL FLOW AND FACILITY OPTIMIZATION TECHNOLOGIES

Several new airspace operations converge in the push to make use of all available capacity in the NAS. Flow and facility optimization technologies have application in the en route airspace, where controller attention load has been implicated in capacity constraints and in terminal operations in which both airspace and ground space are constrained. Technologies are being developed to relax system constraints and integrate flight deck and ground-based operation. These have lead to consideration of new modes of operation and new sources of information to support those modes. Figure 18.4 illustrates an example of tools provided to air traffic to assist in optimization of en route and arrival flows. Figure 18.5 illustrates similar information presented on the flight deck.

These new operating modes affect human performance in two ways. First, the decision-making process becomes distributed, differing from current operations in the number of participants, in the timing of their communication, in their situational awareness, and in the span of their authority. These factors have direct impact on crew and team resource management processes. Second, the dynamic concept of operations provides a new coordination challenge for the human operators of that system. In current operations, the roles of the constituents in airspace operations are relatively fixed; implementation of technologies to share information and authority in ATM relax that stability. The human operators (pilots, air traffic controllers, and airline operations personnel) must monitor and predict any change in the distribution of authority or control that might result as a function of the airspace configuration, aircraft state or equipage, and other operational constraints. The operators are making decisions and sharing information not only about the management of the airspace, but also about the operating state of that airspace and their roles given that operating state. This brings to the fore one of the

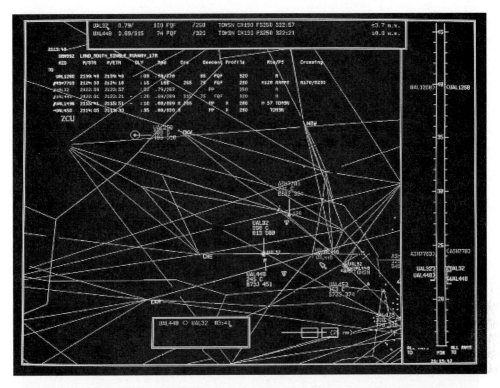

FIG. 18.4. Controller radar display with data link interface (lower box), arrival scheduling (left), and conflict list (upper table), as well as routes, route projections, aircraft data block sector boundaries and conflicts alerts.

most significant issues for cognitive task analysis in the advanced ATM environment: How are the roles and responsibilities in the dynamic system to be assigned, maintained, adapted, and shifted in response to system state? The questions that are raised include the following:

- Who needs to know what, when, and under what operating conditions?
- What are the informational issues associated with cooperative problems solving and distributed team decision making?
- What is the relationship between information and responsibility?
- How do tools (available or new) feed into the information and responsibility issues?
- What are the perceptions (misperceptions) of responsibility; how are they supported?
- How is workload (tasks) distributed, over people and over time?
- Are pilots and controllers cognitively and personally predisposed for their new way of working? Has the selection criteria for past systems provided the right skill mix for new systems?
- Where are the transitions among system operating states, and how are handoffs normally accomplished?
- Are there off-nominal handoff situations?
- What are the issues associated with cooperation (of lack of) among ATM constituents?
- Is there a general perception of fairness? If not, what behaviors ensue?

These issues are shaped by the development of decision-aiding systems and the use to which practitioners put these aids. The decision-aiding systems are focused by the constraints of the target airspace operations, and so we address "terminal" and "en route" operations. The impact

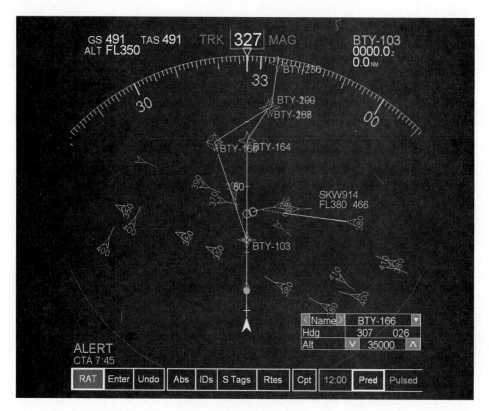

FIG. 18.5.   A cockpit display of traffic information illustrating route and trail routes, potential conflicts, and time to conflict, as well as aircraft trajectory projections.

of selection of certain set points or operational concepts can be approached in support of cognitive task design through a "cognitive walkthrough" technique. Several teams of researchers in the area of structure and analysis of a set of scenarios, based on the operational concept under study, use this approach. The system is then systematically analyzed through several levels of abstraction with specific reference to the information pathways, decision requirements, authority, and responsibility interactions among the cognitive agents in the system. This systematic analysis is conducted among designers, regulators, and practitioners for foreseeable normal and off-nominal operations of the system. The process supports cognitive task design decisions early in the development of system concepts and prototypes.

Automation of flight management operations has had a 20-year development history, and the human factors issues associated with increasingly sophisticated multimodal control operation has received extensive research (Billings, 1997; Wiener, 1989). The issues of automation bias and overreliance, lack of adequate feedback as to operating mode, workload, vigilance decrements, and primary–secondary task inversion need to be investigated in the development of ground-based automation for air traffic control. These issues become even more critical as the link in control tightens between the ground and the air. In related research, alerts and alerting in nominal and off-nominal operations are a significant area of interest and research.

Terminal Operations Decision Aiding

Terminal airspace in particular can exhibit complex operations and structure, with a variety of users and equipage. (Green & Weider, 2000; Leiden & Green, 2000). Tools are being developed to provide optimization support for controllers in the terminal area. The Center TRACON Automation System (CTAS), a prototype system under development for the automated management of arrival traffic, has had several human–system performance investigations (Lee and Davis, 1996; Lee & Sanford, 1998). The three major functional components of CTAS in development are as follows:

- Traffic Management Advisor, which provides the Traffic Management Coordinator with an interactive tool to control aircraft to a specific runway configuration. Traffic Management Advisor introduces the concept of "time-based" scheduling of arrivals.
- The Passive Final Approach Spacing Tool provides the approach controller with aircraft advisories designed to provide optimum sequencing and minimum spacing of final approach. The impact of these functions includes reduction in the length of arrival routes, allowing user-preferred and direct routes to extend closer to the terminal airspace. Another favorable impact would be to sector workload through feedback to the controller on the aircraft's meter time, allowing the controller to manually resequence the meter list to reflect the actual aircraft sequence.
- Finally, "active" advisory functions for Final Approach Spacing Tool are currently under early human factors analysis (Quinn & Robinson, 2000).

These analyses have concentrated on the impact of CTAS on controller operations. Callantine, Prevot, and Palmer (2001) have also made efforts to discern the impact of the controller tools on the flight deck. This kind of systemwide investigation of the impact of individual tools is in keeping with requirements for a thorough cognitive task analysis.

From a cognitive task-analytic perspective, there are several characteristic signatures associated with these terminal decision-aiding systems. First, the system optimization is, by design, less variable and timelier than the controller's performance (although it may be based on heuristics and methods employed by human operators). As such, the standard recommendation of human factors to provide the human the task of accepting or rejecting the option is suspect. On what basis does the flight crew or controller reject the assigned sequence of schedule? This brings up issues of reliance on automation (and overreliance) and trust in automation. These issues have received attention in the human factors research (Mosier & Skitka, 1996; Muir, 1988). The second set of issues has to do with the context in which the operations are performed. The terminal area impresses time constraints on human decision making and action. Time-stressed processes have several possible problematic human performance responses. One is confirmation bias; because of these time constraints the human operator finds only confirmatory data in problem solving. The other is system stability in a high-gain system. The point of automation is to make the system a higher gain receiver and emitter of aircraft. This comes with a requirement for tighter and tighter coupling between air traffic control and the flight deck. When the system stability is disturbed, the high-gain system needs a clearly identified relaxation path. Human performance tends to propagate an error in these structures with tightly coupled operations. Issues associated with automation degradation must be explored.

En Route Operations Decision Aiding

In the future the service providers will maintain the responsibility for separation, except when it is shifted to the Flight deck, under special conditions. The NAS information system will provide timely and consistent information to both users and service providers. Using the data link, the service providers offer the users a variety of information, ranging from routine communications, charts, and accurate weather to clearances, frequency changes, and other flight information.

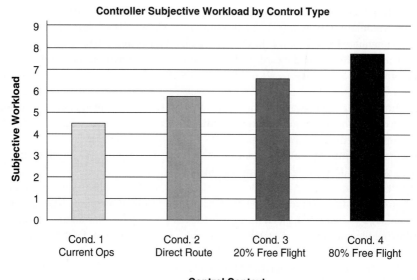

FIG. 18.6. Subjective workload as the control-type employed transitions from (a) current operations with full positive control to (b) direct routing request to (c) 20% self-separating aircraft to (d) 80% self-separating aircraft. Cond = condition; Ops = operations.

Corker et al. (2001) performed and reported a cognitive task analysis and empirical study in which the impact of increasing shifts of separation assurance responsibility from ground to air was examined. The authors found that, in keeping with what the contextual control theory (Hollnagel, 1993) would predict, the burden of control is not simply shed by the air traffic controller, but is represented in a reported increased workload, despite decreasing physical and communicative performance requirements. In the study four, scenarios were presented to the air traffic controllers:

1. Traditional ground-based control
2. Traditional control with all aircraft flying direct
3. All aircraft flying direct with 20% self-separating
4. All aircraft flying direct with 80% self-separating

Each traffic scenario was approximately 1 hour in duration. The experimental group consisted of eight controllers with each subject participating in all four scenarios. The four experimental runs for each subject were presented in the same order as described. The order was kept constant as a conservative approach to moving the controller to the least familiar process of air traffic control in the condition in which they have the most extensive training on the system. In the conditions of self-separation, the appropriate aircraft entered the sector as "self-separating" and so identified themselves to the subject controller at handoff. Figure 18.6 provides a sample of the data, suggesting that controllers respond to shifts in responsibilities by working to maintain a strategic view of operations, despite the fact that the rules of operation reduce their participation to that of an opportunistic respondent.

In the process of delegating responsibilities for separation, the flight deck will be providing separation assurance between self and other obstacles such as aircraft or terrain whenever the responsibility is shifted to them. They will use Automatic Dependent Surveillance—Broadcast, Global Positioning System, data link, and Cockpit Display of Traffic Information (Fig. 18.5) to fulfill this function of self-separation. They will provide the service provider with real-time information on in-flight winds and temperatures aloft for purposes of better forecasting and traffic planning. These communication processes are intended to be supported by a robust digital data communications channel. This channel is said to augment the voice communications channels and provide for routine and repetitive clearances (e.g., voice radio frequency changes at sector boundaries). The claim has been made that digital data link will allow "freeing [of] the voice channels for more time-critical clearances and reducing the voice channel congestion and communication errors, FAA (1995) Air Traffic Services Plan." Despite these pre-operational claims, digital data link operations when subjected to cognitive task analysis do not yield unequivocally positive results, as discussed later.

Airspace Reconfiguration, Dynamic Resectorization

In addition to these communication and separation assurance issues, there is significant FAA investment in developing methods to deal with reconfiguration of the NAS on a timelier basis. This process is dubbed "dynamic resectorization." In this process, the airspace structure would be modified to balance traffic, and presumably controller workload, according to anticipated complexity of operations (calculated as a dynamic density).

Dynamic resectorization is intended to be a requirements-driven process of dynamic re-sectorization based on traffic demand, airspace use complexity, user demand, weather, and manning requirements. It would depend on the development of a measure of sector dynamic density, which is a measure that anticipates the load on the operator of handling such traffic. The definition of dynamic density is intrinsically a human performance issue of workload evaluation based on predicted levels and patterns of traffic. Initially the process is intended to follow a limited application in which additional limited sets of sector boundaries would be developed to address recurring traffic patterns that pose control difficulties within the default configuration. The boundaries would not be completely elastic, but they could be chosen from a set of predefined configurations to best manage the traffic.

It is worth noting that in a study undertaken to access multiple "causal" mechanisms in the development and propagation of "controller error" Connor and Corker (2000) found, as would be anticipated in a analysis of a joint cognitive system, that the controllers relied on the structure of their airspace to support their duties in airspace management. Among the controllers interviewed, which was a subset of all controllers and "selected" to represent controllers with 0, 1, 2, or more operational errors in an 18-month period, it was reported that as the airspace structure became more complex, the complexity of that structure was cited as leading to error (see Fig. 18.7).

Cognitive Analytic Issues: Dynamic Resectorization

Operators' reports and experience in other complex dynamic systems indicates clearly that transition from one operating mode to another in a given operating space potentially leads to loss of situation awareness and to a spiking of workload in the short term, whereas the intent of the process is to reduce workload in the longer term. Three issues come forward:

1. First is the issue of loss of situation awareness during sector transition; which rests on an empirical determination of what constitutes situation awareness for the ATC and how that awareness can be maintained during transition. This traditionally involves display and possibly automation aids.
2. The second issue is the overall cognitive requirement for information maintenance and update as a function of changing airspace configurations for the individual controller. This concerns what information is contributory to the controllers' situation awareness and how that information is encoded.

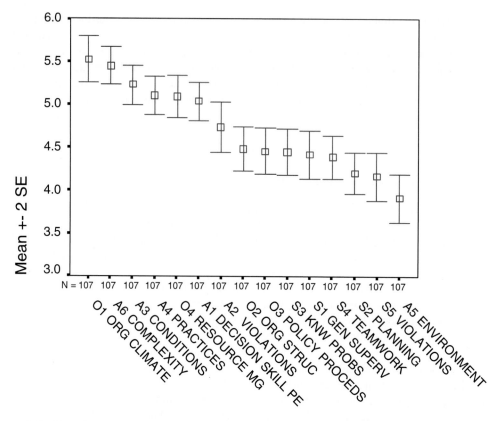

FIG. 18.7. Rank order comparison of all interview items rating on contribution to controller error (±2 standard errors of the mean).

3. The third issue is the requirement for communications among facility collaborators to manage the change and stabilize system boundaries.

Visualization of the airspace is reported to be how the controller currently maintains an airspace picture. This leads to an interesting issue associated with aiding. Do we aid to maintain the current picture-based process, or do we aid to support understanding of required or implemented automation? The problem of maintaining a physical visualization is replaced by the problem of maintaining knowledge of the state of the aiding system in terms of its operating mode and also an understanding of the behavior or the automation in that operating mode. The joint cognitive system implies (as in a human team) that the participants have some expectation of behavior and some model of the basis for behavior observed. Whether that expectation is met as expected reinforces our understanding of our partners' behaviors. In this same manner the controller needs to have some expectation of the behavior of the automation, to predict its expected impact on the system, and, perhaps most critically, to recognize behavior as expected or not and to understand the implication of that observation.

Data Link Communication and Information Exchange: Cognitive Impact

The basic processes of communication among participants in aerospace operations are also undergoing evolution. The primary process to date has been vocal exchange over radio or telephone. The increase in air traffic has led to a problem of "frequency congestion" in which controller–pilot coordination during busy operations is inhibited by overuse of their communication channel. Digital and directed air–ground communication has been targeted as the solution to that problem. The increasingly global scope of air transportation has led to problems in language comprehension and intelligibility. Digital communications have also been posed as a solution to that problem. Finally, throughout the documentation of the NAS development air–air, air–ground, and ground–ground communications efficiencies are cited as critical to the success and timeliness of future NAS operations. The most common assumption in these statements is that some form of digital data link exists among the participants in future operations.

Cognitive Task Analyses and Data Link.   There are a number of human–system performance issues researched with reference to the operation of digital communications between air and ground and among air and ground participants. Lessons learned and reviews of this work can be found in Kerns (1991, 1994) and in Lozito, McGann, and Corker (1993).

Perhaps one of the most interesting issues in data link support NAS modernization is with respect to time. It is postulated that data link communications is a method by which the controller and the flight deck are relieved of tasks, hence incurring a time savings that moves them from tactical to strategic decision making or allows for increasingly complex decisions to be made in a shorter time. In fact, the literature is remarkably consistent about one thing: Data link communications take more time than voice communications, on the order 1 to 2 times as long to complete a transaction. This seems robust across a broad range of display and control configurations and across a broad range of applications for the data-linked information.

With respect to cognitive aspects of the use of directed digital information, it was recognized that when the aviation communication system moved to data link that the so-called party line information that all pilots on a frequency overheard would be significantly reduced. Research was initiated to investigate the impact of that loss on the pilot's situation awareness (Pritchett & Hansman, 1993). The results of these studies indicated a loss of "awareness" in specific operations (e.g., clearance to takeoff) but did not find any "operational impact" of that loss. The lack of operational impact cited may be something of an artifact in the limitation of human in the loop simulation. It may simply be that the coincidence of events in the simulations did not push the joint cognitive system to the point that the requirement for that awareness was manifested. I discuss issue of methodology in the next section.

Because digital communication changes the perceptual mode of the interlocutors (at least in current implementations), several issues associated with the page structure, message log access, and proper cueing and alerting levels for identification of message types (advisory, clearance, or emergency) have been undertaken (Kerns, 1991; Knox & Scanlon, 1991; Lozito et al., 1993; Mackintosh, Lozito, McGann, & Logsdon, 1999) a forward-looking study, Lozito's group performed a cognitive task analysis on the process of using data link communication from the flight deck perspective. They developed an information and procedural flow model examining what the cognitive requirements were for operation in the data link environment. Their analysis anticipated and their empirical work found a significant performance issue with data link operations. Figure 18.8 provides the data of interest. What the researchers found was that under conditions in which there was no loss of information in the initial communication process (i.e., when everything was working smoothly), data link and voice were fairly comparable in terms of the percentage of the communication process that was dedicated to message content clarification and roughly comparable in terms of number of errors in communication that were experienced. If, however, was missing information in the initial communication process, the data link required a significant increase in clarification and there was a significant increase in the number of errors committed in the communication process.

This study provides an excellent example of how cognitive task analysis can support discovery and suggest direction of analyses to reveal the underlying operational issues that are not

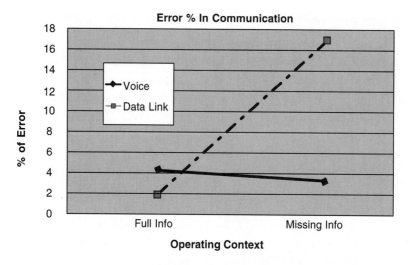

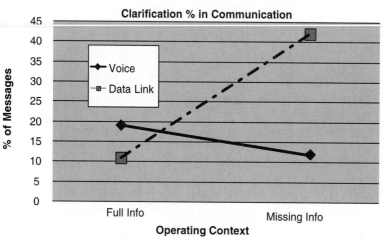

FIG. 18.8.   Data comparing performance between data link and voice communication between air traffic control and flight deck.

immediately manifest if only the first-order effects of technology are investigated. The benefits of this approach are offset by the extent to which a careful cognitive task analysis takes the design's team effort in first the analysis and then in the bookkeeping of the multiple operators' dynamic state.

In the next section, I describe a set of tools to support the process of cognitive task analysis and the application of one of those tools to analyze the impact of context on human performance in aerospace operation.

HUMAN PERFORMANCE MODELS IN SUPPORT OF COGNITIVE TASK ANALYSIS AND COGNITIVE TASK DESIGN

Studies in automation aiding for human operators in complex domains such as distributed battle management, nuclear power and process plant operation, and air traffic control and flight management have concluded that adequate representation of the human–system cognitive interactions in the systems operation are critical to the prediction of its ultimate performance and acceptability (Billings, 1997; Rassmussen & Vicente, 1989; Reason 1990).

General Methodology

How are the human performance factors to be represented? Computational models of the airspace operation have become the common parlance of analysis designers, analysts, and public regulatory agencies. NAS models that account for the movement of airspace assets and predict flow, bottlenecks, and delay are available for all parts of the airspace. However, few include any representation of the human contribution to that system's performance, and fewer still include human performance representation at a level of fidelity that predicts the impact of technology and procedure on the human operator.

I would suggest that human performance models be used in fast-time simulation to pinpoint areas of potential system stress with regard to human performance. The areas of concern could then be the focus of tailored human-in-the-loop studies to identify more clearly the predicted human performance issues. The data from these studies could then be fed back to the model structures to enhance the library of empirically based model performance parameters. Various methods and designs for amelioration of potential problems in airspace operations can then be prototyped in the fast-time modeling environment to search for optimal response. The cycle would then turn again to empirical confirmation of the design and procedure effectiveness.

As an example, human performance models supporting cognitive task analysis and design could be integrated into an airspace sector model and used to address when, how often, and how many sectors of the NAS need to be, or can be, dynamically restructured. Operator loads, communication and decision burdens, working and long-term memory requirements, rules, and knowledge constraints can be explored in various "resectorization schemes." A subset of these schemes could then be subjected to detailed empirical analysis, and techniques for implementation could be suggested. The model can benefit from the human performance data and be used to explore the operational concepts for optimization paths. These can again be refined in a more tightly focused full mission simulation. I suggest that this model and empirical data ratcheting process will enhance efficiency in costly empirical analysis, lend credibility to the necessarily few variations of low statistical power simulation tests, and enhance model fidelity and validation. This process should yield a powerful analysis paradigm for the consideration of human performance and human factors concerns in NAS modernization.

Computational Human Performance Models

There are a number of good reviews of cognitive performance models (cf. AGARD, 1998; Pew & Mavor, 1998) and several reviews of application (cf. Laughery, Archer, & Corker, 2001). My point here is to describe the requirements for these computational models to support cognitive task analysis and make some suggestions as to what further developments might enhance human computational models in this regard.

There are three primary areas of development in computational human performance representation needed to enhance their utility in supporting cognitive task analyses of aerospace technologies on a large scale. Lacking these developments, models of human behavior have failed to meet the challenge of serving in large-scale simulations.

The architectures into which the human performance models are intended to be integrated are, by and large, either time-based discrete event simulations or algorithmic mathematical analyses on system dynamics. The human operator can be represented by discrete event models, but such network models have, until recently (Laughery & Corker, 1997), lacked the capability of emergent behavior that is critical to predict human operation in the cognitive domains of interest in the examination of the NAS operation. This is especially true for the tool developments that are undertaken to improve decision making and offload strategic and tactical workload from the controller and distribute some of the information processing and decision making into a collaborative process. In these operations, the increasingly complex and dynamic NAS representation is critically dependent on higher order cognitive processes and meta-cognitive planning and communication processes (Orasanu, 1994). Network models or discrete event models of human performance tend to rely on a rather rigid network structure to embody such intelligence. This is not an adequately self-referential process to represent the critical impact of aiding on the operator.

Model Scope.   The generative models of human behavior have been developed for the examination of an individual or a small group of individuals in the face of a design optimization task. (These have sufficient complexity in the representation of cognition, planning, decision making, prioritization, and scheduling and parallel concurrent processes in performance. Thus, they support a cognitive task analysis.) The amount of "domain knowledge" that must be integrated into the model's internal reasoning structures presents a considerable challenge to the analysis team. Also, the speed and behavioral granularity at which the models operate (for example, with performance resolutions to 100 msec) is badly mismatched to the coverage and temporal scope of an NAS mode such as Reconfigurable Airspace Mathematical Simulation (RAMS) in which days or weeks of activity across most or all of the NAS are represented in a Monte Carlo simulation environment.

Adaptive and Proactive Behavior.   Both the network and the generative human performance models have tended to represent reactive, event-driven processes in human performance, that is, the kind of behaviors that are scripted in standard operating procedures with triggers that can be associated with a single environmental marker or communication process. In the analyses of the impact of an automated aiding system, the proactive behavior of an operator and a prediction as to the operator's future behavior are critical to the evaluation of the automated aiding system. The scope of the human model process has been limited, however, in the representing levels of abstraction higher than sensory and perceptual input into a propositional and rule-based operation. These shortcomings will need to be addressed and overcome to guarantee the successful integration of human performance models into large-scale system models.

Despite difficulties, attempts are being made to provide at least rudimentary inclusion of some of the critical components of human performance, for example, in the calculation of a dynamic density measure to predict complexity. To capture at least part of the flexibility of discretionary behavior, researchers in Eurocontrol have introduced a rule-based representation of human communication behavior selection into a large-scale airspace representation (RAMS). Additionally, the safety case is being investigated by the active inclusion of human performance characteristics into the process of self-separation in an exploration of free flight. Using dynamic colored Petri net simulation techniques and a model of human–system interaction proposed by Blom, Corker, and Klompstra (2000) are embedding human performance characteristics into analyses of flight deck operations. In these models, the performance of interest is embedded as a contributor to model-output-dependent measures such as "safety" of communications per unit time or implied workload as a weighted contribution of factors.

Taking the prior sections as providing a set of desiderata or criteria for the successful application of human–system performance, I take a closer look at what these requirements imply for human performance representation from the perspective of cognitive task analyses.

Individual Representation.   These decision aids are focused on supporting individual or small-team cognition. As such, the evaluation of the effectiveness of theses technologies must take into account the processes of individual cognition. For a computational representation of those individuals, the representation of cognition, planning, decision making, prioritization, scheduling, and parallel concurrent processes in performance have tended to be developed for the examination of an individual or a small group of individuals. These individuals are interacting with small-scale elements (aircraft and airspace components), but it is the aggregate performance of these elements that is of interest to a NAS-wide analysis.

Aggregate Representation.   In addition to individual performance, there is a requirement to represent behavior in the aggregate to estimate the effect of changes in the large-scale system response. It is still an open question as to whether this aggregate performance representation should be in the same formalism or of the same computational formalism as individual performance. In part, the answer to that question lies in examination of the output or dependent measures required in our analyses.

Output and Measurement Requirements.   The assessment of human performance in the system and of the system's safety risk is ultimately based on behavior (system and human) in time and environmental context. The behaviors represented have safety-neutral, safety-enhancing, or safety-reduction consequences, and it is a measurement of those consequences that predictive models should provide. The dimensions of the output of the models can be scaled from individual and team performance to overall system response to a range of demands, perturbation, and environmental influences. Furthermore, the system and the individual can be characterized as performing under nominal or off-nominal conditions in that range of demands. Measures scaled to individual behavior are necessarily focused at a "human-scale" time representation (seconds, minutes, hours), which can be made stochastic and generated across any number of "runs," but which still provide distributional representation of some relatively constrained epoch of performance. Alternatively, systemwide measures follow the evolution of events and large-scale response to these events over hours, days, months, and years. The measures then are required to be aggregate representations of the physical and dynamic characteristics of the airspace assets in space and time. The behaviors are defined in terms associated with airspace occupancy (rates, trajectories, queues) and dynamics (velocity traces in 4-D airspace and time dimensions). In the evaluation of human–system performance and associated safety-risk or hazard, both these scales (and other intermediate scales) of behavior must be evaluated as to the contribution of those behaviors to some reference level of safety.

Modeling Context.   Another critical dimension for computational support cognitive task analysis is the effective representation of the human operators' response to context in the process of shaping their practice. Concern for context (human, artifactual, social) and its broad influence on the practitioners of work is at the heart of the approach to human system performance that emphasizes understanding the full energetic and informational system in design. The influence of context and its impact on human–system behavior has traditionally been neglected in computational human performance model but is beginning to be undertaken. To illustrate this requirement, I provide an example from my work in modeling human performance. I briefly describe the architecture and functionality of the man–machine integrated design and analysis environment (MIDAS), highlighting its implementation of the contextual control model (Hollnagel, 1993).

MIDAS provides a computational environment to predict human performance in linked human–automation systems. The perceptual and cognitive functions of the human operator in both successful and flawed performances are modeled to gain insight into human automation system integration, training and staffing requirements, and ergonomic and information requirements in design. The system has been used to describe the interplay among the task demands, the characteristics of the operator reacting to those demands, the functions of the equipment with which the operator interacts, and the operational environment (the time course of uncontrolled events in environments of aviation operations, nuclear power plant control, emergency response systems, and military systems).

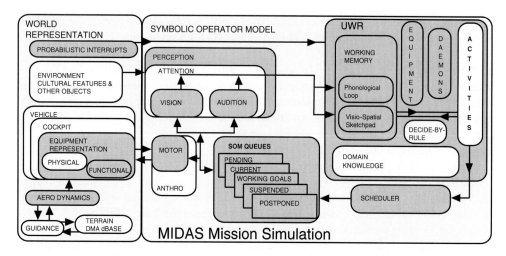

FIG. 18.9. The man–machine integrated design and analysis system (MIDAS) functional architecture. Each element in the diagram represents a function of the model designed to provide psychologically plausible behavior. The addition of context sensitive judgments in task performance is undertaken in the updateable world representation (UWR) and in the scheduling mechanism. A more detailed description of the model and its application to aerospace technologies is available in Sarter and Amalberti (2000).

Human Mental Constructs Represented

Memory.   MIDAS contains representations of human perceptual processes (visual and auditory). I have implemented a working memory, described by Baddeley and Hitch (1974), as comprising a central control processor (of some limited capacity), a "phonological loop" (temporary storage of speech-based information) and a "visuospatial scratch pad" (temporary storage of spatial information), and a long-term memory composed of both episodic and procedural archival structures. Working memory is susceptible to interference and loss in the ongoing task context. We assume a relatively stable long-term memory store in the scenarios we have implemented. In MIDAS, the internal, updateable world representation (UWR) provides a structure whereby simulated operators access and update their own personalized information about the operational world. Data are updated in each operator's UWR as a function of the mediating perceptual and attentional mechanisms.

Activity.   Tasks or activities available to an operator are contained in that operator's UWR and generate a majority of the simulation behavior. Each activity contains slots for attribute values, describing, for example, preconditions, temporal or logical execution constraints, satisfaction conditions, estimated duration, priority, and resource requirements.

Task Scheduling.   Activities, which have their preconditions, met, temporal/logical execution constraints satisfied, and required information retrieved from memory are queued and passed to a model of operator scheduling behavior. Activities are executed in priority order, subject to the availability of required resources. MIDAS contains support for parallel activity execution, the interruption of ongoing activities by those of higher priority, and the resumption of interrupted activities. The MIDAS architecture is illustrated in Fig. 18.9.

MIDAS and Context Modeling.   Within this cognitive architecture structure, MIDAS then implements a computational version of the Hollnagel context control-switching framework. In the original implementation of Air MIDAS, without context switching, the MIDAS model drops or defers task completion as a function of task load. Task load is a computational estimation of the degree to which four "capacities" or resources are used to service a task. The capacities are "visual, auditory, cognitive, and psychomotor" and are assessed prior to the simulation and assigned to the task as attributes. The method of determination of the appropriate levels of load for each task is an extensive Delphi-technique described by Aldrich, Szabo, and Bierbaum (1989). The scheduling mechanism in MIDAS then evaluates the currently active goal or task queue and, according to priority, concurrently performs all tasks with a rating under 7 in any dimension. In the assignment of task load, a scale from 1–7 is used to assign a resource tax on activities. This assignment is provided independently for visual, auditory, cognitive and psychomotor resources. For example, if the pilot, represented by an air-MIDAS simulation agent, wants to access relative position on an approach to landing, then two sets of activities will be need to be completed. One references the primary flight display (inside the cockpit), the other references the external visibility at a "decision height." The visual load associated with these two sets of tasks precludes their simultaneous performance (one out the window, the other inside the flight deck), so they would have to be performed sequentially. If the two visual tasks were both in reference to the primary flight display (e.g., determine altitude and heading), however, the sum of the visual loads may be less than 7, and they could be performed simultaneously. A similar checking is done for all tasks currently active across each dimension of load. This mechanism remains for load determination and task shedding. In addition, a mechanism that changes the set of available actions to be performed is also available. In this control switching is implemented as a function of the number of goals, the decision horizon (time to commitment), and the time available for performance.

The control mode switching module uses a number of operator and task variables to calculate the value of the control switch parameter, which is then mapped onto one of the three possible control models. Specifically, two system variables are used to calculate the control switch parameter value: (a) local past success or failure on a task and (b) the difference between the time required to complete the current goals' activity sets compared with the decision-activity horizon (i.e., how far in the future the task queues needed to be decided to succeed in their performance). This task performance and planning time relationship was defined as "goals/sec" as the control switch metric. This metric is then coupled to as a set of task-specific rules, which determine the control mode switching. The specific switching value is a function of the task domain and is determined through knowledge engineering with experts in the target domain. For example, in the case of en route air traffic control, the event horizon is determined by the number of aircraft in sector and complexity of the maneuvers that these are undertaking. So a rule would relate the complexity metric to a temporal horizon. This would be used with the total number of current goals and compared with time required to perform to arrive at a goals per second estimate. This estimate is then used to make a switch to a control model at, for example, 3 goals/sec in strategic mode or 10 goals/sec in scrambled mode.

The value of the control parameter maps onto differences within the MIDAS goal or task selection process. Specifically, the activity set in each mode is constructed with different tasks to satisfy a goal. In some cases, the control modes also have different goals to perform, so a clearance to taxi, in the strategic mode, would include an activity that provides the flight crew reference to a map of the airport. As the control mode switches to more opportunistic operation, the taxi clearance may be acted on as a gap in surface traffic without prior reference to the map. In an extreme case of scrambled control, the flight crew may service a goal such as "get off active runway" instead of referring to the explicit clearance of the controller.

Verification of the model's performance in switching among control modes as a function of the goal load and time horizon at specific competency levels has been completed. Our research team is currently using the context-sensitive model to evaluate the performance of human flight crews in response to aiding technologies and (under support from NASA and the FAA) examining the secondary and underlying interactions of specific airspace operational concepts. These investigations are also supporting empirical simulation performance, allowing an opportunity to examine the efficacy of the model's prediction with reference to human performance. Focusing on cognitive task analysis and the perspective of human–system work integration, we hope not only to be able to predict the first-order behavior of humans and systems, but also to be able to uncover the secondary and unintended linkages that either impede or contribute to a system's effectiveness and safety.

CONCLUSIONS

The work environment that supports and provides global air transportation of goods and people is going through a significant revolution driven by economic, social, environmental, physical, and human requirements and constraints. The extent of this revolution has impact from manufacture and instrumentation of the world aviation fleet to global air traffic management as a cooperative venture among humans and between the human operators and their far-flung dynamic technological assets.

I have attempted to outline the dimensions of the technosocial systemic changes from system optimization technologies to the human impact of their implementation on work structure, safety, and on the self-regarded roles and responsibilities of the agents in the system. The general, and perhaps competing, trends of increased autonomy and distribution in control, made possible by improved communications and positioning technologies, are implemented in a work context in which increased requirement for system prediction demands the expression of four-dimensional intention across seconds and across months.

The dimensions of implementation of this adaptation to reduce cost, increase capacity, improve efficiency, and enhance safety are the technologies supporting communication, information, and control among the operators and their machines. The developments include increased accuracy for positioning of assets (human and machine) in real time and globally, improved connectivity (via satellite and land line data connections) for dissemination of information among participants, improved control and prediction of aircraft flight in terms of more precise and less variable navigation performance, improved predictivity for environmental effects (weather, turbulence, winds, etc.), and the development of technologies to enhance human performance and surmount constraints in perceptual and cognitive dimensions. At present, technology development is leading the process. It is becoming increasing clear, however, that attention to design is required to understand, for example, how changes at the microlevel of communication protocols produce changes at the macrolevel of airspace capacity how technological support for meteorological predictions change economic and business strategies, and how changes in roles and responsibilities among human and automation elements in the system have an impact on the social and cultural constructs in which these changes are played out.

I have tried to illustrate that, in understanding these multidimensional and propagated effects, effort must be focused on cognitive task design in which humans, the automation, and the environment must be considered in terms of a joint cognitive system. In such design, the tasks are considered in the full context of their implementation, and with direct concern for the informational and procedural requirements of all practitioners and participants in the system. The requirement for such design is illustrated by analysis of systems in which technology led design and then begged an operational concept in which to be placed. The path of interaction among tightly coupled systems of cognitive agents must be explicitly and systematically investigated to understand the systems operation in nominal and off-nominal operation. This same approach is recommended in analysis of points at which the system operation breaks down, in incident and accident analysis.

Finally, I have illustrated some potential for advances in human–system representation in computationally predictive paradigm. The potential value of such an approach to cognitive task design lies in its ability to both explicitly specify and modify the characteristics, roles, and operations of the agents in a complex system and then to manipulate those characteristics, the technologies, and the environments (physical and cultural) in which they are exercised. This method of exploration, I believe, offsets many of the inherent limitations of design in large-scale dynamic systems in which the forces to rapid implementation are strong and persistent. There is an important place for hybrid computational representation and empirical research paradigms in future cognitive task design in aviation systems.

AFTERWORD

I wrote the largest section of this chapter, analyzing the need for cognitive task analysis and design in these large-scale dynamic systems in which the context of operation plays a significant role, prior to September 11, 2001. The events of that day have thrown the use of national and international airspace into a new and irrevocably different context of operations. Although government and industry responses are still being formed, the vector of their response seems clear. The seemingly inevitable addition to airspace operators' duties of security monitoring, conformance monitoring, and enforcement will place demands on flight crews, cabin crews, airline operations centers, and controllers that have never been designed for and for which the information structures and control and command structures are being adapted ad hoc. For instance, to add flight trajectory conformance monitoring to controllers of inbound aircraft in transition airspace without appropriate support (alert, awareness enhancement procedures) seems ill advised, adding yet another class of task burdens to the challenge of air traffic control. Other options under review, including auto land takeover of aircraft operation, are fraught with human–automation interactional and human–system decision-making issues.

The use of the commercial aircraft as a weapon of mass destruction provides a stark reminder of the secondary and unintended consequences of our technical advancement in the concentration of information and power. I am reminded of the words of Norbert Wiener in speaking about what he foresaw as the coming revolution in information sciences to which his work was contributing:

> We have contributed to the initiation of a new science, which, as I have said, embraces technical development with great possibilities for good or for evil. . . . As we have seen, there are those who hope that the good of a better understanding of man and society which is offered by this new field of work may anticipate and outweigh the incidental contribution we are making to the concentration of power (which is always concentrated, by its very conditions of existence, in the hands of the most unscrupulous). I write in 1947, and I am compelled to say it is a very slight hope. (Wiener, 1948, pp. 28–29)

I write in 2001 about the future of aeronautical transportation and its impact on humans and society and am compelled to say that I share that slight hope.

ACKNOWLEDGMENT

K. Corker is supported by research grants from NASA and from the FAA. For further information, please contact the author at kcorker@email.sjsu.edu.

REFERENCES

Advisory Group for Aerospace Research and Development. (1998, December). *A designer's guide to human performance modelling* (Advisory Report 356). NATO Press. Brussels, Belgium.

Aldrich, T. B., Szabo, S. M., & Bierbaum, C. R. (1989). The development and application of models to predict operator workload during system design. In G. MacMillan, D. Beevis, E. Salas, M. Strub, R. Sutton, & L. Van Breda (Eds.), *Applications of human performance models to system design* (pp. 65–80). New York: Plenum Press.

Aviation Space and Environmental Medicine. (1998). *Special edition: Crew factors in flight operations.* Alexandria, VA: Aerospace Medical Association.

Baddeley, A. D., & Hitch, G. J. (1974). Working memory. In G. Bower (Ed.), *Advances in learning and motivation* (Vol. 8, pp. 47–90). New York: Academic Press.

Bailey, L. L., Broach, D. M., & Enos, R. J. (1997). Controller teamwork evaluation and assessment methodology (CTEAM). *Proceedings of the Ninth International Symposium on Aviation Psychology.* Columbus, OH: Ohio State University.

Baker, D. P., & Smith, K. (1997). A methodology for simulation training and assessment of tower cab team skill competencies. *Proceedings of the Ninth International Symposium on Aviation Psychology.* Columbus, OH: Ohio State University.

Billings, C. E. (1997). *Aviation automation: The search for a human-centered approach.* Mahwah, NJ: Lawrence Erlbaum Associates.

Blom, H., Corker, K., & Klompstra, M. B. (2000). *Study on the integration of MIDAS and TOPAZ.* NLR-CR-2000-698. NLR Phase 1 Final Report. Netherlands Aeronautical Research Laboratory.

Boeing Commercial Airplane Group. (1995). *Statistical summary of commercial jet transport accidents.* Seattle, WA: Author.

Bresolle, M. C., Benhacene, R., Boudes, N., & Parise, R. (2000). Advanced decision aids for air traffic controllers: Understanding different working methods from a cognitive point of view. *Proceedings of the 3rd USA/Europe Air Traffic Management R&D Seminar,* Napoli, Italy.

Busquin, P. (2001). *European aeronautics: A vision for 2020.* Brussels, Belgium.

Callentine, T., Prevot, T., & Palmer, E. (2001, December). Flight crew factors under integrated CTAS/FMS operations with controllers in the loop. *Proceedings of the 4th USA–Europe Air Traffic Management R&D Seminar,* Santa Fe, New Mexico.

Cambell, K., Cooper, W., Greenbaum, D., & Wojcik, L. (2000). Modelling distributed decision making in traffic flow management operations. *Proceedings of the 3rd USA/Europe Air Traffic Management R&D Seminar.*

Connor, M., & Corker, K. (2000). *A report of analyses of ATC errors. FAA/NATCA Joint study.* National Institute for Aviation Research. Cooleyvelle, Texas.

Corker, K., Fleming, K., & Lane, J. (2001). Air-ground integration dynamics in exchange of information for control. In (L. Bianco, P. Dell'Olmo, & A. R. Odoni (Eds.), *Transportation Analysis: New Concepts and Methods in Air Traffic Management,* (pp. 125–142). New York: Springer-Verlag.

Dekker, S. W. A. (2001). The disembodiment of data in the analysis of human factors accidents. In *Human factors and aerospace safety* (Vol. 1, No. 1, pp. 39–58).

Dekker, S. W. A. (2002). *Field guide to human error investigations.* Aldershot, England: Cranfield University Press.

Dekker, S., & D. D. Woods (1999). To intervene of not to intervene the dilemma of management by exception. *Cognition, Technology and Work, 1,* 86–96.

Federal Aviation Administration, Human Factors Team. (1996). *The interfaces between flightcrews and modern flight deck systems.* Washington, DC: Author.

Federal Aviation Administration (1995). *Air traffic services plan.* Government Printing Office, Washington, DC.

Green, S., & Weider, T. (2000). Modelling ATM automation, metering conformance benefit. *Proceedings of the 3rd USA/Europe Air Traffic Management R&D Seminar.* Washington, DC: Federal Aviation Administration.

Hancock, P. A. (1997). *Essays on the future of human-machine systems.* Banta Information Services. Madison, Wisconsin.

Hollnagel, E. (1993). *Human reliability analysis context and control.* New York: Academic Press.

Hollnagel, E., & Amalberti, R. (2001). *The emperor's new clothes or whatever happened to "human error."* Presented at the Workshop Proceedings of the Human Error Safety and System Development, Linköping, Sweden.

Jones, S. G. (1993). Human factors information in incident and reports. *Proceedings of the 7th International Symposium on Aviation Psychology.* Columbus, OH: Ohio State University.

Kerns, K. (1991). Data link communication between controllers and pilots: A review and synthesis of the simulation literature. *International Journal of Aviation Psychology, 1,* 181–204.

Kerns, K. (1994). *Human factors in ATC/flight deck integration* (Report MP94W0000098). Fairfax, Virginia. Mitre.

Kirlick, A. (1998). The ecological expert: Acting to create information to guide action. *Proceedings of the 1998 Symposium on Human Interaction with Complex Systems.*

Klein, G. L., & Anton, P. S. (1999). A simulation study of adaptation in traffic management decision making under free scheduling flight operations. *Air Traffic Control Quarterly, 7,* 77–108.

Knox, C., & Scanlon, C. (1991). *Flight tests with a data link used for air traffic information exchange* (NASA Technical Paper 3135). Langley, VA: National Aeronautics and Space Administration.

Laughery, R., Archer, S., and Corker, K. (2001). Computer modelling and simulation of human performance: Technologies and techniques. In G. Solvency (Ed.), *Handbook of human factors and ergonomics* (3rd ed.). New York: Wiley Interscience.

Laughery, K. R., & Corker, K. M. (1997). Computer modelling and simulation. In G. Salvendy (Ed.), *Handbook of human factors and ergonomics.* New York: Wiley.

Lee, K. K., & Davis, T. J. (1996). The development of the final approach spacing tool (FAST): A cooperative controller-engineer design approach. *Control Engineering Practice, 4–8,* 1161–1168.

Lee, K. K., & Sanford, B. D. (1998). *Human factors assessment: The passive final approach spacing tool (pFAST) operational evaluation* (NASA Technical Memorandum 208750). Moffett Field, CA: NASA Ames Research Center.

Leiden, K., & Green, S. (2000, May). *Trajectory orientation: A technology enabled concept requiring a shift in controller roles and responsibilities.* Paper presented at the NASA Distributed Air-Ground-Traffic Management workshop. St. Joseph, Missouri.

Lozito, S., McGann, A., & Corker, K. (1993). Data link air traffic control and flight deck environments: Experiments in flight crew performance. *Proceedings of the 7th International Symposium on Aviation Psychology.* Ohio State University.

Mackintosh, M., Lozito, S., McGann, A., Logsdon, E. (1999). "Designing procedures for controller-pilot data link communications: Effects of textual data link on information transfer." SAE World Aviation Conference Paper 1999-01-5507. Washington, DC. American Institute of Aeronautics and Astronautics.

Mosier, K., & Skitka, L. (1996). Human decision makers and automated aids: Made for each other? In R. Parasuraman & M. Mouloua (Eds.), *Automation and human performance: Theory and application.* Mahwah, NJ: Lawrence Erlbaum Associates.

Muir, B. (1988). Trust Between humans and machines, and the design of decision aids. In *Cognitive Engineering in Complex Dynamic Worlds.* E. Hollnagel, D. Woods and G. Mancini (eds.). Academic Press, London.

National Aeronautics and Space Administration (1999). Airspace Capacity Program Plan NASA AMFS Research Center, Moffett Field, CA.

National Transportation Safety Board. (1986). Engen v. Chambers and Landford W. L. 82575.

Orasanu, J. (1994). Shared problem models and flight crew performance. In N. N. Johnston, N. McDonald, & R. Fuller (Eds.), *Aviation psychology in practice.* Avebury Press. London, England.

Perrow, C. (1984). *Normal accidents: Living with high-risk technologies.* Basic Books, New York.

Pew, R. W., & Mavor, A. S. (1998). *Modelling human and organizational behaviour: Applications to military simulations.* Washington, DC: National Academy Press.

Pritchett, A., & J. Hansman, J. (1993). Preliminary analysis of pilot ratings of "partyline" information importance. In R. Jensen (Ed.), *Proceedings of the 7th International Symposium on Aviation Psychology,* Ohio State University.

Quinn, C., & Robinson, J. (2000). A human factors evaluation of active final approach spacing tool concepts. *Proceedings of the 3rd USA/Europe Air Traffic Management R&D Seminar.*

Rasmussen, J., & Vicente, K. (1989). Coping with human error through system design: implications for ecological interface design. *International Journal of Man-Machine Studies, 31,* 517–534.

Reason, J. T. (1990). *Human error.* Cambridge, England: Cambridge University Press.

RTCA. (1995). *Final report of the RTCA Board of Director's Select Committee on Free Flight.* Washington, DC: Author.

Sarter N., & Amalberti, R. (2000). *Cognitive engineering in the aviation domain.* Lawrence Earlbaum Inc.

Sarter, N., & Woods, D. D. (1995). *Strong silent and out-of-the-loop: properties of advanced (cockpit) automation and their impact on human-automation interaction* (Technical Report CSEL 95-TR-01). Columbus, OH: Cognitive System Engineering Laboratory, Ohio State University.

Smith, P., Billings, C., Woods, D., Mccoy, E., Sarter, N., Denning, R., & Dekker, S. (1997). Can automation enable a cooperative future ATM system? *Proceedings of the Ninth International Symposium on Aviation Psychology.* Columbus, OH: Ohio State University.

Snook, S. A. (2000). *Friendly fire: The accidental shootdown of U.S. blackhawks over northern Iraq.* Princeton NJ: Princeton University Press.

Tenner, E. (1996). *Why things bite back: Technology and the Revenge of Unintended Consequences.* Vintage Books. New York, New York.

Wambsganns, M. (1997). Collaborative decision making through dynamic information transfer. *Air Traffic Control Quarterly, 4,* 109–125.

Wickens, C., Mavor, A., & McGee, J. (1997). *Flight to the future: Human factors in air traffic control.* Washington, DC: National Academy Press.

Wickens, C., Mavor, A., Parasuraman, R., & McGee, J. P. (1998). *The future of air traffic.* Washington, DC: National Academy Press.

Wiener, E. (1989). *Human factors of advanced technology ("glass cockpit") transport aircraft* (NASA Contractor Report 177528). Mt. View CA: NASA Ames Research Center.

Wiener, E. L., Kanki, B. G., and Helmreich, R. L. (1993). *Cockpit resource management.* New York: Academic Press.

Wiener, N. (1948). *Cybernetics; or control and communication in the animal and the machine.* Cambridge, MA: MIT Press.

Woods, D. D. (2001). Presentation to the NASA Design for Safety Workshop. Moffett Field, CA: NASA Ames Research Center.

19

Groupware Task Analysis

Martijn Van Welie and Gerrit C. Van Der Veer
Department of Computer Science, Vrije Universiteit, The Netherlands

Abstract

This chapter describes groupware task analysis, a method for studying the activities of groups or organisations for the purpose of analysis and design. The method uses a conceptual framework comprising the viewpoints agents, work, and situation. We discuss the modelling and analysis steps using those viewpoints. Task design is the next step, which starts with envisioning the future task world. We offer practical representations for the framework, including a discussion of tool support. Experiences with the method are outlined using a case study.

INTRODUCTION

Groupware task analysis (GTA; van der Veer, Lenting, & Bergevoet, 1996) is a recently described design method that combines aspects from several other methods. Systems built today are often used by multiple users, and GTA emphasizes the study of a group or organisation and its activities rather than study of single users at work. Essentially, GTA consists of a conceptual framework specifying the relevant aspects of the task world that need attention when designing groupware.

The broad conceptual framework is based on experiences with a variety of approaches and on an analysis of existing methods for human–computer interaction (HCI) and computer-supported cooperative work (CSCW; van der Veer, van Vliet, & Lenting, 1995). When designing groupware systems, it is necessary to widen the notion of a task model to include descriptions of many more aspects of the task world than just the tasks. GTA makes a distinction between a *descriptive* and a *prescriptive* task model. When starting the design process of a complex interactive system, it is often useful and necessary to understand the current work situation. We label this task world Task Model 1. To design for real-world use, it is also necessary to consider changes in the task world that will result from developing and introducing the envisioned technology. Consequently, GTA stresses the need for designing and, hence, modelling the envisioned future task world, which we label Task Model 2. Task Model 2 may in fact be considered a type of global design that includes not only the artefacts that are being developed but additionally their relation to the users and stakeholders, their goals and tasks, and the context of use.

Analysing the Current Task Situation (Task Model 1)

In many cases the design of a new system is triggered by an existing task situation. Either the current way of performing tasks is not considered optimal or the availability of new technology is expected to allow improvement over current methods. A systematic analysis of the current situation may help formulate design requirements and, at the same time, may later allow evaluation of the design. Whenever it is possible to study the "current" task situation, a thorough analysis is valuable for the design of a new system. In our design practice, we have used a combination of classical HCI techniques such as structured interviews (Sebillotte, 1988), document analysis and observation, and CSCW techniques such as ethnographic studies and interaction analysis (Jordan, 1996). Task Model 1 describes the combined knowledge of the current situation, also referred to as the current task model.

Envisioning the Future Task Situation (Task Model 2)

Many design methods in HCI that start with task modelling are structured in a number of phases. After describing a current situation (Task Model 1), the method requires a redesign of the task structure to include technological solutions for problems and technological answers to requirements. Johnson, Johnson, Waddington, & Shouls, (1988) provided an example of a systematic approach in which a second task model is explicitly defined in the course of design decisions. Task Model 2, the designed future task model, will in general be formulated and structured in the same way as the previous model, but in this case it is not considered a descriptive model of users' knowledge but a prescriptive model of the future task world. Task Model 2 describes the task situation as it should exist when the system has been developed and is in use.

Detail Design

After Task Models 1 and 2 are complete, the details of technology and the basic tasks that involve interaction with the system need to be worked out. This activity consists of subactivities that are strongly interrelated: specifying the functionality, structuring the dialogue between the users and the system, and specifying the way the system is presented to the user. This activity is focussed on a detailed description of the system as far as it is of direct relevance to the end users. We use the term "user virtual machine" (UVM) (Tauber, 1988) to indicate the total of user-relevant knowledge of the technology, including both semantics (what the system offers the user for task delegation) and the syntax (how task delegation to the system has to be expressed by the user). When making the transition from Task Model 2 to the UVM, the users' tasks and the relevant objects in the task world determine the first sketch of the application. The task and object structure is used to create the main displays and navigational structure. From there forward, the iterative refinement process takes off, together with the use of explicit design knowledge such as guidelines and design patterns. This part of design is not covered in detail in this chapter.

The framework that we discuss here is intended to structure Task Models 1 and 2 and, hence, may feature as guidance for choosing techniques for information collection for Task Model 1. For Task Model 2, design decisions have to be made based on problems and conflicts that are represented in Model 1, in combination with requirement specifications as formulated in interaction with the client of the design.

VIEWS ON TASK WORLDS

In GTA, task models for complex situations are composed of three factors: agents, work, and situation. Each describes the task world from a different perspective, and each relates to the others. This allows designers to view and to design from different angles and allows design tools to ensure consistency and completeness. The three perspectives are a superset of the main focal points in the domains of both HCI and CSCW. Both design fields consider agents ("users" vs. "cooperating users" or user groups) and work (activities or tasks, the objectives or the goals of "interaction" and the cooperative work). Moreover, particularly CSCW stresses the situation in which technological support must be incorporated. In HCI, this is only sometimes, and then mostly implicitly, considered. This section discusses the three perspectives of the conceptual framework.

Agents

The first aspect focuses on *agents,* a term that often indicates people, either individuals or groups, but may also refer to systems. Agents are considered in relation to the task world, hence we need to make a distinction between humans, as acting individuals or systems, and the *roles* they play. Moreover, we must consider the concept of agents' organisation. Humans must be described with relevant characteristics (e.g., the language they speak, the amount of typing skill or experience with MS Windows). Roles indicate classes of agents to whom certain subsets of tasks are allocated. By definition, roles are generic for the task world. More than one agent may perform the same role, and a single agent may have several roles at the same time. *Organization* refers to the relation between agents and roles in respect to task allocation. Delegation and mandating responsibilities from one role to another is part of the organisation.

Work

In GTA, both the structural and the dynamic aspects of work are considered, so the *task* is taken as the basic concept and several tasks can share the same *goals*. Additionally a distinction is made between tasks and actions. Tasks can be identified at various levels of complexity. The unit level of tasks needs special attention. A distinction is made between (a) the lowest task level that people want to consider in referring to their work, the "unit task" (Card, Moran, & Newell, 1983); and (b) the atomic level of task delegation that is defined by the tool that is used in performing work, like a single command in command-driven computer applications. The latter type of task is called "Basic task" (Tauber, 1990). Unit tasks are often role related. Complex tasks may be split up between agents or roles. Unit tasks and basic tasks may be decomposed further into user actions and system actions, but these cannot really be understood without a frame of reference created by the corresponding task (i.e., actions derive their meaning from the task). For instance, hitting a return key has a different meaning depending on whether it concludes a command or confirms the specification of a numerical input value in a spreadsheet. The task structure is often at least partially hierarchical. On the other hand, resulting effects of certain tasks may influence the procedures for other tasks (possibly with other roles involved). Therefore, the task flow and data flow over time as well as the relation between several concurrent flows must be understood. A special concept is that of an *event,* indicating a triggering condition for a task, even if the triggering could be caused by something outside the task domain we are considering.

Situation

Analysing a task world from the perspective of the situation means detecting and describing the environment (physical, conceptual, and social) and the *objects* in the environment. Object description includes an analysis of the object structure. Each thing that is relevant to the work in a certain situation is an object in the sense of task analysis; even the environment is an object. In this framework, "objects" are not defined in the sense of "object-oriented" methods including methods, inheritance, and polymorphism. Objects may be physical things or conceptual (nonmaterial) things such as messages, gestures, passwords, stories, or signatures. Objects are described by their structure and attributes. The task environment is the current situation for the performance of a certain task. It includes agents with roles as well as conditions for task performance. The history of past-relevant *events* in the task situation is part of the actual environment if it is involved in conditions for task execution.

MODELLING TASK WORLDS

Work Structure

Work structure modelling is the oldest and most common activity in task analysis. Humans do not think about their work as a collection of tasks, but think in a structured way about their activities (Sebillotte, 1988). This structure can be captured in a task decomposition tree, which forms a hierarchy in which the high-level tasks are found at the top of the tree and the most basic tasks are at the leaf nodes. Such a "classical" task tree is usually enhanced with constructors that indicate time relationships between tasks. Many methods, including hierarchical task analysis (HTA; Annett & Duncan, 1967), méthode analytique de description (MAD, Scapin & Pierret-Golbreich, 1989), and method for usability engineering (MUSE, Lim & Long, 1994), use this kind of task tree.

Work structure is one of the most important aspects in task analysis. A design of an interactive system usually means restructuring the work and adding or removing tasks. A task tree can give an indication of aspects that are considered suboptimal in a given situation. For example, certain subtasks could be part of many complex tasks and could be automated. In other cases, tasks may turn out to be too complex and need to be simplified. When designing for usability, the work structure is important for developing the most appropriate interaction structure and functionality.

Distinguishing Tasks and Goals.   A common definition of a task is an activity performed to reach a certain goal. A goal is then defined as a desired state in the system or task world. Distinguishing between tasks and goals can be useful when analysing a task model. For example, a complex task, which has goal X, may have a subtask with goal Y. In that case, that subtask "belongs" to that task but because it does not have the same goal, one could wonder whether it is really a required task or if it causes problems. An example is a copying task in which the user checks if there is paper in the copier. Checking the paper has "maintenance" as a goal and would ideally be unnecessary. Some methods implicitly presume a one-to-one mapping between tasks and goals; for instance a task knowledge structure (TKS; Johnson et al., 1988) contains only a goal substructure, which would be called a task substructure by others' methods. Other methods, such as GTA (van der Veer et al., 1996) and MAD (Scapin & Pierret-Golbreich, 1989), allow a goal to be reached in several ways. In this way, each task has a goal, and goals can be reached by several tasks. In fact, this is similar to GOMS (goals, operators, methods, selection) in which different methods are selected to reach a goal. One step further is to define a task hierarchy and a goal hierarchy, which occurs only in complex situations. In practice, distinguishing between tasks and goals is not always so easy or clear. Usually it is an iterative process in which the distinction gradually becomes clearer. When describing detailed actions of a user at work, goals often indicate states of the system, but higher level goals are often modelled to states or particular configurations in the task world. Additionally, in complex task trees based on real-life situations, tasks near the leaves in a tree are usually connected with individual goals, and tasks represented by high-level nodes are often closely tied to organisational goals (van der Veer, van Welie, & Thorberg, 1997). When modelling complex situations for which the organisation is of great relevance, it is important to be aware of the difference between individual and organisational goals and the ways that they are related.

Describing Task Detail. Although in most cases a task hierarchy shows a great deal of interesting information, other task details may also be important. For example, conditions describe when exactly this task can be executed and when it ends. Other details may describe the exact transformations that occur in the tasks, the priority of the tasks, or whether they can be interrupted. In highly event-driven tasks it may be vital to know what the priorities of the tasks are and whether tasks can or may be interrupted. Besides such properties, tasks can also be assigned a type. For example, tasks can be characterized as monitoring tasks, decision-making tasks, action tasks, or searching tasks. The interesting aspect of distinguishing task types is that they have characteristic uses of cognitive resources such as perception, attention, memory, and motor capacity. This may be of importance when designing user interfaces because each task type poses constraints on the possibilities. At this point, it is unclear which task typing is needed. A distinction can be made in mental and nonmental tasks, but even for mental tasks, there is no one fixed list of possible types.

Modelling the Work Flow

Another common feature of most task models is task flow, which indicates the order in which tasks are executed. Two forms of flow models can be distinguished: (a) workflow representations, with time on one axis, and (b) task trees enhanced with the "classic" constructors that give a kind of time structure (mixing time and task decomposition). Theoretically they are equally powerful to express any kind of task flow. However, the second type of flow model suffers from the fact that usually many constructors are needed and extra "dummy" tasks need to be added. For example, in Fig. 19.1, two representations are used. The first is a task flow representation with time on the x-axis. The second is a task decomposition with constructors that represent the same task flow as the first representation. Because the constructors scope over all subtasks, the second representation needs an extra task, "Go there," which may not be desired. In Paternò's "ConcurTaskTrees" (Paternò, Mancini, & Meniconi, 1997) these types of tasks are called "abstract tasks." The important point here is that for both representations the specified task flow is basically the same, but the visual representation does not always allow specification of the task structure as desired. Because both representations can be useful, there is a need for an underlying model that allows both.

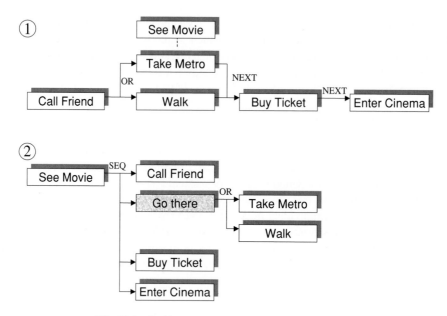

FIG. 19.1. Problems with representation of time in trees.

Events.   Events are used to model external dynamic aspects of the task—things that happen in the task world over which the agent does not always have direct control (e.g., an alarm goes off, a film breaks, a power supply fails, or new mail arrives). Sometimes there may be no need to incorporate the event explicitly in the new design, but in other cases incorporation is important. For example, it may prove useful to model the agent's reaction to an event and how it influences the sequence in which tasks are performed. In complex situations, work can be highly event driven (e.g., in air traffic control positions in which people start and stop doing tasks all the time).

Modelling Work Artefacts

Artefact and object modelling is an addition to task analysis that resembles data structure modelling of the final design implementation. The purpose is to say something about the objects, as they are physically present in the task world or mentally present in the user's mind. Not every object may be directly included in the new design but in the case of models for automated user-interface (UI) generation, there is usually a strong link between objects and UI widgets, such as buttons or menus. The question remains how much object modelling should be in task models. Extensive data modelling does not appear to offer direct help in improving the usability of the product. It also depends on the purpose of the task model; models used as basis for automatic UI generation have different requirements than models used for evaluation. For example, in GTA only the structure the objects and the tasks in which they are used are recorded. Other models such as ConcurTaskTrees (Paternò et al., 1997) and TKS (Johnson et al., 1988) also include actions that are performed on the object. The most important purpose of a task is that it "changes" something; otherwise the task has no purpose. By change we mean any sort of change, including adding information (changing an unknown to a known). Some task analysis methods such as ConcurTaskTrees (Paternò et al., 1997) describe this with task input and task output objects. Another way to describe changes is to specify the initial and final states in terms of object attribute values, as is done in MAD. In this way, the passing of information is achieved indirectly through changes in object attributes. There is no fundamental difference because the list of input and output objects can be generated from the task attributes. It is possible that the changes are not explicitly recorded, however, as in the mental processes involved in human decision making. In models that use object actions, changes are usually defined in terms of the actions instead of the task states.

Modelling the Work Environment

In addition to the work itself, the environment in which it is done is also important to consider. In the past, most methods focused on modelling one user and that user's tasks; in current applications, group factors are becoming more significant. Classic task modelling methods lack the power to deal with these situations, modelling only the static part of the task world by identifying roles. This neglects other parts of the organisation and dynamic aspects of the task world. People rarely perform their work in solitude. They work together with their colleagues and share offices; they help each other and form a social group. Certain aspects such as workplace layout are traditionally the field of ergonomics but are certainly important for UI design as well.

Physical Workplace Layout.   One aspect of the work environment is the actual physical layout. How big is the room? Where are objects positioned, and what are their dimensions? The physical layout can be modelled by assigning a dimension and location attribute to artefacts, but in practice it is usually done by sketching the layout. Usually this is sufficient to gain the required understanding.

People and Organisations.   Modelling the task world means modelling the people that are part of it and modelling its structure, which is often a part of the organisational structure. Although it may be useful to see a model of the "official" organisational structure, for task analysis the structure of how tasks are actually being done is more relevant than how they officially should be done. Specifying roles in the organisation and agents' characteristics gives relevant information that can be used in design. The roles then need to be attributed to agents. In TKS (Johnson et al., 1988), a role is defined to be responsible for performing the tasks it encompasses; for example, a movie projectionist is responsible for starting a movie. In real organisations, however, task responsibilities frequently need to be handled more flexibly, resulting in responsibilities being shifted by delegation or by mandate. The agent playing a role therefore may not perform the task he or she is responsible for; a movie projectionist could have someone from the snack bar push the button to start the movie.

Roles and Actors.   In classic task analysis literature, as well as in ethnography, concepts such as actors and roles are commonly referred to for describing tasks and the task world. Although these terms are intuitively appealing, they can cause confusion when they need to be named during task analysis. A role is defined by the tasks the role is responsible for, for example, a projectionist is responsible for starting and stopping the movie projector as well as setting up the movie projector. Mayhew (1992) defines an actor as a class of humans whereas others consider a particular person an actor. Usually there is no need to consider a particular person and provide a name for an actor (e.g., Chris, Pat) because we are only interested in describing relevant characteristics of the actor. Confusion arises when an actor is to be named and the only sensible name seems to be the role name. For instance, the actor who has the projectionist role is most intuitively called the "projectionist," which is already his or her role name. Therefore, it is usually better to name these actors arbitrarily (A, 123, People having role X) and simply record characteristics such as language, typing skill, computer experience, knows how to use Word, and so on. The important part is their characteristics and their relationships with roles. In other cases, when it does not matter who actually performed the task, it is sometimes more useful to specify that a task was performed by a role rather than by a particular actor. Sometimes even a computer system is the actor of a task (e.g., an automated movie projector).

Work Culture.   Every work environment has its own culture that defines the values, policies, expectations, and the general approach to work. The culture determines how people work together, how they view each other socially, and what they expect from each other. Taking the culture into account for user interface design (UID) may influence decisions on restructuring of work when rearranging roles or their responsibilities. Roles are usually used to describe the formal work structure extended with some "socially defined" roles. In practice, roles such as "management" or "marketing" influence each other and other roles as well. These kinds of influence relationships are part of the work culture. Describing work culture is not straightforward but at least some influence relationships and their relative strengths can be modelled. Other aspects of culture include policies, values, and identity (Beyer & Holtzblatt, 1998).

REPRESENTATIONS FOR TASK MODELLING

The task world is usually complex and simply cannot be captured in a single representation. It is therefore necessary to have a collection of representations that each shows a partial view on the task world. Each view can then focus on one or two concepts with some relationships, and together these views cover all-important aspects of the task world. In this way, each representation can be kept simple and easy to understand while all representations together model a complex task world. In the following sections, we discuss common graphical representations for task modelling. We discuss the strengths and weaknesses of the existing representations and we then propose a collection of improved representations.

Common Representations

Many representations already exist for task modelling as well as other related modelling activities. Not all of them are useful in practice, and the question is what makes a representation useful and usable. One aspect of a representation is that it should be effective. Macinlay (1986) defined the effectiveness of a visual language as "whether a language exploits the capabilities of the output medium and the human visual system." This notion can be expanded to include purpose and audience, that is, what is the representation intended for and who is going to use it, because "visualizations are not useful without insight about their use, about their significance and limitations" (Petre, Blackwell, & Green, 1997). Developing usable diagram techniques is difficult and requires insight into all these aspects. In fact, usability is just as important for graphical representations as it is for UIs, because, both depend strongly on the context of use. Most of the research in this area is the field of visualization (Card, MacKinlay, & Shneiderman, 1999) or visual design (Tufte, 1983, 1990). If we want to compare representations, we must first distinguish several purposes for which they can be used and by whom (Britton & Jones, 1999). Within task analysis the purposes of representations typically include the following:

- To document and to communicate knowledge between designers
- To analyse work and to find bottlenecks and opportunities
- To structure thinking for individual designers
- To discuss aspects of the task world within the design team

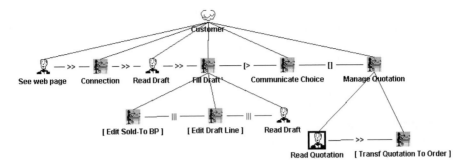

FIG. 19.2.   ConcurTaskTree example.

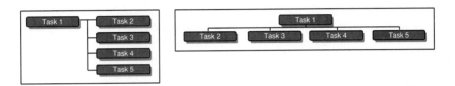

FIG. 19.3.   Depicting left-to-right versus top-to-bottom task tree.

- To propose changes or additions within the design team
- To compare alternatives in the design team or with a client

For the discussion of representations, we take the position that a representation essentially is a mapping of concepts and relationships (and possibly attributes) to the visual domain. Some aspects may concern the concepts and relationships, whereas others concern the appropriateness of the mapping in relation to the purpose and audience.

In the following sections, we discuss the different common representations that are used for task modelling: graphical trees, universal modelling language, and contextual modelling.

Task Trees.   Traditional task models mainly use task tree structures in some flavour. The task trees are intended to show the structure of the work in terms of tasks, goals, and actions. Usually, some time information is added as well. HTA (Annett & Duncan, 1967) uses trees with plans, whereas most others use trees with constructors for time constraints. Some of the older methods (e.g., GOMS; Card et al., 1983) do not even use graphical representations and use long textual descriptions. Such descriptions become highly unreadable when the size of the task model is larger than just a few tasks. The indentation of task is not sufficient for large-scale models, and therefore graphical trees are an effective improvement. In the graphical representation, visual labels (shapes, icons, etc.) add meaning to certain aspects that which make it much easier to distinguish the different parts (goal, task or task type, or action) of the diagram. For example, in ConcurTaskTrees (Paternò et al., 1997), the task trees are combined with icons for task types and LOTOS-based (International Organization for Standardization [ISO], 1988) operators that allow some more elaborate time semantics. In ConcurTaskTrees, tasks are of a certain type (abstract, user, machine), which is reflected in the icon that represents the task (see Fig. 19.2). The figure shows that even when the labels are not readable, the different task types can still be distinguished. Most graphical representations depict a task tree from top to bottom. Drawing the tree from left to right makes better use of the drawing area, however, because trees are usually much wider than they are deep (see Fig. 19.3). This issue becomes relevant when depicting large task models of more than about 25 tasks.

Task trees are generally easy to understand and build, but they are usually built without software tool support. Tool support is slowly beginning to appear, although still not commercially. Using sticky notes on a blackboard, a tree can be (re)structured easily, but for documentation purposes the trees are (re)drawn manually (i.e., using a drawing package).

Task trees are well suited for communication purposes within the design team and, to a certain extent, also for communicating with domain experts. For the latter case, the trees should not be too big. Although task trees can be powerful, they are mainly based on the subtask relationship between tasks. Additionally, some information could be given about the ordering of tasks, but no information about roles, actors, or objects is given.

Templates.   A common way to represent concept properties is a template, a form consisting of fields for every property. Templates typically contain a mixture of properties and some relationships. In MAD and GTA, the template is frequently used to describe the task properties and relationships to mostly the parent task and roles. A template is a simple and easily understood representation, but it is only useful to represent detailed information. A large set of templates is difficult to understand, even by experts. A template focuses on one concept and hence the relationships with other concepts are largely lost. Because of the lack of overview, templates are mostly used as a reference to detailed information of single concepts.

The Universal Modelling Language (UML).   In software engineering, UML (Rumbaugh, Jacobson, & Booch, 1997) is currently one of the most influential modelling languages. It grew out of the intent to standardize the models that were used in object oriented analysis and design. Currently it is widely accepted in industry. UML was not designed with task modelling in mind. In UML, each diagram is defined both syntactically and semantically. Usage of terms between diagrams has been kept as consistent as possible. Although UML was not designed for the purpose of task modelling, several diagrams can certainly be used in a task analysis context. The question is whether it would be useful to standardize on certain UML diagrams for task modelling. Because UML is a standard and many tools exist, there are clearly benefits. If UML is used for task modelling, however, the interpretation needs to be changed slightly. For example, states in an activity model become tasks, and objects become roles. UML has four representations that are directly relevant for task modelling:

- The *activity diagram.* This diagram can be used to describe the task flow in relation to events, roles, and goals. Typically an activity is triggered by an event. As soon as a task is started, a goal consequently gets "activated."
- The *collaboration diagram.* This diagram gives insight into the way different objects work together. The arrows show which roles communicate or work together in the exchange of objects or messages.
- The *sequence diagram.* The sequence diagram can show the sequence of tasks as they are performed between roles. Originally they are used to model method calls to objects but essentially there is no difference. Problems arise when calls are conditional or optional. Parallelism can also be modelled to a certain extent.
- The *Use Case Diagram.* This diagram can be used to describe what is also known as scenarios. The exact difference is an ongoing dispute in the HCI community, but at least they are both used to describe a particular set of tasks in a specified context. This representation is informal.

Considering that the collaboration diagram and activity diagram are so related to each other, usage can effectively be restricted to using the activity view with swim lanes (the dotted vertical lines in Fig. 19.4), that is, no information would be lost because the collaboration diagram

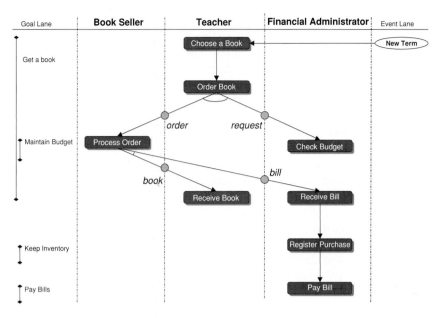

FIG. 19.4.   The workflow model.

contains less information. The sequence diagram is probably less interesting for task modelling because of the problems with conditional and optional paths. Additionally, the method calls that are interesting in the object oriented sense have no important equivalent in task modelling. At most, they could say something about "how" a task is started (by yelling or whispering commands?).

The use case diagram is useful for task modelling, although there is no clear view on the differences between a use case and a scenario. One definition could be that a use case describes a specific "path" through a task tree under specified conditions. A scenario can then be defined as a more general description that also sets the context for a use case.

Using UML diagrams has the advantage that software engineers are familiar with them, but other disciplines in the design team usually do not know them. The diagrams are fairly powerful but would require a small adaptation for task modelling purposes. Additionally, many software tools exist that support the designers to create UML diagrams.

Contextual Modelling.   Contextual modelling (CM) is part of contextual design (Beyer & Holtzblatt, 1998) method and consists of five work models to describe the task world. The models are built to describe work from the point of view of one person, and they are not intended to represent everything that a person does. The five different views are as follows:

- The *flow model* represents the communication and coordination necessary to do the work.
- The *sequence model* shows the detailed work steps necessary to achieve an intent.
- The *artefact model* shows the physical objects created to support the work along with their structure, usage, and intent.
- The *cultural model* represents constraints on the work caused by policy, culture, or values.
- The *physical model* shows the physical structure of the work environment as it affects the work (objects and their locations).

These models as they are introduced are not entirely new. Beyer & Holtzblatt (1998) claim that these representations have been tuned over time and are sufficient in most design cases.

This is questionable because none of the representations allows hierarchical building of representations. For instance, the sequence model is a linear sequence of tasks without the possibility of defining subtasks, choices, plans or strategies. Other representations such as the flow model are almost exactly the same as UML's collaboration diagram, although a different notation is used. The artefact model and the physical model are basically annotated drawings and not structured models. Even though the individual representations are not that new or special, the idea of using these "views" to describe work was not previously stated as such.

The contextual models use a somewhat different terminology than is commonly used. They speak about roles, tasks, and artefacts, but they also use intent to indicate goals. Additionally, they speak about "triggers" as events that start tasks. Events are commonly found in workflow or process modelling but are somehow not often used in task modelling.

Contextual modelling has shown that it is important to look at the work from different perspectives and work that out into practical models. Many have argued for a multiple perspectives view on work, but no one has worked it out in such detail. As Beyer & Holtzblatt (1990) noted in their book, the models usually occupy a whole wall. The problem with the flow model and the sequence model is that they only work for small design cases and do not scale very well to larger cases. Other problems are the undefined semantics. The flow model is actually a renamed UML collaboration model without distinguishing roles and actors. Contextual design also defines the process of gathering data and modelling steps. In time-boxed sessions, the models are created and consolidated in a later session. Only in the consolidated versions are models worked out in detail. This illustrates that designers do not make an exact and consistent task model from the start but rather iterate and slowly consolidate the models.

Choosing a set of Representations for GTA

In the previous sections, several common representations have been discussed. It is clear that some are more useful or usable than others and that improvements can be made. In this section, we define an improved collection of representations that covers the views of GTA. This collection of coherent representations is an attempt to provide a more useful collection of representations for practitioners.

For each of the views, we define one or more representations that form a useful "package" for that view. Together, the representations can form a practical tool set for the designer. The representations are based on existing representations but include some additions or modifications to make them more usable and useful for task modelling. Compared with CM, the main differences are as follows:

- The CM sequence model is replaced by a workflow model similar to the UML activity diagram.
- The CM sequence and CM flow model are combined into one representation.
- Decomposition trees are added.
- The CM cultural model has been redesigned.
- The number of concepts is larger than in CM.

In contrast to UML, we use a modification of the activity model. We have added and event and goals lane and changed representations for parallelism and choice.

Modelling the Work Structure.   The purpose of the work structure model is to represent how people divide their work into smaller meaningful pieces to achieve certain goals. Knowing the structure of work allows the designers to understand how people think about their work and to see where problems arise and how tasks are related to the user's goals. The relation between tasks and goals helps designers choose which tasks need to be supported by the system and why (i.e., which user goals are independent of the technology used). For modelling work structure, the task decomposition tree has proven to be useful and usable in practice. The tree is essentially based on the *subtask relationship* between tasks. Besides tasks, goals can also be incorporated. At the highest level, a tree can start with a goal and subgoals and then proceed with tasks and subtasks. In that case, the *subgoal* and *has relationship* are also used. A task decomposition is modelled from the viewpoint of one role or goal. If complex systems are modelled, several task trees are needed to describe the work for all the roles. It then becomes difficult to see how work is interleaved. Trees normally contain a time ordering using constructors from top to bottom or left to right, depending on the way the tree is drawn. The inclusion of time information can be insightful but often is problematic as we discussed with regard to ConcurTaskTrees, which uses operators based on LOTOS (ISO, 1988), probably the best defined time operators. On the other hand, it is not always necessary to be precise in everything that is modelled. Designers will also typically model that certain tasks occur sometimes or almost never. In our opinion, including some time information is useful, but this kind of information is better represented in a workflow model if precision is required.

If time is included, then a number of time operators are plausible. In our experience, it is useful to have a set of standard operators while allowing designers to create their own operators when needed. For the average usage, the following time relationships have proven sufficient:

- *Concurrent:* The tasks occur concurrently.
- *Choice:* One out of a set of tasks is done.
- *AnyOrder:* All tasks of a set of tasks are done in no fixed order.
- *Successive:* One task is followed by another.
- *Any:* Zero or more tasks of a set of tasks are done in no fixed order.
- * combined with other constructors. Used to express iteration.

In the work structure model, the root of the tree is a goal with possibly some subgoals. Connected to goals are tasks that are represented as rounded rectangles. The tree is drawn from left to right instead of top to bottom for more economical use of space, especially when trees become large. Other aspects of work structure include role structures and the relationships with tasks. For role structure trees can also be used. When used to show goal or role hierarchies, the time constructors are not used.

Modelling the Workflow.   The purpose of the workflow model is to show work in re-lation to time and roles. The model gives the designer insight in the order in which tasks are performed and how different people are involved in them. Additionally, it can show how people work together and communicate by exchanging objects or messages. Typically, a flow model describes a small scenario involving one or more roles. This way, it shows how work is interleaved.

The flow model specified here is a variation on the UML activity graph. We included events and goals to make it more suitable for task analysis. Additionally, the representations of the time operators have been modified to be more appealing. This way the collaboration diagram (or, in contextual design, the flow model) is not necessary because the information has been combined in one representation. Each flow model describes a scenario that is triggered by an event (see Fig. 19.4).

Work usually does not start by itself but instead is often highly event driven (van der Veer et al., 1997). The event is represented by an oval that is connected to the first task. The sequence of tasks is given using a concurrent or a choice operator and not any of the other operators as suggested for the structure model. The concurrent operator is represented by an additional arc,

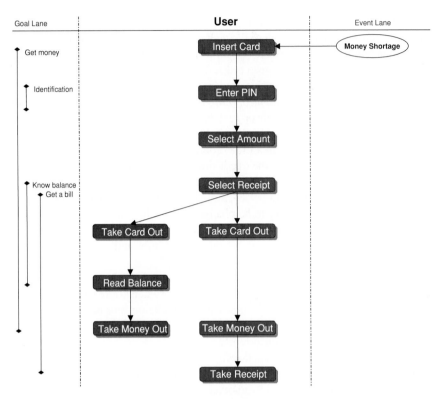

FIG. 19.5.   A flow model of a Dutch automatic teller machine.

and the absence of the arc indicates the choice operator. Tasks can optionally be arranged in swim lanes, one for each role. Objects can be passed between tasks that have different roles and are drawn on the border of the adjacent swim lanes. When needed, goals can also be added to this representation. With a certain task, a new goal can get "activated" until it is "reached" in a later task. The goals are written in the first column with vertical lines to show how long they are activated.

The flow model does not show hierarchical relationships between tasks, and a flow model can only use tasks that are hierarchically on the same level. For subtasks, a new flow model needs to be specified. The addition of the goal lane can show many useful aspects when analysing the work flow. For example, Fig. 19.4 shows that once the "teacher" has received the book his or her goal is achieved but the scenario is not finished yet. Figure 19.5 describes the use of a typical Dutch automatic teller machine (ATM) and shows why people often forget to take their receipts. As soon as the primary goal has been reached, users loose interest in the remaining tasks. In this case, the task of taking the receipt is positioned after the users get the desired money. When ATMs were first introduced, the situation was even worse, because the machines returned the card after the money had been dispensed. Because the user had already achieved the goal, he or she was disinterested in the remaining tasks; users consequently forgot to take their bank card. For objects that are being passed between roles, it holds that each object must be associated with both tasks with the uses relationship. Note that the objects that are only used in one task are not shown in this representation. For example, the card itself is not shown in Fig. 19.5. Iteration is not specified in the flow diagram. If a task is done several times, an asterisk can be used to indicate that the task and its subtasks are done several times. Usually, however, iterations are specified in the work structure model. Iteration is specified on subtasks and not tasks on the same level, which are shown in the work flow model.

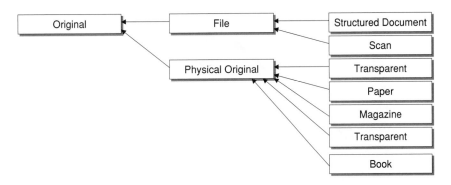

FIG. 19.6. Example of a Universal Modelling Language class diagram.

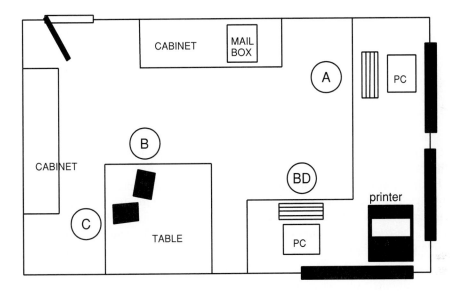

FIG. 19.7. The physical layout model.

Modelling the Work Artefacts. To model work artefacts two relationships between objects need to be represented: the containment and the type relationships. To express containment and type, the UML notation can be used (see Fig. 19.6). The objects themselves can be annotated with their attributes or their visual appearance. It is important to remember, however, that we are only modelling objects and relations that are relevant to the user and not any irrelevant internal system objects. To some this may suggest that the task models describe an object oriented system model, which is not the case.

Modelling the Work Environment. The environment model describes two aspects of the environment: the physical layout of the environment and the culture within the environment. The *physical model* is simply described by one or more annotated "maps" of the environment. The purpose is to show where objects are located in relation to each other. The objects are those that are relevant for the work and also those that are in the same space. Figure 19.7 shows an example of a workplace layout. Such layout diagrams can easily be drawn using commercial drawing software such as Visio (Microsoft).

The other model is the *Culture Model* (Fig. 19.8). The culture model we describe here is an adaptation of the culture model from contextual design, in which the roles are represented

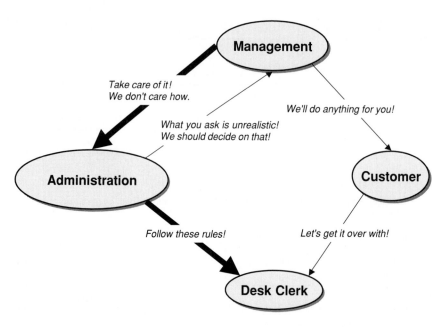

FIG. 19.8.   The culture model.

in overlapping circles. Overlapping of circles has no meaning, however, although it suggests that there is one, and hence we adapted the model.

In Fig. 19.8, we define the culture model as follows: Roles are represented as ovals; the ovals are connected by arrows if there is a force between roles. The relative strength of the force is depicted in the width of the arrow. Forces are annotated with attitudes of the force relationship.

In some cases, a force applies to more than one role. By drawing an extra circle around roles, a force can indicate one to many forces, which can typically be used to describe corporate culture.

Static Versus Dynamic Representations

All of the representations discussed in the previous sections are static. However, representations can also be more dynamic. Traditionally, a representation is static, that is, it does not change after it is drawn and is designed for use on paper. However, it is often convenient to emphasize a certain aspect in the representation. When software is used to draw the representations, they can be changed dynamically. Card et al. (1999) called these active diagrams. For example, one could easily switch between a flow model with or without swim lanes. Alternatively, it could be possible to add some extra information by marking tasks as "problematic" or "uses object X". Designers often make such annotations to explain certain aspects during a presentation or in documentation. In software, we are already familiar with active diagrams and they occur in scrolling, zooming and syntax highlighting. This asks for a more flexible view on what constitutes a representation and when a representation can be modified. Viewers could control dynamic aspects manually but they could also prespecify them using a function, in which case we usually speak of animation. Now that it becomes increasingly easy to create dynamic representations, it is important to understand when and how they could be applied usefully in design.

In task modelling, animation is a way to create more dynamic representations. Animation can be used in simulations of scenarios or task models (Bomsdorf & Szwillus, 1999). Using simulations an analyst can step through a scenario and get a different feel for what takes place. Other uses might be to "debug" a task model, which is particularly useful for envisioned task models.

ANALYSING THE TASK WORLD

Representations help to represent the knowledge that is gained in the process of task modelling. During this process, it is useful to analyse what is actually represented in the specifications. One aspect of this is to see if the specification correctly represents the knowledge; other aspects may focus on seeing the problems in the task world. For envisioned task models, it is important to make sure the specification is correct. One frequent criticism of task analysis has always been that it remained unclear what exactly one should do with the data: "We have the data, now what?" What should be done next is an analysis of the data to uncover both problem areas and opportunities that relieve the problems in the task world. Those results then become the basis for designing an envisioned task model. Representations for envisioned task models can be largely the same as for current task models; only the interpretation is slightly different. Task analysis research has focused on data collection and modelling techniques, but it has neglected research on data analysis. Naturally, during modelling activities some of the data is analysed when models are being constructed and modified, but much more structural analysis is possible. We found that many problems fall into the same categories and have a general and domain independent nature:

- *Problems in individual task structures.* The task structure is suboptimal because too many subtasks need to be performed or certain tasks are too time-consuming or have a high frequency.
- *Differences between the formal and actual task performance.* In cooperative environments, documented regulations and work practices usually exist, for instance, as part of ISO9000 compliance. In reality, tasks are usually not performed exactly as described on paper and that "one way" of how the tasks are done does not exist. When persons in a cooperative environment think differently about what needs to be done, problems arise.
- *Inefficient interaction in the organisation.* Complex tasks usually involve many people who need to communicate and interact for various reasons, such as gathering or dissemenating knowledge or determining responsibility. This can result in time-consuming interactions as well as confict among involved parties.
- *Inconsistencies in tasks.* Tasks are defined but not performed or are executed in contradictory order.
- *People are doing things they are not allowed to do.* In complex environments, people fill a role that makes them responsible for certain tasks; however sometimes other people actually perform these tasks, for which they do not have official permission or are using or changing objects without the authority to do so.

Of course, not all problems can be detected automatically. Nonetheless, by using our model for describing task models, many characteristics can be detected semiautomatically by providing the analyst with a set of analysis primitives. Analysing a work environment can be done when the data present in the model are transformed into qualitative information about the task world. There are two ways to perform qualitative analysis. An analyst may search heuristically by looking at properties of a specification that might point the analyst to problems. Alternatively, the data can be analysed on a logical level by putting some constraints on the model, a form of verification. Constraints that do not hold may show interesting features of the task world. In the next sections, we elaborate and provide examples these forms of analysis.

Heuristic-Model-Based Evaluation

In heuristic evaluation, we try to find out "what is going on" by looking at certain properties of a specification. The goal is to gain an understanding of the task world and to find the nature and causes of problems. Using a standard set of properties, we can increase the chance of a successful analysis. For instance, looking at all tasks in which a certain role is involved may help to gain insight in the involvement of a role in the task structures. Heuristic evaluation is done by checking specification properties that are not objectively right or wrong. It is up to the analyst's interpretation to determine whether there is reason for concern. These properties are more interesting for finding the actual problems in the task world. Possible properties related to problems are as follows:

- The number of roles involved in a task
- The rights roles or agents have for the objects used in the task they are responsible for or perform
- The frequency of tasks
- The frequency of events
- The number of tasks for which a role is responsible
- The number of subroles a role has
- The number of levels in subtasks of a task
- The number of subtasks on the same level of a task
- The objects used in a task
- The roles involved in the task
- The objects that are used by a certain role
- Tasks that are delegated or mandated

Model Verification

Verification concerns only the model as it has been specified. Only a limited degree of verification of a task model can be supported due to the inherent lack of formal foundations for task models; there is no model to verify the task models against. However, it is possible to see if the task model satisfies certain domain-independent constraints. For example, we would prefer that for each task there is at least one responsible role and that each task is actually being performed by an agent. These constraints can be specified as logical predicates and can be checked automatically. Within model verification constraints, we can distinguish constraints on cardinality, type, attributes, and constraints between specifications.

Cardinality Constraints.   These constraints concern the cardinalities of the relationships between concepts; however, they have been defined irrespective of the specific study being done; they should hold in any domain. A task model in which all constraints are obeyed may be considered "better" than one that does not obey all the constraints. In other words, the constraints allow us to denote classes of models, which have an order of preference. Examples include, "Each event should trigger at least one task" or "Each task should be performed by at least one role".

Type Constraints.   These constraints deal with relationships between entities of the same type and are of a general nature, with the possible exceptions of the object constraints. Examples include, "An instance of an object cannot contain itself" and "A task cannot have itself as a subtask".

Attribute Constraints.   Other properties might be related to factors such as missing goal attributes for the tasks that should have a goal. If some of the attributes such as duration and frequency are formally described, other properties could be checked as well. The question is whether these properties make it worthwhile to enforce a more formal specification of the attributes. In such a model, the following properties could be checked:

- The total duration of a task is less than or equal to the sum of the duration of all subtasks. This holds only for sequential tasks that have sequential subtasks.
- It is not desirable to have contradicting task sequence specifications (e.g., A after B after C and A after C after B) at the same time. When analysing a current task model, this may be interesting to detect but is to be avoided in an envisioned task model.

Comparing Two Specifications

The previous sections were concerned with only single specifications. Another option is to compare two specifications, in which case other properties become interesting. There are three ways to compare two specifications:

- Comparing two current task models.
- Comparing two envisioned task models.
- Comparing a current task model with an envisioned task model.

In the last case, comparing specifications may say something about the design decisions taken when redesigning a task world. For example:

- Which tasks were reassigned to different roles?
- Which roles were reassigned to agents?
- Which tasks were removed or added?
- Which objects were added or removed?
- Which events are new and which have been removed?
- Which object rights have changed?
- Which tasks have become less complex?

Model Validation

Validation of current task models means checking if the task model corresponds to the task world it describes. In the process of validation, one may find that certain tasks are missing or that there are more conditions involved in executing a task. Often one finds that there are exceptions that had not been found in earlier knowledge elicitation. Consequently validation needs to be done in cooperation with persons from the task world and cannot be automated. It is possible, however, to assist in the validation process, for instance, by generating scenarios automatically that can be used to confront the person from the task world. Such generated scenarios are in fact simulations of pieces of the task model. Recent work on early task model simulations (Bomsdorf & Szwillus, 1999) has shown promising examples of early simulations based on task models. During the design process, it is difficult to say when to end task analysis. At one stage, it may be considered adequate, but later in the design process new questions may arise that cause task models to be extended or revised (e.g., more information is needed about object attributes, or missing tasks are being discovered). Validation of envisioned task models means checking whether the specified tasks models actually improve the task world. In this case, the model needs to be strict on aspects such as consistency. Techniques such as simulation can be useful in the validation of envisioned task models.

SUPPORTING TASK-BASED DESIGN

With the development of GTA, we developed a design environment or tool called EUTERPE (van Welie, 2001) that is currently available as public domain software (www.cs.vu.nl/~gerrit/gta). The main function of EUTERPE is to support task modelling. Representations include task trees, templates, and other hierarchical representations. Ideally, EUTERPE would be a workbench that supports designers during task-based design. For each activity that could benefit from tool support, a component would be present. Support for task modelling, dialogue modelling, and documentation is present, but many other components could be added (e.g., support for simulation, design rationale, and sketch-based prototyping). In this section, we discuss the functionality that has been implemented in the latest version of EUTERPE (see Fig. 19.9 for an example of a screenshot of the tool in use).

Supporting Task Modelling

EUTERPE provides its most complete support for task modelling. When the tool is started, the designer can choose to create a new task model, after which a new Hierarchy Viewer is shown. This window contains tab sheets for each of the main GTA concepts (tasks, roles, agents, objects, events) except the goal concept. Hierarchical representations of goals have not yet been included in the tool. The first sheet shows a task tree, and the designer can build one by inserting child or sibling nodes. For objects, there are two hierarchies, one for the containment

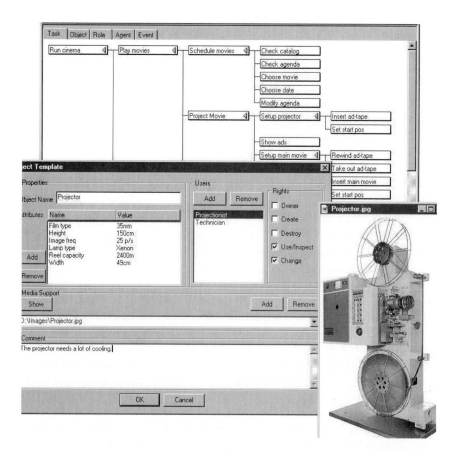

FIG. 19.9.   The Hierarchy Viewer, an object template, and a picture.

hierarchy and one for the type hierarchy. Events and agents are not hierarchically structured and are therefore shown as lists.

For each concept, a template exists that appears when double clicking the node. The task template allows the designer to specify task details such as timing information and task conditions. Additionally, some relationships with other concepts can be established and viewed. The task template has become rather full of fields, and, after user testing, it was decided to initially show only the most used fields and present the others on request. For templates of other concepts, this was not necessary because of the small number of fields.

Figure 19.9 shows the Hierarchy Viewer, an object template, and an image. Task trees tend to become quite large; a hundred tasks or so is not uncommon. Therefore, trees can be (partially) collapsed to give the designer more overview. Additionally, it is possible to zoom in or out so that the visible part of the task model can be optimised. This feature proved very useful during modelling and also during presentations in which low-resolution displays were used. Editing the models has been implemented using the Windows style guidelines and includes cut and paste as well as drag and drop functions. Using the editing function, the designer can move nodes, copy sub trees, delete nodes, and so on. This is actually the tool's most used function. All functions are also accessible through the keyboard. All user actions can be undone using the multilevel undo function.

Supporting Model Analysis

Another activity that is supported by EUTERPE is task model evaluation. As soon as some form of task model exists, the model can be evaluated. Evaluation can be done for several purposes as discussed earlier.

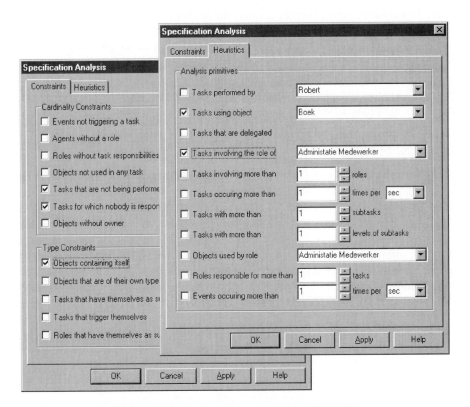

FIG. 19.10. Evaluation using constraints and heuristics.

Figure 19.10 shows the dialogue screens for specifying the questions that the designer wants answered. The questions often need some parameters to be specified, which is done using a form-filling dialogue style. The designer questions have been split into constraints and heuristics. Constraints apply to every specification and should ideally have zero results (see Fig. 19.10). Heuristics can be also used to analyse a specification and to find inconsistencies or problems.

EUTERPE can process several queries in parallel. Each node found by a query is coloured in the Hierarchy Viewer. If a node is selected in more than one query, it takes on the colour belonging to the last query. Although this may not always seem a satisfactory solution, in practice it was never considered a problem.

FROM TASK ANALYSIS TO TASK DESIGN

One of the difficult steps in the user-interface design process is the transition from the analysis phase to the design phase. The results of the analysis phase are a detailed description of the problem domain and the identified areas for improvement that set the design goals (requirements) for the system. The purpose of the detailed design phase is to design a system that meets those design goals. The transition from analysis to an initial design is characterized by a combination of engineering and creativity to incorporate analysis results in a concrete design. This transition can not be done entirely by following a simple set of predefined steps and requires a certain amount of creativity. In UID literature, this transition is called the gap between analysis and design. The gap is concerned with questions such as, what are the main displays? Which data elements need to be represented, and which are merely attributes? Which interaction styles are appropriate? How should the user navigate through the interface structure? How will functionality be accessible? Besides the analysis results, technological constraints and wishes of the client may complicate detailed design even further. In some cases, however, it may even be possible to create new technology that is needed for an optimal design solution. Wood (1997) described a number of methods and techniques that can be used to make this transition. In practice, bridging the gap means coming up with an initial design based on the analysis, which then starts off an iterative development process. Naturally, the goal is to reduce the number of iterations in design by basing the initial design solution directly on the analysis.

Guidelines for Bridging the Gap

Holzblatt described a technique called user environment design that is part of the contextual design method (Beyer & Holtzblatt, 1998). With UED the system is being structured using a mix of functional and object-oriented focus. Using that technique the major displays are identified on the basis of the contextual analysis. From there, the system is again developed with iterative prototypes. All of the guidelines described earlier are based on identification of the interface components, their structure, and the navigational structure. The major interface components are directly derived from the task and object models built during analysis. The way the functionality is distributed over the UI components depends on the type of system being built. The two main types mentioned (product vs. process) offer a broad categorization, but it may not always be easy to classify an application, for example interactive training applications have a bit of both. Although different types of applications can be distinguished, the high-level process can be summarized as follows:

- Develop an essential conceptual model of the new task world. This model describes the task world without any reference to tools and the systems being used.

- Identify the major tasks and objects that need to be part of the system. These will become the high-level interface structure.
- The first two steps result in the design of the future task model, Task Model 2 (see Introduction).
- Depending on the type of application, structure the application based on a process or product metaphor.
- Create navigational paths in the interface structure depending on the task structure.
- Design the presentation using a platform style.

These steps result in the specification of the UVM.

After this short transition process, the iterative design activities are started to mature the system. The actual techniques used in these transition activities are reportedly very low tech (i.e., paper and pencil, sticky notes, and flip charts for making sketches). At this point in the design process, the design solutions have the character of sketches and are hence informal. Nonetheless, using these sketches and paper prototypes, a great deal usability evaluation can be done both internally and with future end users. Constant evaluation drives the design towards a more and more complete specification of a usable system.

Specifying the User's Virtual Machine

User interface design consists of more than just designing some screens. Interfaces can become very complex and once an initial sketch of the interface exists, many aspects need to be worked out in detail, including the interaction, the navigation structure, and the system's behaviour. We now take a closer look at the important aspects of the user interface. For the sake of the discussion, we use a different term that covers a broader range of aspects than is usually thought of when discussing the user interface. Tauber (1988) introduced the term *user's virtual machine* (UVM), and it indicates those aspects of a system that are relevant to the user. The user's attitude typically is, "Who cares what's inside?" To the user, the interface *is* the system. The UVM is a useful concept to show which aspects of the user interface are important and hence need to be covered in the detailed design phase. These facets can be broken down following the Seeheim (Pfaff & ten Hagen 1985) model into following:

- *Functionality design.* The functionality as it is relevant to users. Functionality includes the functional actions and objects that will be available to the user.
- *Dialogue design.* Structure of the interface without any reference to presentational aspects, the navigational structure, and the dynamic behaviour of the interface.
- *Presentation design.* The actual representation of the user interface, including details such as layout, colours, sizes, and typefaces.

All three activities are dependent on each other, and they need to be kept consistent to form a coherent whole. Moreover, from a usability perspective, there is also a forward dependency from functionality to presentation. If the functionality is not designed well enough, the system will not be *useful* to users, and therefore dialogue and presentational aspects are irrelevant. In the same way, the dialogue needs to be sufficiently refined before presentational aspects matter. Nonetheless for each system there will be an emphasis on one of the three aspects because of the nature of the system. It makes a big difference whether it is a safety-critical system or a mass-market consumer application that is being designed. This forward dependency may also offer an explanation for the fact that, in practice, usability aspects are often not discussed until after the software design.

The UVM is user specific, or more precisely, a UVM belongs to *one* role (i.e., a role with the associated tasks and goals). Because systems are usually designed for multiple roles, several UVMs need to be designed. For the final design, these UVMs need to be integrated to design one system for all roles. This means that a design is always a compromise. Besides the different roles, certain user groups have specific needs concerning dialogue and presentation factors as opposed to functionality. For example, elderly users may need larger font sizes, or disabled users might need speech output. This also leads to a need for adaptable or adaptive interfaces. Adaptation should however never be used as an excuse for not making certain design decisions: "Let's just make it configurable".

Task Model and Design Solutions

The relationship between a task model and an interface design is a complex one. Task models are task- and goal-focused, almost by definition, whereas most user interfaces are object oriented. This means that some form of transformation must occur, and both Dayton and Holtzblatt have given guidelines to do so. Although the task structure is central to workflow-type applications, such applications are still very object oriented. The tasks may reappear in menu functions, but there must still be objects to manipulate. It is therefore important to establish a link between the task structure and an object model for use in the interface. Because objects and the operations on them do directly dictate a task flow, the designer must make sure that the original task flow is optimally supported in the detailed design. For example, parallel tasks may require several functions to be accessible at all times, and none of them can be the model. Another example is that certain sequential tasks may be presented sequentially, but in other cases the same result is achieved implicitly by a sequence of operations on objects.

The final detailed design is good when it supports the task model or, even better, improves on it. When design principles are used effectively, the number of steps in the detailed design may be reduced in comparison to the task model. This is where creativity comes in and the designer's expertise is of high importance. However, when the task model is used as a basis for design, it must be able to answer the design questions designers have, such as the following:

- What are the critical tasks?
- How frequently are those tasks performed?
- Are they always performed by the same user?
- Which types of users are there?
- Which roles do they have?
- Which tasks belong to which role?
- Which tasks should be possible to undo?
- Which tasks have effects that cannot be undone?
- What errors can be expected?
- What are the error consequences for users?
- How can prevention be effective?

We believe that the perspectives of GTA and the modelling techniques we use provide sufficient information to answer such design questions. In the process of refining GTA, such questions have led to additions or changes to our techniques. Task modelling for design has requirements different from cognitive task modelling in general. The goal is not to over- or undermodel. In practice, resources are always limited, hence it is of utmost importance that a task analysis can be done efficiently while remaining optimally effective for design.

APPLICATION OF GTA IN INDUSTRY

In this section, we discuss a case in which task-based design was applied in an industrial context. It shows what kind of issues need to be faced when performing a task analysis and the kind of results that can be obtained using a task-based design approach. The case discusses a redesign of an Austrian industrial security system. Other examples of practical applications are several designs created for the Dutch tax office and for Philips Design.

Seibersdorf's Redesign of Security Systems

Our task-based design has been applied at the Austrian Research Centre Seibersdorf (ARCS), mainly for the redesign of a security system produced and marketed by Philips Industry Austria (van Loo, van der Veer, & van Welie, 1999). In its original version, the system has been used at many sites, including banks, railways, and power plants.

The main problem in this case is related to confidentiality of the task domain. The actual security systems in use at these companies are the basis of our knowledge of the task domain. Obviously, none of these companies is eager to have details of its security management and procedures made available to outsiders, even if they are employed by a company that designs their system. Securing large facilities such as factories, museums, banks, or airfields is no small feat. Monitoring and controlling areas in these facilities is accomplished with the help of movement detection systems, video systems, access control systems, fire detector systems, elevator control systems, and so on. In practice, all the information from these (sub-)systems is led to a control room where human operators have to respond appropriately to (alarm) signals. An overview of the status of the building or premises becomes almost unmanageable in complex combinations of subsystems.

To support the operator in monitoring the state of the secured object and to integrate the different subsystems into one system, the sCalable Security System (CSS) was developed. The CSS integrates the information flow among all subsystems in a single central computer system with a generic (graphical) user interface available on several workstations. Little knowledge was available about how the current system is used and what kind of problems the users have with the system user interface in performing their work. To gain more insight into this matter, the first phase of task-based design was performed. The analysis focused on the use of the system by the end users (i.e., operators, gate-keepers, and system managers) in their actual use environments (factories, chemical plants, and office buildings were visited).

The goal of the analysis was to gain insight in the current use of the system and to propose changes and extensions to improve the (practical) usability of the system. In this case study, the physical layout of the control rooms was essential in detecting problems (see Fig. 19.11).

Many systems assumed that an operator would always be within viewing range of the screen, which was not the case. Participating observations and semistructured interviews were first carried out at two sites that seemed particularly relevant. Gaining the necessary cooperation of the managers and employees at the sites took some effort. Explaining the goal of the visits by disclosing some characteristic cases and ensuring that the resulting information remained confidential proved to be helpful in this process. At the end of the visits, most managers and employees were enthusiastic that someone finally took the time to listen to them and took their grievances seriously. Because of the sensitive nature of the work (security control rooms inside the facilities to be secured), camera recordings (which are common in worksite observations) and interviewed tape recordings were (almost) impossible. Crucial to the acceptance of

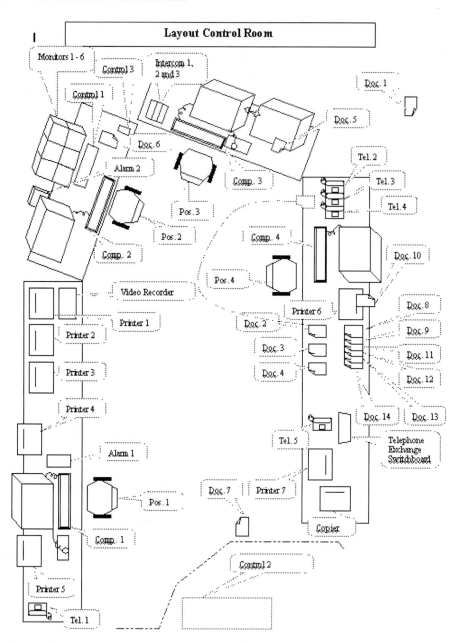

FIG. 19.11.   Layout of a control room.

task-based design in the project organisation were the following factors:

- *External funding.* The analysis of the CSS was carried out in the context of the European ΟΛΟΣ project (OLOS: A holistic approach to the dependability analysis and evaluation of control systems involving hardware, software and human resources, EC Human Capital Programme CHRX-CT94-0577). This made it possible for ARCS to get to know task-based design with little financial risk. The results were the main basis for extending the appointment of the researcher at ARCS, outside European funding, to apply the method on other projects as well.

- *Complementary expertise.* The results of the analysis made apparent a gap in the acquired expertise in the project organisation: the direct translation of context-of-use characteristics in design information.
- *Task-based design (in this case, GTA) as an "official" method.* This offered a more solid base from which to work and communicate. One does not start from scratch. In getting support for gathering data, it was crucial to mention that we were using a specific method. We capitalised on the concepts of "objectivity" and "being systematic," notions that are inherent to GTA and that make a technology-oriented organisation more receptive to analysis.

During analysis, the conceptual framework proved to be the greatest help. It worked as a kind of checklist, focusing attention on things that matter in performing tasks. On the other hand, during information collection the most important characteristic of the analyst is being open to understanding why people do something. One has to be able to "just enter a room" and start observing something, which might turn out to be useless—or quite useful—for the analysis. One has to be able to deal with the open-ended quality associated with a more ethnographic style of analysis. That the method leaves enough room for this without loosing the focus of system design as witnessed in the framework and later stages is regarded as a strong point, not a weakness. In this project, we did not use the complete task-based design approach because we were not involved from the beginning of the project. This is not considered a problem; however; as part of a larger design team and project culture, the results of the analysis stated in the conceptual framework (agent, work, situation) and the translation of these findings into detail design (the UVM) are incorporated in the project. Technical design recommendations were done by iteratively developing sketches of design alternatives and providing these as mock-ups in the real work situation. The "translation" was particularly crucial, and this turned out to be part of the core benefit of using the GTA method in this case. The outcomes of the analysis were translated into concrete, specific recommendations for interface and system design. This way, the results were regarded as beneficial to the overall project. The analyst acts as an intermediary between users and system designers and has to be able to speak both "languages." If the analyst avoids either the conceptual world of the user or of the designer, the method won't work. Reactions to the GTA method are positive, and money has been spent to carry on the work within the framework. What is still lacking, however, is more awareness within the company of what it means to perform task analysis at the location of the customer and how it should be integrated in the existing system design culture. This design case showed that task-based design indeed delivers the kind of results that are expected from such a method.

Application Issues of Task-Based Design

During the application of GTA in industrial contexts, several application issues were encountered, and we discuss these in the next sections. To a certain extent, these issues reoccurred in all cases we have collected so far, and together they provide a critical view on the practical value of task-based design.

Performing Task Analysis. Task analysis and task modelling are at the heart of task-based design. Even though the tools and techniques we developed have improved task analysis, several issues remain regarding practice. For most designers, task analysis is a new technique that needs to be learned. Some problems are related to how designers learn such a technique. Other issues are related to difficulties with the techniques themselves. The issues are as follows:

- *The need for a methodology for practitioners.* Most of our documentation on task analysis is in the form of research publications, and consequently they are written using a certain style required for scientific publications. This is often not the best way to reach practitioners, who may not be interested in the scientific argumentation common to such publications but are more practice oriented. The attitude of designers is focussed towards direct application (i.e., "Ok, but what do I do now?"). If the way practitioners work is to be changed, our "story" needs to be adjusted so that it is more appealing to practitioners. Practitioners need more practical help. It is better to give them diagramming techniques and methodological support that tells them which techniques to use and how to model in the proper way. Techniques need to be outlined with many examples and should discuss the detailed steps of creating models.

- *Task or goal?* Distinguishing between tasks and goals is crucial to good task analysis, but it is difficult to understand for practitioners. During development, exactly what are the goals is not always clear. Some may initially be modelled as tasks (things to do), but it later becomes clear that they are actually goals (states to be reached). It takes time to realize such aspects, and this is one of the most valuable outcomes of the modelling exercise. Practitioners may sometimes be impatient when facing such issues, and their background clutters the issue even more. For engineers who develop the technology, tasks or goals are easily mistaken for system functions. The internals of the system are so familiar to them that it may become difficult to make a shift of thought and 'forget' about the system internals for a moment.

- *Modelling is not easy.* Making and understanding hierarchical models is not always easy. Cooper (1995) reported that end users often have difficulties dealing with hierarchical structures in user interfaces. For some designers, it is natural to use flow diagrams because they are familiar with them. Additionally, translating data into models using observational studies or other sources is not a trivial task. Designers need to learn to recognize the "concepts", the important parts in the raw data, and model them.

- *Task analysis may not be possible.* In some domains, it may turn out to be impossible to do a task analysis simply because it is not permitted. For example, the Dutch company Holland Signaal develops radar systems for naval ships, but because of military security strategies, analysts are not permitted to board the ships to see how people work and to collect information on the context of use. In other cases, as discussed previously, it may not be possible to record interviews or to film on location.

- *Modelling task interruptions.* Often tasks are interrupted and continued later, and it is not possible to model this. Extensions are needed to adequately describe task interruptions and their possible continuation.

- *Modelling conditional task flow.* In some cases, the task flow is determined by conditions. Although these can be modelled using start conditions of tasks, they are not visible enough in the task flow representations. Usually, conditional task flow is related to decision-making tasks.

- *Modelling task strategies.* Task experts often use strategies to deal with their tasks. It allows them to be more efficient and effective. A strategy could be modelled as a specific, separate task flow or tree, but this is not entirely satisfactory. Strategies are often based on implicit knowledge that is difficult to make explicit.

Integration With Current Design Practice.   Currently, most companies are not doing task-based design. Ideally, the design method we developed would be widely used in practice, but making companies switch to a new method is easier said than done. In constructing our method, we design from an idealistic point of view. In practice, there are many other factors to consider. Business goals influence the importance of designing for usability, and most companies already have a design method that may be satisfactory for them. There is a legacy problem here: experienced people who are employed already, as well as the habitual application of current techniques, tend to counteract the acceptance of new techniques even if these could objectively be shown to be advantageous. Integration issues include the following:

- *Task-based design and other software design methodologies.* Most companies already have their own design methods. It is unrealistic to expect them to adopt a new design method easily. For some industries, object-oriented software design is still new, and switching to a task-oriented view for the user interface is difficult. Some wonder whether it is realistic to focus so much on user interface (GUI) when software design is already complicated enough for practitioners! If task-based design is to be integrated in current software design practice, the position of task-based user-interface design activities need to be "positioned" in relation to software design activities. The task modelling and analysis activities can be done before or as part of a requirements phase and should not be problematic. The later activities concerning detailed design and iterative evaluation should at least be done in parallel with internal software design; however, this requires that the parallel tracks are consistent with each other, which is not a trivial accomplishment.

- *Throw-away prototyping.* Many people nowadays use rapid application development (RAD; Martin, 1991) methodology in which the prototype evolves into the final product. This makes it difficult for designers to throw away prototypes because they want to reuse what they have. In some cases, however, it may be advantageous to throw away a prototype because the "main idea" was wrong. From another point of view, prototyping in the sense of task-based design is simply developing a special representation of detailed design specifications for early evaluation. For the sake of task-based design, it is not acceptable if the development of this representation is influenced or constrained by issues such as code reuse.

- *Arrogance or ignorance.* "We know how our products are used or what the user wants to do" is typical attitude in industry, but there is often no data to support such claims. Sales figures seem to be the most important indicator for usability to many companies. Some assume that the usability of their product is effective because they believe they design usable systems, but often they have done no testing to confirm this.

CONCLUSIONS

Groupware Task Analysis is a complete task analysis method that has been built on the experiences of many other methods. The method has a theoretical foundation as well as practical techniques, a set of representations and even tool support. By constantly applying and teaching the method, we have tested and validated the method in practice. Nonetheless, we intend to refine our method in future work, especially in the area of representations and tool support. We believe that task analysis and task modelling should become a practical, efficient and effective activity for any design project that involves complex interactive systems.

REFERENCES

Annett, J., & Duncan, K. (1967). Task analysis and training in design. *Occupational Psychology, 41,* 211–221.

Beyer, H., & Holtzblatt, K. (1998). *Contextual design.* San Francisco, CA: Morgan Kaufmann.

Bomsdorf, B., & Szwillus, G. (1999). Tool support for task-based user interface design. In *Proceedings of CHI 99, Extended Abstracts.* New York: Association for Computing Machinery (pp. 169–170).

Britton, C., & Jones, S. (1999). The untrained eye: How languages for software specification support understanding in untrained users. *Human-Computer Interaction, 14,* 191–244.

Card, S., Mackinlay, J., & Shneiderman, B. (1999). *Readings in information visualization: Using vision to think.* San Francisco, CA: Morgan Kaufmann.

Card, S., Moran, T., & Newell, A. (1983). *The psychology of human–computer interaction.* Mahwah, NJ: Lawrence Erlbaum Associates.

Cooper, A. (1995). *About face, the essentials of user interface design.* Foster City, CA: IDG Books Worldwide.

International Organization for Standardization. (1988). ISO/IS 8807 *Information processing systems—Open systems interconnection—LOTOS—A formal description technique based on temporal ordering of observational behaviour.* Geneva: Author.

Johnson, P., Johnson, H., Waddington, R., & Shouls, A. (1988). Task-related knowledge structures: Analysis, modelling and application. *People and Computers, IV,* 35–62.

Jordan, B. (1996). Ethnographic workplace studies and CSCW. In D. Shapiro, M. Tauber, & R. Traunmuller (Eds.) *The design of computer supported cooperative work and groupware systems* (pp. 17–42), Amsterdam: Elsevier/ North-Holland.

Lim, K., & Long, J. (1994). *The MUSE method for usability engineering.* Cambridge, UK: Cambridge University Press.

Macinlay, J. (1986). Automating the design of graphical presentations of relational information. *ACM Transactions on Graphics, 5,* 110–141.

Martin, J. (1991). *Rapid application development.* New York: MacMillan.

Mayhew, D. (1992). *Principles and guidelines in software user interface design.* Englewood Cliffs, NJ: Prentice-Hall PTR.

Paternò, F. (1999). *Model-based design and evaluation of interactive applications.* London: Springer-Verlag.

Paternò, F., Mancini, C., & Meniconi, S. (1997). ConcurTaskTrees: A diagrammatic notation for specifying task models. In S. Howard, J. Hammond, & G. Lindegaard (Eds.), *Proceedings of INTERACT '97* (pp. 362–369). Sydney: Chapman & Hall.

Petre, M., Blackwell, A., & Green, T. (1997). Cognitive questions in software visualisation. In J. Stasko, J. Domingue, M. Brown and B. Price (Eds.), *Software Visualization: Programming as a Multi-Media Experience* (pp. 453–480). Cambridge, MA: MIT Press.

Pfaff, G., & ten Hagen, P. (1985). *Seeheim Workshop on User Interface Management Systems.* Berlin: Springer Verlag.

Rumbaugh, J., Jacobson, I., & Booch, G. (1997). *Unified modelling language reference manual.* Reading, MA: Addison-Wesley.

Scapin, D., & Pierret-Golbreich, C. (1989). Towards a method for Task Description: MAD. *Work With Display Units, 89,* 371–380.

Sebillotte, S. (1988). Hierarchical planning as method for task analysis: The example of office task analysis. *Behaviour and Information Technology, 7,* 275–293.

Tauber, M. (1988). On mental models and the user interface. In G. C. van der Veer, T. R. G. Green, J.-M. Hoc, & D. Murray (Eds.), *Working with computers: Theory versus outcome.* London: Academic Press.

Tauber, M. (1990). ETAG: Extended Task Action Grammar—a language for the description of the User's Task Language. In D. Diaper, D. Gilmore, G. Cockton, & B. Shackel (Eds.), *Proceedings of INTERACT '90.* Amsterdam: Elsevier Science.

Tufte, E. (1983). *The visual display of quantitative information.* Cheshire, CT: Graphics Press.

Tufte, E. (1990). *Envisioning information.* Cheshire, CT: Graphics Press.

van der Veer, G. C., van Vliet, J. C., & Lenting, B. F. (1995). Designing complex systems—a structured activity. In *Proceedings of Designing Interactive Systems '95.* New York: ACM Press. (pp. 207–217).

van der Veer, G., Lenting, B., & Bergevoet, B. (1996). GTA: Groupware task analysis—modelling complexity. *Acta Psychologica 91,* 297–322.

van der Veer, G., van Welie, M., & Thorborg, D. (1997). Modelling complex processes in GTA. In S. Bagnara, E. Hollnagel, M. Mariani, & L. Norros (Eds.), *Sixth European Conference on Cognitive Science Approaches to Process Control (CSAPC)* (pp. 87–91). Rome, Italy: CNR.

van Loo, R., van der Veer, G., & van Welie, M. (1999). Groupware task analysis in practice: A scientific approach meets security problems. In *Seventh European Conference on Cognitive Science Approaches to Process Control (CSAPC)* (pp. 105–110). Valenciennes, France: Presses Universitaires de Valenciennes

van Welie, M. (2001). *Task-based user interface design.* Unpublished doctoral thesis, Vrije Universiteit, Amsterdam.

Wood, L. (Ed.). (1997). *User interface design: bridging the gap from user requirements to design.* Boca Raton, FL: CRC Press.

20

KOMPASS: A Method for Complementary System Design

Toni Wäfler, Gudela Grote, Anna Windischer,
and Cornelia Ryser
Swiss Federal Institute of Technology (ETH), Switzerland

Abstract

To take into account the complex interplay of human, technology, and organisation, the KOMPASS method provides criteria for task design on three levels. At the level of the *human–machine function allocation,* aspects that are directly influenced by process automation are considered. At the level of the *individual work task,* the whole task to be performed by an individual is taken into account. At the level of the *work system,* tasks of sociotechnical units (e.g., work teams, departments) are the focus. The design criteria on all three levels are deduced from the key notion to provide opportunities for process control as well as to promote respective human motivation and competencies. Furthermore, KOMPASS defines guidelines for a participatory approach to system design.

The chapter has three parts. In the first part the theoretical background of KOMPASS is outlined. Next, the KOMPASS criteria are described. The final section presents the guidelines for the support of the design process.

INTRODUCTION

Task design aims to enable goal-directed performance, allowing for efficient, safe, and sustainable work processes. Goal-directed performance relies on cognitive processes involving the acquisition, the organisation, and the application of knowledge. These cognitive processes are always context-sensitive in two ways. First, the meaning of goals and of goal-directed action is always determined by the context. Second, the context also shapes the cognitive processes behind goal setting and action regulation. Consequently, task design must include design of context. The aim is to create contextual conditions that promote adequate cognitive processes.

In the work world, the actual interplay of humans, organisational structures, and technology influence the task and its context. To a large extent, it is this interplay that enables or restricts certain actions. Consequently, this interplay must be considered when designing tasks. Therefore, a task design that aims at enabling adequate cognitive processes does not only concern the design of an individual's task but also joint cognitive systems in a comprehensive way, be these human–machine systems, sociotechnical units, or entire organisations. Hence, task design must go beyond the design of an individual's task to incorporate different design levels, considering human–machine as well as human–human function allocation.

Furthermore, humans and technical artefacts are different and therefore play different roles within joint cognitive systems. These differences become obvious in information processing, where humans cannot simply be programmed to process information in an algorithmic way. But the differences go far beyond information processing. Human complexity encompasses also emotional and motivational aspects. Moreover, humans—in contrast to artefacts—also live a life outside the work system. They follow their own goals and may even adjust the work system to reach these goals. Task design must take this into consideration as well. This does not mean that task design should aim to control human complexity, which is neither possible nor desirable. But it does mean that task design must also take into consideration aspects that lie beyond cognitive processes. Motivation, especially intrinsic motivation, is central. There is not much use for an optimised cognitive process within a joint cognitive system if the human is not motivated to perform the role the designers envisioned. Therefore, the role that the human is expected to play must be carefully considered before the system is designed, and when designing the system both the cognitive as well as the motivational preconditions that allow the human to contribute to the joint cognitive system's goal-oriented performance must be taken into consideration.

The KOMPASS method was developed to incorporate such factors (KOMPASS stands for complementary analysis and design of production tasks in sociotechnical systems; cf. Grote, Wäfler, Ryser, Weik, Zölch, & Windischer, 1999; Grote, Ryser, Wäfler, Windischer, & Weik, 2000; Wäfler, Windischer, Ryser, Weik, & Grote, 1999). It aims to support analysis and design of work systems in a comprehensive way. It provides task design criteria on the three levels of organisational design, individual task design, and human–machine function allocation, thereby furthering the understanding of joint cognitive systems as being embedded in a broader sociotechnical systems context. Furthermore, KOMPASS defines guidelines for a participatory approach to system design. In the following sections, we present the background of the method and describe the criteria as well as the guidelines.

AIM OF THE KOMPASS METHOD: COMPLEMENTARY SYSTEM DESIGN

The main aim of the KOMPASS-method is to support complementary system design. The concept of complementarity mainly refers to the role envisioned for the human within automated work systems and to the consequences regarding task design.

The human's role in automated work systems is that of supervisory control (Sheridan, 1987). In many systems, it has proven to be impossible for the human to fulfil this role adequately because of a lack of opportunities to develop sufficient knowledge and necessary production skills as well as to maintain adequate situation awareness during operation of the system (Bainbridge, 1982). Consequently, the increasing level of automation has proven to pose a potential danger to efficiency and safety of production processes.

Whether humans are able to fulfil the role of supervisory control does not depend only on their capabilities, but also to a large extent on the task that provides or deprives them of

TABLE 20.1
Criteria for Human–Machine Function Allocation (adapted from Bailey, 1989)

	Implicit Assumptions	
Allocation Criterion	Humans	Technology
Cost efficiency	Cost producing factors	
Leftover	Disturbance and risk factor	Effectiveness and safety factors
Performance comparison	Competing factors	
Humane tasks	Valuable resources	Support for human operator
Flexible allocation	Valuable resources	

corresponding opportunities. Therefore, task design is crucial for effective supervisory control and for system safety and efficiency. A task design that allows for human control over technical systems, based on an understanding of the technical processes and adequate opportunities to influence those processes, is seen as a prerequisite for effective supervisory control (Parasuraman & Riley, 1997). Consequently, task design should further the human's motivation and provide opportunities to develop adequate mental models of the technical system and of its processes and facilitate intervention. Whether the task is designed in this way depends on the function allocation between the human and the technical system, but it also depends on the function allocation among the human and other members of the work system and the function allocation among different organisational units.

Based on different implicit assumptions regarding the roles humans and technology play in automated work systems, five principles for human–machine function allocation can be distinguished (Table 20.1). Following the first three principles has frequently caused failures in system design:

- Cost efficiency: Functions are allocated to the human or machine according to short-term economic considerations. Costs and benefits that are not easily quantifiable (e.g., human capital) are neglected.
- Leftover: Functions are automated as much as possible, assigning to the human operator only those functions that cannot be automated.
- Comparison of performance: A function is allocated to the human if he or she supposedly performs it better than the machine and vice versa.

These three principles are insufficient for adequate function allocation among human operators and technology for a number of reasons (see Bailey, 1989, for a more detailed review). The main criticism is that they do not aim at creating meaningful tasks for humans. Functions are considered singularly and allocated to either humans or technology depending on short-term economic criteria, technical feasibility, a quantitative comparison between the ability of humans and technology, or a combination of these factors. Thus, the set of functions allocated to humans and hence constituting their task is not assorted deliberately. These three principles consequently do not support a well-directed task design, and adhering to them often creates tasks that are impossible for humans to perform.

Almost 40 years ago, as a reaction to the shortcomings of the commonly used allocation principles, Jordan (1963) proposed the principle of *complementarity*. He pointed out that humans and machines are fundamentally different and therefore cannot be compared on a quantitative basis. Instead, they are to be seen as complementing each other in performing a task. Functions are not to be allocated to either humans or machines; rather, the *interaction* of human and machine is to be designed taking into account reciprocal dependencies between task components and the ways they are carried out by human and technical systems. Function allocation should allow support and development of human strengths and compensation of human weaknesses by the technical system to secure efficiency and safety of production as well as humane working conditions. The last two strategies in Table 20.1 both follow the concept of complementarity.

To create a meaningful and challenging task for the human, thereby furthering the overall effectiveness of the work system, a complementary approach may lead to a design strategy that deliberately does not use the full potential of technology for automating a given process. The human as well as the technology are considered valuable resources. A system design is envisioned in which the human can develop specific human potentials as contributions to the system's overall performance. The technology is meant to unfold its own specific contribution. The interaction between human and technology must be designed in a way that strengthens the different capabilities of humans and technologies and compensates for their specific weaknesses. From that point of view, the strengths and weaknesses of humans and technologies differ not primarily quantitatively, but qualitatively. It is not of major concern that humans, for instance, compute more slowly and less reliably than computers that represents the main difference; it is a core assumption of complementary system design that humans possess abilities that machines do not (yet) have (i.e., that humans and machines differ in a qualitative way). Technology, for example, is able to repeat operations constantly and reliable for a long time without fatigue. This is (still) done according to predefined algorithms. But if unforeseeable events occur for which no algorithms exist, technology reaches its limits. Technology does not develop new ideas to face such situations. Also technology is unable to understand what is done in a specific situation and why it is done. In short, technology does not think—it has no imagination, it just performs. On the other hand, humans also have limits. Fatigue, loss of motivation, unreliability, and limited mental capacity are human characteristics. Whereas machines never ask why they are programmed to do something, humans need reasons. This difference can be a human strength. It predisposes them to understand situations, to act in a given way, to learn and to find new solutions for new problems. In short, humans think—they can be creative problem solvers.

To realise a complementary approach in task design, it is essential to allocate functions between the human and the technology in a meaningful way, thereby creating those preconditions that allow the human to develop and to deploy specific human capabilities. Focusing on the human–machine function allocation is not sufficient, however. Because the meaningfulness of a task also depends on the way a task is shared among people, task design must also incorporate the question of human–human function allocation. For instance, the opportunity to program a system may provide deepened insights into the system's working, which may be valuable for effective supervisory control. Allocating programming and control to different people may therefore jeopardise capabilities for supervisory control, and hence, in a complementary task design must also consider the distribution of tasks among humans as well as among organisational units.

The focus of most recent attempts at developing methods supporting function allocation decisions is to operationalise the concept of complementarity. A variety of methods and concepts have been suggested, most of which are rarely applied, however, because of a number of theoretical, methodological, and practical shortcomings (see Older, Waterson, & Clegg, 1997, for a detailed discussion). Against this background, a list of requirements to be met by the KOMPASS method was elaborated (see Grote et al., 2000, for a detailed discussion of the differences between the KOMPASS method and other methods following a comparable approach):

- The method should be applicable to the analysis and redesign of existing work systems and tasks as well as to the prospective analysis and design of planned work systems.
- Three levels of task analysis and design—namely, work system tasks, individual work tasks, and human–machine function allocation—should be integrally taken into account.
- For all three levels, a set of empirically tested criteria for system analysis and design should be provided.
- The application of the method should lead to the deduction of concrete demands for organisational as well as technical design.
- The method should support a design team in pursuing an integral and participative design process.

To meet these requirements, a set of criteria for complementary analysis and design as well as a design heuristic have been developed. Both are presented in the following sections.

KOMPASS CRITERIA FOR COMPLEMENTARY ANALYSIS AND DESIGN

A suitable combination of human and technology on the basis of their complementarity can allow synergetic performance of the joint system. The KOMPASS method provides criteria for complementary analysis and design of human–machine function allocation, the individual work task, and the work system. The criteria are deduced from the key notion to determine explicitly the kinds of uncertainties to be handled in a system and to promote humans motivation to and competency for controlling these uncertainties. The criteria have been operationalised for systems in discrete manufacturing but are considered to be applicable to other types of industrial processes as well.

Human–Machine Function Allocation

Regarding human–machine systems, task design as determined by the human–machine function allocation aims to support humans in their role as supervisors of automated processes (Sheridan, 1987). To empower them adequately, preconditions for human control over technology must be created. Opportunities for process control are not only directly relevant to ensuring efficiency and safety of the process; they are also indirectly relevant because the experience of control is a precondition for unstressed coping with a situation's demands (Karasek & Theorell, 1990). Finally, control is a precondition for a sense of responsibility and meaningful use of autonomy (Grote, 1997).

Three preconditions allow for process control: comprehension of the system to be controlled, predictability of its behaviour, and possibilities to influence the behaviour of the system (Brehmer, 1993; Ganster & Fusilier, 1989; Thompson, 1981). These preconditions are interdependent in such a way that comprehension and predictability are prerequisites for a purposeful use of possibilities to influence, and active influence promotes comprehension and predictability.

Regarding comprehension and predictability of system behaviour, the importance of mental models is generally focused on, i.e., cognitive representations of the system's states and processes, which are relevant for controlling the system (Rogers, Rutherford, & Bibby, 1992).

For the design of human–machine systems, this leads to the question of how the development and the maintenance of mental models adequate to the human operator's control task can be supported. This question refers on one hand to education and training (Sonntag, 1990) and on the other to the actual design of the technology. The latter gains special importance when vocational training is lacking and on-the-job training dominates. With the design of the technical system, possibilities for interaction are determined along with opportunities for practical experience as a prerequisite for learning (Duncan, 1981).

Many aspects of task-adequate mental models seem to be built on implicit and tacit know-how rather than on theory-based knowledge (Bainbridge, 1981). This also shows the importance of technical design in comparison to education and training because this tacit know-how is developed through personal experience gained in a concrete work situation (Böhle & Rose, 1992). Essential aspects in this process are dialogue-oriented, interactive handling; intuitive–associative thinking; and complex sensory perception. Increasing automation impedes these actions because the process becomes separated from the human and can neither be influenced nor directly experienced. Some design approaches aim at enhancing possibilities for sensory experience by providing improved visual or acoustic process information (Kullmann, Pascher, & Rentzsch, 1992). The importance of sensory experience per se, however, especially compared with opportunities for active influence over the process, is considered relatively small (Volpert, 1994). There is still a considerable lack of research regarding new ways to develop tacit knowledge under the condition of technically mediated information.

Against this background, KOMPASS incorporates, at the level of the human–machine system, several criteria for task design. These criteria, described in the following subsections, are particularly relevant to design that aims to further human control over the technical system by creating preconditions for comprehension, predictability, and influence.

Process Transparency.   A crucial prerequisite for fulfilling supervisory control tasks is the transparency of technical processes for the human operator. This enables the formation and maintenance of mental models adequate to the tasks the human performs. Yet although there is agreement about the importance of mental models, there is little common understanding of what exactly mental models are, how they should be empirically studied, what are the requirements for adequate models in the context of human–machine interaction, and how the formation of such models should be supported (Rogers et al., 1992; Wilson & Rutherford, 1989).

For the purpose of KOMPASS, a pragmatic approach was chosen, aimed at identifying potential embedded in a given human–machine system for gaining an understanding of the general nature and temporal structure of the production process and the required process interventions. Based on the assumption that the formation of mental models is supported by active involvement in crucial steps of the production process and direct process feedback, each function is analysed in terms of its relevance to understanding of the production process and the types and extent of available process feedback. Although feedback sources can be determined by means of analysing the production process and the technical systems used to carry out the process, the importance of functions for the formation of mental models cannot be determined solely by process and technical indicators. It has to be evaluated in close cooperation with system operators. This adds a "subjective" component to the analysis, which was accepted as unavoidable when trying to measure a cognitive construct under the time and resource constraints usually encountered in system design processes.

The operationalisation of process transparency is presented in Table 20.2. The logic of the operationalisation's use in task analysis is as follows (the operationalisation's use of the criteria as presented in the subsequent sections follow the same logic). For the set of functions allocated to the human, arguments must be identified that support a high or a low rating of the

TABLE 20.2
Operationalisation of Process Transparency

Low Transparency	High Transparency
There is little interaction between the human operator and the technical process.	The human operator interacts with the process while executing most of his or her subtasks.
The human operator is not given any direct process feedback.	The human operator is given direct feedback from the production process via all sensory channels.
The human operator does not know whether she or he gets any feedback from the process based on computed (simulated) data alone.	Technically mediated feedback from the process based on computed (simulated) data is specifically marked.

different aspects of process transparency. On the basis of these arguments, an overall rating of process transparency must be agreed on. When designing tasks, design possibilities that increase the rating of process transparency have to be identified.

Dynamic Coupling. Coupling is frequently described as a crucial characteristic of human–machine systems (Corbett, 1987a). Usually, it is assumed that a low degree of coupling has to be achieved to ensure high degrees of freedom for the human operator. For human–machine systems requiring constant human input, this objective is not only highly plausible, but also empirically supported by studies showing strong correlations between coupling, job satisfaction, and well-being (Corbett, 1987a). In highly automated systems, however, there may also be too little coupling between human operator and technical system, resulting in very low process involvement with negative effects on cognitive and motivational prerequisites for effective supervisory control. In those systems, technical options should be provided that permit a dynamic coupling. This idea draws on the concept of loose coupling proposed by Weick (1976) regarding the dialectic relationship between autonomy and dependence required for the integral as well as flexible functioning of organisations.

A high degree of dynamic coupling exists if the human operator has a high degree of control with respect to time, place, work procedures, and cognitive effort required in the interaction with the technical system (Table 20.3). Cognitive effort refers to the level of information processing required from the human operator with respect to the technical system, independent of his or her physical proximity to the technical system. It is operationalised in accordance with Rasmussen's (1982) levels of skill-, rule-, and knowledge-based processing.

For all four kinds of coupling, technical options provided to the human for varying the degree of coupling are determined. An example for high dynamic coupling regarding time could be the option either to use a handling robot, thereby reducing the coupling between operator and technical system, or to place work pieces manually, effecting a higher degree of coupling (e.g., when the operator wishes to closely monitor the manufacturing process).

Decision Authority. The distribution of decision authority in a human–machine system determines to what extent the human and the technical system control the actual processes (Kraiss, 1989; Sheridan, 1987). In the KOMPASS operationalisation of the criterion, one distinguishes between decision authority regarding access to process information on one hand and decision authority with respect to controlling the process itself on the other (Table 20.4). The distribution of decision authority can vary between fully manual, manual with automatic support or constraints, automatic with manual influence, and fully automatic. Decisions may

TABLE 20.3
Operationalisation of Dynamic Coupling

Rigid Coupling	Dynamic Coupling
The technical system determines to a large extent at what time the subtasks must be performed.	The human operator can choose when to perform most of the subtasks (without interrupting the process).
The technical system determines to a large extent where the subtasks must be performed.	The technical system provides the human operator with several locations to perform the task for most of the subtasks.
The technical system determines to a large extent how to proceed while performing the subtasks.	The technical system provides the human operator with more than one method to perform the task for most of the subtasks.
Performing the subtasks always requires the same amount of attention from the human operator. The technical system does not provide features that allow the operator to change the required amount of attention.	The technical system offers the human operator the opportunity to freely decide how much attention to devote to the process (without jeopardizing the quality or safety of the process).

TABLE 20.4
Operationalisation of Decision Authority

Automatic	Manual
Decision Authority Regarding Information Access	
The human operator does not have access to all information recorded from the process.	The human operator has access to all information recorded from the process.
All accessible information is automatically presented.	The operator decides which information to retrieve and when.
The operator cannot influence the presentation of this information.	
The operator cannot freely retrieve information.	
Decision Authority Regarding Process Control	
The decision authority regarding process control remains fully with the technical system, which prevents the occurrence of specific system states and proposes actions that the operator can accept or refuse but not change.	The decision authority regarding process control remains with the human operator.
The operator has no influence on the process.	The operator controls the process.
	The technical system does not determine any action restriction.

concern changes in actual processes but also in desired process values. Although the design goal for all other KOMPASS criteria is "the more, the better", the design goal for decision authority is an adequate fit regarding both internal and external aspects. The former refers to a fit of decision authority regarding information access and decision authority regarding process control. High process control authority with little influence over process information, for instance, should be avoided. The latter—the external fit—refers to the match between decision authority on one hand and process responsibility assigned to the human operator as a consequence of the overall task design on the other.

Flexibility. Differing from other conceptualisations, in which flexible function allocation is treated as one form of distributing decision authority between the human operator and the technical system (Kraiss, 1989), flexibility is defined in KOMPASS as a separate criterion because it adds a new quality to a human–machine system. Providing the possibility for flexible and dynamic function allocation does not solve the issue of function allocation, however, but raises new questions regarding which functions should be allocated flexibly and how and by whom this flexibility should be used (Clegg, Ravden, Corbett, & Johnson, 1989; Moray, Inagaki, & Itoh, 2000).

According to the KOMPASS operationalisation, human–machine systems that allow different levels of decision authority provide flexibility. This is the case in aviation, for instance, if flight control can be alternated between the human and the autopilot. The system is *adaptable* when the human decides on the actual allocation of control, whereas it is *adaptive* when the human has no influence. In most cases, adaptable flexibility is the design goal because it allows the human to assign functions to the technical system or to him- or herself based, on for instance, considerations of over- or underload. In human–machine systems with high potential for operator overload adaptive flexibility may be desirable (Rouse, 1988), but this leads to increased complexity of the technical system with higher demands on process transparency. Also, adaptive flexibility fundamentally questions the tenet of human control over technology. Differing from some authors, we argue that only in extreme situations of human incapacitation should human control be abandoned in favour of irreversible automatic decision authority.

Individual Work Tasks

The KOMPASS criteria for task design presented in the previous section refer to human–machine function allocation (i.e., to those aspects of task design that are directly influenced by the fact that a process is automated and that this automation shapes content and requirements of the human's tasks). The criteria presented in this section refer to individual work tasks (i.e., the whole task that is to be performed by one person).

In task design, demands on *the human* are usually specified together with adequate execution support, especially technical support. In most design approaches, the human remains a rather abstract concept and is not concretised as *an individual*. Whether the task is to be executed by an individual or by a group of individuals and how it is distributed within this group remain largely unspecified (Sheridan, 1987). System design cannot proceed from the assumption of an abstract human supervisor, however, if opportunities for human control and supervision over automated processes are a design goal and if the development of mental models is considered to be crucial. The actual execution of both control and supervision is highly dependent on the individual's whole task. An individual who not only supervises the technical system but also programs it, trouble-shoots, and plans the work process has a totally different basis for control and supervision than does an individual who exclusively monitors the process (De Keyser, 1987). Hence, individual work task design cannot be neglected when designing human–machine systems but must be considered explicitly.

Many concepts concerning tasks design in highly automated systems focus on requirements for working conditions that are derived from basic physical and psychological human characteristics. A special focus is vigilance and, derived from that, the notion that the task allocation within the system must provide the human with an active role (Johansson, 1989). Another crucial question is concerned with the problem of providing the human with a meaningful and challenging task despite its reduction due to the automation. Many studies have shown corresponding design options, ranging from task simplification and human dequalification up to a creation of highly qualified jobs (Majchrzak, 1988; Ulich, 1998; Wall et al., 1987). Commonly, it is considered desirable to create conditions and opportunities for more qualification.

TABLE 20.5
Operationalisation of Task Completeness

Low Task Completeness	High Task Completeness
The worker is occupied extensively with one of the following functions: preparation, planning, executing, controlling, finishing, or maintaining and repairing. The worker does not participate or only partially participates in the performance of other functions.	The worker's task includes preparing, planning, executing, controlling, finishing, and maintaining or repairing functions. The worker is strongly involved in the performance of all these functions.

Emery pointed out in 1959 already that highly automated systems are more susceptible to variances and disturbances. Therefore system operation and maintenance for instance should be brought together and allocated to the same people. This increases the humans' autonomy and their opportunities for control. Several studies have shown the positive effects predicted by Emery (Jackson & Wall, 1991; Wall, Jackson, & Davids, 1992). Especially in systems with a high degree of technically determined uncertainty, shorter down times and lower system vulnerability attributable to human operators' preventive interventions have been found. Another possibility to enrich the individual work task is the integration of the human in the process of technical–organisational system development (Friedrich, 1993).

Finally, it must be taken into consideration that a task must not necessarily be divided into individual tasks but can be assigned to work groups in collective responsibility. For highly automated systems, Susman (1976) showed a general shift of work task content away from the transformation process to boundary regulation. He reasoned that boundary regulation, including coping with unpredictable variances and disturbances, can be best mastered by work groups in which parts of boundary regulation are flexibly allocated to individuals who coordinate themselves by mutual agreements. For this, a different understanding of the human role is required. The human is no longer a machine operator but rather a system manager who coordinates him- or herself permanently with the other members of the work group as well as with other organisational units.

At the level of the individual work task, KOMPASS incorporates criteria for analysis and design that aim to create working conditions that enable the human to act as a system manager. These criteria are not directly concerned with the design of the technology, but rather with the allocation of functions between humans.

Task Completeness.   The criterion of task completeness (Table 20.5) measures whether the task of a individual incorporates not only executing functions but also preparation, planning, controlling, finishing, and maintenance or repair (Hacker, Fritsche, Richter, & Iwanowa, 1995; Ulich, 1998). Completeness of work tasks is a precondition for individuals to be able to understand the meaning of what they do as well as for their identification with the result of their work. Furthermore, completeness of work tasks creates independence regarding preceding and succeeding tasks in the production flow and therefore is a basis for local control of variances and disturbances.

Planning and Decision-Making Requirements.   The criterion of planning and decision-making requirements (Table 20.6) measures the complexity of the cognitive processes required for the execution of the task (Oesterreich & Volpert, 1991). Planning and decision-making requirements are high if a task requires self-determined goal setting and planning regarding

TABLE 20.6
Operationalisation of Planning and Decision-Making Requirements

Low Planning and Decision-Making Requirements	*High Planning and Decision-Making Requirements*
The task can be performed without thinking, or The sequence of the processing steps is prescribed, but each step needs to be mentally actualised to guarantee a correct performance of the work task.	Problem solving is required. Subgoals need to be coordinated. After reaching a subgoal, the remaining subgoals are checked, and the procedure must be replanned; generally there are several adjustments.

TABLE 20.7
Operationalisation of Communication Requirements

Low Communication Requirements	*High Communication Requirements*
Communication is limited to information exchange only, or Mutual problem solving is required, in which the sequence of the processing steps is prescribed but a few processing steps need to be actualised mutually and consciously to guarantee a correct performance of the work task.	Mutual problem solving is required. Subgoals need to be jointly coordinated. After reaching a subgoal, the remaining subgoals are jointly checked, and the procedure must be replanned; generally there are several adjustments.

content, means, procedures, and result. Such tasks support the human capability for self-determined goal setting, for the development and flexible adaptation of plans, and for complex problem solving. Furthermore, they serve the maintenance and development of knowledge and know-how.

Communication Requirements. The criterion of communication requirements (Table 20.7) measure the extent to which workers need to mutually coordinate while executing their tasks (Dunckel et al., 1993). The communication can serve the coordination of work steps, the joint planning of subgoals, or the exchange of information. The criterion measures the complexity of the communication in analogy to the criterion planning and decision-making requirements. Consequently the complexity of the communication cannot be higher than the complexity of the individual cognitive processes. Work tasks with a high degree of complexity in communication allow joint problem solving, transfer of knowledge and experience, and hence mutual qualification and support. Furthermore, work-related communication helps prevent or correct variances and disturbances in the workflow.

Opportunities for Learning and Personal Development. The criterion of opportunities for learning and personal development (Table 20.8) measures the ratio between qualification demands inherent to the task and qualification of the worker. Thus, it is a measure for workers' opportunities to apply and to develop qualifications in performing the task (Hacker et al., 1995; Ulich, 1998). When assessing the criterion, the focus is on the frequency of application of the actual qualification as well as on the frequency of needs to increase it. Work tasks that

TABLE 20.8
Operationalisation of Opportunities for Learning and Personal Development

Few Opportunities for Learning and Personal Development	*Frequent Opportunities for Learning and Personal Development*
The full scope of professional qualifications is employed once a year at the most.	The full scope of professional qualifications is employed weekly.
An initial vocational training is sufficient.	Additional vocational training of several days is necessary at least once a year.

TABLE 20.9
Operationalisation of Variety

Low Variety	*High Variety*
The task provides variety regarding fewer than five aspects:	The task provides variety regarding more than eight aspects:

Possible Aspects of Variety

Different production materials:
- Different kinds of the same material (e.g., different types of wood or metals)
- Different materials (e.g., metal or synthetic material)

Different production techniques:
- Different process methods (e.g., drilling, turning, and milling)
- Different types of process control (e.g., conventional or numeric)
- Different operations (e.g., processing, handling, measuring)

Different work procedures:
- Different cycle times per operation
- Different sequences of operations
- Different operations to be performed per job

Different cooperation partners:
- Cooperation with different people of the same organisational unit
- Cooperation with different people of other organisational units (e.g., purchasers, design engineers)
- Cooperation with different people of other organisations (e.g., clients, suppliers)

provide opportunities for learning and personal development serve to maintain and develop qualifications as well as mental flexibility.

Variety. Task variety allows the worker to gain experience with different situations and demands and hence promotes flexible action (Dunckel et al., 1993; Hackman & Oldham, 1976). Variety can be provided by tasks that offer different production materials and techniques, different work procedures, or different cooperation partners (Table 20.9).

Transparency of Workflow. Transparency of workflow is a precondition for anticipatory, preventive, and flexible action and supports short-range adaptation of plans in reaction to changing circumstances (Dunckel et al., 1993). High transparency is provided when the worker is familiar with the actual situation in the organisational unit to which he belongs, as well as with

TABLE 20.10
Operationalisation of Transparency of Workflow

Low Transparency of Workflow	High Transparency of Workflow
Fewer than five aspects of transparency are provided:	More than seven aspects of transparency are provided:

Possible Aspects of Transparency of Workflow

- Knowledge of (external) customers
- Knowledge of jobs or orders of the other workers in the same organisational unit
- Knowledge of number of jobs or orders per week that must be processed by the worker
- Knowledge of time frame within which jobs or orders are to be processed
- Knowing the worker who works upstream in the job or order process
- Knowing the worker who works downstream in the job or order process
- Knowledge of date when jobs or orders will be processed by the subsequent organisational unit
- Knowing the way in which jobs or orders will be processed in subsequent organisational units (processing methods and tolerances)
- Knowing amount and purpose of own contribution to the whole product

the actual situation in units that are upstream and downstream in the workflow (Table 20.10). However, transparency can only support situated action when the worker is provided with an adequate scope of action, especially as determined by the criteria of planning and decision-making requirements and influence over working conditions.

Influence Over Working Conditions. Influence over working conditions is a precondition for a feeling of responsibility as well as for identification with one's own task (Ulich, 1998). It is provided by influence of the worker over the assignment of orders, temporal planning and scheduling, organisational issues, and quantitative or qualitative order characteristics (Table 20.11).

Temporal Flexibility. Temporal flexibility describes the predictability of required interventions of the worker, the amount of externally defined deadlines and time limits, and the amount of time pressure (Table 20.12). The worker's self-determined structuring of orders is desirable. Temporal flexibility allows to balance variances in performance as they occur throughout the workday, it supports task-related communication with coworkers, and it prevents inappropriate work compression (Dunckel et al., 1993; Ulich, 1998).

Work System

The KOMPASS criteria for task design presented in the previous sections refer to human–machine function allocation and to individual work tasks. The criteria presented in this section refer to task design on the level of work systems. A work system is a sociotechnical unit of an organisation (e.g., a work team or a department). Task design on this level must be considered because a work system's task, and with it the distribution of tasks between work systems, limits the possibilities for task design on the level of the individual task as well as on the level of human–machine function allocation. If, for example, programming, maintenance, and executing are assigned to different work systems, then it is difficult or even impossible to enrich individual work tasks.

TABLE 20.11
Operationalisation of Influence Over Working Conditions

Little Influence Over Working Conditions	*Significant Influence Over Working Conditions*
The task offers less than four opportunities of influence:	The task offers more than eight opportunities of influence:

Possible Areas of Influence Over Working Conditions

Organisational issues:	*Assignment of jobs or orders:*
• Job rotation	• The job or order type assigned to the worker
• Definition of operations to be performed (e.g., must there be a quality check?)	• The number of jobs or orders that are assigned to the worker
• The allocation of operations within the organisational unit (e.g., who does the quality check?)	• The machine load within the organisational unit
• The forms of cooperation for order processing (e.g., individual work, teamwork, line work)	• The sequence of job or order processing
• Selection of cooperation partners within the organisational unit	*Temporal planning and scheduling:*
	• Due dates for procurement of materials and resources
Quantitative or qualitative job or order characteristics:	• Due dates for finishing jobs or orders
• Batch size	• Planned cycle time for machining or job or order processing
• Quality standards (e.g., level of tolerance)	• Planning and scheduling of working hours (e.g., shift work, overtime)
• Design improvements	
• Process improvements	

TABLE 20.12
Operationalisation of Temporal Flexibility

Low Temporal Flexibility	*High Temporal Flexibility*
Times for the beginning and end of jobs or orders are prescribed. Jobs or orders cannot be rescheduled.	Neither a time for beginning nor an end of jobs or orders is prescribed. Job or order scheduling is required.
Interventions in the process are unpredictable and require permanent, reactive readiness.	The worker can choose when to intervene in the process.
Backlogs occur permanently.	Backlogs rarely occur.

Corbett (1987b) distinguished between work- and technology-oriented design. This distinction, made on the basis of sociotechnical design principles (Emery, 1959; Susman, 1976), focuses on task demands with which individuals are confronted in automated work systems. Work-oriented design aims at an integration of human resources, technology, and organisation (Ulich, 1998) with the goal to empower the work system to cope locally with variances and disturbances (Grote, 1997). Such variances and disturbances arise from the transformation processes within the work system as well as from the work system's environment. They are

TABLE 20.13
Operationalisation of Task Completeness

Low Task Completeness	*High Task Completeness*
Low vertical range of manufacturing.	High vertical range of manufacturing.
Low product complexity: single-piece products with a simple structure.	High product complexity: multipiece products with a complex structure.
High rate of order repetition.	Low rate of order repetition.
Only one function is performed within the work system.	All functions are performed within the work system.

Functions that can be integrated

- Production planning (technical)
- Order planning (temporal)
- Order scheduling and dispatching (temporal)
- Definition of quality standards
- Machine programming
- Material procurement (raw material, intermediate products)
- Stock administration
- Resource administration

- Manufacturing
- Assembling
- Quality control: interim inspection
- Quality control: final inspection
- Maintenance and servicing
- Repair
- Finished goods inventory
- Shipment

inevitable constituents of complex, highly automated systems, which are integrated in networked production systems. Recent failures in computer-integrated manufacturing showed that technology-oriented design approaches are still followed primarily. These approaches not only hinder adequate coping with variances and disturbances, they also create new sources for such problems because of an unbalanced adjustment of the organisation to the technology (Kidd & Karwowski, 1994; Ulich, 1998).

Many studies have shown that a work-oriented integration of people, technology, and the organisation can be best realised by decentralised organisational structures with a high degree of functional integration combined with self-regulation (Majchrzak, 1988; Schüpbach, 1994; Ulich, 1998; Wall, Clegg, Davies, Kemp, & Mueller, 1987). The KOMPASS criteria for task design at the level of the work system aim at creating corresponding conditions.

Task Completeness. The completeness of a work system's task, and with it the completeness of production processes that are in their complexity transparent for the work system's members, is a precondition for local control of variances and disturbances (Susman, 1976; Ulich, 1998). The completeness of a work system's task can be operationalised by the work system's vertical range of manufacturing, by its functional integration, and by the production's complexity (Table 20.13). Furthermore, the completeness of a work system's task is a precondition for the design of complete individual work tasks that provide opportunities for planning and decision making.

Independence of Work Systems. The criterion of independence of work systems measures the amount of variances and disturbances that affect the work system but are caused outside of it as well as those that are caused by the work system itself but have effects outside of it (Corbett, 1987a; Perrow, 1984; Ulich, 1998). A high degree of independence is characterised

TABLE 20.14

Operationalisation of Independence of Work Systems

Low Independence of Work Systems	*High Independence of Work Systems*
Production delays in the work system have large or rapid effects on other work systems because there are no buffers regarding time, material, etc.	Production delays in the work system have small or slow effects on other work systems because there are buffers regarding time, material etc.
Deviations in quality standards in other work systems cannot be recognised or prevented at their source; they have uncontrollable effects on the work system.	Deviations in quality standards in other work systems can be recognised or prevented at their source; they have no effects on the work system.
Deviations in quality standards in the work system cannot be recognised or prevented at their source, so they spread uncontrolled.	Deviations in quality standards in the work system can be recognised or prevented at their source, so they do not spread uncontrolled.
The workflow between work systems is fixed within strict limits (e.g., processing sequences that are required by technology or prescribed by the organisational structure).	The workflow between work systems can be designed flexibly (e.g., technically possible variants of the processing sequences, organisational scope regarding the planning or handling of work orders).

by local prevention and correction of variances and disturbances (Table 20.14). This is achieved by reducing interfaces and regulation requirements with other work systems. Complete independence, however, would cause a total decoupling of the work system. Other work systems are necessary and desirable to ensure an integration of the work system's activities with the whole organisation's activities. Hence the design goal regarding independence of work systems aims at providing as much independence as possible without jeopardising the functioning of the whole.

Fit Between Regulation Requirements and Regulation Opportunities. The fit between regulation requirements and regulation opportunities is measured by the extent a work system is confronted with uncertainty and whether it has the possibility to adapt itself adequately. The more turbulent the environment is and the more complex the internal processes are, the greater the number of the uncertainties and with these the variances and disturbances to be regulated by the work system. These uncertainties determine the nature and the amount of the regulation requirements a work system has to face (Susman, 1976). The work system's regulation opportunities, on the other hand, are determined by the internal cooperation form that it has adopted (Table 20.15).

The chosen form of cooperation can also be dynamic. This is the case when the work system changes its form of cooperation in dependency of actual production conditions (LaPorte & Consolini, 1991). Whether the regulation requirements and the regulation opportunities fit must be judged for each case independently. The criterion only gives hints and guidelines that can be helpful in the judgement (Table 20.16).

Polyvalence of Work System Members. The criterion of polyvalence of work system members (Table 20.17) measures the amount of qualification of the members of a work system compared with the systems maximal qualification as determined by its tasks (Strohm, 1997). The design goal is that as many members of the system as possible are able to perform as many of the system's tasks as possible. This permits mutual support as well as local compensation for variances and disturbances. Furthermore, the variety of individual work tasks increases

TABLE 20.15
Regulation Opportunities as Determined by the Internal Cooperation Form

Form of Internal Cooperation	Regulation Opportunity
Teamwork: The work system's task is divided into subtasks that are performed in parallel. Everyone works on his or her subtask, yet the individual work performances are interdependent. Fluctuations in individual performance may directly affect the performance of the others.	**Reciprocal coordination:** Coordination of the work processes within the team, perhaps also the coordination with other groups and the allocation of resources, takes place within the work group.
Line work: The work system's task is divided into subtasks that are performed step-by-step according to the principle of flow production. Everyone works alone on his or her subtask, yet fluctuations in individual performance have a direct and sequential effect on the following subtasks.	**Programming:** Coordination of the work processes within and between work systems as well as the allocation of resources is carried out by a superordinate unit (e.g., planning and scheduling department).
Individual work: The work system's task is divided into subtasks that can be performed independently. Fluctuations in individual performance do not affect the work of others. There is, however, a mutual goal.	**Standardisation:** Very often, there is no need for coordinating the work processes. Setting standards concerning work output in relation to the mutual goal and allocation of resources can be carried out by the workers themselves or by a superordinate unit.
Isolated individual work: There is no mutual task. Everyone works on his or her personal task with an individual target and without being dependent on others. A superordinate system guarantees the connection between the tasks.	**Standardisation:** There is no need for coordinating the work processes. Setting standards concerning work output and allocation of resources are carried out by a superordinate unit.

TABLE 20.16
Operationalisation of Fit Between Regulation Requirements and Regulation Opportunities

Low Fit	High Fit
High regulation requirements with few regulation opportunities.	High regulation requirements with high regulation opportunities.
Low regulation requirements with many regulation opportunities.	Low regulation requirements with few regulation opportunities.
Dynamically changing regulation requirements without the possibility of adapting the form of cooperation.	Dynamically changing regulation requirements with the possibility of adapting the form of cooperation.

TABLE 20.17
Operationalisation of Polyvalence of Work System Members

Low Polyvalence	High Polyvalence
Less than half of the work system's members are able to perform more than half of the tasks.	More than half of the work system's members are able to perform more than 75% of the tasks.

TABLE 20.18
Operationalisation of Autonomy of Work Groups

Low Autonomy	*High Autonomy*
Less than four areas of decision making are subjects of collective autonomy.	At least nine areas of decision making are subjects of collective autonomy.
There are no regular group meetings, no meeting rooms used by the group, no group members trained in meeting facilitation.	There are regular group meetings, meeting rooms used by the group, and some group members are trained in meeting facilitation.

Possible Areas of Collective Autonomy

Internal coordination
- Group-internal distribution of work/workplace rotation
- Order scheduling
- Quality control/quality assurance
- Determination of working hours (daily working hours, overtime, planning vacation, etc.)

Internal Staff Affairs
- Election of a spokesperson for the group
- Election of group members
- Training of group members

External coordination
- Order acceptance
- Coordination of the workflow with other work systems
- Coordination with other organisational units

Continuous Improvement Processes
- Improvement of the work organisation
- Improvement/acquisition of resources and machines
- Product improvement (design, manufacturing method, etc.)

TABLE 20.19
Operationalisation of Boundary Regulation by Superiors

Low Regulation	*High Regulation*
The superior makes arrangements with other work systems or external units once a week at the most.	The superior makes arrangements with other work systems or external units several times a day.
The superior is present in the work system more than 2.5 hours a day.	The superior is present in the work system less than 1.5 hours a day.
The superior is occupied mainly with internal coordination.	The superior is occupied mainly with external coordination.

Aspects of internal coordination
- Election of group members
- Qualification of group members
- Group-internal distribution of work/workplace rotation
- Order scheduling (sequencing, resource allocation, etc.)
- Quality control/quality assurance
- Determination of working hours (daily working hours, overtime, planning vacation, etc.)

Aspects of external coordination
- Order acceptance
- Coordination of the workflow with other work systems
- Coordination with superordinate organisational units
- Improvement of the work organisation
- Improvement/acquisition of resources and machines
- Product improvement (design, manufacturing method, etc.)

as well as opportunities for learning and personal development. Moreover, preconditions for an understanding of production processes are created and hence the process transparency in single human–machine systems may be supported.

Autonomy of Work Groups.   Generally, autonomy is characterised by the opportunity for self-determination regarding the setting of goals as well as the definition of rules and procedures for reaching these goals (Grote, 1997). Autonomy makes possible flexible and situated adaptation of goals and procedures and hence also furthers active and comprehensive consideration of the production process. Restrictions regarding individual autonomy due to the complexity of production processes can be overcome by collective autonomy. It is important, however, that potential conflicts between individual and collective autonomy are taken into account and handled explicitly by the group. A high degree of collective autonomy is provided when organisation and coordination can be determined collectively by the work system's members (Table 20.18).

Boundary Regulation by Superiors.   On one hand, boundary regulation incorporates coordination requirements as determined by demands emerging from a work system's environment. On the other hand, it incorporates also the communication and possibly the assertion of demands the system poses on its environment (Susman, 1976). Supervisors of work systems should mainly take care of boundary regulation, leaving internal coordination as a subject of collective autonomy as much as possible to the work system's members (Table 20.19).

KOMPASS DESIGN HEURISTIC

The KOMPASS method provides a design heuristic that supports interdisciplinary teams in the design of automated systems, based on the criteria presented in this chapter (for a detailed description, see Wäfler et al., 1999). In drawing up the design process, one crucial objective was to achieve a balance between participation and expert-driven design. This overriding principle is described before the concrete steps of the heuristic itself are presented.

Participation in System Design

There is widespread agreement that system design should follow a participative strategy to improve design quality, user satisfaction, and user acceptance (i.e., actual use) of the new system (Greenbaum & Kyng, 1991; Majchrzak, 1988; Ulich, Rauterberg, Moll, Greutmann, & Strohm, 1991). It has been found, however, that user involvement in the actual design of technology, as compared with system implementation and optimisation, is rather rare and often does not extend beyond information distribution (Scarbrough & Corbett, 1992). Hornby and Clegg (1992) also provided some evidence for the relationship between general organisational characteristics and the form of participation chosen or rather the form evolving in the course of system design and implementation. In their case study, the mechanistic structure and unstable processes of the organisation fostered "arbitrary" participation instead of the open participation to which the design team had explicitly committed itself.

In the KOMPASS method, formal and direct forms of participation are aimed for as early in the design process as possible, allowing for at least systematic consultation of all parties involved and potentially also common decision making, especially where direct effects on everyday work practices are concerned. At the same time, the design team should take into account the nature of the organisation, thereby potentially and deliberately reducing participation so as not to create expectations that cannot be adequately fulfilled within the actual organisational context.

The Role of the KOMPASS Design Expert

Hirschheim and Klein (1989) described how implicit and explicit assumptions of system developers and managers about basic topics such as the nature of social organisation, the nature of technology, and their own role in the development process will greatly influence the design and implementation process. The role of the KOMPASS design expert was defined in view of two extreme positions on organisational decision making: rationality versus self-organisation. The first position assumes that experts determine and implement the "what" and "how" of changes founded on a model of rational behaviour in and by organisations, which argues that expert knowledge is heard and followed when it is in the organisation's best interest (cf. the empirical–rational strategy in Chin and Benne's [1976] account of change strategies). The second position stresses the self-organizing qualities of organisations, which restrict the influence of experts to providing ideas and initiatives mainly regarding the "how" of a decision process, the course of which is only partially predictable and, to an even lesser degree, controllable (Baitsch & Alioth, 1990).

From the at times heated discussion between these two positions, it can be concluded that both positions need to be taken into consideration when acting as an expert in a design process. Knowledge of both the "how" and "what" of a design process should be provided, even though different contexts will require balancing inputs differently. In particular, normative design assumptions should always be handled with great care, because they can easily create insurmountable resistance and severely disturb a democratic dicussion process. So, in many cases, designing the design process to help participants bring together their own knowledge may be the best strategy in line with the assumption that asking the right questions sometimes may be more important than giving the right answers (Klein, 1993).

Besides providing the organisation with his or her knowledge on the "what" and "how" of good system design, the expert's role includes as a second and at least equally important element to be able to switch between involvement in and distance to the design process. This allows the expert to understand the contextual nature of his or her as well as everybody else's design inputs and to transfer this understanding back into the design process (Clegg, 1994; Winograd & Flores, 1986).

In the KOMPASS method, expert knowledge on both the "what" and "how" of design is incorporated, but with the important addition that the design team is instructed to reflect explicitly and continuously the experts' role and the relevance and status of their knowledge during the design process.

Modules of the KOMPASS Design Heuristic

KOMPASS covers three phases of design processes, providing an initial phase concerning project management and decisions on system definition and form and degree of participation of future system users: In Phase 1 an expert analysis of existing work systems is carried out. In Phase 2, a design team, composed of experts from different disciplines involved in the design of automated systems as well as future system users, is guided through a reflection process of the design approach. In Phase 3, concrete design requirements are derived. The sequence of these phases is not mandatory. In some cases, it might be desirable only to carry out the analysis and evaluation of an existing system to identify needs as well as potential for change without intending a major technical or organisational design project. An automation project might also be carried out without reference to a specific existing work system, in which case the analysis

TABLE 20.20
The Phases Supported by the KOMPASS Method (from Grote et al., 2000)

Phase	Goal
1. Expert analysis of existing work systems	Evaluation of current system design with respect to complementarity and control of variances and disturbances
2. Discussion and formulation of design philosophy	
2.1. Definition of the primary task and the functions of the planned work system	Definition of primary task and functions of the system Collection of potential variances and disturbances
2.2. Definition of a shared evaluation concept	Definition of goals for the system differentiating between successful and unsuccessful work systems
2.3. Derivation of the main potentials for improvement	Identification of potentials for improvement on the basis of variances and disturbances as well as of system goals
2.4. Identification of the potential contributions to a successful work system by human operator, technical system and organisational conditions	Definition of potentially promoting and hindering contributions of people, technology, and organisation to successful system performance according to potentials for improvement (cf. 2.3)
2.5. Specification of the working conditions required for human operators to make their specific contributions	Specification of technical and organisational prerequisites for specific furthering contributions of the human operator
2.6. Usefulness of the KOMPASS criteria for the analysis, evaluation and design of work systems	Critical reflection of the KOMPASS criteria and of the system to be (re)designed in light of those criteria Validation of results from expert analysis (cf. 1) Agreement on complementary design approach
3. Derivation of concrete design requirements	
3.1. Derivation of requirements for system design	Definition of demands on system design according to KOMPASS criteria in relation to potential for improvement (cf. 2.3)
3.2. Definition of work packages	Clustering of design requirements Formulation of work packages to be assigned to specialists for realisation

phase will be omitted. The three phases included in the KOMPASS method and the goals of each phase and step within the phases are illustrated in Table 20.20.

Phase 1: Expert Analysis of Existing Work Systems

In this phase, existing systems are analysed with respect to the criteria on all three levels of analysis. Guidelines are provided that comprise detailed instructions for data collection by means of workplace observation and semistructured interviews as well as for data reduction and interpretation with respect to the criteria. Particular attention is also given to variances and disturbances occurring in the system. If no systems similar to the one to be (re)designed are available, this phase is omitted.

Phase 2: Discussion and Formulation of Design Philosophy

The rationale of Phase 2 is to support the design team in following an integral design approach by developing a shared understanding of the system's objectives and the contributions of technical, organisational, and human aspects to successful system performance and by reaching an agreement on adopting a complementary design philosophy. The main goal of this phase is to establish a common learning process that enables the team to apply the concept of complementary system design to its concrete problems by reflecting implicit assumptions and theories about the roles of people, technology, and organisation in work systems. To achieve this, a facilitator leads the design team (in a 2-day workshop) through five steps. These steps are not necessarily intended to be followed in sequential order, but they contain all the aspects relevant to the explicit reflection of the design team's automation philosophy.

Step 2.1: Definition of the Primary Task and the Functions of the Planned Work System. Because the outcomes of a work system always depend on the interplay of people, technology, and organisation, the aim of this first step is to further a view on the system to be designed as a sociotechnical, not just a technical, system. To achieve this goal, the design team first defines the primary task of the system as well as the functions of the work system. For each function, potential variances and disturbances are collected to identify the functions with high potentials for improvement (cf. Step 2.3).

Step 2.2: Definition of a Shared Evaluation Concept to Differentiate Between Successful and Unsuccessful Work Systems. In this step, the design team defines a set of design goals with regard to the overall quality of the system. The different expectations by the members of the design team can be made explicit and integrated into a common evaluation concept. This also serves as a basis for discussion and comparison of different design scenarios.

Step 2.3: Identification of the Main Potentials for Improvement. From the collected variances and disturbances as well as from the system goals defined in the previous step, the design team delineates the main problem areas (i.e., the system's characteristics containing the most potential for improvement).

Step 2.4: Identification of the Potential Contributions to a Successful Work System by Human Operator, Technical System, and Organisational Conditions. The principle of complementarity is the focus of Step 2.4. The design team has to identify and discuss the different contributions of people, technical systems, and organisation that further or hinder the reaching of the design goals and improvements defined in the previous steps. The explicit discussion of different contributions of people, technical systems, and organisation is meant to make their qualitative differences and their complementarity evident.

Step 2.5: Specification of the Working Conditions Required for Human Operators to Make Their Specific Contributions. In this step the focus is shifted to humans' working conditions. Technical as well as organisational prerequisites that enable humans to provide specific contributions to the success of the overall system are discussed. If, for example, experience was defined as one of the potential specific contributions of the human operator, the question is which technical and organisational prerequisites are necessary to allow the operator to actually gain that experience. The members of the design team must recognise that their decisions regarding technical and organisational aspects substantially affect the operator's possibility for contributing to the system's efficiency.

Step 2.6: Usefulness of the KOMPASS Criteria for the Analysis, Evaluation, and Design of Work Systems. In the last step of Phase 1, a facilitator introduces the design team to the KOMPASS criteria for complementary system design. To initiate a critical reflection on the criteria and the system to be (re)designed in relation to those criteria, an analysis of an existing system (if possible, the one chosen in Phase 1) is carried out together with all members of the design team. Following this common analysis, the results of the expert analysis carried out in Phase 1 are presented by one facilitator and discussed in the group. The overall goal of this step is to enhance the acceptance of the KOMPASS criteria as a chance to structure the design process and to agree on the complementary design approach as the basis for the further design process.

Phase 3: Derivation of Concrete Design Requirements.   In Phase 3 concrete requirements for system design—and consequently specific demands on the function allocation in the human–machine system—are derived on the basis of the KOMPASS criteria. The KOMPASS method supports the design team in the definition of these requirements as well as in the definition of concrete work packages to be assigned to the various design experts for their realization.

Step 3.1: Derivation of Requirements for System Design.   On the basis of the KOMPASS criteria, the team formulates—in another 2-day workshop with the whole design team—design requirements to realise each potential for improvement (as identified in Phase 2, Step 3). In this way, the problem areas can be systematically analysed with respect to their potential for complementary design. The goal of this step, however, is not to produce the ultimate technical solutions with all the details at this point. Instead, the aim is to develop design requirements that will have to be examined with respect to their technical feasibility and actually realised in a later phase by specialists in each technical domain.

Step 3.2: Definition of Work Packages.   After the workshop, members of the design team are charged with the formulation of design work packages. The aim of this step is to gather the interdependent design requirements were collected in the previous step for each potential for improvement and to combine them in concrete work packages that can be assigned to the system experts for realisation.

KOMPASS does not provide specific guidelines for the technical realisation of the design requirements. Rather, the realisation takes place within the enterprise's usual methodologies for organisational and technical development.

OUTLOOK

The KOMPASS method has been applied in several projects in the field of manufacturing (Grote et al., 2000). In these projects, requirements for organisational as well as for technical design have been fruitfully developed. However, one of the main difficulties when applying the method was the rather high number of design criteria. Because most of these criteria are not independent but interrelated, it turned out to be difficult to maintain an overview of the requirements specification process. The discussions oscillated between details (e.g., when discussing the process transparency regarding one subfunction of the system) and a holistic perspective (e.g., when figuring out trade-offs between the different criteria and the subfunctions and among criteria and subfunctions). Another problem concerns the tractability of design decisions, as has also been pointed out by Boy (1998) in the context of function allocation decisions in aircraft design.

To improve the support that KOMPASS offers to system designers, a computer-based software tool is currently being developed that incorporates the KOMPASS criteria as well as the design heuristic. Specifically, the tool users are instructed to discuss interrelations of the design objectives and the KOMPASS criteria. They have to define for every criterion their relevance for reaching each of the objectives. In this way, a commonly developed network between the objectives and the criteria is established. With this network's help, the tool users can explore and evaluate the dependencies between the objectives and criteria in an interactive way. In this process, the tool offers three specific benefits: (a) visualisation of assumptions regarding interrelations between design criteria and design objectives as well as among criteria and objectives, (b) visualisation of the dynamics resulting from complex interrelations between and among design criteria and design objectives, and (c) support of "virtual workshopping" distributed in time and place by means of the possibility for tool users to individually provide their inputs and check on other's inputs.

In a first test of the tool, the procedure and the support for developing and evaluating solutions offered by the tool was regarded "interesting, challenging, and supportive" (Little et al., 2001). Further experience with the tool will determine whether it can truly achieve a better handling of the complexity created by the integrative system design approach intended by the KOMPASS method.

ACKNOWLEDGEMENTS

The KOMPASS method has been developed at the Institute of Work Psychology at ETH Zurich. The project has been funded by the Center of Integrated Production Systems (ZIP) at ETH Zurich as well as by the Commission of Technology and Innovation (KTI) of the Swiss Federation. The research in the further developments as described in the outline has been carried out in the EU IST-IMS PSIM Project (Participative Simulation environment for Integral Manufacturing enterprise renewal; IMS 1999-00004). This project was funded by the European Commission and the Swiss Federation. PSIM is part of HUMACS, a project within the international IMS research program.

REFERENCES

Bailey, R. W. (1989). *Human performance engineering* (2nd ed.). London: Prentice-Hall International.

Bainbridge, L. (1981). Mathematical equations or processing routines? In J. Rasmussen & W. B. Rouse (Eds.), *Human detection and diagnosis of system failures* (pp. 259–286). New York: Plenum.

Bainbridge, L. (1982). Ironies of automation. In G. Johannsen & J. E. Rijnsdorp (Eds.), *Analysis, design and evaluation of man-machine systems* (pp. 129–135). Oxford: Pergamon.

Baitsch, C., & Alitoth, A. (1990). Entwicklung von Organisationen—Vom Umgang mit Widersprüchen. In F. Frei & I. Udris (Eds.), *Das Bild der Arbeit* (pp. 244–257). Bern: Huber.

Böhle, F., & Rose, H. (1992). *Technik und Erfahrung. Arbeit in hochautomatisierten Systemen.* Frankfurt/Main: Campus.

Boy, G. (1998). Cognitive function analysis for human-centred automation of safety-critical systems. *Proceedings of CHI'98* (pp. 1–9). New York: Addison Wesley, pp. 265–272.

Brehmer, B. (1993). Cognitive aspects of safety. In B. Wilpert & T. Qvale (Eds.), *Reliability and safety in hazardous work systems* (pp. 23–42). Hove: Lawrence Erlbaum Associates.

Chin, R., & Benne, K. D. (1976). General strategies for effecting changes in human systems. In W. G. Bennis, K. D. Benne, & R. Chin (Eds.), *The planning of change* (3rd ed., pp. 22–45). New York: Holt, Rinehart & Winston.

Clegg, C. (1994). Psychology and information technology: The study of cognition in organizations. *British Journal of Psychology, 85,* 449–477.

Clegg, C., Ravden, S., Corbett, M., & Johnson, G. (1989). Allocating functions in computer integrated manufacturing: A review and a new method. *Behaviour & Information Technology, 8,* 175–190.

Corbett, J. M. (1987a). A psychological study of advanced manufacturing technology: The concept of coupling. *Behaviour & Information Technology, 6,* 441–453.

Corbett, J. M. (1987b). Computer aided manufacturing and the design of shopfloor jobs. In M. Frese, E. Ulich, & W. Dzida (Eds.), *Psychological issues of human-computer interaction in the work place* (pp. 23–40). Amsterdam: Elsevier.

De Keyser, V. (1987). Structuring of knowledge of operators in continuous processes: Case study of a continuous casting plant start-up. In J. Rasmussen, K. D. Duncan, & J. Leplat (Eds.), *New technology and human error* (pp. 247–259). Chichester: Wiley.

Duncan, C. D. (1981). Training for fault diagnosis in industrial process plants. In J. Rasmussen & W. B. Rouse (Eds.), *Human detection and diagnosis of system failures* (pp. 553–573). New York: Plenum.

Dunckel, H., Volpert, W., Zölch, M., Kreutner, U., Pleiss, C., & Hennes, K. (1993). *Kontrastive Aufgabenanalyse im Büro. Der Kaba Leitfaden.* Zürich: vdf.

Emery, F. E. (1959). *Characteristics of socio-technical systems.* Tavistock Document No. 527, London.

Friedrich, P. (1993). Technische Veränderungstätigkeit. In G. Cyranek & E. Ulich (Eds.), *CIM—Herausforderung an Mensch, Technik, Organisation* (pp. 167–194). Zürich: vdf.

Ganster, D. C., & Fusilier, M. R. (1989). Control in the workplace. In C. L. Cooper & I. Robertson (Eds.), *International review of industrial and organizational psychology* (Vol. 4, pp. 235–280). New York: Wiley.

Greenbaum, J., & Kyng, M. (1991). *Design at work: Cooperative design of computer systems.* Hillsdale, NJ: Lawrence Erlbaum Associates.

Grote, G. (1997). *Autonomie und Kontrolle—Zur Gestaltung automatisierter und risikoreicher Systeme.* Zürich: vdf.

Grote, G., Wäfler, T., Ryser, C., Weik, S., Zölch, M., & Windischer, A. (1999). *Wie sich Mensch und Maschine sinnvoll ergänzen. Die Analyse automatisierter Produktionssysteme mit KOMPASS.* Zürich: vdf.

Grote, G., Ryser, C., Wäfler, T., Windischer, A., & Weik, S. (2000). KOMPASS: A method for complementary function allocation in automated work systems. *International Journal of Human-Computer Studies, 52,* 267–287.

Hacker, W., Fritsche, B., Richter, P., & Iwanowa, A. (1995). *Tätigkeitsbewertungssystem (TBS). Verfahren zur Analyse, Bewertung und Gestaltung von Arbeitstätigkeiten.* Zurich: vdf.

Hackman, J. R., & Oldham, G. R. (1976). Motivation through the design of work: Test of a theory. *Organizational Behavior and Human Performance, 16,* 250–279.

Hirschheim, R., & Klein, H. K. (1989). Four paradigms of information systems development. *Communications of the ACM, 32,* 1199–1216.

Hornby, P., & Clegg, C. (1992). User participation in context: A case study. *Behaviour & Information Technology, 11,* 293–307.

Jackson, P. R., & Wall, T. D. (1991). How does operator control enhance performance of advanced manufacturing technology? *Ergonomics, 34,* 1301–1311.

Johansson, G. (1989). Stress, autonomy, and the maintenance of skill in supervisory control of automated systems. *Applied Psychology, 38,* 45–56.

Jordan, N. (1963). Allocation of functions between man and machines in automated systems. *Journal of Applied Psychology, 47,* 161–165.

Karasek, R. A., & Theorell, T. (1990). *Healthy work: Stress, productivity and the reconstruction of working life.* New York: Basic Books.

Kidd, P. T., & Karwowski, W. (1994). *Advances in agile manufacturing. Integrating technology, organisation and people.* Amsterdam: IOS Press.

Klein, L. (1993). On the collaboration between social scientists and engineers. In E. Trist & H. Murray (Eds.), *The social engagement of social science, Vol. II: The socio-technical perspective* (pp. 369–384). Philadelphia: University of Pennsylvania Press.

Kraiss, K.-F. (1989). Autoritäts- und Aufgabenteilung Mensch-Rechner in Leitwarten. In G. Daimler- and K. Benz-Stiftung (Eds.), *2. Internationales Kolloquium Leitwarten* (pp. 55–67). Köln: Verlag TÜV Rheinland.

Kullmann, G., Pascher, G., & Rentzsch, M. (1992). Erhöhung der Prozesstransparenz durch differenzierte akustische Rückkopplung bei CNC-Maschinen. *Zeitschrift für Arbeitswissenschaft, 46,* 219–223.

LaPorte, T. R., & Consolini, P. M. (1991). Working in practice but not in theory: Theoretical challenges of "high-reliability organizations". *Journal of Public Administration Research and Theory, 1,* 19–47.

Little, S., Bovenkamp, M. van de, Jongkind, R., Wäfler, T., Eijnatten, F., & Grote, G. (2001). The STSD-tool—IT support for sociotechnical system design. In G. Johannsen (Ed.), Analysis, design, and evaluation of human-machine systems. *Proceedings of the 8th IFAC/IFIP/IFORS/IEA Symposium* (pp. 409–414). Kassel (Germany).

Majchrzak, A. (1988). *The human side of Factory automation.* San Francisco: Jossey-Bass.

Moray, N., Inagaki, T., & Itoh, M. (2000). Adaptive automation, trust, and self-confidence in fault management of time-critical tasks. *Journal of Experimental Psychology, 6,* 44–58.

Oesterreich, R., & Volpert, W. (1991). *VERA Version 2. Arbeitsanalyseverfahren zur Ermittlung von Planungs- und Denkanforderungen im Rahmen der RHIA-Anwendung.* Berlin: Technische Universität.

Older, M. T., Waterson, P. E., & Clegg, C. (1997). A critical assessment of task allocation methods and their applicability. *Ergonomics, 40,* 151–171.

Parasuraman, R., & Riley, V. (1997). Humans and automation: Use, misuse, disuse, abuse. *Human Factors, 39,* 230–253.

Perrow, C. (1984). Normal accidents. *Living with high-risk technologies.* New York: Basic Books.

Rasmussen, J. (1982). Human errors: A taxonomy for describing human malfunctions in industrial installations. *Journal of Occupational Accidents, 4,* 311–333.

Rogers, Y., Rutherford, A., & Bibby, P. A. (1992). *Models in the mind. Theory, perspective & application.* London: Academic Press.

Rouse, W. B. (1988). Adaptive aiding for human-computer control. *Human Factors, 30,* 431–443.

Scarbrough, H., & Corbett, J. M. (1992). *Technology and organisation.* London: Routledge.

Schüpbach, H. (1994). *Prozessregulation in rechnerunterstützten Fertigungssystemen.* Zürich: vdf.

Sheridan, T. B. (1987). Supervisory control. In G. Salvendy (Ed.), *Handbook of human factors* (pp. 1243–1268). New York: Wiley.

Sonntag, K. (1990). Qualifikation und Qualifizierung bei komplexen Arbeitstätigkeiten. In C. Graf Hoyos & B. Zimolong (Eds.), *Ingenieurpsychologie, Enzyklopädie der Psychologie,* Themenbereich D, Serie 3. (Vol. 2, pp. 536–571). Göttingen: Hogrefe.

Strohm, O. (1997). Analyse und Bewertung von Arbeitssystemen. In O. Strohm & E. Ulich (Eds.), *Unternehmen arbeitspsychologisch bewerten* (pp. 135–166). Zürich: vdf.

Susman, G. I. (1976). *Autonomy at work. A sociotechnical analysis of participative management.* New York: Praeger.

Thompson, S. C. (1981). Will it hurt less if I can control it? A complex answer to a simple question. *Psychological Bulletin, 90,* 89–101.

Ulich, E. (1998). *Arbeitspsychologie* (4th ed.). Zürich: vdf.

Ulich, E., Rauterberg, M., Moll, T., Greutmann, T., & Strohm, O. (1991). Task orientation and user-oriented dialogue design. *International Journal of Human-Computer Interaction, 3,* 117–144.

Volpert, W. (1994). *Wider die Maschinenmodelle des Handelns. Aufsätze zur Handlungsregulationstheorie.* Pabst: Lengerich.

Wäfler, T., Windischer, A., Ryser, C., Weik, S., & Grote, G. (1999). *Wie sich Mensch und Technik sinnvoll ergänzen— Die Gestaltung automatisierter Produktionssysteme mit KOMPASS.* Zürich: vdf.

Wall, T. D., Clegg, C. W., Davies, R. T., Kemp, N. J., & Mueller, W. S. (1987). Advanced manufacturing technology and work simplification: an empirical study. *Journal of Occupational Behaviour, 8,* 233–250.

Wall, T. D., Jackson, P. R., & Davids, K. (1992). Operator work design and robotics system performance: A serendipitous field study. *Journal of Applied Psychology, 77,* 353–362.

Weick, K. E. (1976). Educational organizations as loosely coupled systems. *Administrative Science Quarterly, 21,* 1–19.

Wilson, J. R., & Rutherford, A. (1989). Mental models: Theory and application in human factors. *Human Factors, 31,* 617–634.

Winograd, T., & Flores, F. (1986). *Understanding computers and cognition: A new foundation for design.* Norwood, NJ: Ablex.

21

Relating the Automation of Functions in Multiagent Control Systems to a System Engineering Representation

Michael D. Harrison, Philip D. Johnson,
and Peter C. Wright
University of York, United Kingdom

Abstract

Engineers of safety-critical systems are beginning to recognise that human issues are critical to their safe automation and that appropriate techniques for taking into account the people in the system should be integrated into the design process. This chapter gives a brief introduction to a two-step decision procedure that can be used to help decide how to automate an interactive system. The procedure is intended for use early in the development of systems in larger scale collaborative settings with the aim of improving their safety and performance. Two issues are particularly important. The first concerns the appropriate choice of automation so that the tasks designed for the different roles satisfy criteria that have significance from a cognitive perspective. The second is to understand the mapping from concepts of function allocation to notations that are meaningful and usable by system engineers. The method has received preliminary evaluation in aviation and naval contexts.

INTRODUCTION

As the automation of complex processes becomes more achievable, the need for system engineering procedures that help decide how to automate becomes more important to the safety and flexibility of automation use. Work systems are often complex interactive systems involving many people and many technology components. The work involved must be implemented in a way that is most compatible with roles that are designed for the people in the system. The implementation must satisfy general criteria such as minimising workload, maximising awareness of what is going on, and reducing the number of errors. The basic problem, therefore, is to reduce the cognitive demands of the tasks being performed by the people involved in the system while maintaining fully their ability to function within their given roles. This chapter is concerned with these procedures, collected together in a process known as function allocation. It is also concerned with how these procedures can be integrated effectively with methods and representations that are used by system engineers. Although many function allocation methods have been designed to be used by human factors experts, relatively little attention has been paid to the linkage of these methods to system engineering.

Allocation of function has evolved since the early 1950s. Methods that have been developed are problematic for a number of reasons:

- *Context:* Most available methods fail to take proper account of the context in which the functions are to be automated. Functions are often considered in isolation using general capability lists describing what people do better and what machines do better (Fitts, 1951). If context is to be considered, it is presumed to be done implicitly by the appliers of the method. The cognitive task is therefore considered in isolation from the environment in which it is to be performed.
- *Comprehension:* Most available methods fail to be comprehensible and easily applied by engineers and depend on a firm understanding of human factors. For example, the KOMPASS method (Grote, Ruser, Wäfler, Windischer, & Weik, 2000) described criteria for complementary system analysis and design: process transparency, dynamic coupling, decision authority, and flexibility. These criteria are difficult to interpret. Methods do not use design representations that are familiar to system engineers. Allocation decisions are usually binary and, as a result, functions are identified as to be automated or not. This means that they must be described at a low level of granularity.
- *Collaboration:* Most available methods fail to recognise that the system is more than a single human and a single device. They do not take into account the broader collaborative system, the roles defined therein or the design of cognitive tasks that bridge these roles. They offer no guidance about how dynamic allocation of function should be implemented. This problem becomes more relevant as the possibility of automating a system dynamically, when necessary or appropriate, becomes more realistic.

Recent methods of function allocation have been developed to overcome some of these problems. In particular, one developed at the Institute of Work Psychology, University of Sheffield (Older, Waterson, & Clegg, 1997) aims to overcome context problems while taking account of collaborative aspects of the organisation. Another method developed at the Department of Computer Science, University of York (Dearden, Harrison, & Wright, 2000) makes the decision procedures more explicit while also having a strong emphasis on context. This method also provides a clear representation for implementations of functions that automate some parts but not all of the functions (hereafter called partially automated functions).

Another approach that, although relatively weak at supporting the decision process or representing the context in which the functions are performed, has a strong representation of the output of the method in terms of a classification of levels of automation. These methods of automation classification, particularly that of Parasuraman, Sheridan, & Wickens (2000), are useful for introducing the important automation categories for different phases of the function. They work well in aiding the production of an output format. They are relatively difficult to convert into implementation for any given situation however.

This chapter proposes a modest extension to the method described in (Dearden et al., 2000) aimed at addressing some of the concerns about collaboration and also ties more effectively to a format that can be understood by system engineers. In this method, roles can be assigned to parts of a function, thereby defining some of the collaborative characteristics required in performing the function. To link with the requirements of system engineers, the system, and modelling language UML (Unified Modelling Language) is used to represent elements required in the process of function allocation.

UML is a notation for describing different views of a system using different kinds of diagrams (Rumbaugh, Jacobson, & Booch, 1999a). It presumes an object-orientated design approach but, within this approach, provides a range of techniques for modelling important features of the design in a standardised notation. UML is of considerable interest in industry, particularly to those companies that are attempting to satisfy external requirements for human factors integration such as military organisations. The representations of UML can be used to describe the inputs and outputs to the method. If function allocation is to be made an explicit step in the system engineering development process, then it must offer solutions to problems faced by engineers in the design of highly automated control systems. In practice, function allocation is not often performed as an explicit step because results are perceived to be imprecise and benefit is unclear. Engineers recognise that human operators are a source of error and attempt to solve this problem by minimising their role. The assumption is that doing this will have the effect of reducing these risks. Inevitably, situations arise that the automation cannot handle. In these circumstances, the operators are expected to step in and resolve the situation. Because operators have been "out of the loop", their ability is impaired. The view of allocation of function is to keep operators in the loop in relation to their roles and to view automation as assisting them; it is therefore a problem of cognitive task design. The *cognitive task design* goal is to take advantage of the benefits offered by automation, but to do so in a way that does not impede the operators' abilities to perform their roles. The achievement of this goal is difficult because it requires the combination of two disciplines, system engineering and human factors engineering, neither of which can provide the solution alone.

This chapter describes the proposed method in two passes: The first (the next two sections) describes the essentials of the method; the second describes how the method provides support for the system engineering process. The next section gives a brief description of the method and describes in more detail what is required as input and what is produced as a result. This is followed by a more detailed description of the method, in particular, the decision steps that are taken. The link between the allocation of function method and UML is then considered. The final section contains a brief description of possible further extensions to the method. It is possible to get an appreciation of how the method works without getting into the detail of UML. The process is fully described without reading the UML sections.

THE INGREDIENTS OF THE METHOD

Introduction

The problem is to take a set of functions, describing the work that the system is to do in the contexts in which the work is to be carried out, and to decide how to automate them. The procedure for automating functions involves two decision steps. First, functions are matched to defined roles that capture high-level functions of the agents. In practice, it is often difficult to be clear about these roles at the outset. Initial definitions will be refined iteratively, therefore through successive applications of the method. The aim is to decide how closely these functions fit the roles in a set of given scenarios. Functions that are entirely subsumed within a role are proposed to be "totally manual within the role"—to automate would in effect remove part of the agent's role. Functions that can be separated entirely from any of the roles and for which it is possible to automate are designated as "to be automated". In practice few functions fit either of these categories. The functions that are left are therefore considered in more detail to decide which aspects should be automated. By this means, automation can be designed to support specified roles most effectively. The effect therefore is to design cognitive tasks for the different roles specified and to render them in a form that can be employed by a system engineer.

In the following subsections, we describe function, role, and scenario in more detail. Other inputs to the method (e.g., description of the technology baseline, mandatory constraints, and evaluation criteria) are also required as inputs. We leave a discussion of some of these issues until the section in which we consider UML aspects in more detail and others we ignore to maintain a clear description of the method. Function, role, and scenario represent the minimum information required to gain a reasonable understanding of the method. The output of the method is a set of functions that can either be described as manual, totally automated, or partially automated. The format for partially automated functions is indicative of how they should be implemented.

Role

The aim is to consider the match between function and role as well as whether functions contain aspects that are most appropriate for humans or for machines. The machine or technology part of the system will be described generically as a *device* in what follows. In practice, the design of roles, whether of a human or a device, takes account of human considerations such as capability and training as well as the overall balance of the work design. Often, whether a function should be automated and how much it should be automated, depends on the context in which the function is performed. It makes more sense to be specific about how compatible the tasks are with the various roles specified for the various personnel in the particular contexts than to ask in isolation the question, "Would it be sensible for a human to do this?"

In practice, role is difficult to define. It is an activity that can be performed either by a human or a device. Normally it is not necessary to produce a statement for a device's role, but there are circumstances in which doing so is helpful in providing scope for the definition of a technology component. For example, the envelope protection provided in a fly-by-wire aircraft (in other words, the boundaries beyond which pilot manoeuvres are judged to jeopardise the safety of the aircraft) could be clarified through an explicit role statement. Doing so would help to ensure that functions are not allocated in a way that prevents the device from being able to keep the plane within the safety envelope.

An example of a role statement for the captain of a civil transport aircraft might be: *The captain is the final authority for the operation of the airplane. The captain must prioritise tasks according to the following hierarchy: safety, passenger comfort, and flight efficiency.*

In the context of a single seat military aircraft an example role statement for the pilot might be: *The pilot is the final authority for the use of offensive weapons and for evasive manoeuvres. The pilot is responsible for mission planning, tactical and strategic decision making, and coordination with other aircraft. The pilot is responsible for the management of systems to maximise the probability of successful mission completion.*

These roles may be taken as a starting position subject to refinement as the allocation process unfolds. Ambiguities may be removed, assumptions may be seen as inappropriate, incoherence between roles may be resolved, and role definitions changed to establish function integrity and functions changed to support roles provided.

Functions

A function or unit of work (an activity that the system—people and devices—is required to be capable of performing to achieve some result in the domain) might include "finding the current position of the vehicle", "withdrawing cash", or "detecting a fire". Although in practice, identifying and determining a function is a matter of expert judgement, some characteristics are important to prevent premature commitment to a means of automation. A function does not

TABLE 21.1
Fitts's List

Humans Are Better at . . .	Machines Are Better at . . .
Detecting small amounts of visual, auditory, or chemical energy	Responding quickly to control signals
Perceiving patterns of light or sound	Applying great force smoothly and precisely
Improving and using flexible procedures	Storing information briefly, then erasing it completely
Storing information for long periods of time and recalling appropriate parts	Reasoning deductively
Reasoning inductively	Doing many complex operations at once
Exercising judgement	

contain any indication of who or what performs it (Cook & Corbridge, 1997). For example, "key in way point" is not a function because it implies that the operator enters the waypoint manually, whereas "set way point" does not. Functions are related to, and can be derived from, the functional requirements used in system engineering. The assumption made in this process of function allocation is that, unlike typical practice, indications about who or what should perform the function are removed at the requirements stage.

Functions can be defined at a variety of levels of granularity and can be arranged hierarchically. In the same way as task analysis, a top-down hierarchical approach can be a useful aid to function elicitation and is also helpful in deciding what level of granularity of function can be used most effectively in the decision process. Descending the hierarchy reveals both a decrease in the complexity of the function and in the size of the subsystem required to perform it. At the top of the hierarchy, functions may be performed by a team or department, whereas at the bottom by a single operator with the aid of automation.

Scenarios

Function allocation methods such as the "Humans are better at, Machines are better at" approach presume that the suitability of a function for automation is based on an individual function. They use capability lists usually based on the so-called Fitts's List described in Table 21.1 (Fitts, 1951). Using this approach, functions are matched against the lists and, on the basis of the match, a decision is made whether to automate the function.

As has been noted already, these capability lists ignore the complex interactions and dependencies between activities of work. To overcome this, the method we describe here considers groups of functions structured in the context of a scenario. This allows the designer to appreciate the interactions between the functions and therefore to understand more effectively those contextual factors that might influence the design of the cognitive tasks entailed by function allocation decisions. Scenarios focus on situations relevant to the decision criteria that are used to decide what level of automation is appropriate Examples of such criteria include workload and situation awareness. Scenarios aim to represent the functions under consideration in a range of contexts. For example, in the field of civil aviation where workload has been identified as a decision criterion, takeoff and landing may be considered important as a basis for scenarios (among others, such as emergency conditions) because these periods are recognised as producing high workload.

TABLE 21.2
Scenario Input Template

Use-Case Number	Name of Use-Case

	Scenario 1 - Each use-case can have a number of scenarios
Environment	This is a description of the environment within which the system is operating when the scenario occurs.
Situation	This is a description of the state of the system at the start of the scenario. Are all the operators on duty, are there any known or unknown faults in the system etc.?
Sequence of events	*Step* *Event* 1 The main sequence of events that take place during the scenario. This includes events that happen in the environment, events that cause changes to the system and the actions that the system must perform. . . .
Event extensions	*Step* *Event* 1 Variations on the main sequence of events are recorded as event extensions. . . .
	Scenario 2

The method only considers those functions used within the current scenario, and therefore the analyst must ensure that every function occurs in one scenario and ideally several. It is important to ensure that a variety of scenarios are used, covering the range of activities that the functions will engage in. To achieve this, scenarios should be selected that cover all the normal operating conditions of the system. There is a wide range of sources for possible candidate scenarios, for example:

- The experience of practitioners in previous systems
- Incident and accident reports for previous systems
- Scenarios developed during the business modelling stage of the system life cycle
- Use cases and scenarios in the previous system's development documentation or training manuals.

Scenarios often entail a detailed account of what happened in a set of circumstances. They are therefore concrete in terms of the actions that are described. If the scenario is describing some current possibility with the baseline architecture, these events will be expressed in implementation-dependent terms. Once a scenario is elicited, therefore, the description must be re-expressed neutrally in terms of functions (using the initial list of functions) rather than the baseline actions. The format for representing scenarios is based on a modified version of the scenario template for Technique for Human Error Assessment (Pocock, Harrison, Wright, & Johnson, 2001) and employs the UML format for scenarios described in Cockburn, 2001; see Table 21.2.

This scenario description is then transformed. The events and event extensions headings are changed to functions and function extensions. The account of what happens in the scenario is re-expressed using the set of functions rather than the event descriptions. This process can be a useful check that the functions are expressed at an appropriate level and whether they incorporate too much implementation bias. The functional scenarios are then used as the basis for the two decision procedures.

The Nature of the Output

Function allocation methods typically consider functions at a very low level and provide two allocation options H (human) or M (machine). In many cases, the designer's interest lies with those higher level functions that require some form of collaboration between operators and the devices in the system. The process is therefore concerned with how to implement these higher level functions in terms of how the collaboration or automation boundaries work. By this means, the cognitive tasks that are relevant to the potentially various human roles may be understood. Rather than working with numerous low-level functions to determine this boundary, it is easier to work with the parent functions and to declare them as partially automated. Of course there are many ways in which the operators and devices can interact to execute a function, and there is a need for the designer to specify how the collaboration will be achieved.

One possible approach is to use a classification that defines levels of automation with the aim that engineers recognise how to implement the particular level. The allocation of function method then provides advice about appropriate automation in terms of level. A number of such classification schemes have been suggested. For example, Sheridan and Verplanck (1978) suggest a classification of levels in which decisions and control pass progressively from the human to the device. Later authors have produced alternatives. Kaber and Endsley (1997) provided 10 levels of automation. Billings (1991) suggested seven levels of management automation. For each function that the designer defines as being partially automated, there would be an indication of what the level of automation would be. Parasuraman et al. (2000) went further and suggested guiding the decomposition of functions into four broad classes: information acquisition, information analysis, decision and action selection, and action implementation. These functions are then automated according to Sheridan and Verplanck's scale. In summary, all these approaches have in common two difficulties:

1. They assume a spectrum of levels of automation between human and device. In practice, a number of different roles may carry out different functions.
2. They provide solutions that are not sufficiently clear to form a basis for engineering implementation of the functions.

The approach used here, called IDA-S (Table 21.3; Dearden, 2001), provides more hooks for thinking about partial automation. It is based on Malinkowski, Kuhme, & Schneider-Hufschmidt's (1992) framework for adaptive systems and is capable of expressing all types of automation provided by the various classifications, splitting into four components that have

TABLE 21.3
The IDA-S Template

	Information	Decision	Action
Planning the response	Collect Integrate Configure Initiate response	Propose Evaluate Modify Select	Approve
Supervise ongoing execution	Monitor progress	Identify exceptions	Revoke authority
Supervise termination	Determine output content	Identify completion	Stop process
Action	Execute actions		

TABLE 21.4
Solution to Plan Route

Function: F1 Plan route
Solution: Sol6 Plot way points
Design solution: The navigator plots a number of way points describing the destination and the route required. The electronic chart evaluates the proposed route based on its knowledge of navigation and sailing, proposing any changes or conflicts it identifies. The navigator can modify the route as required and approve the route. The electronic chart then calculates the distances and bearing between the points. The navigation officer supervises the entire process.

	Information		*Decision*		*Action*	
Planning the response	Collect	N	Propose	N	Approve	
	Integrate	N	Evaluate	E		
	Configure	N	Modify	N		
	Initiate response	N	Select	N		
Supervise ongoing execution	Monitor progress	C	Identify exceptions	C	Revoke authority	C
Supervise termination	Determine output content	E	Identify completion	N	Stop process	C
Action	Execute actions	N				

Note. N = Navigator; E = electronic chart; C = navigation officer or command and control.

common features with Parasuraman et al. (2000). The top-level components are information, decision, action, and supervision (hence IDA-S). Each component is further split into a number of elements, each describing a particular aspect of the function. The designer specifies which role is responsible for performing that aspect of the function. If the element is not applicable within the context of the function, then it is marked not applicable (n/a).

The elements in the information component cover issues such as which role integrates the information required to carry out the function and which role is responsible for initiating a response. The decision component covers such issues as which role proposes what plan or action to take, evaluates it, modifies it, and selects one if there is more than one possibility. The action component covers which role carries out the action. The supervision component covers such issues as which role monitors the performance of the action, identifies exceptions, and revokes the action if necessary. It allows the designer to express precisely how the function is implemented in terms of the various roles that are responsible for these aspects of the function. The given solution is indicated by placing the role identifiers in appropriate cells of the template.

Consider, for example, the automation of a function used in scenarios related to ship navigation, for example, "plan route". One solution is described in Table 21.4 in the format required by the method. This function allocation assumes the roles: navigator, electronic chart, and the navigation officer or command and control.

The IDA-S definition clarifies how to develop an implementation that satisfies the requirements. The informal description, under the heading "Design solution" in Table 21.4, states in English which roles are responsible for what aspects of the function. The IDA-S representation invites the analyst to consider how to decompose the function and to consider which role should be responsible for what aspect of the function.

TABLE 21.5
The First Trade-Off

State of Automation Research vs. Relation to Role	Role	Existing with Immediate Access	Existing in Competitor Systems	Low-risk or Low-cost R&D	High-risk or High-cost R&D	Unfeasible
Separable	ALL		Sol 1			
Role related information or control	R1 R3					
Role critical information or control	R1 R3					
Central to role	R1 R3					

Note. R&D = research and development; R = role; Sol = solution.

THE METHOD: TWO DECISION STEPS AND CONSOLIDATION

The approach contains two decisions. The first provides an initial view at how much automation to provide and hence prevents the system from being under- or overautomated. The second decision refines the situations where some level of automation is appropriate for a function. This step makes it easier for the analyst to work towards tasks that optimise performance, workload or situation awareness for example in the context of the different scenarios. The procedures aid the process of deciding what tasks are to be designed for what roles.

Can It Be Matched Entirely to Role?

Once roles, functions, and scenarios have been defined, the first decision step concerns which functions can be totally allocated to one of the roles (these roles may be device or human). Each scenario is considered in turn. The functions that are employed within the scenario are identified. The designer bases decisions about automation of these functions on their use in the context of the scenario under consideration. Suitability for total automation is not based solely on the technical feasibility of a solution; it is also based on the function's relation to the roles. If a function is not seen to be separable from an operator's role, then it cannot be totally automated, because it would interfere with the operator's ability to do the job effectively. There are two dimensions to the trade-off (Table 21.5).

Firstly, each function is considered in relation to the feasibility of automating it. The concern here is with the cost and technical possibility (1st row, Table 21.5). The system engineer must consider how feasible it would be to automate the function along a spectrum of "already possible" to "unfeasible." The engineer therefore uses expert judgement to decide where it best fits. Second, the function is matched against the set of roles. The roles are likely to continue to be refined as the design evolves (First column in Table 21.5). In the example, two roles, R1 and R3, are relevant to the scenario. The functions are entered into the cells of the matrix as possible solutions or implementations of the function. Hence, Sol1 is a possible implementation of a function that is separable from all the roles that have been defined and can be automated using technology that exists in existing competitor systems.

Two classes of functions can be distinguished using this trade-off. If a function can be separated from all the defined roles and is feasible (for example, cost-effective) to automate, then it makes sense to automate it totally. On the other hand, if the function is totally subsumed within one of the human roles, whether or not it can be automated feasibly, it makes sense to consider it as totally manual within the role. These functions can be identified by finding the functions that are "central to role" (Rows 3 and 4 in Table 21.5) and appear in the high-risk or unfeasible columns (Columns 6 and 7 in Table 21.5). It is also likely, however that other functions central to one role may be considered "manual" because it is important to that role's activity that they perform the function manually, however easy it is to automate. The class of functions that contains both the "wholly separable" and "entirely within role" types is not considered further in the method. If the function is to be automated, then the means of automation is dealt with in some other component of the general development process. The interface to this functionality is of no concern here. Notice that a function may appear in more than one row because it relates to several roles but may only appear in one column because feasibility to automate is invariant. This leaves functions that are to be "partially automated". There are usually a number of ways in which partial automation may be achieved. Choosing the most appropriate is the subject of the second trade-off that will be considered next.

The function "plan route" discussed earlier is critical to the role "navigator" (N) and related to the role "navigation officer" (C). In addition, it might be reasonable to assume, depending on system engineering judgement, that the automation can be achieved through low-risk, low-cost research and development. Hence, a potential solution to this function fits into the "partially automated" category and must be considered further. On the other hand, in the context of the definition of the "pilot" role defined earlier for a military aircraft, a function such as "fire missile" is central to role and, although it is technically feasible to automate, would be defined as manual even if there were no mandatory requirement to ensure that this should happen.

Candidates for the Second Step

The second decision procedure is concerned with comparing alternative IDA-S solutions defining partial automation possibilities with a "baseline" solution. The aim is to obtain, in the first instance, a set of most favoured IDA-S representations for each scenario in the sense that these function implementations have the most beneficial effect under criteria such as workload or situation awareness in the context of the scenarios. When this has been done for each scenario, the choices are consolidated into a set of choices for the system as a whole. The result of this analysis will be to produce an implementation reflecting the function allocation decisions. Hence, implementations will be produced that support most effectively the roles that are engaged in the scenario. Tasks will, in effect, be designed so that function implementations gather information, support decisions, and support the most appropriate mechanisms for action and supervision optimally in the context of a set of criteria. In each scenario, comparison is made with the baseline design. Often the baseline is the existing design that is currently the subject of modification, but there are circumstances in which a new concept is being designed and the most conservative of the design alternatives may be considered.

The Rating Process

Having constructed a number of alternative possibilities for each function listed from the scenario, these candidates are rated using a second matrix. An example of possible alternatives in the case of "calculate point to point information" would be the following: The navigation system might propose alternative routes from which the navigator selects the most appropriate or no such choice is provided. The reason for choosing particular representations may be

TABLE 21.6
Identify Partially Automated Functions

Primary Concern	Performance			
State of automation vs. level of improvement	Suggestion is immediately available	Available on competitor systems	Low-risk or low-cost R&D	High-risk or high-cost R&D
Large improvement in primary concern, no deterioration in secondary concerns				
Improvement in primary concern, no deterioration in secondary concerns		F1.2.4 Sol 2		
Improvement in primary concern minor deterioration in secondary concerns		F1.2.4 Sol 1		
Improvement in primary concern, large deterioration in secondary concerns				
No significant improvement in primary concern				
Large deterioration in primary concern				

Note. R&D = research and development.

random or based on some assumptions about the abilities of the off-the-shelf technologies available to the project. The options for all these functions are then rated in relation to a criterion such as workload or performance in comparison with the "baseline" design.

In practice, a collection of criteria will most likely be relevant to the current scenario; for example, it may be appropriate to consider workload but not to the detriment of situation awareness. It is possible that the relation between the criteria will be uncertain, so, for example, improving workload may have a negative effect on situation awareness. The second decision is therefore made in relation to a "primary concern" (the most important criterion) in the context of the potential parallel effect on the other relevant criteria.

This decision step, therefore, takes all the options that have been produced for all the functions relevant to the scenario and the primary concern identified and enters them into a second matrix (see Table 21.6). The design options used in the baseline should also be included in the cell that is labelled "no significant improvement in primary concern". The alternative solutions are then placed in the table. Two criteria decide where the solution should fit. The first depends again on the feasibility of the particular solution—how easy will it be to implement with achievable technology? The second requires a judgement about the effect of the solution in terms of the criteria (workload, performance, situation awareness, etc.). The judgement here is whether the solution causes an improvement or deterioration to the primary criterion and what the consequent effect will be on the other criteria. This process is therefore significant in understanding cognitive load.

All these judgements are carried out in comparison to the baseline design. It would be expected that some solutions do better and others worse. It may be that the analyst uses expert judgement, but it could be that the situation requires a more careful human factors analysis of these decisions. In this case, it may make sense for a team to be involved and for the workload analyses or situation awareness calculations to be performed by appropriate experts.

In the case of the function "calculate point to point information", if workload were the primary criterion and situation awareness the secondary criterion, the process might be as follows.

If the baseline assumption were to be that the function was performed entirely manually, then Solution 1 in which optional routes are presented to the navigator would improve workload with no deterioration to situation awareness (because alternative route information is judged to be key to situation awareness by human factors experts). Solution 2, in which no options are presented, would also improve workload but would, in the judgement of human factors experts, cause minor deterioration to situation awareness.

Choosing the Best Candidates

Each potential solution is placed in the matrix. At the end of the process, it will be possible to derive a set of best candidates, solutions that are most favourable in relation to the criteria and are technically feasible. By this means tasks will be designed in the scenario contexts. This process is achieved by searching from the top left of the matrix and selecting new design options. If a design option for a function is selected, then all other options for that function are deleted from the table. If a design option is selected from the "high-risk research and development" column, then an alternative, low-risk solution should also be considered as a "fall-back" position. In Table 21.6, two options are provided for a function F.1.2.4. Both are possible to implement because they are available on competitor systems. However, the second solution is preferable because, although it results in an improvement in performance compared with the baseline solution, it has no negative effect on any of the secondary concerns such as workload.

After a number of options for functions have been selected, the designers should re-evaluate the scenario and consider whether the primary concern should be changed as a result of the decisions made so far. For instance, consider a scenario in which high workload is the primary concern. If new, partially automated solutions are selected that significantly reduce the expected workload, then a different concern, such as performance or situation awareness, may now be more significant. If the primary concern is changed, then options for the remaining functions are rearranged in a new matrix reflecting the changed priorities. Option selection then proceeds as before, starting with the options that provide the greatest improvement for the new primary concern. This procedure iterates until one design option has been selected for every function.

One possible outcome of the procedure is that some functions cannot be successfully allocated without making use of options from the "high-risk research and development" column or from a row involving a "large deterioration" with respect to a secondary concern within the scenario. If this occurs frequently and cannot be solved by generating alternative design options, it may indicate a need to review the system requirements or to review assumptions about the number and role of human operators in the system.

Emerging Functions

When a design option for each function has been selected, the scenario is reanalysed using the proposed allocations as the set of baseline designs. The purpose of this reanalysis is to identify any new functions that may be an emergent consequence of the new design. Such functions could include for example a requirement to coordinate two separate functions that control the same system resources (for example, in one domain we considered, we recognised that both "terrain following" and "missile evasion" had similar IDA-S characterisations and could be combined as variants of the same function). Design proposals for the partial automation of these functions are made. Hence, the task is redesigned in the light of the analysis.

If new functions are identified, then designers must consider whether their impact on criteria such as performance, workload, or situation awareness is acceptable. If the emergent functions create an unacceptable situation, then the selection matrix is revisited to consider any options that might improve outcomes for the emergent functions. This may result in changing the level of automation or may result in changed selections for the original functions. Hence, if emergent functions are identified, then steps of the process dealing with these are repeated, that is, feasible design options are suggested for partially automating the emergent functions, and the optimum choice is selected. These new functions may have unexpected effects on the work, and therefore the whole process of function allocation must be considered again using these new functions. In situations in which these functions might be particularly critical, it may be necessary to find new scenarios in which combinations of features may be considered.

Consolidation

Once functions within each scenario have been allocated, any contradictions of allocation across scenarios are resolved. This is done either by changing one of the allocation decisions so as to resolve the conflict or by allowing automation levels to change across scenarios. Components of IDA-S allocations can be transferred from one role to another during the activity supported by the system. This redistribution typically occurs in response to a change in the environment or a change in the state of one of the agents. The designer must decide how the allocation of function changes and the extent to which the operator is involved in this process. In practice, the changeover can be seen as another function that can be refined in the same way as any other function that emerges during the process.

The following section describes in more detail how the method is integrated with UML. A reader interested only in the method may skip to the final section (Extension and Conclusions).

INTEGRATION WITH UML

Two further developments of the method make it more accessible to system engineers. The first reformulates the representations that are input and output for the method into UML (Rumbaugh et al., 1999a). The second produces a representation of the mapping between the roles and system components. The aim of the first step is that the method can be more easily assimilated into existing system engineering practice and of the second that the implementation of roles may be more easily visualised. In other words, an additional dimension is introduced, that of recognising how a role is implemented either as a human or a device within the system. In earlier sections issues of feasibility of automation were discussed without considering the architecture of the system.

Integration of Input

The previous discussion of scenarios and functions has already indicated that they can be represented using UML's use-case model. The advantage of this approach is that system engineers may use a UML supported design process (or profile) to develop the system's design model based upon the use-case model. Function allocation can therefore be inserted into the development process at the point during which the use-case model (a representation of requirements) is transformed into the design model.

A UML use-case describes a collection of scenarios related to an actor's goal (Jacobson, 1995). The use-case is the general goal and scenarios are a sample set of means by which the goal can be achieved. This has a strong similarity with the goals, subgoals, actions, and plans of hierarchical task analysis (Kirwan & Ainsworth, 1992). Each scenario contains a different description of who does what to fulfil the goal. At least one scenario must describe the normal set of steps taken to fulfil the goal successfully. The other scenarios can describe alternative ways of fulfilling the goal or ways in which the goal may fail to be fulfilled. Failure may occur because of operator error, mechanically induced faults, or unexpected events in the environment. It must be made clear that any erroneous steps are not the required functionality of the system. Use-cases also describe what guarantees the system provides for the other stakeholders. Together the use-cases form a model of the system's behaviour known as the use-case model.

Use-Case Template

Cockburn (2001) suggested a more detailed description of use-cases than proposed in the standard (Rumbaugh et al., 1999a). He proposed that the sequential description of a use-case should consist of a sequence of steps each taking the form 'Subject . . . verb . . . direct object . . . prepositional phrase' (Cockburn, 2001). The description can also include statements that control the flow, for example, REPEAT $<x_1 - x_x>$ UNTIL $<condition>$ or STEPS $<x_1 - x_x>$ ANY ORDER. Because a specific instance of behaviour in scenarios is being described, conditional branches such as the "if" statement can be avoided.

The scenario should describe the functions to be allocated but should not preempt allocation; therefore, the system element (to be discussed in more detail later) responsible for performing the function should be unspecified unless it has already been formally decided. The scenarios associated with allocation of function are therefore grouped according to goal as defined by the use-case. The set of scenarios covering all the use-cases should cover all the functions that are candidates for function allocation.

An example of a use-case scenario is shown in Table 21.7. The example is based on a ship navigating from open sea along a familiar channel into harbour. Only the main success scenario is shown. The description has three columns: step number; the role or roles that carry out the functions; the functions themselves in the order that they are to be performed. As the allocation of function is completed then the roles responsible for providing each function, extracted from IDA-S descriptions, can be inserted into the middle column (see Table 21.7).

In one of the preferred UML design processes, the so-called rational unified process (Rumbaugh, Jacobson, & Booch, 1999b), the use-case model is created during the "requirements" workflow and is realised as an analysis model during the "analysis and design" workflow. The analysis model is a design model that ignores any specific implementation issues and may be used to implement an object-oriented software analysis design using any

TABLE 21.7
A Use Case With Roles

Step	Role(s)	Function Description
1	Captain	Orders the ship into harbour
2	Navigator/Electronic chart	Plans the route into harbour
3	GPS subsystem	Obtains the ship's current fix
4	Electronic chart	Plots the ship's current fix
5	Navigator/Electronic chart	Calculates direction, time and appropriate speed to next checkpoint
6	Navigator	Sends calculated information to helmsman as required
7	Repeat Steps 3–6 until ship is docked	

Note. GPS = global positioning system.

TABLE 21.8
Characteristic Information

Goal in Context	The goal is to navigate the ship from the open sea into and along a channel so that it safely comes into harbour.
Scope	System
Level I	User goal
Pre condition	1. Ship is in open sea close to harbour (less than 1 hour away).
Rationale for Scenario	Approaching harbour is recognised as a high workload period for the navigation system. It is also a period during which the fixes and projected fixes must be highly accurate.
Success End Condition	1. The ship is stationary. 2. It is alongside the correct point on the quay. 3. It has not hit anything.
Failed End Condition	1. The ship hits an object.
Minimal guarantees	1. The ship will not be navigated into an obstruction.
Primary Role	Captain
Stakeholder Roles	1. Crew: safety. 2. Helmsman: requires navigation information to steer the ship.
Trigger Event	The order is given to sail into harbour

TABLE 21.9
Main Success Scenario

Situation and environment	This scenario assumes that the ship is arriving at a familiar harbour for which an up to date chart is available. The approach takes place during daytime and in good weather. The ship is not obstructed by any other vessels and there are no technical or human failures.

Step	*Role(s)*	*Function Description*
1	Captain	Orders the ship into harbour.
2	N/A	Plans the route into harbour.
3	N/A	Obtains the ship's current fix.
4	N/A	Plot the ship's current fix.
5	N/A	Calculates direction, time, and appropriate speed to next checkpoint.
6	N/A	Sends the calculated information to the helmsman as required.
7	Repeat steps 3–6 until ship is docked.	

Note. N/A = not applicable.

object-oriented language. The diagrammatic notations available to the analysis and design models can be used to produce views of the allocation of function.

These tables provide the developer with a fuller description of the distribution of functionality than is possible using the use-cases. Tables 21.8, 21.9, and 21.10 give a full representation of the scenario in this format. Table 21.8 provides the "characteristic information", that is,

TABLE 21.10
Related Role Information

Schedule	Date/build the use case can be tested.
Priority	Must
Performance Target	If applicable, how fast should this use case proceed?
Frequency	Twice a month
Super Use Case	Navigate
Sub Use Case(s)	
Channel to Primary Role	Navigation officer accountable for navigation
Secondary Role(s)	Secondary roles that are controlled by the primary roles are described.
Channel(s) to Secondary Role(s)	How do we get to the secondary roles?

information about the general circumstances in which the scenario takes place. Table 21.9 provides the main success scenario, similar to that described in Table 21.7. Table 21.10 describes related information that is valuable in understanding the circumstances in which the scenario takes place.

Characteristic Information

The characteristic information (Table 21.8) includes a description of the overall goal to which the scenario relates, the pre- and postconditions that govern the specific circumstances of the scenario and the actors that are involved. It also describes what a successful end condition would be as well as possible failure situations. Some of the information contained here may be used in formulating the extent to which different criteria are relevant to the scenario during the process of deciding the effect of alternative IDA-S solutions in terms of the criteria within the scenario. Human factors experts may use this part of the scenario to place their assumptions about the situation.

Main Success Scenario

Table 21.9 describes the main success scenario. This is the central scenario that will form the basis for function allocation in normal circumstances. There may be other scenarios that describe extreme or exceptional behaviour that may also be taken into account, and these may be used to consider circumstances in which dynamic function allocation may be appropriate. Such decisions will mean that these scenarios will require modification to include functions that are concerned with the decision about which function to perform.

In this particular example, it is imagined that the function allocation process is in progress. The first step, that is the decision about whether functions should be totally manual or totally automated has been completed and the function in Step 1 is entirely manual, performed by the captain role. The remaining steps involve functions for which appropriate IDA-S representations must be selected.

Related Information

Related information that may be used by both system engineers and analysts in the allocation of function process provide the last element in the scenario description (Table 21.10). This information includes version and project schedule information as well as information about the frequency of the scenario, the roles that are involved, and issues of accountability.

Integration of Output

The output from the allocation of function method is not easily translated into UML. Two types of information have to be expressed. The first is the representation of the partially automated functions (the IDA-S representations defined in UML terms). The second issue has not yet been dealt with in the description of the method. A set of representations is required that describes how the roles are implemented in terms of the elements of the system. The IDA-S representations are expressed as UML activity diagrams. An activity diagram can be divided into a number of columns (called swim lanes in UML), each of which is assigned a particular role. Any activities that lie within a column are the responsibility of that role. A partially automated function is expressed by creating a column for each role involved and by representing each element of the IDA-S as an activity placed in the appropriate columns (see Fig. 21.1). This representation of partially automated functions requires a little more design commitment than the IDA-S matrix representation. In particular, it requires the analyst not only to suggest partially automated solutions, in which the different roles perform different parts of the function, but also to decide the order in which these function components are performed. In practice, it appears that to understand which role performs what part of the function some idea of order is required. This extra detail provides important information that can be used in the assessment of workload or situation awareness.

Hence, in Fig. 21.1, it is stated that collection, integration, configuration, and proposing will be done in sequence by the device for the function set and adjust aimpoint. Evaluation, however, which is done next will, be shared between the human and the device. The pilot will modify the information and select and authorise in the context of execution.

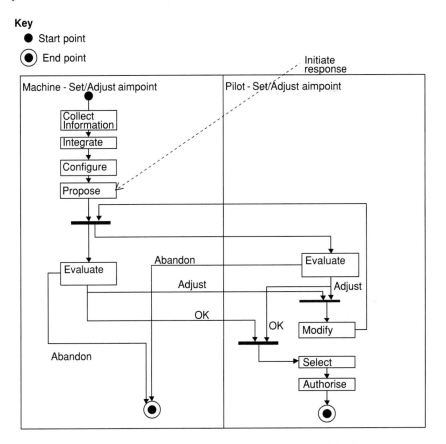

FIG. 21.1.   Representation of a partial function using an activity diagram.

TABLE 21.11
Mapping Function Allocation to UML

FA Model Element		Meta-Element	Stereotypes		
			Profile for Software Development	*Profile for Business Modelling*	*Customised UML profile for Function Allocation*
Function		Op			Function
Role		C			Role
System element class	Component	P	Subsystem		
	Human	C		Worker	
	Device	C			Device
System element	Component	O	Subsystem		
	Human	O		Worker	
	Device	O			Device

Note. Op = operation; C = class; P = package; O = object; UML = Unified Modelling Language.

Apart from brief mention of baseline architectures, little has been said about the system architecture in terms of which the function allocation is conceived. Which aspects of the system will be feasible to automate, for example? An important aspect of this description is to find a means of representing roles in terms of system components. The UML proposal provides tags or stereotypes for describing the system elements. These are the basic ingredients that engineers use UML to represent. Because UML is intended to be used generically, it provides a notion of profile as a means of supporting a particular development process directly. Profiles therefore involve a method and a collection of defined elements (described by stereotypes) to support a particular development process. Two standard profiles are relevant to the process that surrounds allocation of function: the profile for software development and the profile for business modelling. These profiles contain some of the elements that are required to support the method envisaged here, but other components that have been described in the chapter are not included. More detail on the UML extension mechanisms is contained in the OMG [Object Management Group] Unified Modeling Language Specification available at the OMG Web site (OMG, 1999).

The first distinction required in describing architectures is between system element *classes* and system element *instantiations*. A class, "naval rating" for example, will have as instantiation a particular naval rating. System element classes required by the allocation of function method must distinguish between *human* and *device* as well as describe groupings of such elements, *teams,* for example. A team may consist of human and device components. In addition, neither of the profiles that are presumed in this method include stereotypes associated with function or role. For this reason stereotypes <<Function>> and <<Role>> are included.

Therefore, in addition to the stereotypes required by the software development and business modelling profiles, further stereotypes are included as part of the allocation of function profile to support the allocation of function method. These stereotypes are: <<Subsystem>>, <<Worker>>, <<Device>>. It is assumed therefore that system engineers are already fluent with UML software development and business process profiles, and therefore little additional overhead is involved in using these additional tags. Table 21.11 describes the new customised profile for function allocation along with the stereotypes used from business modelling and software development. Note that underlining the stereotype implies instantiating the class.

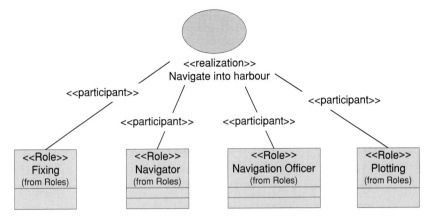

FIG. 21.2.   Role participation diagram.

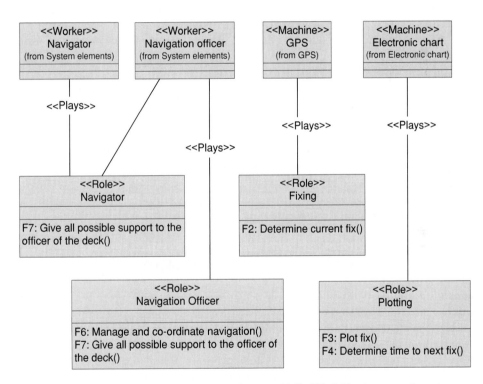

FIG. 21.3.   Allocation of roles to system elements: Unified Modelling Language format.

Hence, the function allocation method uses two existing stereotypes concerned with software development and business modelling and adds to them the extra stereotypes defined. The aim is that the process subsumed within the UML software development and business modelling profiles will form the basis of the function allocation approach described in earlier sections of this chapter.

The final representations for use by system engineers indicate (a) how roles participate in scenarios, and the nature of the participation (Fig. 21.2); (b) which functions relate to which roles (Table 21.11), and (c) how the roles are implemented in terms of system elements (Fig. 21.3). The role participation diagram shows which roles participate in the scenario. In

TABLE 21.12
Allocation of Function—Unified Modelling Language Format

<<Role>> Navigator	F7: Give all possible support to the officer on the deck.
<<Role>> Plotting	F3: Plot fix. F4: Determine time to next fix.
<<Role>> Navigation Officer	F6: Manage and coordinate navigation. F7: Give all possible support to the officer of the deck.
<<Role>> Fixing	F2: Determine current fix

the example in Fig. 21.2, the "Navigate into harbour" use-case is associated with four roles (each tagged by a <<Role>> class) involved in the sequence that describes the achievement of the main success scenario. The fact that the description of the relation with the scenario is <<participant>> indicates that this same diagram may be used to indicate other relationships between roles and scenarios. Other types of stakeholder, outside the scope of the function allocation method, may be described in these diagrams.

Table 21.12 shows the relation between functions and roles. It summarises the mapping between functions and role without specifying the functions. If further details are required, this may be obtained using the activity diagram given in Fig. 21.1. In Table 21.12, the plotting <<Role>> has two operations representing the functions "Plot fix" and "Determine time to next fix". The function "Give all possible support to the officer of the deck" is shared between two roles, namely, the navigation officer and the navigator.

Finally, roles and system elements are connected using stereotypes. Figure 21.3 gives an example of roles being allocated to system elements. The navigation officer system element can play two roles: navigation officer role and navigator role.

EXTENSIONS AND CONCLUSIONS

This allocation of function method has been designed to be sufficiently procedural to be usable in practice in a straightforward way. The process can be easily documented and therefore made traceable. The method has been confined therefore to two decision steps that may be applied by a team involving system engineers and human factors specialists. It is reasonable to expect that the process could be performed by system engineers if they subcontract the process of assessing the criteria related questions in the decision step to human factors experts. The UML representations have been introduced to provide formulations of inputs and outputs that are directly usable by system engineers and have indicated proposed extensions to standard UML profiles.

The method has been used in a realistic case study within QinetiQ (formerly the UK Defence Evaluation Research Agency, or DERA) based on a ship-based fire emergency system. Information about whether the proposed function allocations influenced the design of the system are not available.

The application of the method was reviewed by practitioners (system engineers, human factors experts, and domain experts) in the context of a 2-day workshop at QinetiQ. During the workshop, results from the case study were presented, and participants were invited to consider a specific scenario: to apply the decision procedures and produce the appropriate representations. The participants were divided into teams involving one naval officer, system engineers, and human factors experts. Useful information about the method came out of the review. It was felt that the IDA-S template was too complicated. The purpose behind its design was that it provided a representation that would help engineers produce solutions based on proposals made by human factors experts. It was aimed at helping human factors experts to consider key aspects of the automation of the interface as it relates to user tasks. The problem with it was associated with seeing how all the elements specified in the IDA-S template related to a particular function. As a result of this feedback, alternatives to IDA-S are being considered. In particular, we are concerned with attributes of a solution that are simpler to understand by all participants in the allocation of function process. The second problem concerned the initial proposal for mapping roles to system architectures. It was felt that the connection between roles and system elements was too complicated, and as a result the method presented in this chapter is a simplification. This continues to be an issue that is under exploration.

A further area where the method is perhaps too simplistic and untried is that of adaptive automation. A simplifying presumption is that automation adapts by function substitution and that in all cases the result is the same. In practice, both assumptions may be false. As noted by Sperandio (1978) in discussions of air traffic control, different strategies are adopted depending on the number of aircraft in the airspace. Sequences of functions making up procedures rather than individual functions are substituted. It may also be appropriate that dynamic mechanisms should be prepared to shed certain less critical functions in the face of hard deadlines. Both of these issues are discussed in more detail in Hildebrandt and Harrison (2002).

ACKNOWLEDGEMENTS

We would like to acknowledge the contributions of Andrew Dearden, Colin Corbridge, and Kate Cook to this work. It has been funded by two companies, BAE SYSTEMS and QinetiQ (DERA CHS).

REFERENCES

Billings, C. E. (1991). *Human-centered aircraft automation*. Technical report number 103885. Moffet Field, CA: NASA AMES Research Center.

Cockburn, A. (2001).*Writing effective use cases*. Reading, MA: Addison-Wesley.

Cook, C. A., & Corbridge, C. (1997). Tasks or functions: What are we allocating? In E. Fallon, L. Bannon, & J. McCarthy (Eds.), *ALLFN'97 Revisiting the Allocation of Function Issue: New Perspectives* (pp. 115–124). Louisville, KY: IEA Press.

Dearden, A. M. (2001). IDA-S: A conceptual framework for partial automation. In A. Blandford, J. Vanderdonckt, and P. Gray (Eds.), *People and Computers XV—Interaction Without Frontiers. Proceedings of IHM-HCI 2001* (pp. 213–228). Berlin: Springer.

Dearden, A., Harrison, M. D., & Wright, P. C. (2000). Allocation of function: Scenarios, context and the economics of effort. *International Journal of Human-Computer Studies, 52*, 289–318.

Fitts, P. M. (1951). Human engineering for an effective air navigation and air traffic control system. In D. Beevis, P. Essens, & H. Schuffel (Eds.), *Improving function allocation for integrated systems design* (CSERIAC SOAR 96-01). Wright Patterson Air Force Base, OH.

Grote, G., Ryser, C., Wäfler, T., Windischer, A., & Weik, S. (2000). KOMPASS: A method for complementary function allocation in automated work systems. *International Journal of Human Computer-Studies, 52*, 267–288.

Hildebrandt, M., & Harrison, M. D. (2002, September). *The temporal dimension of dynamic function allocation*. In S. Bagnara, S. Pozzi, A. Rizzo & P. Wright (Eds.), Proceedings of the 11th European Conference on Cognitive Ergonomics Instituto Di Scienze e Technologie della Cognizione Consiglio Nazionale delle Ricerche. Viale Marx, 15-00137 Rome, Italy, pp. 283–292.

Jacobson, I. (1995). Use cases and scenarios. In J. M. Carroll (Ed.), *Scenario-based design: Envisioning work and technology*. New York: Wiley.

Kaber, D. B., & Endsley, M. R. (1997). The combined effect of level of automation and adaptive automation on human performance with complex, dynamic control systems. In *Proceedings of the Human Factors and Ergonomics Society 41st Annual Meeting* (pp. 205–209). Santa Monica, CA: Human Factors and Ergonomics Society.

Kirwan, B., & Ainsworth, L. K. (1992). *A guide to task analysis.* London, England: Taylor and Francis.

Malinkowski, U., Kuhme, D. H., & Schneider-Hufschmidt, M. (1992). A taxonomy of adaptive user interfaces. In A. Monk, D. Diaper, & M. D. Harrison (Eds.), *People and Computers VII, Proceedings of HCI'92.* Cambridge, England: Cambridge University Press (pp. 391–414).

Older, M. T., Waterson, P. E., & Clegg, C. W. (1997). A critical assessment of task allocation methods and their applicability. *Ergonomics, 40,* 151–171.

Object Management Group. (1999). OMG Unified Modeling Language Specification, ver. 1.3. Retrieved from http://www.omg.org.

Parasuraman, R., & Mouloua, M. (Eds). (1996). *Automation and human performance: Theory and applications.* Hillsdale, NJ: Lawrence Erlbaum Associates.

Parasuraman, R., Sheridan, T. B., & Wickens, C. D. (2000). A model for types and levels of human interaction with automation. *IEEE Transactions on Systems, Man and Cybernetics—Part A: Systems and Humans, 30,* 286–296.

Pocock, S., Harrison, M. D., Wright, P. C., & Johnson, P. D. (2001). THEA: A technique for human error assessment early in design. In M. Hirose (Ed.), *IFIP TC 13 International Conference on Human-Computer Interaction* (pp. 247–254). Amsterdam, The Netherlands: IOS Press.

Rumbaugh, J., Jacobson, I., & Booch, G. (1999a). *The unified modelling language reference manual.* Reading, MA: Addison Wesley.

Rumbaugh, J., Jacobson, I., & Booch, G. (1999b). *The unified software development process.* Reading, MA: Addison Wesley.

Sheridan, T. B., & Verplanck, W. L. (1978). *Human and computer control of undersea teleoperators* (technical report). Cambridge, MA: Man-Machine Systems Lab, Dept. of Mechanical Engineering, MIT.

Sperandio, J.-C. (1978). The regulation of working methods as a function of workload among air traffic controllers. *Ergonomics, 21,* 195–202.

22

Product Design to Support User Abstractions

Alan F. Blackwell and Rachel L. Hewson
University of Cambridge, United Kingdom

Thomas R. G. Green
University of Leeds, United Kingdom

Abstract

Both software products and new consumer products mask substantial abstractions behind apparently simple button-pressing interfaces. The simplicity of the interface belies the complexity of the device, however. We analyse a range of new product types, including a survey of retail consumer devices and a series of interviews with the designers of simple home automation devices. We find that there is little support for usable access to abstraction facilities. A series of interviews with office workers confirms that they are able to create abstractions in their working environment but are inhibited from doing so when using software products. We propose an alternative analysis of product abstractions in which the presentation of the abstraction is considered as a notational system. A case study of a well-established and sophisticated abstract notational system provides a test case for this analysis technique.

INTRODUCTION

Even simple domestic automation tasks and products are rapidly developing abstract capabilities, largely due to the excess computational capacity of microcontrollers originally used only to implement basic control functions in the product. These abstract capabilities are manifested in new cognitive demands on the user. Abstraction can result a loss of direct manipulation feedback, because the user is defining actions that should occur once or many times in the future (abstraction over time). Direct feedback can also be lost because the user is describing actions that should apply to a range of objects or situations (abstraction over a class). Abstract interaction can require that the user define categories or maintain a hierarchy of type relations. Finally, abstract interaction is characterised by syntactic relations defining the behaviour of some notation, rather than concrete cause–effect relations within a physical system.

Similar trends have already been noted and analysed in the domain of professional practice, for example, by McCullough in his book *Abstracting Craft* (1996). In this analysis, the physical medium of work is giving way to the abstract media of notation. As a result, the craft skills of continuous interaction within a syntactically dense medium (Goodman, 1969) are being replaced by discrete and discontinuous styles of interaction. This trend has clear implications for cognitive task design. If we consider those situations in which the cognitive task is grounded in the physical world, human perception and motion capacities constitute significant mediating constraints between cognitive processing and physical context. By contrast, notational interaction involves a cognitive artefact, the external representation of which is principally applied as a facilitator for the cognitive manipulation of abstraction. The structure of the notation is not inherent in the situation, but designed to facilitate cognitive tasks. In cognitive task analysis, this is specifically recognised in the consideration of representations such as diagrams employed by the user (Seamster, Redding, & Kaempf, 1997).

Our analysis of notational systems—combinations of a notation and an environment for manipulating that notation—concentrates on the ways in which the properties of notational systems impact the cognitive tasks of their users. The cognitive dimensions of notations (CDs) framework has been applied to complex notational systems such as programming languages (Green & Petre, 1996), as well as to a wider range of information artefacts (Green & Blackwell, 1998). CDs provide a discussion vocabulary for designers, as well as a framework in which to consider the activities and needs of the user. The aim is to identify and highlight the properties of notational systems that may either facilitate or impede general classes of user activity. Those classes include generic types of notational activity such as transcription, search and structural modification. Individual dimensions allow designers to identify and discuss the tradeoffs that result from features such as hidden dependencies between the components of the abstract structure represented in the notation.

In this chapter, we investigate the newly abstract demands placed on users by new generation products. The fundamental operation of these products is relatively simple in terms of cognitive task design, and applying classical techniques such as cognitive systems engineering (Rasmussen, Pejtersen, & Goodstein, 1994) would suggest well-defined design solutions (see, for example, chapter 16 by Elm et al.). In fact, the reality of commercial product design is that even potentially well-defined analyses are not incorporated into the design process. The abstract functionality of the products is expressed through notational conventions, but these notations are adopted on an ad hoc basis, often by imitation either of similar products or of user interface conventions from computational domains. Finally, we propose an appropriate analytic level at which these notational considerations can be incorporated into the design process alongside other concerns of cognitive task design.

ABSTRACTION IN DOMESTIC AUDIO APPLIANCES

The user interfaces of domestic audio products provide an ideal case study to investigate the issues raised above. The history of domestic audio appliances shows that they have always incorporated relatively advanced contemporary technologies (mechanical, electrical, and electronic) compared with other domestic appliances. The proliferation of microprocessor controls in domestic products, and the resulting increase of abstraction load for users, is also evident in domestic audio products.

Previous generations of domestic audio products have provided a relatively simple range of user controls corresponding to relatively concrete functions, as follows:

- Power supply—usually a simple on-off switch
- Media control—specific to media types, such as disc rotation, pickup position, or tape spooling

- Receiver tuning—selection of modulation circuits and frequency adjustment
- Sound quality adjustment—usually limited to volume control and a variety of frequency-band filtering

The current generation of microprocessor-controlled audio equipment has introduced a number of kinds of abstraction that are presented to users. These fall into two classes: design abstractions are created by the product designer and are simply presented for the user's consumption. In contrast, abstraction managers are facilities that the user is supposed to employ to create his or her own abstractions.

Design Abstractions

A common design abstraction is the identification of related system behaviours that can be integrated into a single user control. For example, a combined compact disc and minidisc player might have a single "stop" button or a single "pause" button that can be used to control whichever of the two media is currently rotating.

A second type of design abstraction is the overlay of multiple functions onto a single physical control, with behaviour that is dependent on the current system mode. An example is a product incorporating a radio tuner, compact disk player, and cassette tape player. This product has three buttons to select one of the three sound sources. Pressing one of those buttons a single time starts music playing from that source, operating transport motors as required. Pressing the button a second time produces a range of mode-dependent behaviours. Pressing the compact disk button a second time pauses the playback. Pressing the cassette button a second time reverses the tape direction. Pressing the radio button a second time toggles between AM and FM tuning. More extreme examples of mode-dependent functions mapped onto controls include a single button that resets the time of day clock or pauses a compact disk, depending on mode.

A third type of design abstraction is the use of notational conventions to indicate related functions. One integrated audio system includes seven "eject" buttons, five of them relating to compact disk trays, and two relating to cassette drives. These are distributed around the control panel, sometimes in proximity to the opening where the relevant media should be inserted. All seven are marked with a common symbol. Similar notational abstractions are applied to multiple "play", "pause", "stop", and "search" controls.

A fourth type of design abstraction is the use of graphical and physical design elements to express overall system function through visual cues. Some systems arrange groups of corresponding buttons in a layout that expresses their relationship to physical functions or to other groups. Some systems apply arbitrary or unrelated graphic vocabulary to create abstractions that are irrelevant and confusing.

A fifth type of design abstraction anticipates the user's requirement by hiding technical detail that can be described by a single abstraction. One common example is the replacement of tone filter controls (either bass and treble adjustment knobs or graphic equalisers) with predefined filters anticipating users' needs. These include tone control buttons labelled "rock", "pop" and "jazz", or "rock", "pop", and "classical" (purchasers who wish to listen to both jazz and classical music would be required to buy two separate products in this particular shop).

User Abstractions

In contrast to these design abstractions, a recent, and perhaps surprising, innovation in microprocessor-controlled domestic audio devices is the extent to which users can create and maintain data abstractions that are stored within the system memory. The longest standing example has been the definition of tuning presets for radios. In fact, these have been implemented using mechanical techniques for many years. More recently, the random access facilities of digital media such as compact disks and minidiscs have allowed users to define playlist abstractions, although these are often transient with only a single list stored and overwritten when a new one is defined. Recent systems are able to store information related to a single compact disk, such as tracks that should not be played, or data that should be displayed when that compact disk is inserted. Recordable digital media such as minidiscs allow the user to define directory information such as track positions and titles or disc names. Further user abstractions include the ability to define programmed events that should occur at future times.

ABSTRACTION IN DOMESTIC AUDIO APPLIANCES

To quantify the incidence of these types of abstraction, we conducted a comprehensive survey of a consumer electronics retail store. We treated the products on display over a specific 3-day period as a representative sample of current-generation domestic audio technology. We surveyed 114 domestic audio and video appliances, analysing only the labelling and arrangement of front panel controls.

This therefore constitutes a conservative measure of the incidence of abstraction management functions in domestic appliances, as many classes of appliance provide further levels of abstraction (perhaps including menu navigation) via a remote control. The advantage of concentrating on front panel controls is that these are most directly analogous to the ideal of "direct manipulation" in software products.

Of the 114 appliances surveyed, 65 fall into the class of "integrated systems", with multiple controls that exhibit some or all of the five types of *design abstraction* discussed earlier. Many of these also support a range of *user abstraction* facilities with which users can create, manage, and maintain their own abstractions for future use.

We separate these 65 integrated systems into three general classes within which products have similar price and form factors. Form factor is particularly relevant, because it constrains not only the number of sound sources, but also the size of the user interface and hence the number of controls. Within each form factor class, the total number of user interface features is similar, whereas there is more variation in distribution of features between the classes. The first of the three classes is that of "mini systems", which generally have two cassette transports, multiple compact disk drives, and are around 40 to 50 cm high and 30 cm wide (not including speakers). A typical example is shown in Fig. 22.1. The second class, "micro systems", generally have a single cassette transport (or minidisc), a single compact disk drive,

FIG. 22.1.   A typical audio mini system.

TABLE 22.1

Aggregate Number of Controls Found in Each Abstraction Class (and Average
Number per Product)

	Mini Systems (24 Systems)	Micro Systems (17 Systems)	Portables (24 Systems)
Distributed functions	22 (0.92)	14 (0.82)	N/A
Overlaid function	42 (1.75)	36 (2.12)	59 (2.46)
Signal filtering abstractions	29 (1.21)	10 (0.59)	13 (0.54)
Predefined control abstractions	50 (2.08)	11 (0.65)	11 (0.46)
Time abstractions	26 (1.08)	26 (1.53)	17 (0.71)
Memory abstractions	57 (2.38)	14 (0.82)	23 (0.96)
Interface abstractions	32 (1.33)	9 (0.53)	4 (0.17)
Total	258 (10.75)	120 (7.06)	127 (5.29)

Note. N/A = not applicable.

and are around 30 cm high and 15 cm wide (not including speakers). The third class, "portable systems" may have several cassette transports and multiple compact disk drives but also have integrated speakers that constrain the available space for user interface controls.

The total number of abstraction features that were found in these systems is aggregated in Table 22.1. Features are grouped into the following categories:

- Distributed functions: multiple controls performing the same function for different sources, with functional commonality indicated by notational convention, such as multiple "eject" or "play" buttons
- Overlaid function: single controls that perform different but conceptually related functions, such as a single button that controls "play" and "change direction"
- Signal filtering abstractions: filtering behaviour that anticipates the user's needs, such as "Jazz" or "Pop"
- Predefined control abstractions: control behaviour that anticipates some sequence of control actions, such as "Shuffle" or "Rev mode"
- Time abstractions: programmable timer and clock controls
- Memory abstractions: controls for manipulating memory values, including tone presets, tuning presets, or directory data
- Interface abstractions: controls that change the interface mode of either the display or the other controls (such as "function" or "menu"), as well as navigation controls such as cursor buttons

The results of this survey indicate that, within the constraints of form factor, these types of microprocessor-controlled consumer products do include a substantial number of design abstractions. They also include significant scope for users to define and manipulate their own abstractions for storage within system memory.

The fact that smaller systems include fewer abstractions of both types suggests that manufacturers are restricted only by price and form factor considerations. Even the larger mini systems are marketed very much at the bottom end of the market for domestic audio equipment, so it seems unlikely that this level of user abstraction has been provided with an especially sophisticated user in mind. In fact, as is well known, premium audio products generally have fewer user controls. A special case may be that of more complex digital media such as recordable compact discs, which require special directory management functions in the user interface.

The aspect of product design that has not been addressed in this section is that of communicating abstractions through graphical design language. This is far more difficult to quantify and has been addressed in a separate study, described in the next section. That study considers a mundane utility product—domestic heating controls. Heating controls also incorporate electronic abstractions, but their physical appearance is low-key and restrained, in contrast to that of current audio systems, which possess a deliberately striking visual style, ranging from *Star Wars* to *Star Trek,* by way of the fifties' American automobile—shiny chrome, tail fins, and so on. Although the audio systems' internal complexity is in no way disguised by their external extravagance, the inner workings of the heating controls are almost equally baroque, despite their meeker-appearing bodies. In both cases, the abstractions within the interface designs that exist to provide ease of use through simplicity actually generate greater levels of complexity.

Abstraction in Domestic Heating Controls

To investigate the design approaches that are taken by industrial designers to packaging abstraction handling facilities, we chose one very well-constrained class of abstraction management device: time controllers for domestic central heating systems. The circulated-water heating systems commonly used in British houses generally include a thermostat and time controller that are manufactured independently of the main boiler unit. The time controller itself is a fairly simple device; it includes an electronic clock and allows the user to program start and stop times during the day. More advanced models support further abstractions, such as varying start and stop times for the 7 days of the week (for two or three cycles a day).

This style of controller is not used in many locations outside the United Kingdom, and they are therefore designed and manufactured locally. In this study, we contacted four manufacturers of heating time controllers and identified the designer in each company who had direct responsibility for the design of the product. We then conducted an in-depth telephone interview (about 1 hour in each case) with each designer, following a structured questionnaire.

The questionnaire was structured to address questions of user abstraction through a discussion of the manufacturer's current product range, followed by questions about user studies and then plans for future products. Questions about the current product range included asking designers to classify the user features, say which of these presents the greatest difficulties for users, and identify any particularly interesting or poor designs by their competitors. Questions about users included the kind of people they have in mind as typical users, how they analyse the needs of these users, how they evaluate their designs, what users are thinking while using the product, and whether usability is more affected by the electronics, the user manual, or the case design and graphics. Questions about future development asked the designers to anticipate user needs for abstractions, such as separate controls for different zones, coping with a part-time worker, personalisation for different users, ability to take more factors than time of day into account, or ability to control other programmable devices, using a generic programming language. We also asked about plans for technical facilities such as advanced touchscreen interfaces, control from mobile phones or the Internet, and industry standardisation.

The range of devices that was described in these interviews was quite limited. Figure 22.2 shows two typical timer products, one with a digital display that can be used for programming a 7-day cycle and one with a rotary display that programs a 24-hour cycle.

From the engineers' perspective, the 'user', whether that refers to the plumber who installs the heating controller, or the home owner, is not trusted to have an accurate picture of the

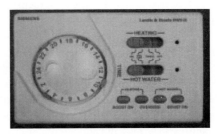

FIG. 22.2.   Typical central heating time controllers. Left: timer product with a digital display and 7-day-cycle programming capabilities. Right: timer product with a rotary display and 24-hour programming.

abstraction as intended by the engineer. End users were variously described as task driven, not interested in programming per se, and as thinking about 'very little' when attempting to use the heating controller. These seem accurate descriptions (if somewhat disrespectful). The misalignment between engineer and user occurs because the engineer believes that the user, to make full use of the heating controller, ought to be thinking about abstraction, whereas the average user only wishes to achieve a simple end. Of course, ideally the user should be able to do complex things, but in such a way that they are not conscious of how complex the actions are, because the interface is so skillfully designed that using it is completely intuitive after a brief and shallow learning curve.

There is little incentive for the user to attempt to conquer a complex and counterintuitive interface for the supposed advantages of manipulating such a system, because the overheads are so high and the user's perception of the positive trade-offs is so limited. As one of the heating controller engineers observed, the public is more motivated by entertainment, for example, to work out how to use video recorders than to work a mundane domestic heating system. Another engineer observed that relative age and familiarity with an interface also contribute to ease of learning and use. He asserted that the menu-driven interface of a particular controller was easier for young mobile-phone users and that, even so, in comparison to the mobile phone or video recorder, a heating controller is set infrequently, and this low-volume usage–unfamiliarity cycle is never broken.

The notion of familiarity raises interesting issues because there are competing forces at work. For the domestic audio unit, a novel interface that may even contrive to be obscure as part of its appeal, may be commercially successful because the target audience is willing and able to invest time to learn a new interface—there is both some pleasure to be derived from that activity and peer kudos from becoming a fluent user of an opaque but trendy system. Even in these interfaces, however, familiarity resides in the now-standardised names and symbols for particular operations, although they may be applied in confusing ways (as considered in the context of audio products in the previous section).

By contrast, the heating controllers that we considered resemble heating controllers, but, despite the familiarity of their genre, they offer little affordance for executing what should be straightforward tasks. This applies even to devices, which incorporate a circular dial that resembles a clock, for setting times, but is emphasised in devices with a liquid crystal display (LCD) that shows both time and other operations. The low level of affordance is partly due to buttons and displays being labelled with obscure or misleading terms (e.g., 'Boost', which implies an increase in boiler power, but actually has the effect of extending the period of boiler operation by an extra hour on a cold day), and partly because the LCDs try to provide feedback about complex actions within a confined space. The size of the units, constrained by cost, dictates that any textual display will be laconic at best. This degree of abstraction introduced as a direct consequence of 'lack of screen real estate' coupled with the engineers' desire to incorporate greater functionality, enabling users to program the units to perform more complex tasks, suggests that this issue will become even more pressing. One engineer observed that elderly customers prefer the clocklike interface because of its familiarity but claimed "they never set it correctly anyway". This is probably because, although part of the external aspect of the interface has a reassuring familiarity, its working differs from its precedent.

The connections between what is represented, how it is represented, and how it actually works are at the heart of abstraction in interface design. Using a 'familiar' interface to perform an unfamiliar task may introduce greater confusion by false mapping than a well-designed, novel interface. Designing the interface is a matter of balancing technical product capability with human usability. The engineers perceive what the technology allows, and wish to push it to its limits, partly through attempting to achieve commercial advantage and partly through intrinsic satisfaction of developing the technology.

But an engineering achievement is not necessarily a step of progress in the wider world. For the advance in engineering to be beneficial to the general public, the development must be applied to an appropriate niche in an appropriate way. This includes iterative testing premarket. One of the heating control engineers interviewed, on being asked how the designs were evaluated, was honest enough to admit that "once it works [presumably in the most basic, technical sense] we go straight to market, and then get feedback from the customers". This kind of usability beta testing at the customers' expense is not restricted to the heating controller market but is indicative of why so many poor user interfaces make it into the marketplace. The hidden cost to the manufacturers who operate like this is indicated in the observation of an engineer from another company: "some people hate it [this company's heating controller] and give up—I never hear of them, but we get customers migrating from other manufacturers for this reason".

The possibilities of programmable heating control systems may be of genuine benefit to a much broader constituency than heating control engineers (who, presumably, are fluent programmers of their own heating systems at home). However, the physical interface to such systems needs to cater for members of the public as diverse as highly educated, white-collar workers with perfect vision, illiterate adults, and those suffering from sight and motor-sensory disabilities.

In contrast, the users of the audio products surveyed in the previous study are a predominantly young, techno-fluent, self-selecting minority (teenagers, early-twenty-somethings, and disk jockeys). Thus, designers of audio interfaces may be able to indulge in conspicuous rampant featurism; indeed, this may even be a desirable quality of the product, given the fashion-consciousness of the narrow target market. Users of heating controllers, by contrast, are the majority of householders in this country, comprising all ages, but typically considerably older than the teenage audio system user, and of all educational levels.

Designers of heating controllers, therefore, are designing equipment for a much wider range of users, to carry out a mundane activity. In this situation, particularly taking into account the age range of the users, the degree of abstraction that is both desirable and possible is much more limited. The question of how users can be characterised and what range of abstractions they can be expected to maintain is addressed in a later study.

A COMPARISON TO ABSTRACTION IN SOFTWARE PRODUCTS

First, let us compare this development of abstract facilities in physical products to the more familiar domain of software product interfaces. The ideal of direct manipulation is that "users think about the screen objects much as they would think about real world objects" (Hackos & Redish, 1998, p. 355). The use of a graphical "desktop" to represent file operations has provided clear usability advantages over command line interfaces. The desktop enforces some real-world constraints that do not necessarily apply to file systems but assist users in reasoning about what will happen when they act (Johnson et al., 1989). Examples of these direct manipulation constraints include the fact that an object must have a location, that when it is moved to a new location it is no longer in its old location, and that it cannot be in two locations at the same time. Some of these constraints are in fact a little too restrictive—for example, special "shortcut" or "alias" objects have been introduced so that an object may appear to be in more than one place. Facilities such as this introduce data abstractions that are superimposed on the "real" world.

Most graphical user interfaces, however, do not provide direct manipulation facilities. A user interface constructed using the standard widgets of the Java Swing or Windows MFC library consists of pseudo-realistic controls that only marginally resemble real-world objects. Buttons and sliders appear and operate by analogy to the controls on electronic appliances, but the shape of the button and slider has no relationship to the actual operation that the button controls—it is standardised according to the interface style guide of the operating system vendor. The cognitive advantages of the desktop, referring to objects in terms of spatial locations chosen by the user, are not available in dialogue boxes, where the controls cannot be moved around.

The most powerful varieties of user interaction with software are not restricted by the widget sets of a minimal window-icon-menu-pointer interface. Rather than mimicking the controls of electronic appliances (push buttons, radio buttons, etc.), powerful software tools provide new visual formalisms that the user can employ to describe and reason about his or her task (Nardi, 1993).

It is clearly more difficult to design and implement a radically new visual formalism than it is to assemble a collection of widgets into a dialogue. Moreover, new visual formalisms also provide substantial cognitive challenges for the user. The diagrammatic elements of the formalism may be unfamiliar or be combined in novel ways. The application of the formalism may require the user to define complex information structures or extend the standard vocabulary. These challenges are far greater obstacles to usability than more mechanical issues such as the length of time it takes to locate and select an item on a menu or move the mouse pointer from one button to another.

All of these are general characteristics of notational systems. Learning and applying new notations often involves cognitive effort that is far greater than that devoted to button-pushing interfaces. In fact, research into the cognitive problem of learning and applying new notations bears more similarity to the psychology of programming than it does to conventional usability theories. Green's cognitive dimensions of notations framework (Green & Blackwell, 1998; Green & Petre, 1996) provides one technique, developed from the study of programming languages, for analysing this type of usability issue.

Many product designers simply abandon any hope of making abstractions available to the average user of their product. To take one example, Microsoft Word paragraph styles provide a flexible and sophisticated type hierarchy for defining the appearance of documents in a structured manner. The definition of type inheritance hierarchies is not easy, as is clear to anyone who has taught (or recently learned) a first object-oriented programming language. Most object-oriented programming environments include special tools to help the programmer visualise and refine the inheritance hierarchy, usually in the form of a graphical tree. Unfortunately, it is not possible (so far as we can tell) for the user to view the inheritance tree in the version of Word that we are using for this chapter; it is only possible to see the immediate parent of the type currently being viewed. Word provides less user assistance than tools for professional programmers, even though one might think that Word users would need more assistance. As a result, few Word users have any idea of how to employ paragraph styles efficiently. One of the studies described later in this chapter includes further investigation of Word paragraph styles.

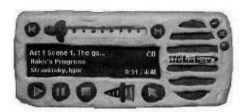

FIG. 22.3.   A selection of skins for Real Jukebox.

Skins as Abstractions in Software Audio Products

Increasing abstraction in user interfaces is also becoming evident in the popularity of "skinnable" applications in which window borders and backgrounds, button styles, or complete user-interface metaphors can be modified by the user. A vast number of alternative skins for popular skinnable applications are available on the Web from sites such as www.skinz.org. Users of these applications are encouraged to create novel skins for the benefit or entertainment of other users, in return for appreciation of their creative design work. As an example, a small selection of available skins for the RealJukebox MP3 player is shown in Fig. 22.3. These provide an instructive contrast to the abstractions discussed earlier in the context of physical audio devices.

Skinnable software provides an ideal opportunity for a software publisher to capture larger market share. The time that users spend creating and exchanging skins reinforces their allegiance to the product. Meanwhile, the time that might otherwise be spent evaluating alternative products from competitors is instead spent in locating and experimenting with alternative skins. For these reasons, it seems that skinnable software is likely to become far more common.

At present many skinnable products allow the user only to change the appearance of the window background. Some also allow the user to define new appearances and arrangements of the control widgets, or even remove some widgets altogether. We are not aware at present of skinnable products that provide scripting facilities or application programmer interfaces (APIs) such that the user can add completely new functionality to the product, but we believe that these facilities will also become available in the near future.

The overall effect of the growth in skinnable products is that the product vendor is no longer providing the user with a preconfigured user interface package. Existing techniques for measuring usability are hardly relevant to a product of this type. If the users are able to radically change the appearance of the product, even to the extent of superimposing a new user-interface metaphor, how can we analyse its usability?

We propose the same approach as for other visual formalisms that depart from the desktop paradigm. Skinnable products should be regarded as notational systems in which the user is free to create an interface from the vocabulary and syntax of a new product-specific notation. Even the "end user" of the skins may face an increased level of abstraction, because he or she must be able to recognise the control abstractions that underlie the various "concrete", but transient, skins that have been created by the skin designers. It is therefore essential that new approaches to usability be able to measure the quality of this abstract notation rather than concentrating on the specifics of any given skin or set of skins.

RELATING ABSTRACTION IN HARDWARE AND SOFTWARE PRODUCTS

The ideal of direct manipulation and user-interface metaphors has been to create software products that look and behave like physical objects. In many cases, this has resulted in software products that look and behave like electronic appliances, insofar as they are constructed from a small selection of buttonlike objects and single or multiline text displays, arranged among static text and graphic elements. Of course these "physical" front panels are only a front end to far more complex software behaviour running behind the user interface. As a result, the effect of pushing the buttons may be far more substantial than the effect of pushing a button on an electronic appliance. Arguably, this is what makes a software application more difficult to use than (for example) a microwave oven.

It is ironic that the inclusion of microprocessors in consumer products is actually making them behave more like software applications, just at the time that software applications have acquired graphical widgets that make them look like consumer products. The control panel of a microwave oven might now be a front end to a recipe database stored in the oven memory. The control panel may have been simplified to a set of navigation buttons that scroll through hierarchical menus on the display. The display may even include a touchscreen interface so that the user can select menu items directly. In a final irony, some consumer appliances have touchscreens with metaphorical buttons displayed on them, so that the user knows where to push.

As consumer products get more software, the user interface is more likely to offer abstractions. It is not yet possible for the user to change the whole appearance of a microwave oven in the way that he or she can install a new skin on an MP3 player, but product designers have certainly proposed this type of facility. Salomon, Wong, Erickson, & Kemink (1994) described a conceptual design for a personal wayfinding device that can be fitted with a variety of "clothing" making it suitable for use by populations as diverse as adult professionals, children, the visually impaired, and the elderly. Of course this study does not propose that users should define their own interfaces, in the sense of the community creating new software skins. But from the point of view of the user, the product offers abstractions that must be recognisable when clothed in their different forms.

Finally, many microprocessor-based consumer products also offer users the opportunity to define and manipulate their own abstractions. If a microwave oven allows the creation of new preset cooking profiles for favourite recipes, the user must be able to create, name, find, and manipulate those profiles as abstract entities in the memory of the oven. Where a telephone allows the definition of a directory of quick-dial codes, the user must be given facilities to create and maintain the database entries. When recording music on minidiscs, the user has the opportunity to define track boundaries and provide names that will help future navigation through the contents of the disk. Even the controls for central heating boilers may allow the user to define and store abstract descriptions of a weekly schedule, including definitions of holiday requirements.

The usability issues that are raised by these kinds of facilities are not addressed by conventional training or commercial methods in product design. As with new generation software products, the key issue here is that product users are being expected to recognise, maintain, and manipulate abstractions. Our earlier discussion of software product design also raised this issue, and we believe that the correct approach to usability analysis is not to attempt to hide the abstraction behind a layer of physical metaphor, but to recognise and criticise the characteristics of the notational system that is being created.

THE RANGE OF HUMAN CAPACITY FOR ABSTRACTION USE

What benefit do people derive from learning, creating, and maintaining abstractions in their environments? Abstraction is a costly tool. It requires time and mental effort to learn or create abstractions, and this effort may not always be recouped during the period that the abstraction remains relevant to the user's requirements. We have in the past described the "ironies of abstraction" (Green & Blackwell, 1996)—the fact that abstraction is risky, often bringing added labour in the pursuit of labour saved. More recently, this has been formalised in a theoretical description of attention economics (Blackwell, 2001).

To estimate the extent of public abstraction usage, we conducted a small-scale study of abstractions created by office workers. We considered one particular class of abstraction: the creation of named organisational categories for documents and other items of information. The goal of the study was to distinguish between users' willingness and capacity to invest in creating new abstractions, and the effect of usability obstacles posed by abstraction-management facilities in their software environment. The professional office environment is an ideal investigation domain, because abstractions are almost always created for pragmatic, labour-saving reasons. Several respondents in the study mentioned that at home they were more likely to "fool around" with software without having any specific goal in mind.

Carroll's (1982) study of computer file naming suggested a suitable methodology for this task. That study used annotated listings of personal file spaces provided by 22 staff members at a scientific research centre. The results suggested that in-depth interviews with a smaller number of subjects would have provided the opportunity for subjects to reveal more about their naming behaviour. We took the approach of using in-depth interviews, inviting office workers to volunteer information about the organisational strategies that they employed when naming and categorising in both physical and software contexts.

We conducted in-depth interviews with 12 respondents. These were selected from two populations to gauge the magnitude of the effect that computing and programming research expertise could have on habits of abstraction use. Six respondents were drawn from among secretarial and administrative staff at Cambridge University, and six were drawn from among doctoral students and research and teaching staff at the university.

Interviews took place in the respondents' offices. In the course of the interview, the respondents were asked to provide names for all categories of document organisation that could be found in the office. These included the contents of desk drawers, shelves, baskets, paper spikes, diaries and address books, and so on. The interviewer also counted existing labels on all folders within filing cabinet drawers, on the spines of ringbinders, and on file boxes. In all cases, the respondents were asked to point out only labels that they had created (rather than being inherited from previous occupants of the office or holders of their position). Where organisational collections were not labelled, the respondents were asked to provide a name. If they were not able to name the collection, it was not counted. Some collections spanned a number of folders or boxes, as in the case of chronological series or alphabetical series. These were counted as a single abstraction.

In the second phase of the interview, respondents were asked to provide details of their electronic environment. Working with the interviewer, they identified all directories in their personal filespace that they had created themselves. Within these directories, the interviewer then ran a simple script to count nested directories and the number of files in each. Several

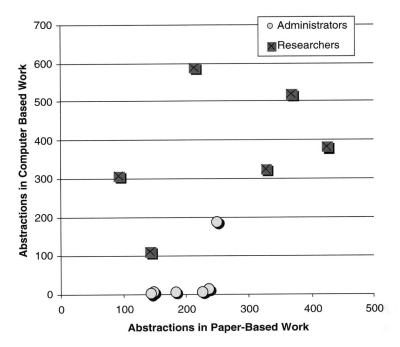

FIG. 22.4.   Comparison of abstractions created in computer- and paper-based work by administrators and researchers.

other types of categorisation in the software environment were also measured. These included the number of categories used to organise archived e-mail messages, the number of categories used to organise Web bookmarks, the number of e-mail aliases that had been created, and the number of quick-dial codes that had been programmed into the respondent's telephone. Finally, the interviewer measured a particularly demanding abstraction task; the creation and use of styles in Microsoft Word. All respondents used Microsoft Word regularly. They were asked whether they knew how to create or apply paragraph styles. If they were aware of the paragraph styles feature, the interviewer then selected a random sample of 10 Microsoft Word documents within their filespace and counted the number of styles that had been used in each.

The two data sets from these interviews, comparing paper abstractions and computer abstractions, were compared by normalising each measure of paper abstraction and computer abstraction and adding the normalised values in each category to obtain a weighted average. As shown in the scatter plot of Fig. 22.4, the two groups are not clearly separated in their creation of paper abstractions but are very clearly separated in their use of computer abstractions. We can conclude from this that administrative and secretarial workers are willing and able to create and manage data abstractions where the data is in paper form but are unwilling or unable to create and maintain data abstractions in a computer environment.

In fact, anecdotal evidence from the interviews strongly supports this conclusion. Only one of the six administrative workers interviewed ever created any directories on the computer hard disk, and this was a computer enthusiast who had multiple computers at home. Another of the administrative workers did not archive any old e-mail messages, because any message that was important would be printed out and the hardcopy filed into a paper filing system. Others achieved impressive mnemonic feats rather than experiment with unfamiliar facilities of the computer; one secretary kept 358 working files in a single directory, and was able immediately to find the file she needed, despite the historical use of eight-character filenames.

This study gives a clear impression of the magnitude of the problem in expecting product users to create and manage abstractions. The problem is not one of abstraction capability, but of the unsuitability of computer user interfaces for abstraction management. It seems most unlikely that the increasing number of abstraction facilities in domestic appliances will be used any more enthusiastically. The challenge for designers is to enable users who do have abstraction management skills to employ those in situations in which the abstractions will be of value to the user. The home central heating application is one clear example where this could be beneficial.

COGNITIVE DESIGN FOR ABSTRACT NOTATIONAL SYSTEMS

If the abstraction processing capabilities of software and hardware products are to be regarded as notational systems, how should we analyse them? The CD framework provides a powerful vocabulary for discussing the usability of information artefacts, especially programming languages. It was originally proposed as a broad-brush discussion tool, offering a vocabulary with which to discuss the usability trade-offs that occur when designing programming environments (Green, 1989; Green & Blackwell, 1998; Green & Petre, 1996). This is an alternative to cognitive analysis methods that use knowledge elicitation techniques to analyse the representation skills of potential users (e.g., Seamster et al., 1997).

The CDs all describe ways in which a notational system can directly affect the cognitive tasks faced by the user. A notational system consists of some notation by which abstract structures and relations are represented by the system and user and also a set of manipulation facilities provided by the system. The potential implications of the chosen notation and environment can be discussed in terms of dimensions such as that highlighted in the following bulleted list. The description of dimensions in this list is the one evaluated in the study described in the next section, although further dimensions have been added, as identified at a recent research workshop (Blackwell et al., 2001).

- *Premature commitment:* constraints on the order of doing things. When working with the notation, can the user go about the job in any order, or does the system force the user to think ahead and make certain decisions first? If so, what decisions does the user need to make in advance? What sort of problems can this cause?
- *Hidden dependencies:* important links between entities are not visible. If the structure of the product means some parts are closely related to other parts, and changes to one may affect the other, are those dependencies visible? What kind of dependencies are hidden? In what ways can it get worse when creating a particularly large description? Do these dependencies stay the same, or are there some actions that cause them to get frozen? If so, what are they?
- *Secondary notation:* extra information in means other than formal syntax. Can users make notes to themselves, or express information that is not really recognised as part of the notation? If it was printed on a piece of paper on which one could annotate or scribble, what would users write or draw? Do users ever add extra marks (or colours or format choices) to clarify, emphasise, or repeat what is there already? If so, this may constitute a helper device with its own notation.
- *Viscosity:* resistance to change. When users need to make changes to previous work, how easy is it to make those change? Why? Are there particular changes that are especially difficult to make? Which ones?
- *Visibility:* ability to view components easily. How easy is it to see or find the various parts of the notation while it is being created or changed? Why? What kind of things are difficult to see or find? If users need to compare or combine different parts, can they see them at the same time? If not, why not?

- *Closeness of mapping:* closeness of representation to domain. How closely related is the notation to the result that the user is describing? Why? (Note that if this is a subdevice, the result may be part of another notation, not the end product). Which parts seem to be a particularly strange way of doing or describing something?

- *Consistency:* similar semantics are expressed in similar syntactic forms. When there are different parts of the notation that mean similar things, is the similarity clear from their appearance? Are there cases in which some things ought to be similar but the notation makes them different? What are they?

- *Diffuseness:* verbosity of language. Does the notation (a) let users say what they want reasonably succinctly or (b) is it long-winded? Why? What sorts of things take more space to describe?

- *Error-proneness:* the notation invites mistakes. Do some kinds of mistake seem particularly common or easy to make? Which ones? Do users often find themselves making small slips that irritate them or make them feel stupid? What are some examples?

- *Hard mental operations:* high demand on cognitive resources. What kind of things require the most mental effort with this notation? Do some things seem especially complex or difficult to work out in one's head (e.g., when combining several things)? What are they?

- *Progressive evaluation:* work completed to date can be checked at any time. How easy is it to stop in the middle of creating some notation and check work so far? Can users do this any time? If not, why not? Can users find out how much progress they have made or check what stage in their work they have achieved? If not, why not? Can users try out partially completed versions of the product? If not, why not?

- *Provisionality:* degree of commitment to actions or marks. Is it possible to sketch things out when the user is trying different ideas or when he or she is unsure which way to proceed? What features of the notation help the user to do this? What sort of things can the user do when he or she doesn't want to be too precise about arriving at an exact result?

- *Role-expressiveness:* the purpose of a component is readily inferred. When reading the notation, is it easy to tell what each part is for? Why? Are there some parts that are particularly difficult to interpret? Which ones? Are there parts that users employ solely out of habit, without really understanding what they are for? What are they?

- *Abstraction:* types and availability of abstraction mechanisms. Does the system give users any way of defining new facilities or terms within the notation, so that they can extend it to describe new things or to express ideas more clearly or succinctly? What are they? Does the system insist that users start by defining new terms before they can do anything else? What sort of things? These facilities are provided by an abstraction manager—a redefinition device. It will have its own notation and set of dimensions.

- *Creative ambiguity:* potential for multiple interpretations. Does the notation encourage or enable users to see something different when looking at it a second time? Is it useful to create an initial version that might turn out to mean something else later?

- *Specificity:* intentional restriction in the power of abstract interpretation. Do the elements of the notation have a limited number of potential meanings (irrespective of their defined meaning in this notation)? Or do they already have a wide range of conventional uses?

- *Detail in context:* ability to see both complete descriptions of local information and their relation to a wider picture. It is possible to see how elements relate to others within the same notational layer (rather than to elements in other layers, which is role expressiveness)? Is it possible to move between them with sensible transitions?

- *Indexing:* references to structure for navigational use. Does the notation include elements to help users find specific parts when they need them?

- *Synopsis:* support for holistic or Gestalt views. Does the system provide an understanding of the whole structure when the user "stands back and looks"?

- *Free rides:* new information is generated as a result of following the notational rules. Can users read new information off, as a result of making measurements and observations of the things they put there previously?
- *Useful awkwardness:* cognitive advantages of restricted fluency. It is not always desirable to be able to do things easily. Does the system sometimes help users by not letting them do things too easily, thereby forcing them to reflect on the task? Can this result in an overall gain in efficiency (perhaps despite initial impressions)?
- *Unevenness:* bias toward specific solutions or actions. Does the system push users' ideas in a certain direction because certain things are easier to do?
- *Permissiveness:* alternative representations or actions. Does the system allow users to do things in multiple ways? Can users choose alternative expressions of the same thing?

ANALYTIC CHARACTERISATION OF COGNITIVE ACTIVITIES

Individual dimensions are seldom good, bad, or even important in all contexts. The CD framework therefore characterises broad categories of use of notational systems to establish usability profiles relevant to some range of cognitive tasks that have a shared notational abstract structure. The profiles correspond to activities including the following.

- *Search:* finding information by navigating through the notational structure, using the facilities provided by the environment (e.g., finding a specific value in a spreadsheet). Visibility and consistency are critical elements of the profile.
- *Exploratory understanding:* acquiring an understanding of the structure and capabilities presented by some information artefact through exploration of the notation. Role expressiveness and synopsie are critical elements of the profile.
- *Incrementation:* adding further information to a notation without altering the structure in any way (e.g., adding a new formula to a spreadsheet). Premature commitment and error proneness are critical elements of the profile.
- *Modification:* changing an existing notational structure, possibly without adding new content (e.g., changing a spreadsheet for use with a different problem). Hidden dependencies can be critical.
- *Transcription:* copying content from one structure or notation to another notation (e.g., reading an equation out of a textbook and converting it into a spreadsheet formula). Juxtaposability is essential.
- *Exploratory design:* combining incrementation and modification, with the further characteristic that the desired end state is not known in advance (e.g., programming a spreadsheet "on the fly" or "hacking"). Viscosity and premature commitment are critical elements of the profile.

These indications of profiles are only preliminary; in any cognitive task domain, this initial characterisation can be greatly extended. They are sufficient, however, to show the way in which user activity can be classed according to the consequences for usability of abstract notational design decisions.

SAMPLE EVALUATION OF A COMPLEX ABSTRACT NOTATION

As originally conceived it was expected that CD evaluation would be performed by somebody who understood the framework well and also understood the product well. That person might be the product designer (although it has to be admitted that designers usually have priorities in design rather than evaluation), or the evaluator might be an human–computer interaction (HCI) expert (which is how most HCI evaluation schemes work).

Kadoda, Stone, and Diaper (1999) introduced a questionnaire approach in which the evaluation was performed by system users rather than by designers or HCI specialists. That has to be a good idea. Not only does it give users a new set of ideas in such a way that they will be motivated to grasp them and use them, but it also means the users do all the work! We have extended that approach. Kadoda et al.'s questionnaire only presented those dimensions that they thought were relevant to the system under consideration, and to make it easier for the users to pick up the ideas, they paraphrased those CDs in terms of the system under consideration. There may be a problem here: Filtering the CDs constrains the respondents to commenting on the dimensions that the researcher or designer has already identified and thereby mixes the HCI expert's evaluation (in choosing the subset) with the users' evaluations, so the data will not be a pure account of what the users thought. This is particularly dangerous when the questionnaire designer is also the designer of the system, who may completely overlook aspects that are important to the users.

Of course, if a different questionnaire has to be designed for each system, then the evaluators still have to do some of the work, rather than leaving it all to the users. So we have set out to develop a questionnaire that presents all the CDs and lets the users decide which are relevant. We have also attempted to present the CDs in general terms, applicable to all information artefacts, rather than presenting descriptions specialised to the system under consideration. On the plus side, this means (a) the users do all the work, (b) the data only reflects their opinions, and (c) the same questionnaire can be used for all information artefacts. On the down side, it means that the resulting questionnaire is longer (because it has to present all the dimensions, not just a subset) and possibly harder to understand, because the CDs are presented in general terms.

The resulting questionnaire was described by Blackwell and Green (2000). In this study, we evaluated the application of the questionnaire to the problem of user abstractions in product design. We chose a specialised notational system and conducted a trial study with the questionnaire to find out what insight expert product users can develop into the nature of their abstract tasks. Previous sections of this chapter have described notational conventions for playing and archiving music, in the context both of domestic music products and software products for playing and archiving music. In this section we consider the most detailed abstract level of description of musical structure: musical performance notation.

The respondents for our trial of the CDs questionnaire were eight staff recruited from the Music Department of Glasgow University. Their interest in notation usability results from their use of the music-typesetting package Calliope, developed by William Clocksin of the Cambridge Computer Laboratory. It soon became clear that for our sample of musicians, drawn from the academic world, the staff notation of music is a given for describing their specific abstractions. Music is staff notation, and vice versa. Some respondents used slight variations on the standard notation (e.g., Gregorian chant notation, which has four lines rather than five, and diamond-shaped neumes rather than note-heads), but musicologists clearly found it more difficult than programmers might to accept the possibility that the mapping between notation and product could be completely different. Of course, musicologists are also far more expert in their notation than any programmer; academic musicologists will typically have been reading music notation since childhood. Even if the users of programmable devices had been programming them since childhood, there is no programming notation that has remained the same for the last 40 years.

For the purposes of the current investigation, these users might therefore be regarded as representing a potential future ideal in the description of abstract behaviour. Music notation is one of the most highly developed specialised notations that we have. It has evolved to fit its purpose rather precisely, and the users are practised and accomplished. It would be a great achievement if any software product or domestic product presented abstractions to its users in such a refined and appropriate manner. By studying the users of a software package for music notation, we therefore gain insight into the usability factors that will potentially be of interest when we have formulated other successful notations for user abstractions.

The first part of the CD questionnaire characterises the types of activity that the user carries out with the notational system. Our analysis framework considered five activity types:

1. Searching for information within the notation.
2. Translating substantial amounts of information from some other source into the system.
3. Adding small bits of information to a previously created description.
4. Reorganising and restructuring previously created descriptions.
5. Playing around with new ideas in the notation, without being sure what will result.

All of these clearly correspond to activities that might be required when creating user abstractions in other contexts. In the case of musicians, their familiarity with the notation means that they spend much of their time (they estimated 49% on average) transcribing music from paper sources or from notes. If they need to experiment with ideas, they tended to do it on paper. Only one composer regularly experimented with ideas directly on the screen; the others estimated that they spent only 2% of their time experimenting. Respondents did spend a substantial amount of time modifying existing representations, however—21% of their time adding new information to an existing score and 19% modifying the structure of information that was already in place. Because they tended to work on one piece of music for a period of time and then move to another, they spent relatively little time searching through the notation for specific information—about 8%. In the case of an abstract notation that users keep for a long period of time, returning to modify it at irregular intervals (such as a central heating controller), it may be the case that users would spend the majority of their time searching for the information that they had previously entered.

The main part of the questionnaire asked users to reflect on specific aspects of system usability in order to gain insight into the way that the facilities for manipulating the notation interacted with their abstract tasks. The CDs describe specific localised scenarios affecting notational usability, but for most respondents these provided initial cues that prompted broader reflection about the way that they used the notation. It is these aspects of the responses that are of most relevance to the general question of supporting user abstraction with notational devices.

One significant observation is the importance of layout details in expert notations. Just as many programmers have strong opinions about details of indenting or placement of punctuation, musicians interpret many aspects of the notation that does not have formal meaning. In the CDs framework, this is described as secondary notation. In the domain of programming languages, secondary notation also refers to extra annotation in the program source code (comments); in musical notation, even the annotations are largely formalised (as in the use of Italian vocabulary of terms for volume and tempo). As a result, layout conventions are even more meaningful, and expert users have precise requirements with little room for flexibility. Twelve responses mentioned this factor, although they did so under the headings of eight CDs when they recognised ways in which the tool can frustrate their notational expression.

A related observation is that expert notations provide richer expressiveness than the editing or interpreting environment recognises. If a system takes a heavily rule-driven approach to assisting the user construct abstractions, it can become difficult for the user to step outside the boundary of the system's "understanding" of his or her intentions. In the case of these expert users, their intentions are generally fixed, so they find themselves fighting the system or second-guessing it. Nine responses mentioned this factor, once again under the headings of five CDs.

Several respondents described the fact that they use different aspects of the notation to construct skeletal representations. This allows them to describe relevant structure that can later be augmented with specific detail. For different respondents, the skeleton exploited different aspects of the representation: One simply created bars in an appropriate time signature, one entered notes without accidentals, one wrote in the key of C for later transposition, others included only one voice of the harmony or started from spoken word rhythm or a single verse of the word underlay. Six responses mentioned this factor, under the headings of five CDs.

Regarding the usability of Calliope itself, most respondents were using this system (which is itself the result of an extended research exercise) because they considered it to be the best available product for their requirements. The specific usability problems that they recognised are therefore of less immediate interest within the context of the current project. All eight respondents stated that Calliope provided the closest possible mapping to their use of music notation on paper—far closer, as one claimed, than is provided between a word processor and the paper document that is being produced. This very close relationship between notation and product is much closer than has been achieved with more abstract software systems such as programming languages, and it is likely to be some time before similar results are achieved with the other forms of user abstraction discussed in this paper.

Nevertheless, it is noteworthy that some of the specific usability issues that the CDs recognise in programming environments do also occur in Calliope. We believe that these general issues are likely to arise in any abstract notation. They include premature commitment and viscosity—the need to think carefully about abstract structure early in the task because it can be difficult to make fundamental changes later (13 comments were made along these lines). They also included the problem of visibility—navigating through a project that has been divided into multiple parts (five responses commented on this) and local errors that are a function of notational details that appear disconcerting to experts (such as accidentally placing a note at the wrong pitch; five responses commented on this). It is clear that these types of usability issues are completely relevant to the more mundane or less evolved abstract notations described earlier in this chapter. Studies of expert notation users who have substantial insight into the abstractions that they use are a valuable source of design guidelines for new abstract notations.

SUMMARY

We have outlined an approach to cognitive task design that spans both conventional software products and information-handling consumer products. This approach emphasises the cognitive task of abstraction management, and especially the characteristics of notational abstraction managers—facilities within a user interface that form subdevices for creating, inspecting, and modifying abstractions. New generations of products are incorporating ever-increasing complexity of abstraction management, to the extent that this is the primary usability problem facing designers.

Our interviews with users, product designers, and expert notation users indicate that there are well-established techniques for managing information abstraction and that these are accessible to most people in the form of conventional abstract notations and paper-based information management systems. However, both software devices and microprocessor controlled domestic products are lagging far behind in their support for usable abstraction. We believe that cognitive task design can be applied in this broad class of cognitive tasks through the use of the cognitive dimensions framework. This is just as true for software products that have developed baroque programmable facilities with usability that cannot be accounted for by conventional analyses of dialogues, buttons, and direct manipulation interfaces.

Analytic Procedure

Although CDs provide a design perspective and vocabulary that can be applied throughout the design process by expert analysts, we have found that initial critique of a design, when CDs are first introduced to a project team, can be conducted in the following stages:

- *Identify notational formalisms within the system:* There is likely to be one "main notation", perhaps closely related to the final output of the system (in the case of mapping software, the output is a map, and the main notation is likely to be a map). There may be more than one main notation (e.g., architectural software that offers both orthogonal views and perspective views). There will also be conventional notations adopted from other products (menus, dialogues etc.) that may be applied in helper devices or abstraction managers.
- *Characterise the kinds of activity for which the system will be used:* Based on this mix of activities, it will often be possible to anticipate dimensional profiles—sets of dimensions that should be prioritised in this product.
- *Analyse the various devices and notations within the product:* Assess their characteristics with respect to each of the dimensions, and in particular identify dimensions on which that notation represents some extreme (for example, having no provision whatsoever for secondary notation).

In cases when particular notational characteristics are inappropriate for the activities that are to be supported, investigate design manoeuvres that can be applied to adjust them. The trade-offs associated with these changes will usually result in (potentially undesirable) changes on other dimensions, and these consequences must be analysed before finalising design changes.

An alternative approach is for expert users of the system to act as critical informants, with their input to the design process structured using the CDs questionnaire. The questionnaire effectively asks users to follow the analytic steps just listed, reflecting on the notations they use, the nature of their activity, and the ways in which the system facilitates or hinders that activity. This requires that informants have a high degree of expertise and familiarity with the system (so the design must be based on an existing tool or area of practice) and also requires that they be comfortable with a rather abstract analytic vocabulary. We have successfully found these attributes among academic users (not necessarily scientists). The questionnaire can be administered in an interview format, although this must be done with caution because it may potentially result in the interviewer directing informants toward a limited range of notational attributes, as has occurred in previous questionnaire studies conducted before the current questionnaire was designed.

Future Work

This work is part of an active and continuing research programme, in the course of which we are experimenting with deploying new abstract notational systems for use by users without computer programming experience (e.g., Blackwell & Hague, 2001). This is directly relevant to the new challenges of programmable ubiquitous computing systems. In pursuing novel application domains, we are also continuing to develop the cognitive dimensions of notations framework as a means of understanding the issues of cognitive task design that will be raised in future generations of intelligent products.

ACKNOWLEDGEMENTS

The authors would like to thank the manager and staff of Dixon's electrical retailer in Lion Yard, Cambridge, and Andrew McNeil, who assisted with the survey of domestic audio appliances. We are grateful to Dr William Clocksin and the Music Department of Glasgow University for assistance with evaluation of the Cognitive Dimensions questionnaire. This research is funded by the Engineering and Physical Sciences Research Council under EPSRC grant GR/M16924 "New paradigms for visual interaction".

REFERENCES

Blackwell, A. F. (2001). See what you need: Helping end users to build abstractions. *Journal of Visual Languages and Computing, 12,* 475–499.

Blackwell, A. F., Britton, C., Cox, A., Green, T. R. G., Gurr, C. A., Kadoda, G. F., Kutar, M., Loomes, M., Nehaniv, C. L., Petre, M., Roast, C., & Young, R. M. (2001). Cognitive dimensions of notations: Design tools for cognitive technology. In M. Beynon, C. L. Nehaniv, & K. Dautenhahn (Eds.), *Cognitive Technology 2001* (LNAI 2117) (pp. 325–341). Berlin: Springer-Verlag.

Blackwell, A. F., & Green, T. R. G. (2000). A cognitive dimensions questionnaire optimised for users. In A. F. Blackwell & E. Bilotta (Eds.), *Proceedings of the Twelfth Annual Meeting of the Psychology of Programming Interest Group* (pp. 137–152). Cosenza, Italy: Edizioni Memoria.

Blackwell, A. F., & Hague, R. (2001). AutoHAN: An architecture for programming the home. In *Proceedings of the IEEE Symposia on Human-Centric Computing Languages and Environments* (pp. 150–157). Los Alamitos, CA: IEEE Computer Society.

Carroll, J. M. (1982). Creative names for personal files in an interactive computing environment. *International Journal of Man-Machine Studies, 16,* 405–438.

Goodman, N. (1969). *Languages of art: An approach to a theory of symbols.* London: Oxford University Press

Green, T. R. G. (1989). Cognitive dimensions of notations. In A. Sutcliffe & L. Macaulay (Eds.), *People and computers V.* Cambridge, UK: Cambridge University Press.

Green, T. R. G., & Blackwell, A. F. (1996). Ironies of abstraction. In *Proceedings 3rd International Conference on Thinking.* London: British Psychological Society.

Green, T. R. G., & Blackwell, A. F. (1998, October). *Cognitive dimensions of information artefacts: A tutorial* (ver. 1.2). (An earlier version was presented as a tutorial at HCI'98 under the title *Cognitive Dimensions of notations and other Information Artefacts.*) Retrieved from http://www.cl.cam.ac.uk/~afb21/CognitiveDimensions/

Green, T. R. G., & Petre, M. (1996). Usability analysis of visual programming environments: A "Cognitive Dimensions" approach. *Journal of Visual Languages and Computing, 7,* 131–174.

Hackos, J. T., & Redish, J. C. (1998). *User and task analysis for interface design.* New York: Wiley.

Johnson, J., Roberts, T. L., Verplank, W., Smith, D. C., Irby, C. H., Beard, M., & Mackey, K. (1989). The Xerox Star: A retrospective. *IEEE Computer, 22*(9), 11–26.

Kadoda, G., Stone, R., & Diaper, D. (1999). Desirable features of educational theorem provers—a cognitive dimensions viewpoint. In T. R. G. Green, R. Abdullah, & P. Brna (Eds.), *Collected papers of the 11th Annual Workshop of the Psychology of Programming Interest Group (PPIG-11)* (pp. 18–23).

McCullough, M. (1996). *Abstracting craft: The practiced digital hand.* Cambridge, MA: MIT Press.

Nardi, B. A. (1993). *A small matter of programming: Perspectives on end user computing.* Cambridge, MA: MIT Press.

Rasmussen, J., Pejtersen, A. M., & Goodstein, L. P. (1994). *Cognitive systems engineering.* New York: Wiley.

Salomon, G., Wong, Y. Y., Erickson, T., & Kemink, J. (1994). Interfaces for adaptive systems: Design of a personal wayfinder. *American Center for Design Journal, 8,* 62–75.

Seamster, T. L., Redding, R. E., & Kaempf, G. L. (1997). Applied cognitive task analysis in aviation. Aldershot, England: Avebury Aviation.

PART III
Field Studies

23

Mapping the Design Space for Socio-Cognitive Task Design

Alistair Sutcliffe
*University of Manchester Institute of Science and Technology,
United Kingdom*

Abstract

This chapter describes a framework for cognitive task design that focuses on social and communication issues. It is intended to complement the more traditional task-oriented views. The chapter starts by describing a process model for cognitive task design then introduces cognitive and social pathologies: heuristics for analysing potential causes of failure in sociotechnical systems. This is followed by an introduction to generic tasks as reusable models for specifying human action, to which requirements for computer system task support, as well as the pathologies to warn the designer about potential failure points, can be attached. A general model of discourse is described, which is also associated with pathologies and generic requirements to guide the designer towards reusable knowledge at appropriate stages in the communication process. The discourse model is supplemented with patterns for specific types of communication such as command and control, proposals and liaison, and so forth. The next section proposes models for analysing power and trust in human–human and interorganisational relationships. Once again, models are associated with pathologies and countermeasures. Use of the models is illustrated by analysis of command and control tasks in the combat subsystem of a navy frigate. A scenario-based approach is taken to walk through the models using operation scenarios of a missile attack. The design implications for cognitive task support and communication are assessed, then two alternative designs for intelligent computerised radar and standard radar are compared using the generic models for analysing communication, perceptions of power and trust. The chapter concludes by reviewing the lessons learned for functional allocation and future work in providing more effective decision support for operators in sociotechnical systems.

INTRODUCTION

Hollnagel (1998), Vicente (1999), and others have pointed out that cognitive task design takes place in a context. Although considerable attention has been paid to the environmental context in terms of affordances and constraints on action, less research has been carried out on the social implications in task design. The importance of social context has been pointed out by researchers in distributed cognition (Hutchins, 1995), who described how work is often achieved by groups of individuals interacting via shared artefacts. Distributed cognition and ethnographic studies of work (Ball & Ormerod, 2000; Luff & Heath, 2000) illustrate the influences of complex social factors on cognitive task design, but these literatures are less forthcoming about how social factors should be incorporated within the task design process. Furthermore, communication between people, and between people and computers, is an important component of work systems; however, the implications for designing human (and human–machine) discourse are not made explicit from ethnographic analysis.

This chapter treats cognitive task design as "design of goal-directed human activity and the machines people interact with to take account of the abilities and limitations of human psychology". A familiar tension in cognitive task design is the trade-off between the detail of modelling and the benefits or insights gained in design from a model-based approach. At one end of the extreme lie cognitive task analysis methods, based on formal theories of cognition such as interacting cognitive subsystems (ICS; Barnard, 1985) or executive process interaction control (EPIC; Kieras & Meyer, 1997). Barnard's cognitive task analysis uses his ICS model to reason about human operational effectiveness and limitations; unfortunately, it requires a considerable modelling effort to create descriptions of task and domain knowledge to enable formal reasoning (Barnard, May, Duke, & Duce, 2000), and even then only small fragments of interaction can be considered (Sutcliffe, 2000). Clearly, this approach has scaling problems for design in the large. At the other extreme, approaches founded on simple taxonomies of influencing factors (e.g., Chemical Industries Association, 1993; technique for human error rate prediction, Swain & Weston, 1988) have been criticised for their poor psychological validity (Hollnagel, 1993). The compromise in the middle ground points to approximate cognitive models, as argued by Hollnagel, Vicente, and others. Although I agree with this approach, these models are still general in their recommendations.

Models can be used in cognitive task design in two ways: first as representations that help analyse design problems (the Bayesian belief networks model of human performance, trust and power models) and second as generic models describing tasks and discourse patterns that can be used as templates for design. Generic task models are proposed as families of domain-oriented cognitive models that can be applied as templates or as tools for thought in task design. Generic task models occupy a higher level of abstraction than would normally be achieved by generalising from a specific task analysis (e.g., task analysis for knowledge description, task-related knowledge structures, Johnson, Johnson, Waddington, & Shouls, 1988). Generic task models are reusable abstract task models that are annotated with design advice. The models might not have the predictive precision of ICS, but they do encapsulate valid psychological knowledge to inform design decisions. Cognitive models and methods usually provide guidelines, modelling techniques, and advice that emphasise the positive aspects of design, that is, they indicate functional requirements that support users. Cognitive task design also has to anticipate human error and failure, however. This approach is common in safety engineering, and I follow it in this chapter to focus on the possible causes of human failure and how to design to avoid them. The design approach is to propose countermeasures for pathologies in human behaviour, communication, and social interaction as *generic requirements* for task design.

Taxonomic approaches to safety engineering have developed a layered series of influences on human reliability ranging from social, managerial, and cultural influences, through an environmental layer that accounted for geographic and climatic effects, to properties of human agents, types of task, and properties of supporting computer technology (Sutcliffe & Ennis, 1998; Sutcliffe & Maiden, 1998; Sutcliffe, Maiden, Minocha, & Manuel, 1998; Sutcliffe & Rugg, 1998); however, taxonomic approaches do not address social influences on human behaviour. To scale cognitive task design up to the design of complex sociotechnical systems, we need models that address phenomena at both the cognitive and social levels. Although some would argue that the variability and complexity of interactions at the social level defeats a model theoretic approach, without models to inform design we are left with nothing more than examples and anecdote. In this chapter, I introduce models for trust and power in interpersonal and organisational relationships to examine social influences on task design.

The chapter investigates three problem areas rarely addressed in other cognitive task design approaches. These areas form the structure for the chapter, which is organised first into the following section, which introduces causal models of human error using Bayesian belief networks to incorporate influences from the environment with a model of cognition. This builds on the taxonomic approaches to cognitive task design. The next section sets the scene for a different design approach by introducing a generic task model for command and control that will be used in the case study, as well as introducing cognitive reasoning pathologies that are a particular concern in knowledge-intensive tasks. The fourth section develops the design picture from a single-user cognitive perspective to collaborative systems by describing a theory of discourse modelling approach for investigating communication between people and automated agents. A second set of pathologies is given for common causes of communication failure. The fifth section investigates social phenomena that affect the design of collaborative systems, focusing on how trust, power, and motivation can affect the design of computer supported collaborative work. A sociocognitive model of trust is introduced, and pathologies at the social level are explained. The design approach is applied in a case study that addresses the issues of introducing intelligent agent technology into a command and control domain. The chapter concludes by returning to reflect on the merits of the different modelling approaches that have been described.

OVERVIEW OF THE DESIGN APPROACH

It is not the intention of this chapter to describe a method for cognitive task design in detail, but instead to introduce a set of discrete design techniques that address specific issues. To place the techniques in perspective, however, this section presents a model of the design process to point out where contributions made by the subsequent sections fit in. The process model also indicates where contributions in this chapter can be integrated with other topics in this volume, and more generally how the cognitive design process fits within a wider perspective of software engineering.

The design approach follows two concurrent strands (see Fig. 23.1). First is cognitive task analysis that fits with the tradition of classic task analysis (Annett, 1996; Johnson et al., 1988) or more ecological methods for task modelling (Vicente, 1999). The next section provides generic task models that are reusable templates for assisting task analysis as well as cognitive pathologies that help the next stage of analysing human reasoning. Whereas task analysis describes human activity, both physical and mental, human reasoning analysis examines mental activity implied by tasks at a more fundamental cognitive level. This is carried out by applying Rasmussen's rules, skills, knowledge framework (Rasmussen, 1986) to each task to specify whether it is usually a trained skill or requires more conscious rule- or knowledge-based processing. Cognitive pathologies are applied to the resulting model to anticipate typical causes of suboptimal human performance. Treatments to avoid the pathologies are imported into the nascent cognitive task design.

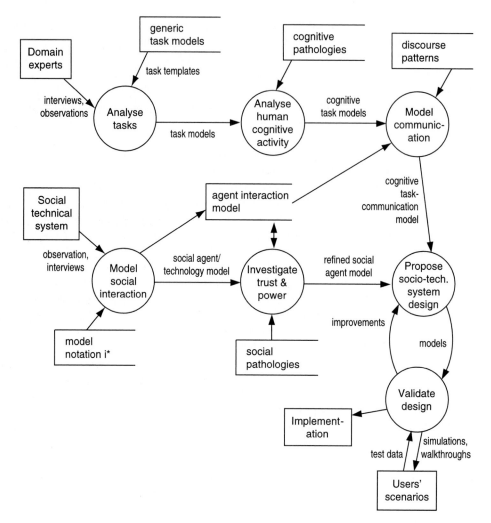

FIG. 23.1.   Stages in the cognitive task design process, cross-referenced to models in the following sections.

The next stage is modelling communication between people and between humans and machines (usually computers). Communication analysis is not dealt with in the traditional method, so a separate section is devoted to this stage. Reusable templates are provided following the same motivations as in task analysis, as guides to help analysis while design knowledge in the form of "generic requirements" can be imported into the system design. Generic requirements are abstract design advice that can be applied to several domains. For example, the diagnosis generic task is associated with generic requirements pointing to information needs (e.g., fault location, problem symptom) and is of use in supporting the human operator (e.g., provide suggestions for possible causes given a set of observed symptoms). Generic requirements are refined into specific functional requirements for software or become inputs for training and design of human operational procedures. Communication pathologies are also applied to add defences as functional or operational requirements against potential human errors leading to communication failures.

The second method strand models the social system structure, composed of agents, technology, and the communication–information flows between them. This chapter borrows a modelling approach from requirements engineering for this purpose (using the i* notation; Yu & Mylopoulos, 1995). The next stage is to investigate the power and trust relationships between all of the agents or a key subset thereof. This stage requires input from communication analysis so both analytic strands naturally converge from this stage. The following section describes power and trust models that help analyse interagent relationships, with pathologies that point out potential pitfalls and remedial treatments in sociotechnical systems design. Both strands lead to a sociotechnical system design being proposed that describes agents, their tasks, and communication and information flows between agents and technology. Technology components and affordances are included, following the descriptive conventions of the designer's choice (e.g., i*, ecological task design, or Unified Modelling Language [UML]) for software engineers.

Validation involves testing the design with users and running the design against a set of operation scenarios. In this process, the scenario is used as a walkthrough to trace the task steps implied by the scenario narrative. Then questions are asked to identify if the human operator has the necessary skills, time, and information to carry out the task when supported by functions provided by available equipment. For instance, in a scenario used in the case study, a hostile target (an enemy missile) is detected by the ship's radar. This implies the need for a human operator to detect and interpret the radar image. Validation questions focus on the operator's knowledge, training, the desired operational time, and so forth. Equipment questions concern fidelity and range of the radar, clarity of the display, adequacy of alerting signals, and so forth. Space precludes a detailed description of scenario-based validation, although this theme is introduced. More detailed descriptions of this approach can be found in Sutcliffe (1998); further treatment of the communication analysis is given in Sutcliffe (1999, 2002).

GENERIC TASKS AND COGNITIVE PATHOLOGIES

In this section, two new concepts are introduced. First are generic tasks, which are reusable templates for design of human activity systems. Imagine you are interested in designing a system (and hence a task) to support fault finding in computers. The essence of fault finding is diagnosis (discovering what's wrong) and repair. It might help if you had a general model of diagnosis tasks to help your analysis and subsequent task design. This is where generic tasks fit in. They are reusable knowledge that can be applied in many domains; for instance, diagnosis is practised by doctors as well as by computer-repair engineers. Description of a library of generic tasks covering a wide range of domains can be found in Sutcliffe (2002). The second concept is a pathology. This is the cause of suboptimal behaviour or system performance, following the medical analogy of pathology. In this case, pathologies are described for human reasoning and communication. They act as warnings for designers in the form "this is something that probably could go wrong" and are associated with design treatments that might avoid the pathology.

Cognitive pathologies are introduced within Rasmussen's (1986) framework of reasoning. Some pathologies are familiar in the cognitive design literature, such as slips, lapses, and mistakes (Norman, 1988; Reason, 1990); however, the causes of poor reasoning at the knowledge-based level have been less well documented. When tasks involve new problems, rule- and knowledge-based errors will be more frequent than slips (i.e., failures of attention and action in highly trained operators; Reason, 1990). At the skill level, errors are caused by failure in attention, when we miss a cue for action or when we omit an action in an otherwise correct sequence. There are also failures of memory when an incorrect skill routine is used in the right context and vice versa. Skill-based errors have been extensively documented and discussed elsewhere (Norman, 1986; Hollnagel, 1993).

Rule-based errors result from memory recall failure and from inadequate learning in the first place. The right rule may be applied in the wrong context or the wrong rule used in the right

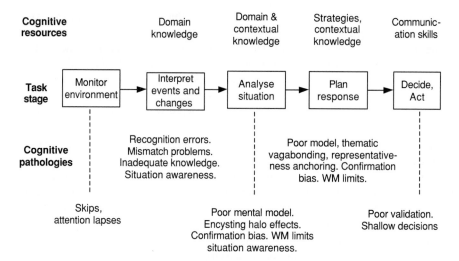

FIG. 23.2.   Generalised task model for command and control.

context. This gives the "strong but wrong" mistakes in which people carry out actions convinced they are right even though the environment clearly contains cues suggesting otherwise. Frequency and recency gambling errors (Reason, 1990) affect rule-based processing in which operators apply the most recent or frequently used memory that approximately matches the input events. Several nuclear power plant operator failures fall into this category (see Reason, 1990, for case histories). Other mismatches between triggering conditions and the appropriate rules can arise from biases towards particular rules and policies that are adopted even if they are suboptimal choices (Payne, Bettman, & Johnson, 1993). Rules also may be incomplete or malformed so even if triggering is correct the action is inadequate.

At the knowledge-based level, people suffer from several limitations (Gardiner & Christie 1987; Johnson-Laird 1983, 1988). First, models of the problem domain are often incomplete, may be biased in scope, and contain inconsistent facts. This tendency leads to many reasoning errors because facts are not available or are incorrect. A further class of errors are related more closely to the problem-solving process, such as thematic vagabonding (a "butterfly mind" that does not concentrate on any one topic long enough to solve the problem) and encysting (getting bogged down in detail and not being able to see the forest for the trees).

The link between these pathologies and more fundamental limitations of human cognitive processing is illustrated in Fig. 23.2, which shows the generic model of command and control. In the safety-critical military applications literature, generic command and control models are commonplace, for instance, "observe, orient-assess, decide, and act". Generic tasks produced by a comprehensive domain analysis have been specified for other areas (e.g., information retrieval, Kuhlthau, 1991; Sutcliffe & Ennis, 1998). In this chapter, a more sophisticated generalised command and control task is described that is based on the domain theory generic task library (Sutcliffe, 2002). Command and control is composed of five subtasks that have separate abstract goals:

- Monitor (for new events)
- Interpret (events)
- Analyse (the event in context)
- Plan (the response)
- Decide/act.

TABLE 23.1
Summary of Cognitive Pathologies and Countermeasures for the Knowledge-Based
Level of Reasoning

Pathology	*Countermeasure/Task-Support Requirement*
Thematic vagabonding (butterfly mind)	Task lists, checklists, aide-memoires
Encysting (can't see forest for the trees)	Conceptual maps, issue hierarchies, priority lists, detail filters
Halo effects	Exception reports, history browsers
Confirmation bias	Challenge assumptions, alternative scenarios
Poor mental model of problem space	Problem simulations, design rationale, decision tables/trees, concept maps
Lack of situation awareness: wrong model or no model	Training, role variety, simulations, embed regular update tasks
No/poor validation	Model-checkers, testing tools, what-if simulations
Representativeness/availability of data	History lists, hypertext, tracing tools
Belief in small numbers	Data sampling/mining tools, visualisations
Anchoring	History lists, tracing tools, reminder agents

While specific instances of command and control tasks will exhibit considerable variation, even in written procedures, generalised task models are useful as templates for specialisation. In addition, they can provide a locus for considering reasoning pathologies and other potential causes of failure such as working memory limitations (Baddeley, 1990).

Although such models are only approximations of human cognition, they do help situate how human error relates to fundamental properties of our mental machinery (see also Hollnagel, 1993; Reason, 1990). Working memory is one such limitation. Only a small number of facts can be held in the "foreground" at any one time, so we tend to form incomplete and partial models of problem domains. We tend to find positive facts easier to hold in working memory than negative facts; similarly, exceptions give more memory problems than facts that fit into a pattern (Johnson-Laird, 1983). This leads to confirmation biases (looking only for positive evidence), halo effects (ignoring exceptions), and poor causal reasoning.

The more specialised sets of biases, which affect mental model formation and reasoning with numerical estimation problems, have been described by Kahneman and Tversky (1982) and Gigerenzer (1991) among others. These include, for example, the following:

- *Availability of data:* The frequency of an event is judged by the ease of recall. Recall is affected by frequency and recency of memory use (Reason, 1990), so a biased data set is the norm rather than an exception. Memory of a recent aircraft accident may bias our estimate of how safe flying is.
- *Belief in small numbers:* Small samples are judged to represent faithfully the variation in a larger population.
- *Adjustment and anchoring:* Once an estimate has been made, we tend to be conservative in revising it, a consequence of "stick to the known hypothesis".
- *Representativeness:* The probability of an object or event belonging to a particular class depends on how prototypical it is or how well it represents the overall class.

Countermeasures to prevent or at least ameliorate some of these pathologies are given in Table 23.1. In some cases, personnel selection and training people to think effectively are the appropriate remedies, whereas computer system support can counteract mental model errors through information visualisation and validation functions.

To summarise, generalised task models can be productive in reasoning about design in three ways. First, the model describes the process of human reasoning and action related to a general goal. Second, generic requirements for computer support systems can be attached to the models, so designers can consider the functionality and dialogue structure that are necessary to support decision making. Finally, cognitive pathologies can be added to analytic models to suggest defences and countermeasures against errors that people might make.

Case Study: Combat Operational Subsystem in a Navy Frigate

Before proceeding with a description of models for task design, I introduce a case study that links each analysis and design treatment. The problem is designing the next generation of military systems with intelligent technology. With the need to reduce costs in defence systems, customers and suppliers are looking to reduce human resource levels. Increased use of intelligent technology is one obvious way of maintaining capability while saving costs by eliminating human roles. Unfortunately, this raises complex questions about how human interaction with intelligent machines should be structured. Furthermore, increased automation changes roles, responsibilities, and lines of communications between agents, both human and machine. I focus on one aspect of the problem: increasing automation in radar systems to identify potential targets and threats, referred to as IFF (identify friend or foe). The motivation for increased automation comes from the "faster, better, cheaper" mantra that procurers of defence systems hope to achieve. The danger in saving money by reducing human resources, that is, by increasing automation, is that increased automation might not be "better" or "faster" if it increases errors.

The setting is the control room in a navy frigate with antiaircraft and anti–surface ship capability. The ship has a set of radars, sonars, and other devices that detect events and threats in the environment. The control personnel are responsible for interpreting and analysing these threats (e.g., a missile attack, a suicide bomber in a speed boat) and taking countermeasures to ensure the ship's safety (e.g., launch decoys to confuse the missile, take evasive action to avoid the suicide bomber). The system description has been sanitised for security considerations; nevertheless, the dimensions of the problem are realistic and anchored in specific experience.

Three personnel roles in the current, unautomated design are analysed. The *radar operator* monitors radar contacts, identifies contacts as friendly, hostile, or unknown, and relays this information to the *tactical assessor,* who interprets radar contacts in terms of the wider context. Warships have a variety of radar and other sensing devices for surface, aerial, and subsea object detection. The tactical assessor integrates these information feeds and interprets their implications. This information is passed on to the *combat control officer,* who analyses the situation, plans a response, and gives orders to junior officers who control weapons systems (e.g., missiles, decoy launchers, and electronic jamming devices). The combat control officer also coordinates action with the captain and other officers who are responsible for navigation and manoeuvring. The roles and their responsibilities are summarised in Fig. 23.3, which also illustrates the assignment of roles for the generic command and control task. The case study only considers the first three phases for monitoring and interpreting data on the radar monitor, followed by a situation analysis that involves interaction between the radar operator and tactical assessor roles.

The interpretation of the command and control task, shown in Fig. 23.3, breaks the sequence down into one skilled but repetitive subtask of monitoring and three judgement-related tasks that require reasoning at the rule-based level with knowledge-based processes for analysis of unfamiliar situations and planning responses. The generic requirements attached to the task

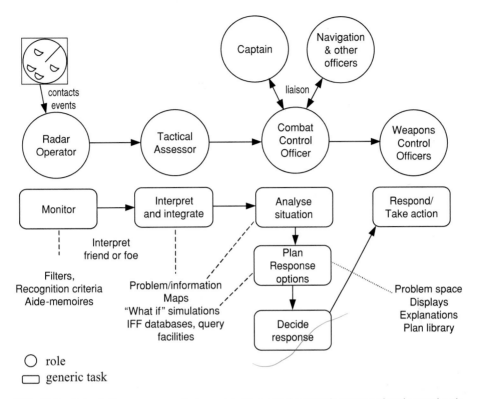

FIG. 23.3.   Roles in the sensor monitoring subsystem and the generic command and control task.
IFF = identify friend or foe.

model suggest that simulations and displays of the problem space are necessary to help the user form a detailed mental model. Other task support requirements are query and explanation functions, with filters to select different views of the problem space and facilities for interacting with the model to explore different "what-if" scenarios. The model also highlights the critical problem of situation awareness and decision making under time pressure and stress (i.e., frequency gambling pathology, Reason, 1990).

Situation awareness entails keeping the operator aware of changes in the environment while carrying out a skilled but repetitive task that inevitably becomes boring. Countermeasures such as placing training exercises in operational systems to increase awareness and swapping operator roles are recommended as generic requirements to counteract this problem. Countermeasures suggested in Table 23.1 are added to the system requirements.

MODELLING COMMUNICATION

Communication is modelled in this section first by using a generic model to analyse the process and then by applying communication pathologies at each stage. This first part focuses on human cognitive activity (and failings therein) for the sender and receiver. The second part focuses on the interaction or discourse exchanges between the communicating agents. In this case, discourse patterns are proposed as templates for analysis and design via their associated generic requirements.

Cognitive task design involves communication in two ways. First is design of communication channels and procedures between people; second is communication between people and

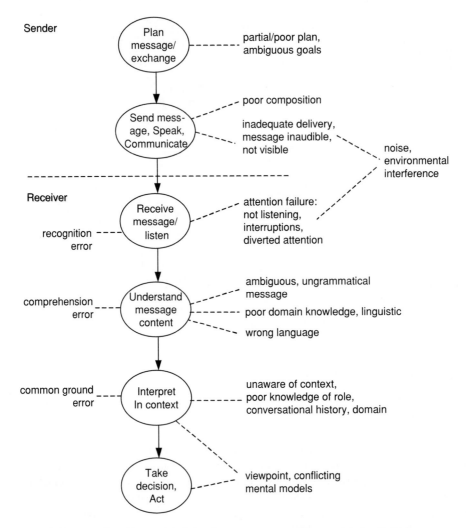

FIG. 23.4.   Cognitive model of the discourse process with associated pathologies.

computers. Although the second concern has been traditionally treated as straightforward input commands and output messages in simple models of interaction (Norman, 1988; Sutcliffe et al., 2000), as intelligent agents become increasingly prevalent the design of human–computer communication will require richer models of discourse. This section describes a cognitive model of communication from a generic task perspective, with associated pathologies. A taxonomy of discourse acts is then used to specify generic patterns motivated by the communication goals of clarifying understanding, information gathering, justifying arguments, and so on. The grounding theory in this case is Clark's (1996) theory of language.

Clark argued that people communicate to establish and achieve shared goals. Language is therefore a prelude to action, and conversations are motivated by a purpose. Meaning develops through a cycle of exchanges between the conversees that establish a *common ground* of shared understanding. Clark proposed a *ladder of action* that moves from initial recognition to comprehension of intent, elaboration of implications, planning actions, and finally to executing plans. Meaning is construed not only through language, however, but also via reference to mutually accessible artefacts (e.g., pointing to diagrams or objects in a room) and via shared knowledge of the communication context and the roles of conversees. Space precludes an

TABLE 23.2
Communication Pathologies and Remedial Treatments

Pathology	Remedial Treatment	Generic Requirements: Task Support
Inaudible message: communication not audible or corrupt	Request resend: improve communication channel, reduce ambient noise	Amplification, dedicated communication channels, noise filters
Not listening: receiver not paying attention	Request attention, test understanding	Alerting, store and repeat message
Semantic misinterpretation: words or grammatical construct not understood	Simplify language, eliminate ambiguity, reduce jargon, acronyms, check linguistic competence of receiver	Thesauri and lexicons, parsers/interpreters, explain symbols and diagrams, query facilities
Role confusion: interpretation mistake because intent attributed to incorrect agent	Make implications and role identification clear; provide background knowledge of roles	Image of sender integrated with speech, nonverbal communication, role information displays
Pragmatic misinterpretation: mistake caused by failure to appreciate the location and setting of the conversation	Provide better explanation of background and domain knowledge	Information/multimedia integration, cross-referencing, pointing, history and contact info.
Plan failure: utterance not generated or incomplete	Clarify, request repeat	Reuse library previous discourse
Plan failure: conversation not linked to action or does not make sense	Clarify viewpoint, restart conversation to establish common ground	Explanation and query facilities for context/role information

elaborate treatment of Clark's theory; however, even this précis can be used productively for reasoning about design. As before, the approach is to locate pathologies within a model and to suggest guidelines to remedy such problems. A generalised task model of communication based on Clark's theory is illustrated in Fig. 23.4, and the associated pathologies and design guidelines are explained in Table 23.2.

The communication model follows the traditional descriptions of cognitive processes for natural language understanding that commence with the lexical level of phoneme and word recognition, progress to interpretation of the utterance using the rules of grammar (syntax), and knowledge of word meanings (semantics), with final understanding being created by integrating input within contextual knowledge of the conversation setting, participant roles, the conversation history, and domain knowledge (pragmatics). In Rasmussen's (1986) framework, communication is generally an automatic and hence skilled process, so it is prone to slip and lapse errors in failure to listen and poor generation. Indeed, spoken speech is full of grammatical and lexical slips; however, the semantic and pragmatic levels of understanding are prone to many of the previously described rule- and knowledge-based pathologies. For instance, we often fail to understand the implications of communication because of halo effects, confirmation biases, encysting, and mismatch errors. The communication pathologies and remedial treatments are described in Table 23.2. The general cognitive pathologies (see Table 23.1) have been omitted for simplicity; however, skill-level problems apply to the send–receive stages, whereas rule- and knowledge-based pathologies are associated with the remaining stages.

A second approach to modelling communication is to use generic patterns of discourse acts to model how the exchange of messages should be structured between people to achieve a shared task goal. Primitive conversation patterns were originally described by Sacks, Schegloff, & Jefferson (1974), and larger patterns associated with tasks were specified in the conversation for action research by Winograd and Flores (1986). Although natural human discourse is variable, when associated with specific task goals such as buying and selling goods, natural regularities are observed; hence, patterns have utility for planning task-related conversations.

Discourse acts describe the intent of a communication (be that a sentence, phrase, voice mail, etc.) rather than the content. The following patterns are based on the classification of Sutcliffe and Maiden (1991), which elaborated the taxonomy of Austin (1962):

- Request (question)
- Inform (provide information, answer)
- Acknowledge (communication received)
- Check (belief)
- Confirm
- Propose (action)
- Command
- Agree
- Disagree
- Correct (i.e., amend).

The discourse acts impose different degrees of force or obligation on the receiver. *Request, check, inform,* and *report* generally impose only the obligation to provide or process information, whereas *disagree* and *correct* necessitate reasoning. Proposals require the receiver to consider the proposition and then *agree* or *disagree.* Commands have the highest degree of force because they necessitate reasoning and action; furthermore, the receiver has less choice, and hence both agents have a higher mutual dependency. Communication is planned using these acts to amplify the task model. For example, when a decision task is completed, there is usually a need to communicate the results to another party for action. This implies a command in hierarchically organised systems or a propose in more democratic networks of collaborating agents. Alternatively, communication can be analysed by applying generic patterns that can be specialised to specify communication exchanges between agents. Three patterns germane to the case study are presented; further details of other patterns can be found in Sutcliffe (2002).

Check or Clarify

This pattern tests understanding about beliefs or facts, so Agent A can ascertain that he or she shares knowledge with Agent B. Clarification or checking understanding may be by overt questions or by statements that assert a belief to be confirmed or denied; for example, "so patterns can be used to construct user computer dialogues", is a statement that one might either agree or disagree with.

A: *Check* (assert belief); may be a *request* or *inform* act.
B: *Confirm* or *disagree,* with a possible *correct.*

Propose Action

Proposals are suggestions for action in collaboration with another party. The initial proposal invites agreement or disagreement. In the case of agreement, there should be a commitment to action, either immediately or in the future, thus building an action ladder from communication

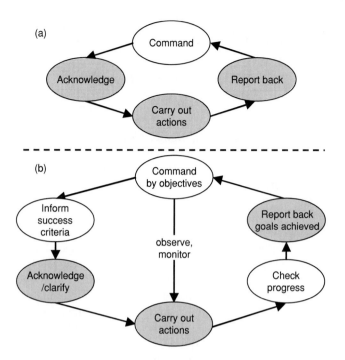

FIG. 23.5.   Command patterns: (a) strict hierarchy and (b) decentralised.

to intent to behaviour (Clark, 1996). The pattern structure is as follows:

A: *Propose* (action, strategy, or goal).
B: *Agree* or *disagree*.
IF agree THEN B is committed to action
ELSE B: *Clarify* or *propose* (alternative)

This pattern is related to *command* but has less binding force on the recipient to act.

Command

Two variations of the command pattern are illustrated in Fig. 23.5. In strict hierarchies, commands should be followed to the letter of the law, and the only response expected is acknowledgement of the command and reporting back on completion. A more loosely coupled command pattern gives subordinates more freedom of action and implies more trust in their abilities.

Command patterns not only have different forces and levels of trust in the subordinate but also vary in their constraints on action. The recipient may be told either to carry out an action in some detail or may be given a higher level goal with considerable freedom to interpret the plan of action on his or her own.

Patterns have mappings to generic tasks that depend on role allocation and where communication has to occur. For multiagent command and control, shared monitor and interpret stages involve clarify and check patterns, whereas at the analyse situation and plan response stages, patterns of propose, check and clarify, and negotiate (not illustrated) are expected. The command pattern is associated with the final stage of decide and act. The association of patterns and generic tasks is not always so simple, however. As we shall see later, the social setting influences communication, so in command and control with nonhierarchical relationships, communication involves proposals and liaison rather than command; furthermore, the task changes from command and control to judgement and decision making.

Case Study: Communication Analysis of Human Operator or Intelligent Agent

The discourse patterns are used to analyse the communication requirements following the scenario of replacing the radar operator role with an intelligent IFF analyser. The original system model has a reporting sequence and information flow from the radar to the radar operator to the tactical assessor with one check or clarify pattern. First we can apply the pathologies to the original system to vet the communication structure and facilities. Communication between the radar and the radar operator is visual via a video display unit–style display overlaid with text and graphic codes that represent known features (e.g., friendly aircraft, other ships, etc.). The other human roles are all located in the same operations room, so voice communication is the normal channel. Operators have microphones and voice messages are broadcast to officers in other locations over a loudspeaker network. Hence there is potential interference from several operators speaking at once. This is controlled by training operators to report only important events. The communication pathologies raise design issues for the communication channel as well as planning how dialogues should be structured. Semantic misinterpretation and role confusion pathologies are less probable in military domains with highly trained personnel because sublanguages have evolved to express ideas concisely, and role allocation does not usually vary during task execution. Interpretation in context errors may arise, however, because the listener may have to integrate spoken communication with other information in visual media. The data fusion problem and reference resolution in language (e.g., unknown contact bearing 070, 55 miles—just which one is this on a cluttered display?) can be addressed by pointing and highlighting the object being referred to by the speaker. Visual information needs to be shared to alleviate this problem. A use case that illustrates the discourse pattern and task stages for the radar system is given in Fig. 23.6a (manual system) and 23.6b (automated system). The discourse acts are shown on the connections between the agents (vertical bars), while the command and control task stages are shown as vertical boxes on each agent. The sequence reads from top to bottom and illustrates how communication is integrated with task action.

In the intelligent radar model (Fig. 23.6b), the computer identifies all radar contacts, interprets them as possible friend or foe, and notifies the tactical assessor (TA) of any significant contacts. The monitor and interpret tasks are automated, so the TA now has a human–computer interface. Given that the identification of friend or foe is never 100% reliable, complete trust (see Social Models of Interaction) is unlikely.

When intelligent technology is introduced, the communication load of the TA could actually increase because of the need to check and clarify the interpretations made by the intelligent IFF interpreter; furthermore, the TA's task of interpreting input data does not go away. The radar officer role is eliminated, so the running cost in personnel is reduced. Unfortunately the TA role could become more complex unless absolute trust is placed in the intelligent radar. This increases the dependency of the TA on the accuracy and reliability of the device. References have already been made to social influences on design of communication; this theme is taken up in the next section, which describes models for considering social factors. Increasing automation has a potential impact on system flexibility. Computer natural language understanding and generation capabilities are limited; even if they were not, the limited knowledge and inferential abilities of current intelligent technology would place limits on the flexibility of response. Human–machine dialogues therefore have to be largely preplanned. Exchanges have to conform to patterns (see Fig. 23.5) with limited adaptability through branch points and changing patterns. Human dialogues, in contrast, are much more flexible and the pattern is

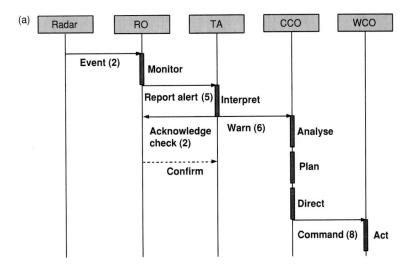

FIG. 23.6a. Discourse patterns for original manual system. RO = radar operator; TA = tactical assessor; CCO = combat control officer; WCO = weapons control officer.

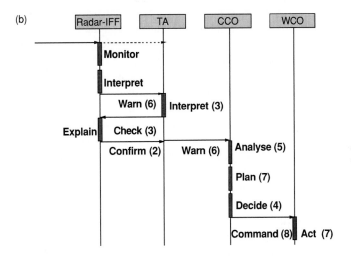

FIG. 23.6b. Discourse patterns for the automated system, intelligent identify friend or foe.

only a general model that has a wide variety of specialisations. Automation therefore imposes inflexible communication on the human–operator who has to converse within the machine's limited capabilities. Automation also has an impact on flexible role allocation. Although people can change roles, given sufficient training, changing the extent of automation is difficult when specialist hardware is involved; furthermore, the dialogue exchange noted earlier presents further difficulties.

SOCIAL MODELS OF INTERACTION

In the last section, models and patterns for communication were proposed. Communication takes place in a social context in which the relationship between individuals and groups colours action and intent. This section examines two key social influences: power and trust.

Power

Power is determined by access to resources, authority, and the influence of control. Sociological views of power emphasise resource dependencies in which one organisation needs goods, services, or information from another (Emerson, 1962). Psychologists, in contrast, have modelled power from the viewpoint of personality and dominance: subordinance relationships (Ellyson & Dovidio, 1985; McCrae & John, 1992). Power and trust are interrelated, because power assumes the ability to influence the policy or behaviour of others, which may be exercised benevolently or to the advantage of the more powerful party (Gans et al., 2001). Trust implies the assumption that power will not be used to the disadvantage of the less powerful party. In interindividual settings, power constrains choice and intent and influences communication by altering the roles and setting of conversation. For instance, we generally make assumptions of politeness and deference when speaking to people in authority (Brown & Levinson, 1997; Clark, 1996). The more powerful participant has fewer constraints, more opportunities for action, and greater leverage for influencing the decision making of another party. The less powerful party, in contrast, encounters more constraints, has fewer opportunities for action, and will have less leverage in influencing others. Power is acquired by assignment, authority, or experience. Societal roles have power connotations; for instance, judges and policemen have positions of authority and assumed power. Few people question their judgement and most will be influenced by their proposals. Power may also be attributed by membership of organisations, such as learned societies or secret societies. The experiential aspect of power is a consequence of our perception of actions carried out by other agents and the effects of those actions. The more frequently actions of an agent affect others, and the greater the number of agents affected, so our perception of an agent's power is increased.

Power influences our judgement by introducing a bias towards making a choice that we assume would meet with a more powerful agent's approval. In conversations, power influences the role we ascribe to other parties and our reaction to discourse acts. An intuitive example is our reaction to commands or perlocutary acts (Austin, 1962). In a power-asymmetric relationship, the less powerful listener will be disposed to obey commands from the speaker. In contrast, a more powerful listener in unlikely to obey a command from a subordinate party, while in equal-power relationships commands will be obeyed if they seem to be reasonable; otherwise they may be questioned. Clearly other factors enter into reactions, such as personality, the force with which the command was expressed, and the contextual setting. Highly skewed power relationships can have pathological influences on communication by preventing questioning and clarification when one party's command or course of action is perceived to be wrong. Other consequences can be biased judgements and ineffective collaboration

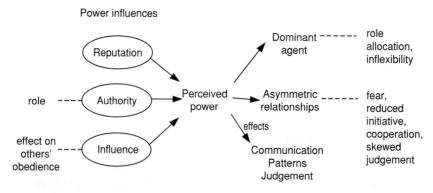

FIG. 23.7.   Model of influences on and consequence of power relationships.

TABLE 23.3
Pathologies of Asymmetric Power Relationships

Symptoms and Implications	Countermeasures
Commands not questioned even when wrong	Provide anonymous channels to challenge superiors
Decisions biased to appease the powerful	Establish consensus decision procedures, encourage viewpoints
Dangerous events, procedures, and systems not challenged	Provide anonymous reporting
Poor collaborative task performance: fear of superior	Reduce fear, improve feedback, praise good performance
Responsibilities abrogated leading to task failure: fear of criticism	Encourage and reward delegation

in tasks, when fear of superiors inhibits a subordinate's performance. The merits of power distribution also depend on context, however. In some domains, power asymmetries are a necessary part of the culture (e.g., military forces or the police). In these circumstances, equal power relations could be pathological. Some power asymmetry is a necessary consequence of leadership. A map of influences of power and other social relationships on design is illustrated in Fig. 23.7, and power-related pathologies and their design implications are given in Table 23.3.

In human–computer interaction, it might be assumed that the computer is always in the subordinate role; but this may not be true. People ascribe human qualities to computers when speech or human images are portrayed in multimedia (Reeves & Nass, 1996). Indirect power can be exercised via projection of dominant personalities using human images and forceful speech. A more direct manifestation of computer power comes from overdependence on automated systems for control or providing information; the computer can thus have considerable perceived power in organisations. Such perceptions can lead to pathological effects of stress when computers are designed to monitor, check up on, or control human operators.

Trust

Trust is an important mediator of relationships. It has the beneficial effect of reducing the need for governance and legal contracts in managing relationships and is therefore a powerful means of reducing cost overheads in relationship formation and maintenance (Sutcliffe & Lammont, 2001). Doney, Cannon, and Mullen (1998) defined trust as "a willingness to rely on another party and to take action in circumstances where such action makes one vulnerable to the other party". This definition incorporates the notion of risk as a precondition for trust and includes both the belief and behavioural components of trust. Different disciplines emphasise different aspects in definitions of trust, however. For example, economists emphasise costs and benefits and define trust as the expectation that an exchange partner will not indulge in opportunistic behaviour (Bradach & Eccles, 1989). Psychologists focus on consistent and benevolent behaviour as components of trust (Larzelere & Huston, 1980). Sociologists consider aspects of society, viewing trust as the expectation of regular, honest, and cooperative behaviour based on commonly shared norms and values (Fukuyama, 1995). I have synthesised a common schema from these different views that models trust between individuals, organisations, and also the resources and artefacts on which they rely.

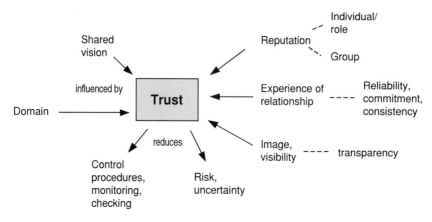

FIG. 23.8. Components of the PLANET (PLANning Enterprise Technology) trust model and their interaction.

As with power, trust has an assigned, prerelationship component: reputation, followed by an experiential component. Trust is assessed initially from information gathered about the other party. Reputation information may come from references supplied by third parties: general publicity, corporate image and brand visibility for organisations, or role assignment for individuals such as professional society memberships. A model of trust, developed from Sutcliffe and Lammont (2001), is illustrated in Fig. 23.8.

An initial level of trust is derived from reputation. This is then modified by experience, which may reinforce the trust if it is positive or lead to mistrust when negative experience in encountered (Gans et al., 2001). Positive experience results from evidence that our trust in another party has been justified, for instance, by the other party's support for one's actions, help in achieving a shared goal, altruistic action, or simple communicative acts such as praise and supporting one's case in arguments (Reeves & Nass, 1996). In brief, trust is based on an expectancy of another party that has to be fulfilled; if it is not, then mistrust can quickly set in. Mistrust is not a linear function of bad experience; instead, it can accelerate as cycles of mistrust feeding on suspicion, which lead one party to question the other's motives in more detail and find criticism where previously there was none. The trust-related pathologies and countermeasures are summarised in Table 23.4.

The need for trust is also domain dependent. In some systems a moderate level of mistrust can be beneficial to promote mutual awareness and performance monitoring (e.g., rival design teams, internally competitive teams).

The HCI implications for trust can be partitioned into trust engendered by illusion and conscious trust in the whole system. Strangely, when human images and speech are used in multimedia communication, we attribute computers with human qualities and react accordingly (Reeves & Nass, 1996). As a consequence, the variables that influence human trust and perceptions of power can be used to engender trust with the computer (e.g., use of praise, polite attitudes, friendly conversational style, smiling, paying attention, supporting suggestions, and proposals). Projections of power can be achieved by choice of the male authoritarian voice, conversational initiative, image of mature male leaders, and social indicators of power and authority. These effects can be thought of as within-dialogue trust, because as soon as the computer is turned off the attribution is unlikely to persist. Longer range, conscious trust will be built by reputation and experience, that is, the utility and validity of the computer-provided information (e.g., timeliness, accuracy, comprehensibility), the usefulness of its advice, and its transparency (e.g., explanation facilities, queries, ability to inspect reasons for decisions, etc.).

TABLE 23.4

Trust-Related Pathologies and Countermeasures

Mistrust Pathology: Symptoms	*Countermeasures*
Excessive legal governance needed to control relationship	Promote arbitration, clear agreements, third-party regulation of relations
Suspicion about motives of other party: ineffective collaboration	Improve negotiation, clarify shared goals, create agreement for confidence-building measures
Failure to accept proposals and commands from other party	Clarify relationships of proposals to shared goals, explain motivation
Suspicion about information supplied by other party, failure to act on basis of such information	Clarify information, invite third party to validate information
Excessive checking and monitoring other party	Encourage transparency, free information exchange

Case Study: Social Implications for Intelligent Agent Solutions

Increased automation of the radar raises questions about how well the operator will trust identification of friend and foe by the intelligent agent. As alluded to in the previous section, lack of trust will have an impact on communication because more cycles of clarification and explanation will be required if mistrust exists. Mistrust will diminish the operator's faith in the machine's proposals, thus increasing the tendency to check on the identification and acquire justification and evidence to understand the automated decision making. This will defeat the object of increased automation by imposing a greater workload on the human operator. The design question is how trust might be increased in the first place.

Using the model in Fig. 23.8, we need to investigate the reputation component and how experience might be positively affected. Reputation will depend on knowledge of similar designs and the equipment manufacturer. In this case, there are no similar artefacts for ref-erence because the design is new; hence, as with all innovative products, its reputation rests with the manufacturer. In defence equipment no strong contribution from the reputation of the organisation can be assumed. This leaves experience. Unfortunately, experience that is likely to engender trust is rare (i.e., real combat situations in which correct performance by the radar led to a successful outcome). Confidence-building measures can be designed into the equipment, however. Use of the radar in training exercises and being able to simulate a variety of known and unknown threats in scenarios may help to build confidence in the in-telligent radar's ability to identify contacts correctly. Building in explanation facilities will improve transparency and faith in the intelligent agent's decisions, as will an audit trail of its decision making. A final possibility is to design the operator–system dialogue so it engages the user periodically to maintain situation awareness and involve them in system-initiated decisions.

Increased automation will skew the balance of power from the operator to the intelligent IFF because the operator has a greater dependency on the computer for information and decisions. Although this change in power might not present problems in normal operation, in emergencies, when operators are under time pressure, a bimodel reaction is likely. Either people will ignore the computer's advice, possibly to their detriment, or they will accept it without question and assume a subordinate position. This presents a dilemma for communication when increased automation uses a variation of the command pattern in Fig. 23.5(a) for time-dependent decisions (e.g., command [proposed action], offer override [agree or disagree], take action by default if no reply within *x* seconds). The opportunity of human operators to use an override effectively is unlikely, given increased automation and resource dependency for information.

EVALUATING ALTERNATIVE DESIGNS

So far, the models and pathologies have been used to discover and refine functional requirements for one design that has assumed increased automation to replace a human role. In this section, the models are used to assess two sociotechnical designs that take the automation debate a little further. Some of the interdependencies between the social influences and structure of communication and task design are summarised in Fig. 23.9.

Power relationships will affect the structure of communication because discourse exchanges are partially determined by the roles of the participants. Higher ranking individuals can use

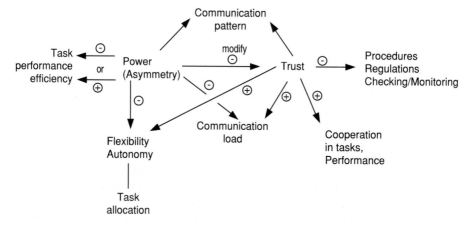

FIG. 23.9. Relationships between power, trust, communication, and task performance. A plus sign (+) indicates increase in variable, and minus sign (−) indicates decrease.

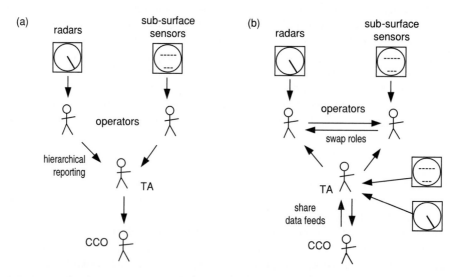

FIG. 23.10. Alternative role allocations for the radar identify friend or foe subsystem: (a) hierarchy mode and (b) overseer mode. TA = tactical assessor; CCO = combat control officer.

TABLE 23.5
Comparison of Two Role Allocation Designs for the radar identify friend or foe subsystem

	Hierarchy Mode	Overseer Mode
Power	Authority clear	Shared partially
Trust	Assumed in responsibility	Personal trust important
Communication	Command pattern	Proposal liaison patterns
Task design flexibility	Preset	More adaptable
Situation awareness	Could wane	Change should improve*
Training requirements	No change economical	Higher
Intelligent user interfaces	Possible	Difficult

Note. *Indicates where situation awareness may be prone to mental model pathologies and slips when roles are swapped.

commands, whereas subordinates cannot. Trust affects communication in a more subtle manner. Trusting relationships require fewer check or clarify acts; proposals are more likely to be accepted, and negotiation over conflicts should be easier to achieve. The two designs to be compared are summarised in Fig. 23.10.

The traditional, hierarchical mode has clear role separation between the human operators of sensing equipment (various radar and subsurface sensors), and the TA who collates and integrates the information before passing it on to the CCO (combat control officer). This model is prone to slip errors when operators are fatigued or bored (see Reason, 1990) and may also be prone to situation awareness problems if the TA and CCO roles do not maintain an up-to-date mental model of the ship's environment. Increased automation could exacerbate the lack of situation awareness by the TA and CCO if they are left out of the loop of monitoring and interpreting radar and sonar contacts. The alternative role allocation "overseer mode" has the operators periodically swapping roles, which also makes the role responsibility boundaries deliberately less rigid. The TA and CCO are explicitly given the task of overseeing the operators' roles and consequently share the same data feeds. The motivation for this organisation is to increase situation awareness by periodic rotation of tasks. Variety in job design tends to improve alertness and satisfaction (Bailey, 1982; Wickens, 1992). Furthermore, the less rigid role boundaries should increase communication because the operator and officers are required to problem solve cooperatively. This in turn should improve situation awareness. The trade-offs between the alternative designs are summarised in Table 23.5.

The traditional model (hierarchy mode) has clear lines of authority. Trust is assumed in the lines of command, although experience and individual relationships built over time are still important. Communication tends to be structured (see Fig. 23.10) so that interaction is efficient, although it may pay a penalty in inflexibility. Tasks are determined by role allocation, which economises on training but does not allow for flexibility. Intelligent operator interfaces are possible in this model, but as the previous analyses demonstrated, automation will make the system less flexible.

The alternative overseer model has power shared more equitably among the roles, although lines of command are still present. Trust at a personal level acquired through experience will be more important because individuals must have confidence that they can cooperate flexibly to solve problems under pressure. Communication will be less structured, with liaison and proposal–negotiation exchanges more common than commands. The change of role and higher incidence of communication should improve situation awareness; however, there is a penalty in higher training costs because individuals may possess more diverse skills as well as being trained in collaborative problem solving. Furthermore, longer conversational exchanges may lead to longer task completion times, so there is a trade-off between these sociotechnical system designs in terms of efficiency (i.e., time) and overall effectiveness (i.e., decision quality). The dangers of increased automation in this organisational design are twofold. First there is reduced awareness for the TA and CCO because role sharing or rotation among the individuals who carry out knowledge-intensive task stages will require more training and will be vulnerable to handover errors (e.g., mental model acquisition pathologies). Second, the increased automation will tend to impair maintenance of accurate mental models by the TA and CCO, who have to interpret machine-based proposals and decisions. The psychology of judgement and decision making can be used to illuminate the automation problem a more depth.

The elaboration likelihood model (Cacioppo, Petty, & Kao, 1984) stresses the importance of attitudes and beliefs and proposes a dual route for the decision-making process. This theory predicts that in many choice situations people engage in peripheral processing (i.e., memory intensive, rule- and skill-based, fast) rather than central route processing (i.e., knowledge-based, deliberate, slow). Adaptive decision making (Payne et al., 1993) argues that the decision-making strategies adopted by people are a trade-off between the perceived seriousness of the problem and the cognitive effort involved in the strategy. This means that people will only tend to reason in depth (i.e., slow route) about problems they consider serious. Finally, the theory of planned behaviour (Ajzen & Fishbein, 1980) emphasises the role of attitude in decision making and contributes the concept of the controllability of risk (i.e., people's behaviour depends on their ability to take action to avoid a risk). Unfortunately, as Reason (1990) demonstrated, many critical decisions have to be made under time pressure, when people exhibit frequency or recency gambling pathologies.

The situation awareness dilemma can be interpreted in terms of how to influence human judgement towards a fast or slow cognitive route. Fast-route processing is characterised by decisions driven from memory and attitude, so decisions will be made without in-depth reasoning. In contrast, the slow route involves in-depth reasoning and forming a mental model of the problem to arrive at a rational decision. The fast route is equivalent to skilled reasoning in Rasmussen's concept because the decision is already preformed so the only mental processing is mapping between the input variables and memories that indicate the choice. Slow-route decision-making is equivalent to Rasmussen's (1986) knowledge-based reasoning, but this is only feasible if a mental model of the decision space is developed. The design problem is how to encourage slow-route reasoning by encouraging the operator to maintain an active mental model while avoiding the tendency to lapse into fast-route frequency gambling. Task swapping is one strategy; however, when automation is increased, mental model maintenance becomes more difficult, hence the "designing-the-operator-out-of-the-loop" problem.

As a result, if increased automation is to succeed, human operators will have to function in overseer mode for the intelligent agents. The design implication is for two concurrent user system dialogues. One presents a channel of interpreted information, but in the context of basic input data to enable the human operator to make his or her own interpretation. Visual summarisation of information will be necessary to prevent information overload (Card, MacKinlay, & Shneiderman, 1999). The second dialogue presents a separate channel of interpreted information produced by the intelligent processing. How this information is represented in diagrams, speech, or text requires skills in multimedia design (Sutcliffe, 1999), coupled with explanation and simulation facilities. The final design challenge will be integrating the two (or possibly more) information channels to enable the user to check machine-based advice by validating proposals with information input and their models of contextual knowledge.

Sample storyboards for the radar-intelligent IFF subsystem are shown in Fig. 23.11. Slow-route processing can be encouraged by continuous user engagement with the intelligent IFF

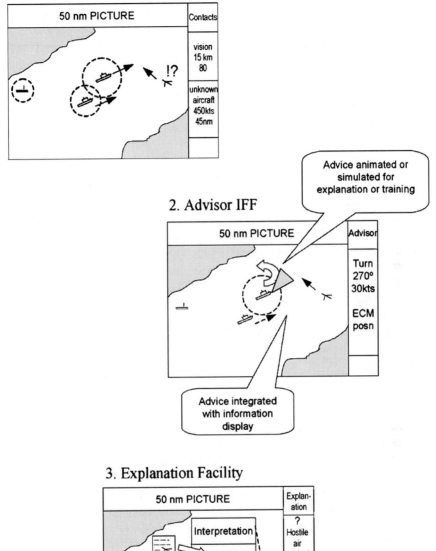

FIG. 23.11. Storyboards for the intelligent advisor user interface for the radar identify friend or foe (IFF) subsystem.

system to maintain situation awareness. This might be achieved by showing explanations when identifying threats, by ensuring the user is in the loop by check–confirm dialogue patterns that ask for the user's opinion on interpretation, and by identifying proposals or trade-offs so that active judgement by the operator is solicited.

These trade-offs and other aspects of the prototype design are still under debate. Studies on the implications of reducing manpower in navy ships have pointed out the need for increased levels of technical skills and more flexible role allocation (Strain & Eason, 2000). Other arguments in favour of the more flexible overseer mode come from evidence of collaborative problem solving and dynamic allocation of roles in work practices (Luff & Heath, 2000; Rognin, Salembier, & Zouinar, 2000). Human operators often have to respond flexibly to events and partition complex tasks between themselves; however, mutual situation awareness is important not only between the human operators but also between the operators and automated systems (Hollnagel & Bye, 2000). This chapter has illustrated how different types of model can contribute to an understanding of functional allocation and task design problems and, in particular, how trust and human reactions to automated systems need to be considered. Functional allocation has been driven from a task-goal perspective, although this view has evolved to account for the importance of information, artefacts and resources in effective operation (Hollnagel & Bye, 2000; Vicente, 1999). The THEA method for safety critical system design (Fenton et al., 1998) and the IDAS (information-decision-action-supervision) framework (Dearden, Harrison, & Wright, 2000) emphasise the importance of contextual information and the scenario-based approach to analysing functional allocation; while Clegg has argued that sociotechnical issues such as ownership, power, and responsibility need to be components of the task allocation process (Clegg, Axtell, Damodaran, & Farbey, 1997). These issues have been addressed by the models used in this case study with a scenario-based approach.

CONCLUSIONS

This chapter has proposed a different approach to design that has been partially motivated by the patterns movement (Alexander, Ishikawa, & Silverstein, 1977). Generic models provide ready made and reusable models of knowledge that can be applied to design. Rather than analyse everything de novo, designers only have to interpret which generic models apply to a new application, then specialise the models with domain specific information. An additional benefit is that design knowledge in the form of generic requirements can be attached to the model, thereby reusing knowledge as a form of organisational learning. The second contribution has been a taxonomy of pathologies linked to particular design issues and models. Cognitive task analysis and design in general have a tendency to model the positive, normal course of events, whereas the real world is full of errors and potential hazards. The pathologies encourage design thinking to anticipate errors as an important aspect of safe system design.

One omission from this chapter is an explicit cognitive model of human error; however, elaborate taxonomies of the causal influences on human error have been available for some time (management oversight and risk tree taxonomy; Johnson, 1980; Hollnagel, 1993), and these have been developed into more predictive models such as COCOM (Hollnagel, 1993, 1998) and probabilistic models represented in Bayesian belief nets (Galliers, Sutcliffe, & Minocha, 1999). However, it has to be acknowledged that social variables can have important influences on human cognition, so Bayesian belief network models could be elaborated to include the influences of trust and power on human error and judgement. I doubt this would be a profitable way to proceed because it compounds the validation problem. Cognitive models are difficult to validate because of the difficulty in collecting large complex data sets; furthermore, even at the level of the individual, cognitive models can only make coarse-grained predictions. Although we know social influences are important, predicting the exact effect and force of such influences requires many more years of social-psychological research.

Greater immediate utility can be found by treating the social modelling approach as tools for thought. Although the models cannot predict the design solution, they do at least point out the dimensions of the design problem. It is then part of the designer's expertise to interpret the key variables that impinge upon the problem in hand. Interpreting general models with contextual knowledge specialises them for a given domain. A useful supplement to the causal modelling approach is to reuse generic task models with associated advice, either in terms of generic requirements for task support or pathologies to avoid or counteract through user interface design. This approach is as yet in its infancy, and considerable research is necessary to build up a library of reusable task models, although generalised tasks have been researched in several areas (Feiner & McKeown, 1990; Sutcliffe, 2002; Wehrend & Lewis, 1990).

One danger in providing several models for interpreting design problems is that, while giving the designer choice, it may also lead to confusion. The obvious question is how to integrate generic task models with pathologies, with cognitive and social models. As yet I have no direct answer, although I have found that applying generic cognitive and social models to tasks helps to generate new reusable generalised task models with associated requirements and pathologies, as demonstrated in information retrieval, command, and control and other generalised tasks (Sutcliffe & Ennis, 1998; Sutcliffe, 2002). Design in many cases concerns trade-offs and taking different viewpoints, so the identification of a single magic bullet solution will probably always be an illusion. We will have to learn to tolerate complexity of different views and design approach when dealing with complex sociotechnical systems. The family of models described in this chapter is one small step towards dealing with such complexities.

ACKNOWLEDGEMENTS

This work was partially supported by the EPSRC (Engineering, Physics and Science Research Council) projects DATUM and SIMP (Systems Integration for Major Projects). It is a pleasure to acknowledge the help of colleagues on the SIMP project and BAE Systems. However, the content of this chapter is the author's sole responsibility.

REFERENCES

Ajzen, I., & Fishbein, M. (1980). *Understanding attitudes and predicting social behaviour*. Englewood Cliffs, NJ: Prentice-Hall.

Alexander, C., Ishikawa, S., & Silverstein, M. (1977). *A pattern language*. London: Oxford University Press.

Annett, J. (1996). Recent developments in hierarchical task analysis. In S. A. Robertson (Ed.), *Contemporary ergonomics 1996*. London: Taylor and Francis.

Austin, J. L. (1962). *How to do things with words*. London: Oxford University Press.

Baddeley, A. D. (1990). *Human memory: Theory and practice*. Hillsdale, NJ: Lawrence Erlbaum Associates.

Bailey, R. W. (1982). *Human performance engineering: A guide for system designers*. Englewood Cliffs, NJ: Prentice-Hall.

Ball L. J., & Ormerod, T. C. (2000). Putting ethnography to work: The case for a cognitive ethnography of design. *International Journal of Human-Computer Studies, 53*, 147–168.

Barnard, P. (1985). Interacting cognitive subsystems: A psycholinguistic approach to short term memory. In A. Ellis (Ed.), *Progress in psychology of language* (Vol. 2, pp. 197–258). London: Lawrence Erlbaum Associates.

Barnard, P., May, J., Duke, D., & Duce, D. (2000). Systems, interactions and macrotheory. *ACM Transactions on Computer-Human Interaction, 7*, 222–262.

Bradach, J. L., & Eccles, R. G. (1989). Price, authority and trust: From ideal types to plural forms. *Annual Review of Sociology, 15*, 97–118.

Brown, P., & Levinson, S. C. (1987). *Politeness: Some universals in language usage*. Cambridge, England: Cambridge University Press.

Cacioppo, J. T., Petty, R. E., & Kao, C. F. (1984). The efficient assessment of need for cognition. *Journal of Personality Assessment, 48,* 306–307.

Card, S. K., Mackinlay, J. D., & Shneiderman, B. (1999). Information visualization. In S. K. Card, J. D. Mackinlay, & B. Shneiderman (Eds.), *Readings in information visualization: Using vision to think* (pp. 1–34). San Francisco: Morgan Kaufmann.

Chemical Industries Association. (1993). *A guide to hazard and operability studies*. London: Author.

Clark, H. H. (1996). *Using language*. Cambridge, England: Cambridge University Press.

Clegg, C., Axtell, C., Damodaran, L., & Farbey, B. (1997). Information technology: A study of performance and the role of human and organisational factors. *Ergonomics, 40,* 851–871.

Dearden, A., Harrison, M., & Wright, P. (2000). Allocation of function: Scenarios, context and the economies of effort. *International Journal of Human-Computer Studies, 52,* 289–318.

Doney, P. M., Cannon, J. P., & Mullen, M. R. (1998). Understanding the influence of national culture on the development of trust. *Academy of Management Review, 23,* 601–621.

Ellyson, S. L., & Dovidio, J. F. (1985). *Power dominance and non-verbal behaviour*. New York: Springer-Verlag.

Emerson, R. (1962). Power dependence relations. *American Sociological Review, 27,* 31–41.

Feiner, S., & McKeown, K. R. (1990). Generating coordinated multimedia explanations. *Proceedings: 6th IEEE Conference on Artificial Intelligence Applications* (pp. 290–296). Los Alamitos, CA: IEEE Computer Society Press.

Fenton, N., Littlewood, B., Neil, M., Strigini, L., Sutcliffe, A. G., & Wright, D. (1998). Assessing dependability of safety critical systems using diverse evidence. *IEE Proceedings: Software, 145,* 35–39.

Fukuyama, F. (1995). Trust: The social virtues and the creation of prosperity. New York: Free Press.

Galliers, J., Sutcliffe, A. G., & Minocha, S. (1999). An impact analysis method for safety-critical user interface design. *ACM Transactions on Computer-Human Interaction, 6,* 341–369.

Gans, G., Jarke, M., Kethers, S., Lakemeyer, G., Ellrich, L., Funken, G., & Meister, M. (2001). Requirements modelling for organisation networks: A (dis)trust-based approach. In S. Easterbrook & B. Nuseibeh (Eds.), *Proceedings: RE 01: 5th IEEE International Symposium on Requirements Engineering* (pp. 154–163). Los Alamitos, CA: IEEE Computer Society Press.

Gardiner, M., & Christie, B. (Eds.). (1987). *Applying cognitive psychology to user interface design*. Chichester, England: Wiley.

Gigerenzer, G. (1991). How to make cognitive illusions disappear: Beyond heuristics and biases. In W. Stroebe & M. Hewstone (Eds.), *European review of psychology (Vol. 2)*. Chichester, England: Wiley.

Hollnagel, E. (1993). *Human reliability analysis: Context and control*. London: Academic Press.

Hollnagel, E. (1998). Cognitive reliability and error analysis method: CREAM. Oxford, England: Elsevier.

Hollnagel, E., & Bye, A. (2000). Principles for modelling function allocation. *International Journal of Human-Computer Studies, 52,* 253–265.

Hutchins, E. (1995). *Cognition in the wild*. Cambridge, MA: MIT Press.

Johnson-Laird, P. N. (1983). Mental models: Towards a cognitive science of language, inference and consciousness. Cambridge, England: Cambridge University Press.

Johnson-Laird, P. N. (1988). *The computer and the mind: An introduction to cognitive science*. Cambridge, MA: Harvard University Press.

Johnson, P., Johnson, H., Waddington, R., & Shouls, R. (1988). Task-related knowledge structures: Analysis, modelling and application. In D. M. Jones & R. Winder (Eds.), *Proceedings: HCI '88* (pp. 35–61). Cambridge, England: Cambridge University Press.

Johnson, W. G. (1980). MORT (Management Oversight and Risk Tree): Safety assurance systems. New York: Marcel Dekker.

Kahneman, D., & Tversky, A. (1982). Intuitive prediction: Biases and corrective procedures. In D. Kahneman, P. Slovic, & A. Tversky (Eds.), *Judgement under uncertainty: Heuristics and biases*. Cambridge, England: Cambridge University Press.

Kieras, D. E., & Meyer, D. E. (1997). An overview of the EPIC architecture for cognition and performance with application to human computer interaction. *Human-Computer Interaction, 12,* 391–438.

Kuhlthau, C. (1991). Inside the search process: Information seeking from the user's perspective. *Journal of the American Society for Information Science, 42,* 361–371.

Larzelere, R. J., & Huston, T. L. (1980). The Dyadic trust scale: Toward understanding the interpersonal trust in close relationships. *Journal of Marriage and the Family, 42,* 595–604.

Luff, P., & Heath, C. (2000). The collaborative production of computer commands in command and control. *International Journal of Human-Computer Studies, 52,* 669–700.

McCrae, R. R., & John, O. P. (1992). An introduction to the five factor model and its applications. *Journal of Personality, 60,* 175–215.

Norman, D. A. (1986). Cognitive engineering. In D. A. Norman & S. W. Draper (Eds.), *User-centred system design: New perspectives on human-computer interaction*. Hillsdale, NJ: Lawrence Erlbaum Associates.

Norman, D. A. (1988). *The psychology of everyday things*. New York: Basic Books.

Payne, J. W., Bettman, J. R., & Johnson, E. J. (1993). *The adaptive decision maker*. Cambridge, England: Cambridge University Press.

Rasmussen, J. (1986). *Information processing in human computer interaction: An approach to cognitive engineering*. Amsterdam: North Holland.

Reason, J. (1990). *Human error*. Cambridge, England: Cambridge University Press.

Reeves, B., & Nass, C. (1996). The media equation: How people treat computers, television and new media like real people and places. Stanford, CA/Cambridge, Englaand: CLSI/Cambridge University Press.

Rognin, L., Salembier, P., & Zouinar, M. (2000). Cooperation, reliability of socio-technical systems and allocation of function. *International Journal of Human-Computer Studies, 52*, 357–379.

Sacks, H., Schegloff, E. A., & Jefferson, G. (1974). A simple systematics for the organization of turn-taking in conversation. *Language, 50*, 696–735.

Strain, J., & Eason, K. D. (2000). Exploring the implications of allocation of function for human resource management in the Royal Navy. *International Journal of Human-Computer Studies, 52*, 319–334.

Sutcliffe, A. G. (1998). Scenario-based requirements analysis. *Requirements Engineering, 3*, 48–65.

Sutcliffe, A. G. (1999). A design method for effective information delivery in multimedia presentations. *New Review of Hypermedia & Multimedia, Applications & Research, 5*, 29–58.

Sutcliffe, A. G. (2000). On the effective use and reuse of HCI knowledge. *ACM Transactions on Computer-Human Interaction, 7*, 197–221.

Sutcliffe, A. G. (2002). *The Domain Theory: Patterns for knowledge and software reuse*. Mahwah, NJ: Lawrence Erlbaum Associates.

Sutcliffe, A. G., & Ennis, M. (1998). Towards a cognitive theory of information retrieval. *Interacting with Computers, 10*, 321–351.

Sutcliffe, A. G., & Lammont, N. (2001, December/January). Business and IT requirements for B2B e-commerce. *New Product Development and Innovation Management*, 353–370.

Sutcliffe, A. G., & Maiden, N. A. M. (1991). Analogical software reuse: Empirical investigations of analogy-based reuse and software engineering practices. Special issue: Cognitive ergonomics: Contributions from experimental psychology. *Acta Psychologica, 78*(1/3), 173–197.

Sutcliffe, A. G., & Maiden, N. A. M. (1998). The Domain Theory for requirements engineering. *IEEE Transactions on Software Engineering, 24*, 174–196.

Sutcliffe, A. G., Maiden, N. A. M., Minocha, S., & Manuel, D. (1998). Supporting scenario-based requirements engineering. *IEEE Transactions on Software Engineering, 24*, 1072–1088.

Sutcliffe, A. G., & Rugg, G. (1998). A taxonomy of error types for failure analysis and risk assessment. *International Journal of Human-Computer Interaction, 10*, 381–405.

Swain, A. D., & Weston, L. M. (1988). An approach to the diagnosis and mis-diagnosis of abnormal conditions in post-accident sequences in complex man machine systems. In L. Goodstein, H. Andersen, & S. Olson (Eds.), *Tasks, errors and mental models*. London: Taylor and Francis.

Vicente, K. J. (1999). Cognitive work analysis: Toward safe, productive, and healthy computer-based work. Mahwah, NJ: Lawrence Erlbaum Associates.

Wehrend, R., & Lewis, C. (1990). A problem-oriented classification of visualization techniques. *Proceedings: First IEEE Conference on Visualization: Visualization '90, October 1990* (pp. 139–143). Los Alamitos, CA: IEEE Computer Society Press.

Wickens, C. D. (1992). *Engineering psychology and human performance* (2nd ed.). New York: HarperCollins.

Winograd, T., & Flores, F. (1986). *Understanding computers and cognition: A new foundation for design*. Reading, MA: Addison Wesley.

Yu, E. S. K., & Mylopoulos, J. (1995). From E-R to "A-R": Modelling strategic actor relationships for business process reengineering. *International Journal of Cooperative Information Systems, 4*(2/3), 125–144.

24

The Design of POGO World

Antonio Rizzo
and Patrizia Marti
University of Siena, Italy

Françoise Decortis
University of Liège, Belgium

Claudio Moderini
Domus Academy, Milano, Italy

Job Rutgers
*Philips Design, Eindhoven,
The Netherlands*

. . . a creative personality is prepared through the creative imagination that in the present expresses itself and becomes concrete.

—L. Vygotsky

Abstract

In this chapter, we describe the design process that led us to the development of an innovative narrative environment for children at school and home, POGO world. The design process was iterative and co-evolutionary among four design components: user study, concept design, content design, and technology design. Grounded on a shared vision, this design process allowed us to use principles and concepts of the cultural psychology theoretical approach to drive the design concepts and development of the tools of POGO world. The process we describe shows how cognitive task design can envision and guide technological development and innovation. To this aim, we first present the pedagogical principles derived from the work of Lev Vygotsky and Jerome Bruner that became the conceptual basis for assessing the results of the design process. Subsequently, we present an analysis of the pedagogical activity involving narratives carried out in two European schools and the development of the design concepts that were embodied in mock-up and working prototypes. Finally, we briefly report the testing at school of the mock-ups and prototypes and conclude with the implication of the described design process for extending cognitive design to more formal working context.

INTRODUCTION

The design experience of POGO was carried out within the European Experimental School Environment initiative. POGO addresses story building as a key concept to drive technology development: a virtual and real environment for supporting the unfolding of narrative competence in children aged 6 to 8 years (primary school), a world where children create, explore, and develop narrative language and social skills.

Since the start of the project, it was considered essential to build and share a cognitive vision of the project among the multinational and multidisciplinary design team and to anchor the design process in psychological theories of narrative.

The design team, although multidisciplinary and with different theoretical backgrounds concerning design (engineering, architecture, industrial design, interaction design, literature), had the common sense that any artefact had a dual nature—a material one and a conceptual one—and that both, but especially the conceptual, give shape to the human cognitive process when people interact with that specific artefact. An interesting difference was related to the aspects that the different professionals reported as example of cognitive process shaped by the artefact: for the architects and industrial designers, they were mostly related to aesthetics and identity; for engineers, they were the new problem-solving strategies offered to the user; for the interaction designers, they were the way in which intentions are stated, actions carried out, and expectations about feedback put forward; for the writer, they primarily related to the way by which meaning is built and transformed.

These positions were synthesized in the theoretical framework of cultural psychology, which, resting on the seminal work of Vygotsky, deserves particular attention to the role of artefact in human cognition. According to Vygotsky (1978), a human never reacts directly (or merely with inborn reflects) to environment. Instead, tools and signs mediate the relationship between human agent and objects of environment; thus, for Vygotsky, mental process can be understood only if we understand the tools and signs that mediate them (the law of semiotic mediation.) Vygotsky named the process of mediation as the principle of extracortical organization of complex mental function. He claimed that the development and the use of tools extend cognitive processing beyond the biological dimension of the nervous system, giving a crucial role to artificial stimuli in psychological activity. One corollary of the semiotic law was that the embodiment of external representations in psychological activity permits completely new and unpredictable behavioral patterns.

From such assumptions, it was clear that the vision would orient the design process toward a progressive definition of the POGO world, where the conceptual level of the artefact would step by step be embodied in the physical and functional form of the artefact and in cognitive activity it would mediate.

For POGO we also deployed the theoretical framework of cultural psychology to define the nature of narration following another of the leading figures of the discipline, Bruner, who paid special attention to narratives as a fundamental tool mediating human communication.

Bruner (1990) considered narrative a primitive function of human psychology, being at the heart of human thought. The representation of experience in narratives provides a frame that enables humans to interpret their own experiences and one another. For Bruner, narrative is a fundamental aspect of meaning construction. If it were not for such narrative framing, 'we would be lost in a murk of chaotic experience and probably would not have survived as a species in any case' (Bruner, 1990, p. 56).

In becoming a higher psychological function, narrative, as a primitive function, undergoes transformations according to the mediation role of cultural artefacts. Writing, printing, and broadcasting have allowed the evolution of the narrative process (Ong, 1985). Each of these new artefacts has not dramatically changed the previous one; writing and printing have been de facto currency on information and culture and are destined to remain so for the foreseeable future, even though broadcasting has become the dominant media in many social contexts. Computing, the most recent communication artefact, has a property that earlier communication artefacts did not have: It can change previous ones. Computers now control almost all printing and so encompass many aspects of the media; likewise, computing increasingly mediates writing, drawing, and painting.

TABLE 24.1
Summary of the Objective of POGO Vision for Each of the Design Components

User Research	We will investigate how narrative mediates construction of meaning and the child's' organization of knowledge. The description of how narrative as an artefact interacts with other physical and cultural artefacts is seen as equally important. These research questions will be addressed by the pedagogical objectives that will guide the POGO project.
Concept Development	In order to create "an experience" it is necessary to develop a system that supports the creation and management of behaviours, meaning and senses. In order to create "one collective story" it is necessary to add the elements of cooperation, rules, language and acting. In order to create a story world, it is necessary to add the elements of reporting, memory, interpretation and representation.
Content Development	We will provide a virtual world organized by story structures that will facilitate children's individual, collaborative and intercultural narrative creation processes. An open research question is at which level the POGO story world should be pre-programmed and how much space is left for children to create their own stories.
Enabling Technology	The POGO application will be a computer program using video features (MPEG4) in a 3D environment (VRML) in a networked context. Interactive devices will be linked to on-screen VRML/MPEG4 objects. Multi user technology will provide a collaborative environment supporting children in their manipulation of narrative content.

Computing is on the way to transforming again the modalities of narration, allowing completely new and unpredictable human behavioral patterns. POGO aimed at envisioning, experimenting, and testing new modalities, with the idea that they could be, at least partially, deliberately designed as cognitive activities and tasks.

From these ideas, we developed our POGO Vision in plenary meetings and articulated it according to four design components: user research; concept development, content development, and enabling technology. Table 24.1 summarizes the stated objectives for the vision of each component.

In the following sections we present the design process of the POGO project, starting with the definition of the pedagogical principles and objectives and proceeding through (a) the analysis of the pedagogical activity concerning narratives carried out in two European schools, (b) the development of the design concepts, (c) the interaction between narrative activities at school and design concepts, (d) the development of mock-ups and prototypes and their testing at school, and (e) concluding with the results of evaluation of the tools and the implication of the proposed cognitive task design approach for designing human activities.

PEDAGOGICAL OBJECTIVES

The development of the POGO environment was driven by a set of pedagogical objectives derived from the theoretical framework of cultural psychology. These objectives stem from the original work of Vygotsky and Bruner and are related to fundamental characteristics of the individual's ontogenetic development.

The first three objectives of POGO are rooted in the relations Vygotsky (1998) identified between reality and imagination, which are at the basis of the creative process. Vygotsky maintained that the notions of reality and imagination are not as separate as they appear to be in everyday life. To understand the psychological mechanisms of imagination and creative thinking, Vygotsky argued that the creative activity of imagination directly depends on the richness and variety of previous experiences made by an individual, because fantasy's constructions are composed of the material supplied by experience. Fantasy is not antithetical to memory of real events, but supported by it; fantasy disposes traces of events in new forms. This constitutes the first relation between fantasy and reality.

The second relation is more complex, because it does not consist of a relation between the elements of fantastic construction and reality but between the product of fantasy and any complex phenomenon of reality. This form of relation is only possible because of social experience. For instance, the French Revolution, which is a historical event, can be known only through social experience and interactions with others.

These two relations are at the core of the first pedagogical objective. The POGO environment should (a) expand as much as possible the sensorial experience of children within each type of relationships—the imaginary relationship between real elements (first relation) and real relationships between unexperienced elements (second relation); and (b) allow comparison and experimentation among both relationships, stressing the social origin of the second relationship (cf. the forth pedagogical objective).

The third relation concerns emotion. Here, Vygotsky distinguished two aspects: the law of the common emotional sign and the law of reality of the imagination. In the first law, the author claims that fantasy's images supply an inner language to our feelings. Our feelings select elements that are isolated from reality and combine them together in a relationship that is internally conditioned by our state of mood, instead of by temporal or logical relationship among images.

On the other hand, in the second law, Vygotsky stated that every form of creative imagination includes affective elements. This means that each construction of fantasy influences our feelings, so even if it is not a construction that corresponds to reality, the feeling it evokes is effective and actually experienced.

This emotional relation is the central issue of the second pedagogical objective: *The POGO environment should support children in developing emotional knowledge (e.g., empathy) through the laws of the common emotional sign and of reality of the imagination.*

The law of reality of emotion is a bridge toward the fourth relation between reality and imagination. A construction of fantasy may constitute something effectively new, something that has never existed before in the experience of man and which does not correspond to any object or concept that actually exists. Yet once this crystallized image of imagination is externally embodied, once it is made concrete, once it has become a thing among other things, it begins to exist in the world and to affect other things. This holds true for physical objects (e.g., devices) as well as for concepts (e.g., metaphor, conceptual blending).

This brings us to the third pedagogical objective: *The POGO environment should help children to complete the circle of creative imagination that starts from sensorial knowledge of reality and return to reality through active modification of the environment produced by the embodiment of the imagination. The process of embodiment concerns not only the material and technical aspects but also emotional and conceptual ones.*

The fourth and fifth pedagogical objectives adopted in POGO come from the notion of intersubjectivity and the role of cooperation and social interaction in the development of metacognitive consciousness put forward by Bruner.

For Bruner (1996), intersubjectivity is the way humans come to know each other's minds. Human beings have an innate propensity to intersubjectivity that manifests itself in children's prelinguistic phase of development and evolves through social experience. Intersubjectivity becomes in Bruner's pedagogical framework, both the object and the instrument; at the same time, it is an epistemological and a methodological issue. Bruner related intersubjectivity to our interpretative activity. We do not use analytical tools to describe our own and other's mind; our mental activity is a "mystery", not accessible to introspection. Through collaboration children can negotiate these interpretations with others "to achieve not unanimity but more consciousness". Meaning construction is an activity of negotiation that starts in early childhood and characterizes the whole of human life (Bruner, 1990). Even if he did not attribute to the prelinguistic child a conscious intentionality, Bruner (1990) considered our early ability to have meaningful exchanges with other people as a "primitive form of folk psychology", "a set of predisposition to construct the social world in a particular way and to act upon our construal" (p. 72).

Hence we arrive at our fourth objective: *The POGO environment should support the child in building his or her fantasy, world model and understanding the process through discussion and cooperation. The POGO environment should sustain and enhance this discussion among children and among children and adults and should enable cooperative mechanisms and reflection.*

Metacognition is a mental activity in which the object of thinking is thinking itself. The main objective of metacognition is the creation of different ways of building reality. Metacognition offers a basis for the interpersonal negotiation of meaning, a unique way to understand each other even when the negotiation process does not succeed in producing a shared meaning. Metacognition can relate to several thinking processes, but in POGO we focused on metacognition of narrative processes, not only because narrative was the focus of the project but because stories are always expressions of a point of view—one of the possible interpretations of reality. Educational programs devote a great attention to the scientific paradigm, but the interpretative one is often neglected. One of Bruner's most important pedagogical claims is to devote more attention, more time, and more effort to the development of children's narrative competence through collaboration

Our fifth objective is as follows: *POGO should support the understanding and the management of different ways to organize narrative processes (i.e., narrative structures) to enable a richer way to build meaning out of our experience and, consequently, to become conscious of what the narrative interpretation imposes on the reality it builds.*

This set of pedagogical objectives represented an operational way to allow the principles of cultural psychology to guide the design process. They were proposed within the user research design component but soon become part of the project's shared vision. Both the concept envisioning activity, carried out by the architect and industrial designers, and the analysis of narrative practice at school were oriented by the pedagogical objective, but in different ways.

ANALYSIS OF NARRATIVE PRACTICES AT SCHOOL

We observed and analyzed more than 30 narrative activities in two schools, one located in Siena, Italy, the other in Brussels, Belgium. With the collaboration of the teachers, we tried to understand why and how narrative activities are specific and how these activities could be empowered by POGO world.

Narrative activities at school are part of a pedagogical practice; they aim to mediate the learning of various skills, including the proper use of narrative. They are at the same time an instrument and an object of pedagogical practice. For each activity observed in class, a narrative activity model was constructed. The model includes indications on the following variables:

- The main structural phases of the activity
- Their temporal sequences
- The artefacts used
- The type of activity (individual, collective, cooperative)
- The role of the teacher(s)
- The relation with the POGO pedagogical objectives

The analysis was carried out with the aim of (a) identifying the principal phases of the narrative activity in class, (b) identifying narrative practices requiring further investigation, and (c) supporting the mapping of concepts design onto educational narrative practices. All three objectives of the analysis were instrumental to producing heuristic guidance for evolving narrative practices by exploiting the knowledge produced by the practitioners and for preventing the disruption of effective and successful practices.

Modelling Through the Cycle of Creative Imagination

In describing the narrative activities, we observed that the cycle of creative imagination proposed by Vygotsky (1998), as a psychological process to account for the creation of knowledge that occurs within the zone of proximal development, could be used to represent the different chronological and structural phases of a narrative activity as they occur in school.

The cycle of creative imagination is a process that develops in four phases: exploration, inspiration, production, and sharing. It describes how the individual experiences the external world, elaborates the impressions received, assembles them in a novel way, and shares this production with others. What is a seamless and continuous psychological process occurring as a whole in the activity of a child could be broken down—not in the children's mind but in the focus of an articulated pedagogical activity. The narrative activities at school included a focus on all of these phases, often in linear sequences, sometimes with small loops or repetition, sometimes with a leap. This is how the four phases of the cycle of creative imagination can capture the different chronological and structural phases of a narrative activity.

The first phase of the cycle of creative imagination, *exploration,* consists of the interactions with the real world, which can be either direct or mediated by social relations. For instance, at this stage of activity, children manipulate various kinds of objects and materials, discover sounds and natural elements, listen to stories, and meet other people.

The impressions, received during the exploration of the real world, are complex wholes resulting from a large amount of different parts. For additional activities of fantasy, the dissociation or the division of the complete whole into components and features is fundamental. Dissociation also implies granting relevance to some characteristics and leaving out others. Thus, the second phase of the cycle of creative imagination corresponds to a phase of reflection and an analysis of the experience (*inspiration*). As we observed in narrative activities in both schools, teachers often dedicate time to return to and reflect on an experience, asking children to think about what was relevant for them after the exploration phase.

The next moment in the complex process of creative imagination is *production,* which means the reunion of the dissociated and transformed elements. In other words, this phase of the activity corresponds to the point at which children produce new contents, usually through a variety of media, based on selections and choices of previously self-made elements. This allows children to practice different means of expression and to find the ones that better fit their expressive needs. During this phase, children write narratives telling their emotions or produce oral stories of particular characters (puppet) or cartoonlike plots on posters.

The cycle of creative imagination is not complete after the production phase, but when imagination is embodied or crystallized in external images shared with others. Children externalize the product of their imagination during the production phase, but *sharing* adds another dimension, that is, when children's externalised productions start to exist in the world, children can experience the effects of their imagination on the others (peers, teachers, and parents). During the sharing phase, children may present their performance to their parents, explain their drawing to the classroom and teacher, or hang a poster in a hall. This last phase of the cycle closes the circle opened by the first phase: from real world, to imaginary world, back to the real world.

The models we produced were the result of the following methodological process. From the pool of observed activities (scenarios) we selected, together with the teachers, the most successful ones, that is, the ones that allowed the best achievement of pedagogical aims and, at the same time, that allowed the best engagement for children (it is worthwhile to note that even though in one of the schools advanced digital technologies were available, and although there were several activities that included the use of computers, not one of these activities turned out to be among the most successful). Next, this subset of narrative activities was represented in narrative activity models (NAMs).

In the models, we described features and details that were relevant from a theoretical viewpoint (tools, physical space, social relations, pedagogical objectives, etc.) and from the teacher's viewpoint (chronological sequence, macroclustering of some basic tasks, educational aims, etc.). These models were iteratively developed with the teachers and licensed after their approval. The emerging NAMs were finally assessed for their ability to represent the salient components of the narrative activities conducted at school (i.e., there were no other successful activities concerning narration with features that were not present in the NAMs produced thus far). We ended up with seven NAMs describing seven successful pedagogical activities in the two schools.

The NAMs were used to define the users' requirements and to relate them to the POGO concept and enabling technologies. In particular, the requirements mainly emerged by trying to answer the following questions:

- What are the components of the activities performed in the classroom that make them effective and successful in achieving the pedagogical objectives?
- In what ways could we hinder the success of these activities?
- How could we move functionalities and properties from one artefact to another?

The set of requirements were produced and assessed for their heuristic capacity to foster the concept to be expressed in mock-ups. In addition, the representation of examples of educational practices in the NAMs associated to the five pedagogical objectives was particularly useful for the design team for sharing the understanding of the pedagogical objective.

In what follows, we present a summary of a NAM that reflects the complexity of the use of narration at school and to provide an example of the activity models.

NAM, an Example: "Element Festival"

A NAM is composed of two parts: a textual description of the events and a descriptive model that represents the main features of NAMs in a chronological sequence of events. The model highlights the mediation role of the artefacts in children's experiences. The elements of the model are as follows:

- The main structural phases of the activity related to the cycle of creative imagination
- Their temporal sequences
- The artefacts used (pencil, whiteboard, scissors, paper, etc.)
- The type of activity (individual, collective, cooperative)
- The space where the activity took place: the classroom, a dedicated space (library, theatre, museum, etc.), outside the school (e.g., court yard)
- The physical setting describing the way teachers and pupils interact in the space (Are pupils at their desk, with teachers moving around the room, or is the teacher also seated at her desk? Are the children free to move in the space, or do they stand in front of the other pupils?)
- The supported POGO pedagogical objective

The following is a reconstruction of the salient phases of the element's festival activity.

The Sensorial Experience

During the school year, children met five times with Rosa, an external educator who proposed activities and games about the four natural elements: earth, air, water, and fire. In each of these activities, children manipulated the elements and expressed (or externalised) their feelings and emotions both through their body (movement or dance) and products (drawing on posters using different colours or collage).

Remembering Activities Through Expression of Favourite Games.   The teacher asks the children to remember the activities with Rosa. The teacher stimulates the children asking many questions, such as, "What did you do with Rosa? Did you remember the activity concerning Air? Did you like it? Why did you like it? Which game did you prefer?".
    Example (speaking about air):

    Child: You gave us a bag and then we had to dance and we caught the air.
The teacher: How did you feel?
    Child: Like a feather!

Choosing Favourite Element Through a Ritual of Identification.   The teacher asks the children to choose individually, without sharing this choice with the classmates, one of the elements they would like to become for the final festival (the last meeting with Rosa). Through a simple ritual proposed by the teacher, children "become" that chosen element. The ritual is based on the use of simple artefacts, such as spangles, glue, and teacher's use of voice and prosody to make children believe in the magic.
    Example: The teacher goes to Chiara, puts a stick in the spiral of her notebook, throws some spangles (shining stars) on it and in Chiara's hair, saying "magic, magic". Then the teacher says, "Chiara will become, will become the . . . ." Chiara says to the teacher the element she wants to be and the teacher continues saying loudly, "FIRE" From this moment on, Chiara is the fire, and she behaves like fire.

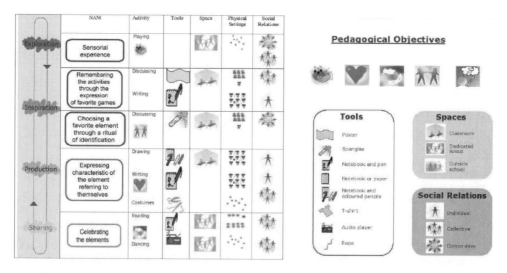

FIG. 24.1.   Summary of a narrative activity model. On the right, the legend presents the items for each class of the activity model. An exception is made for the setting and teacher role; this class would require a richer description.

TABLE 24.2
Summary of an Example of a Requirement Extrapolated from the Narrative Activity Model

Requirement	Provide teachers with facilities to acquire content and to structure that content to facilitate their retrieval by children.
Rationale	In many activities, the teachers set up ahead of time the material that children would use; they collect and arrange the content that will support narration according to their pedagogical aims.
Example	In the "Little Red Riding Hood" activity, children built a story starting from abstract drawings of the fairy tale's elements that a teacher brought from home (see video clip associated).

Expressing Characteristics of the Element Referring to Themselves.   Children define the emotional and behavioural characteristics of the elements they will perform by writing, drawing, and by creating a costume. For instance, children write in a notebook why they would like to be that element, what they would do if they were that element, and how they feel.
Example:

Child: I would like to be water. I've become water, I am very powerful, and I am of many colours. Inside me there are fish. I am waves. I am not bad. I don't get on well with fire. I am the cascade, which goes down very fast, fresh, with plenty of foam. This is me!

Celebrating the Elements.   Children celebrate the elements by reading and dancing. Rosa calls out an element (water, for instance), and all children embodying that element stand up and go to the stage. One at a time, they read the narratives they wrote in their notebooks ("Here is what I am like and what I can do"). When all the children have finished reading their compositions, the music (especially composed for each element) starts, and Rosa invites the children to dance as if they were the element they embody.

By developing a model of each activity (UniSiena & UniLiegi, 1999; see Fig. 24.1 for a schematic summary), we were able to (a) identify the requirements for empowering successful narrative activities, (b) identify the requirements for avoiding to chain the activity with the technology, and (c) guide the evolution of design concepts.

The requirements were defined according to the NAM providing a rationale for each requirement and an example from the documented scenarios. Thus, for example, for each phase of the NAM we had a set of requirements articulated according to interaction with tool, social relations, space, and settings and teacher role. The user requirements were not presented as a list, but as a map projected on the seven activity models. Table 24.2 is an example for a requirement concerning the production of content in the early phase of interaction between the children and teacher.

There were two peculiarities in the way requirements were proposed: (a) Video clips offered examples of each requirement, providing contextual evidence of the activity in which the users were actually involved. For example, in the case reported in Table 24.2, a video clip showed the way in which a teacher presented, distributed, and used the material she brought from home. This was important for the whole design team to have a better understanding of the conceptual use of the tools by the teacher and by the children. (b) Requirements were stated at a conceptual level with the specific aim of merging with the concept design phase and with the idea that only after their match with the concepts, they would be refined in functional terms. For example, in Table 24.2, a conceptual aspect of the requirement concerns the way in which the content should be classified by the teacher to allow children to retrieve it. This could become an aspect of the artefact that mediates the memorization of the content and even the operation that one could perform on that content (e.g., content is memorized according to the activity in which it was produced, to the group of people who joined the production, and to its relationship with a special event such as Christmas or activities at the school).

The basic idea in specifying requirements was that to properly design cognitive activities and tasks, one needs first to define the conceptual nature of the artefact, explore it in an empirical way, and then specify its material shape and its interaction modalities (see the functional activity model matrix in Fig. 24.3, and the section, The Concept–Mock–Up–Test–Prototype–Concept Cycle). Thus, in the POGO design process, in parallel with the definition of the pedagogical objectives and the user study, there was an open concepts generation phase.

CONCEPT DEVELOPMENT

One of the design partners of the project, Domus Academy, partly inspired by user observations, produced 14 ideas of narrative environment described in a technical document (Domus Academy, 1999). These design concepts represent different ways in which enabling technologies can facilitate narrative processes and mediate the development of child communication. In a next stage, the participants in the project (researchers, users, designers, and technologists) were asked to evaluate the design concepts. This evaluation was performed first by filtering the concepts through the defined pedagogical objectives and then through the comments of the teachers (described later in the chapter). The main question was how these concepts would allow us to achieve the defined objectives. The comments collected were structured in different sections: One was directly related to the objectives and responds to the question stated earlier, a second section grouped general comments, and a third represented the teachers' comments.

We describe these parts of the design process (concept generation and evaluation) by briefly describing two concepts and their respective evaluations.

The design theme of the concept *TheatReality* (after the combination of the words theatre and reality) is related to how to "visualize" the emotional aspects of story narration by emphasizing the elements of expression of the characters and how to integrate in the performance the reactions of the public. The concepts outline has been described as an environment that emphasizes the physical (i.e., voice) and behavioral characteristics (i.e., expression) of children performing or telling stories. The narrator can choose, modulate and emphasize his or her emotions during the narration by adding visual and aural effects to the performance. The voice can be altered and transformed into music, facial expressions can be modified and highlighted, and gestures can be exaggerated for a more effective communication. The same can be applied to virtual characters that can be used to convey emotional aspects of interpersonal communication. The public, both real and virtual, intervenes, expressing consensus by way of enhanced reactions. The combination of the choices of the different actors participating in the building of the stories, both the narrator and the public, determines the appearance and behaviour of the POGO world.

Theatreality

Consistency With Pedagogical Objectives.   Theatreality appears to be consistent with the first objective because it is based on rich and various visual and auditory experiences. This concept is also coherent with the objective of developing a language of emotions through "images". Indeed, after a free and informal exploration of the possibilities offered by the instrument, children could use these possibilities to better express their emotions and to build new meanings through an "augmented" dramatization. The third objective should be emphasized further. Children have to be able to understand clearly the effects that the changes they introduce produce on the public. It should be clarified how the public interacts.

In general, we can say that this concept is positive, but it has to provide stronger and more emphasized moments of socialization and negotiation. For instance, more actors could collaborate. It would also be useful to explore some elements belonging to the children's sociocultural context (i.e., a Sicilian accent or a French one). When thinking aloud about this concept, we imagined the following situation: Mario wants to make an angry face, so he pushes the button "angry face", and his own face is projected on a screen, but it is not his normal expression; instead, it has been transformed into an angry one. According to our vision, the child should not passively ask POGO to transform a normal expression to an angry one; instead, POGO must invite him to learn *how* to change his appearance. Moreover, POGO may suggest the way different cultures represent an angry face (Chinese, European, African, etc.).

Teachers' Remarks. The proposal about voice is interesting; not only can the voice be altered, it can be transformed into music, too. We find this opportunity amazing particularly and are intrigued by the possibility of exaggerating movements and altering facial expressions. Indeed, the more children can modify themselves, the more they can identify with the embodied characters and the more their capacities to develop stories may be stimulated and augmented.

We aren't in favor of the idea that the public's reactions can be altered and emphasized. Observing an exaggerated reaction could be stimulating for some children, but it could be negative input for others (shyer and more insecure children) who could be inhibited by strong reactions.

When we think about the POGO world, we imagine a room where children can have many educational opportunities. POGO world could offer the possibility of projecting scenarios on three sides of that room. These scenarios would be significant enough to make children feel themselves inside the projected environment (a forest, for example). We think it would be better to use scenes children have drawn, so that they forest be inside a forest exactly as they imagined it. This way, the child feels completely engrossed in the scenery and in the story. This active immersion into the story allows an extended comprehension of the story. The more the child feels immersed in it, the more he or she will be able to create and invent new events. Indeed, if children feel as if they are really in a forest or at the bottom of the ocean, they will automatically be stimulated by particular inputs and continue in with the story.

Having the possibility of projecting children's drawings or those taken from books is also useful because it takes less time than preparing them on big posters. Preparing handmade scenes takes so much time that often children have the opportunity of making a good dramatization only once a year. We think that dramatization is of fundamental importance in education, but preparing scenes takes too much time, and the long period that passes before the final product is finished leads one to lose the sense of immediateness of the moment. The POGO world would allow children to make dramatizations more frequently.

Redrum

A second concept, *Redrum,* after the movie *The Shining,* was designed on a different theme: How can a virtual partner guide the narrator in the discovery of the story world? How can POGO support children's curiosity in informal ways? The concept outline is the following: A virtual partner introduces and guides children into a story. The child can create or find a virtual friend, who has his or her own identity and a personal story. The "ghost friend" can help the child solve problems, but at the same time, it never stops browsing through the possible story world. The ghost friend identity is affected by its experience and by the interaction with the narrator and other characters. Ghost friends learn from each other and grow along with the child that has to take care of them.

Consistency With Pedagogical Objectives.   This concept does not seem to meet the first objective related to the expansion of sensorial experience. Indeed, there aren't particularly stimulating sensorial activities. Moreover, with such a concept the child cannot properly explore the distinction between real and virtual. For the same reasons, we think that this concept doesn't support the second objective. To allow children to verify the effects of their own action on the social world, the activities proposed in "Redrum" are inappropriate because they are totally individual and therefore also do not meet the third pedagogical objective. We do not think it is possible for children to develop "one's mind theory", through interactions between themselves and "virtual friends". In the educational development of a child, the activity proposed in this concept might become ambiguous and dangerous; we don't think the metaphors used (e.g., Tamagotchi) are particularly stimulating or interesting.

Teachers' Remarks.   Teachers commented that if the virtual friend is already predeter- mined, if it has its own story already, it is too cold and somewhat appropriate. On the other hand, if children can create it as they want (a person or an animal, big or small), as they create the virtual friend, it becomes a part of them; it is not something strange from the external world. If it comes from his inner world, the child likely will build a story made of things and events coming from his unconscious that couldn't be externalized in an objective way without the support of the virtual friend. The concept might be interesting and avoid emotional distress. Children would build another ego, which has another name and another form, that could be used to project their inner feelings. The virtual friend becomes a kind of inner mirror; but it can be used only at an individual level, only for children who have problems or who are unable to express their feelings.

These two sketched design concepts and the associated summarized comments provide a simple but representative example of the cooperative strategy adopted to design POGO world. It was an effective way to allow different stakeholders, such us teachers, social scientists, and industrial designers to join the design process and to promote the mutual awareness of theoretical issues, pedagogical practices, and design practices. In particular, one can observe how both the social scientists' and the teachers' comments concern cognitive aspects of the envisioned concepts and not technical or functional ones. Indeed, through their comments teachers contributed to the conceptual design of the POGO environments, and their remarks were projected on the requirements map inside the NAMs.

A similar process was carried out for the evaluation of the enabling technologies (virtual reality modeling language, Moving Picture Expect Group version 4). The potentialities of such technologies were presented to the teachers, and they were asked to project such potentialities on the NAMs or onto the "what if" concepts. Again, their remarks were connected to the requirements map.

INTERACTION BETWEEN NAM AND CONCEPT DESIGN: ACTIVE TOOLS

The basic idea that was communicated to designers through NAM was that narrative practices at school are more than just storytelling and story creation in the mind. They are articulated social practices that make the child the fundamental actor in a close relationship with peers and, foremost, with the teacher(s). These practices span from sensorial experience, at the base of each narrative activity (exploration), to the expression of children's production (sharing), going through a rich phase of inspiration and creation. POGO world aims at supporting all the phases of the activity.

Thus, design concepts evolved to envision the different ways in which children, supported by enabling technologies, can create new narratives interacting with the physical and virtual

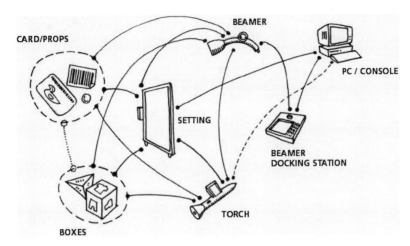

FIG. 24.2.   The basic POGO tools as first sketched.

POGO world. From the first interaction among design concepts and narrative activity models, Domus Academy developed a set of Active Tools to support children in their story-creation processes (Domus Academy, 2000). This design activity was mainly intended as an input to the prototype development, but at the same time had the role of envisioning future scenarios that overcome technological and conceptual limitations of early "what-if" concepts. The tools were conceptualized as cooperating devices for the construction of a narrative experience based on situated story editing, supporting the entire process from exploration to inspiration, to production, and finally to sharing. The active tools are the mediating artefacts for the narrative process both in the classroom and between the classrooms.

The functionality of the tools span many areas, from gestural (live performances), visual (manipulation of images and drawings), to aural (sounds and atmospheres), to manipulative (physical feedback, kinematics), to material (surface and texture, weight, etc.). Fourteen active tools were proposed, and the basic tools, the ones that embody the more peculiar functions are briefly illustrated here. These reflect also a propaedeutic choice that sees the pedagogical Objectives I (sensorial experience), III (embodiment of imagination), and IV (intersubjectivity) as preliminary to the II (emotional knowledge) and V (metacognition of narrative structures). That is, the POGO world should first of all be a rich sensorial environments in which social activities that produce sharable outputs can be carried out. (See also Fig. 24.2.). Following are six of the proposed active tools.

- *POGOsettings.* POGOsettings are basically windows that open onto the virtual story world and permit screening from the POGO world. They are therefore mobile video walls that configure the POGO world scenery. Because each scene is site related, the main goal of the POGOsettings is to define the place where the episodes of each scene take place. Changing scene, the POGOsettings configuration updates either automatically or by selection. By positioning the POGOsettings in the POGO world, a virtual story map is defined. This process can also occur in reverse, selecting a given virtual story map and placing the POGOsettings in the appropriate location. The combination of the POGOsettings affects the virtual relation between the story elements, triggering different combinations.

- *POGOtorch.* POGOtorch is a portable device to explore and manipulate the virtual story world. It can be used both to interact with the virtual elements displayed on the POGOsettings and with the augmented object (tags, cards/props) within the POGO world. The manipulative functionalities related to the POGOtorch allow children to point to story elements and to select, move, or capture them to share these items later with their classmates.

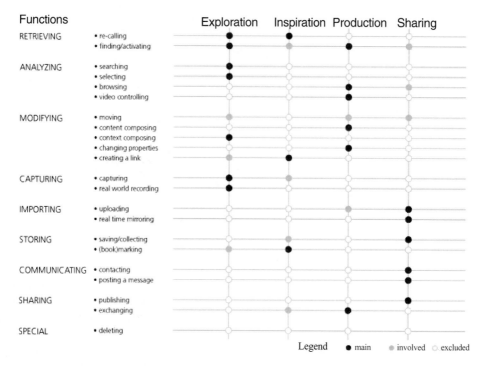

FIG. 24.3.   Mapping functionality on the cycle of creative imagination. On the left, functions are spit into macro and micro functionalities.

- *POGObeamer.* POGObeamer is a specific threshold tool that connects the real and the virtual environment by allowing physical things to pass into the virtual story world. Raw materials (produced or found by children) and affective objects (such as a puppet or a flower from the school's garden) can be imported into the virtual story world using the POGObeamer. During theatrical narration the POGObeamer can act as real time mirroring and performance enhancer.

- *POGOcard, POGOprops.* The POGOcard and-props relate to physical story components: children, common objects, props ad hoc. With a POGOtag, a child can be recognized by the system when entering in a cover area (e.g., by proximity with a POGOsetting or using the POGObeamer). Common objects can be defined and associated to virtual correspondences by attaching to them a POGOtag. Ad hoc objects can be found within the POGOworld. These objects are archetypal story elements that update their behaviors depending on the context.

- *POGOboxes.* The POGOboxes are the story elements modifier tools. They allow the children to change the story elements' properties and attributes. They are a physical palette for modifying the selected virtual story elements.

- *POGOPC/console.* The POGOPCconsole is the tool used to browse, collect, partially edit, and publish stories; it is mainly by the teacher.

The envisioned Active Tools were linked to the narrative model based on exploration, inspiration, production, and sharing to define for each phase the combination of functionality that best supported the process. This process can be represented by a three-dimensional matrix depicting the relation between functionality, the four stages of the cycle of creative imagination (see Fig. 24.3), and POGO tools (Fig. 24.4).

This exercise allowed exploration of the potential role of the artefacts in the narrative activities at school and preparation of their testing in the field, on the basis of which precise

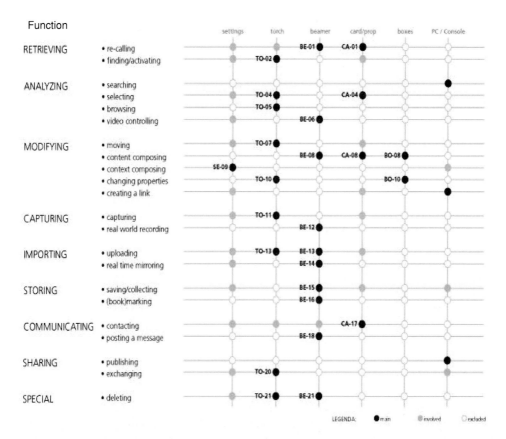

FIG. 24.4. Mapping functionality on POGO tools. The main functionalities of each tool are coded in bold and tagged to identify the priority in the following phase of the interaction design.

functional specifications have been further elaborated and translated into detailed interface descriptions.

The proposed tools were designed by the logic of not reducing the complexity and the richness of existing activities as represented in the NAMs, but rather from the perspective of increasing the narrative expression and communication through technology. Thus, in POGO the dedicated tools for supporting the different phases of narrative activities are also integrated tools for collecting, producing, and editing story elements and story structures.

At the end of the concept design process, we developed prototypes of POGO tools by producing in parallel "role" prototypes and "look and feel" prototypes (Houde & Hill, 1997). Role refers to questions about the function that an artefact serves in a user's life—the way in which it is useful to them. Role requires the context of the artefact's use to be established. Look and feel denotes questions about the concrete sensory experience of using an artefact—what the user looks at, feels, and hears while using it. The role prototypes where built as rudimentary mock-ups and tested in their proper context in real activities. The look and feel prototypes were built as three-dimensional models to be presented to the whole design team and to the teachers after the mock-up testing in the school. The role mock-ups were build in a "quick and dirty" way because the aim was just to explore how the envisioned tools produced by the interaction of NAMs and concept design could allow the user to mediate their cognitive process. That is, what was the heuristic power of these conceptual tools when projected in their everyday practice? How these conceptual tools would mediate their thinking and communication?

CONCEPT TESTING: MOCK-UPS AT WORK

The prototyping phase of the POGO project was carried out through "Wizard of Oz" testing sessions. The Wizard of Oz is a technique used to present advanced concepts of interaction to users (Erdmann & Neal, 1971). An expert (the wizard), possibly located behind a screen or hidden to the subject of the test, processes inputs from a user and emulates system outputs. The aim is to show and explore with the users future system capabilities that are not yet implemented. This method allows user requirements and creative design to be explored at an early stage in the design process, particularly for systems that go beyond readily available technology. Furthermore, designers who play the "wizard" can gain valuable insight from the close involvement in the users' activity.

The prototyping phase was directly embedded in a real activity proposed by teachers with whom we hypothesized, on the basis of the model, how to improve their everyday activity through the use of POGO tools. The prototyping we illustrate here as an example was inserted in an activity proposed by an Italian teacher whose objective was to allow children to learn the structure of story and enable them to develop their creative imagination and emotional expression through the construction of stories (UniSiena & UniLiegi, 2000).

The teacher started by telling children an unknown tale (exploration) that children analyzed in terms of temporal sequence and narrative elements (inspiration). The teacher suggested that children could use POGObeamer to keep track of their production that they could reuse it in a next phase of the activity. So, after the children had drawn the main narrative elements of four familiar tales (inspiration), they recorded drawings and voices of the tales' characters in the POGO room (Day 1). The next phase was concerned the production of the story (Day 2). Teachers organized the dramatization of "Little Red Riding Hood", suggesting the use of POGOsetting to project scenes (the kitchen at the little girl's house, the forest, and the grandmother's bedroom) accompanied by sounds of ambience and children's voices previously recorded by POGObeamer to place them in a self-constructed environment. Children invented a new version of the classic tale when the teacher, by means of POGOtorch, provided them with new story elements (e.g., the little girl was meeting an invisible dwarf instead of the wolf, represented by a recorded voice and then embodied by a child playing the role).

Setting and Material

For the POGObeamer prototyping (see Fig. 24.5), an acquisition room was set with a lamp and a box with three buttons for the acquisition of drawings, sounds, and video. There was also a microphone on the control box. Children could monitor the images they wanted to capture on a small projection positioned in front of them. Another room next to the acquisition room was set up to allow a "wizard" to simulate the functionalities of the tools. To record sounds and images,

FIG. 24.5.   The early POGO mock-ups. Left: POGObeamer. Right: POGOsetting and POGOtorch.

the wizard manipulated an iMac Digital Video plugged into a camera (Sony Handycam) and to a projector.

The POGOsetting prototyping took place in a room equipped with curtains on the windows, necessary to allow projections on one wall of the room. The space in front of the wall was dedicated to the actors' play, and the space ahead was for the public (children not participating directly in the dramatization). This test required the use of two computers: one to manage the sounds (iMac DV connected to baffles directed to the stage) and one for the projections (Apple PowerBook 3400 connected to a video projector).

Qualitative Indications

From the observation, the activity of "Little Red Riding Hood" enabled by the beamer, torch, and setting tools, was generally successful. This role prototyping phase allowed us to provide designers with first comments of concepts refinements and tool functionality. For example, teachers proposed using the projection on the floor to facilitate children entering the scene and to stimulate them to move more freely in the space around them.

At a pedagogical level, we noticed other interesting effects produced by projections and sounds. For instance, children who actively participated in the story building knocked on the virtual door when they heard the bell or performed movement pretending to be cooking in the virtual kitchen. These few elements showed how much the children felt involved in the context (images, sounds, and lightening) and, by extension, story building was facilitated. Children wanted to play the character, stayed longer on the stage, and did not want the activity to come to an end (compared with other children who did the same activity without the POGO-tools). Moreover, children in the audience paid particular attention to what was happening in the scene; they wanted to participate in the story. According to the teacher responsible for the activity, this greater degree of motivation from the students had positive consequences at the pedagogical level. For instance, back in the classroom children were willing to write their story and tended to write more than usual with more descriptions elaborated more spontaneously. At the level of creative imagination, it was clear that sounds and images inspired the narrative and expression process because children were using them when they lacked inspiration. For the teacher, the reuse of children's material (drawing, voices) and projections represent technology that can enrich their everyday practice by providing an opportunity to dramatize stories quickly without the time and effort required for the organization of a theater play. Moreover, even the rough voices and sound used to dramatize characters were effective in facilitating children's emotional expression.

TABLE 24.3
Summary of Requirements in the Early Interaction Phase

Requirement	Allow autonomous production of content starting from scratch, facilitating seamless integration with existing tools and physical settings. Enhance the feeling of control of content production and manipulation by teacher. Results in inexpensive and economical content productions and facilitates the chance to share best practices among teachers.
Rationale	The teachers feel more in control of content production if they can build the resources along the pedagogical activity. An autonomous production of content (instead of using a library of resources) has individualized pedagogical value. Producing content is also economical and allows the involvements of available tools not used in pedagogical activities. The modular availability of the content produced allows teachers to exchange their best practices easily.
Example	The clothes, their colour and patterns, and even eyeglasses were used in producing a scene (video-clip associated).

Seven role mock-up testing sessions were performed, all using normal pedagogical activities scheduled in our two schools. The results of this testing were used to refine the requirements and to specify the basic set of working prototypes to be developed in the first version of POGO world. For example, Table 24.3 describes refinement of the production of content in the early phase of interaction between children and teacher.

POGO WORLD: WORKING PROTOTYPES

The first generations of Active Tools and virtual story POGO World are illustrated in Figs. 24.6 and 24.7. The environment is composed of the following elements: physical walls where the virtual story is developed and four active tools: Cards, Torch, Bucket, and Camera Table (Fig. 24.6). In this prototype, the physical walls are simple screens where story elements are displayed.

The cards are media to exchange physical story elements such sounds, pictures, and video clips. They are a "memory" of story elements that can be associated to real-world objects by physically attaching the card on drawings, clay models, or toys. Cards contain a unique ID tag and are used as physical pointers to virtual story POGO World elements. In the first release, cards are simple stand-alone elements, but a number of attachment mechanisms have been envisioned to integrate them into clothing, accessories, and other artefacts.

The bucket is a container (or alternatively, a support if inverted) that reads background story elements stored within the cards. Dropping cards into the bucket (or locating cards on the top of the inverted bucket) will activate background images and related sounds in mixed media combinations. Children can create background images (drawings, collages, and compositions of elements picked from the real world), or they can be selected from a database. The bucket provides continuous output, so even if there is no card in it, a live video image is shown as background. In a sense, the live image allows children to "perform" a story in the real world on a virtual background.

The camera table captures new story elements such as real-world objects and people (including the children) or live video. It is a base unit integrating a video camera, a card reader, and a composition area. In the composition area, drawings can be positioned, objects can be placed, and collages can be created. The tray that is used as container of the composition area can be used to collect real objects such as sand, leaves, and shells. The camera can be used either to capture these elements as a simple static shot or in live video mode, in which the images are directly projected on the walls.

The torch reads foreground story elements by layering them on the virtual story background. It is a wireless device enabling children (and teachers) to explore the elements associated to the cards. Using the torch, foreground elements are activated and brought to life (sounds are triggered, and images are displayed on the screen).

FIG. 24.6.   From left, the cards, the torch, the bucket, and the beamer.

FIG. 24.7. The POGO World.

THE CONCEPT–MOCK-UP–TEST–PROTOTYPE–CONCEPT CYCLE

The POGO prototype was tested in both schools, in Siena and Bruxelles; in these tests, we considered not only the role but also the look and feel of the program. We began to consider the actual interaction modalities with the POGO tools. The testing was organized using macro and micro scenarios. The macro scenarios were constructed with the teachers according to their scheduled didactic activities; in essence, a macro scenarios was a planned activity carried out using POGO. The micro scenarios were determined primarily by the design team and focused on specific aspects of interaction; these were carried out with a sample of children, teachers, and researchers. Whereas in the macro scenario the role aspect of the prototypes is dominant, in the micro the interaction modalities are the focus. The results of the testing were reported both as edited video clips of the activity analysis and as further specification of requirements.

From this test, we did not move on toward a refinement of the prototype but returned to the concept design and the mock-up activities. That is, instead of focusing our attention on improving the interaction modalities (which, by the way, badly needed some improvement; for example, there was no way to know what was on a card until it was in the bucket or in contact with the torch), our effort was directed toward the further exploration of the conceptual design because from the testing it was clear that the narratives activities and the related cognitive process could be extended and transformed more then was expected. This was considered a clear sign that the designed POGO tools were not just the rough implementation of some interesting functions, but were embodied conceptual tools that were mediating the cognitive activity of teachers, children, and even designers. Thus, a new set of concepts and related active tools was envisioned and the related mock-up produced and tested again in the school (Figs. 24.8 and 24.9). Again, from the results of the mock-up testing, we developed a second set of POGO prototypes (Fig. 24.10) that were tested at schools. In this second set of prototypes, we collected evidence that there was room for extending the conceptual design. It was clear to the design team, however, that it was more important to consolidate the new cognitive practices

FIG. 24.8.   Children and a teacher playing with POGO world.

FIG. 24.9.   Left: mock-ups of the mambotool and the carpettool. Right: mock-ups of the serpentool and the new cards.

FIG. 24.10.   Left: a general view of POGO world. Right: a new version of POGObeamer.

by designing more robust and reliable interactions then to follow the heuristic promoted by the new conceptual tools. The teachers and the children wanted to carry out their narrative activity through the mediating role of POGO.

CONCLUSIONS AND FUTURE DIRECTIONS

Our challenge in creating POGO was to design innovative technologies for children that would be attractive, fun, and long-lasting, yet offer sound pedagogical learning opportunities. In other words, POGO should foster and mediate human cognitive activities inspired by well-established pedagogical principles. The aim was not to fix or repair inefficient school practices;

indeed, we did not focus our activity analysis on problematic processes. We focused on the most successful practices that teachers experienced in the classroom. The results were encouraging. POGO world does not replace any of the current tools that teachers successfully use in their narrative practices; instead, it empowers some of these tools and integrates them with new opportunities.

POGO world is evolving as a way to support the five pedagogical objectives, yet improvements are still possible, as was evident from the testing. Likewise, important issues remain that to be addressed. But most of these are at the interaction level, that is, at the level for example, where intentions are mapped into action, where actions take on shapes that evoke their meaning on the interface, where the interface reacts promptly and consistently to the action, and where the modality and form of the feedback is meaningful for the user. But POGO had the primary goal of becoming a new semiotic tool; it aims to foster new modalities to organize and communicate knowledge among children involved in narrative activities. In this sense, POGO world was a successful environment both for the teachers and for the students.

POGO tools allow rich sensorial interaction where reality can be explored, decomposed, built, and analyzed. The existing objects or the new ones produced by children can be captured and edited in real time through simple artefacts. What a child builds or brings in can be combined with the products of other children in a continuous negotiation process in which the evolution or transformation of the objects is recorded and moves back and forth along this process of meaning construction to be used as a way to understand the other point's of view. The product of a single child or the output of cooperation among children becomes real in several ways. They become real by being embodied in the settings, but also in the card; children can be immersed in that reality together. Moreover, the physical objects that were produced along the way during the iterative and combinatory activity remain living features of the product and can be used as the physical address for the articulated production of creative activity.

We now provide an outline of the lessons learned from this design experience and point out how the same guidelines that we propose are currently applied in a more critical context, Air Traffic Management.

Lessons Learned

Vision.   Nurture the building and sharing of a vision among members of the design team. Continue building the vision during concept design and activity analysis, but be sure that the vision is sound before the implementation phase. It is difficult to overestimate the importance of creating and sharing a sound vision of the project, especially when a team wants to promote a cognitive vision of the design process. The effort dedicated to the vision is compensated for with better communication among the diverse experts on the team and with more effective results produced by the directions of the project leader, to cite the more practical effects.

Theoretical Framework.   Try to ground the vision in a theoretical framework that can provide heuristic guidance to the concept generation phase and to the evaluation of the project's products. A theoretical framework allows the team (a) to be explicit concerning the potential cognitive implication of the artefacts that are produced in the design process and (b) to keep the main cognitive assumptions that inform the project clear and stable.

Concurrent and Interactive Concept Design.   Begin the concept design as soon as the vision starts to form. Use the user research to inform the concept generation, and use the concept generation to promote the vision of the project. Bring the vision to the users. Produce enough concepts so that most of them can be discarded without problems. It is hard to overestimate the importance of producing plenty of concepts. If there are only few concepts, critiquing and modifying them is difficult and time-consuming because the author will defend them beyond reason according to the well-known principle that any concept is better then no concept. At the same time, criteria for accepting or discarding concept need to be established; the more salient one should come from user research and the related activity models.

Activity Model of Both Best and Worst Activities.   Produce activity models of human practice that will be mediated by future designed artefacts. We stress the need to produce activity models rather than user models, because the latter are too narrow in perspective and too time-consuming to be of any realistic value in design. We want to describe a task not to conserve what users do but because we want to change what they do. The description of task and user should be heuristically positive so as to inform the concept generation phase and to define heuristic requirements. We proposed a drastic, "quick and dirty" approach to the description of users and tasks. We included in the description goals, artefacts (materials such as tools, space, etc., and concepts as procedures, practice, ethics, etc.), and people. The key for accepting a description of an activity as a model is its capacity to cluster scenarios and facilitate communication among the design team. If the proposed model description of the activity is effective in providing guidance in collecting data from the field (scenario) and in clustering the observation, it is a good candidate for an activity model.

Progressive Requirements Definition Expressed in Relation to Activity Models. State requirements as soon as activity models are available and relate requirements to specific aspects of a scenario. Focus first on conceptual requirements, those directly related to the cognitive processes, but consider worthwhile only requirements that can be presented in (or at least explored through) mock-ups. Discuss and refine requirements with people on the design team involved in concept and implementation, then submit the requirement to users. Learn from this what aspect of the activity description should be better specified in the activity model. Loop in requirement definitions until they can be expressed immediately as mock-ups.

Parallel Mock-Up of Role and Look and Feel.   Produce in parallel mock-ups of both role and look and feel. Do not worry about prototyping the same concept or requirement; role prototypes address aspects completely different from look and feel prototypes. The more mock-ups you have of different aspects, the easier it will be to find a sensible mapping between activity model and concept—working on possible solution is better than working on problems. Do not worry about implementation in the early phase of mock-up but start benchmark analysis of enabling technologies. According to the selected functionalities of the mock-up explore what kind of technological solutions are available and report them in the concept, activity model, and functionality matrix.

Testing Sessions Involving the Whole Design Team.   Carry out some of the testing sessions with the whole design team using mock-ups and keep the session informal. Do not worry about formal testing until reaching advanced versions of the prototypes; instead carry out "quick and dirty" testing session, but do them directly in the field with the whole design team or at least with representatives of the different competencies (industrial designer, computer scientist, electronic engineers, content designer, interaction designer, etc.). What is really important is the feeling of development in the way in which the activity is carried out by the users or by the designers. Remember, the embodiment of external representations in psychological activity permits completely new and unpredictable behavioral patterns.

Testing With Macro and Micro Scenarios. Organize your testing session with respect to macro and micro scenarios. Sketch the micro scenarios after the macro scenarios. Macro scenarios should represent whole, meaningful activities such as the one selected to build the activity model. They should include the entire range of tools, actors, and settings that the activity demands. You can do this both early in testing using mock-ups and methods like the Wizard of Oz or later using working prototypes. In this testing, be concerned first with envisioning the power the mock-ups and prototypes have. The core issue is how well they enable the evolution of the activity.

Micro scenarios should concentrate on specific aspects of the interaction in a specific context. Micro scenarios can be used to assess solutions that concern similar interactions in different macro scenarios (activities). This will allow refining of the interaction modalities and of the look and feel of the prototypes.

Creative Research Environment for ATM

Even if our design experience was carried out for the design of an educational environment, we believe the proposed guidelines and principles can be applied in many contexts and application domains. Indeed, the proposed approach is currently applied in the air traffic management (ATM) domain. Some of the authors are involved in CREA! (Creative Research Environment for ATM), a project funded by Eurocontrol within the CARE Innovative Action Programme.

The project aims to develop innovative concepts for ATM, ranging from the current air traffic control scenario, in which control is centralized and managed by control units on ground, to a more futuristic scenario, the free flight, in which traffic management is decentralized and control is moved on board to the aircraft. The approach adopted is challenging because the exploitation of design practices and methods from disciplines different from engineering and human factors is largely unexplored in ATM. All innovation in such domains is constrained by a problem-solving view, neglecting other factors such as aesthetic, affective, cultural, and emotional aspects of human cognition. But the design of cognitive tasks and human activity can introduce a new way of approaching even safety-critical systems.

The Crea! project was recently started, and we are moving through the initial steps of the cognitive task design process. In the following, we report how the activity in the process was articulated along the guidelines for cognitive task design illustrated in this chapter.

We started by building a clear vision of the process to reorient design choices whenever the contributions coming from the different disciplines involved in the process tended to develop only partial aspects of cognitive task design. The vision plays a fundamental role in maintaining the focus of the project and in exalting the centrality of the human activity in cognitive task design. The vision was articulated along three hypotheses of transition from the current ATM scenario to more visionary and futuristic scenarios:

1. The control scenario: the current situation in which controllers act mainly as air traffic managers.
2. The advisory scenario: the future free-flight situation in which the responsibility of traffic management is mainly moved on board and the controllers play an advisory and support role for the pilots.
3. The service scenario: a completely futuristic scenario in which traffic management is decentralized and totally moved on board. In this scenario, the controllers play the role of connecting air and ground services. They becomes "service and information brokers".

The vision focuses on solutions that can be appropriate in the first two situations (control and advisory scenarios) and that can be partially implemented in a short term and partially developed further in the medium term.

As an integral part of our vision, we decided that our solutions should have addressed the following objectives:

- To support personalization and individual styles of work (individual differences and cultural differences are fundamental cognitive properties of human beings)
- To guarantee a certain level of transparency of the processes and of the related activities (construction is a nonstop cognitive process)
- To use multiple channels and reduce the information overload on the visual interfaces (human communication is mediated by multiple and parallel sensorial modalities)

With the aim of building a theoretical framework for inspiring the cognitive task design process, we decided to start to create a collaborative environment in which artists, designers, and technologists could work as equal partners in the exploration and development of innovative concepts and media for ATM. Each participant brought to the group his or her own expertise, both in terms of ideas and of theoretical background. For example, the interaction designer defined a conceptual framework based on the ideas of knowledge exchange and identity recognition, sensorial experience based on active manipulation of the working tools, and a mixed reality, reactive environment. The framework was articulated around key issues, namely, the physical experience of the space, memory creation, awareness mechanisms, multiple point of views of reality, and situated information.

The architect set up a framework in which the characteristics of the space were defined in the following terms:

- Openness and horizontality, to facilitate the visual interaction among the different people
- Flexibility and adaptability, to permit interoperability and task-driven configuration of the environment
- "Theatre setting", so that the disposition of the different elements on the scene and the role of people contribute to the fluidity of the processes.

All these views are currently in the process to be harmonised into a single theoretical model that will be consolidated as soon as activity models are defined. It is important to state here that CREA! adopted a full co-evolutionary process of cognitive task design in which concept, technology, and cognitive activity design are carried out in parallel so that each strand of the process can inform the others. CREA! interpreted the guidelines for cognitive task design defining a co-evolutionary process articulated in two main phases: divergence and convergence.

Divergence included inspiration to gain insight from the application domain (user and domain understanding) and elaboration to develop concepts from the perspective of single disciplines (music, architecture, interaction design, visual design, human factors). Convergence included sharing to present and confront concepts elaborated separately (concept testing) and production to evolve single concepts in integrated "concept scenarios" (scenario testing and briefing).

At the moment, the convergence phase has just begun, so we do not have consolidated, detailed concept scenarios to test with the users. However, the choice of implementing this particular co-evolutionary process demonstrates that guidelines for cognitive task design are flexible and that the principles can support the requirements of specific projects like CREA!

The preliminary outcomes of this cognitive task design process produced a set of concepts that are briefly sketched in the list that follows. We develop five new elements/tools (Marti and Moderini, 2002):

- The flag, a public display that can be used to extend the action and communication area of a single user and support awareness by visualizing the processes

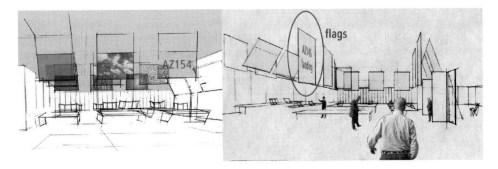

FIG. 24.11. Two views of the flags concept.

- The trittico, a multiple display tool integrated in a table/desktop that constitutes the main interface for the operational work
- The hard book, a personal tool that permits both a focus on specific processes and the sharing of information and activities with others
- The cameo, a personal ID tool that allows both set up of the system according to the individual preferences and management of the communication processes
- The osmosis skin, flexible and adaptable partitions to organize working and resting areas

As soon as the process define ATM activity models, they will be mapped on the proposed tools to define progressively both functional requirements and system specifications (see Figs. 24.3 and 24.4 for reference).

We started by communicating our five concepts to air traffic controllers. Again following the cognitive task design process, we are developing simple mock-ups that represent only the vision and the main characteristics of the concepts without being overly concerned with the look and feel of the tools. This allows us to concentrate on the role of our concepts without constraining air traffic controllers to detailed solutions at the level of look and feel. Examples of mock-ups are illustrated by the Fig. 24.11, which shows the sketches we used to present the concepts of "flags" to air traffic controllers.

We believe that there are enormous opportunities for designers to amplify the creative design process. Thus far the design process has not explicitly recognized the implications of new conceptual and physical tools for the shaping of the human mind, but today the design disciplines must design not only for aesthetics and functionality but also for the full range of human experience.

REFERENCES

Bruner, J. S. (1990). *Acts of meaning*. Cambridge, MA: Harvard University Press.

Bruner, J. S. (1996). *The culture of education*. Cambridge, MA: Harvard University Press.

Domus Academy. (1999). *Concept design*. POGO deliverable no. 99007/v.1.

Domus Academy (2000). *Specifications of the functional components*. POGO deliverable no. 00003/v.1.

Erdmann, R. L., & Neal, A. S. (1971). Laboratory vs. field experimentation in human factors. An evaluation of an experimental self-service airline ticket vendor. *Human Factors, 13,* 521–531.

Houde, S., & Hill, C. (1997). What do prototypes prototype? In M. Helander, T. Ê. Landauer, & P. Prabhu (Eds.), *Handbook of human–computer interaction (2nd ed.)*. Amsterdam: Elsevier Science.

Marti, P., & Moderini, C. (2002). Creative design in safety critical systems: A methodological proposal. *Proceedings of European Conference on Cognitive Ergonomics*. S. Bagnara, S. Pozzi, A. Rizzo, P. Wright (Eds.), Design, Cognition, and Culture, Consiglio Nazionale delle Ricerche, Istituto di Scienze e Tecnologie della Cogniziome. Roma, Italy, ISBN: 88-85059-13-9. 11.

Ong, W. (1985). *Orality and literacy*. London: Routledge.

UniSiena & UniLiegi. (1999). *Narrative and learning: School studies*. POGO deliverable no. 00001/v.1.

UniSiena & UniLiegi. (2000). *Concept testing on site*. POGO deliverable no. 00004/v.1.

Vygotsky, L. S. (1978). *Mind in society: The development of higher psychological processes*. M. Cole, V. John-Steiner, S. Scribner, & E. Souberman (Eds.), Cambridge, MA: Harvard University Press.

Vygotsky, L. S. (1998). Imagination and creativity in childhood. In R. W. Rieber (Ed.), *The collected works of L. S. Vygotsky*. New York: Plenum.

25

Approaches to Transportation Safety: Methods and Case Studies Applying to Track Maintenance Train Operations

Kenji Itoh
Tokyo Institute of Technology, Japan

Masaki Seki
Central Japan Railway Company, Japan

Henning Boje Andersen
Risø National Laboratory, Denmark

Abstract

In this chapter, our objective is to describe several methods for addressing transportation safety with an illustration of case studies that have been applied to the operations and organisation of track maintenance trains for a Japanese high-speed railway. We discuss the following six methods: experiments, observation of operator tasks, probabilistic risk assessment and human reliability analysis, cognitive simulation, accident–incident analysis, and questionnaire-based surveys. For each method, we offer a general description of its purpose, main characteristics, benefits, limitations, and the steps involved in its application. Then we describe three case studies of track maintenance train operations and their associated organisations. Throughout the review of the case studies, we seek to bring out implications from each of the methods to the design of cognitive tasks, including safety management issues in organisations.

INTRODUCTION

Background and Objectives

Human factors have become of increasing concern in the modern technical workplace, such as in production facilities, power plants, airplane cockpits, train cockpits, or ship bridges, where large-scale, high-tech human–machine systems are operated. This is mainly because *human error* is the single most common cause of accidents in modern technical work environments. The rate of human error is known to be greatly affected by human factors such as skills, knowledge, experience, fatigue, stress level, and vigilance, as well as by task factors and design of the work environment, including the human–machine interface. Thus, there is extensive evidence that human errors contribute massively to the rate of incidents–accidents in industrial applications.

Regarding the potential risks of human errors resulting in accidents, Hollnagel (1993, 1998) reported that the proportion of cases in which human actions are the causes of accidents has been increasing over the past 15 years. He suggested two major reasons for the growing rate of accidents caused by human errors. One reason is that human error is now focused on as a salient source of accidents to a much greater extent than before. An event or phenomenon that was previously considered to be a "purely" technical fault is now seen as the result of a failure in the maintenance and design of the system and is therefore now categorised as a human error made by maintenance personnel or designers (Reason, 1997). The other reason is related to advances in engineering and technology. Technical components and subsystems have gradually attained very high levels of reliability, and whole systems are designed to be fault tolerant. Thus, the accident rate caused by human error is becoming relatively higher in comparison with that caused by system failure. This is particularly apparent in aviation, where the rate of accidents has dropped dramatically over the past 50 years—as a result of the increased reliability of modern aircraft—but where the proportion of accidents in which "crew factors" or maintenance errors are the major cause has increased (e.g., Amalberti, 2001).

However, not only are modern human–machine systems vulnerable to human error, but *organisational factors* also play a crucial role for their safe operation. Organisational problems are frequently latent causal factors that contribute to the occurrence of human error made by frontline personnel. Thus, it has been pointed out that the majority of causes contributing to major accidents may be attributed to the organisations themselves that shape the *safety culture* or climate within which the employees operate (Hee, Pickrell, Bea, Roberts, & Williamson, 1999; Reason, 1997). For example, it was reported that 40% of the incidents in the Dutch steel industry were caused by organisational failures (van Vuuren, 2000).

There are many proposed definitions of safety culture or climate (e.g., Advisory Committee on the Safety of Nuclear Installations [ACSNI], 1993; Flin, Mearns, O'Connor, & Bryden, 2000; Pidgeon & O'Learry, 1994; Zohar, 1980), and it is difficult to define or explain this concept precisely. One of the more succinct definitions was presented in a report by ACSNI (1993), which stated "it is the product of individual and group values, attitudes, perceptions, competencies and patterns of behaviour that determine the commitment to and the style and proficiency of an organisation's health and safety management" (p. 23). According to this definition, safety culture is coupled not only to management's commitment to safety, its communication style, and the overt rules for reporting errors, but also to employees' motivation, morale, their attitudes towards management, and their perception of factors that have an impact on safety, such as fatigue, checklists, and violations of procedures (Andersen, 2002). This suggests that a healthy safety culture not only will contribute to a good working system but also will bolster positive human factors elements such as morale and motivation. In turn, this is likely to lead to more reliable operations in which human errors are captured before they have an impact on safe operations.

Our objective in this chapter is to review a set of methods for tackling safety issues in transportation and to illustrate their use in terms of case studies. We aim at describing each method—so that readers can use it for their own applications—by describing its procedures, its main advantages and limitations, and its implications for cognitive task design. In each case study, different but interrelated methods are applied. In other words, to improve the safety of operations in our target domain, we applied different methods from various points of view to tackle either single issues or a set of related ones. We assume, from a practical point of view, that cognitive task design focuses on methods to support the design of social and technological artefacts—such as interfaces, devices, human–machine systems, rules, procedures, training programmes and organisations—all of which have to have knowledge acquired, organised, and put into use to ensure efficient and safe operation. The design activity is supposed to be composed of planning, analysis, formulation or composition of functions and structures (design in the narrow sense), evaluation or test, and improvement or maintenance, directed at one's target artefact. Thus, as an application field of the cognitive task design, this chapter focuses on track maintenance trains operated to maintain tracks and rails for the Japanese high-speed train.

Approach to Railway Safety

As we already mentioned, human errors are involved in a large proportion of accidents in industrial and transportation domains. However, it is well known from accident cases in aviation, shipping, and nuclear accidents that a single machine or system failure will rarely by itself cause a catastrophic outcome (e.g., Pidgeon, 1991). Nor will a single human error by itself result in a catastrophic or critical accident. That is, an accident is typically caused by a series of independent multiple human errors, and the accident would not have occurred had any one of such errors been captured or avoided. This mechanism of error causation is the same in railway operations.

It has also been widely realised that it is impossible to get rid of human errors completely, although we can reduce their rate of occurrence: "to err is human" (to echo both Seneca and the title of a recent influential report on medical error—see Kohn, Corrigan, & Donaldson, 1999). Therefore, it is of primary importance to establish mechanisms that allow us to capture or to escape from a series of human errors in a given work system, for example, by means of foolproof and fail-safe functions. Thus, we must develop work systems that will work safely even though operators make mistakes.

In addition to the design of error-resistant mechanisms, management efforts are required to improve or reinforce the safety culture of the surrounding organisation because it is impossible to "design away" all the potential causes of accidents in advance. Nobody can create a foolproof system unless he or she knows the cause–effect relations of all possible accidents. Even so, safety depends not just on the hardware and the written procedures, but also on the actual practice of operators and management. A healthy safety culture will include practical actions for reducing and capturing human errors and will encourage operators to establish and maintain a high professional morale and a strong motivation. Such actions may involve the provision of good working conditions, leadership, training, and manuals and checklists, thereby facilitating effective cooperation and communication to ensure a high quality in the usability of human–machine interfaces and equipment and tools.

Methods Applied to the Establishment of Safe Operations

There are three aspects involved in the design of cognitive work for enhancing safety and productivity: cognitive, operational, and organisational aspects. The cognitive aspect focuses on an operator's *individual* cognitive functions and processes for the purpose of analysing a task, working out a plan or diagnosis, or devising a measure for improving operational safety. In contrast, operational safety is analysed and evaluated, and possible suggestions are made for its improvement as a *global* task carried out in terms of *teamwork* and through cooperation and allocation of tasks among team members. In the following paragraphs we analyse work systems and discuss actions and activities for enhancing operational safety by focusing on the organisational aspect. Application of this aspect to work and organisation design includes issues such as training, manning, team organisation, work planning and scheduling, work rules, working procedures, incident and error reporting, manuals and checklists, personal feedback (evaluation), and management communication (to mention just a few).

TABLE 25.1
Approaches Applicable to Transportation Safety

Methods	Cognitive	Operational	Organisational
Experiment	YY	Y	
Observation	YY	YY	(Y)
PRA–HRA	Y	YY	(Y)
Cognitive simulation	Y	YY	
Accident–incident analysis		YY	Y
Questionnaire	(Y)	Y	YY

Note. YY = best match; Y = applicable; (Y) = partly applicable.

There are several methods and techniques that apply to analysis and intervention measures targeted at safety critical domains such as railway operations. The following are the ones that have been most frequently applied to the safety issues of human–machine operations: controlled experiments, field observations, probabilistic risk assessment (PRA) and human reliability analysis (HRA), cognitive (modelling and) simulation, accident–incident analysis, and questionnaire surveys. Each method has its own suitable and efficient application(s) of analysis and practical use. A comparison between these methods and the aforementioned aspects is roughly summarised in Table 25.1.

GENERAL DESCRIPTION OF METHODS

Method 1: Accident–Incident Analysis

An intensive study of a single case can yield valuable information about the circumstances leading up to critical accidents. Where sufficient evidence is available about both the antecedent and the prevailing circumstances of a particular adverse event or accident, the interaction of the various causal factors can be investigated over an extended time scale in a way that would be difficult to achieve by other means. In addition, by combining the knowledge obtained from accident analysis with a more adequate theory of human performance, we can extend our knowledge of the human error mechanisms in performing cognitive functions required for accomplishing the target tasks. As mentioned previously, accidents are typically preceded by the unfortunate conjunction of several distinct single errors. The precise effects of a particular combination of human errors teach us something about the limits of human performance. Thus, the lessons learned from a specific *accident–incident analysis* can be expected, if implemented, to reduce either the occurrence of errors or their harmful consequences.

To manage this analysis efficiently when—and preferably immediately after—an accident has taken place or an *incident report* has been submitted, an organisation must have an effective incident-reporting system. This is because the report will always contain less information than was potentially available. A written account will contain a series of selected, often simple, discrete samples from the original, complex, and continuous series of events. The reporting system prescribes a procedure for submitting a report when an accident or incident has occurred, including the type and format of information to be reported and the feedback to the reporting employee. The need for further analysis is determined for each reported case on the basis of its severity and its potential for serving as an alert for further potential accidents. If the case is investigated, a process of detailed analysis is commenced. To perform this process efficiently, it is useful to have procedures for obtaining additional, detailed data sources and for analysing causal relations between the accident–incident, any possible human errors, system failures, and other relevant factors. Analysis results applying such procedures may allow domain experts to develop countermeasures to eliminate a specific accident potential in the work system (e.g., an improper procedure or an ill-designed task) or to implement measures that reduce or capture human errors that may lead to a similar accident. The resulting investigation report will include, in addition to details about where and when and persons involved, factors directly and indirectly linked to the accident, a timeline of actions and events leading to the accident, and, finally and most importantly, a summary of the "lessons learned," including recommendations for changes to prevent the same kind of accident from happening again.

Like any other method, the accident analysis approach has some limitations and prerequisites for its application. A major source of data for this analysis is typically the statements of the operators involved in the adverse event, as well as logs and physical data. In relation to accident analysis, it is often mentioned that management style will influence the quality and quantity of the verbal data that are included in incident reporting. There are two extreme styles: "management by freedom" versus "management by punishment." In the former management style, operators are encouraged to tell the truth about human errors that they may have made, even if they did cause an adverse event. In contrast, in the latter style, workers or operators will typically be blamed and sometimes sanctioned or punished for having been causally involved in an accident or incident. Therefore, they are likely to tell a story that may be inaccurate or incomplete, even when a report is prepared by experienced and open-minded investigators.

Method 2: Observation

It is common to obtain many varieties of data through the *observation* of an actual task. In applying observational methods, it is required to clearly describe a task scenario. The scenario will include descriptions of operators, dates and times, tasks, working conditions and work setting, equipment, recording data, and so forth. The data that are most frequently recorded by observation are human actions and behaviours, verbal protocols, eye movement and fixation, system log, and operators' psychological and physiological data during task performance. Recorded data are analysed, for example, to uncover operators' cognitive processes, interaction between human and machine or between humans, to examine critical factors affecting task performance and safety, or to obtain insights about the human–machine interface. Several modalities of data are often recorded at the same time by using video recorders and additional equipment for analysis of cognitive operations in a modern human–machine system. Exploratory sequential data analysis tools such as MACSHAPA (Sanderson & Fisher, 1994) are useful for data recording and analysis of multimodal synchronised data from various sources (Andersen & Hansen, 1995).

In safety research into human–machine operations, it is of critical importance to analyse cognitive tasks performed in high-risk or emergency situations. It is scarcely possible to observe a task preformed in such a situation in an actual work setting, and it is unethical to stage an emergency; therefore, simulators are highly useful for data recording under high-risk conditions. In most safety critical technical domains there is a growing use of simulators—such as flight, ship, car, train, nuclear power plant control, and anaesthesia simulators—for the purpose of training and investigation of system safety (e.g., Itoh, Hansen, & Nielsen, 1998).

In observing tasks performed in a simulator, however, one needs to be aware of potential problems of fidelity. Thus, operators will very seldom become incapacitated by fear and panic in a simulator—though there are some relatively rare cases—but they do so in real life.

Method 3: Experiment

The most powerful and reliable method in examining the effects of particular factors on human errors as well as on human performance is an *experimental study,* carried out in a real work setting or under controlled laboratory conditions. This approach is very effective in trying to uncover causal relations between experimental factors and human errors. In particular, and in this respect similar to observations as just described, simulator experiments are of considerable help for human error studies.

In contrast to these advantages, experiments are usually very expensive to perform in terms of time and costs. As researchers we must prepare, sometimes laboriously, an experimental setting to mimic the actual work environment. In addition, repetitive trials are required with a number of subjects to obtain reliable and comparable data. There may also be limitations in the extent to which task conditions can be controlled in terms of scale and complexity. With regard to these difficulties and limitations, Reason (1990) pointed out the following two problems associated with experimental investigation of human error in work settings. First, it is necessary to establish precise control over the possible determinants of the error. This will often force investigators to focus on rather trivial phenomena. Second, the greater the measure of control achieved by the experimenter, the more artificial and unnatural are the conditions under which the error is elicited.

Method 4: PRA–HRA

The most common method for estimating accident risks and human error rates is the *PRA* or the *HRA*. With this method, one models a task sequence, usually by using a fault tree, an event tree, or a similar form. Subsequently, using predefined human error–reliability data, one combines individual subtasks or component tasks according to the task sequence model to estimate the error probability for the entire task. But one of the biggest problems with this method is the lack of databases on human reliability or human error probabilities. To make the method useful for a wider application, one must collect and record human error data for many varieties of tasks in various industries. Major steps of the HRA for the purpose of performance (risk) prediction are described as follows (e.g., Reason, 1990):

1. Define the system failures that may be influenced by human errors and for which probabilities are to be estimated.
2. Identify, list, and analyse human operations performed and their relationships to system tasks and functions of interest.
3. Estimate the relevant human error probabilities by using a combination of available data and expert judgment on performance shaping factors.
4. Determine the effects of the human errors on the system failure events.
5. Recommend changes to the system for reducing the system failure rate to an acceptable level.
6. Steps 2–4 should be repeated to evaluate the changes.

One of the most famous and the most widely used first-generation HRA techniques is the *technique for human error rate prediction (THERP),* developed by Swain and his colleagues (e.g., Swain & Guttman, 1983). This technique can be used to generate quantitative estimates of task reliability, interdependence of human activities, the effects of performance shaping factors, equipment performance, and other system influences. Like other first-generation techniques, THERP uses an event tree representation, called a "probability tree diagram," as a modelling tool of an accident sequence. This diagram is described in a binary tree in which two branches at each decision point indicate the correct and the wrong actions. In addition, labels of human activities as well as their conditions, that is, levels of relevant performance shaping factors, can be attached to the branches. The THERP provides 27 tables concerning nominal human error probabilities of specific human activities (Swain & Guttman, 1983).

Besides the lack of human error data, it is widely recognised that the first-generation HRA techniques have several shortcomings, such as inadequate treatment of performance shaping factors, no account of human performance in context, insufficient classification scheme of erroneous actions, and no operator model (e.g., Hollnagel, 1998; Swain, 1990). Thus, second-generation HRA techniques have been developed to overcome these criticisms. For example, in one of the representative second-generation HRA methods, CREAM (cognitive reliability and error analysis; Hollnagel, 1998), the classification scheme describes both effects, that is, error modes, and causes of erroneous actions. CREAM defines eight general categories of error modes: timing, duration, sequence, force, distance–magnitude, speed, direction, and wrong object, and it largely classifies antecedents of erroneous actions into three types, that is, person-, technology-, and organisation-related genotypes, each of which comprises several categories. In addition, the context is analysed by using the concept of "common performance conditions" that were also described by their relationships with main genotypes (cf. Hollnagel, 1998, for details).

When these features extended from the first generation are applied, it is of great importance to emphasise the qualitative analysis of cause–effect relationships in the HRA approach. The qualitative analysis is composed of the following steps (Hollnagel, 1998):

- Describe the context or common task condition. A detailed analysis may be required to determine the context.
- Describe the general possible error modes. The description of the error mode should be provided for all possible actions.
- Describe the probable causes. With reference to expert knowledge about the context, categories of causes could be identified.
- Perform a detailed analysis of main task steps. This process is carried out to identify the possible consequent–antecedent links.

Method 5: Cognitive Simulation

A fairly recent major development with applying the *simulation* approach has been promoted for its advantages in cost and time savings when compared with the experimental study. This approach requires modelling either the human cognitive performance for the task under study or a causal relation between human errors and error-relating factors (Cacciabue, 1998). Once such a model is built up, a person-in-the-loop simulation can be performed. A series of simulation runs allows us to apply, for example, predictions of error rate or accident occurrence under any given condition to perform a risk analysis. The great appeal of this approach is that computer-driven data collection and analysis may be made without any intervention by analysts. Batch processing of computer simulation allows us to estimate operator behaviours and system states under various conditions relatively quickly.

In the simulation approach, the modelling of human behaviour and cognition is of primary interest. The main focus is on developing an appropriate model that represents relevant behaviour and process for each component of the human–machine system, that is, the operator's actions and activities, system behaviours, and the interaction between these two parts at a specific level of description, depending on the purpose of the model. One of the most critical problems in this approach is the difficulty in building a model of human cognition involved in task performance. Only a good and accurate model can produce correct results. However, little is known about human cognitive processes compared with motor behaviour. In addition, many human–machine operations are characterised as skill-based and rule-based performance (Rasmussen, 1986). Therefore, their performance is automated or semiautomated and thus executed unconsciously. This in turn makes it difficult to elicit or infer the hidden cognitive process of the operator. For this reason, an in-depth thorough task analysis is usually required, which in turn requires support from observational and other methods, in order for a cognitive model for the task domain under study to be established.

Method 6: Questionnaire-Based Survey

Questionnaire-based surveys are utilized in a wide range of application areas to assess operators' perceptions of and attitudes to their organisation, management, human errors, and so on. This can be a useful and time-effective method for collecting subjective but uniform data from a great number of respondents. As mentioned at the beginning of this chapter in relation to safety culture, this method has been widely applied to safety analysis of human–machine systems and their organisational aspects.

In general, long questionnaires tend to discourage respondents, so the longer it takes to fill out a questionnaire, the lower its expected response rate will be. As a general rule, questionnaires that are to be filled out on a voluntary and unpaid basis should require no more than approximately a 30-min. effort by respondents. Depending on how much the operators are used to process written material, it may be desirable to restrict a questionnaire to 15 min. or even less, according to the authors' experience. The drawback of achieving a low response rate is not the lack of data per se, but rather that there may now be a systematic bias in the data collected. Thus, with a low response rate (~50% or less), there is a risk that the sample is biased so that responses are mainly drawn from employees who are more likely to respond in a certain way. Thus, those who have not responded may tend to have different perceptions of and attitudes toward their jobs, their organisation, and management. If this is true, the sample is biased and conclusions are therefore likely to be misleading.

Appropriate questions must be prepared to meet the goals of the survey. Each question item should be described so that all respondents will understand it in the same way. It is always a good idea to carry out a preliminary survey with relatively small samples for the purpose of evaluating the questionnaire and developing its revision. Techniques often employed in developing a questionnaire are the semantic differential (SD) method and the subjective rating of questions on the Likert scale. In a questionnaire applying the former technique, each question is represented as a pair of adjectives with opposite meanings, such as democratic versus autocratic. A respondent rates each question on a 5-, 7- or 9-point scale between these two extremes. With the latter technique, each question is rated on a 5-, 7- or 9-point scale referring to levels of agreement and disagreement.

CASE STUDIES IN TRACK MAINTENANCE TRAIN OPERATIONS

Track Maintenance Organisation

There are several high-speed railways in Japan; the major line is Tokaido Shinkansen, which started to run between Tokyo and Osaka in 1964. Today, the maximum speed of the bullet train, as it is called, has increased to 270 km/hr, and the bullet train now covers the distance in

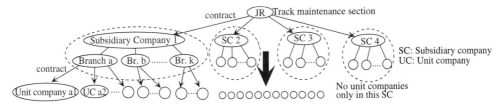

FIG. 25.1. Organisational structure of track maintenance operations.

approximately 2.5 hr. There has been no loss of life in operations on the high-speed rails ever since the start of operations. The railway safety of the bullet train is supported by an efficient organisation and by the high quality of their tracks and rails, which are maintained everyday after the last train and before the start of the first passenger-carrying morning train.

Today, the track maintenance of the high-speed railway is conducted in cooperation with the bullet train company (Japan Railway; JR). There is a four-level hierarchy of track maintenance organisations operating for the high-speed railway. The hierarchy is structured in an organisational tree: the JR is at the top of hierarchy, followed by four subsidiary companies, their branches, and finally a number of unit companies (see Fig. 25.1). The JR contracts track maintenance operations in a specific area to one of its subsidiary track maintenance companies.

The entire line is divided locally into 20 areas between Tokyo and Osaka. Each area has a track maintenance base and its operations are managed by a branch of one of these subsidiary companies. Each subsidiary company has several branches, each of which is responsible for contracted operations, including the driving of track maintenance trains, in a local area. Within each branch of a subsidiary company, there are one or several small independent operational units, called unit companies, each of which employs 10 to 50 operators. Most of the operators are employed by unit companies, but a small number of operators (less than 3% of all operators) are employees of the subsidiaries (and these operators thus work in particular branches but not in unit companies).

Track Maintenance Trains and Their Operations

Although there are several types of track maintenance trains, for example, general purpose trains for transporting materials and equipment (see Fig. 25.2), and track inspection trains, they have almost identical driving operations. These trains are operated under more changeable and stressful conditions than those of regular passenger trains. For example, the track maintenance train has no traffic signals available when it is operated during the night. Therefore, this requires the train crew to make go–stop decisions on the basis of only their own perception and judgment. Unlike the bullet train, there have been accidents and incidents with track maintenance trains, though the accident–incident rate is actually very low in terms of absolute numbers. It is the goal of management to further reduce the number of accidents and incidents of these trains and, by pursuing various approaches to safety, to reach a level matching that of the bullet train.

Driving operations in track maintenance trains are performed by a team of two operators, a driver and a supervisor, in a train cockpit following a prescribed plan and directives and control by station-based traffic controllers. In this task, the driver controls the train within safety bounds, usually just manipulating acceleration and brake levers according to the time schedule and the supervisor's command. In addition to supervising the driver, the supervisor communicates with the traffic controller and personnel in other trains. During most of the time, the driver and the supervisor monitor the tracks and their surroundings. Therefore, the driver and the supervisor rely almost exclusively on their visual perception and attention allocation.

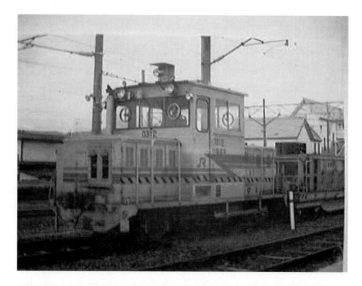

FIG. 25.2.   The most typical, general purpose track maintenance train.

Case A: Analysing a Derailing Risk From Cognitive Aspects

Description of the Case.   In the first case study, the work and performance of track maintenance train operators were investigated from the cognitive point of view (Itoh, Tanaka, & Seki, 2000b). In particular, we analysed their track-monitoring process because this task is one of the most important and common activities for all types of track maintenance trains. For this purpose, we focused on a special train called the "track inspection train," in which a supervisor performs a track inspection task while a driver is taking care of the driving. The major purpose of this task is to ensure that there are no obstacles on the track so that bullet trains can run safely on the rails. When the supervisor finds something on the track during his monitoring process, he orders the driver to stop the train and he retrieves the object.

Using the results of the task analysis, we discussed the risk of derailing when an obstacle on the rails escapes detection. In addition, we constructed a train operator's cognitive-perceptual model of track inspection, based on the task process elicited. We also examined other risks associated with failed obstacle detection, which may lead to derailing of the bullet train, by using a computer simulation of the model, including variations of individual operator factors and operating conditions (Itoh, Seki, & Hoshino, 2001).

Applied Methods to the Case.   We first performed a cognitive task analysis, applying an observational method to the track inspection task. This task, which is performed by operators in a skill-based manner, depends very much on the operators' visual perception and attention to the tracks and their surroundings. Therefore, supervisors' eye-tracking data were recorded to enable an in-depth analysis of the track inspection process. We observed four track inspection sessions, each of which was carried out in the same driving area on a different day with a different scenario. There are actually very few occasions when something is found on the track. Therefore, in the last session, four boxes (15 × 25 × 8.5 cm) were put at different locations dispersed throughout the driving area, and the supervisor's eye-tracking data were recorded to analyse the supervisor's actual monitoring process in detecting an obstacle.

On the basis of the supervisor's monitoring process elicited through the task analysis, we built a cognitive-perceptual model for this task. The model reproduces the operator's perceptual and cognitive processes of obstacle detection with respect to a specific object under given

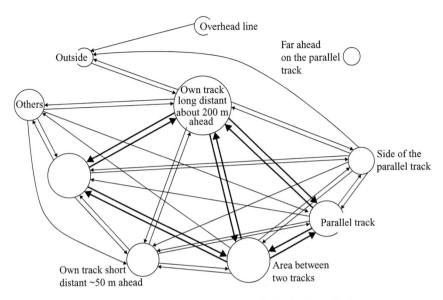

FIG. 25.3.   Attention transition pattern during track monitoring.

conditions. To accommodate several factors that affect the risk of missing (failing to detect) an obstacle, we estimated the detection rates with a number of simulation runs; we applied this model to various operational scenarios, such as change of driving speed, size of obstacle, and its location on the track.

For the sake of generating simulated behaviour of the obstacle detection, the model required a database of the operator's visual perception functions, that is, a detection rate for a different-sized obstacle with a different eccentricity of the visual field. To produce the database, we carried out an experimental study with 10 track maintenance train operators: Each subject identified a visual target from the cockpit of a track maintenance train at a level of luminance identical to the actual operation condition on the track. All the visual targets were rectangle shaped and covered with fluorescent paint on their surfaces similar to that attached on tools used for usual track maintenance work. Four different-sized targets were presented at five different locations 100 m ahead of the subject, that is, at 0, 2, 4, 6, and 8 deg. in eccentricity from the fixation point, ordered randomly. Each subject was asked to report whether he could identify a target or not, while he fixated at a mark 100 m straight ahead between the rails.

General Pattern of Track Monitoring.   We analysed the track-monitoring process by using eye-tracking data from three sessions involving a supervisor who had geographical familiarity with the area—the "home area" supervisor (Itoh et al., 2000b)—and its pattern is depicted in Fig. 25.3. This figure indicates the transition network of the supervisor's attention allocation during the track inspection task. In this figure, the size of each circle represents the percentage of total gaze time at each location on the tracks and surrounding environment. The bigger a circle is, the more attention was paid to that location during the task. The thickness of the arrowed arc between two locations indicates the frequency of attention transition in terms of the relative percentage over the total number of eye movements. The thickest lines in this figure, namely, the arcs between the distant position on the (currently) running track and the side of the running track, and those between the distant position and the parallel track, represent 5% to 10% of the transition over all attention shifts. The thinnest lines represent 0.5% to 1.0% of transition. No line is provided for a smaller percentage of transition than 0.5% in the network.

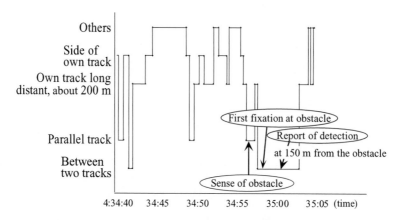

FIG. 25.4. Train operator's eye movement in detecting an obstacle.

The most frequently attended location was the distant region on the running track (~200 m ahead of the cockpit) on which a headlight of the train focused. The supervisor gazed at this location for approximately 30% of the time during the task. Therefore, we call this location the *home base* of the supervisor's track-monitoring attention. Besides this location, the supervisor also frequently looked at the side of the running track, spending approximately 20% of the time on this. His attention was also shifted frequently to the area between the two tracks (9.3%), the parallel track (8.9%), and a nearby point (~50 m ahead) on the supervisor's own running track (7.1%). From these results, the supervisor's track inspection can be conjectured to be driven by applying a home base monitoring strategy. In this strategy, the operator monitors the states on his tracks around his home base, that is, the distant region on the running track approximately 200 m ahead, by a short look at each location (taking less than a few seconds). His attention shifts from the home base to another location such as the side of his own track or the parallel track and then returns to the home base of his monitoring.

Process of Detecting an Obstacle. Using the eye-tracking data of the last session, in which four obstacles were intentionally placed on the tracks, the supervisor's actual detection process of an obstacle was analysed. The transition of fixation points is shown in Fig. 25.4 for one of the four boxes left on the track. This figure describes the case in which a box was put at the most difficult place to detect, that is, on the centre pathway between two tracks (1 m below the track level). Until approximately 10 s before his report of the obstacle detection, the supervisor was inspecting the states on his track and its surroundings in the usual home base monitoring style as just mentioned (also see Fig. 25.3). However, approximately 5 s before detection, the supervisor got a sense of something left on the centre pathway when he attended to the parallel track. He later said, during debriefing, that when he gazed at the parallel track he obtained a visual image of its surrounding in his peripheral vision. He performed a pattern matching of his acquired visual image with the visual template stored as geographic knowledge. As a result of the pattern matching, he supposed there was a discrepancy between his visual acquisition and his geographic knowledge. Then, he shifted his attention to the place where he sensed the discrepancy, that is, to the centre pathway, to test his hypothesis about the obstacle. After fixating properly at the box, he reported its detection approximately 150 m before the obstacle.

Detection Rate in Peripheral Vision. A train operator's visual detection rate obtained by an experimental study is shown in Fig. 25.5(a) for each combination of target size and its location. This index was calculated over data from all 10 subjects, although there were

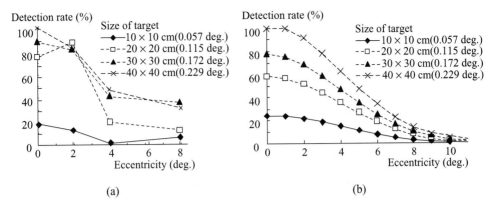

FIG. 25.5.   Obstacle detection rates with eccentricity from the fixation point: (a) experimental results; (b) approximated function.

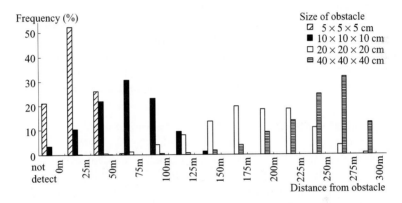

FIG. 25.6.   Simulated train position when an obstacle is detected (speed: 60 km/hr).

individual differences in the detection rate. As we can see in this figure, the detection rate was reduced with an increase of eccentricity from the operator's fixation point and with a decrease of target size, but not linearly for both factors. To generate input data to the cognitive-perceptual model, we approximated the detection rate as a function of the eccentricity and the visual angle of a target (both in degrees), as shown in Fig. 25.5(b) for several example conditions.

Simulated Detection Process.   The operator's cognitive-perceptual process was modelled for track monitoring and obstacle detection on the basis of the task analysis results (Itoh et al., 2001). As a result of the simulation study, Fig. 25.6 indicates a distribution of the train position (the distance between the train and the obstacle) where the simulated supervisor detects a different-sized obstacle left in the centre of the running rails while the train is running at the maximum speed limit, which is 60 km/hr. This figure was obtained by 1,000 simulation trials with different seeds of random digit for each scenario. As we can see in this figure, the detection rate for a not-so-small obstacle placed in the centre of one's own rails is estimated at 100% in driving at the present speed limit. Even for smaller obstacles, there were not many cases of missing detection—21% and 3.5% for the 5 × 5 × 5 cm and 10 × 10 × 10 cm boxes, respectively. However, as we can easily guess, the size of an obstacle is found to affect the timing critically when the operator detects it. For example, the operator was expected to detect a 40 × 40 × 40 cm obstacle more than 200 m ahead in approximately 90% of the cases. This means that the train can always stop before the obstacle. In contrast, for the smallest ($5 \times 5 \times 5$ cm) box, the simulated operator detected an obstacle at a distance shorter than 50 m in all the cases. In these cases, it is impossible for the operator to stop the train before the obstacle, but it is so small that the train may pass it safely.

Case B: Devising a Foolproof Method for Avoiding Derailing

Description of the Case.   In the second case study, we first illustrate an analysis of an actual derailing case of a track maintenance train (Itoh, 1999). The accident analysis allowed us not only to uncover the causes of the accident but also to suggest some countermeasures for avoiding similar potential accidents and incidents. The accident case occurred when a train passed points (the switch for the tracks) that had failed to change to the correct position, thus resulting in a derailing when the train approached a station approximately 30 min. before the end of that workday.

We estimated the probability of this kind of derailing accident by using the PRA–HRA approach, developing an accident sequence model derived from the accident analysis. As mentioned in the general description of this method, however, there are not enough data about human error probabilities and performance data relevant to this accident case. Therefore, we also performed a task analysis employing an observational approach to generate such kinds of data sets. By applying the PRA approach, we also examined the effect of a countermeasure— which was implemented in the railway after the analysis—to reduce the accident rate.

Methods Applied to the Case.   The accident analysis of this case study focused on a timeline analysis of operators' actions and events leading to the accident. The original report of the accident analysis included five A4 pages of timeline accident sequence (originally in Japanese; Itoh, 1999). Because of the task characteristics—work carried out in the middle of the night—the original analysis also included the work schedules for the previous 7 days of the operators involved as well as the timeline activities preceding their working hours of the day. Accident histories of the train as well as those of the operators were also included in the original report.

For the estimate of accident probability using the HRA approach, we first applied the THERP to this derailing accident. However, because we had some difficulties with the application of this technique to our accident case, we modelled the accident sequence by adopting a flow-type representation of actions and events with state descriptions to estimate the accident probability more adaptively. We made use of this model for risk prediction.

For the purpose of generating the PRA human error and performance databases relevant to this accident case, we recorded large volumes of data on the train operators' activities in a train cockpit. The observations were carried out almost everyday for approximately half a year with two video cameras fixed in one of the train cockpits. One camera focused on the driver's face while the other recorded the view out of the front windscreen. Video recordings were made on either the outbound trip from or the return trip to a track maintenance base, lasting between 20 min. and 1 hr. Finally, verbal protocols were recorded of both the driver and the supervisor during their tasks.

During the first 3 months or so, the operators showed that they were quite conscious of the video camera and their behaviours were thought to be somewhat different from their usual patterns. However, after approximately 2.5 months, their behaviour changed considerably compared with what it was in the first month. Therefore, we decided that the video recordings for the first 3 months should be excluded from analysis data because they were nonrepresentative of the operators' usual style of work.

Analysis of the Accident.   As results of the timeline analysis, the immediate cause of this case was identified as an omission made by a points manipulation operator when he neglected to switch the points. However, additional errors made by the supervisor and the driver were also closely related to the occurrence of this accident. These errors occurred one after another—independently—during the last few minutes just before the accident. Both the driver and the supervisor missed checking the position of the second points successively followed by the first points. When the supervisor requested the points manipulation operator to switch the points, the latter acknowledged the request very promptly. However, the points manipulation operator did not in fact actually switch the points. Moreover, the supervisor did not in fact expect that they would be switched within such a short time frame. Nevertheless, the supervisor did not express his suspicion to the points manipulation operator. In addition, the driver did not reduce the train speed to the required upper limit when passing the points. As pointed out, it is a typical, but not an invariable, feature of accidents that several errors have occurred, each being necessary for the accident. This is true of this accident as well: if any one of these human errors had not taken place, the accident would not have happened.

Suggestions Obtained by Accident Analysis.   Several potential risk factors could be identified in relation to the human errors that caused this accident. Some of these errors are related to cognitive and operational aspects during task performance, whereas others fall within the realm of organisational factors. On the basis of an analysis of these risk factors, we can obtain some ideas about countermeasures that will eliminate or diminish the risk factors by means of organisational support and actions.

In this accident case, the failure of the driver and the supervisor to detect the wrong points setting is to a large extent explained by the fact that it was psychologically difficult for them to keep their attention continuously on a successive set of points. As already mentioned, there were two sets of points within a very short interval—approximately 100 m—and it took just a few seconds to run between these sets. In this situation, the cockpit crew were led to perceive the two sets of points either as a single set or in the same way. So, when the first set of points was in the correct position, they were not unlikely to believe "the next one is also OK." Therefore, they did not shift their attention to the second points. In addition, it was highly unusual and had not happened in ages that points were in the wrong position; therefore, the cockpit team took it for granted that the points would be in the right position. Finally, as the driver and the supervisor stated during the accident investigation, their vigilance level was rather low as a result of fatigue and their circadian rhythm. In such situations, it is impossible to ensure with 100% confidence that operators pay sharp attention to points whenever they approach them.

Accordingly, a major proposal from this analysis was the suggestion that a *foolproof* devise, to be called a "blocking system against wrongly set points," be built into the work system to ensure that a train stops before it encounters points in the wrong position. This system would trigger an inspection terminal at a location just before points by taking the braking distance into account. When a train arrives at the terminal, the position of the points would be examined, and if they were in the wrong position, the train would automatically be made to stop.

Besides this design proposal, some other important suggestions were obtained from this accident case. An organisational problem, a so-called invisible high wall between departments, was found. This is not unusual and is actually the case with nearly every Japanese organisation. The train operators involved in the accident were employed by a unit company that had one of the lowest positions in the organisational hierarchy (see Fig. 25.1). So, they worked under the administration of the track maintenance department of the JR, whereas the points manipulation operator was an employee of the JR and belonged to the signal department. As the supervisor stated afterwards, it is difficult for an operator to doubt or question a decision made by a person in another department, particularly at a higher position or in a superordinate company.

This suggested that a system should be developed for facilitating mutual understanding and communication among operators on the one hand, and between operators, staff, and managers beyond their organisations, departments, and positions on the other.

Generation of Performance Indices for PRA.   Similar to the callouts that pilots are required to make when performing checklists, the driver or supervisor is obliged to make a callout and state the position of the points—for example, "position of points engaged" (the practice is that the driver will do this)—thereby indicating to his colleague that the points position has been checked. After the initial callout, his colleague is required to repeat the callout, which—comparing once again with aviation procedures—is similar to the required "read back" that a pilot will make when receiving an instruction over the radio from an air traffic controller.

From the observed video and protocol data of operators' actions, it was established that the driver made the callout and the pointing gesture in 81% of the cases (830 out of 1031) while the supervisor did so in 16% of the cases. In most cases (87%), the supervisor or the driver made the prescribed "read back" of his colleague's callout, whereas he failed to make the proper callout in 3.4% of the cases. In addition, the driver or the supervisor failed to "read back" his colleague's callout in 4.7% and 5.4% of the cases, respectively. On some occasions of these missing "read backs," the vigilance level of the operator skipping his "read back" might be lower than usual.

On the basis of the observed data, the train speed at passing the points varied, depending on the running area. The speed was low, ranging from 7 to 13 km/hr, when passing the points outside the main line—not least because the train stopped near the points before starting to move again on these tracks. In contrast, the train passed the points at a speed of more than 15 km/hr when running on the main line. This indicates a similar tendency as in the accident case just mentioned, and it supports the suggestion that train operators are under a great time pressure to comply with their time schedule in order not to cause delays to the first morning bullet train. This may suggest the need for providing a new means by which the train can be stopped before the points, compensating for the increased speed. An example of a countermeasure could be the installation of a clearly visible indicator (e.g., a coloured on–off lamp) showing the points status at a long distance.

Risk Analysis Applying the Observed Data.   When the train driver detects the wrong position of points, as in the accident case just described, he will immediately pull a brake lever to stop the train. When we combine the observed case data, that is, speed and distance for each case of passing the points, with the braking performance of the train, we can derive an estimate of a derailing risk by the wrong position of points. That is, if we assume that the driver and the supervisor make no "missing callout" errors, we derive an estimate of the risk by examining whether the train can be stopped safely before the points.

Figure 25.7 shows a graph that plots all the observed video cases in terms of their actual braking distance at the time when the points are checked and that also contains the performance lines of braking distance depending on speed, slope, and state of rails. So, a mark plotted over a braking distance line indicates that the train can be stopped safely before the points. As we can see in this figure, there is almost no derailing risk (0.003%; 2 out of 647 cases) if the points were to be in the wrong position on a level track, but there is approximately a 1% likelihood of a slight incident on the steepest tracks (25/1, 000) at the time when points are checked *if* they were to be in the wrong position.

Simple Application of THERP.   When we apply the THERP to the analysis of the accident case just considered, we derive a probability tree diagram as depicted in Fig. 25.8. In estimating

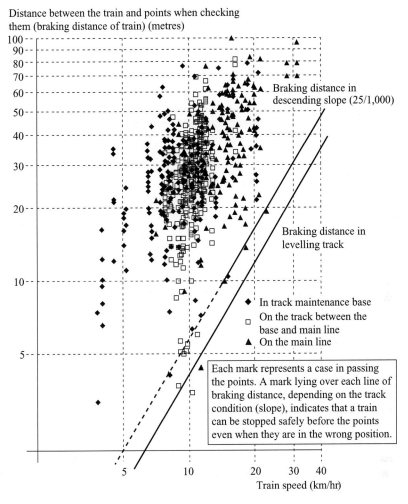

FIG. 25.7. Derailing risk caused by the wrong position of points.

an accident rate, we need data about the human reliability–error probabilities that may be used for the THERP analysis. In this case the calculation of the observed data mentioned in the previous paragraphs yields the following results:

- The probability of a low level of operator vigilance (almost sleeping) during the task: 0.1%.
- The probability of an adequate level of vigilance to perform the task: 99.9%.
- The probability of neglecting a points status check: 3.0%. Here we assume that one third of the total cases, 3.4%, of missed callouts and the missed "read backs," 5.4% and 4.7%, respectively, are cases of neglected points status checks; so for the supervisor it is (3.4% + 5.4%)/3 = 2.9%; for the driver, (3.4% + 4.7%)/3 = 2.7%; hence, the average is approximately 3.0%.
- The probability that a points status check is appropriately made: 97.0%.
- The probability of a wrong position of points: 0.34% (averaged rate over all the opportunities of passing points; no cases of wrong position on the main line were revealed in the video observations, but a few cases are held in the track maintenance base).
- The probability of correct position of points: 99.66%.

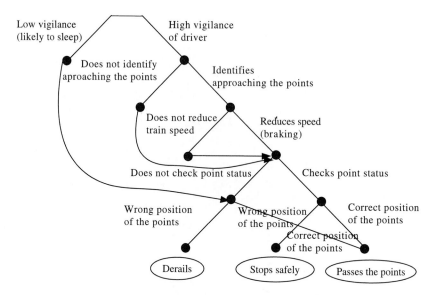

FIG. 25.8.   THERP model of derailing accident when the points are passed.

Applying these probabilities to the accident sequence model shown in Fig. 25.8, we calculate that the accident rate of derailing caused by a wrong position of points is

$$0.001 \times 0.0034 + 0.999 \times 0.03 \times 0.006 = 1.06 \times 10^{-4} \approx 0.01\%.$$

Currently, approximately 50 track maintenance trains are routinely operated to maintain tracks and rails for the high-speed railway everyday. If we assume that each train passes two sets of points per day on average, the rate just calculated is almost equivalent to five accident–incident cases per year. This estimate is, however, much higher than the actual rate. As mentioned previously (see Fig. 25.7), whether a train will be derailed as a result of the wrong position of points will depend on its braking distance. The accident sequence model in the form of a THERP-type binary tree did not include a detailed relationship between work conditions and the possible events. Therefore, the model will predict a derailing whenever an operator fails to check the points that are set in the wrong position. This is the major reason why the model estimates a higher rate of derailing than is actually observed.

Estimation of Derailing Risks by Action Flow Modelling.   As explained earlier, to produce a reliable accident rate estimate we must describe in the accident sequence model the details about cause–effect relations, including the interactions between the cockpit crew. Thus, we show in Fig. 25.9 a flow-type model with state description of each activity for this derailing case. In this action flow model, human actions and system behaviour and events are described as nodes and their transitions are represented as arcs. Human actions and internal cognitive states are represented inside single-lined rectangle nodes, whereas double-lined nodes contain either conditional branches or descriptions of system and machine behaviours and states. Each rectangle node referring to a human action contains a brief description, an identification code, and probabilities of success and as well as the present states of the system. Finally, circle nodes describe conditional decisions.

Applying the data set mentioned in the previous section, we calculated the risk of derailing. When the points are correctly positioned, the model indicates, as it should, that the train passes safely, regardless of the internal states and actions of the operators. As described in the bottom node of the action flow model, there are three situations to be distinguished when the points

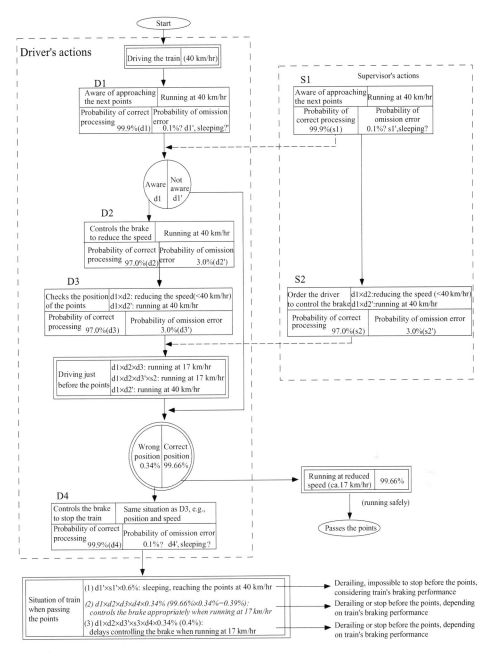

FIG. 25.9. Action flow modelling of passing the points.

are in the wrong position. In the first situation, where vigilance levels of both train operators are very low (almost sleeping), a severe derailing accident may occur if the train reaches the wrongly positioned points at relatively high speed. This adverse event is estimated to occur at the probability of 3.4×10^{-9} ($0.1\% \times 0.1\% \times 0.34\%$).

In the second situation ($99.9\% \times 99.7\% \times 97.0\% \times 99.9\% \times 0.34\% = 0.33\%$), the driver's vigilance level is sufficiently high and he inspects points in the usual way. In this situation, when he detects the wrong points, the driver immediately pulls the brake lever to stop the train. According to the data obtained by the video observation, the driver checked the points status on average 9.7 s prior to passing (standard deviation: 3.6 s) on the main line. The mean train speed when the points were checked was approximately 17 km/hr. Applying these data as well as track and train conditions to the function of braking performance shown in Fig. 25.6, we calculate the probability that a train will not stop before the wrongly set points to be 0.62%. Thus, multiplying this probability with that of the reaction time of a vigilant driver, we estimate the derailing risk in this situation at 2.11×10^{-5} ($0.62\% \times 0.34\%$). Even in this case, the derailing effect may be rather small because of low speed at the points.

The derailing probability can be calculated in the same way with reference to the action flow model for the last situation. Summing up all three situations, regardless of the severity level of the possible accident–incident, we estimate the summed-up probability of a derailing caused by wrongly positioned points to be 2.11×10^{-5} per case of passing points. This rate means that 0.75 derailing accident–incident of track maintenance trains is likely to occur on the entire line each year on average. This estimate is almost identical to the actual mean occurrence rate over the past 10 years.

Evaluation of a Planned Foolproof System.   As a way to examine the effects of a foolproof system proposed by the accident analysis—a so-called "blocking system against wrongly set points"—the action flow model depicted in Fig. 25.8 was adapted. The adapted model resulted from adding nodes incorporating the foolproof function to the original flow model, followed by a conditional branching according to the state of the points.

Depending on the reliability of the foolproof devise, that is, how correctly it detects position of points, the accident rate can now be estimated by the adapted action flow model in the same way as just described. Assuming a probability of 0.1% of devise malfunctioning, for example, we estimate the derailing rate at 3.64×10^{-12} per case. This means statistically that almost no derailing accident will take place in practical terms of operation, namely, a single accident for every 1,336 years.

Case C: Surveying Operator Attitudes—Accident Rates and Safety Culture

Objectives of a Safety Culture Survey.   A questionnaire-based study was carried out for the purpose of identifying issues and points that (a) might be improved to enhance the safety culture of the track maintenance organisations and (b) might be influenced by management decisions (Itoh & Andersen, 1999; Itoh, Andersen, Tanaka, & Seki, 2000a). The questionnaire-based study aimed at (a) eliciting and analysing track maintenance train operators' perceptions of factors that have an impact on safety, including their views of morale and motivation, and (b) demonstrating correlations between operator perceptions and attitudes and the rates of accidents and incidents of different parts of the organisation. The method of analysis applied to the questionnaire data involved, as a first step, a principal components analysis of responses, allowing us to identify a set of attitudinal factors. The factors involved operators' attitudes to and views of, inter alia, their jobs, management style, stress factors, teamwork, and organisational issues. The second step consisted of calculating an objective risk index derived from the company's accident and incident records; and finally, as a third step, we analysed the correlation between the qualitative measures of attitudinal factors and the quantitative risk index by applying data of all five branches within a single track maintenance company.

Questionnaire and Respondents.   The questionnaire used for this case study was based on the widely known aviation-oriented questionnaire, the FMAQ (Flight Management Attitudes Questionnaire; Helmreich, 1984; Helmreich, Merritt, Sherman, Gregorich, & Wiener, 1993), and its derivative applied to the maritime domain, the SMAQ (Ship Management Attitudes Questionnaire; Andersen, Garay, & Itoh, 1999). From these predecessors, the train version

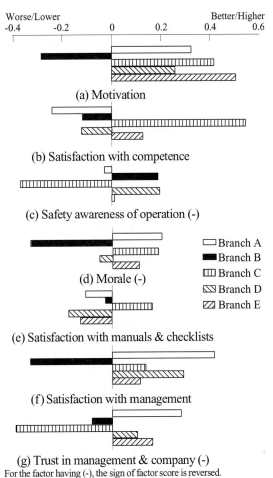

FIG. 25.10. Attitudinal factor scores for each branch.

of the questionnaire was termed the Train Management Attitudes Questionnaire, or TMAQ. Adapting the questionnaire from other domains allows us to compare the results across domains. The original version of the TMAQ (Itoh & Andersen, 1999) has approximately 100 questions, grouped into five sections: general questions concerning training and schedules, followed by questions regarding cooperation and teamwork, organisational issues, train management issues, and train automation issues. Respondents were asked to rate each question on a 5-point Likert scale between 1 and 5 (from "strongly disagree" to "strongly agree").

The questionnaire was distributed in the autumn of 1997 to all drivers and supervisors of one of the four track maintenance companies. This company, which has five branches spread out between Tokyo and Osaka, operates several types of special-purpose track maintenance trains for the entire line. A total of 291 response samples were collected; the response rate was almost 100% because of the strong organisational support of the survey.

Attitudinal Factors of Operators. As an outcome of the principal component analysis applied to 222 valid responses of the TMAQ, seven attitudinal factors were extracted accounting for 60% of the cumulative variance (Itoh et al., 2000a). These factors were interpreted on the basis of highly loaded question items, and they were labelled as shown in Fig. 25.10. This figure shows averaged factor scores for each attitudinal factor across the branches of the company surveyed. As we can see in this figure, operators in Branch B had relatively negative attitudes for most factors compared with the other branches. In particular, their motivation, morale, and satisfaction with management were much lower than those in the other branches. In contrast, Branch A and E operators, which were quite similar in their attitudes, showed particularly high motivation, morale, satisfaction with management, trust in management, and satisfaction with company. In Branch C, operators showed a pattern somewhat similar to the two "positive" branches, as they scored high on motivation and morale.

Effects of Attitudinal Factors on Railway Safety.   As a way to explore the relationship between train operators' morale and motivation data and the safety levels of train operations, branch-based accident–incident statistics were collected of the track maintenance company surveyed for the previous 5 years (1994–1998). In the JR, accidents and incidents of track maintenance trains are classified into three types in terms of lost money and the possible delay of the first morning bullet train: big, small, and no-loss accidents–incidents. A big accident is defined as one that either involves a loss of more than 500,000 Japanese yen (approximately $4,000) or causes a delay of more than 10 min. to the first bullet train. A small incident involves a delay and a loss that are shorter and smaller, whereas a no-loss incident is a very minor event that causes no loss and no delay to the bullet train, and because they nearly always result from technical faults, they are excluded from the statistical analysis.

Actually, very few accidents and incidents were produced by the track maintenance company surveyed. However, there was a difference in their occurrence across the five branches: Branch B, which was the lowest in motivation and morale (see Fig. 25.10), was alone in having produced big accidents during the 5-year period, although its rate was not high (approximately 0.4 accident/100 km of territory a year on average). The other branches had no big accidents. The top two branches in operators' morale and motivation, that is, Branches C and E, did not produce even a small incident at all over the 5-year period.

For a one-dimensional index of railway safety to be produced, 24 staff and managers in the track maintenance department of the JR were asked to indicate the relative severity of big and small accidents–incidents. The 24 respondents were asked the following question: "In your estimate, one 'big' accident is the equivalent of how many 'small' incidents?" As a result of this rating, the severity of a big accident was perceived on average to be 5.5 times that of a small incident (median 5). Accordingly, using a weight of 0.2 for a small incident, we define a weighted total accident–incident rate.

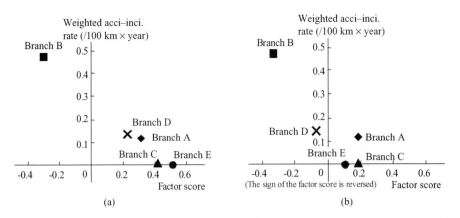

FIG. 25.11.   Relations between attitudinal factors and accident–incident rate: (a) motivation, (b) morale.

For the effects of the operators' attitudes on railway safety to be represented quantitatively, each branch is mapped on the geometric plane of (a) motivation and (b) morale and the weighted total accident–incident rate, as shown in Fig. 25.11. As we can see from this figure, both motivation and morale are negatively correlated with the accident–incident rate. That is, the higher the motivation and morale of operators within a given branch, the lower the accident–incident risk of that branch.

IMPLICATIONS FOR COGNITIVE TASK DESIGN

Implications From the Accident–Incident Analysis Approach

In our experience with the accident–incident analysis, as mentioned earlier, the most critical issue is to obtain a reliable and informative report about the accident or incident from the operators involved in the adverse event that does not contain any misleading descriptions. Fear of sanctions will often be justified and will prompt operators to be less than candid when they are questioned about their own involvement. In turn, this may make it difficult or even impossible to gain a clear picture of and a valid analysis of events. A model for introducing people to tell investigators candidly what they think happened is to change a management style to a "blame-free" style or "management by freedom." In some reporting systems, employees are granted immunity against sanctions if they inform of their role in the accident; in others there may be a conditional immunity (in which acts of gross negligence or criminal acts are exempted); and in others there may be no immunity granted (Orlady & Orlady, 1999). Following this style of management, an operator involved in an incident will be granted immunity against sanctions (except for cases of gross negligence or for cases in which others are injured—in which case it is usually a matter dealt with by the police or the courts), but will receive strong incentives for reporting his or her own errors and the events that led up to them. In contrast, the "management by punishment" style will instruct operators to tell the truth at their own peril.

Another key to success of an incident reporting system is management's commitment to the system and their attitudes to applying reported cases to the improvement of safety. Such a commitment will allow operators to realise that their reported cases are useful for improving the safety of their operations and will boost their motivation and morale. In turn, this may be expected to contribute to a reduction of the human error rate during operations.

In addition to the derailing case mentioned in this chapter, some 30 accident and incident cases of track maintenance trains that were reported during the previous 15 years were also analysed. On the basis of this analysis, several issues were suggested for improving the operational safety of the track maintenance trains. Most of these suggestions were subsequently implemented. A few of these are listed here.

One human–machine factor, for example, was that there were several varieties of almost identical specifications of different tools and pieces of equipment. Operators sometimes confused a particular tool with another tool and then engaged in the wrong type of skill-based behaviour when manipulating it, resulting in an accident or incident. Today, this problem is largely solved and each tool and piece of equipment has its own unique specification. Regarding the work factors, it was found that operators nearly always worked under a strong time pressure and that this in turn was a potential risk factor indirectly involved in almost every accident and incident. In addition, in some cases, poor communication between operators and managers was found to be a background for adverse events. Prompted by this analysis, the track maintenance department has therefore been engaged in continuous discussions with employees to have formal and informal systems for facilitating communication between operators and management as well as organisations at different levels.

Implications From the Observation Approach

An observational study follows the descriptive approach by which we seek to understand how operators actually behave in practice (Vicente, 1999). Therefore, the most important thing in observing actual tasks is to record operators' behaviours and actions performed in their usual way. Data recorded in this way can be also analysed as a preparatory step to applying another approach, as presented in our case studies.

Although the primary purpose of observation is indeed to get data for a cognitive analysis of the operator's tasks, this method can also be applied to get suggestions for designing or improving the task. For example, in our first case study, the effects of the operator's geographic knowledge of the driving area were found to be a critical element for track monitoring when we compare the performance of the "home area" supervisor with that of a supervisor who had no geographic knowledge of this area, but who was an expert in this task in his usual and familiar area. Besides this factor, we identified effects of fatigue on the operating processes. When an operator became tired some hours after the start of his shift, his attention allocation was much more concentrated on his "home base" of monitoring. A similar phenomenon was seen when the driving speed was increased. When the speed increased, there was an increase in task difficulty in general and the operator's monitoring strategy became more conservative, concentrating more on the home base, to compensate for the difficulties in maintaining task performance. Such outcomes from observations can be utilised to create a new procedure and task improvement as well as to develop appropriate tools for supporting operators, depending on task difficulty and complexity.

Implications From the Experimental Approach

The experimental method was applied to generating a data set about the visual functions of operators in one of the case studies of this chapter. The experiment in the case study was typical of its kind in that we spent several weeks from the planning of the experimental design to its execution and analysis. A large number of staff in the track maintenance department of the JR were involved at various stages of the experiment, as were train operators acting as subjects. Still, the number of subjects was not enough to derive a statistically reliable data set as a result of large individual differences involved in the study. Therefore, as stated earlier in the general description of the method, this experiment is not an exceptional case in terms of cost and time. Nevertheless, an experimental study is the only means to obtain this kind of functional data of human processing with a satisfactory level of reliability. It might be helpful to reinforce the results derived by such a relatively small number of subjects by seeking to generalise from physiological or psychological data, for example, visual acuity or distribution of visual cells on the retina—in this case, in relation to eccentricity.

As mentioned in the case study, there are individual differences among train operators in their visual detection performance. From the aspect of human performance in designing or improving a cognitive task, this visual function was just one example of skills that were required for track maintenance train operators. However, individual data on human characteristics elicited by an experiment can be effectively utilised for cognitive task design in areas such as selection, manning, skills and competence training, and development of a training programme.

Implications From the PRA–HRA Approach

A key condition of success of the HRA–PRA approach is that an error process be built appropriately into an accident sequence model. An estimate was made of the effects of installing the aforementioned "blocking system against wrongly set points" on the likelihood of having zero derailing accidents. However, this does not necessarily mean, of course, that no derailing accident will ever happen with this system. The result of the PRA—that virtually no derailing incident will take place over a reasonable time span—is limited in its validity to the known factors and the known logic of mechanisms described in the model. There will be other causes of derailing accidents that are not yet known. Therefore, it is natural to consider two approaches to risk analysis, one referring to causes that are already known and the other to accident causes that are largely unknown. In the first approach, processes contributing to and causes of accidents should be uncovered by applying various methods mentioned in this chapter, yielding an increasingly accurate and dedicated model of the HRA. The other approach tackles potential causal factors with the aim of improving performance shaping factors or common performance conditions to reduce human errors as potential accident causes.

In addition to a probability estimate of an accident, the HRA has the advantage that a causal relation between a specific accident and human errors can be visualised in the accident sequence model and the operator model, particularly in the second-generation HRA methods. As mentioned previously, one of the most critical limitations of the PRA and the HRA methods is the lack of relevant human error data. Even if we have such error data available for accident modelling, their level of accuracy may not necessarily be high enough simply because our data are too scarce, as in our case study. However, we believe it is more important for a safety analysis to (a) understand the nature of the accident-generating process and its structure in relation to human errors, the task context, or the performance shaping factors and (b) establish a comprehensive accident sequence model than to (c) estimate an accident probability that by itself is derived from possibly not so reliable data. This is actually one of the most important purposes of the second-generation HRA approach (Hollnagel, 1998). Such understanding of the error-generating process is useful for different stages of the cognitive task design from "analysis" and "test", for example, planning. One idea for compensating for less reliable data is to perform a sensitivity analysis, repeatedly changing the human error probability in a small interval to evaluate its effect on the accident rate. We can also use this "generate (hypothetical error rate) and test (its effect on accident occurrence)" approach for the purpose of goal setting for improvement of safety procedures.

Implications From the Simulation Approach

As mentioned earlier, the cognitive simulation may have potential abilities for large-scale risk analysis of human–machine operations because of its advantages in cost and time saving. One idea supporting this is to prepare various kinds of human performance databases available to cognitive modelling. In our case study (not included here for space reasons), we conducted many additional simulation studies by applying different databases of operator's visual functions (Itoh et al., 2001). For example, applying the attention allocation database of the operator' having no geographic knowledge, we estimated his detection rate, and we found it to be lower for obstacles lying far from the operator's own running rails than that of the operator who does have geographic knowledge.

Implications of the Questionnaire-Based Survey Approach

As mentioned in the general description, a key point for leading a survey to success is to prepare a questionnaire having a small number of questions that can effectively elicit respondents' opinions matched to the purpose of the survey. In our case study, we wished to obtain confirmation (or disconfirmation) of the correlation between motivation and morale and accident risk obtained, originally derived from a survey of a single company, and thus obtain a more robust and generalisable result. Therefore, we carried out a similar survey for all the track maintenance train operators working for the JR by using a revised version of the TMAQ that had a smaller number of questions. We adapted this shortened questionnaire of 30 questions by selecting items from the original TMAQ that had a reasonably strong relation to attitudinal factors and risk anticipation based on results of the statistical analyses. When using the shortened questionnaire, we obtained results that were almost identical to the five-branch study described earlier in terms of the effects of the operators' attitudinal factors on railway safety. That is, companies employing operators who have higher morale and motivation exhibit lower accident–incident rates. This suggests that the shortened, "easy-to-use" questionnaire is just as powerful as the original and three times longer version in extracting meaningful and diagnostic results. In addition, using this type of shortened questionnaire, we collected responses twice from the track maintenance company surveyed in the case study, separated by a 2-year interval, in 1997 and 1999. This repeated survey also allowed us to detect and analyse a change in attitude that took place over the time interval within the company that we had sampled twice.

Finally, it is also important to compare results obtained from respondents' subjective replies to questionnaire items with facts that characterise the organisation and the specific work setting. For example, in our case study, the variation in the absolute levels of responses between the branches is quite interesting and provides us with some suggestions about organisational issues because the branches were, in the formal sense, entirely alike: They were alike in terms of their tasks, their employment system, management system, operating procedures, training system, and manuals and checklists. Therefore, given the differences between the branches that we uncovered, it is reasonable to conjecture that each branch may have developed its own local ways and informal culture or "safety climate," and that such an informal system with its tacit canons of conduct may vary in an independent manner from the formal system and the overall management system. It might reveal important clues to establishing means of shaping the informal elements of safety culture if it were possible to identify the causal factors behind the generation and upholding of such safety cultural differences.

From this case study we have obtained what we believe are interesting results regarding organisational aspects of operational safety. We found that operators' motivation and morale are key factors for operational safety and, therefore, are potential risk factors that may be identified before accidents happen. We also saw that the two attitudinal factors, morale and motivation, are strongly related to operators' satisfaction with management and their trust in their management and company. These organisational and managerial factors include, for example, issues concerning training, manuals and checklists, and work schedule and procedures, as well as management style and organisational rules. Therefore, these findings suggest that it is of great importance to change or adapt training programmes, working procedures, and management style, with a consideration of both formal and informal ways in which safety culture is shaped, and also to improve manuals and checklists in order to enhance morale and motivation and, in turn, operational safety.

CONCLUSION

In this chapter, we mentioned an approach to tackling safety issues in modern, complex, large-scaled human–machine systems, which was based on the widely acknowledged recognition that most accidents in technical workplaces are caused by human errors. As we mentioned earlier, it is impossible to eliminate human error during operations. Therefore, we must address these issues on the assumption that humans will inevitably make errors during their tasks. However, this does not mean it is impossible to prevent an accident: errors may be captured and the causation of accidents will typically involve multiple human errors. Therefore, it is a key point in safety advances to establish mechanisms that allow us either to avoid or to capture human

TABLE 25.2
Relative Strengths and Weaknesses of Each Method

Methods	*Cost*	*Effectiveness*		
	Time & Econ.	*Certainty*	*Reliability*	*Variety*
Experiment	N	YY	YY	N
Observation	(Y)	Y	Y	(Y)
PRA–HRA	YY	YY	(Y)	N
Cognitive simulation	Y	Y	(Y)	(Y)
Accident–incident analysis	YY	Y	Y	Y
Questionnaire	Y	(Y)	(Y)	YY

Note. YY = always good; Y = frequently good; (Y) = sometimes good; N = bad.

errors or to build safeguards against errors in a work system, such as foolproof protections, checklists, and dual or multiple check systems.

This chapter has reviewed methods involved in observational studies, experimental studies, accident–incident analysis, the PRA–HRA approach, cognitive simulation, and questionnaire-based surveys by using illustrations of three case studies that applied to track maintenance train operations and organisations. Using the outcomes from the case studies and their implications, Table 25.2 summarises the relative strengths and weaknesses of these methods in general in terms of cost and effectiveness of their applications. The cost attribute is meant to include both the time and the expense or economy required to apply a given method. Effectiveness is divided into three indices: certainty, reliability, and variety. The first index, certainty, relates to the question of how a method can generate appropriate information matched to the purpose of its application; the last index, variety, indicates how many varieties of information can be submitted. Reliability is a measure of how reliable or credible the information generated by a method is. For example, an experimental study is relatively expensive in terms of the cost, compared with the other methods. It consumes plenty of time and expenses to apply this method to a real-scale complex task. However, it can also yield reliable results on the effects of focused factors—indeed, it is the very purpose of this method to investigate their impact, and therefore it has comparable strengths in the certainty index. However, as we can see from this statement, the method is not useful if we want to generate a great variety of information. In contrast, it typically does not cost very much to apply a questionnaire-based survey, though it is costly to collect a large sample from quite different strata of a population. This method allows us to obtain a rich amount of information from various aspects—strength in variety—but its reliability may not always be high because of low response rates, problems of confidentiality, and problems in eliciting responses to complex and possibly dynamic issues.

Considering their strengths and weaknesses, Table 25.3 shows a matching of these methods with each design activity mentioned in the Introduction, that is, planning, analysis, designing, test, and maintenance—in terms of balance between cost and effectiveness. It may be impossible to apply these methods to design, that is, to the formulation of functions and structures, and therefore support to this activity was interpreted as idea seeking for designing. Most of the methods are well matched to analysing existing artefacts such as tasks, procedures, interfaces, systems, and organisations. A questionnaire-based survey, for example, can support the planning of a new artefact or improvements of an existing system or organisation; and it may also

TABLE 25.3

Methods Suitable to Support Design Activities for Operational Safety: Balance Between
Cost and Effectiveness

| Methods | Design Activities | | | | |
	Plan	Analyse	Idea Seek	Test	Maintain
Experiment		(Y)		(Y)	
Observation	Y	YY	(Y)		(Y)
PRA–HRA	(Y)	Y		Y	(Y)
Cognitive simulation		(Y)	Y	Y	
Accident–incident analysis	Y	Y	(Y)		
Questionnaire	YY	Y	(Y)	(Y)	Y

Note. YY = always suitable; Y = application dependent; (Y) = sometimes applicable.

allow us to obtain suggestions on specifications or functions for new designs as well as ideas for improving an existing system. An accident–incident analysis as well as an observational study may also be applied to get hints for plans for new design or system improvement. Cognitive simulation can be applied to support the test or evaluation of existing and alternative systems, but it may also be used to select design parameters of a system—for example, for the purpose of idea seeking for design—if a reliable, appropriate cognitive model is possible to establish. The PRA–HRA approach is also available to test alternative designs by developing an updated accident sequence model.

For the purpose of maintaining a safe organisation, for example, a questionnaire-based survey can be periodically applied to identify changes. In addition, it may be used to measure the effects of safety enhancement programmes and initiatives—to the extent that other working factors, such as tasks, environment, and staff—are not changed. In particular, from the case study applying this method as well as from general recognition that the quality and safety of tasks that operators accomplish are affected not only by their professional technical competence and skills but also by their attitudes, motivation, and morale (e.g., Andersen et al., 1999; Davies, Spencer, & Dooley, 2001; Helmreich & Merritt, 1998), the latter issues have been surveyed applying this technique as important indices to measure safety levels of organisations. It is important to notice that these indices can be measured *before* an actual accident takes place, and therefore a method of measuring safety attitudes by questionnaires seems to be very useful, especially in domains or companies where incidents and accidents rarely occur.

In many cases, only one of these methods is applied individually, but some of these may well be combined to tackle a single problem. For example, when the safety level of some target operations requires improvement, as illustrated in the case studies of this chapter, we can identify causal relations leading to an accident by applying either one or several methods, such as observation, PRA–HRA, an accident–incident analysis, or a questionnaire-based survey. We devise and implement one or more countermeasures on the basis of these analysis results. During this process, we can evaluate them by using the PRA–HRA or cognitive simulation. In contrast, for the need of designing a novel task, we can apply an observational study to similar tasks to a to-be-designed task to support planning of its design. It will be quite unusual if there are no tasks similar to or relating to the one that must be designed.

One of the important issues in operational safety is the level and style of automation. Some people take an approach to automation that is based on the premise that humans inevitably make errors. However, this is not necessarily a good idea. There are several modes or phases in operations of human–machine systems. Complete automation is impossible to develop over all the functions or tasks that require a human–machine system. It is true that functions required for a human–machine system are increasingly being allocated to automation, and therefore human tasks have been shifted to supervisory duties. However, not all the functions required can be performed by automation or machines alone, and therefore human operators must be responsible for difficult and complicated tasks such as intervention in abnormal and emergency situations. To design, say, a well-organised procedure, we should investigate these tasks in detail from various aspects by applying some of the methods reviewed in this chapter. Operators must be prepared for abnormal situations, which seldom occur, while they are only monitoring states in a system. This is bound to lure an operator into a lower state of vigilance, and it is therefore difficult for him or her to deal with an abnormal situation when it happens. In addition, as we mentioned earlier, it is known that the likelihood of human error becomes much higher in such situations. Therefore, the functions required of a human–machine system should not necessarily be locally or individually optimised with respect to cost and efficiency, say. Instead, the functions must be distributed between human and machine with consideration of the full range of working situations to achieve a global optimisation.

Finally, we stress the importance of participation and involvement of all members of an organisation in safety activities, including operators, middle management, and senior management. We do so in part on the basis of the outcomes of the case studies, particularly the questionnaire-based survey and the cases of accident analysis. Through such participation, it is possible to tackle safety activities continuously, to build on the experience of such initiatives, their outcomes, the knowledge and techniques involved, and the repeated applications to new cases, and, finally, to help in reinforcing safety awareness of operations with front line staff and management. This allows us to develop a sound and robust safety culture within an organisation.

ACKNOWLEDGEMENTS

The authors acknowledge Hiromasa Tanaka, Kunihiro Kondo, Masahiro Kawagoe, Takehiro Hoshino, Ken-ichiro Hisanaga, Isami Nakura, Suguru Ushirosako, Minoru Hiranaga, Toshiaki Urano, and Kimihiro Ichikawa, all with the Central Japan Railway Company, as well as Tokuhei Sugiyama, Fumio Kitagawa, and Shigemi Tani of the CN Construction Company, Ltd., Akikazu Umoto and Norimasa Nogawa of the Futaba Tetsudo Kogyo Company, Ltd., Toshitatsu Ishii and Jushiro Takahashi of Nihon Kikai Hosen Company, Ltd., and Shigeyasu Hara of Meiko Construction Company, Ltd. for their cooperation in the railway safety project of track maintenance trains. We also thank Hirotaka Aoki, Shinji Akiyama, Yoshiaki Nakai, Hironao Tanaka, and Satoru Mashino of the Tokyo Institute of Technology; Robert L. Helmreich, University of Texas at Austin; John Paulin Hansen, the IT University of Copenhagen; and Thomas Bove and Gunnar Hauland, with Risø National Laboratory.

REFERENCES

Advisory Committee on the Safety of Nuclear Installations. (1993). *Human Factors Study Group Third Report: Organising for safety.* Suffolk, UK: HSE Books.

Amalberti, R. (2001). The paradoxes of almost totally safe transportation systems. *Safety Science, 37*(2/3), 109–126.

Andersen, H. B. (2002). *Assessing safety culture* (Tech. Rep. R-1459). Roskilde, Denmark: Risø National Laboratory.

Andersen, H. B., Garay, G., & Itoh, K. (1999). *Survey data on mariners: Attitudes to safety issues.* (Tech. Rep. No. I-1388). Roskilde, Denmark: Risø National Laboratory, Systems Analysis Department.

Andersen, H. B., & Hansen, J. P. (1995). Multi-modal recording and analysis of operator data. Paper presented at the Fifth International Conference on Human-Machine Interaction and Artificial Intelligence in Aerospace, HMI-AI-AS '95, Toulouse, France.

Cacciabue, P. C. (1998). *Modelling and simulation of human behaviour in system control.* London: Springer.

Davies, F., Spencer, R., & Dooley, K. (2001). *Summary guide to safety climate tools.* Suffolk, UK: HSE Books.

Flin, R., Mearns, K., O'Connor, P., & Bryden, R. (2000). Measuring safety climate: Identifying the common features. *Safety Science, 34*(1–3), 177–192.

Hee, D. D., Pickrell, B. D., Bea, R. G., Roberts, K. H., & Williamson, R. B. (1999). Safety management assessment system (SMAS): A process for identification and evaluating human and organization factors in marine system operations with field test results. *Reliability Engineering and System Safety, 65,* 125–140.

Helmreich, R. L. (1984). Cockpit management attitudes. *Human Factors, 26,* 63–72.

Helmreich, R. L., & Merrit, A. C. (1998). *Culture at work in aviation and medicine: National, organizational and professional influences.* Aldershot, UK: Ashgate.

Helmreich, R. L., Merritt, A. C., Sherman, P. J., Gregorich, S. E., & Wiener, E. L. (1993). *The Flight Management Attitudes Questionnaire (FMAQ)* (Tech. Rep. No. 93-4). Austin: University of Texas, Aerospace Crew Research Project.

Hollnagel, E. (1993). *Human reliability analysis: Context and control.* London: Academic Press.

Hollnagel, E. (1998). *Cognitive reliability and error analysis method (CREAM).* London: Elsevier Science.

Itoh, K. (1999). Approaches to the prevention of accidents and the analysis of psychological factors behind their occurrence. In Y. Uehara et al. (Eds.), *Handbook of safety engineering* (Chap. 11.6.1, pp. 765–769). Tokyo: Corona. (in Japanese)

Itoh, K., & Andersen, H. B. (1999). Motivation and morale of night train drivers correlated with accident rates [CD-ROM]. Paper presented at the International Conference on Computer-Aided Ergonomics and Safety, Barcelona, Spain.

Itoh, K., Andersen, H. B., Tanaka, H., & Seki, M. (2000a). Attitudinal factors of night train operators and their correlation with accident/incident statistics. In Cacciabue, P. C. (Ed.). *Proceedings of the 19th European Annual Conference on Human Decision Making and Manual Control* (pp. 87–96). Ispra; Italy: Joint Research Centre.

Itoh, K., Hansen, J. P., & Nielsen, F. R. (1998). Cognitive modelling of a ship navigator based on protocols and eye-movement analysis. *Le Travail Humain, 61*(2), 99–127.

Itoh, K., Seki, M., & Hoshino, T. (2001). A simulation-based study on night train operator's track inspection performance by use of cognitive-perceptual model. In Onken, R. (Ed.). *Proceedings of the 8th European Conference on Cognitive Science Approaches to Process Control* (pp. 103–109). Bonn, Germany: Deutsche Gesellschaft für Luft- und Raumfahrt-Lillienthal-Oberth e.V.

Itoh, K., Tanaka, H., & Seki, M. (2000b). Eye-movement analysis of track monitoring patterns of night train operators: effects of geographic knowledge and fatigue. In *Proceedings of the 14th Triennial Congress of the International Ergonomics Association* (Vol. 4, pp. 360–363). Santa Monica, CA: Human Factors and Ergonomic Association.

Kohn, L. T., Corrigan, J. M., & Donaldson, M. S. (Eds.). (1999). *To err is human: Building a safer health system.* Washington, DC: National Academy Press.

Orlady, H. W., & Orlady, L. M. (1999). *Human factors in multi-crew flight operations.* Aldershot, UK: Ashgate.

Pidgeon, N. F. (1991). Safety culture and risk management in organizations. *Journal of Cross-Cultural Psychology, 22*(1), 129–140.

Pidgeon, N. F., & O'Learry, M. (1994). Organizational safety culture: Implications for aviation practice. In N. A. Johnston, N. McDonald, & R. Fuller (Eds.), *Aviation psychology in practice* (pp. 21–43). Aldershot, UK: Avebury Technical Press.

Rasmussen, J. (1986). *Information processing and human-machine interaction: An approach to cognitive engineering.* New York: North-Holland.

Reason, J. (1990). *Human error.* New York: Cambridge University Press.

Reason, J. (1997). *Managing the risk of organizational accidents.* Aldershot, UK: Ashgate.

Sanderson, P. M., & Fisher, C. (1994). Exploratory sequential data analysis: Foundations. *Human-Computer Interaction, 9,* 251–317.

Swain, A. D. (1990). Human reliability analysis: Need, status, trends and limitations. *Reliability Engineering and System Safety, 29,* 301–313.

Swain, A. D. & Guttman, H. E. (1983). *Handbook of human reliability analysis with emphasis on nuclear power plant applications* (NUREG/CR 1278). Albuquerque, NM: Sandia National Laboratories.

van Vuuren, W. (2000). Cultural influences on risks and risk management: Six case studies. *Safety Science, 34*(1–3), 31–45.

Vicente, K. J. (1999). *Cognitive work analysis: Toward safe, productive, and healthy computer-based work.* Mahwah, NJ: Lawrence Erlbaum Associates.

Zohar, D. (1980). Safety climate in industrial organizations: Theoretical and applied implications. *Journal of Applied Psychology, 65,* 96–101.

26

Lessons From a Focus on Artefacts and Implicit Theories: Case Studies in Analysis and Design

Ann M. Bisantz
University at Buffalo, The State University of New York, USA

Jennifer J. Ockerman
Georgia Institute of Technology, USA

Abstract

A claim in human–computer interaction is that technological artefacts embody implicit theories about people, including how they will use the artefacts. More broadly, artefacts embody theories of the work setting in which the artefact will be implemented. By identifying these embedded theories and examining the use of a built artefact in a real life setting, one can assess the adequacy of those theories in describing how humans will interact with the technology and how work in the setting is conducted. This chapter describes such an investigation of the use of decision support technology in a quick-service restaurant environment. An analysis indicated that the decision support technology was not used as intended, in part because the theories embodied in the technology did not represent conditions in the restaurant. Specifically, the system embodied theories regarding human decision making, the use of decision aids, and the way work was organized that ran counter to those descriptive of the way these activities were actually conducted. This contrast illustrates differences between competing theories of decision making, decision support, and work organization: the study serves as a falsifying case to refute those theories that are highly prescriptive in nature. The chapter describes the work environment and methods of data collection. Subsequently, evidence is provided from field observations that support accounts of human decision making and work organization that emphasize the flexible, distributed, and continuous nature of those activities. In addition, the analysis provides information regarding the types of decision aids that are more appropriate for intentional work environments. Finally, the role of similar embedded theories in informing design are explored through a second case study in which computerized technologies were being introduced into a different work domain.

INTRODUCTION

Computerization and advanced technologies are being applied to aid and support an ever-expanding set of work activities, spanning technologies from advanced cockpit automation (Pritchett, 2001; Wiener, 1985), to decision aids for military command and control (Cannon-Bowers & Salas, 1998; Martinez, Bennett, & Shattuck, 2001; Morrison, Kelly, Moore, & Hutchins, 1998), to technologies for retail establishments and light manufacturing. Of issue in the design of such advanced technologies is the role that theories of human–system interaction can play in their development, how such theories can be developed, and the role that technology itself can play in illuminating important theoretical issues (Carroll & Campbell, 1989; Dowell & Long, 1998; Karwowski, 2000; Woods, 1998).

For example, Carroll and Campbell (1989) describe a task–artefact cycle, in which technological artefacts are developed to support existing tasks, which in turn are shaped by the new technology, resulting in the need for redesigned artefacts, and so on. By designing artefacts to support specific tasks, designers are necessarily making an argument that certain features of the artefact will enable performance of the task. In their analysis of psychological research and human–computer interaction, Carroll and Campbell (1989) asserted that "artefacts embody implicit theories of human-computer interaction" (p. 250). Specifically, artefacts embody theories regarding the psychological characteristics of users of the artefact, and the work that the artefact was designed to support. That is, by creating a computer artefact, the designer is—explicitly or implicitly—considering theories of the user's needs and how people will use the artefact. By extension, one can claim that artefacts (including noncomputerized artefacts) embody theories not only of the psychological characteristics of users, but theories of their physical characteristics, along with the workplaces and social systems in which the artefacts will be used. It follows that the success of the artefact will depend on the adequacy of these embedded theories (Kirlik, 1995).

Identification and consideration of the theories implicit in an artefact can be used for analysis, to understand difficulties in system performance in terms of the inadequacy of the embedded theories in describing user characteristics (Bisantz, Cohen, & Gravelle, 1996). More fundamentally, artefacts can serve as theories of human–computer interaction, in place of explicit theoretical claims (Carroll & Campbell, 1989). As noted by Kirlik (1995), every instance of human interaction with an artefact acts as an experiment in which the theories implicit in the artefact are tested. It has been argued that design activities in human–computer interaction, through the production of technology, contribute to, rather than rely on, a body of theoretical knowledge (Dowell & Long, 1998). Instead of moving from scientific theories and knowledge to the design of computer artefacts based on that knowledge, the design and subsequent study of the computer artefacts themselves result in a greater theoretical understanding of human–computer interaction, and, more generally, greater understanding of human cognition. Through the study of technological artefacts in situ, the theories implicit in the artefacts can be tested in a real life laboratory. That is, the successes (or failures) of the artefact will provide information that can support or counter the implicit theories of human–computer, or, more broadly, human–system interaction embedded in the artefact. Taken even further, theories of human–system interaction can be tested explicitly by implementing and testing systems whose elements are based on those theories (Woods, 1998).

For example, Payne (1993) studied properties of memory and prospective remembering through an analysis of relevant artefacts and task properties. Payne analyzed interviews from paper and electronic calendar users and attempted to identify the psychologically relevant design characteristics of the calendar, the artefact of interest. Payne examined the psychological tasks supported by the artefact (e.g., prospective memory, or remembering to do things) and showed how aspects of the tasks were affected by aspects of the artefact (e.g., recording an upcoming event necessarily schedules it, because of the time and date layout of calendars). Payne also examined how conceptual entities, such as the events to be remembered, were represented by the artefact. Other researchers have focused on artefacts to understand properties of work and interactive systems with a focus beyond individual users and their cognitive characteristics. For instance, Suchman and Trigg (1991) analyzed the use of a paper grid sheet detailing luggage and passenger transfers between flights and how that sheet along with its annotations over time served to support communication between personnel at different locations. Pascoe, Ryan, and Morse (2000) considered physical characteristics of the work environment in their evaluation of the suitability of hardware and software designed to support field research tasks. Characteristics, such as the fact that users may be in a variety of dynamic postures during data collection, serve as theories of the field researchers and their activities in a similar fashion to theories of work activities and users' cognitive characteristics.

Hutchins (1995) studied theories of distributed cognition through a description of how artefacts supported the team task of navigating a merchant ship close to land. In particular, Hutchins studied how the artefacts used by the crew members supported the distributed nature of the work. The artefacts supported the distribution of information among crew members and its simultaneous transformation into forms of information that allowed the location of the ship to be determined. For example, specialized navigational tools transformed bearing information sighted by one crew member into location information plotted by another crew member on a navigational chart. The open nature of the information transformation and recording processes supported both the effective coordination of tasks between individuals and also the process by which more junior crew members learned the skills needed to perform more advanced duties. These artefacts were successful in part because they correctly captured and embodied the information transformation and crew coordination tasks that had to be performed to accomplish the system goals. In this case, the artefacts embodied theories not only of individual cognitive characteristics (e.g., the necessary knowledge to perform crew duties) but also of the manner in which information flowed and tasks were coordinated, and the social structure of the system (e.g., how junior officers were trained).

In the following case study, observations were made regarding the use of decision support technology in a quick-service restaurant environment. An analysis indicated that the decision support technology was not used as intended, in part because the theories embodied in the technology did not represent conditions in the restaurant. In this case, the decision support technology embodied normative, top-down theories of how decisions are made, decision aids are used, and work is organized. The analysis provides information contrary to those theories, and instead it supports alternative theories of human cognition, decision making, and work organization that are more descriptive of human performance.

CASE STUDY: FOCUSING ON ARTEFACTS AND EMBEDDED THEORIES

Environment Description

A major segment of the food service industry is composed of fast-food or quick-service restaurants. Determining how much food to cook is a ubiquitous problem in these environments. It is critical that the appropriate amount of food be prepared to serve customers in a timely manner, without cooking so much food that it is either thrown away or served to customers past its expiration time. These decisions are difficult for several reasons. First, they are affected by a number of variables including cooking times, product hold times, cooking capacity, holding capacity, and customer demand, which is to some extent uncertain. For example, demand has predictable trends based on the time of day, but unexpected rushes can occur as a result of community events or busloads of travelers pulling off the highway. Second, these decision tasks embody the sometimes conflicting goals of customer service and cost containment. Because of the competitive nature of the business, it is necessary to provide good customer service, which encompasses quality food and rapid service. Service times are affected not only by the speed of ordering and packing items, but also by the immediate availability of the items ordered, because cooking times are high relative to acceptable service times. However, the need to control costs and serve quality products can limit the ability to prepare enough food to meet these service times in the face of unpredictable demand, because excessive waste is costly and serving customers expired product is problematic.

These stores can be characterized as a system of production and service equipment staffed by a team of individuals with varying skill levels. This team works in concert to achieve the restaurants' goals of satisfying customers and controlling costs. The team includes supervisory and nonsupervisory personnel and has a mix of experience levels, from no experience to many years of experience. The positions in the store are differentiated (e.g., cooking, preparing food, taking orders, and packing food), and people are assigned to tasks in different ways. Tasks may be assigned on a daily basis or may change throughout the day as demands change. Some positions may be assigned permanently, either formally or informally. For instance, someone might be hired and trained to fill a position such as a cook, whereas certain other food preparation activities might be "claimed" by an employee who would always perform that activity.

Management of cooked food inventory in these restaurants is a cooperative task that requires information and actions from team members in various restaurant positions. For example, team members serving customers should ideally monitor customer flow to assess demand, and team members that pack food should arrange and restock products so they are readily available to be served, and they should also alert management when product levels are low. These activities affect the decision of determining the correct amount of food to cook. Thus, information about the state of the store (e.g., amount of product cooked, amount of raw product, state of in-process food) and the activities associated with the cooking decisions are distributed across the individuals working in the store.

To assist store personnel with this task, restaurant management in one chain installed a computerized decision aid to provide minute-by-minute cooking instructions based on current sales, historic sales data, and cooking parameters (e.g., cooking times). This aid was intended to support the activities of cooking, monitoring the amount of food available to be sold, and disposing of cooked product in order to maintain appropriate levels of quality cooked product available for sale. It was believed that this aid would improve store performance, because cooks and managers no longer had to consider any factors in the cooking decision or rely on order-taking personnel to monitor product levels, but instead could just act based on computer instructions. The Cook Decision System (CDS) provided minute-by-minute cooking instructions based on predicted sales, actual current sales, and the amount of product on hand. The system also computed the amount of on-hand product that had expired based on product hold times. The system consisted of one monitor with an alphanumeric display mounted in the kitchen, along with a numeric, membrane key data entry device, which allowed the cooks to select a product and enter the number of units they had just cooked. For each product, a number indicating the amount to cook, the amount currently being cooked, the cooked amount on hand, and the amount no longer fresh enough to be sold were shown in a matrix format. This information was also displayed on a computer in the manager's office.

Method

Data regarding the use of the CDS were collected during a comprehensive study of operations at several restaurant chains, which comprised interviews with upper management, focus groups with restaurant managers, hands-on training, observations, and videotaping (see

Cohen, Gravelle, Wilson, & Bisantz, 1996 for a more complete description of the methods). Eleven interviews were conducted with upper management personnel from the areas of field operations and information technology. Interview topics included corporate goals, functions of the subjects' departments, problems with store operations, and information technology use. During the interviews, one interviewer drew concept maps (McFarren, 1987) of the subjects' responses. These maps presented the concepts mentioned in a response, and their relationship to each other, in a pictorial format. The maps were reviewed with the subject during the interview, allowing the subjects to modify or augment their responses. Later, a content analysis was performed on the concept maps.

Four focus groups were conducted with groups of 5 to 10 restaurant managers or assistant managers. Topics of discussion included store operations and performance measures, product and inventory management, and labor issues. A content analysis on the transcribed focus group responses was performed by using the tool MacSHAPA, which was developed to support analysis of sequential data streams such as verbal protocols or video data (see Cohen et al., 1996 for a more complete description of the analysis, and Sanderson et al., 1994 for a description of MacSHAPA).

Several phases of observations were conducted in this study. First, two of the analysts participated in a week of hands-on training in the restaurants, learning team member and management activities. Subsequently, several days of observation were conducted at six different locations, which spanned both geographic regions and sales patterns. During this time, we observed activities, participated in certain activities (e.g., packing food), and asked questions of the management and team members. Finally, multiple stationary cameras were used in two restaurants to capture activities ranging from raw product management, cooking activities, and food preparation and packing to customer service activities. Tapes were analyzed to determine the activities performed and their duration over the course of 1 complete day at each restaurant.

Unlike a laboratory experiment in which the experimental conditions are both created and controlled as part of the research, researchers performing field studies of real work environments cannot typically intervene in the work process in order to create the conditions of observation. However, field researchers can take advantage of naturally occurring "experimental conditions" in order to exert control over the conditions of observation. Given the characteristics of the work environment, it may be possible to selectively sample from different locations (e.g., assembly lines with different equipment or experience levels) or across time periods in which relevant variables, such as production demands, differ (e.g., across shifts, across days or weeks, or across time periods within a shift or day). Additionally, observations can be collected and compared across individuals with different responsibilities or levels of experience, or across types of tasks.

In this study, several steps were taken to control the conditions of observation. Restaurants from different geographic and demographic regions were selected for study in order to compare findings across different levels of automation (which varied regionally) and different sales and demand patterns (which varied as a result of demographics). Activities under different sales patterns were also captured by performing observations on selected days of the week, and at different times of day. Furthermore, personnel from different levels of the organization, from management to employees, were interviewed or observed.

Analysis

Extensive observation and subsequent data analysis indicated that far from being an aid to real time decision making in the restaurants, the CDS was at best largely irrelevant to the decision-making process, and was at worst a burden by imposing activities on managers that did not contribute to the tasks at hand. This conclusion was substantiated by documenting the information provided by the CDS and the cooking activities that occurred over 1 complete day at one of the restaurants, using videotapes, observations, and computer logs. Attempts to use the system were made for only two of eight products, and cooks did not cook the amount indicated on the screen nor consistently enter the correct amount cooked in a timely manner. Instead, observations indicated that the system was used more as a post hoc accounting of how much product was available: the on-hand count would become incorrect because of missing, incorrect, or delayed inputs, and the managers would manually count the number of product units on hand and enter this correction into the system.

The reasons for this "use" of the CDS were due in large part to a mismatch between the assumptions regarding work practices in the restaurants and the actual practices as they were shaped by decision-making methods, organizational constraints, store layout, and the tasks themselves. For instance, at a very basic level, the system required cooks to input information by using a membrane keypad: this task was difficult because the cooks' hands were often messy from the cooking process, making it difficult for them to use the entry device and to keep it clean. From an organizational perspective, the CDS failed because it assumed managers could stay in their offices and monitor the information provided by the aid on their computer screen. In contrast, managers did not typically spend time in the office except to handle cash during shift changes. The typically young, inexperienced team in the store needed close supervision, and managers had to take on store positions (i.e., serve customers or cook food) in order to keep labor costs within limits set by upper management.

More substantially, the system embodied theories regarding human decision making, the use of decision aids, and the way work was organized that ran counter to those descriptive of the way these activities were actually conducted. This contrast illustrates differences between competing theories of decision making, decision support, and work organization: the study serves as a falsifying case to refute those theories, which are highly prescriptive in nature. The analysis also provides information regarding the types of decision aids that are more appropriate for particular forms of work environments.

Decision-Making Analysis

Research on human decision making has spanned more than half a century, and it has encompassed multiple theoretical viewpoints. Notably, traditional theories of decision making considered this activity to be a mathematically defensible process with well-defined stages of enumeration and comparison of alternative courses of action (Von Neumann & Morgenstern, 1944) and selection of a single alternative as the effective endpoint to the process. These theories were offered as a prescription for normatively correct decision making (e.g., Lindley, 1985; Savage, 1954), and any deviations from this pattern were typically considered to be nonnormative biases, resulting from heuristics developed to cope with information-processing demands (Kahneman, Slovic, & Tversky, 1982; Tversky & Kahneman, 1982), rather than appropriate adaptations to the demands of a complex and dynamic environment. Essentially, normative decision making has been modeled as a process based on an essentially feedforward strategy. Globally optimal decisions are made without a need for further refinement on the basis of additional information or changing circumstances.

The CDS, as designed, was consistent with this more normative paradigm. Its design rested on an assumption that, during its development, designers could enumerate, compare, and select from all potential circumstances, and thus compute algorithmically what decision was appropriate at any point in time. The data collected during the field study regarding the actual use of the CDS provided information which discounted this method as appropriate for decision making in dynamic, complex environments. In particular, the design failed to account for the dynamic, unpredictable nature of the environment, which made it difficult for decision outcomes to be predicted. That is, it was difficult to predict whether or not an amount of food would be sufficient to meet demand. Instead, it was more appropriate to make decisions serially in response to feedback about the changing demands for products. In this way, the results suggest that the normative models of decision making just described are not just inappropriate descriptions of human decision making, but are in fact inappropriate theories for expert systems and decision support technologies to embody in environments that are dynamic and uncertain.

Instead, on the basis of information collected during the field study, we concluded that managers and cooks made cooking decisions by using a combination of feedforward and feedback strategies and based their decisions both on experience and through observation of dynamic variables (demand) in their environment. In fact, in the case studied here, a number of strategies that attempt to maintain appropriate levels of cooked products can be considered with regard to their effectiveness. A simple strategy would be to cook all food to order: however, although this strategy would minimize waste, it is precluded by a requirement for quick service times. Alternatively, food could be cooked in large batches at the beginning of the day, where batch size is based on past experiences in daily sales. However, this strategy could result in either poor quality products (caused by long holding times for the cooked food) or excessive waste. A more sensitive feedforward strategy based on experience could also be utilized: given knowledge of likely demand for a particular time of day, managers and cooks could cook appropriate amounts ahead of time to ensure an adequate, but not excessive, supply is on hand. Finally, personnel could use a more feedback-oriented strategy, by monitoring the amount of prepared product in the store and cooking more when the amount drops below some trigger level.

We found that strategies used by store personnel corresponded to the latter two cases. Cooking decisions were made on the basis of experience with typical sales patterns and the size of food-holding equipment available. In particular, personnel would compare the current day's sales demand with typical demand and adjust the amount they cooked accordingly. Instead of relying on information provided by the CDS, managers would position themselves in the restaurant so they could observe both the customer flow into the store and the amount of unsold product, and they would also check sales figures for the day to see if demand was typical. Personnel serving customers also alerted cooks as to the amount of food left in the warmers, and they would call back to the kitchen if they served an unusually large order.

These strategies, in addition to refuting earlier, normative theories of decision making, are consistent with and thus provide support for more current theories of how people make decisions in dynamic, uncertain environments. In particular, store personnel made decisions in a way consistent with theories claiming that decision making in dynamic environments is continuous and cyclical in nature. Several researchers (Brehmer, 1990; Brehmer & Allard, 1991; Connolly, 1988; Hogarth, 1981) have considered the continuous, cyclical nature of judgment and action in the natural environment. In these conditions, people receive immediate feedback about the effects their judgments and subsequent actions have had on the environment. The availability of this information makes it possible for people to act without making highly accurate initial judgments, because they can take remedial action based on feedback.

For example, because of the continuous, cyclical nature of judgment and action in the natural environment, Hogarth (1981) claimed that in a continuous environment, people can reduce effort and risk by making locally sound choices and then adjusting their actions over time on the basis of both feedback and the new state of the world. In addition, the relations between environmental cues and the criterion to be judged may change, or new cues might become available. Connolly (1988) described how the characteristics of continuous decision tasks support different decision-making strategies than the normative paradigm of selection of a single, globally optimal choice. For instance, a situation in which actions have low cost, low consequence, and a range of acceptable solutions (e.g., trimming a hedge) supports a trial-and-error approach to decision making. A gardener can make small, local cuts to shape a hedge until it is an acceptable size, or a consumer can purchase and return a series of items until he or she finds one that is suitable. This can be contrasted to the decisions involved in removing a tree, or purchasing a home, in which the costs and consequences of acting are much higher, and the reversibility of the decisions are low.

Brehmer (1990) and Brehmer and Allard (1991) also considered factors that influence selection of feedforward versus feedback decision strategies. They found that participants adopted a combination of feedforward and feedback strategies to control the deployment of firefighting equipment in an experimental microworld, which provided participants with real-time feedback about the results of their actions. Participants monitored the spread of the fires and dispatched firefighting units. Brehmer and Allard (1991) found that participants tended to use a feedback strategy, waiting before moving firefighting units, when they could not predict system delays. In contrast, when system delays were more predictable (e.g., the delays in the fire spreading), participants used a feedforward strategy, responding quickly and with many firefighting units even before a fire had spread. Thus, participants' decision strategies were dependent on their understanding of the system dynamics. Feedforward control was possible in circumstances in which experienced decision makers had a good, predictive system model, whereas feedback strategies were necessary when system behavior was unpredictable.

Decision makers in the restaurants were able to use their experience to make general predictions about daily demand and begin cooking products, and thus had some understanding of system behavior. Their feedforward approach was necessary because of the speed at which customers demanded services (e.g., cooking times precluded a solely feedback strategy). These cooking decisions were then fine tuned throughout the day by using feedback about the unpredictable level of demand.

Such feedforward, experience-based strategies are also consistent with another theory of decision making, recognition-primed decision making (RPD) (Klein, 1993), which proposes that decision makers attempt to recognize the current situation and identify actions on the basis of their past experience with the situation. In this case, store personnel used experience to identify likely demand throughout the day, and they actively sought out information so they could identify, or recognize, the current system state.

Decision Aid Design

In addition to providing a means to explore theories of decision making, the CDS allowed the analysis of human behavior with respect to the utilization of decision-aiding technology. The CDS required cooks to enter information accurately and in a timely manner, and then cook product according to its instructions. In essence, this required store personnel to behave as data gatherers for the system and take the advice of the system. This type of aid can be categorized as a cognitive prosthesis, or as a tool developed primarily to remedy a human deficiency (Woods, Roth, and Bennett, 1990). The observations collected in this study provide evidence supporting claims of the limited value of prosthesis-style aids. Woods et al. (1990) describe that a primary example of a cognitive prosthesis is an expert system that produces problem solutions. Human interactions with such systems serve primarily to provide an interface between the environment and the expert system. Humans are reduced to data gatherers for the systems, and they are expected to accept and implement solutions provided, as long as the solution is correct. Thus, human operators, who are often ultimately responsible for the joint human–system performance, must serve to assess and filter the solutions provided by the expert system.

In this case, the CDS functioned as a system, which "remedied" the restaurant personnel's inability to compute detailed, minute-by-minute sales projections. The CDS offered a problem solution in the form of a detailed instruction regarding how much to cook. However, these projections failed to take into account time-based and restaurant-based sources of variability (e.g., increases in sales caused by payday, or after the end of a nearby baseball game), as well as unpredictable demand fluctuations (e.g., busses pulling off of a highway). Such sources of variability would be difficult or impossible to enumerate across hundreds of restaurants with different sales patterns, in order to include them in the CDS algorithms during system development.

As already noted, to function, the CDS required cooks to enter information into the CDS and follow its instructions. This was problematic because, as described by Roth, Bennett, and Woods (1987), difficulties arise when humans are placed in the role of data gatherers for machines, and also when instructions are provided with no explanation of their derivation, as was the case here. Similarly, Mackay (1999) described how operators, in this case air traffic controllers, were resistant to technology that required them to input data to a computerized system with unknown benefits to them. As an alternative, Woods et al. (1990) proposed a paradigm of cognitive instruments, or tools that support the activities of experienced practitioners. Rather than force a human into a role as a passive data gatherer and solution implementer, a cognitive instrument is used, as needed, by operators actively engaged in pursuing a goal. Correspondingly, it is informative to examine the "low-tech" aids that cooks and managers actually used in the restaurants.

For example, decision making was simplified through the reliance on and manipulation of aspects of the physical environment. Cooks looked into holding equipment where cooked product was kept to assess the amount of product available and told us they cooked to "keep the warmers full." In observations made in this study, and across other restaurant chains, employees reported heuristics for making cooking decisions that depended on the configuration of holding equipment. They claimed to cook to keep shelves in warming cabinets or chutes for sandwiches full. The employees would change the number of chutes to keep full at different parts of the day based on expected demand, and employees were observed removing shelves from warming cabinets for cleaning at the end of the day, effectively reducing the holding space. Although the latter example was driven by another store process (i.e., cleaning before the restaurant closed), the timing of the activities were enabled by the reduced demand at the end of the day, and they served as a concrete reminder of the reduced demand.

The conclusion that the physical environment served as a decision-making aid was reinforced by problems observed in maintaining appropriate levels of cooked product at another chain, which had only a large "slide" to store prepared products. Employees attempted to create structure by lining up products in rows. This method failed, however, because of the lack of physical constraint, and employees were therefore unable to adequately assess the current state of cooked product. Other researchers have indicated how the physical structure of the environment can be used to support memory recall tasks (Beach, 1988) or manipulated to reduce uncertainty about the state of certain variables (Kirlik, 1995).

Thus, in this case, the physical environment itself can be considered a cognitive instrument. The physical environment was used by restaurant personnel to support their own, experienced methods of decision making, by flexibly supporting the task of identifying how much cooked product was available. Such methods are consistent with the decision-making strategies used by restaurant personnel, as already described: they allowed personnel to identify the system state and take action on the basis of their recognition of the situation.

Work Organization

Finally, the implementation of the CDS provides a means to highlight the insufficiency of theories of work organization that assume work to be conducted in a highly planned fashion corresponding to normative procedures, and instead provides support for theories of work as a set of highly situated activities. Numerous methods and theories regarding the description of work activities assume those activities to occur in a very structured and often sequential fashion that is consistent across individuals and circumstances, and therefore that optimal sequences of activities can be identified and prescribed. The principles of scientific management developed by Taylor (1911), for example, in which work was decomposed into essential elements, reorganized to eliminate unnecessary elements or inefficiencies, and then linked to optimal completion times, motivated many of the task analytic methods used to describe and improve work (see Kirwan & Ainsworth, 1992, or Luczak, 1997, for a review). Task models based on such rationalized descriptions or prescribed sequences of actions have been used as the basis for interface design and analysis (e.g., Gray, John, and Atwood, 1993; Mitchell & Miller, 1986), whereas in the fields of cognitive science and artificial intelligence, theories that human behavior is structured by detailed, goal-oriented plans (e.g., Newell & Simon, 1972; Schank & Abelson, 1977) have influenced the development of cognitive and computational models.

Opposing theories, however, have suggested that such proceduralized accounts of work fail to account for the actual methods by which human activity in general, and work activities in particular, are organized (e.g., Adler & Winograd, 1992; Button & Harper, 1993; Sachs, 1995; Suchman, 1987). For example, Suchman (1987), through a field study of a photocopier help system, determined that the failure of the help system to assist users was not due to the lack of any particular help function or interface component, but rather the underlying assumption that user activities regarding the photocopier could be interpreted on the basis of a predetermined plan of how users would act. Instead, Suchman claimed that user activities were situated or determined on the basis of the moment-by-moment, changing activities and characteristics of the users, the device, the situation, and their interaction. Sachs (1995) also described how a system based on a top-down rationalized model of work activities failed to support system troubleshooting tasks, because the model failed to recognize the value of ad hoc communication between workers.

As designed, the CDS was an example of a support system developed on the basis of proceduralized notions of work organization, product flow, and information flow. For instance, the only personnel who could easily view the CDS were cooks, who had a screen in their working area, and managers, who had a screen in the office. From a normative work organization standpoint this was rational: cooks prepared the food, in a specified area of the restaurant, and managers performed tasks in their office and could therefore monitor the food inventory through the CDS.

This conflicted, however, with actual work practices. First, because of labor and organizational constraints, managers did not typically spend time in the office except to handle cash during shift changes. Because the typically young, inexperienced team in the store needed close supervision, managers circulated throughout the food preparation and serving areas. Managers also regularly took on store positions (i.e., acted to serve customers or cook food) as needed in order to keep labor costs within limits set by upper management.

In addition, food preparation activities were flexible in terms of their location, time, and assignment to employees. Cooks did not prepare all products: some were prepared in a different area by other employees, including cashiers when they were not serving customers. Minute-by-minute instructions provided on the CDS for some items were unnecessary because some products were prepared in bulk, during slow parts of the day, so that labor was utilized efficiently. The heating of some products that could not be prepared ahead of time was managed through communication among the employees and manager, based on their task demands at the time. For example, the drive-through operator would prepare some products at the same time she was listening to and talking to customers, before she assembled the order. For the CDS system to support this allocation of labor, cooks (who were in a position to view the information on the CDS screen) would have to communicate the need for food to be prepared (as indicated by the CDS) to other employees such as the drive-through operators and cashiers, because they were located in the front of the store where they could not see the CDS. In fact, cashiers communicated among themselves regarding the amount of certain items that were needed, and also communicated to the cooks about other products that were running low. Thus, rather than flowing from the CDS to the cooks and managers, to other personnel in the restaurant, information flowed within the front customer service area from one cashier to another, or from the service area to the cooks in the kitchen area.

Finally, the CDS was based on the model that cooked products were moved through the restaurant in an ordered fashion so that the portion of the product units that had expired could be identified. That is, in order for store personnel to dispose of expired product as indicated by the system, they must be able to differentiate expired from nonexpired product. The theoretical path for cooked items was from the kitchen area to holding warmers to a front area from which it was served. However, this did not reflect actual store practice. Storage racks held different types of food, and store personnel often took products directly from the warmers if the type they needed was not in the front area, leaving partial racks in the warmers and front case. Cooks would consolidate racks from the warmers and the front case when they ran out of racks to use. Thus, products with different expiration times could be mixed on the same racks.

In summary, the design of the CDS assumed predictable work activities in terms of their location, timing, and assignment to personnel, and it was based on normative flows of information and product through the restaurant. These rationalized models of restaurant activities did not account for the situated methods that store personnel actually used to communicate information, allocate, and perform tasks, and therefore the CDS as designed did not adequately support those activities.

CASE STUDY: IMPLICATIONS FOR COGNITIVE TASK DESIGN

As described earlier, identification of theories of decision making, cognition, communication, and work organization implicit in technological artefacts provides a measure for testing or evaluating those embedded theories in a real world laboratory. Taking such an approach, either with an existing artefact or by identifying theories or work activities during design, can also be informative for design of environments, which support necessary activities. As illustrated in the aforementioned descriptions, and as implied by the task–artefact cycle, the design of artefacts (and by extension work environments) affects the type of tasks possible, and the manner of their performance. Identifying relevant embedded theories and applying them to design provides some specification of the opportunities for, and constraints on, work activities, in essence "designing" aspects of the cognitive tasks. Thus, making explicit the theories embedded in the CDS, and contrasting those theories to the actual situations in the restaurant, could provide a basis for the redesign of existing artefacts. For example, the physical design of the artefact should be such that it can provide information regarding cooking needs and projections at multiple locations throughout the facility, where all personnel who make cooking decisions can access the information and communicate regarding the decision recommendations. Information relevant to demand projections and cooked food available should be collected automatically, where possible, instead of relying on data entry by store personnel.

Likewise, when a new artefact is designed, the theories in the work environment that should be embodied have to be articulated and related to higher level goals to ensure that they support the purpose of work. To illustrate this, we present a case study from a different work environment—the inspection of poultry products in a poultry processing plant.

Environment Description

As a result of increasing competition, demand for high-quality poultry products, and governmental regulations, the poultry processing industry is faced with increasing requirements to collect and store process data. Currently, in the poultry processing industry, quality assurance inspectors collect a variety of data on paper and then, at the end of their shift, type some of that data into the company computer system to be used in daily, weekly, and monthly reports and charts. This process, which leaves a majority of the data on paper, means that data cannot be evaluated on a real-time basis and is of little use in proactive control of the poultry plant processes. In addition, the process of transcribing the data from paper into the computer leads to an unknown number of errors in the entered data that are due to typographical errors. Finally, all of the collected data must be saved for several years to meet government regulations. Retaining this information in paper format has become increasingly cumbersome, as collection requirements have increased. In response to these pressures, the industry is moving toward electronic data capture and storage. The poultry industry has identified three benefits to electronic data collection and storage: reducing the need for physical data storage space and making it easier to retrieve stored data; supporting data analysis and reporting; and providing real-time data for process control. All three reasons may result in significant economic benefit, particularly from the ability to better understand and control the production process through the use of real-time data. For these reasons, a poultry processing company asked us to design a computerized system to aid in the collection, storage, and distribution of product quality data on a real-time basis. To meet their needs, we determined that a field study would be necessary for us to understand the current situation and work practices.

Method

To gather information relevant to design, we conducted open-ended interviews with both managers and inspectors to understand the current process and identify perceived problems with the current method of collecting, storing, and disseminating process and quality data. In addition, we observed several inspectors doing a variety of tasks (videotaping was not allowed), and we recorded some verbal protocols of inspectors performing their tasks. Two tasks with very different characteristics were selected for detailed study and comparison. The first task had a variety of steps and took several hours to complete. This task occurred once per day and was completed by a single inspector. The second task comprised just a few activities and took only minutes. This task was completed every half hour throughout a shift, by several inspectors working in parallel throughout the processing facility. Finally, the approximately 50 different data collection forms used by inspectors were collected and examined in detail, as were the few reports and charts that were generated from the data. Observations were made, and data was collected, at intervals over a 2-year period. On the basis of these data, we developed theories about the work environment. Aspects of the environment and work activities led to specific theories that would have to be accounted for, or included, in the new data collection system. Aspects of the environment and the theories these suggest are discussed in the paragraphs that follow.

Embedded Theory Development

Theories addressing work organization, communication, cognition, and physical constraints were developed after an analysis of the poultry processing work environment, and they are described in the following sections. Although these theories were identified heuristically, and do not necessarily represent a complete set of design requirements, developing design solutions that integrate or embed those theories would shape the resultant work activities in a manner consistent with the structure of the environment.

Theories Regarding Work Organization

Three theories concerning the work organization existing in the plant were identified. These theories would have to be accommodated by a new system. First, *inspectors required the ability to collect and share any type of data or information about organizational responsibilities at any time.* To meet the changing needs of the facility and because of a high turnover rate, inspection activities could change on a monthly, weekly, daily, and even hourly basis. Tasks were flexibly and dynamically assigned to inspectors either by managers or through communication among the inspectors themselves. This need to support flexibly assigned work activities in order to most effectively use the available workforce constrains the system's design. Designing a data collection system that integrates this theory supports an important quality assurance objective: collecting all the necessary inspection data.

Second, *inspectors required access to the electronic data collection system in variable locations throughout the plant.* This theory reveals aspects of the organization of the inspector's routine work. Inspectors were responsible for collecting different data at various sites throughout the plant. To ensure that inspections remained irregular in terms of product, location, and time (e.g., product *x* should not always be checked at location *y* at time *z*, so as not to bias the data with some unknown correspondence between these aspects), the inspectors were encouraged to do random inspections of different products in different areas of the facility. In poultry processing plants the distance between potential data collection sites can range from a few feet to hundreds of feet. Thus, this aspect of the environment imposed a physical constraint on the design of a new artefact: it had to be either portable or accessible from many locations. However, because one of the goals was random inspection, a portable device best met this work requirement. Thus, the nonprocedural aspects of the data collection process called for a flexible hardware component to the entire data collection system. Producing a new system design, which embedded this theory, was necessary to provide accurate and unbiased inspection data.

Third, *the forms and procedures were frequently revised.* As a result of externally and internally driven changes in quality control procedures and constantly changing manufacturing procedures, the data collection forms and procedures changed frequently. For example, in hotter weather, the frequency of temperature checks might be increased from two to four times per hour so that product temperatures could be monitored more closely. Forms and the tasks had to be responsive to the changing processing environment. This characteristic placed constraints on how a new design must support changing data collection procedures and inspection practices in order to keep up with dynamic processing procedures, environmental conditions, and raw product characteristics.

More fundamentally, on the basis of these three theories of the work organization, this work environment can be classified as an intentional work environment. In intentional work environments, human and organizational intentionality provide the primary constraints on human activity (Rasmussen, Pejtersen, & Goodstein, 1994). In such environments, although there are some technological constraints, work practice itself is the primary organizing feature. This characterization of the work organization is based on several factors. First, the actions of the inspectors and their managers were determined primarily by the current situation and skill levels available, rather than a set of standard operating procedures tied to the state of processing facility or natural laws. The inspectors were flexibly assigned to different data collection procedures in order to best utilize the variable workforce in the face of changing production requirements and environmental conditions. Under most circumstances, the inspection manager made the assignments ahead of time on the basis of the information that was currently available. However, as the plan was implemented, it was up to the inspection supervisors and actual inspectors to ensure that all the quality assurance goals were met. As with the restaurant case described previously, it was sometimes necessary for the supervisors and manager to take on one or more of the roles of the inspectors to ensure that the quality goals were met. With the high turnover rate in poultry processing plant employment, this was the case more frequently than the supervisors or manager desired.

Second, as already noted, data collection tasks deliberately included random elements of time and place. The goal of some of the inspections was to incorporate elements of random sampling, in order to produce a more thorough, comprehensive, and valid inspection that would not be biased by unknown, systematic factors in the production process. The unpredictability of the inspections relied on the self-directed actions of the inspectors rather than a preplanned schedule of inspections. This method of determining inspection sites depended on opportunistic behavior by the inspectors and relied on inspectors' knowledge of the process to investigate possible problems more closely during a shift.

Third, there were frequent changes resulting in new operating procedures. Changes were implemented to meet evolving production processes, quality assurance objectives, and environmental conditions. For instance, the poultry processing plants would make changes to production processes over time to improve productivity and meet the changing needs of their customers. These changes in turn drove the need to change the quality inspections that take place in the plant. In addition, over a shorter time frame, the quality objectives would change to ensure that the product leaving the plant was safe and of good quality. For example, at various times, plant and quality managers would identify quality data that they thought would help them to better understand the production processes and ensure higher quality products. These needs required the data collection forms to be continually modified to collect and generate the new data that were required. Finally, on an even shorter time scale, changes in environmental conditions might require changes in the data collection procedures. These changes were usually short lived and did not result in changes in the data collection forms; instead they were simply communicated to the inspectors.

This field study identified how important the intentional nature of the work organization was to meeting the goals regarding product quality. There was some desire during the design of the new data collection system to formally describe inspectors' activities in order to develop prescriptive, and ostensibly more efficient, inspection procedures. This would constrain the activities of the inspectors to a more predictable set of actions in order to facilitate the design and development of the new data collection system. Sachs (1995) referred to this tendency as the "rationalization" of the work. However, as evidenced in the theories just outlined, the ability of inspectors to flexibly and quickly respond to unanticipated changes in processing was essential to the activity of inspection. This flexibility must be supported by any new technology. Such conclusions parallel those made by other researchers regarding the need to consider the methods by which people accomplish their work, rather than relying on a rationalized or "organizational" view of the work process in designing new technology (Sachs, 1995; Zuboff, 1987). Instead, a recommendation for the new electronic data collection system included making the data collection software easy for the plant personnel to change in the face of new requirements and supporting the flexible allocation of work among inspectors.

Theories Regarding Communication

Because of the need for flexibility and quick reactions in this work environment, there were no formal communication channels. In more prescriptive environments, one would expect the formal communication channel to follow the hierarchy of positions, from the inspection manager, to the inspection supervisors, to the inspectors, and back up. However, the actual quality assurance work was self-organizing because the inspection employees as a whole had responsibility for the satisfactory completion of the inspections. Thus, it was the responsibility of each inspection employee to have another employee take over his or her normal activities if the inspection employee was called into a meeting or had to do further sampling at one site because of a failure. Thus, in the new design, it was necessary to provide for a means of communication among the inspectors as well as along the expected channels of communication, so that the work could continue to be done in a flexible manner that met the needs of the quality objectives. In addition, it was possible to enhance the means of communication because all the inspection employees (i.e., manager, supervisors, and inspectors) would have electronic access to the same system.

Theories Regarding Cognition

In addition to theories of work organization, several theories of the cognitive activities occurring in the work environment would have to be embedded in a new system to adequately support the inspectors tasks. For example, *inspectors required assistance to conduct error-proof calculations.* Many of the data collection tasks required arithmetic to determine if conditions warranted additional quality control checks. Although these calculations were simple, consisting of addition, multiplication, and division, they were tedious and often involved a large amount of data. In addition, these calculations were important in ensuring the safety and quality of product because they determined if a sample met the necessary requirements or further sampling of product was required. Small calculation errors often occurred. This characteristic of the work environment placed constraints on the type of support a new artefact must provide for data calculation activities, in order to meet reporting accuracy and quality shipping standards.

Additionally, *inspectors required reminders of when to do conditional samples and how to accomplish them.* Many of the forms that were used by the inspectors contained instructions regarding when and how to perform additional quality control checks. For example, if a sample had an acceptance rate of less than 95%, then it was necessary for the inspector to take additional samples in a prescribed manner from following batches of product. Samples did not often fail to meet inspection criteria. Although this is a positive circumstance with respect to product quality, it also meant that even experienced inspectors had little experience in the supplemental data collection procedures that were required when a sample failed, or the background to understand the statistical underpinnings of the additional data collection requirements. In the paper-based system, these additional requirements were listed at the bottom of the data collection forms. This characteristic of the environment puts constraints on the information that the new artefact had to provide to inspectors.

Furthermore, *inspectors needed a method of scheduling and reminding themselves of tasks that had to be done.* Inspectors were required to record the time at which they took a sample. Often, instead of doing this at the time they did the test, the inspectors would record the times they planned to do the tests. They then relied on the paper artefact to remind them of the need to perform an inspection, by periodically glancing through the paper forms. It appeared during the field study that the inspectors decided on what product to inspect at which location serendipitously as they were walking around the facility observing production. For example, if a type of inspection they had previously "scheduled" on their forms came due, they would stop and inspect the product where they were currently located.

Design recommendations made on the basis of these theories included providing automatic calculations and alerting inspectors to samples that failed the inspection criteria, rather than relying on inspectors to perform these tasks. This would prevent calculation errors allowing substandard product to be shipped, and it would free the inspectors to concentrate on the inspection task. Second, the system had to provide information to inspectors of when and how to complete additional inspections when a sample did not meet the necessary criteria. Third, the system could benefit the inspection process by providing methods for the inspectors to schedule different inspections throughout the day, and to prompt inspectors to perform inspections at the correct time. In addition, it is possible that the inspectors might improve their thoroughness and randomness with some form of feedback describing the inspections and sites they had already completed in the previous day or week.

Theories Regarding Physical Constraints

Finally, theories about physical aspects of the work environment were used to provide recommendations regarding the new system design. Inspectors required a method of inputting data as it is collected, without using their hands. Inspectors often used both hands to take temperatures of chicken pieces and examine chicken pieces for bones, skin, fat, and foreign material. The current data collection method required the inspectors to remember collected data until they had chance to wipe their hands off and then write the data down. Additionally, inspectors often had to collect temperature data in locations of the plant accessible by ladders: their hands would be occupied in holding on, as well as taking the temperature of the poultry. These aspects of the work environment imposed physical constraints on the design of the inspectors' portion of the data collection system, because a system that was physically inconvenient to use would either not be used or be used on a limited basis. In the case of this data collection system, if the artefact designed for the inspectors was not physically easy to use while both their hands were busy or dirty, then it would be a limited improvement over the usability of the current paper-based systems from the inspectors' point of view. In fact, a hands free, voice-recognition data collection system was recommended and prototyped. This design provided additional positive impacts on quality, because it reduced the possibility of cross-contamination that resulted from inspectors' handling of pens and clipboards as well as fresh product.

DISCUSSION

The case studies just illustrated demonstrate how focusing on the theories of users and work implicitly embedded in technology, or identifying those theories in a work context prior to design, can provide input to a design process involving redesigned or new systems, ultimately affecting the design of tasks performed by using those systems. This approach may provide particular benefits for design in domains where work activities are less predictable than domains typically studied in human–machine systems. Numerous theories and models of human–system interaction have been proposed and used to structure and analyze field observations in order to provide support to system designers through the identification of information and training needs. For instance, normative models of operator functions, based on prescribed functions and sequences of actions, have been used to predict operator actions or infer operators' information needs (e.g., Jones, Chu, & Mitchell, 1995; Mitchell, 1987). Sundström (1993) developed a model of goals and information needs, on the basis of knowledge of the goals and activities an operator should be pursuing given a particular system state. This model has been used to develop a field study methodology in which user's information needs are collected and organized based on a structured model of information processing goals (Sundström & Salvador, 1995). However, such models and the efforts based on them rely on the identification of normative procedures and activities. Such an approach is more suited to highly proceduralized domains, where the high levels of cost and risk have resulted in the development of standard operating procedures that restrict many human operator actions.

Typically, these environments can be characterized as highly *causal* work domains. Causal work domains have been defined as domains in which human activities are constrained by tightly coupled, technological systems governed by the laws of nature, the system requirements and processes defined by those laws, and the typically high levels of risk and cost associated with the systems (Rasmussen et al., 1994). In these work environments, such as aviation or nuclear power plants, normative procedures are often implemented in order to manage the assignment and completion of anticipated tasks (e.g., Degani & Wiener, 1997; Roth, 1994).

Modeling techniques used in these environments can more appropriately leverage the explicit procedural structure of the tasks in understanding and predicting tasks and information needs.

In contrast, the application of strongly proceduralized models of operator activity is less appropriate for other classes of work environments in which human and organizational intentionality rather than the laws of nature provide constraints on human activity (Rasmussen et al., 1994). Examples of these environments, termed *intentional work domains,* include the retail–restaurant and manufacturing environments discussed here. In these environments, although there are some technological constraints, work practice itself is the primary organizing feature. Such practice may include structured procedures determined at high levels of an organization, in addition to activities determined at lower levels. Alternatively, work practice may be primarily guided by actors' intentions. In these situations, higher level management goals are implemented at lower levels, shaped not by strict procedures but by situational context (Rasmussen et al., 1994). Workers may have a variety of skill and experience levels and may lack the training or motivation to follow detailed procedures. Instead, the timing, ordering, and location of work activities are flexible. Tasks are dynamically assigned to personnel based on available skill level, worker preference, and situational demands. The normative task models described above are not as well suited to focus and analyze observations from field studies in these more flexible situations, in which well-defined normative procedures are not used to manage tasks. Instead, as is demonstrated in this chapter, a focus on theories of users and work implicitly embedded in technology, or identifying those theories in a work context, may provide insight for the design of systems, which ultimately influence the tasks that are performed.

CONCLUSIONS

Through an analysis of theories of people, work systems, and the use of technology embodied in a decision-aiding system in a real life environment, conclusions were drawn regarding the appropriateness of competing theories of decision making, effective decision-aid design, and work organization. Specifically, the decision-aiding system embodied theories that ran counter to those descriptive of the way these activities were actually conducted, and it serves as a falsifying case for theories of humans and work that are highly prescriptive in nature. Instead, evidence based on field studies of the decision aid in a work setting supported accounts of human decision making, decision aiding, and work organization that emphasize the flexible, distributed, and continuous nature of those activities. A similar approach was used in a somewhat different domain in order to articulate the theories of users and work practices that should be embodied in the design of a new artefact. These analyses emphasize the role of system and environment design on the ultimate conduct of tasks, and the need to identify appropriate embedded theories to design systems that support successful task performance. In both cases, the focus on the theories of users and work implicitly embedded in technology was more suited to providing design insights than more normative models of human–system interaction, because of the intentional nature of the environments.

ACKNOWLEDGMENTS

The restaurant study was performed with and supported by Sally Cohen, Michael Gravelle, and Karen Wilson of the Human Factors Group of NCR Corporation, and supported by that organization. The inspection study was performed with Larry Najjar and Chris Thompson of the Georgia Tech Research Institute. Funding was provided through the Agricultural Technology Research Program, directed by Craig Wyville. We thank Alex Kirlik for his suggestions on framing this work.

REFERENCES

Adler, P., & Winograd, T. (1992). The usability challenge. In P. Adler & T. Winograd (Eds.), *Usability: Turning technologies into tools* (pp. 3–14). New York: Oxford University Press.

Beach, K. (1988). The role of external mnemonic symbols in acquiring an occupation. In M. M. Gruneberg, D. E. Morris, & R. N. Sykes (Eds.), *Practical aspects of memory: Current research and issues* (Vol. 1, pp. 342–346). New York: Wiley.

Bisantz, A. M., Cohen, S. M., & Gravelle, M. D. (1996). *To cook or not to cook: A case study of decision-aiding in quick-service restaurant environments* (Tech. Rep. GIT-CS-96/03). Atlanta: Georgia Institute of Technology. Cognitive Science Program, College of Computing.

Brehmer, B. (1990). Strategies in real-time, dynamic decision making. In R. M. Hogarth (Ed.), *Insights in decision making* (pp. 262–279). Chicago: University of Chicago.

Brehmer, B., & Allard, R. (1991). Dynamic decision making: The effects of task complexity and feedback delay. In J. Rasmussen, B. Brehmer, & J. Leplat (Eds.), *Distributed decision making: Cognitive models for cooperative work* (pp. 319–334). New York: Wiley.

Button, G., & Harper, R. H. (1993). Taking the organisation into accounts. In G. Button (Ed.), *Technology in working order: Studies of work, interaction, and technology* (pp. 98–107). London: Routledge.

Cannon-Bowers, J. A., & Salas, E. (1998). *Making decisions under stress.* Washington, DC: American Psychological Association.

Carroll, J. M., & Campbell, R. L. (1989). Artefacts as psychological theories: The case of human-computer interaction. *Behavior and Information Technology, 8*(4), 247–256.

Cohen, S., Gravelle, M., Wilson, K., & Bisantz, A. M. (1996). Analysis of interview and focus group data for characterizing environments. In *Proceedings of the 1996 Human Factors and Ergonomics Society Annual Meeting* (pp. 957–961). Santa Monica, CA: Human Factors and Ergonomics Society.

Connolly, T. (1988). Hedge-clipping, tree-felling, and the management of ambiguity: The need for new images of decision-making. In L. R. Pondy, R. J. J. Boland, & H. Thomas (Eds.), *Managing ambiguity and change* (pp. 37–50). New York: Wiley.

Degani, A., & Weiner, E. (1997). Procedures in complex systems: The airline cockpit. *IEEE Transactions on Systems, Man, and Cybernetics—Part A, 27*(3), 302–312.

Dowell, J., & Long, J. (1998). Conception of the cognitive engineering design problem. *Ergonomics, 41*(2), 126–139.

Gray, W. D., John, B. E., & Atwood, M. E. (1993). Project Ernestine: Validating a GOMS analysis for predicting and explaining real world task performance. *Human Computer Interaction, 8,* 237–309.

Hutchins, E. (1995). *Cognition in the wild.* Cambridge, MA: MIT Press.

Hogarth, R. M. (1981). Beyond discrete biases: Functional and dysfunctional aspects of judgmental heuristics. *Psychological Bulletin, 90,* 197–217.

Jones, P. M., Chu, R. W., & Mitchell, C. M. (1995). A methodology for human-machine systems research: Knowledge engineering, modeling, and simulation. *IEEE Transactions on Systems, Man, and Cybernetics, 25*(7), 1025–1039.

Kahneman, D., Slovic, P., & Tversky, A. (Eds.). (1982). *Judgment under uncertainty: Heuristics and biases.* Cambridge, England: Cambridge University Press.

Karwowski, W. (2000). Symratology: The science of an artifact–human compatibility. *Theoretical Issues in Ergonomics Science, 1*(1), 76–91.

Kirlik, A. (1995). Requirements for psychological models to support design: Toward an ecological task analysis. In J. Flach, P. Hancock, J. Caird, & K. Vicente (Eds.), *Global perspectives on the ecology of human machine systems* (Vol. 1, pp. 68–120). Hillsdale, NJ: Lawrence Erlbaum Associates.

Kirwan, B., & Ainsworth, L. K. (1992). *A guide to task analysis.* Taylor and Francis.

Klein, G. A. (1993). A recognition-primed decision (RPD) model of rapid decision making. In G. A. Klein, J. Orasanu, R. Calderwood, & C. E. Zsambok (Eds.), *Decision making in action: Models and methods* (pp. 138–147). Norwood, NJ: Ablex.

Lindley, D. V. (1985). *Making decisions.* London: Wiley.

Luczak, H. (1997). Task analysis. In G. Salvendy (Ed.), *Handbook of human factors and ergonomics* (pp. 340–416). New York: Wiley.

Mackay, W. E. (1999). Is paper safer? The role of paper flight strips in air traffic control. *ACM Transactions on Computer-Human Interaction, 6*(4), 311–340.

Martinez, S. G., Bennett, K. B., & Shattuck, L. (2001). Cognitive systems engineering analyses for army tactical operations. In *Proceedings of the Human Factors and Ergonomics Society Annual Meeting* (pp. 525–526). Santa Monica, CA: Human Factors and Ergonomics Society.

McFarren, M. R. (1987). *Using concept mapping to define problems and identify key kernels during the development of a decision support system* (Tech. Rep. AFIT-GST-ENS-87J-12). Alexandria, VA: Defense Technical Information Center, Cameron Station.

Mitchell, C. M. (1987). GT-MSOCC: A domain for research on human-computer interaction and decision aiding in supervisory control systems. *IEEE Transactions on Systems, Man, and Cybernetics, 17*(4), 553–572.

Mitchell, C. M., & Miller, R. A. (1986). A discrete control model of operator function: A methodology for information display design. *IEEE Transactions on Systems, Man, and Cybernetics, 16*(3), 343–357.

Morrison, J. G., Kelly, R. T., Moore, R. A., & Hutchins, S. G. (1998). Implications of decision-making research for decision support and displays. In J. A. Cannon-Bowers & E. Salas (Eds.), *Making decisions under stress* (pp. 375–407). Washington, DC: American Psychological Association.

Newell, A., & Simon, H. A. (1972). *Human problem solving.* Englewood Cliffs, NJ: Prentice-Hall.

Payne, S. J. (1993). Understanding calendar use. *Human-Computer Interaction, 8,* 83–100.

Pascoe, J., Ryan, N., & Morse, D. (2000). Using while moving: HCI issues in fieldwork environments. *ACM Transactions on Computer-Human Interaction, 7*(3), 417–437.

Pritchett, A. R. (2001). Reviewing the role of cockpit alerting systems. *Human Factors and Aerospace Safety, 1*(1), 5–38.

Rasmussen, J., Pejtersen, A. M., & Goodstein, L. P. (1994). *Cognitive systems engineering.* New York: Wiley.

Roth, E. M. (1994). Operator performance in cognitively complex simulated emergencies: Implications for computer-based support systems. In *Proceedings of the Human Factors and Ergonomics Society 38th Annual Meeting* (pp. 200–204). Santa Monica, CA: Human Factors and Ergonomics Society.

Roth, E. M., Bennet K. B., & Woods, D. D. (1987). Human interaction with an "intelligent" machine. *International Journal of Man-Machine Studies, 2*(7), 479–525.

Sachs, P. (1995). Transforming work: Collaboration, learning, and design. *Communications of the ACM, 38*(9), 36–44.

Sanderson, P., Scott, J., Johnston, T., Mainzer, J., Watanabe, L., & James, J. (1994). MacSHAPA and the enterprise of exploratory sequential data analysis (ESDA). *International Journal of Human-Computer Studies, 41*(5), 633–681.

Savage, L. (1954). *The foundation of statistics.* New York: Wiley.

Schank, R. C., & Abelson, R. P. (1977). *Scripts, plans, goals and understanding.* Hillsdale, NJ: Lawrence Erlbaum Associates.

Suchman, L. (1987). *Plans and situated action.* Cambridge, England: Cambridge University Press.

Suchman, L., & Trigg, R. H. (1991). Understanding practice: Video as a medium for reflection and design. In J. Greenbaum & M. Kyng (Eds.), *Design at work: Cooperative design of computer systems* (pp. 65–89). Hillsdale, NJ: Lawrence Erlbaum Associates.

Sundström, G. A. (1993). Towards models of tasks and task complexity in supervisory control applications. *Ergonomics, 36*(11), 1413–1423.

Sundström, G. A., & Salvador, A. C. (1995). Integrating field work in system design: A methodology and two case studies. *IEEE Transactions on Systems, Man, and Cybernetics, 25*(3), 385–399.

Taylor, F. W. (1911). *The principles of scientific management.* New York: Norton.

Tversky, A., & Kahneman, D. (1982). Judgment under uncertainty: Heuristics and biases. In D. Kahneman, P. Slovic, & A. Tversky (Eds.), *Judgment under uncertainty: Heuristics and biases* (pp. 3–22). Cambridge, England: Cambridge University Press.

Von Neumann, J., & Morgenstern, O. (1944). *Theory of games and economic behavior.* Princeton, NJ: Princeton University Press.

Wiener, E. (1985). Beyond the sterile cockpit. *Human Factors, 27*(1), 75–90.

Woods, D. D. (1998). Designs are hypotheses about how artefacts shape cognition and collaboration. *Ergonomics, 41*(2), 168–173.

Woods, D. D., Roth, E. M., & Bennett, K. B. (1990). Explorations in joint human-machine cognitive systems. In S. Robertson, W. Zachary, & J. Black (Eds.), *Cognition, computing, and cooperation* (pp. 123–158). Stanford, CA: Ablex.

Zuboff, S. (1987). *In the age of the smart machine.* New York: Basic Books.

27

Case Studies: Applied Cognitive Work Analysis in the Design of Innovative Decision Support

Scott S. Potter, James W. Gualtieri, and William C. Elm
Cognitive Systems Engineering Center,
ManTech Aegis Research Corporation, USA

Abstract

There have been a growing number of applications of cognitive analysis techniques to understand a work domain and the cognitive demands it imposes on practitioners in order to provide a foundation for the design of support for these cognitive tasks. Although many techniques have proved successful in illuminating the sources of cognitive complexity and in explicating the basis of expertise, the results are often only weakly coupled to the design of support tools. A critical gap occurs at the transition from cognitive analysis to system design. In this chapter we briefly discuss two recent examples in which the Applied Cognitive Work Analysis approach was utilized to design innovative decision support concepts.

INTRODUCTION

The purpose of this chapter is to provide two case studies to illustrate the results of the Applied Cognitive Work Analysis (ACWA) process. As noted in the companion chapter describing the ACWA methodology (Elm, Potter, Gualtieri, Roth, & Easter, Chap. 16, this volume), ACWA is a structured, principled methodology to systematically transform an analysis of the demands of a domain into decision-aiding concepts through the construction of a continuous design thread from domain analysis to decision support concept. The ACWA approach is predicated on the premise that in order to develop revolutionary decision support systems (not just an evolutionary change in the existing system), the support concepts must be based on the fundamentals of the domain of practice and the demands it imposes on domain practitioners. This enables the cognitive task designer to deliver the transparent, effortless decision support needed to provide revolutionary advances in net decision effectiveness. As described in Elm et al. (Chap. 16, this volume), the steps in this process include:

- Using a *functional abstraction network* (FAN) model to capture the essential domain concepts and relationships that define the problem space confronting the domain practitioners;
- Overlaying *cognitive work requirements* (CWRs) on the functional model as a way of identifying the cognitive demands, tasks, or decisions that arise in the domain and require support;
- Identifying the *information–relationship requirements* (IRRs) for successful execution of these cognitive work requirements;
- Specifying the *representation design requirements* (RDRs) to define the shaping and processing of display surfaces so as to exhibit the information requirements for domain practitioner(s);
- Developing *presentation design concepts* (PDCs) to explore techniques to implement these representation requirements into the syntax and dynamics of presentation forms in order to produce the information transfer to the practitioners given their task and goal context.

The analysis of the cognitive work domain starts with the building of a function-based goal–means decomposition of the problem space. This provides a representation of the cognitive work domain as an abstraction hierarchy (e.g., Lind, 1991, 1993; Rasmussen, 1986; Rasmussen, Pejtersen, & Goodstein, 1994; Roth & Mumaw, 1995; Vicente, 1999; Vicente & Rasmussen, 1992; Woods & Hollnagel, 1987). A work domain analysis is conducted to understand and document the goals to be achieved in the domain and the means available for achieving them (Vicente, 1999). The objective of performing this analysis is to develop a structure that links the purpose(s) of individual controllable entities with the overall purpose of the system. This includes knowledge of the system's characteristics, and the purposes or functions of the specific entities. We refer to the goal–means decomposition as a FAN to emphasize the levels of abstraction as a fundamental characteristic of the model as well as the rich, network topology inherent in the model.

The ACWA approach creates design artifacts that capture the results of each of these intermediate stages in the design process. These cognitive task design artifacts form a continuous design thread that provides a principled, traceable link from cognitive analysis to the final system design. However, the process of generating these artifacts, not the artifacts themselves, constructs the spans of the bridge. The artifacts serve as a mechanism to record the results of the process in the form of intermediate artifacts. These intermediate artifacts provide multiple opportunities to evaluate the completeness and quality of the analysis–design effort, enabling modifications to be made early in the process. The linkage between artifacts also ensures an integrative process; changes in one artifact cascades along the design thread, necessitating changes to all. These features of the ACWA process provide a rich scaffolding that is both supportive and highly adaptable to a variety of work domains.

Because of the complexity inherent in the work domains in which we have used the ACWA methodology, the case studies described in this chapter emphasize a single, relatively thin design thread that, while providing a complete walk from initial cognitive analysis to final visualization design, is necessarily lacking in breadth. The first case study examines the development of a single presentation design concept for a system used to train concepts in the area information operations. The second case study examines the ACWA process used to make available a decision aid to support the understanding of the complex abstract concept of combat power for military commanders. For an in-depth description of each step of the process, see the companion chapter (Elm et al., Chap. 4, this volume).

CASE STUDY 1: INFOCHESS® TEMPORAL OVERVIEW DISPLAY

Domain Background

The target domain for this case study is the online version of ManTech Aegis Research Corporation's Information Operations (IO) training environment, InfoChess (ManTech Aegis Research Corp., Falls Church, VA, USA). The online version of InfoChess was developed as part of an internal research and development effort with a rapid development cycle. This affected the level of analysis conducted at each step in the development process. Thus this case study represents the effectiveness of the ACWA methodology given a relatively small budget and short time frame. Despite the abbreviated analysis, the ACWA methodology provided critical insights from the work domain analysis for system design.

InfoChess uses the ancient war game of chess as the context in which to learn IO and information warfare. InfoChess deviates from traditional chess in that games state (i.e., situation awareness) is *not* directly perceived. All information exchanged between the two sides of an InfoChess game requires intelligence collection and is therefore subject to deception and denial. This is accomplished by a set of InfoChess rules (hereafter referred to as InfoMoves) that mimic IO and are used in conjunction with traditional chess rules associated with physical movement. All of these InfoMoves have been faithfully modeled to reflect joint doctrine for IO. Like real world warfare, there are limits to the number of IO or InfoMoves that an InfoChess player can make on the basis of available resources. The types of InfoMoves available in the basic version of InfoChess include several moves:

- Operations Security—the suppression of information about own forces from being conveyed to the opponent;
- Deception—the portrayal of false movement information;
- Psychological Operations—the influence of opponent pieces to resist movement or engagement;
- Electronic Warfare—the jamming of the transmission of opponent movement orders as well as the protection of own transmissions;
- Physical Destruction—the rescue or mobilization of own forces;
- Intelligence—the active collection of information about opponent operations; and
- Counterintelligence—the detection and reporting of opponent intelligence activity.

The key winning strategy for InfoChess becomes very different from chess. The cunning integration of conventional chess moves (physical) and InfoMoves is much more important than the most effective chess strategy. It is this synergy and use of the limited IO resources that is essential for victory and is the focus of the training. This translates into a need to show how to integrate the player's various InfoMoves into a unified plan.

Cognitive Task Design Challenges

The focus of this case study is on the development of the presentation design concepts that supported the development of this unified IO plan. This is an extremely challenging task because of the complex temporal aspects of the different InfoMove operations. Just as in real world IO, there are delays, limited durations, and stagnancies associated with each IO. Without a display to represent these temporal relationships, it would be necessary for the player to remember when the various InfoMoves started, stopped, and interacted with one another.

In fact, given the complexity of the decision making, InfoChess is most often used in a team environment in order to spread the cognitive demands associated with planning and

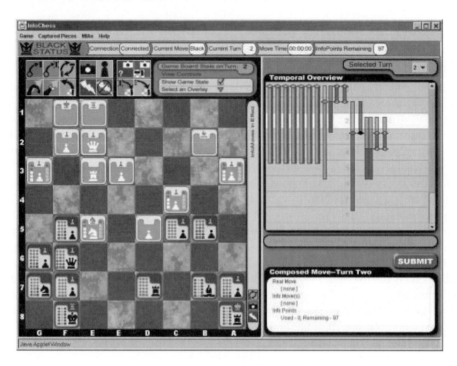

FIG. 27.1.   InfoChess® interface.

executing InfoMoves. The cognitive work associated with the management of the temporal issues associated with InfoChess also manifested itself in the time it took for controllers to adjudicate the game turns, the transfer of training issues, and the use of self-constructed external aids by InfoChess players to augment the chess board to track InfoMoves.

In addition to the temporal management challenges associated with the play of InfoChess, there was also customer pressure to field an online version of InfoChess. An online version of InfoChess would enable distributed play, reduce variability in rule interpretation by controllers, shorten game time, and increase training by providing better integration between physical chess moves and InfoMoves. Beyond the desire to address these demands was the desire to produce improvements in the InfoChess cognitive work artifacts used for training IO, using a relatively small budget and short time frame.

A screen capture of an InfoChess game in progress can be seen in Fig. 27.1. The focus of this case study is the temporal overview display, which is in the upper right-hand portion of the figure.

Modeling the Work Domain

Our ACWA approach begins with the creation of a "road map" of the cognitive work domain's problem space, using the notions of abstraction and aggregation. To maintain focus for this case study, we show a simplified version of the FAN developed for InfoChess in Fig. 27.2. It served as the basis for the design of the online InfoChess training environment.

This case demonstrates several characteristics of a pragmatic application of an approach based on cognitive work analysis (CWA) as the basis for designing decision support concepts:

- The FAN is a tightly coupled network of related functional nodes;
- It demonstrates a recursive application of Rasmussen's abstraction levels;

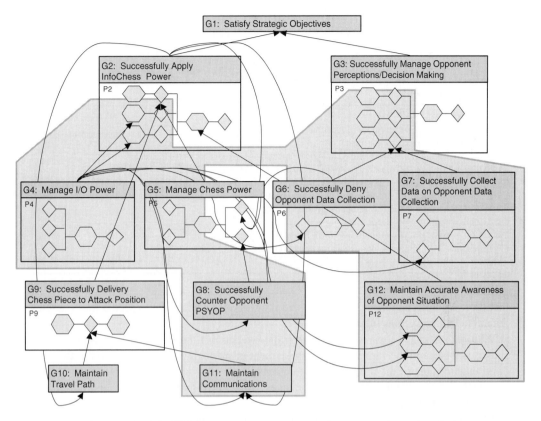

FIG. 27.2. InfoChess® FAN with shaded area covered by a temporal overview display.

- There is a many-to-one mapping between the nodes of the FAN and the presentation design concept;
- The FAN has circular dependencies—supporting functions have an impact on the goals of their supported functions.

As indicated by the top-level goal–means nodes in the FAN, winning in InfoChess is the balance of applying InfoChess power (G2; see Fig. 27.2) with managing opponent perceptions and decision making (G3), which is the essential aspect of satisfying the strategic objectives (G1). It is important to note that each of the goal–means nodes is success oriented. This explicit representation of goals within constraints or limits provides a purposeful basis on which to build.

The goal of satisfying the strategic objectives (G1) is dependent on the victory conditions established at the beginning of the InfoChess game (note: unlike traditional chess, InfoChess supports asymmetric victory conditions and force structures). To successfully apply InfoChess power (G2), players must integrate physical chess power (G5) with their information operation (InfoMove) power (G4). A similar decomposition is seen for the successful management of opponent perceptions and decision making (G3).

As you move down the FAN between the goal–means nodes, there is a shift in the commodities. This is an indication of a well-formed FAN. Another indicator of an accurate functional representation is the anchors of the support–supported links. Each supporting goal should be linked to the component of the process that it supports. This linkage enables the required level of precision of the support–supported link. Examination of the support–supported links shows

TABLE 27.1
Sample CWRs Within Scope of Region

CWRs	Included in TOD
CWR G4-1—"Monitor the inventory of remaining IO power."	
CWR-P2-1—"Choose–control the impact mechanism on opponent status (targets)."	X
CWR-P2-2—"Monitor the impact of Psychological Operations on opponent status (targets)."	
CWR-P2-3—"Monitor the impact of jamming on opponent status (targets)."	
CWR-P3-1—"Choose–control the information mechanism affecting the opponent."	X
CWR-P3-2—"Monitor the impact of displays on opponent."	
CWR-P3-3—"Monitor the impact of opponent collection efforts."	
CWR-P3-4—"Monitor the state of display operations."	X
CWR-G6-1—"Monitor the success of information denial."	
CWR-P6-1—"Monitor the state of information denial (OPSEC) operations."	X

that they exhibit functional circularities and recursions. This is to be expected in a complex dynamic domain such as InfoChess.

Although it is not apparent in this simplified version of the FAN (but is revealed in the subsequent steps in the case study), a key element of managing InfoChess power is the temporal management of the information operations. If players wait until they realize they need intelligence, for example, it is too late (as it takes several turns for the collection and subsequent reporting of intelligence).

This case study focuses the need to build support for this temporal management of IO as one component of the overall decision support environment. Thus, this display covers only a portion of the total functional space. The overall functional scope of that focus is shown overlaid on the FAN. It is important to note this display works in concert with several others, which together cover the complete set of CWRs associated with this cognitive work domain.

Identifying CWRs

The next step in our process is the identification of CWRs for each of the goal–means nodes. These decisions or cognitive demands are derived from an analysis of the domain (using the FAN) as well as through an analysis of various types of knowledge elicitation outputs.

Within the functional scope assigned to the temporal overview display (TOD) of the InfoChess training environment, a subset of the decisions within that region are "assigned" to the TOD. That is, within the shaded region shown in Fig. 27.2, a portion of them relate to decisions associated with the temporal application of InfoMoves. Information necessary to make other decisions within this region are included in separate presentation design concepts. Although a primary cognitive task design principle is to achieve the most integrated visualization possible by constructing support for the largest number of decisions into the most integrated context possible, representing the temporal issues was given its own display region to most effectively support the CWRs imposed by the domain associated with time management.

Table 27.1 is a small subset of both the overall set of CWRs and the subset assigned to the TOD. The numbering refers to the node in the FAN, and there is a unique number for each decision at that node. Decisions not marked with an "X" were assigned to companion displays. The term *decision* is used in a very broad sense. It is not for only "either–or" decisions; it is also used to denote all cognitive activities associated with the functional process.

The CWR number starts with "CWR". The next letter, "G" or "P," denotes whether the decision is associated with the goal or the process portion of the functional nodes that make up the FAN. The next column in the character string identifies which goal–means node in the FAN the decision is associated with. The final column after the dash is simply a running count of the number of decisions associated with a particular goal–means node. This character string serves as a unique identifier for the decision and enables a traceable link to be established between cognitive task design and system software development.

When examining the list of CWRs in Table 27.1, note that each contains an action verb related to decision making. Of the four TOD-specific decisions listed in the table, two (CWR-P2-1 and CWR-P3-1) are associated with strategic decision making and the control of information within InfoChess through the coordination of multiple InfoMoves, and two (CWR-P3-4 and CWR-P6-1) are associated with monitoring the status of individual InfoMoves. The TOD is more than just an information display; it enables InfoChess players to develop a coordinated attack on their adversary.

Choose–Control the Impact Mechanism on Opponent (targets), CWR-P2-1, captures the ability of players to decide which InfoMove to apply against their opponent. Choose–Control the Information Mechanism Affecting the Opponent (CWR-P3-1) enables the InfoChess player to determine how much information will be provided about that InfoMove. Thus the TOD supports not only the display of information but also the control of the game.

Monitor the State of Display Operations (CWR-P3-4) and Monitor the State of Information Denial (OPSEC) Operations (CWR-P6-1) represent two examples of the decisions associated with each of the InfoMoves that enable InfoChess players to keep track of their InfoMoves within the context of the game. Similar decisions exist for Psychological Operations, Electronic Warfare, and Intelligence. The objective of the TOD was to support an InfoChess player in monitoring the application (state) of his or her InfoMoves. In this context, the defeating of an opponent is a function of a player's ability to coordinate active InfoMoves with his or her physical moves.

As mentioned earlier, providing support for questions such as, When will this IO move go into effect? Which IO moves are active this turn? When will this IO move expire? and When will data be reported back to me? has been observed to be a particularly challenging area, as users struggle to manage the temporal aspects of lags, durations, and reporting latencies.

Supporting Information–Relationship Requirements

The next step is to identify and document the information–relationship requirements (IRRs) to support the CWRs. The FAN provides the context in which these CWRs reside and furnishes the framework for determining what information is needed. A small example of the IRRs needed by the user to support two of the decisions in Table 27.1 is shown in Table 27.2. The IRRs often dictate significant transformational processing that must occur within the system architecture's middle tiers to convert available data to the information needed. In the case of InfoChess, this is not necessary because the board is "perfectly instrumented." However, in many domains, this transformation becomes problematic. Thus, identified IRRs may serve as the impetus for further sensor placement, instrumentation, or algorithm development in order to provide the decision maker with the information needed to reach a decision.

In the case of the InfoChess TOD, the processing demands involved an expanded database design to fully satisfy the IRRs. An examination of the six IRRs in Table 27.2 finds that no "sensors" are listed. That is, IRRs describe what is needed to make the decision rather than list physical data that are available. For example, rather than just listing InfoMoves and when they start and end, the IRRs for decision CWR-P6-1 include issues associated with synchronization, move states, and temporal considerations. Nor do the IRRs simply list data sources; they clearly

TABLE 27.2
Supporting IRRs Associated With the TOD

Sample Information Relationship Requirements

CWR-P2-1

IRR-P2-1.1—"States of Psy Op and jamming moves over time (preactivation lag; active; ending; defeated (consumed) by opponent."

IRR-P2-1.2—"Synchronization (temporal relationship) of Psy Op and jamming moves (any activation state) with other IO and physical chess moves."

CWR-P6-1

IRR-P6-1.1—"IO & physical moves that have been protected with Information Denial Operations."

IRR-P6-1.2—"States of Information Denial Operations over time (preactivation lag; active; ending; defeated (consumed) by opponent."

IRR-P6-1.3—"Relative "intensity" of the Information Denial Operations."

IRR-P6-1.4—"Synchronization of Info Denial Ops with other IO and physical chess moves."

---

describe what is needed in the form of a phrase to convey the intent of the IRRs. Because these IRRs are information centered rather than data centered, they are capable of conveying the complex constructs necessary for supporting decision making in a naturalistic environment.

Representation Design Requirements

Once the decisions and supporting information requirements have been identified and anchored to the FAN, attention can shift to the design of support tools. Because the ultimate goal of the cognitive task design effort is to reveal the critical information requirements for effective decision-making performance, it is necessary to define the goals and scope of a display in terms of the cognitive demands it is intended to support as well as the supporting information requirements it must portray. As noted in the companion methodology chapter (Elm, et al., Chap. 16 this volume), this is the role of the representation design requirements (RDRs). The key aspect of a RDR is that it requires designers to be explicit about the specific cognitive work that a given representation is intended to support—prior to defining specific presentation concepts to achieve this goal. Thus RDRs they constitute explicit hypotheses—a model of support—that can be empirically evaluated. As a consequence, RDRs enable more informed and pointed testing of the effectiveness of the proposed aiding concepts. The RDR exists as a specification of the task meaningful domain semantics to be mapped into the structure and behavior of the representation. These serve as a checklist to ensure that the completed design will provide effective support for the decision maker, who in this case is an InfoChess player.

Table 27.3 provides a subset of the information contained within the RDR for the TOD. The TOD was designed to provide an integration of temporal information related to each information operation in a single display component. In this way, temporal synchronization and coordination of InfoMoves (a key aspect of the problem-solving environment) becomes a prominent element in the IO training environment. In addition, this representation is designed to complement a physical–spatial display that in itself is designed to support other CWRs. Note that each of the IRRs identified in the previous design artifact are included in the RDR. This traceable design thread is critical to the continuity from analysis to design to development.

The RDR starts with a general description of the objective(s) of the display concept that specifies its primary use. In the case of the TOD, its primary function is to provide a single source

TABLE 27.3
RDRs for the TOD

Decision and Information Requirements	Visualization Requirements

CWR-P2-1
    IRR-P2-1.1—"States of Psy Op and jamming move over time (preactivation lag; active; ending; defeated (consumed) by opponent."
    IRR-P2-1.2—"Synchronization (temporal relationship) of Psy Op and jamming moves (any activation state) with other IO and physical chess moves."
CWR-P6-1
    IRR-P6-1.1—"IO & physical moves that have been protected with Information Denial operations."
    IRR-P6-1.2—"States of Information Denial operations over time (preactivation lag; active; ending; defeated (consumed) by opponent."
    IRR-P6-1.3—"Relative 'intensity' of the Information Denial operations."
    IRR-P6-1.4—"Synchronization of Info Denial Ops with other IO and physical chess moves."
	1. Provide a depiction of purchase time, activation time, end time, and activity (i.e., intel collected, reported) of each InfoMove in order to convey the full life cycle of each IO.
	2. Provide an organization of InfoMoves to include friendly turns in order to synchronize different types of IO (offensive and defensive).
	3. Depict an indication of IO to convey intensity of the operation–single or double.
	4. Information Denial operations will be indicated by qualifiers to the individual IO in order to integrate the two operations.
	5. Indication of Information Denial operations will convey full life cycle of these operations—preactivation lag; active; ended; defeated.
	6. Indication of Information Denial operations will convey intensity of the operation—single or double.

Note. Context: The TOD will integrate temporal information from InfoMoves into a single, unified temporal display that will complement the physical–spatial display.

display that fuses the temporal information associated with the InfoMoves. It is important to note that the RDR does not specify the visual form that the presentation design concept should take (e.g., list, pie chart, or bar graph), but rather the syntax and dynamics of the subsequent presentations to achieve the desired information transfer. Issues of readability, aesthetics, consistency, and the like are still important to the interface design, but they are a consideration in the finalization of the presentation design effort.

After the general context under which a presentation design concept will be used has been set, the specific representation requirements are specified in "shall–will" language. In the case of the TOD, the RDRs are concerned with both temporal relationships among information (indication of information denial operations will convey the full life cycle of these operations—preactivation lag; active; ended; defeated) as well as dependencies between InfoMoves (information denial operations will be indicated by qualifiers to the individual information operations in order to integrate the two operations).

Presentation Design Concepts

Figure 27.3 shows the storyboard version of the presentation design concept. The fundamental aspect of the TOD is that the entire life cycle (purchase, activation, and expiration) of each IO move is depicted by a single but multidimensional indicator (vertical bars with diamonds) on the display as required by RDR 1. This indicator conveys lag (by a less saturated region of the

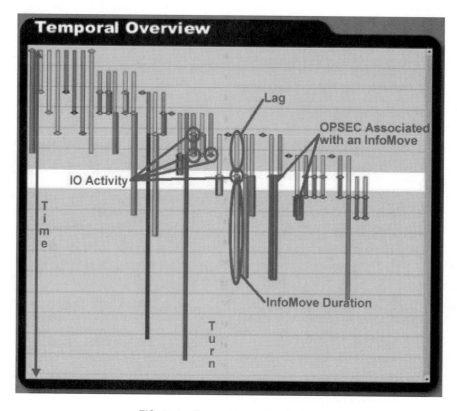

FIG. 27.3. Temporal overview display.

bar), IO activity (diamonds on the bar), and duration (fully saturated region of the bar). The integration of this information into a single indicator reduces the cognitive demand placed on the players of InfoChess.

On this basis, synchronization between different InfoMoves is achieved by plotting these multidimensional indicators against a single temporal axis (the y axis depicts time in units of turns in the game, because all activity is turn based). The different InfoMoves are grouped temporally by turns (horizontally). In this manner, the temporal sequence of InfoMoves can be easily seen by the horizontal arrangement of the different indicators. The association of information denial operations (i.e., operations security) with an InfoMove (such as protecting a feint from intelligence collection) is shown as adjoining bars attached to the left and right of the InfoMove. Intensity is indicated digitally within the state change indicators. The relationship to current time (turn) is the highlighted temporal region in the background of the TOD.

Although this description conveys the basic properties of the TOD, it does not provide a direct link between the RDRs and the presentation concept. Figure 27.4 provides a direct link between the elements of the RDR and the TOD. Each vertical bar in the TOD provides an indication of a depiction of purchase time, activation time, end time, and activity (RDR 1). The second RDR provides for a turn-based display in order to synchronize InfoMoves (RDR 2). In the display in the figure, the current turn is 6. Two of the RDRs (RDR 3 and RDR 6) relate to the intensity of the InfoMove. The presentation concepts within the TOD that satisfies this requirement are the diamonds on the display. Within these diamonds the number "1" or "2" is displayed to indicate the intensity of the InfoMove.

The bars on either side of the InfoMove life-cycle bar indicate the two types of information denial operations used to mask other InfoMoves (RDR 4). Whereas RDR 1 sets the requirement

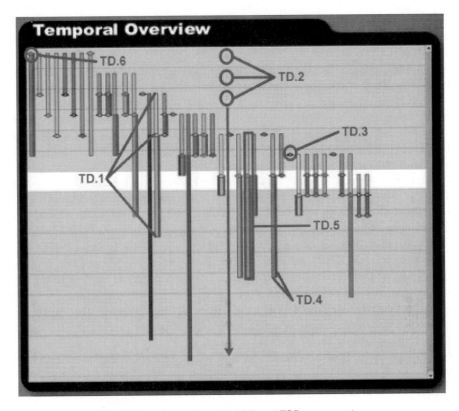

FIG. 27.4.   Linking between RDR and TOD components.

for displaying the full life cycle for all InfoMoves, RDR 5 is specific to information denial operations. This subset of RDRs along with those for the full InfoChess display fully specify the presentation design concepts that have to be conveyed to a user of the system.

Over the course of the development of the system, various versions of the presentation design concepts were developed and "tuned" against the RDR. This tuning often results in modifications to all the earlier artifacts, as creating the visualizations enables insights not revealed during the initial analytical steps. The variety of InfoMove behaviors, when they are employed in the game, and the variety of temporal effects—some instantaneous, some long in duration—are all made "perceptually transparent" by the resulting visualization.

Discussion

This first design case study provides a concrete illustration of how the continuous nature of the ACWA process can be used to provide a principled, traceable cognitive task design process from domain description to visualization and decision support concepts for effective decision-making performance by a domain practitioner. Consistent with the pragmatic approach presented here, the process of developing the display concept began with a FAN, which captured the goals in the domain, the means available for achieving them, and the constraints and interactions inherent in the domain. From the FAN it was only a short step to the identification of CWRs to which they are integrally linked. IRRs then provided the next piece for linking the domain and the user, by making explicit what needed to be displayed to the domain practitioner to make the decisions transparent. Then, specifying the RDRs defined the transformations required in order to achieve the information requirements for domain practitioner(s). Finally, perceptual and presentation theories guided the presentation and form of the presentation design concepts (PDCs) to close the gap between the machine's representation and the user's understanding.

Through this process, the temporal-based information requirements were identified that laid the foundation for the TOD. The resulting visualization has provided a powerful tool for effective decision making in InfoChess. One of the most challenging and error-inducing aspects of the game (managing the temporal aspects of InfoMoves) has been transformed into a manageable and well-understood concept. Players are now able to stay "ahead of the game" with respect to synchronization of InfoMoves and anticipating IO activity rather than reacting (poorly, most of the time) to the complex temporal dynamics.

The TOD provides a means for integrating various Information Operations to (a) ensure that one InfoMove does not "step on" or negate another InfoMove, and (b) facilitate the integration of InfoMoves so that the movement of physical chess pieces has a greater impact. Prior to the development of the TOD, InfoChess players routinely were unable to develop a cohesive plan for integrating their InfoMoves into their scheme of maneuver (physical chess piece movement).

The identification of emergent properties based on this integration of physical chess piece movement (maneuver) with IO provides the InfoChess player with a new functional perspective. This in turn facilitates the development of new strategies for the application of both components as well as a deeper understanding of how the two components interact. This deeper understanding is characteristic of domain experts.

With this expertise, InfoChess players are able to innovate and think "outside of the box." Thus, with the help of the TOD, InfoChess succeeds in providing a rich training environment and supports a rapid transition from novice to expert in the area of IO.

CASE STUDY 2: COMMAND AND CONTROL DECISION MAKING

Domain Background

This section describes the design of a second decision aid in order to illustrate the breakthroughs that can be achieved from following the ACWA-based methodology (Elm et al., Chap. 16, this volume) to support effective decision making in a very different work domain—military command and control. Unlike the unique nature of the first case study, this example is a more traditional problem domain. Specifically, the focus is on the planning of military forces to achieve a decisive victory against the enemy at a specific point in time and space in the future. Achieving victory in this situation requires assembling the right combination of forces at the right time (referred to as "husbanding combat power for the culminating point in the fight") in order to overtake the enemy. This is an extremely challenging task because of the temporal complexities, military doctrine, and less tangible factors related to a unit's effectiveness.

The specific example from this domain used in this case study was a small-scale simulation presented in a distributed manner over the Internet as a training test bed for interface design concepts. As such, much of the uncertainty, dynamism, and realism were reduced. However, there is sufficient complexity to create a cognitively challenging problem in need of effective decision support.

Cognitive Task Design Challenge

The cognitive task design challenge in this example was to design a decision support system concept that would assist military commanders in this complex decision-making situation.

The challenge was in escaping from the physical presentations and designing a tool that would support problem solving based on the key abstract concepts in the domain. This included integrating disparate data into a unified representation of the problem, supporting "what if" analyses for different scenarios, and maintaining an awareness of the inherent uncertainty in the domain.

Modeling the Work Domain

For the FAN model to be constructed, a variety of knowledge elicitation techniques were used to develop an initial understanding of the decision problems faced by military commanders, how military commanders conceptualize the problem space, and the complications that arise in the domain that increase the difficulty of decisions. These knowledge elicitation–acquisition activities included a fairly typical set of activities, including:

- Reviewing military documents;
- Conducting structured interviews with military commanders at different levels in the command chain, including recently retired General Officers with combat experience; and
- Observing decision-making behavior in the simulated "microworld" exercises that were conducted as part of the overall program.

One of the fundamental benefits of FAN-based modeling is that it provides the underlying structural framework that can then serve as the organizing model through which additional knowledge elicitation data can be resolved. Thus, the FAN was used as the mechanism for integrating complementary as well as seemingly conflicting information generated during the knowledge elicitation process. Although this can be done in a behind the scenes manner, in this effort it was an explicit part of the knowledge elicitation process. We presented draft versions of the FAN to military commanders and revised the model on the basis of their feedback. This was an iterative process that occurred over several cycles.

One of the fundamental findings of the work domain modeling effort was that military commanders think about military units in terms of the abstract concept "combat power" that military units possess and are able to generate. This abstract concept includes not only the number and type of military equipment that the unit possesses, but also less tangible factors such as the unit's morale and fatigue, as well as a number of external factors such as terrain. In Fig. 27.5, this concept has been generalized to "apply military (combat) power" (shown in more detail in Fig. 27.6) to reflect the fact that the modern-day military is often asked to engage in noncombat operations (e.g., providing humanitarian aid).

The function of applying combat power (or more generally military power) is performed in the context of meeting several higher level goals. The most direct higher level goal is to satisfy mission objectives. There are other high level goals, however, that must be taken into account. These include complying with military law, minimizing collateral damage, complying with local laws and cultures, and a need to attain positive public perception of the operation. These additional goals place constraints on applying military power. The fact that high-level goals can compete with each other in this way is reflected in the FAN by the multiple support–supported links entering that node.

The Military Command and Control FAN illustrates one of the important benefits of developing a functional abstraction hierarchy representation of the domain. The exercise of developing a FAN enables the cognitive analyst to see beyond the physical level of description of the domain and to begin to understand and represent the domain at higher levels of abstraction. The concept of combat power that is represented in the Military Command FAN is a good example. Combat power can be thought of as a commodity. The commander is given resources

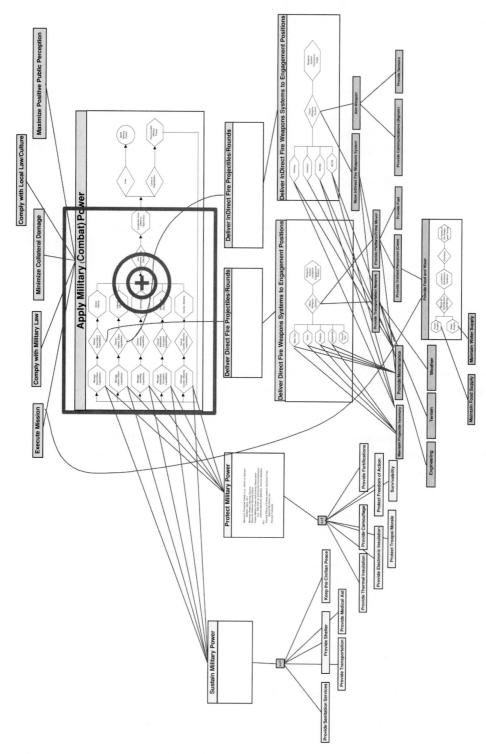

FIG. 27.5. FAN model of military command and control with the "apply military (combat) power" portion highlighted.

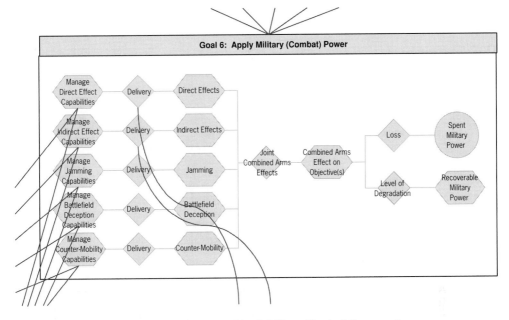

FIG. 27.6.   Detailed view of Apply Military (Combat) Power node.

(troops, planes, tanks, and logistics support) that have many complex functional properties and interdependencies that affect the amount of combat power that can be delivered at a given point in space and time. Thus, the commander can be thought of as the manager of a very precious commodity, seeking ways to apply the combat power that can be brought to bear to satisfy the set of constraints from the higher level goals.

Military commanders routinely think in terms of abstract concepts such as "relative combat power" in making decisions about movement of troops and equipment to achieve a mission objective. Yet the current tools available to them (e.g., physical maps of the terrain with icons representing placement of troops and equipment) provide a much more physical representation of the problem space, making the estimation of relative combat power difficult and prone to inaccuracies and error.

A second important attribute illustrated by the Military Command and Control FAN is the value of the conceptual structure it provides. The focus of effort in developing a FAN was on constructing a representation that captures the important concepts and interrelationships in the domain at appropriate levels of abstraction and makes the goals, means, relationships, and constraints explicit. Initially, the emphasis was placed on shaping the representation around key concepts (goals and processes) and relationships. Once the right structure was in place, then supporting details (i.e., less abstract concepts, e.g., procedures and functional capabilities) were added to the FAN in order to flesh out the model.

It has often been our experience that when we finally get the functional modeling 'right' (after many iterations), the resulting FAN looks simple and is readily understood and accepted by domain subject matter experts. This is a typical reaction by domain practitioners to a FAN that successfully crystallizes the most important abstract concepts and interrelationships in the domain. It should be pointed out, however, that although the FAN may look simple and obviously true to the subject matter experts once it is presented to them, the conceptual structure represented in the FAN is not something that the domain experts could have spontaneously generated on their own. Often the concepts represented in the FAN are issues that the domain experts understand implicitly, and readily resonate to once shown, but could not generate themselves.

TABLE 27.4
CWRs Associated With the Apply Military (Combat) Power Functional Node

---

CWRs for Apply Military (Combat) Power

CWR P6-1—"Determine the point in time when the enemy will reach a specified point in space, and monitor the enemy's combat power over time." (This decision is driven by the need to establish the mission requirements, i.e., evaluate enemy combat power, so as to understand the friendly combat power that will be required to achieve mission goals, i.e., defeat the enemy.)

CWR P6-2—"Choose among the friendly combat resources that can bring their combat power to bear at the specific point in space and time." (This decision considers the availability of resources to achieve mission goals.)

CWR P6-3—"Estimate the potential to defeat the enemy after the application of the chosen power at the specific point in time and space." (This decision is related to the choice among the various alternatives.)

CWR P6-4—"Determine the impact of moving combat power to a specific point in space and time; i.e., what else were these resources committed to and what will happen if they change their assignments–commitments?" (This decision considers the side effects of choosing various alternatives.)

The power of this abstraction hierarchy is that it allows the cognitive analysts to recognize the need to create decision aids that allow the commander to visualize and control the domain at this higher, more abstract level of goal achievement. The FAN provided the starting point for developing the visualization to support commanders in choosing the appropriate combat power to carry out the mission objective(s). For this case study, we will focus on the 'Apply Military (Combat) Power' node in the FAN, as that was the focus area in one of the microworld exercises. This entails (as shown in Fig. 27.6) a process of choosing the combination of resources that the commander believes will effectively carry out the mission.

Identifying CWRs

Once the framework provided by the FAN had been constructed, the next step in the analysis and design process was to utilize the structure of the problem space map (i.e., the FAN) and derive the supporting decisions–cognitive tasks for accomplishing the objective in question. In this specific context, the 'Apply Military (Combat) Power' node in the FAN requires the decision of "Choose combined combat power to achieve the objectives at a specific point in time and space." Note that the decision is written in a form that is independent of the particular battle or terrain of any single situation that originally exposed the need for the decision. This is part of the process of making the decision support system not situationally brittle. Then, the specific set of decision requirements were derived by further knowledge elicitation as well as by imposition of a template of generic decisions (adapted from Roth & Mumaw, 1995). The resulting set of cognitive requirements is shown in Table 27.4.

It is important to note that there were other CWRs associated with the 'Apply Military (Combat) Power' goal-process node. However, because this case study is on the analysis and design of a particular component of a decision aid, the cognitive requirements not directly related to this component are omitted in order to maintain focus.

Supporting IRRs

Associated with each of the CWRs identified in the previous step in the process is a requisite domain-specific IRR. A description of the type of information necessary to support the "Choose

TABLE 27.5

Supporting IRRs Associated With the Apply Military (Combat) Power CWRs

---

Information Relationship Requirements

CWR P6-1

IRR P6-1.1—"Expected arrival time of enemy combat resources at the specified point in space," i.e., the lead unit, as well as other follow-on units.

IRR P6-1.2—"Estimated measure of combined enemy combat power at the specified point in space, beginning at the arrival time of the first enemy unit and extending through follow-on units."

CWR P6-2

IRR P6-2.1—"The time required for selected friendly combat resources to reach the specified point in space."

IRR P6-2.2—"Estimated measure of combined combat power of the selected friendly combat power resources once they reach the specified point in space."

CWR P6-3

IRR P6-3.1—"Measure of combat power ratio of friendly to enemy combat power, beginning with the arrival of the first unit (friendly or enemy) over time."

IRR P6-3.2—"Indication of combat power ratios required to defeat the enemy under different battle conditions (i.e., doctrinal–procedural referent information)."

IRR P6-3.3—"Location of alternative resources of both friendly and enemy combat power that could be brought to bear."

IRR P6-3.4—"The time required to bring to bear the combat power of these alternative friendly and enemy combat resources."

IRR P6-3.5—"Measures of cumulative combat power of both friendly and enemy resources as additional friendly and enemy resources are selected (over a specified window in time)."

---

combat power" decision is given in Table 27.5. As in the previous case study, the identification numbers for IRRs cross-reference the CWRs they support.

It is interesting to note that the IRRs for these decisions included "relative combat power over time," which is a very untraditional concept that created significant excitement from the subject matter experts because it solidified some doctrinal writings about controlling the "tempo" of the conflict (but this is precisely the intent of building the FAN as the framework for decisions—to identify those key abstract concepts within the work domain). This also provides an example of the typical nontrivial data-to-information transformation that is often required to satisfy the information requirements identified as part of the analysis. As part of the system development process, the definition of these transformations would have to be specified. However, that level of detail is outside the scope of this case study.

Representation Design Requirements

On the basis of the FAN, the CWRs, and the IRRs, a set of RDRs was specified to define how these IRRs should be mapped into the structure and behavior of the specific decision support concept. The RDR provides a specification of the specific information requirements that must be satisfied by the resulting visualization and decision-aiding concepts to support the target cognitive tasks or decisions. If the graphic elements and visualizations effectively convey the information transfer requirements identified during the information requirements analysis, we have successfully bridged the gap between the essential demands of the work domain and the resulting decision support concepts. In most cases, this also requires consideration of allocation of the functional–decision–information scope for each display. However, for the present

TABLE 27.6
RDRs for the "Choose Combat Power" Decision Support Concept

Decision and Information Requirements	Visualization Requirements
CWR-P6-1—"Determine the point in time when the enemy will reach a specified point in space, and monitor the enemy's combat power over time."	1. Visualize the time required for enemy unit(s) to reach location designated by commander, and visualize cumulative enemy combat power at that location as a function of time.
IRR-P6-1.1—"Expected arrival time of enemy combat resources at the specified point in space (i.e., the lead unit, as well as other follow-on units)."	
IRR-P6-1.2—"Estimated measure of combined enemy combat power at the specified point in space, beginning at the arrival time of the first enemy unit and extending through follow-on units."	
CWR-P6-2—"Choose among the friendly combat resources that can bring their combat power to bear at the specific point in space and time."	2. Visualize the time required for friendly "combat power" forces to bring their combat power to bear on designated point (in space and time) and visualize cumulative friendly combat power at that location as a function of time.
IRR-P6-2.1—"The time required for selected friendly combat resources to reach the specified point in space."	
IRR-P6-2.2—"Estimated measure of combined combat power of the selected friendly combat power resources once they reach the specified point in space."	
CWR-P6-3—"Estimate the potential to defeat the enemy after the application of the chosen power at the specific point in time and space."	3. Visualize relative friendly to enemy combat power ratio at the designated point in space as a function of time, and compare to combat power ratio required to defeat enemy under different battle conditions.
IRR-P6-3.1—"Measure of combat power ratio of friendly to enemy combat power beginning with the arrival of the first unit (friendly or enemy) over time."	4. Visualize uncertainty of estimates of combat power ratio.
IRR-P6-3.2—"Indication of combat power ratios required to defeat the enemy under different battle conditions (i.e., doctrinal–procedural referent information)."	5. Visualize changes in enemy and friendly combat power and combat power ratio as selected enemy and/or friendly units are added or removed.
IRR-P6-3.3—"Location of alternative resources of both friendly and enemy combat power that could be brought to bear."	
IRR-P6-3.4—"The time required to bring to bear the combat power of these alternative friendly and enemy combat resources."	
IRR-P6-3.5—"Measures of cumulative combat power of both friendly and enemy resources as additional friendly and enemy resources are selected (over a specified window in time)."	

Note. Context: This display is intended for combat applications where the mission objective is to engage and defeat an enemy. It should assist the commander in choosing combined combat power and managing the space–time trade-off to control the battle space. Specifically it is to assist in deciding (a) at what location to engage the enemy; (b) when to engage the enemy; and (c) what combat resources to deploy in order to maximize the potential to defeat the enemy.

example, we have not included that broader scope issue in this discussion. Table 27.6 presents the RDR for the "Chose combat power" display.

Presentation Design Concepts

A prototype of the "Choose combat power" display was developed to provide a concrete instantiation of the specifications embodied in the RDR. The objective of this particular display is to support commanders in choosing the appropriate military power to achieve mission objectives (the fundamental CWR) through effectively presenting the IRRs. The display is targeted at combat applications where the mission objective is to engage and defeat an enemy at a particular geographic location. In this combat context, the probability of meeting mission objectives is a function of the ratio of friendly to enemy combat power at the point in time and space where the engagement is to take place. Thus, the "Choose combat power" display is designed to support commanders in deciding (a) at what location to engage the enemy; (b) when to engage the enemy; and (c) what combat resources to deploy in order to achieve an acceptable potential to defeat the enemy. It should be noted that, although this discussion focuses on the "Choose combat power" display, the display is intended to be only one of a suite of displays that would encompass a complete battle command decision aid.

Figure 27.7 presents a scenario that illustrates the context in which the visualization concept would be used. The commander would begin by viewing a physical representation of the terrain with enemy and friendly units and their positions identified. This is similar to the types of geographic displays that commanders use today. This scenario starts with the Red (enemy) forces attempting to deny the Blue (friendly) forces bridgehead at Weston and Easton bridges. Initially, Blue had decided to use the Second and Third Battalions to support the advantage gained at Easton Bridge. This would allow Blue to capture Fairview and possibly Clifton. However, at that point, Red's Fifth Battalion breaks through Blue's position at Weston Bridge and threatens a flanking attack from the west. At this point the Blue Commander recognizes the need to interdict the Red Fifth Battalion that is pursuing the Blue Fourth Battalion. The commander considers using the Second Armor to interdict the Red Fifth Battalion at the "choke point" south of the Burke Bridge.

To evaluate whether the Second Armor is sufficient to achieve the mission goal, the commander brings up the "Choose combat power" display. The commander selects the choke point south of the Burke Bridge as the designated location of the engagement and the Second and Fourth Blue Battalions and the Fifth Red Battalion as the units involved in the engagement. The system computes the time it will take for these units to reach this designated location and the combat power that the Red and Blue sides will be able to bring to bear at that location over time.

Figure 27.8 presents the resulting "Choose combat power" presentation design concept. The display has three major areas. The top third of the display shows all the Blue units that could potentially be deployed, their current status, the time at which they would reach the designated location, and the combat power they could bring to bear over time (IRR-G6-2.1, G6-2.2, G6-2.3, and G6-2.4). The shaded areas represent amount of combat power as a function of time. The horizontal axis represents time, with the leftmost end of the axis representing the current time. The heights of the shaded areas represent the amount of combat power at a given point in time. The units that are highlighted (appear darker) are the units that the commander has selected. Similarly, the bottom third of the display shows the enemy (Red) units and their combat power (IRR-G6-1.1, G6-1.2, G6-3.3, and G6-3.4), with the highlighted unit (Fifth Battalion) indicating the Red unit selected by the enemy commander. The middle third of the display is used to present the cumulative combat power of the combined Red and (separately) combined Blue units selected as a function of time (IRR-G6-3.1, G6-3.2, and G6-3.5). Thus,

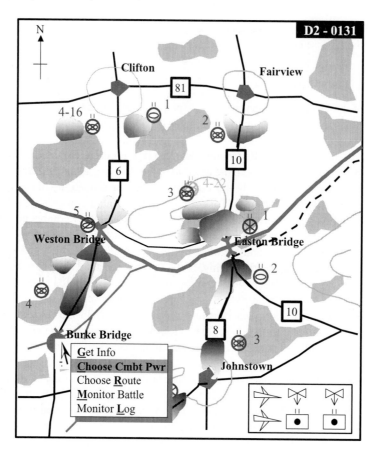

FIG. 27.7.   Geographic map providing the scenario context for the "Choose combat power" visualization. This is classically the form of decision support a commander uses (along with some outboard math) to make the decision.

the top curve in the center area represents the combined combat power of the Second and Fourth Blue Battalions over time. Combat power increases significantly when the Second Battalion reaches the designated location. The vertical line indicates the estimated time of arrival of the first Red unit. These three major areas map directly to the decision and information requirements contained within the RDR, as indicated in Table 27.7.

The "Choose combat power" display can be used to visually compare the combat power of the Blue and Red units at the time that the Red unit first reaches the designated location. The graphic presentation of combat power as a function of time makes it visually apparent that the Red Fifth Battalion will reach the designated point before the Blue Second Battalion can arrive. As a result, the difference in combat power between the Blue and Red units at the time of the initial engagement will be small. Therefore the goal of defeating the Red units is unlikely to be achieved.

Figure 27.9 illustrates two additional powerful features of this decision aid. First, it can be used to explore alternative combat resource choices by adding or removing enemy and friendly units to assess the impact they would have on combat power at the specified point in time and space. In this case, the commander decides that the Red First Armor Battalion is also likely to join the engagement. The commander decides that additional Blue resources (artillery and helicopters) will need to be brought in to affect the temporal dimension of the battle. The commander selects these additional resources and the display is updated to reflect the changes

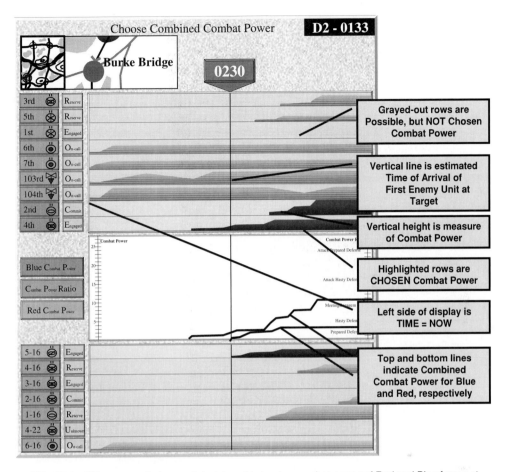

FIG. 27.8. "Choose combat power" decision aid showing combat power of Red and Blue forces at the time that the Second and Fourth Blue Battalions and Fifth Red Battalion reach the choke point south of the Burke bridge.

in Red and Blue combat power that would occur as a result of introducing these additional units.

In the middle region of the display, the commander can display the combat power ratio of friendly to enemy forces (estimated, with uncertainty bands). The highlighted line is an estimate of the combat power ratio (Blue to Red). The bands around the ratio line represent the uncertainty band around the estimate of the combat power ratio. A combat power ratio scale appears on the right with indications of combat power ratios recommended for different types of engagements based on conventional military guidance (this "referent information" would typically be included as an IRR but has been omitted here for simplicity). As is visually apparent, with the new set of choices, there is enough Blue combat power brought to bear to make it highly probable that the Blue forces will achieve their mission.

As mentioned earlier, although this discussion has focused on the "Choose combat power" display, the display is intended to illustrate one of a suite of displays that would encompass a complete Battle Command decision support system. As is suggested by the menu that appears in Fig. 27.7, additional displays would address other aspects of battle command such as selecting routes and monitoring the battle. In addition, display navigation mechanisms would be provided to enable the commander to more fully understand the basis for the combat power values. For example, the commander would be able to view the factors and values that contributed to

TABLE 27.7
Mapping of Graphical Elements to Visualization Requirements for
the "Choose Combat Power" Display

Visualization Requirements	*Graphical Elements*

Visualize the time required for enemy unit(s) to reach location designated by commander, and visualize cumulative enemy combat power at that location as a function of time.

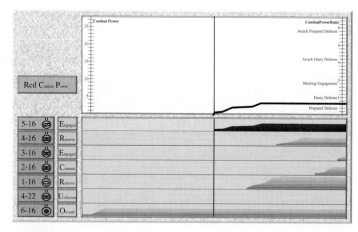

Visualize the time required for friendly "combat power" unit to bring their combat power to bear on the designated point (in space and time) and visualize cumulative friendly combat power at that location as a function of time.

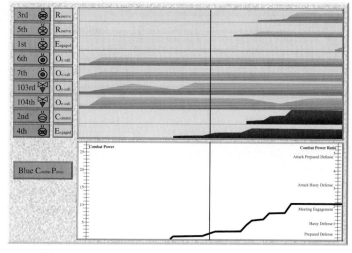

Visualize relative friendly to enemy combat power ratio at the designated point in space as a function of time, and compare with combat power ratio required to defeat enemy under different battle conditions.

Visualize uncertainty of estimates of combat power ratio.

Visualize changes in enemy and friendly combat power and combat power ratio as selected enemy and/or friendly units are added or removed.

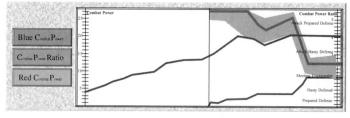

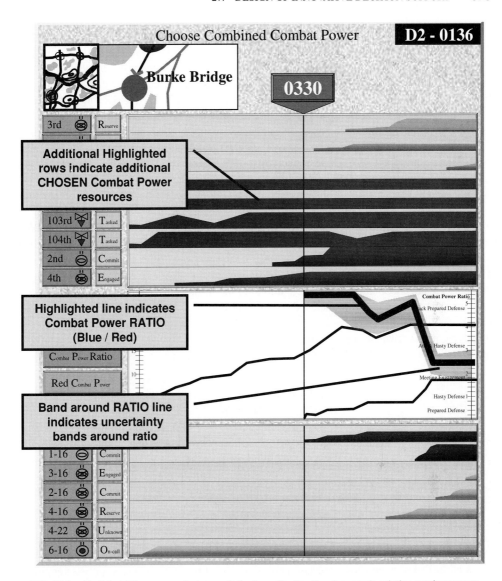

FIG. 27.9.   Updated "Choose combat power" display reflecting the changes in relative combat power based on additional units selected. The highlighted line with light bands around it portrays estimated combat power ratio (Blue to Red) with associated uncertainty around the estimate.

combat power calculations and make changes to the values and factor weightings as judged appropriate. This is consistent with principles for design of effective decision aids that include the importance of making the basis of recommendations transparent to the user (the principle of decomposability) and the importance of enabling the user to direct the decision aid (the principle of directability) described in Roth, Malin, and Schreckenghost (1997). In addition, mechanisms would also be provided to enable the commander to assess the impact of selecting particular units to join an engagement on the achievement of other goals. Specifically, the commander would be able to select a unit being considered for a particular engagement, and be shown the current (or other planned) commitments for this unit, and the potential impact on achieving those goals if the unit were to be reassigned. This supports the CWR to be able to assess the side effects of decisions that was identified in Fig. 27.6.

Discussion

This discussion of the design of the "Choose combat power" decision support concept is intended to provide a concrete illustration as well as demonstrate the value of how the application of the structured, principled ACWA methodology can systematically transform the problem from an analysis of the demands of a domain to the design of effective decision-aiding concepts. The process of developing the FAN representation was central to the identification of the key abstract concept of combat power. This concept would not have been recognized without developing the FAN model, as evidenced by typical physical map-based displays in widespread existence (despite the fact that the concept of relative combat power is taught as an integral part of military decision making). In addition, the importance of representing combat power as a function of time and location emerged out of the analysis. It takes time for resources to be assembled and moved to a specified location. One of the breakthrough insights in developing the "Choose combat power" display was the realization that such a visualization could allow commanders to explicitly manage the *time* of the culminating point of the engagement. Without this type of visualization, estimates about units and power over time are typically inaccurate. These insights provided the basis for development of the "Choose combat power" display concept. This decision aid would enable domain practitioners to visualize combat power as a function of time and to manipulate it directly. It provides a clear example of an "ecological" interface (Vicente & Rasmussen, 1992) in that it translates an abstract, functional concept (relative combat power as a function of time) into a concrete visualization that can be directly perceived—providing effective support for commander's cognitive tasks.

SUMMARY

In the development of an innovative decision support environment that enables revolutionary changes in users' cognitive tasks, one main bottleneck is the struggle to discriminate between aspects of task performance that result from "surface" characteristics of the current work environment and fundamental cognitive task demands that will carry over to the new environment. To address this issue, we find it critical to have a robust approach that explicitly models the fundamental goals, processes, and relationships within the domain and the cognitive demands and strategies of the operators in response to these functional properties (see Roth & Woods, 1989). As a whole, the generation of intermediate artifacts that model the structure of the work domain, the demands and cognitive task activities to be supported, the IRRs for these cognitive tasks, and decision support concepts designed to provide this support are needed to provide a traceable link from analysis, to design requirement, to decision support concept. The key is to use a model of the underlying, fundamental behavioral characteristics of the work domain in a principled manner to generate well-grounded decision support concepts for the cognitive task demands facing the human–machine decision-making team.

The steps in these two applications of the process were presented as if they were performed in a strictly sequential order. However, it is presented this way for expository simplicity. In practice, the process is much more parallel, opportunistic, and iterative in nature than presented here. For example, it is not unusual for an initial presentation design concept to emerge before the complete CWRs it is intended to support or the supporting IRRs have been clearly articulated. The order in which the artifacts are produced is not as important as the fact that all artifacts are eventually produced that provide a functional description of the cognitive tasks that the display is intended to support, and the representation requirements for the required support. The process of generating these artifacts, not the artifacts themselves, constructs the end-to-end design thread. In generating these artifacts, the cognitive analysts are forced to think about the problem in a systematically transformed manner and capture the evolving requirements in a manageable sequence of steps. The analysts must revisit the functional model and consider issues possibly overlooked, be explicit about the cognitive tasks associated with the nodes in the model, and consider the often-complex transformation between data and information. This type of approach is essential to the development of innovative decision aids to provide highly effective and robust decision-making performance. With this design thread as an explicit part of the prototype development, each assessment of the prototype can provide insights into the effectiveness of the presentation design concepts and, what is more important, additional understanding of the underlying requirements or basis for effective support. It is only with this type of iterative, closed-loop approach that the gap between cognitive analysis and decision support system development can be narrowed.

REFERENCES

Elm, W. C., Potter, S. S., Gualtieri, J. W., Roth, E. M., & Easter, J. R. (2003). Applied cognitive work analysis: A pragmatic methodology for designing revolutionary cognitive affordances (this volume).

Lind, M. (1991). Representations and abstractions for interface design using multilevel flow modelling. In G. R. S. Weir & J. L. Alty (Eds.), *human-computer interaction and complex systems*. London: Academic Press.

Lind, M. (1993, July). Multilevel flow modelling. Paper presented at the AAI93 Workshop on Reasoning about Function, Washington, DC.

Rasmussen, J. (1986). *Information processing and human-machine interaction: An approach to cognitive engineering* (North Holland Series in System Science and Engineering). New York: Elsevier Science.

Rasmussen, J., Pejtersen, A. M., & Goodstein, L. P. (1994). *Cognitive systems engineering*. New York: Wiley.

Roth, E. M., Malin, J. T., & Schreckenghost, D. L. (1997). Paradigms for intelligent interface design. In M. Helander, T. Landauer, & P. Prabhu (Eds.), *Handbook of Human-Computer Interaction* (2nd ed., pp. 1177–1201). Amsterdam: North-Holland.

Roth, E. M., & Mumaw, R. J. (1995). Using cognitive task analysis to define human interface requirements for first-of-a-kind systems. In *Proceedings of the Human Factors and Ergonomics Society 39th Annual Meeting* (pp. 520–524). Santa Monica, CA: Human Factors and Ergonomics Society.

Roth, E. M., & Woods, D. D. (1989). Cognitive task analysis: An approach to knowledge acquisition for intelligent system design. In G. Guida, & C. Tasso, (Eds.), *Topics in Expert System Design* (pp. 233–264). North Holland: Elsevier.

Vicente, K. (1999). *Cognitive work Analysis*. Mahwah, NJ: Lawrence Erlbaum Associates.

Vicente, K., & Rasmussen, J. (1992). Ecological interface design: Theoretical foundations. *IEEE Transactions on Systems Man and Cybernetics, 22,* 589–606.

Woods, D. D., & Hollnagel, E. (1987). Mapping cognitive demands in complex problem-solving worlds. *International Journal of Man-Machine Studies, 26,* 257–275.

28

Analysis of Procedure Following
as Concerned Work

Peter Wright
University of York, United Kingdom

John McCarthy
University College, Cork, Ireland

Abstract

In this chapter our concern is with the design and use of operating procedures. Operating procedures feature strongly in safety-critical work domains, where it is argued they reduce the likelihood of human error. Procedures can be viewed as the result of cognitive task design. Information processing analyses of procedure following have shown how the design of procedures can be improved. Here we explore an alternative to the prevalent information processing approach to take a more critical view of the proceduralisation of the workplace. Our approach takes as its starting point ideas from the field of literary studies and narrative rather than information processing psychology. The analysis focuses on autobiographical data. Unlike a more traditional cognitive analysis, it highlights the concerns facing operators who have to make procedures work in practice, the paradoxes and dilemmas they face in doing this, and the amount of intelligent effort that goes into making sense of procedures. Our analysis leads us to conclude that the gap between procedures and practice is filled by the creative work of the operator. Such creative work is based on a history of experiences both inside and outside the workplace. Currently, there are very few ways in which this experience can be given a legitimate voice in the process of procedure design.

INTRODUCTION

Operating procedures feature strongly in many work domains, either as informal but routine ways of coordinating work or as documented prescriptions about how operators must carry out tasks. In the latter, more formal, manifestation procedures may be found embedded in computational artefacts, such as electronic checklists or in paper-based documents designed to be referenced by operators when they carry out procedures. In some situations, procedures take the form of checklists, which the operator has to check off and sign to show that all the required steps have been taken. In many safety-critical work settings, operators can be held responsible for failing to follow a written procedure.

Orr (1996) reminds us that procedures come out of the scientific management tradition that tries to rationalise human work. This can be seen in aviation, where it is argued that procedures should be used to minimise the consequences of human variability (Hawkins, 1987). It can also be seen in the directive documentation provided to photocopier technicians in the belief that they will solve problems more quickly (Orr, 1996). Berg (1997) features a number of similar justifications in his account of the introduction of decision support tools into medical practice. Advocates argue that, in the face of people's limited capacity for remembering important but infrequently used information, procedural artefacts can equalise workers who would otherwise perform quite differently. In short, procedures are provided to rationalise performance that might otherwise depend on "unreliable" individual experience.

When viewed as prescriptions of how to carry out work, procedures are of course highly relevant to cognitive task design. Indeed, procedures can be seen as an explicit expression of cognitive task design. Such a view is in line with recent accounts of cognition as distributed between people and artefacts. However, as Hutchins (1995) has pointed out, a procedure is not simply an externalised cognitive task. Following a procedure itself requires cognitive tasks not specified in the procedure. To quote from Hutchins (1995), "in order to use a written procedure as a guide to action, the task performer must coordinate with both the written procedure and the environment in which the actions are to be taken" (p. 295). According to Hutchins, transforming the written procedure into activity requires a series of cognitive tasks: reading and comprehending the words, understanding what they mean, and turning that understanding into some notion of appropriate action. DeBrito (2000) has carried out a more formal analysis of the cognitive tasks involved in procedure following. She identified nine cognitive components, including, for example, *detection of triggering conditions, searching for procedure, reading and understanding instructions,* and *planning action.*

There is evidence to suggest that badly designed procedures can impose excessive cognitive demands on operators and lead to errors in procedure following. Orr's technicians reported that the procedures they were required to use were sometimes circular and others were easy to misread. They also warned that without understanding the intent behind a procedure, one could easily get lost trying to follow it. DeBrito identified a number of processing difficulties, such as failure to detect a warning and unsuitable formats for displays, which led to processing difficulties and errors in procedure following. Thus it would seem that documenting a procedure and requiring an operator to follow it does not simply externalise the cognitive task and thus reduce cognitive demand. Rather, it can create new cognitive tasks that are themselves prone to different forms of error. DeBrito's analysis of the information processing steps involved in procedure following draws our attention precisely to the information processing requirements of task following and the new forms of error that poorly designed procedures invite.

However, even when procedures are well designed, their value as a means of controlling and rationalising human activity has been brought into question. In her analysis of instruction following, Suchman (1987, 1993) argues that procedures are inevitably incomplete specifications of action. They underspecify the activity that they purport to prescribe because they contain abstract descriptions of objects and actions that relate only loosely to the particular objects and actions that are encountered in an actual situation. Procedures are designed to be contextually independent, yet, like utterances in a conversation, procedures have to be interpreted with respect to the particulars of the situation in which they are used. The problem of mapping abstract procedures to particular situations is solved by the person using the procedures on each occasion of their use. This means that, in practice, procedures are often adapted, modified, and circumvented on the basis of the operator's appraisal of the local situation. There are a number of studies that illustrate Suchman's point.

Button and Harper (1993) describe a case study in which workers often ignored the normal accounting procedures associated with a manufacturing process in order to meet the contingencies presented by rush orders. Button and Sharrock (1994) observed how methodologies imposed on design teams are "made to work" in practice by those using them. Such *making to work* involved shortcutting steps in the sequential process and patching up the product afterwards. Symon, Long, and Ellis (1996) describe the dilemma faced by staff in a radiography department who have concerns both for following procedures and for the care of their patients. Their analysis demonstrates a subtle decision process in which at times procedures are followed and at others *workarounds* are used in order to resolve the conflicting concerns. These authors do not interpret such behaviours as bad practice or procedural violation. Instead they argue that flexibility in the light of local contingencies is in the nature of human work, and it is in the detail of this kind of engagement with methods, procedures, and other formal artefacts of work that the intelligence of human-involved work resides.

Most of these examples come from domains that are not safety related. In safety-critical settings where violation of a procedure might have hazardous consequences, we might expect to find a stricter adherence to mandatory procedures. However, studies of procedure following in these situations reveal similar gaps between procedures and practice, especially in emergency situations where procedure following is most strongly mandated. In the preface to his book, Vicente (1999) describes in detail the procedure violations that nuclear power plant workers have to perform on a routine basis in order to make the procedures with which they are provided work. Yet, at the same time, procedure violation is an oft-cited cause of accidents (e.g., Civil Aviation Authority, 1998). It seems clear that the difference between the procedures and practice is as important, if not more important, in these safety-critical settings.

What Suchman and others achieved by their critique of procedure following is an exposure of the limits of a purely information processing account of procedure. Suchman turned our attention away from the idea that procedures are rather like programs in a computer that the operator can follow without intelligence. Instead she turned our attention to the idea that it is only through the intelligent, pragmatic sensemaking of the operator that the procedure can be made to work in the diverse situations in which it is required to work. For Suchman, meaning and sensemaking, not information processing, are at the heart of procedure following. Procedures are not context-free algorithms that can be followed by an unintelligent agent; they are resources for action that have to be made sense of afresh in each new situation.

There is a need, then, to extend our information processing approaches to task design. In particular, we need to find a way of analysing the processes involved in making sense of procedures in a situation and the conflicting concerns this often reflects. In this chapter we attempt to do this by developing an approach to the analysis of procedure following that emphasises a particular individual's experiences of using procedures and the processes he has gone through in order to make meaningful use of them in his work as an airline pilot. This complementary perspective on cognition and cognitive task design provides us with a quite different perspective on the problems of procedure following.

FROM INFORMATION PROCESSING TO SENSEMAKING

Bruner was among the first psychologists to take experimental cognitive psychology seriously (Bruner, Goodnow, & Austin, 1956). In his groundbreaking book, he identified differences in the cognitive strategies that were used to carry out a concept formation task, and he related these differences to personality factors. Yet, in a more recent book (Bruner, 1990), he argued that the information processing and computational metaphors implied by such cognitive tasks analysis are inappropriate for analysing human activity. Indeed, he argued that the obvious success of such computational models has been gained only at the expense of dehumanising the very concept of mind that cognitive science, as a new discipline, sought to reestablish.

For Bruner it is the construction of meaning and not the processing of information that is central to cognition as a human activity. Information processing involves the assignment of meaning to data by virtue of preexisting codes. Creative choice is reduced to the selection among precomputed alternatives. In contrast, in Bruner's contructivist account of meaning making, meaning is the outcome of a computation, not its input, and choice is about the construction of alternatives, not just selection among them. Perhaps most importantly for Bruner, such meaning making is a cultural process in which individuals take part. It is through our participation in culture that meanings are made public, shared, evaluated, and changed; through their sharing, we come to make sense of ourselves, others, and activities.

Bruner's analysis of cognition as sensemaking is centred on the idea of narrative. People make sense of their activities in terms of unfolding events that have storylike structures and offer explanations of activity in terms of agency, needs, desires, commitments, and concerns. This folk psychological vocabulary serves as a means of explaining our activities and the activities of others in a coherent way. A historical perspective is part of such a narrative account. Bruner demonstrates how autobiographies—the stories that people tell about themselves—can be used to analyse why a person does what he or she does, how he or she accounts for the behaviour of others in his or her life, and how these understandings change over time.

Bruner is not particularly concerned with analysing work activity or technology, but others have used a narrative approach to sensemaking in such settings. The most notable of these is Orr (1996). He studied photocopier repair technicians and identified the telling of "war stories" as a central part of their activities. These stories of previous experiences with repair problems are frequently told during diagnosis but also in purely social situations. These stories have a discernible narrative structure and are designed not only to inform problem solving but also to amuse, instruct, and celebrate an individual's identity as a member of the technician community. The stories describe particular experiences and often identify particular technicians, particular machines, and particular occasions. What Bruner's and Orr's analyses have in common is the way in which they show the kinds of concerns that people have, what motivates them to do what they do the way they do, and how they justify their actions to others.

Bruner's perspective on sensemaking as narrative suggests a methodological approach that focuses on how individuals reflect and recount their experiences. It also suggests that such a focus should look beyond immediate ongoing activities towards individual life histories of experience. Bruner's emphasis on intentional states as explanations of human action also suggests a methodological approach that focuses on the concerns that motivate a person's activities and how these activities are explained in terms of such motivations. This point echoes Rosaldo's approach to the analysis of activity as experience (Rosaldo, 1986). In our study of procedures, therefore, we have collected autobiographical data looking at an individual's reflections on his career as an airline pilot, his experiences of procedures over that career, and the concerns engendered by their use. However, in our approach to cognition as sensemaking, we go beyond Bruner's narrative account to look at sensemaking as *dialogical*.

Within cognitive psychology, the term *dialogue* is often used to refer to exchange of information between humans or between a human and a machine, and quite often this assumes a simple coding–decoding model. The information requirements of the sender are coded as some form of utterance or sequence of keyboard inputs; these are then transmitted to the receiver and decoded so that the original information requirements are retrieved and an appropriate response is similarly coded for transmission. Knowledge of the common code is all that is required for successful dialogue in this model.

Our approach to sensemaking as dialogue moves away from such traditional information processing. It takes as its starting point the ideas of Bakhtin and others (Bakhtin, 1981, 1986; Morson & Emerson, 1990). In this approach, dialogue is less about the exchange of information and more about the idea of an author desiring to draw out a response from the reader. For Bakhtin, all utterances are made with the expectation of a particular response and carry an evaluative weight. Moreover, utterances are made in the knowledge of the many ways in which the words and gestures comprising it have been used on past occasions. Thus an utterance is neither neutral nor unitary with respect to its meaning. In this view, an utterance carries the meanings of many different authors and thus the potential for many different interpretations. Bakhtin (1981) put it thus:

> There are no neutral words and forms—words and forms belong to "no-one"; language has been completely taken over, shot through with intentions and accents. . . . All words have the "taste" of a profession, a genre, a tendency, a party, a particular work, a particular person, a generation, an age group, the day and hour. (p. 293)

Such utterances, then, cannot be simply decoded because there is no unitary meaning. Rather the reader or receiver must actively construct meaning in order to make an evaluative response. Bakhtin (1986) put it thus:

> when the listener perceives and understands the meaning (language meaning) of speech, he simultaneously takes an active, responsive attitude toward it. He either agrees or disagrees with it (completely or partially), augments it, applies it, prepares for its execution, and so on. . . . (p. 68)

This active process of a reader forging a sense of an utterance is much the same as the active process of authoring an utterance. It involves interpreting the utterance in the knowledge of who uttered it, a history of other possible meanings, and with a view to responding evaluatively. This process involves more than just the meanings of individual words. It also involves understanding the perspective of the individual who made the utterance and the tone or weight with which it was made (e.g., sincere, sarcastic, or authoritative). The symmetry has led some to refer to the "reader as writer" (Shusterman, 2000). Others have referred to the process as the cocreation of meaning (Morson & Emerson, 1990). The point here is that it is only in the dialogical interaction between reader and writer that utterances have any sense. In collapsing the more traditional distinction between sender and receiver or reader and writer, a dialogical perspective on sensemaking orients us to the idea that meaning making is a process of bringing together different perspectives, and it is in this creative bringing together that understanding is forged. Bakhtin referred to this as *creative understanding*. Morson and Emerson (1990) summarised the position thus:

> Rather than merely understand a text as the author himself understood it we should seek something "better" (ibid.). True understanding both recognises the integrity of the text and seeks to "supplement" it. Such understanding is active and creative by nature. Creative understanding continues creativity, and multiplies the artistic wealth of humanity. We must therefore value "the cocreativity of those who understand (N70-71, p. 142)." (p. 56)

As we can see from a careful reading of this quote, bringing together different perspectives, motivations, and concerns does not result in the determination of the "correct" perspective; nor does it result in the "synthesis" of perspectives into a single view. Rather, it results in the overlaying of one perspective on another in order to see their relationships and create a way of going on.

In taking a dialogical approach to sensemaking in this chapter, we explore the relationships between the concerns of a number of individuals or collectives who author procedural documents. We then go on to look at how individual readers or users of these documents seek to make sense of them in the light of their own concerns and their understanding of the concerns of the documents' authors. We reveal the reader as a writer creatively relating perspectives by examining the annotations that one individual makes on a procedural document. More than this, we seek to understand how this individual's motivations and concerns relate to those of the authors. In making this comparison, we are able to reflect back on issues of authorship and offer alternative visions for the use of such documents.

THE STUDY

The data we report here were collected as part of a research project exploring the design and use of procedures in commercial aircraft. As part of this project we interviewed commercial airline pilots about the documents they use when following procedures and the problems they encounter. One key document used by pilots is something referred to as a Quick Reference Handbook (QRH). This document is intended for use in the cockpit as a reference book for procedures. In an interview with one particular pilot about his use of the QRH, a fascinating and complex story emerged. The pilot provided us with a copy of a QRH that he owned. The QRH contained numerous pencilled annotations. An analysis of these annotations has been reported elsewhere (McCarthy, Wright, Monk, & Watts, 1998; Wright, Pocock, & Fields, 1998). We also interviewed the pilot about his life as a commercial airline pilot. In this context, he gave us an account of where his annotations came from, why they were made, how they feature in his ongoing work as an airline pilot, and what he intends for them in the future. It is this interview that forms the central part of the analysis of concerns. However, before we report this analysis, we provide some necessary background about procedures and the QRH in the modern commercial cockpit.

Some Background to Procedure Following in the Cockpit

There have been many technological advances in the commercial aviation industry. A modern aircraft is now capable of automatic flight practically from takeoff to landing. For the humans on the flight deck, this technological advance has been associated with dramatic changes in the nature of their work and, in particular, an increasing proceduralisation of their work practice. Much of the human work on a flight deck is prescribed by so-called standard operating procedures (SOPs), which are typically defined by aircraft manufacturers or operators. As we have already argued, these procedures are designed to reduce the cognitive demands made on pilots in both routine and emergency tasks. According to Degani and Wiener (1990), the "received" aims of introducing such procedures include:

- Act as a memory guide;
- Ensure all critical actions are taken;
- Reduce variability between pilots; and
- Enhance coordination during high workload and stressful conditions.

On all flight decks the QRH is used to provide quick and easy documentary access to SOPs, and, in the more modern aircraft, they are also supplemented by electronic displays. A QRH will include procedures for carrying out tasks conceived by the designers, including both normal procedures such as the before-start checklist (see Fig. 28.1) and procedures especially designed for dealing with emergencies.

757 NORMAL CHECKLIST
BEFORE START

GEAR PINS .... 3 ON BOARD
OXYGEN .... CHECKED
PASSENGER SIGNS .... AUTO & ON
FLIGHT INSTRUMENTS .... SET
PARKING BRAKE .... SET ON
FUEL CONTROL .... CUT OFF
ALTIMETERS QNH .... SET-X-CHECKED
TAKEOFF BRIEFING .... COMPLETED.

CLEARED FOR START

FUEL .... CONFIRMED
AIRSPEED BUGS .... SET
CDU .... SET
TRIMS .... ZERO-ZERO
FLIGHT CONTROLS .... CHECKED

FIG. 28.1. A page from a QRH showing the preengine start checks.

The QRH is not the only document in which the defined operating procedures can be found. Normal checklists are often printed on cards. The cockpit also contains an expanded checklist and technical manuals. This holds more elaborate explanations of both normal and emergency procedures. However, although it is carried on the flight deck, it is not intended for use during the execution of a procedure.

Our Approach to the Analysis of Concerns

Little has been said in the literature about dialogical analysis methods. Instead, the approach is to sensitise the analyst to dialogical ways of looking at artefacts, activities, and texts. We have adopted a qualitative case study method of the sort advocated by Robson (2002) and consistent with the grounded theory approach to research and data analysis (Miles & Huberman, 1994). These methods have been used to analyse a variety of qualitative data sources, including documents, transcriptions of semistructured interview data, and field notes.

In adapting this approach to carry out a concerns analysis of our interview data, we first identified five features of the dialogical perspective most relevant for this study. These have already been detailed and can be summarised here as follows:

- Look for evidence of readers and writers as coconstructors of meaning;
- Identify the multiple voices to be found in utterances;
- Explore the multiple perspectives and concerns of these voices;
- Identify to whom reader and writer are answerable for responses; and
- Explore the dilemmas of participants and their creative responses to them.

The analysis proceeded by first identifying segments of the interview that were considered relevant to the analysis of concerns. These were then grouped together into clusters that related specifically to the particular dialogical features listed here. These clusters were then iteratively refined. When the clusters had stabilised, each was given a code to reflect the reason for grouping those segments together in that cluster. These codes were then grouped together into clusters and coded to identify themes relevant to the features we were seeking to exemplify. Although done electronically, this process is equivalent to the process of affinity diagramming described by Beyer and Holtzblatt (1998), in which segments of interview data are posted onto the walls of a room and juxtaposed in order to create insightful affinities between ideas. In some cases this process can be repeated in order to generate different possibly orthogonal themes and patterns. We report the results of our analysis in terms of the following themes:

- On authoring a QRH;
- On the reader as writer and enactor; and
- On the enactment of procedures.

In each of these, we have drawn out the concerns of these multiple voices and focussed on how these multiple concerns create dilemmas for the pilot required to enact the procedures. We also tease out how the activity of procedure following extends far beyond the job of flying a real aircraft from one destination to another on a daily basis.

On Authoring a QRH

SOPs express the voices of many authors. Their utterances are also addressed to many different readers. In identifying the authors, the most obvious place to start is with the team of technical document writers at the sharp end of text production (see Orr, 1996, for such an analysis). However, these individuals are not the originators of the procedures themselves. Engineers and test pilots in dialogue within the manufacturing company are most often responsible for the actual procedures and the actions that constitute them. Operating companies (i.e., organisations such as British Airways) may add their voice to the text, as may the people they employ to train and check the competencies of their pilots. If we extend our analysis of authoring over larger units of time and space, we also find institutions such as air accident investigation branches and regulatory authorities, for example the Civil Aviation Authority (CAA) and the Federal Aviation Authority (FAA), contributing to the text in a number of ways. Thus an analysis of authors' concerns could be quite complex. In this chapter, we confine ourselves to exploring three of these collective voices: the manufacturers, the operating company, and the regulatory authority.

The Concerns of the Manufacturing Company.   Engineers, test pilots, and technical writers must demonstrate to the regulatory authorities that the aircraft they have designed can be flown safely. They must also demonstrate to the operating companies that it is efficient and economical. How the aircraft is automated and how the pilots' procedures have been designed are parts of this demonstration. The SOPs are central to this process. The engineers speak for the safety procedures in the context of engineering hazards, and the test pilots speak for the capability of a pilot with the required common sense and airmanship to understand and follow the procedures appropriately. The types of mission the aircraft will fly will have been thoroughly analysed and, for any failure that could reasonably be expected to occur, a procedure will have been devised. The procedures will also have been analysed to ensure that certain kinds of deviations from their intended execution are both unlikely and do not in themselves lead to hazards.

As Degani and Wiener (1990) have pointed out, the manufacturing company must first meet the concerns of the regulatory authorities by persuading them to provide a certification for flight. In the United States, this is achieved by proving that the manufacturers have complied with the Federal Aviation Regulations (FAR) Part 25 document. Such certification allows the manufacturer to fly the aircraft. It does not, however, allow the aircraft to be operated by the manufacturer's customers, the operating company.

The Concerns of the Operating Companies and Their Training Officers.   Operating companies must prove to regulatory authorities that they are capable of flying their aircraft.

TABLE 28.1
Extract from FAR 121.315

FAR 121.315

Each certificate holder shall provide an approved cockpit check procedure for each type of aircraft.
The approved procedures must include each item necessary for flight crew to check safety before starting engines, taking off, or landing, and in engine and system emergencies. The procedure must be designed so that the flight crew will not need to rely upon their memory for items to be checked.
The approved procedures must be readily usable in the cockpit of each aircraft and the flight crew shall follow them when operating the aircraft.

Note. From Degani and Wiener (1990).

In the United States, this involves a second certification process to prove compliance with FAR Part 121 regulations. As part of this process, companies may change the manufacturer's procedures to conform more closely with company philosophy and practice regarding such things as, for example, the role of automation in the cockpit. They may also change procedures to conform more closely to particular types of operations, for example, short haul versus long haul. In doing this, these companies are concerned that their pilots receive procedures that reflect the way they want their aircraft to be operated, and what they feel is best for pilots given the type of flying they are required to do. Of course, if they undertake to extend, revise, or otherwise interfere with the certified operating procedures, they must do so in dialogue with the manufacturers and the regulatory authorities.

Standardisation is also a concern of operating companies. Most companies operate more than one aircraft type, and pilots are usually licensed to fly more than one type. Differences in procedures across cockpits can lead to pilot error. A conscientious management will thus endeavour to minimise the differences in procedures across types flown by the same people. Consistency across aircraft types is also one of the selling points of many aircraft manufacturers, who argue that it reduces the cost of training and the likelihood of errors. However, an overemphasis on consistency can also result in problems, such as when items are added to a checklist in the interests of consistency.

The Concerns of the Regulatory Authorities. Regulatory authorities contribute other texts of significance to the authoring of procedures. Table 28.1, for example, presents extracts from the FAA regulations on the design and use of cockpit procedures to which operating companies have to comply. These regulations speak both to other authors of procedures and user–readers, and their tone is one of compliance and accountability.

Regulatory authorities can also contribute to the emergence of new procedures following accidents and incidents. For example, total engine failure after the ingestion of volcanic ash seems an unlikely failure mode for a commercial aircraft, but a procedure for dealing with just such an emergency exists following incidents over Alaska. When a Boeing 747 encountered volcanic ash around Mount Redoubt in Alaska in 1989, it suffered a total engine failure. The aircraft lost 14,000 feet in altitude before the pilots could restart the engines (Beaty, 1995).

The Concerns of Authors Extended in Time and Space. As has already been illustrated, even before an aircraft flies, the authoring of the cockpit procedures has already been

TABLE 28.2
Analysis of Annotations to a QRH

Type of Annotation	Frequency	Percentage
1. Clarifying the reason for the procedure	63	19.5
2. Addition of extra actions or procedure	61	18.9
3. Additional relevant background systems information	48	14.9
4. Alternative procedure or uncertainty about the effectiveness of the current procedure	38	11.8
5. Consequence of taking an action	32	9.9
6. Delimiting the scope of a condition or procedure	26	8.0
7. Cause of warning or condition	21	6.5
8. Consequence of warning or condition	20	6.2
9. Highlighting importance of particular steps	14	4.3
Total	323	100

stretched across numerous individuals and across both time and location. In one sense there is closure to this process when the manufacturer and, later, the operating company receive certification to fly. However, there is also a sense of chronic openness to the process of authoring. This arises from many sources. For example, manufacturers gain experience from flight operations that lead to modifications. To some extent, this is to be expected, but numerous and frequent changes are considered by regulatory authorities, companies, and pilots alike to be unsatisfactory, because they can lead to pilot confusion. Other changes occur when a company changes its philosophy or introduces new technology onto the flight deck. Changes in regulations concerning flight operations can also affect procedures in unexpected ways. An example offered by Degani and Wiener concerns the introduction of nonsmoking flights. Previously, the automatic activation of the no smoking sign was taken as a cue to cabin staff to prepare for landing. Once no smoking throughout the flight was mandated, a procedure had to be modified to remind flight deck crew to warn cabin staff to take their seats for landing.

On the Reader as Writer

Having looked in a general way at the authors' concerns that motivate procedure design and documentation, we now move to look at how one reader makes sense of a QRH. In the interests of confidentiality, we refer to him as Captain Roy. As part of a larger project, Captain Roy was kind enough to allow us to observe him on a number of European flights. In preparation for these, Captain Roy had shown us a copy of a QRH that he had annotated extensively. At that time we were concerned with the usability of the QRH. We thus carried out analyses of the annotations, the primary purpose of which was to categorise them in terms of the types of information they suggested were missing from the QRH.

A total of 323 annotations were present in Captain Roy's QRH. Randomly chosen segments of the QRH were used to generate a classification scheme, which was then validated against the remaining segments of the QRH. Captain Roy was subsequently asked to comment on the classification scheme, which he confirmed as being an accurate representation of the intent behind his annotations. The classification scheme was then further validated by asking a volunteer researcher, who was knowledgeable in the field, to use the classification scheme to classify a sample set of just over 12% of all QRH annotations. Full agreement was reached in

TABLE 28.3
Examples of the Four Most Frequent Annotation Types

Type of Annotation	Example
Why the procedure is the way it is	
1. Clarifying the reason for the procedure	"De-energises DC fuel pump" [following procedure "APU selector . . . Off" position]
3. Additional relevant background systems information	"Deploys automatically when both engines fail" [referring to ram air turbine in loss of engine thrust section]
Modifications to actions of the procedure	
2. Addition of extra actions or procedure	"Try another autopilot" [when procedure calls for autopilot disengage if inoperative]
4. Alternative procedure or uncertainty about the effectiveness of the current procedure	"No mention of using APU at low level to start engines" [when attempting engine restart after loss of thrust]

62.5% of cases, while partial agreement occurred in 27.5% of cases, yielding a combined full or partial agreement rate of 90%.

The results of the analysis are summarised in Table 28.2. As we can see, the modifications to the written procedures take a number of different forms reflecting different purposes and concerns. Although there are nine categories in total, the four highest scoring categories cover over 65% of the annotations. Categories 1 and 3 cover over 38% of annotations and are concerned with providing a rationale for *why the procedure is the way it is*. Categories 2 and 4, covering 26% of the annotations, can be seen as concerned with *modifications to the actions in the procedure*.

Table 28.3 presents some examples of the kinds of annotations analysed in Table 28.2. Captain Roy's annotations reflect his concerns about the effectiveness of the procedures, understanding how and when to use them, and why the procedures are the way they are. They can be seen as an attempt to bridge the gap between work as represented and the work in practice. In bridging the gap, Captain Roy renders the procedures meaningful to himself.

The content analysis of Captain Roy's annotations focuses our attention on the QRH as a statement of procedures and highlights the information that is missing from the procedures as stated. However, it gives no insight into where these annotations came from, nor why they were made, nor what concerns were foremost in Captain Roy's mind when he was motivated to make them.

Therefore we interviewed Captain Roy to talk about the annotated QRH. This interview, lasting approximately 2 hours, took place in Captain Roy's home. The interview took the form of a semistructured interview (Robson, 2002) around the following topics:

- Why were the annotations made?
- What sources of information were used to make the annotations?
- When were the annotations made?
- How is the annotated QRH used?
- Is annotating QRHs common practice?

The interview was audiotaped and transcribed for analysis. The interview data made it clear that, in Captain Roy's eyes, there is a far greater set of voices, concerns, commitments, and values residing in what he has written than is revealed by our earlier analysis of its information content. This lead us to reorient our analysis towards examining the concerns and dilemmas experienced by Captain Roy and to which the annotated QRH is a partial solution.

On the Enactment of Procedures

Captain Roy's interview data were analysed from the dialogical perspective, using the grounded theory method already described. It quickly became clear that limiting our analysis to the task of procedure following in a real in-flight emergency would not only fail to reflect the richness of the interview data but also misrepresent important parts of the experience of procedure following. Procedure following as an experience involves learning about procedures in the classroom and in private study, practising them in simulators, and gathering experience from other members of the aviation community. It seldom involves following a procedure during a real in-flight emergency. Thus we sought a metaphor for procedure following that would allow us to make visible the experience of this larger activity. We chose the metaphor of *enactment.*

In choosing the term *enactment,* we are making explicit reference to Turner and Bruner's *Anthropology as Experience* (Turner & Bruner, 1986). The term is used by them to draw attention to a context of performing a script that is larger than the individual on an isolated occasion of its performance. Turner and Bruner are concerned primarily with drama as enactment and see researching and negotiating the script of the play, rehearsals, the selection of actors, and the actual performance as composite and equal parts of the experience of the dramatic production. Such large-scale enactments also make visible the fact that the final performance is a product of many voices, and, as any one involved in drama knows, it also always unfinalised and creative. Captain Roy's story about his experiences with procedures fits neatly into this metaphor. Our analysis is described in four parts. In *researching the text,* we describe the diversity of information sources that contribute to Captain Roy's interpretation of the QRH. In *rehearsal,* we highlight the fact that in the day-to-day world of airline pilots, they are more likely to encounter emergency procedures in a training simulator than on a fee-paying flight. Thus much of their experience of emergency procedure following is as regular practice for an event that never occurs. *Between rehearsal and performance* captures Captain Roy's reflections on the times when he has actually needed to use an emergency procedure or his anticipation of possible future occasions of use. Finally, *putting experience into circulation* describes Captain Roy's efforts at sharing his experiences of the previous three stages.

Researching the Text.   In dealing with an emergency situation, pilots do not simply pick up the QRH and read it as if for the first time. An enormous amount of work goes into the preparation for such an event. This includes extensive research of numerous other documents and, at least in the case of Captain Roy, the approaches taken by other companies flying the same type of aircraft. In this way, he situates what he learns from the QRH in a much broader experience of use.

The QRH is not the only source of information about procedures to which pilots have access. There are full technical manuals, extended checklists, and notes from lectures and training sessions. Some of these sources are available on the flight deck in addition to the QRH, but the information is sometimes hard to find and widely dispersed. This is one of the reasons for the QRH. As its name implies, it is intended to be a quick and easy source of information.

The authors of the QRH, in distilling this large and dispersed body of information into something manageable, necessarily have to leave some information out. Captain Roy's annotations are to some extent a result of his research into these other sources of information:

> All the annotations are to be found in the manuals somewhere—it's a sort of collation exercise to bring it all together. . . . I just tried to get lots of bits of the manuals all together in one

book. . . . There's too much stuff scattered all around in different parts of the books which is why
I did it in the first place.

However, this is not just a passive collation exercise; it is a learning experience. For Captain
Roy, the research highlights not only what he sees as mistakes or omissions in the QRH, but
also bits of information that are hard to find or poorly classified. It also provides, for him, an
understanding of the intentions that lay behind some of the actions specified in the QRH:

When you're reading it through, you want to know why things are being done this way, and if you
know why a certain check is being done, it makes more sense. For example there's a "cargo fire"
checklist and if you work your way down this it says to turn an air conditioning pack off. The first
time you read this you think "why on earth are you doing that—what's that got to do with it?" and if
you go into the [expanded checklist] books it says you're reducing the mass flow of air, oxygen, into
the cargo hold . . . so that makes sense, but just knowing that just makes the checks more sensible.

The concern voiced here is one of wanting to know why things are the way they are in order
to make sense of the text as part of an activity. Another important concern for Captain Roy is
to become familiar not only with the procedures he is required to perform, but also with the
QRH as a document:

You know you get so many mistakes, and really there are some things that you've got to find quite
quickly, smoke problems, major electrical problems—yeah, you want to be onto the checklist
pretty quick, and most people I've seen in simulators fumble about trying to find the right drill and
come up with the wrong drill.

There are many similarities here with an actor preparing his part in a performance. Captain
Roy is primarily concerned to know his part well, but more than this to understand why his part
has been written the way it has. There is also a major difference, of course. Captain Roy, unlike
an actor in a play, will not be required to act from memory. On the contrary, most companies
require the opposite, that pilots always read the procedures step by step from the QRH. With
only a few exceptions, they do not allow pilots to act from memory. The authorial concern here
is that pilots acting from memory may perform poorly because of memory lapses.

Dispersed technical manuals are not the only documentary source that informs Captain Roy's
annotations. He has been a pilot with a number of airlines in his career and has experienced many
versions of the QRH. Some companies, like his current one, use the manufacturer's version.
Other companies have developed their own versions. This is a lengthy process, based on a
company's experience both of real incidents and accidents and of training pilots in simulators.
Changes are made to reflect company policy or philosophy in a number of areas. Captain Roy's
experience of these other company QRHs is a valuable source of additional information:

When I joined *Company Z,* I was dismayed to find that they used the *manufacturer's* basic checklist,
whereas I had been used to the *Company X* expanded one which had a lot more information on it.
What the [*Company Z*] flight deck needs is an expanded QRH like the *Company X* one, which is
a bit more helpful—more user friendly. That's what you need on the flight deck.

In this comment we also sense frustration and dismay at this situation. This tension between
what Captain Roy sees as best practice and what is required in his own company takes on
a clearer expression in the following quote. After being asked again whether his company
(*Company Z*) used the manufacturer's QRH, he searches for an explanation:

Yes [we use the *manufacturer's* QRH] and I think that's criminal. It may be because we're a brand
new company (we've only been going a year and a half, and we've merged with two airlines since

we started) and the training department and powers that be are probably well overloaded anyway. But I know *Company Y* [a parent company of *Company Z*] use this [the *manufacturer's* QRH] as well, and they've got no excuse because they've been going twenty years or so.

In the following extract he identifies specific deficiencies with the Manufacturer's QRH:

On the *manufacturer's* checklist for example, "smoke" is not itemised and you have to know from experience that they quantify smoke as "pressurisation/air conditioning smoke" or as "electrical smoke." And that's ridiculous really, its very difficult to know what sort of smoke you've got in the cockpit. I mean you're asking people to sniff and go "yeah that's electrical smoke." . . . When you've got smoke coming out you want this drill quickly and its not itemised. . . . The *Company X* one is much better, its actually got smoke itemised as "smoke or internal fire in the air." On the first page of the index . . . a great improvement I would say.

The sarcastic humour expressed here serves to emphasise what Captain Roy sees as both the frustration and the ridiculousness of the situation. He goes on to point out how relatively inexperienced, qualified pilots, new to the aircraft, might be caught out by the manufacturer's QRH, in contrast with the *Company X* version:

You've got to imagine this scenario where you've got this first officer who's been on one aircraft type a long time, he's got at least 5000 hours which qualifies him to be a Captain. Now he's been on another aircraft type for six months so he knows how to operate the aircraft, he's done, he's done maybe, one or two simulator checks, and in theory, he should know all there is to know about the 757, but he's going to get caught with this [*manufacturer's*] QRH—either for real or in the simulator because he won't know that some of these checks are all over the place and "smoke" isn't there when you need it.

What is being expressed here is Captain Roy's concern to bring together in one place all of the information he considers necessary to enact a procedure. This includes not only information about what to do and when, but also information about why things are done in certain ways and what the authors intended. Another thing we see is a concern on the part of Captain Roy to become familiar with the QRH as a document, so that, in practice, the document itself is not an impediment to performance. However, there is a tension here. Pilots are not provided with personal copies of the QRH for the purposes of annotation or revision. Captain Roy does not disclose from where he got his copy. In discussion, when asked why pilots do not get their own copy, Captain Roy responds as follows: "Expense I think. I think it's a false economy, isn't it?" When asked whether it would be a good idea for pilots to have their own copy, he replies, "Oh God yes. . . . Not so much annotate (they could if they wanted to) but just familiarity with the book itself."

Another thing we see is a concern not only to bring what Captain Roy sees as best practice into a younger company with less experience, but also for Captain Roy himself not to lose valuable knowledge he has gained from his previous experiences.

There is also a tension here that becomes more palpable as we move towards rehearsal and performance. The tension is that because the company authorises only one version of the QRH, the knowledge gleaned by Captain Roy from other companies' checklists has an ambiguous status—something akin to "not invented here" or "unauthorised." This tension serves, as we shall see later, to suppress this knowledge.

Rehearsal.   It might be considered that the only concern about an emergency procedure in a pilot's mind is "getting it right on the night." But, of course, an in-flight emergency, particularly of significant proportions, is a rare event in the actual practice of flying passengers around the world. A much more significant and recurring event for pilots is the regular simulator training and assessment they undergo. Pilots must undergo such training at regular intervals and must perform acceptably well. The training, as described by Captain Roy, is quite adversarial. Training officers from the company present the pilot with a progressive unfolding of ever more difficult problems. Captain Roy reports how the training officers never give you all of the problems straight off. They keep some hard ones in reserve so that, when you just think you are doing well, they will throw something at you that the procedures won't deal with. They call these the "gotcha" problems:

> I've gleaned it is from being put in this situation in the simulator where some instructor has realised that this is a bit of a "gotcha"—with this failure you don't get this, for instance, and I've been caught on it and I've come home and written it up, and I can't for the life of me understand why they don't publish all this sort of stuff, and I've said this to trainers over the years and they go a bit quiet, and they don't seem to understand what your talking about. You know, if I was a Chief Pilot (perhaps in an idealised world), I'd like my pilots to know all the "gotchas" from day one.

His main concern with this approach to training is what happens if relatively inexperienced pilots who have not been through as much simulator training encounter one of these "gotchas" in real life:

> Just about every time I go in the simulator I learn something which I think is wrong. You know, why was I not given that information on day one? I feel sorry for people that have just come onto the aircraft. I was flying with a guy the other day who's just come off the Airbus. He's done six months on the 757, and he's going for a command course next week and I was chatting to him on the flight deck about certain things that I've learned over the years, and he hadn't a clue what I was talking about, and he's going to be a captain next week, you know. You might not get that problem in a simulator, and these are not things that are written down in the books anywhere, but if it happens for real he won't know about it.

One of the chief sources of the QRH annotations is experience of the "gotchas" in simulator training. As Captain Roy comments, "after a simulator session I always write up my notes on the simulator and add anything that I've learned."

Together these extracts are revealing for our analysis. The QRH, as a distillation of one set of experiences and concerns, those of the manufacturers, are overlaid here with the concerns of training officers to make clear the shortcomings of the QRH as a received text. Overlaid again by Captain Roy himself is the concern for people who have not had access to the experiences of simulator training.

There is a sense in which the annotations represent expression of experience, not present in the QRH as a received document; but these experiences are not just those of Captain Roy. They are retellings of experience from training officers, which in turn may be retellings of others' experience and so on. The annotations express not only the voice of Captain Roy but also his reading of the voices of his training officers. Their concerns are for the particularly hard-to-solve problems that are not captured in the QRH by the generalisations of the engineers and test pilots. The annotations are also shaped by a desire to make the "gotchas" tractable for his colleagues. This is the multiperspectival character of dialogical sensemaking.

As far as rehearsal is concerned, the annotations serve as a record of many people's experiences in the simulator but also as a source of preparation for simulator sessions to come. When asked why he did the annotation in the first case, he answers, "initially it was for simulator revision."

Yet there is in this reading an interesting tension. Captain Roy is not allowed to use his annotated version in the simulator, or for that matter on the flight deck. When asked when he used the annotated QRH, he replied:

> Well simulator revision—its brilliant for that. It jogs your memory, as I say, I try to memorise most of it anyway before I go in the simulator. The actual pilots with the training department would be horrified if they thought I'd get that out during a flight, obviously, because its not an approved book at all."

So Captain Roy is faced with the ironic and paradoxical situation. He has to memorise his annotations to the QRH in order for them to be of use in the simulator, yet they are representations of things learnt in the simulator. Furthermore, having committed these things to memory he is required to read the official procedures from the official book.

From Rehearsal to Performance.   If the annotations to the QRH began as a recording of, and revision for, rehearsal in the simulator, what role does the QRH have in the actual performance of a procedure in the cockpit when dealing with a real emergency?

Captain Roy has developed his annotations further. He has produced what he himself describes as a "QRH for Dummies." This is an attractive and well-presented, word-processed document that captures, in a more usable form, all of the information in the annotations and other sources of information: "What I've done now is put it all on computer now in a nicely coloured book [indicating a book entitled QRH for Dummies]."

His concern here was to produce a personalised version of the QRH, which distills everything he has gleaned of best practice, and present it in a way that is more usable than "*the manufacturer's* QRH" he is officially required to use. He is quite proud of its usable design: "My customised QRH has a cross referenced index at the back with page numbers, so you can find things quickly. I've noticed in the simulator people scratching around looking for these drills all over the place."

Captain Roy carries a copy of the "QRH for Dummies" with him on the flight deck and has had occasion to use it: "I would not feel very comfortable having left it at home. I feel very comfortable having all that extra information sitting next to me on the flight deck, and I've used it on quite a few occasions and it's been very useful."

Captain Roy's concern here is to have, at his fingertips, a usable expression of his collected experiences of enacting procedures and his readings of other people's enactments. His concern is to bring them together in the cockpit, in the event that they will be required, possibly to save his own life and that of others. But it is in this concern that the most profound of all tensions begin to emerge. Although Captain Roy argues that most pilots may have ad hoc notes, few he knows have put in the time and effort to produce what is essentially an alternative QRH:

> Pilots do make all sorts of notes. Most pilots write notes after simulator sessions so that they don't make the same mistakes again. I think most pilots will have a folder in their flight bag with various bits of notes. But I don't think many will have a QRH like mine. . . . I've organised mine into something a bit more useful. But I can't start distributing my unofficial QRH around the place. Its got to be approved, read, updated and verified by the manufacturer that what I am saying is correct, so I fully see their point of view.

Although he believes in the value of his own QRH, he is also aware that he should not use it: "But you've got to be careful not to start distributing copies of the thing because the training department would go bananas because its not approved etc., etc. It's very much a personal thing that I don't actually spread around."

Given the tension resulting from the presence of the alternative QRH on the flight deck, how is it actually used in relation to the Manufacturer's QRH? When asked whether he would use the *manufacturer's* QRH, Captain Roy replies, "Oh yes, I'd always do the company book—the proper QRH first." However, he also says, "But what I actually do is I take it with me, and if I have a problem (I had one the other day) I do the proper drills from the normal QRH, and then I would get mine out just for further reading." He is quite insistent on this point:

> If you've got a problem that is no great rush, then I might have mine out at the same time, or perhaps for further reading, but you must do the drill from the *manufacturer's* QRH, so I would do that and probably use the other one for further reading, but you MUST do the drill from the *manufacturer's* QRH.

His tone is at times both defensive and ambiguous:

> There's nothing in mine which conflicts . . . although actually . . . something like a loss of thrust on both engines is an interesting one. Because the *manufacturer's* checklist assumes that their drill which is given starts one or both engines, but there's nothing there to say what to do if you don't get the engines re-started. The *Company Y* QRH has a procedure for starting the engines using the APU when you get to low level, and I've got that in my book, so in that case, having done the *manufacturer's* checklist, I'd probably get out mine and do the drill for that.

We see in these extracts a number of attempts at diffusion of the tension of actually using the "QRH for Dummies" on the flight deck. We see, for example, how the alternative QRH is used "just for further reading" or as an addition to, or duplication of, official procedures. This provides an ambiguity to its role as alternative. Furthermore, we see how it is claimed that there are no contradictions in the alternative QRH, though this claim is heavily qualified and the assertion that the official drill must be done first is strongly asserted. In the following extended extract, which ostensibly is about procedures that have to be done from memory versus procedures that have to be read from the official QRH, we see a clear conflict of concerns. This conflict is born of the tension between what the official QRH says and what has been learnt in simulator training:

> You can get into a bit of a conflict here because there are checks that appear quite serious that you would think should be done from memory, but which the QRH says NOT to do by memory. Occasionally, when you're in the simulator (not with this company but with a previous company) you can do something from the QRH and the instructor will say "no—I'd do that by recall." Now this is not standard but happens occasionally, and you're in a bit of a quandary because you're not following standard procedures . . . [describes at length a particular procedure] . . . So it's checks like this where you think "God, I wish this was a memory item—I could just do it." But you have to think that well, the *manufacturer* must have thought of this and they've not made it a memory item, therefore it isn't doing that much damage. So you do the standard procedure and clean the aircraft up. Get the wheels up etc., etc. . . . You're five minutes down the line from the problem and, for real, I'd be awfully tempted to switch it off even though I know I'm violating a procedure. If I did it in the simulator I'd get shot! I perhaps shouldn't have said that."

There are several points worth drawing out here. The first is that Captain Roy's concern to avoid a procedure violation (or even talk about entertaining such a thing) is partly because disciplinary action might follow ("I'd get shot," and "I perhaps shouldn't have said that"). However, it is also due to the knowledge he attributes to the authors of the procedure ("*the manufacturer* must have thought of that"). We also see something of the frustration of being required to follow procedures when what needs to be done is both self-evident and straightforward.

Putting Experience into Circulation.   We have seen throughout the aforementioned analysis that one of Captain Roy's concerns in writing the annotations to his QRH was to record his own experiences for use both in simulator training and assessment and also on the flight deck.

However, this preparation for an enactment goes beyond a concern for self in activity. Captain Roy is also concerned to record these experiences for use by others, what Turner (1986) refers to as putting expressions into circulation. This is expressed directly by Captain Roy:

> I'd like my pilots to know all the "gotchas" from day one. OK, they may not have time to read all this stuff, but I've always quite fancied writing a book on the "gotchas." I mean, the training department would kill me, but I can't understand why someone doesn't.

Part of his motivation for the "QRH for Dummies" is to encapsulate what he has learnt in a readable form. He further adds, "any pilots who have seen it have wanted copies."

His choice of title is also an allusion to the numerous alternative manuals for everyday computer applications such as Microsoft Office products, but it also reminds us of other "do-it-yourself" documents circulated by pilots (e.g., Bulfer & Gifford's 1996 *FMC User Guide*). There is, however, a tension in Captain Roy's concern to put these experiences into circulation. As we have already seen, operating companies do not condone personal versions of the QRH. This reflects a concern on the part of operating companies to ensure that everyone is "singing from the same hymn sheet." This in part protects the company from accusations of non-uniform practice and also provides a source of quality assurance.

For Captain Roy, the considerable time and energy he has voluntarily put into his project reflects his concern that there is something paradoxical about a document like the manufacturer's QRH, which claims to make procedures usable by average pilots and falls so far short of the mark. The title of his book implies that one day he will make this available to other pilots, but at the same time he voices considerable reservation about this, as the following quote makes clear: "I've organised mine into something a bit more useful. But I can't start distributing my unofficial QRH around the place. Its got to be approved, read, updated and verified by *the manufacturer* that what I am saying is correct, so I fully see their point of view."

So there is a final paradox. The QRH for Dummies represents the aggregation of the considerable experience of many people: Captain Roy, the manufacturers, other operating companies, and simulator training officers. Yet the very same people whose collective experience Captain Roy attempts to express deny him the opportunity of sharing it with other pilots.

Summary of Concerns Analysis: The Paradoxes of Procedures

From the authors' perspective alone, we have seen that there are many voices in this dialogue. There are the voices of the manufacturers and the operating companies, the training officers, and the regulatory authorities. We have also seen that these authors do not write at a single moment for a single reader, nor is it the case that the full story they wish tell is contained in a single document, the QRH. Rather their writing is spread across texts and over time and space. Their addressees are not only the pilots who will use the QRH in simulators and in the air, but also regulatory authorities and operating companies. Some of what is being voiced is an understanding of the pilot-reader as a fallible human that detracts from an otherwise ideal rational system. Yet in this understanding, the authors see both an inherent unfinalisability of the procedures as a means of controlling what pilots do and, at the same time, the need to use the procedures to confirm that the pilots will do the right thing.

Focussing on the voicings directed to the pilot-reader as addressee, we see voicing of an authoritative tone declaring that the only things the pilot needs to know is written in the QRH. At the same time the very presence of extended checklists on the flight deck indicates that this is not so. There is also a voicing that sets up the QRH as something to be read and followed during enactment. This, it is understood, is in order to compensate for the pilot's fallibility. Yet at the same time the pilot is not given access to the document for the purposes of familiarisation or revision, for fear that it will lead to memorisation, customisation, and nonstandard enactment.

The analysis of Captain Roy's annotations and his story about whence they came and whither they will go provided us with a rich account of the reader as writer and enactor. Here we see responsivity to the official QRH in both moral and material form. Captain Roy has invested considerable energy in bringing together his experiences of enactment both within his present company and from previous companies. His experiences involve extensive document research, simulator training, and flight operations. These are distilled by Captain Roy as author, into the annotated QRH, a response overlaid on the QRH itself. They are also articulated in his QRH for Dummies, reflecting a concern to put these experiences into a more usable expression for himself and others.

Captain Roy's experience of using the official QRH in simulator and line operations is that the QRH does not contain all the information he would like to have. This awareness most significantly arises from Captain Roy's experiences of company simulator training, which is actually geared to highlighting the shortcomings of following the QRH procedures. Indeed the original motivation for the annotated QRH was as revision for simulator assessments. It is here that we first encounter the most fundamental paradox that surrounds Captain Roy's activity. Although the annotations are the result of (among other things) these simulator experiences, the annotated QRH (and the QRH for Dummies) cannot be used in the simulator. Neither can they officially be used on the flight deck for which the simulator training is supposed to be a rehearsal. We see here multiple and ambiguous answerabilites. Answerable to the operating companies, Captain Roy should not use the QRH for Dummies. Answerable to the training officers, he should not *be seen* to use the QRH for Dummies. Answerable to himself, he should both use and be seen to use the QRH for Dummies in the simulator and in flight.

As well as his own enactment of procedures on the flight deck and in the air, it is clear from Captain Roy's story that he is concerned, too, for other pilots who face similar deficiencies in the use of the official QRH. He is strongly motivated to share these experiences through his composition of the QRH for Dummies. Although he is convinced of the value of this sharing, he is prevented from doing so by his concerns for what the training officers and operating company would say, as well as by the fact that he is aware of the importance of having the document authorised. This leads to an ambiguous position in which he both sees the document as something he ought to make available and at the same time as something he ought to keep to himself. The concerns summarised here are listed in what follows.

1. The manufacturer:
   - You (the pilot-reader) should be able to understand the procedures because we have designed them with you in mind.
   - We believe we have anticipated every problem and provided an easy-to-follow solution.
   - We know that you (the pilot) are the weak link in the chain. Consequently your actions must be controlled and reduced to simple instructions and you must follow them.
2. The operating company:
   - We have benefited from a lot of experience of operational work, which we pass on to you (the pilot-reader) in our version of the QRH.
   - We do things differently from other places you (the pilot-reader) may have worked.
3. The training officers:
   - They (company, manufacturer, or regulators) say that you (the pilot) must have written procedures and you must follow them; otherwise you will go wrong.
   - However, there are things we can show you where doing what they say will lead you astray.
   - These things must not be written down.

4. The pilot:
   - There are things that are not written down that I have learnt and need to remember.
   - I have written them down for myself and for others.
   - But you (the training officers, etc.) say I mustn't use these things.
   - But I am not prepared to do without this information when it really matters (in flight).
   - You (other pilots) ought to have this information, so I have written it down for you.
   - But I can't give it to you.

CONCLUSIONS

DeBrito and others, by analysing procedure following as a cognitive task, have shown how the usability of procedures could be improved. In our analysis, too, we have seen how Captain Roy's annotations highlight some very fundamental ergonomic design failings of the QRH, such as poor indexing. Some of our findings, particularly Captain Roy's reports of the difficulties of navigating paper documents and the desire to know why certain procedural actions are deemed necessary, reinforce those of DeBrito. However, in this chapter, we have attempted to paint a much broader picture of the activity of procedure enactment extended in time and space through an individual's career as an airline pilot. This kind of experiential analysis focuses on emotional or motivational concerns about the activity of procedure following. This leads us to consider not only the design of a QRH per se, but rather the organisation of the sensemaking activities that surround a pilot's performance. In this regard, the analysis leads us to a quite different quality of "design" consideration.

Increasing attention has been paid in recent years to the concept of *organisational learning*. In the safety-critical systems world, this has been taken to mean learning from accidents or incidents (see, e.g., Hale, Wilpert, & Freitag, 1997). This learning is often expressed as changes to operating procedures or piecemeal modifications to system components. However, the analysis reported in this chapter speaks to a much broader set of experiences that could inform a more open learning process. The analysis points to the way in which the QRH as an authored document engenders a creative response from a number of quarters. Operating companies interpret the procedures inscribed in the manufacturing company's official documents against their own experience and concerns. Some operating companies rewrite the document as part of that creative response. Pilots working for these companies come to understand that a range of company interpretations of the manufacturer's document exist, and they also assimilate more of these interpretations as their career progresses. They use them as a means of evaluating their current company's interpretation of the document. Trainers use the same documents not so much as statements of what to do but as means of creating challenging learning experiences for those under their guidance. In so doing, they come to critique and make manifest the limitations of the document both for themselves and for the pilots they train. All of these experiences speak to the potential for learning that exists within the complex organisation of commercial aviation. Yet this learning is not being exploited, as is made evident by the deep paradoxes apparent in Captain Roy's story.

A procedural document such as the QRH is an attempt to finalise a solution about "what to do." But, as this analysis points out, and most other research on procedure following has shown, the openness of life defies such finalisation. The paradoxicality of the learning experience that we describe in this chapter arises from the fact that those involved in it "pretend" that things are finalisable, that there can be a last word on what to do in the event of an emergency. And hence *responsibility* for knowing what to do can be removed from those who do it. The philosophy here is a philosophy of "one best way" (Vicente, 1999), and there is no room for a creative response. If we are to capitalise on the learning that takes place in organisations, then a philosophy of coconstruction, what Vicente refers to as "let the worker finish the design" (Vicente, 1999), has to be considered.

It is not that the learning experiences we have reported here lead us to conclude that procedures are a bad thing or that procedural documents are, in principle, unworkable. Captain Roy's annotations and his carefully designed QRH for Dummies tend to make the opposite point. They are proof positive that workers can and will finish the design in a way that is quite consistent with proceduralised working. The results also lead us to conclude that the workers are the nexus for experience of use but, currently, there are very few ways in which this experience can be given a legitimate voice in the process of procedure design.

So what would happen if regulatory authorities and manufacturers declared an end to their monopoly on experience and allowed Captain Roy to publish his QRH for Dummies, perhaps as a Web-based resource? What would happen if pilots started taking versions of the document into the simulator with them and testing out their efficacy? What if trainers used it as source for generating even harder "gotchas"? Perhaps the individuals would engage in open dialogue and critique, and a wider sharing of experiences would be made possible. A utopian vision, perhaps, but one worth considering.

From a conceptual and methodological point of view, the dialogical analysis of concerns reported in this chapter presents a new way of looking at the activity of procedure following. The emphasis is on the individuals' experience of using procedures rather than on their physical behaviour or their information processing task. What this amounts to is an analysis of how people reflect on and make sense of their own activities. This includes what problems they see with designed artefacts and how well their own and other people's concerns are mediated by such artefacts. Such reflection necessarily involves them in recounting their past experiences in the broader sociocultural context in which they find themselves. We have tried to argue in this chapter that such a perspective on activity is a valuable one, both as an approach to work analysis and as a way of informing design.

The dialogical analysis of concerns that we have presented draws our attention to the emotional and motivational aspects of cognition and the everyday ethical dilemmas faced by individuals in making decisions about how to proceed. It might be argued that such an analysis is outside of the scope of cognitive task design because changes to the design of procedures per se will not contribute to resolution of such dilemmas. In part this is true. Note, however, that the concerns analysis presented here highlights a number of usability problems that could be addressed by better task design. More importantly, in our view, such an analysis leads us to explore better ways of allowing operators to become involved in procedure design so that experience can be more effectively utilised.

It could also be argued that a dialogical analysis of concerns would be of little value in other, less regulated, domains of work. However, there is much recent interest in the issues of trust, loyalty, and user experience in domains as mundane as Internet shopping. Website designers are concerned to understand how trust can be engendered through mediated interaction. Such questions are hard to address in traditional information processing terms because their solution requires an understanding of how people make sense of themselves and their relationships to other people. Such an understanding is at the heart of a dialogical approach to sensemaking.

REFERENCES

Bakhtin, M. (1981). Discourse in the novel. In M. Holquist (Ed.), *The dialogic imagination: Four essays by M. M. Bakhtin* (pp. 259–422). Austin: University of Texas Press.

Bakhtin, M. (1986). *Speech genres and other late essays.* Austin: University of Texas Press.

Beaty, D. (1995). *The naked pilot.* Shrewsbury: Airlife Publishing.

Berg, M. (1997). *Rationalising medical work: Decision-support techniques and medical practices*. Cambridge, MA: MIT Press.

Beyer, H., & Holtzblatt, K. (1998). *Contextual design: Defining customer centred systems*. San Francisco, CA: Kaufmann.

Bruner, J. (1990). *Acts of meaning*. Cambridge, MA: Harvard University Press.

Bruner, J. S., Goodnow, J. J., & Austin, G. A. (1956). *A study of thinking*. New York: Wiley.

Bulfer, B., & Gifford, S. (1996). *FMC User's guide*. Kingwood, TX: Authors.

Button, G., & Harper, R. H. R. (1993). Taking the organization into accounts. In G. Button (Ed.), *Technology in working order: Studies of work, interaction and technology*. London: Routledge.

Button, G., & Sharrock, W. (1994). Occasioned practices in the work of software engineers. In M. Jirotka and J. Goguen (Eds.), *Requirements engineering: Social and technical issues* (pp. 217–240). London: Academic Press.

Civil Aviation Authority (1998). *Global fatal accident review 1980–1986* (CAP 681). Cheltenham: Westward Digital.

DeBrito, G. (2000). Experience-based cognitive modelling of procedure following. In P. Wright, S. Dekker, and C. P. Warren (Eds.), *Confronting reality: ECCE-10, Proceedings of the Tenth European Conference on Cognitive Ergonomics* (pp. 86–96). Linkoping: Linkoping University Press.

Degani, A., & Wiener, E. L. (1990). *Human factors of flight-deck checklists: The normal checklist* (Con. Rep. No. 177549). Moffet Field, CA: NASA–Ames.

Hale, A., Wilpert, B., & Freitag, M. (1997). *After the event: From accident to organisational learning*. Oxford: Elsevier.

Hawkins, F. H. (1987). *Human factors in flight*. Guildford, UK: Gower Technical Press.

Hutchins, E. (1995). *Cognition in the Wild*. Cambridge, MA: MIT Press.

McCarthy, J., Wright, P. C., Monk, A. F., & Watts, L. (1998). Concerns at work: Designing practically useful procedures. *Human-Computer Interaction, 14,* 433–458.

Miles, M. B., & Huberman, A. M. (1994). *Qualitative data analysis: An expanded source book*. London: Sage.

Morson, G. S., & Emerson, C. (1990). *Mikhail Bakhtin: Creation of a prosaics*. Stanford, CA: Stanford University Press.

Orr, J. (1996). *Talking about machines: An ethnography of a modern job*. New York: Cornell University Press.

Robson, C. (2002). *Real world research: A resource for social scientists and practitioner-researchers* (2nd ed.). London: Blackwell.

Rosaldo, R. (1986). Ilongot hunting as story and experience. In V. W. Turner & E. M. Bruner (Eds.), *The anthropology of experience* (pp. 97–138). Urbana: University of Illinois Press.

Shusterman, R. (2000). *Pragmatist aesthetics: Living beauty, rethinking art* (2nd ed.). Boston: Rowman and Littlefield.

Suchman, L. (1987). *Plans and situated actions: The problem of human-machine communication*. Cambridge, England: Cambridge University Press.

Suchman, L. (1993). Response to Vera and Simon's situated action: A symbolic interpretation. *Cognitive Science, 17,* 71–75.

Symon, G., Long, K., & Ellis, J. (1996). The co-ordination of work activities: Cooperation and conflict in a hospital context. *Computer Supported Cooperative Work, 5,* 1–31.

Turner, V. W. (1986). Dewey, Dilthey, and drama: An essay in the anthropology of experience. In V. W. Turner & E. M. Bruner (Eds.), *The anthropology of experience* (pp. 33–44). Urbana: University of Illinois Press.

Turner, V. W., & Bruner, E. M. (1986). *The anthropology of experience*. Urbana: University of Illinois Press.

Vicente, K. J. (1999). *Cognitive work analysis: Toward safe, productive and healthy computer-based work*. Mahwah, NJ: Lawrence Erlbaum Associates.

Wright, P. C., Pocock, S., & Fields, R. E. (1998). The prescription and practice of work on the flight deck. In T. R. G. Green, L. Bannon, C. P. Warren, & J. Buckley (Eds.), *ECCE9, Ninth European Conference on Cognitive Ergonomics* (pp. 37–42). Limerick: Limerick University Press.

29

Representations for Distributed Planning

Valerie L. Shalin
Wright State University, USA

Pamela M. McCraw
Johnson Space Center, NASA, USA

Abstract

When representations are used to represent hypothetical conditions in planning or design, current world conditions do not completely determine the represented content. By definition, this content must reflect desired conditions that do not currently exist, created by actions that have not yet occurred. Human operators, rather than sensors, configure these representations, which are used to examine the consequences of possible actions. This chapter describes a cycle of domain apprehension, analysis, and design that resulted in new displays to address three issues that arise in creating and modifying representations for planning: (a) ambiguity in represented plans, (b) the relationship between representations for alternative plans, and (c) the history of changes to representations.

INTRODUCTION

The Flight dynamics officer (FDO), accompanied by a team of assistants, is responsible for the trajectory of objects in orbit during a flight of the U.S. Space Shuttle. Space shuttle engines can alter this trajectory, for example, to enable the deployment of a satellite in a particular orbit, or to rendezvous with the International Space Station. To alter this trajectory, flight controllers determine engine parameter values in a series of exploratory, real-time-distributed planning activities. Our view of planning is consistent with that of Hoc (1988), in which participants develop schematic representations to guide future activity. Although the chapter focuses on a particular work domain, the conclusions and recommendations pertain to any synthesis task (Clancey, 1985) with a sufficient cycle time to permit the development, exchange, and evaluation of schematic plans. Example domains similar to trajectory planning in the requirement for representations of hypothetical conditions include manufacturing, construction, and the coordination of emergency responses.

The study of computer-based representations for hypothetical conditions contrasts with the more typical study of displays in human factors, addressing computer-based representations slaved to the external world through sensors. We call such displays "sensor based." Sensors respond to current world conditions in a generally predictable fashion. Humans may influence the contents of sensor-based representations, but only by changing the conditions of the world to which the sensors respond. In contrast, when representations are used for hypothetical conditions, current world conditions do not completely determine the represented content. For example, in designing the actions that will create desired conditions, flight controllers configure the parameter values for a simulation that permits an evaluation of the consequences.

The resulting representations retain a degree of ambiguity not present in sensor-based representations. First, multiple representations may exist simultaneously, with implicit differences in their assumptions and minor (sometimes barely detectable) differences in appearance. Second, the meaning of a representation may depend on its history of changes. For example, identical display values mean something different depending on whether they have been updated for the latest information on resource availability, a process that is at the operator's discretion with no indication of currency on the representation itself. The process that results in a particular representation is not software or hardware, but variable (and somewhat unpredictable) human reasoning. Together, these properties result in a representation that is not self-contained (Bødker, 1998). If one individual conducted all trajectory planning activities, the implicit nature of the representation would simply create an additional demand on working memory. However, in distributed work, ambiguity threatens portability between individuals. A possible consequence of an ambiguous representation is misunderstanding or surprise, not unlike the confusion that complex automation sometimes evokes from human observers (Sarter & Woods, 1994). In a distributed work environment, the demands on working memory for an individual also result in additional demand for coordination and communication between individuals who use ambiguous representations.

The project reported in the paragraphs that follow had its origins in a wider research effort to understand how work groups use their computational tools. The chosen work group (Orbit Flight Dynamics group) at NASA—JSC Mission Control had a reputation for forward thinking, encouraged by several champions within the group (Brulé & Blount, 1989), including at the outset the second author. The group had expressed general concern regarding the existing displays, particularly with respect to information overload and legibility.

This chapter first identifies the methods for apprehending and analyzing task content, providing an understanding of the domain sufficient to support recommendations for new displays. Next we present the results from these methods and introduce the legacy display used to represent and evaluate plans. In the body of the chapter we indicate some of the consequences to work procedures resulting from the legacy display and describe features of a new representation designed to remediate the ambiguity just noted.

METHODS

Apprehending Domain Content

Whether applications or research goals motivate the analysis of a work domain, an investigator faces the problem of developing an understanding of a new domain without benefit of the years of formal training and experience that shape a domain expert. This section places participant observation (used primarily in the present research) in the context of some alternative methods for apprehending domain content.

Alternatives.   Investigators may attempt to acquire domain content in more or less structured fashion, imposing a corresponding filter on the gathering process. Interviews provide an efficient alternative when the interviewer has specific questions. However, interviews are typically not part of an informant's job and therefore detract time from it. Worse, Forsythe (1999) provides examples of persistent self-editing in open-ended interviews. Other specialized tasks may communicate the investigators' interests, which can influence responses. The critical decision method (Klein, Calderwood, & MacGregor, 1989) communicates a theoretical focus on decisions. Specialized tasks that relate to a target task representation (Hall, Gott, & Pokorny, 1995; Shute, Torreano, & Willis, 2000) may communicate a technical focus. Social scientists argue that such representation-directed efforts gain efficiency only through assumptions that can exclude the social facets of expertise (Forsythe, 1993; LaFrance, 1997).

An alternative to these focused methods is observation, which permits the work environment to structure apprehension and is suitable when the investigator has open-ended goals regarding an existing work system. Opportunities for gaining insight include conversation between members of the target group, as well as the more explicit guidance of apprenticeship training that makes tacit knowledge explicit for the newcomer. Button and Harper (1996) note that the scope of observation remains important, determining the success of systems designed from those observations. If investigators observe only documentation and displays, then they risk adopting a formal account of work practice that excludes the "real" work required (Star & Strauss, 1999). Within social science, ethnographers developed *participant* observation, which "combines participation in the lives of the people under study with maintenance of a professional distance that allows adequate observation and recording of data" (Fetterman, 1998, p. 34). It involves living and working "in the community for 6 months to 1 year or more learning the language and seeing patterns of behavior over time. Long-term residence helps the investigator internalize the basic beliefs, fears, hopes and expectations of the people under study" (p. 35).

Analysis

The results from observational study require an interpretation from an analyst who embodies certain preconceptions concerning the important properties of work (Dourish & Button, 1998). A social scientist would likely offer a description of living and working in an observed culture as experienced by the people comprising that culture (Harper, 2000). Descriptions will include the observed environment and its artifacts (e.g., Cicourel, 1988; Hutchins, 1993; Suchman & Trigg, 1993), including mechanical and computational tools (Cole, 1995). In contrast to the sort of quantitative description that an ergonomist might produce related to physiological and muscular-skeletal function, the required record of the task environment and artifacts is related to social and behavioral function. For example, Suchman and Trigg identify a prominent, broadly visible whiteboard in an artificial intelligence laboratory not for its accessibility, but for its potential as the host of a set of diverse representations.

In cognitive engineering, the analysis of work includes the creation of a task representation. Ideally, the representation permits inferences, concerning opportunities for error (Hollnagel, 1991), information and control requirements (McDonald & Schvaneveldt, 1988; Mitchell & Miller, 1986; Roske-Hofstrand & Paap, 1986), the contents of training (Shute et al., 2000) and testing (Dubois & Shalin, 1995, 1997/1998, 2000), and so on. Alternative representation frameworks emphasize different aspects of the work, such as its physical execution (e.g., Card, Moran, & Newell, 1983; Payne & Greene, 1986), its different levels of abstraction (Annet & Duncan, 1967 and Gagné, 1962, as well as Rasmussen, 1983; Rubin, Jones, & Mitchell, 1988), or an operator's context-sensitive choice (Johnson, Johnson, & Wilson, 1995). These frameworks enable an explicit account, but they do not ensure sensitivity to all task content. For example, Sebillote (1995) represented a distributed task but did not identify knowing the whereabouts of team members, or team member progress as a content requirement for displays. Sebillote might have legitimately argued that these properties of the task were not of interest. Our point is that, independent of representation, the analyst always exercises discretion in determining the scope of interest in creating the representation and inferring its implications.

A risk in analysis is that the sifting, summarizing, and explanation excludes certain types of work, particularly hidden work (LaFrance, 1997; Star & Strauss, 1999), that is, work required for a functional system that is not acknowledged in engineering-oriented technical description. To guard against such exclusions, social scientists recommend deliberate comparison between actual work and recorded procedures (Forsythe, 1999; Symon, Long, & Ellis, 1996). Consistent with the human factors tradition, comparison in what follows involves the match between actual work and the representations of information. This is important for two reasons. First, emphasized by a social science perspective, mismatches reveal hidden system dependencies on human capability. Second, good displays support the cognitive demands of work by serving as an external memory that offloads human short-term memory (Hess, Detweiler, & Ellis, 1999; Zhang, 1997; Zhang & Norman, 1994). Mismatches between work and displays suggest opportunities to better support the work.

Design

Shalin and Geddes (1994, 1998) conduct design in three phases. First, they identify information requirements based on the tasks. Display design proceeds with the assignment of information to display elements and controls. A final step is to lay these out on a screen.

We agree with Nielsen's (1994) design goal to indicate the status of a system in interface design, but expand the notion of system beyond the technical or computational aspects to include the status of the work system. Consistent with Eckert (2001), we focus on ambiguity in boundary objects (Star, 1989), that is, representations created by one operator and passed along to another operator who typically has different responsibilities. Drawing on research in computer-supported cooperative work, particularly in software development, we suggest that operator information requirements include the following: owners or creators of these representations (Grinter, 1996; Teege, 1995), changes in the status of intermediate work products (Agostini, DeMichelis, Grasso, Prinz, & Syri, 1996), and implications of these changes, concerning not only the scope of responsibilities under a given operator's control (Watts-Perotti & Woods, 1999) but also the responsibilities of other operators (Button & Sharrock, 1996).

A large research base addresses the realization of particular display elements, and the formatting, organization, and labeling of content. This research has lead to a number of design guidelines, such as:

- Integrate multiple displays (Watts-Perotti & Woods, 1999);
- Reduce unnecessary inference (Gerhardt-Powals, 1996);
- Provide overviews and long shots (Watts-Perotti & Woods, 1999);
- Combine pictures and words (Vaughn, 1998); and
- Eliminate opportunities for error and confusion (Nielsen, 1994).

Although we adopt these and similar guidelines, we list them here to distinguish between these guidelines and the content-oriented recommendations we emphasize here. A prototype display for orbit flight dynamics followed from the particular cycle of apprehension, analysis, and design that is described in the paragraphs that follow.

APPREHENDING ORBIT FLIGHT DYNAMICS

Readers will note that the coauthor on this chapter is a member of the Mission Operations Directorate. She is an expert in the task domain and has used the plan representation discussed in the paragraphs that follow. However, at the outset of this effort, her expertise did not encompass the general issues governing the influence of representations on cognitive work.

The first author acquired an understanding of the work over more than 2,000 hours of observational study, conducted over a period of more than $2\frac{1}{2}$ years. The majority of this observation occurred in conjunction with real shuttle flights. Observation of flights was supplemented by observation of work in some simulations, preflight and postflight meetings, ongoing software development meetings, and access to available documentation. When not present on site, the first author participated in nearly daily exchange with the work group by means of telephone and e-mail, including personally directed messages as well as messages distributed to various work-related mailing lists.

A schematic for the location of participants in orbit flight dynamics appears in Fig. 29.1. The FDO sits in the front row of Mission Control, assisted by the trajectory officer (Traj). They work with support personnel located in the backroom. This distributed work environment facilitates remote observation, because much of the equipment for observation is already in place for normal operations. Cameras already mounted in the front room of Mission Control provide a video feed for public affairs purposes. Backroom personnel view this feed to assist in coordination across rooms. Front room personnel communicate with other front room and backroom personnel through a system of voice loops. An additional ambient microphone added for project purposes captures face-to-face conversation between FDO and Traj. Occasionally, other cameras and microphones were placed in the backroom. All of these resources were routed to the backroom for observation and recording on a set of VCRs.

During flights, the first author was seated at the Landing Support workstation, covering approximately half of the 24-hour work day, scheduled (like the flight controllers) according to the astronauts' cycle of activity rather than local day shifts. The observer listened to front room activity on both the voice loops and the ambient microphone while viewing

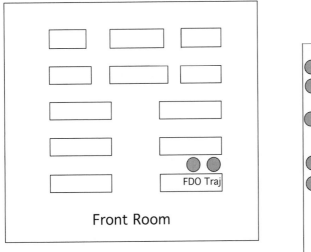
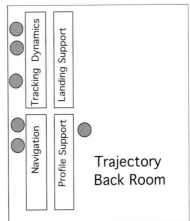

FIG. 29.1. Schematic for the location of participants in trajectory operations.

the front room on a closed circuit broadcast. Incoming flight controllers often passed by to chat, offer explanations, and, occasionally, ask a question about the status of an ongoing flight.

ANALYSIS

Description

Task Overview.   Orbit flight dynamics includes a core task of planning, refining, and scheduling the shuttle engine burns that mission objectives such as satellite deployment or a rendezvous require. The FDO's task rests on a mathematical model or representation of objects in orbit and the effects of velocity changes on this orbit. A computer-based representation for the position of an object in orbit over time is called an *ephemeris*. Because the ephemeris contains hypothetical vehicle maneuvers (engine burns), we also refer to an ephemeris as a representation of a plan.

Engine burns appear in an existing schedule, which provides a guide for the mission-related events that will occur. However, because events take place in a dynamic and uncertain environment, assumptions behind the original schedule may change. In addition, the outcome of execution is not completely predictable, for example because of mechanical failure, or, on occasion, because of a lack of experience maneuvering under novel shuttle configurations. Moreover, a sequence of burns cannot be executed exactly as planned because of differences between the fidelity of the underlying model and the real environment. Although these normal deviations do not affect the feasibility of mission goals, such deviations must be considered in refining the plans for future burns.

Consequently, although prespecified plans serve as resources for action, they still require revision. In this case, a critical resource in revising the burn plan is the propellant budget. Existing automation cannot optimize burn plans for propellant consumption within the available response time. Instead, revision consists of a process of human-guided simulation and evaluation. FDOs consider current properties of the environment and properties of the shuttle to select input parameters for specialized burn simulation software that predicts trajectory results. Relevant properties of the environment include the position of the sun and moon, communication satellites, debris, and targets. Relevant properties of the shuttle include weight (determined by available fuel) and center of gravity, and a number of modifiable parameters related to an engine burn, including the three axes of shuttle attitude, the engines to be utilized, and the ignition time of the burn. Because appropriate responses to anomalies take some time to derive, the FDO is responsible for developing several different contingency plans in advance, so that good options are readily available should the need arise.

Team Member Responsibilities.   As Fig. 29.1 implies, a team of flight control officers and assistants contribute to the burn planning process. The FDO verbally delegates tasks to the Traj as needed, which may include contingency planning and emergency landing site evaluation. The backroom plays a key role in ephemeris maintenance, concerning the accuracy and currency of the several ephemerides typically in use, as burns are executed and as tracking data are obtained. One set of individuals (Navigation) tracks all objects; another set (Dynamics) changes all models upon front room request. Tracking, also shown in Fig. 29.1, assists in the management of the ground facilities for tracking. Profile Support assists the front room during periods of high workload, real-time planning. One other group, whose location is not shown in the figure, also influences the planning process. This group projects the future attitude of the orbiter, represented in an attitude time line (ATL). Navigation performs a similar function, projecting the future introduction of accelerations apart from engine burns, for example as a

FIG. 29.2.   The Traj's workstation.

result of dumping water overboard. Such accelerations are called vents, and they are represented in the vent time line (VTL).

The FDO, Traj, and Profile Support work in parallel on different contingency plans, or different phases of the same plane. In most (if not all) cases, they should be working with the same general assumptions concerning future attitudes and vents, and any other aspect of context that might influence the coupled plans they are formulating.

Workstation and Displays.   In general, the front room workstations for each discipline contain three 38-cm-wide by 29-cm-tall video terminals designed for a single flight controller. However, the trajectory operations workstation was retrofit with a fourth screen to provide both the Traj and the FDO with two screens each. The trajectory workstations in the front room appear in Fig. 29.2. A 66-cm desk separates the operator from the screens. This distance requires at least a 12-pt font (Sanders & McCormick, 1993) and limits the amount of legible content that can appear on one screen. The distance and small font make it difficult for two controllers to view the same display hosted on a single terminal.

The trajectory team uses approximately 100 different windows at different points in the flight, with a similar appearance to the tabular display in Fig. 29.3. The legacy display Trajectory Profile Status (TPS) in Fig. 29.3 appears on all trajectory team members' workstation screens. TPS provides a representation (ephemeris) for seven objects in orbit. The three panels at the bottom of the screen correspond to TDRS (Tracking and Data Relay System) satellites that provide communication between Mission Control and the orbiter. The remaining four reconfigurable panels may correspond to models of the shuttle (called orbiter), the International Space Station (ISS), rendezvous targets, deployed satellites, or debris. When used to model

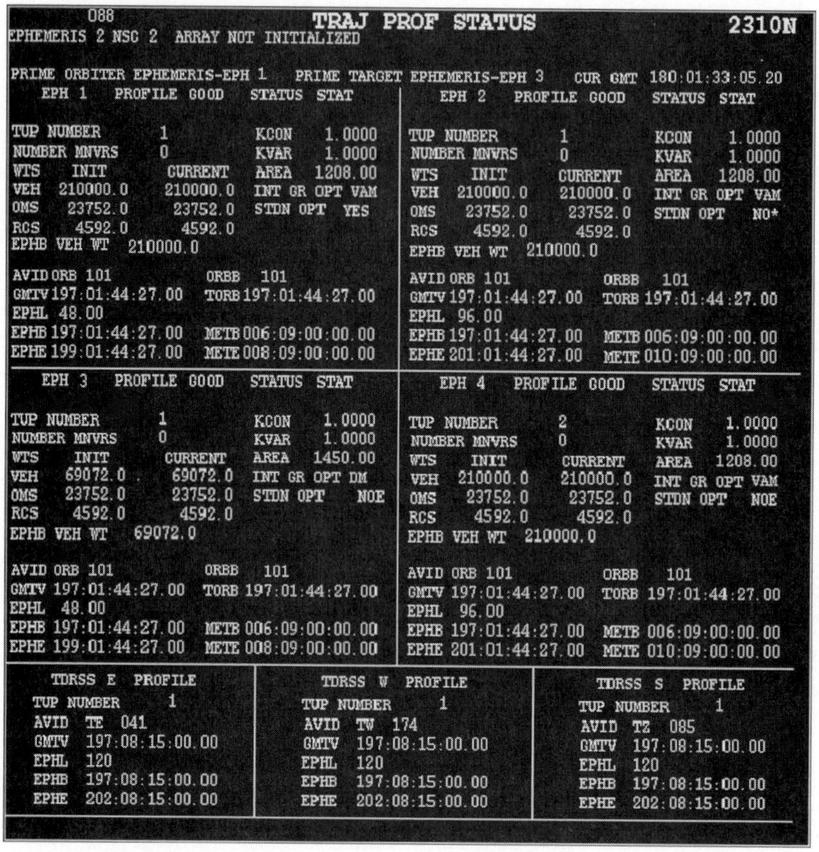

```
  088                      TRAJ PROF STATUS                        2310N
EPHEMERIS 2 NSC 2  ARRAY NOT INITIALIZED

PRIME ORBITER EPHEMERIS-EPH 1    PRIME TARGET EPHEMERIS-EPH 3   CUR GMT 180:01:33:05.20
   EPH 1    PROFILE GOOD    STATUS STAT          EPH 2    PROFILE GOOD    STATUS STAT

TUP NUMBER     1         KCON   1.0000      TUP NUMBER     1         KCON   1.0000
NUMBER MNVRS   0         KVAR   1.0000      NUMBER MNVRS   0         KVAR   1.0000
WTS    INIT      CURRENT AREA   1208.00     WTS    INIT      CURRENT AREA   1208.00
VEH  210000.0   210000.0 INT GR OPT VAM     VEH  210000.0   210000.0 INT GR OPT VAM
OMS   23752.0    23752.0 STDN OPT  YES      OMS   23752.0    23752.0 STDN OPT  NO*
RCS    4592.0     4592.0                    RCS    4592.0     4592.0
EPHB VEH WT  210000.0                       EPHB VEH WT  210000.0

AVID ORB 101           ORBB  101           AVID ORB 101           ORBB  101
GMTV 197:01:44:27.00  TORB 197:01:44:27.00 GMTV 197:01:44:27.00  TORB 197:01:44:27.00
EPHL 48.00                                 EPHL 96.00
EPHB 197:01:44:27.00  METB 006:09:00:00.00 EPHB 197:01:44:27.00  METB 006:09:00:00.00
EPHE 199:01:44:27.00  METE 008:09:00:00.00 EPHE 201:01:44:27.00  METE 010:09:00:00.00

   EPH 3    PROFILE GOOD    STATUS STAT          EPH 4    PROFILE GOOD    STATUS STAT

TUP NUMBER     1         KCON   1.0000      TUP NUMBER     2         KCON   1.0000
NUMBER MNVRS   0         KVAR   1.0000      NUMBER MNVRS   0         KVAR   1.0000
WTS    INIT      CURRENT AREA   1450.00     WTS    INIT      CURRENT AREA   1208.00
VEH   69072.0    69072.0 INT GR OPT DM      VEH  210000.0   210000.0 INT GR OPT VAM
OMS   23752.0    23752.0 STDN OPT  NOE      OMS   23752.0    23752.0 STDN OPT  NOE
RCS    4592.0     4592.0                    RCS    4592.0     4592.0
EPHB VEH WT   69072.0                       EPHB VEH WT  210000.0

AVID ORB 101           ORBB  101           AVID ORB 101           ORBB  101
GMTV 197:01:44:27.00  TORB 197:01:44:27.00 GMTV 197:01:44:27.00  TORB 197:01:44:27.00
EPHL 48.00                                 EPHL 96.00
EPHB 197:01:44:27.00  METB 006:09:00:00.00 EPHB 197:01:44:27.00  METB 006:09:00:00.00
EPHE 199:01:44:27.00  METE 008:09:00:00.00 EPHE 201:01:44:27.00  METE 010:09:00:00.00

   TDRSS E  PROFILE            TDRSS W  PROFILE            TDRSS S  PROFILE
 TUP NUMBER  1              TUP NUMBER  1              TUP NUMBER  1
 AVID TE 041               AVID TW 174               AVID TZ 085
 GMTV 197:08:15:00.00      GMTV 197:08:15:00.00      GMTV 197:08:15:00.00
 EPHL 120                  EPHL 120                  EPHL 120
 EPHB 197:08:15:00.00      EPHB 197:08:15:00.00      EPHB 197:08:15:00.00
 EPHE 202:08:15:00.00      EPHE 202:08:15:00.00      EPHE 202:08:15:00.00
```

FIG. 29.3. The legacy TPS display.

the orbiter, or the orbiter docked with the space station, ephemerides include the affects of maneuvers and comprise a plan. When used to model rendezvous targets, deployed satellites, or debris, ephemerides generally do not contain maneuvers. The four ephemeris addresses on the legacy display comprise a limited resource in trajectory planning. The limited number of these ephemerides requires the team to swap alternative plans for the orbiter in and out of TPS.

Each of the top four ephemeris panels contains information concerning the anticipated behavior of an object in orbit and the values of parameters that the FDO selected to generate the ephemeris. Some of these parameters correspond to states of the world such as object weights (WTS) decomposed into various propellant sources (OMS, RCS), object area, object drag (KCON, KVAR, corresponding to constants and variables that may be used), and the number of hypothetical maneuvers incorporated. Other parameters correspond to properties of the model, such as the trajectory propagation options (D, V, A, and M), the vector that anchors the ephemeris (AVID) and its Greenwich Mean Time (GMTV), orbit number at the beginning of the ephemeris (ORBB), the length (EPHL) and the beginning and ending times of the ephemeris in GMT and Mission Elapsed Time (MET).

In a typical alteration cycle for an orbiter ephemeris, the FDO will first ask Dynamics to input specific modifications to the features of a burn, and then to run the trajectory propagation software to incorporate the newly modified burn, generating a new ephemeris. Using tools that display relative motion, the FDO can examine the new ephemeris and repeat the modification cycle. Not every operation on the ephemeris will result in a detectable change to the ephemeris as it is represented in the TPS display. The TUP numbers in the upper left corner of each ephemeris compensate for the lack of observable change. TUP number stands for trajectory update number. It allows viewers to determine whether a change they requested has in fact been executed (on the proper ephemeris), by recalling the old count and comparing it to the new value that appears on an update. TUP numbers exist solely to assist in the coordination of work with the computer.

Candidate Foci.   After approximately 6 months of elapsed calendar time, 10 issues were suggested as possible foci for future work. Two of the suggestions concerned general properties of the work: facilitating the reporting of software anomalies and the accessibility of documentation. Several other suggestions related to fairly standard human factors recommendations and converged with initial interest in displays: Reducing the high frequency of data entry, enhancing the legibility and formatting of numerous tabular displays, reducing display content and managing limited display real estate, enhancing limited feedback regarding the selection of parameter values used in processing, and ameliorating lags in system response time to parameter selection.

The remaining suggestions depended on a detailed understanding of the domain. One of these concerned the integration of multiple sources of information. In some cases, the organization of the display content reflected the technology source rather than the task. For example, one of the displays grouped together parameters originating from the orbiter. However, it was common for most of this display to be covered up, leaving visible just that portion that revealed the status of the orbiter's onboard models for objects in orbit.

Two other suggestions for future work concerned the function of the discipline-specific display TPS used to model the trajectory of objects in orbit. Although the available information is sufficient to generate a unique ephemeris, this information is not sufficient to constrain the meaning of the ephemeris within the planning process. For example, the available information is not sufficient to communicate whether an ephemeris incorporates a delayed burn, or some other contingency plan such as a breakaway from a rendezvous target. Therefore, one suggested focus of future analysis concerned the absence of explicit indications of model meaning in an ephemeris panel. A second, related suggestion concerned the limited number of ephemeris slots. This required team members working in parallel on different nominal and contingency plans to reconfigure ephemerides in order to study more than four plans. The frequent reconfiguration left open ample opportunity for error and confusion. And yet, the work system functions quite well, with no missed mission objectives caused by errors in trajectory operations since study of this work group was initiated.

A rehosting of the trajectory software further encouraged us to examine the work practices surrounding the successful use of TPS for two reasons. First, the rehost will provide 40 ephemerides instead of 4. Work methods appropriate for the management of 4 ephemerides may not scale to 40. Second, the rehost will relocate TPS and associated software from a mainframe computer with a command line interface to a distributed workstation environment with a windows interface. This will enable multiple individuals to alter ephemeris configuration without mediation from Dynamics.

The focus on the limitations of TPS earned the immediate interest of the work group, and eventually operations resources to design and build a replacement for TPS. In retrospect, the selection of this display for focused analysis appears obvious. No one works in the domain without bringing up TPS on the workstation. The integrity of the ephemerides is essential for safe, effective, and efficient trajectory operations. When a group member developed a diagram depicting the relationship between various software elements, this display appeared in the center of the diagram. Though obvious in retrospect, no one thought to focus this completely open-ended project on this display at the outset.

TABLE 29.1
Partial Task List in Matrix Form

Number	Task
1.0	Monitor world
1.1	Monitor orbiter
1.1.1	Monitor orbiter trajectory
1.1.1.1	Monitor orbiter drag (solar activity)
1.1.1.2	Monitor other orbiter vector properties (e.g., vents, attitude)
1.2	Monitor debris
1.3	Monitor target
1.4	Monitor TDRS system—satellites, network, etc.
1.5	Monitor payload
1.6	Monitor weather
1.7	Monitor shuttle facilities
1.8	Shipping info to Goddard
1.9	Monitor mission control Moscow
2.0	Task supervision
2.1	Assign global ephemerides
2.2	Monitor task progress
2.3	Copy ephemerides
2.4	Manage computational resources
3.0	Model world
3.1	Model orbiter
3.1.1	Setting–changing ephemeris parameters
3.1.1.1	Creating ephemeris
3.1.1.2	Update anchor vector (VAT)
3.1.1.3	Change maneuver properties
3.1.1.4	Change drag properties (e.g., ATL, KCON)
3.1.1.5	Change vents (VTL)
3.1.1.6	Eph init weight and WGL table
3.1.1.7	F10 & KP
3.1.1.8	Changing eph length
3.1.2	Set–change dependant processes
3.1.2.1	Auxiliary processes (acquisitions & automated mission operations computation)
3.1.2.2	Auxiliary process flags (STDN options)

Task Representation.   Several hours of the available audiotapes and videotapes were transcribed, and, with the available documentation, served as the point of departure for creating an initial hierarchical task decomposition, intended to center on a portion of the domain associated with ephemeris maintenance. Individual members of the work group provided some initial feedback. As the representation approached maturity, several group meetings were held to gather a broader range of suggestions across the different specialties participating in trajectory operations. The backroom specialties added whole branches that front room participants overlooked, particularly concerning troubleshooting the ephemeris models.

A portion of the representation for ephemeris maintenance is shown in Table 29.1. It begins with the identification of six abstract goals (monitoring, task supervision, and modeling, shown in part in the table, and planning, executing, and troubleshooting, not shown). In a typical cycle, the orbiter and other relevant objects are modeled. Monitoring these objects results in the need to update these models. Depending on the phase of the mission, the models are distributed

TABLE 29.2
Sample Evidence for Task Supervision

2.1 Assign global ephemerides	FDO: I'll put the de-orbit there. This maneuver one ephemeris two, do a delta V override the cut-off without safe HP. (technical property of altitude) Dyn: Did you want to do that in one or just take the same solutions and put it in two? FDO: Put it in two.
2.2 Monitor task progress	FDO: Yeah, do four and then do maneuver one ephemeris one, and then when you are done, give me a holler, and we'll uhm, we'll proceed....
2.3 Copy ephemerides	FDO: TUP four out of one at zero days, twenty-one hours, twenty-five minutes, thirty-four seconds. Dyn: OK. FDO: The maneuver will be the same time.
2.4 Manage computational resources	FDO: And let's go, uhm, twenty four hours, uh, and ephemerides one, two and four. That might help a little . . . Dyn: OK. FDO: To speed things up.

across the team to construct plans for future engine burns and contingency conditions. Although the crew executes the engine burns, the FDO has execution tasks as well, tidying up the models to reflect the actual (rather than planned) burns. Troubleshooting refers to the models and occurs whenever an observer determines an error in configuring the model.

The full representation has over 100 entries. Although this partition of abstract goals is meaningful to the work group, evidenced by their ability to critique the representation, this partition is not likely one that members of the work group would have generated themselves and therefore required explanation. The distinction between monitoring changes in the world and modeling these changes in a computer-based representation reflects the fact that different work groups contribute to these goals. The distinction is also consistent with the artificial intelligence planning literature and as well as Norman's (1991) conceptualization of computer-mediated action. However, in the present case, the configuration of the model is not done one time by a software developer, but repeatedly, as part of the work process.

The task supervision branch emerges from an appreciation for the distributed nature of work. Although the research literature recognizes the role of cooperation in planning (Hoc, 1988, 2001) prior to the present project, from the perspective of the work domain, managing the access and distribution of models was a tacit responsibility. Documentation, for example, does not address the work practice surrounding the distribution of ephemerides across the team, which is currently governed by courtesy and negotiated over the voice loops. Table 29.2 presents sample evidence for four subtasks under supervision. Similar evidence or documentation supports the other entries into the task decomposition.

Design

The task list was used in conjunction with design to audit and refine recommendations. Consistent with the design methods that Shalin and Geddes (1994, 1998) developed, each node in the task representation was annotated for information requirements by interviewing the

TABLE 29.3
Further Decomposition of Tasks, With Information Sources Annotated

Tasks	Task Details	Information Requirements–Sources
3.1.1.2 Update Anchor vector (VAT)	• Update per rule of thumb criteria or flight rule annex criteria if applicable • Nav shows FDO latest tracking solution via vector compare (MSK 0372) and makes recommendation on update • On FDO "Go," Nav moves solution to V39 & FDO tells Dyn	• Nav tells FDO new vector ID • Vector compare (MSK 372) shows 0s for move to V39 vs. solution • TPS flashes after update then has new vector ID • DDD's (digital displays) flash yellow, then go green again • Vector compare shows 0s for new ephemeris
3.1.1.3 Change maneuver properties (OMP, DMT, RET, RELMO)	• Use PMP software for latest mass props & update engine characteristics => MOC • OMP: Transfer a rndz plan into eph in TSU/MOC or TSA • Direct input: FDO DI into TSA; Dyn DI into MOC/TSU • DVSensor: FDO can confirm (and input) maneuver in MOC/TSU via DV sensor • DMP: FDO/Dyn can run	• Engine Char table; latest mass props • RCT, RET, MCT, DMTs for ephemeris & maneuvers (detailed tables displaying rendezvous maneuver and engine characteristics) • DMTs, Deorbit Summary
3.1.1.4 Change drag properties (e.g., ATL as well as CBD status, area, KCON, KVAR, and integrator parameters)	• MOC—Because only 1 ATL and only in MOC, ATL will be nominal plan; changes to ATL will need TUPs into the ephemeris; DDDs show ATL update waiting for; ATL only applicable for AM or VAM ephemeris	• Cargo bay door status, TPS • ATL DDDs & AM or VAM (modeling options) in TPS for the ephemeris

second author and then a member of the Orbit Flight Dynamics section and one of the project champions. A portion of the task list and corresponding information requirements is presented in Table 29.3. The information requirements serve not only to audit display contents, but also as notes to the software development team concerning the current location of information that will appear in the new displays. A separate frequency list was used to ensure that the displays maintained easy access to frequently required information. The matrix form of Table 29.1 provided support for estimating and recording task concurrences to ensure the proximity of related information.

The second author leads the design of the new TPS. Design work was initiated well before the task representation stabilized, but it was suspended when early efforts failed to suggest improvements. Design resumed after the task representation began to stabilize, serving primarily as a template for auditing display content to ensure the availability of all necessary information. However, the design process still suggested features that prompted an addition to the task representation. In this sense the representation and the display are refined simultaneously, with the goal of maintaining consistency between the final display and task representation. An agreement to maintain this consistency enables a judicious response to new suggestions for additions to the displays.

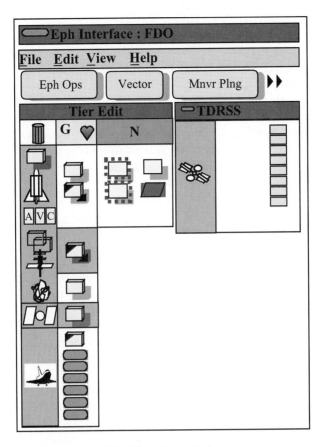

FIG. 29.4.   Tier 0 display.

The redesign of TPS focuses on the semantics (meaning) of models it contains. Though a recent concern within cognitive engineering (e.g., Flach, 1999), the semantics of a representation or model has been the focus of philosophical debate for decades (Carnap, 1947), discussed in cognitive psychology at least as far back as Greeno (1978) and Johnson-Laird (1983). Our design encompasses two views of model semantics: (a) the correspondence between model parameters and the world and (b) relations between the model parameters and other computational processes. The first view is the classical extensional theory of semantics. The second view is inspired by the alternatives (procedural semantics and intensional semantics), generally acknowledging the processes that use the models and interrelationships between models as important contributions to their meaning. In describing the new representations for plans, we discuss three issues: the ambiguity in plan meaning, the relationship between plans and the rest of the work system, and the history of changes to these representations.

Tier 0 TPS.   Tier 0 (shown in Fig. 29.4) provides the overview that Watts-Perotti & Woods (1999) recommend. Task supervision (2.0) and distributed model manipulation (3.1) suggest the need for a display that permits supervisory control. The increased numerosity in available ephemerides drove a design that represented all models simultaneously. For example, the new representation includes access to models that are stored on board the orbiter (represented as cubes above the orbiter and station icons in the leftmost column), and not available through the legacy TPS. In addition, limited display real estate and a substantial viewing distance required an overview with low resolution. In this section, we discuss the properties of Tier 0 that address the three aspects of model meaning just identified, and we compare these with work practice while using the legacy TPS.

One facet of plan meaning concerns the mapping between the plan and an object in orbit. In the legacy TPS, the third line from the top indicates which ephemeris is prime for the orbiter and rendezvous target. However, it does not identify the type of target, and it does not indicate the roles of ephemerides 2 and 4. An observer may infer whether an ephemeris refers to the orbiter, space station, or a satellite by using the weight parameter. Small weights suggest either satellites or debris. Large weights suggest the space station or the orbiter docked to the space station. The area parameter also provides an indirect indicator for the content of the object being modeled, along with the vector identifier.

To eliminate the inference of object identity based on weight and area, in the new representation models addressing the same object (orbiter, space station, debris, or satellites) appear in a row labeled with an icon, consistent with Gerhardt-Powals (1996). An object parameter is easily set when an ephemeris is initialized, and in most cases poses no maintenance issue.

A second aspect of model meaning concerns its role in the work system. Tier 0 portrays three pertinent properties of this role discussed next: whether it is an argument for other computational processes, whether its access is restricted, and whether someone is presently using the model. Acquisition software generates predictions concerning the future relationships between orbiter ephemerides and communication satellites. Other groups in mission control use these predictions, and they must access them through the ephemeris that is configured to participate in the acquisition processing. However, acquisition processing is expensive and must be recomputed every time the ephemeris changes. Consequently, acquisition processing is typically set up for a single orbiter ephemeris. Acquisition processing is sometimes suspended briefly in order to update the models of the orbiter and the communications satellites. The FDO must resume acquisition processing, ideally invoking it for the same ephemeris. Both the old and new representations required some means for identifying and managing these relationships. The legacy TPS provides a text indication (STDN options ON) to indicate that an ephemeris participates in acquisitions. Instead of requiring viewers to search text fields serially to identify the ephemeris participating in options, the new representation provides an indication of participation in acquisition with darkened corners.

Tier 0 allows the supervisor (FDO) and the team members to determine which ephemerides are in use (2.1). To assist in the distribution of ephemerides across the team, the work group has agreed to a convention, delegating a subset of models to Dynamics, and requiring them to input all changes dictated over the voice loop. These special ephemerides are called "golden" and appear in the leftmost column of the new representation. Dynamics remains present to ensure the integrity of the ephemeris, and traceability in the event of any error in configuration. Tier 0 distinguishes between models that are restricted (under the G column, for golden) and available for allocation to different team members (under the N for nonGolden). Dashed outlines indicate the delegation of a nongolden ephemeris to a team member. The coordination that this display supports corresponds to verbal exchange over the voice loops that previously had no other persisting visual support.

The use of models in hypothetical reasoning acknowledges a distinction between the world and the model representing that world. Hence, one facet of model meaning concerns the health and currency of the model. The legacy representation indicates health with a text box. The need to update the model for changing assumptions concerning future world conditions (emanating from Tasks 1.0 and 3.0) is implicit in changes to the attitude and vent time lines (ATL and VTL, respectively) represented elsewhere. A time line change does not affect an ephemeris until the point at which the ephemeris is updated. At this point, the ephemeris will incorporate the change automatically. Because the legacy TPS ephemeris panel has no indications of changes to the ATL or VTL, the FDO relies on the operator making these changes to provide a verbal notification of such changes, a small warning button located on another display, and memory to recall the notification.

The new representation indicates health with a heartbeat and status indicators. Consistent with work by Agostini et al. (1996), the new TPS contains icons that indicate when changes to the ATL and VTL have been made, thereby reducing the dependence on verbal warning from the operator initiating the change and providing a persistent reminder that a change has occurred. These flash on update, but they also provide access to more detailed information regarding the change.

Capturing the history of such purposeful changes to the model is discussed in detail in a later subsection. Here we note a subtle difference in managing change enabled by the increase in the number of models in the new display. With the use of the old representation, four model slots were reused. For example, at one point in time the model number might refer to the space station, and at another time it might be adapted to refer to a piece of debris. Because of the number of models available in the new representation, models no longer in use can be thrown away, marking the next round of changes as properties of a new model rather than modifications on an old one. Hence, the new representation includes a trash can.

Tier 1 TPS.   Tier 1 (shown in Fig. 29.5) duplicates much of the information in the original TPS panels, except in icon form. These permit the reduction of text fields and decrease the real estate required for an individual ephemeris. Relative to Tier 0, Tier 1 ephemerides have comparable topological relationships with each other. We discuss Tier 1 with reference to the same issues that drove the discussion of Tier 0.

Both the legacy representation and Tier 1 include a reference to the anchor vector, as a partial constraint on the meaning of the model in questions and consistent with Grinter (1996) and Teege (1995). Relative to Tier 0 (and the legacy TPS as well), Tier 1 has more features to establish model meaning. The legacy TPS ephemeris panel indicates the number of maneuvers (or engine burns) incorporated into an ephemeris, suggesting something about the purpose of an ephemeris. A maneuver count of zero in an orbiter ephemeris suggests that no burns are being planned in that ephemeris. However, maneuver count provides a very limited account of the purpose of a maneuver. For example, a model with two maneuvers might correspond to a nominal rendezvous plan. It might also correspond to a planned delay in reaching a waypoint. Moreover, replacing an existing maneuver will not change the maneuver count.

Further, because limitations in the number of ephemeris models require flight controllers to swap models in and out of ephemerides, they must *remember* which of the two-maneuver plans is actually being modeled. In contrast, observers must *infer* the identity of an ephemeris without having participated in its configuration. Provided observers are listening, they may infer the purpose of the maneuvers being altered in the voice loop exchange between the FDO and Dynamics. For example, the repeated calls to Dynamics for planning a collision avoidance burn have a distinct pattern and pace. However, there is no persisting indication of maneuver content in the visible, legacy representation.

The Tier 1 representation has several compensatory features to indicate the maneuver contents in an ephemeris. First, the reentry ephemerides with specialized engine burns now appear in a dedicated row (at the bottom of Fig. 29.3, indicated by the orbiter with visible landing gear). Text-based time information from the legacy display appears in Tier 1 in time-line form, with a tape moving behind a fixed window. Second, the length of an ephemeris is indicated by a tape with a fixed resolution and field of view, providing a visual location for maneuvers. Filled tape corresponds to the past. Hatched tape indicates remaining ephemeris length. Stippled tape corresponds to the endpoints of ephemeris. An unfilled triangle indicates burns executed in the past. A filled triangle indicates burns to be executed in the future. The position of a maneuver on

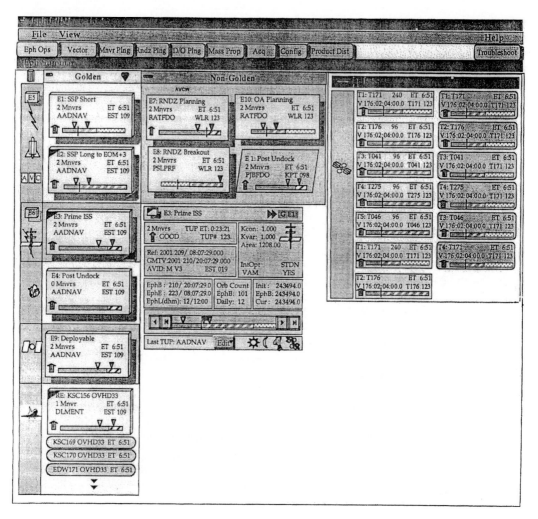

FIG. 29.5.   Tier 1 display.

this tape communicates some implicit information about its purpose because some maneuver differences may appear as different placements on the tape. For example, the maneuvers in orbiter ephemeris 7 and 8 cannot be identical, because they do not occur at the same point in time. A new field indicates the initials and position of the last person who updated an ephemeris (eg., AADNAV, RATFDO, etc.)

Before starting work in an ephemeris, a team member must determine the role of the ephemeris in question across the team. The two alternative representations provide different indications of an ephemeris in use. In the legacy display, the TUP number already noted increments with each TUP. The detection of use with TUP number can proceed in two ways, both of which impose cognitive demand. An observer might watch the TUP counter increment, although this requires extended constant attention and limits parallel attention to other work. Alternatively, the observer may memorize the TUP number for an ephemeris, work elsewhere while a change occurs, and confirm the incremented TUP number. Third an operator may hear an ephemeris being discussed over the voice loops, indicating that change is occurring, and who is initiating the change. When these fail, and someone asks the Dynamics operators to initiate a change on an ephemeris in use, the Dynamics operators can declare that someone else is using the ephemeris. Dynamics will know of the conflict, because they execute all requested changes.

Dynamics will no longer be involved in every ephemeris change and cannot warn of interference. We have added two features to Tier 1 in the new display to counteract this potential conflict. First, an elapsed time (ET) since TUP will complement the TUP number and indicate that an ephemeris is in active use. ET, unlike the TUP number, does not require users to remember previous TUP numbers to detect change. The ET counter allows viewers to confirm that expected updates are occurring without requiring constant attention to watch the TUP counter increment, and it decreases memory load associated with remembering TUP counts for multiple ephemerides. Tier 1 also provides a comment line at the top of each ephemeris. This permits operators to express the purpose of an ephemeris. Ephemeris 3, for example, is the prime ephemeris for the International Space Station. Ephemeris 10 includes an orbit adjust, conducted in two maneuvers. At present we are concerned about the maintenance of this comment line. In the absence of an automated link between this comment and changes to ephemeris contents, nothing prevents the persistence of stale, misleading comments.

The resulting individual ephemeris panel within Tier 1 takes up less than half of the space of the legacy TPS panel. However, additional detail on an ephemeris, corresponding to the digital information on the legacy display, can be requested. This detail is shown for E3 (in the center of the figure) and contains much of the information from the legacy display, though slightly rearranged. Although this panel is larger than the comparable panel on the legacy display, it contains more information and it can be dismissed when not needed. The original TUP counter is available here. Additional information not previously available includes an icon indicator of the object being modeled, more detail on the vector, and explicit indications of relationships with other ephemerides with linking icons (chains) and copying icons (arrows). The detailed panel contains a higher resolution, interactive tape with a distinctive icon corresponding to closely spaced burns characteristic of an altitude reboost.

Part of the ephemeris configuration process may involve copying one ephemeris to another. For example, the FDO might ask the Traj or Profile Support to plan a contingency breakaway by using a copy of the ephemeris for a nominal rendezvous. Using a copy ensures that the nominal and breakaway plans are based on the same underlying assumptions. The FDO might make one ephemeris a subset of another, simply to speed the simulation computations (Ephemeris 1 in Fig. 29.5 is a subset of ephemeris 2). If the source ephemeris is updated, for example, as a result of new tracking information, the copies or subsets also require an update. Because the legacy TPS does not represent relationships between ephemerides, the flight controllers must remember which ephemerides are related, and to update the multiple copies as appropriate.

These relationships may be indicated during the process of creating a new ephemeris. The double arrows on E3 in Fig. 29.3 indicate that E3 is the parent of E1. Detail for E1 would show that it is the child of E3. These relationships allow for automated reminders regarding the propagation of change from one ephemeris to another. Tracking the relationship among ephemerides is an example of difficulty created by the availability of more ephemerides in the rehosted trajectory processors. More ephemerides can result in more relationships to remember.

Tier 2 TPS.   Tier 2 (shown in Fig 29.6) supports ephemeris troubleshooting (Task 6.0), a task that was not identified until fairly late in the task analysis process, and only with the guidance of the backroom support personnel who perform it. The vast majority of ephemeris changes are executed correctly and correspond to correct intentions. In a small minority of cases, a change is executed incorrectly, or mistaken intentions are executed. If these are not caught immediately, someone must review the history of ephemeris interactions in an attempt to restore ephemeris integrity.

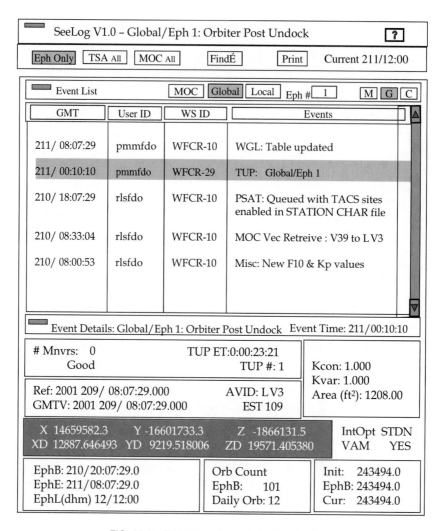

FIG. 29.6.   Tier 2 for ephemeris troubleshooting.

The Dynamics operator was largely responsible for ephemeris troubleshooting, and, as a dedicated resource, would not have to manage interruptions from other disciplines if troubleshooting were required. When required, the primary tool provides a merged running account of commands for all ephemerides. This makes it difficult to track the changes to individual ephemerides or determine the values of parameters that have not changed.

The new software allows many different individuals to interact with the nongolden ephemerides. It may be difficult, at best, to ask the Dynamics operators to troubleshoot the ephemerides that they did not alter. However, the other operators are not completely dedicated to ephemeris maintenance, and they will need better tools to take on troubleshooting for the ephemerides they change. An entirely new representation has been designed to support ephemeris troubleshooting (in Fig 29.6). The new prototype allows viewers to obtain sorted logs of ephemeris interaction, formatted to indicate event time, user, workstation, and event. Placing the cursor on a line of the log causes a detailed view to appear in the bottom third of the display, illustrating *all* of the parameter values after the highlighted event. Changes resulting from the event are highlighted in the detailed display. Providing the values of unchanged parameters eliminates the need for the troubleshooter to hold these in memory while trying to reconstruct the problematic sequence of events.

Project Status

Implementation has been in progress for more than 1 year, challenged in part by the availability of the lead software developer. However, the nature of the project has contributed to an extended development effort, and readers who attempt similar projects may benefit from the identification of some of these issues. At present a member of the Flight Dynamics Section is testing the interface for accurate functionality. This can only be done when all the complementary trajectory processing software is running, typically in conjunction with the testing of other software development activities. Although users have informally and positively reviewed the representation, performance testing must wait until all of these functionality issues have been identified and resolved. Only then can we determine whether or not the design benefits performance as anticipated.

Thus far, 57 anomalies have been identified, with four of these remaining unresolved and the rest being closed with successful modifications. No member of the target work group is likely to view these anomalies as anything but uncontroversial oversights. We distinguish among three types of anomalies that occur with approximately equal frequency: design errors, internal errors, and linking errors. Design errors refer to features that the design team should have realized when they specified the design for implementation. For example, the ephemerides for TDRS did not include a field for solar radiation pressure, and they had unnecessary fields for drag-related parameters. Internal errors refer to interrelationships between components of the new representation. For example, changing the status of an entry ephemeris on Tier 0 resulted in bogus text on the Tier 1 display. Linking errors refer to the relationship between the new display and other existing software. For example, the display will not start in the absence of certain trajectory software. Unlike the other two classes of error, linking errors (or successful links, for that matter) require the expertise of scarcely available specialists in other aspects of system software. Similarly, aspects of the full design have not yet been implemented. User identification is one example. Linking with the user name is not a trivial matter in a software system that was not designed to make user name available to the applications. This example illustrates the type of difficulty that arises when displays do more than provide illustrations for physical sensors' but rather reflect manipulated parameters embedded in other software. One third of the difficulties in developing this display arise from this type of difficulty.

CONCLUSIONS AND IMPLICATIONS

In this section we identify features of the project that we believe are responsible for the results, in an effort to guide other development activities. Because the display has not been evaluated in performance testing, we cannot ensure a relationship between the methods used here and a quantitatively characterized positive result. Moreover, only a controlled experiment varying the methods of apprehension, analysis, and design could establish a causal link between methods and outcomes. Pending such a study, we can only speculate on the relationship between the methods used and specific outcomes.

Apprehension

The access to the work environment was exceptional, and largely the result of the project champions. Brulé and Blount (1989) pointed to the role of such a champion, though it merits mention that the primary champion was a member of the target work group, and not merely an administrative or research point of contact. Combined with participation in other work group activities, notably extensive discussions of anticipated software changes and future work practices, the observer obtained a surprisingly detailed understanding of the work domain.

Although the degree of site presence is consistent with participant observation, the period of observation was extensive, and therefore expensive and slow. Elapsed calendar time reflects the broader goals of the project, as well as the prevailing rate of shuttle flights and other demands on project champions. Without such influences, elapsed calendar time might have been shortened with little project consequence. However, calendar time proved helpful because it encompassed organizational change, including the software development that converged with the current project. Although the duration of site presence is worrisome, the observer's time trades off against domain expert time participating in interviews. In many domains, experts are too busy for extensive interviews. Further, extensive observation contributes to the credibility of recommendations (Harper, 2000).

Analysis

Analysis is reflected in two aspects of the project: the selection of a problem for focused study and the creation of a task representation.

Problem Selection.   The selected problem did not emerge from standard human factors guidelines for display formatting. Instead, the selected problem focused on ambiguity in a boundary object (Bødker, 1998; Star, 1989), in this case, ephemeris representations created by one operator but viewed by others. The broader (but not deeper) background and experience of an outside observer led to an appreciation for distributed nature of the work, and the centrality of the ephemeris as a model of hypothetical conditions that operators configure, refine, and share. This supports the view that operators are not necessarily the best source of insight into system design issues that influence their efficiency.

The changing trajectory software provided the impetus for this analysis. Sound motives underlie the shift from an aging mainframe to a set of modern desktop workstations enabling parallel work. However, as Hutchins (1993) suggested, we must not to eliminate the very features of work that are responsible for its success in the name of technological progress. The concern here is that side effects of mainframe operation rendered work intelligible to the team. The need to instruct Dynamics to make changes over the voice loops kept the rest of the team aware of the work occurring with an ephemeris. The inability for multiple team members to change multiple ephemerides simultaneously prevented multiple operators from effecting change on the same ephemeris. As the sole point of contact with an ephemeris, Dynamics was also in a position to prevent multiple operators from effecting change on the same ephemeris, and incidentally troubleshoot errant ephemerides.

Increasing a limited resource required for planning, ephemeris slots, would seem to relieve the burden of coordination. However, the relief is only superficial. First, more ephemerides translate into more representations to understand and more relationships to remember. The plans, and therefore the ephemerides, share some common but changing assumptions that require coordinated updating (e.g., the ATL and VTL). This coordination becomes an important consideration when new technology intervenes in the work practice. A number of the new display features simply provide persisting, visual support for the coordination functions of transient voice instruction in the legacy system.

Task Representation.   Like the selection of a problem, the task representation reflects the observer's interpretation, guided by experience in the domain for terminology and goals, and by the scientific literature for major distinctions between monitoring and modeling, and such recognizable engineering-oriented activities versus hidden supervisory work. Group review and endorsement substantiated the representation, along with the experience of modifying the representation during design. Nothing like Woods' (1993) thorough process tracing occurred for the applications focus of this project. Moreover, the representation does not play a completely a priori, prescriptive role in the present project as it has in previous research projects built on similar foundations (DuBois & Shalin, 1995; Shalin & Geddes, 1994; 1998). Readers familiar with this previous work may note the difference in task representations. Typically, the first author employs a hierarchical decomposition that distinguishes plans from goals. This allows for the representation of different plans to address the same goals. This representation was not employed here because there were very few goals that could be addressed with different plans.

The representation proved a surprisingly helpful role in gathering the opinions of the work group related to the displays and work practices. In particular, Tier 2 arose precisely from a work group review of the information in Table 29.1. Like the project of Paris, Balbo, and Ozkan (2000), the present project suggests that the target community may benefit from a new ability to represent their work for themselves. It also merits mention that both authors worked directly with the representation as part of the design process. That is, just as the first author acquired an understanding of the work, the second author acquired an understanding of how to represent work.

Design

Identification of Information Requirements. Having identified the task, the first author's standard methods led to the identification of information requirements from the task representation, largely concerning the meaning behind configured ephemerides. The coordination of models previously supported only by means of voice-loop communication now has enduring visual support. Several features reduce dependence on human memory. The TUP counter, an ingenious parameter for enhancing human–computer interaction, has been replaced with an elapsed time counter.

The same methods also comprised an analysis of the legacy TPS display, identifying information requirements for hidden work concerning the supervision of a distributed team drawing on shared computational resources. Thus methods developed for design also constitute a content-oriented approach to the analytic evaluation of existing displays.

Analyses of team performance oriented to human factors emphasize the need for communication, concerning roles and responsibilities (Blickensderfer, Cannon-Bowers, Salas, & Baker, 2000), reconciliation of alternative approaches and views (Klein, 2000; Orasanu, 1991), and the negotiation among team members with different functions (Essens, Post, & Rasker, 2000). With the exception of Zachary, Ryder, and Hicinbothom (2000), who specifically note the need to identify competition for shared resources, these human factors oriented analyses do not clarify the implications of collaborative work for the design of computer-based support. Consistent with research in computer-supported cooperative work, we supported information requirements that indicate the owners or creators of these representations (Grinter, 1996; Teege, 1995), changes in the status of intermediate work products (with the A and V buttons; Agostini et al., 1996), and an appreciation for the work that other operators will conduct (Button & Sharrock, 1996) by marking ephemerides that participate in acquisition.

Consistent with work by Zhang and Norman (1994), the new representation distributes the memory burden between operators and computers. As in many problem solving tasks, the issue is one of preserving a representation of work in progress while it is being expanded and refined (Larkin & Simon, 1987). This is simply not an issue in the analysis tasks (diagnosis and monitoring) that have captured much of human factors. Most of the new features (e.g., indications of object type, maneuver contents, linked ephemerides, and user) enable the coordination of distributed work on parallel problems by using shared tools, creating a representation that was more complete for those who did not participate in its configuration. Some of the features (e.g., links between ephemerides, iconic indicators for the role of an ephemeris in other processes, and the ET indication) appear in response to the increased number of ephemerides in the rehosted software.

Layout.   Apart from the identification of new information requirements that clarify the meaning of ephemerides, three features are reflected in the new design: (a) a hierarchical display provides scope by locating different levels of detail in different displays; (b) similar-appearing text fields have been replaced or augmented with icons; and (c) information spread across displays in the legacy system is now integrated with the new display. Unfortunately, even full performance testing will not be able to isolate the contributions of each of these features because we will not be able to develop displays that vary these features independently.

On the basis of what we have understood here, how might we approach the problem of understanding representations used, for example, in manufacturing design? The implications of this study for work design in other synthesis domains consist of a list of issues that must be considered, in examining an existing pairing of technology and work practice, and in pondering a modification to this pairing. We would investigate the explicitness of the representation for the designed object, and the representation of the processes that the design implied. We would identify how changes in assumptions occur, how they are incorporated into the design, and how change is indicated. We would determine how members of the design team become aware that someone else is working on a part of the design or using the design as a foundation for other tasks such as planning manufacturing or advertising. Undetected, either of these activities have the potential to undermine coordination. We would find out how relationships between related designs are captured. We would search for indications of design age, and the means for capturing the content of revisions and thinking that led to a current design. For all of these questions, we would seek to uncover the role of representing technology, particularly regarding limited design resources and the controls on making changes to designs, and the manner in which hidden human coordination takes advantage of and compensates for these properties. Those issues managed purely with verbal exchange become candidates for persisting display so that human intent is as apparent as technical detail in visual representations for work.

ACKNOWLEDGMENTS

This project has been supported under cooperative agreement NAG2-1237 with the Work Systems Design and Evaluation program in the Information Sciences Division at NASA Ames Research Center. The authors are grateful for invaluable opportunities and guidance under this agreement from the program director, William J. Clancey. Several students participated in setting up equipment, organizing recordings, and transcribing some of the audio: Scott Bachmann, Rodney Halgren, Judith Isaacson, Paul Jacques, James Kondash, Alexa Painter, Mark Palumbo, Louise Rasmussen, and Brian Simpson. We are profoundly grateful for the exceptional cooperation and contributions of the Orbit Flight Dynamics group in the Mission Operations Directorate at NASA's Johnson Space Center, particularly Joseph Williams. The views and conclusions presented in this paper are those of the authors and do not represent an official opinion, expressed or implied, of the sponsoring or cooperating NASA agencies.

REFERENCES

Agostini, A., DeMichelis, G., Grasso, M. A., Prinz, W., & Syri, A. (1996). Contexts, work processes and workspaces. *Computer Supported Cooperative Work, 5,* 223–250.

Annet, J., & Duncan, K. D. (1967). Task analysis and training design. *Occupational Psychology, 41,* 211–221.

Blickensderfer, E., Cannon-Bowers, J. A., Salas, E., & Baker, D. P. (2000). Analysing knowledge requirements in team tasks. In J.-M. Schraagen, S. Chipman, & V. L. Shalin (Eds.), *Cognitive task analysis* (pp. 431–447). Mahwah, NJ: Lawrence Erlbaum Associates.

Bødker, S. (1998). Understanding representation in design. *Human-Computer Interaction, 13,* 107–125.

Brulé, J. F., & Blount, A. (1989). *Knowledge acquisition.* New York: McGraw-Hill.

Button, G., & Harper, R. (1996). The relevance of "work-practice" for design. *Computer Supported Cooperative Work, 4,* 263–280.

Button, G., & Sharrock, W. (1996). Project work: The organisation of collaborative design and development in software engineering. *Computer Supported Cooperative Work, 5,* 369–386.

Card, S. K., Moran, T. P., & Newell, A. (1983). *The psychology of human-computer interaction.* Hillsdale, NJ: Lawrence Erlbaum Associates.

Carnap, R. (1947). *Meaning and necessity: A study in semantics and modal logic.* Chicago: University of Chicago Press.

Cicourel, A. V. (1988). The integration of distributed knowledge in collaborative medical diagnosis. In J. Galegher, R. E. Kraut, & C. Egido (Eds.), *Intellectual teamwork: Social and technological foundations of cooperative work* (pp. 221–242). Hillsdale, NJ: Lawrence Erlbaum Associates.

Clancey, W. J. (1985). Heuristic classification. *Artificial Intelligence, 27,* 289–350.

Cole, M. (1995). Socio-cultural-historical psychology. In J. V. Wertsch, P. Del Rio, & A. Alvarez (Eds.), *Sociocultural studies of mind* (pp. 187–214). Cambridge, England: Cambridge University Press.

Dourish, P., & Button, G. (1998). On "technomethodology": Foundational relationships between ethnomethodology and system design. *Human-Computer Interaction, 13,* 395–432.

DuBois, D., & Shalin, V. L. (1995) Job knowledge test design: Cognitive contributions to content oriented methods. In P. Nichols, S. Chipman, & R. Brennan (Eds.), *Alternative diagnostic assessment* (pp. 189–220). Hillsdale, NJ: Lawrence Erlbaum Associates.

DuBois, D., & Shalin, V. L. (1997/1998). A cognitively-oriented approach to task analysis and test development. *Training Research Journal, 3,* 103–142.

DuBois, D., & Shalin, V. L. (2000). Essential properties of cognitive task analysis for applications goals. In J.-M. Schraagen, S. Chipman, & V. L. Shalin (Eds.), *Cognitive task analysis* (pp. 41–55). Mahwah, NJ: Lawrence Erlbaum Associates.

Eckert, C. (2001). The communication bottleneck in knitwear design: Analysis and computing solutions. *Computer Supported Cooperative Work, 10,* 29–74.

Essens, P. J. M. D., Post, W. M., & Rasker, P. C. (2000). Modelling a command centre. In J.-M. Schraagen, S. Chipman, & V. L. Shalin (Eds.), *Cognitive task analysis* (pp. 385–399). Mahwah, NJ: Lawrence Erlbaum Associates.

Fetterman, D. M. (1998). *Ethnography* (2nd ed.). Thousand Oaks, CA: Sage.

Flach, J. M. (1999). Ready, fire, aim: Towards a theory of meaning processing systems. In D. Gapher & A. Koriat (Eds.), *Attention and Performance XVII* (pp. 187–221). Cambridge, MA: MIT Press.

Forsythe, D. E. (1993). Engineering knowledge: The construction of knowledge in artificial intelligence. *Social Studies of Science, 23,* 445–477.

Forsythe, D. E. (1999). "Its just a matter of common sense": Ethnography as invisible work. *Computer Supported Cooperative Work, 8,* 127–145.

Gagné, R. M. (1962). The acquisition of knowledge. *Psychological Review, 69,* 355–365.

Gerhardt-Powals, J. (1996). Cognitive engineering principles for enhancing human-computer performance. *International Journal of Human-Computer Interaction, 8,* 189–211.

Greeno, J. G. (1978). Understanding and procedural knowledge in mathematics instruction. *Educational Psychologist, 12,* 262–283.

Grinter, R. E. (1996). Supporting articulation work using software configuration management systems. *Computer Supported Cooperative Work, 5,* 447–465.

Hall, E. M., Gott, S. P., & Pokorny, R. A. (1995). A procedural guide to cognitive task analysis: The PARI methodology (Tech. Rep. No. AL/HF-TR-1995-0108). Brooks Air Force Base, TX: Air Force Materiel Command.

Harper, R. H. R. (2000). The organisation of ethnography: A discussion of ethnographic fieldwork programs in CSCW. *Computer Supported Cooperative Work, 9,* 239–264.

Hess S. M, Detweiler, M. C., & Ellis, R. D. (1999). The utility of display space in keeping track of rapidly changing information. *Human Factors, 41,* 257–281.

Hoc, J. M. (1988). *Cognitive psychology of planning.* London: Academic Press.

Hoc, J. M. (2001). Towards a cognitive approach to human-machine cooperation in dynamic situations. *International Journal of Human Computer Studies, 54,* 509–540.

Hollnagel, E. (1991). The phenotype of erroneous actions. Implications for HCI design. In G. R. S. Weir & J. L. Alty (Eds.), *Human computer interaction and complex systems* (pp. 73–121). London: Academic Press.

Hutchins, E. F. (1993). Learning to navigate. In S. Chaiklin & J. Lave (Eds.), *Understanding practice* (pp. 35–63). Cambridge, England: Cambridge University Press.

Johnson, P., Johnson, H., & Wilson, S. (1995). Rapid prototyping of user interfaces driven by task models. In J. Carroll (Ed.), *Scenario-based design* (pp. 209–246). New York: Wiley.

Johnson-Laird, P. (1983). *Mental models: Towards a cognitive science of language, inference and consciousness.* Cambridge, England: Cambridge University Press.

Klein, G. A. (2000). Cognitive task analysis of teams. In J.-M. Schraagen, S. Chipman, & V. L. Shalin (Eds.), *Cognitive task analysis* (pp. 417–429). Mahwah, NJ: Lawrence Erlbaum Associates.

Klein, G. A., Calderwood, R., & MacGregor, D. (1989). Critical decision method for eliciting knowledge. *IEEE Transactions on Systems Man and Cybernetics, 19,* 462–472.

LaFrance, M. (1997). Metaphors for expertise: How knowledge engineers picture human expertise. In P. J. Feltovich, K. M. Ford, & R. R. Hoffman (Eds.), *Expertise in context: Human and machine* (pp. 163–180). Menlo Park, CA: American Association for Artificial Intelligence.

Larkin, J. H., & Simon, H. A. (1987). Why a diagram is (sometimes) worth ten thousand words. *Cognitive Science, 11,* 65–100.

McDonald, J. E., & Schvaneveldt, R. W. (1988). The application of user knowledge to user interface design. In R. Guindon (Ed.), *Cognitive science and its applications for human-computer interaction* (pp. 289–338). Hillsdale, NJ: Lawrence Erlbaum Associates.

Mitchell, C. M., & Miller, R. A. (1986). A discrete control model of operator function: A methodology for display design. *IEEE Transactions on Systems, Man and Cybernetics, 16,* 353–369.

Nielsen (1994). Heuristic evaluation. In J. Nielsen & R. L. Mack (Eds.), *Usability inspection methods* (pp. 25–62). New York: Wiley.

Norman, D. A. (1991). Cognitive artefacts. In J. M. Carroll (Ed.), *Designing interaction: Psychology at the human-computer interface* (pp. 17–38). New York: Cambridge University Press.

Orasanu, J. (1991) Individual differences in airline captains' personalities, communication strategies, and crew performance. In *Proceedings of the Human Factors Society 35th Annual Meeting* (pp. 991–995). Santa Monica, CA: Human Factors Society.

Paris, C., Balbo, S., & Ozkan, N. (2000). Novel uses of task models: Two case studies. In J.-M. Schraagen, S. Chipman, & V. L. Shalin (Eds.), *Cognitive task analysis* (pp. 261–274). Mahwah, NJ: Lawrence Erlbaum Associates.

Payne, S. J., & Greene, T. R. G. (1986). Task action grammars: A model of the mental representation of task languages. *Human-Computer Interaction, 2,* 93–133.

Rasmussen, J. (1983). Skills rules knowledge; signals, signs and symbols; and other distinctions in human performance models. *IEEE Transactions on Systems Man and Cybernetics, 13,* 257–266.

Roske-Hofstrand, R. J., & Paap, K. R. (1986). Cognitive networks as a guide to menu organization: An application in the automated cockpit. *Ergonomics, 29,* 1301–1311.

Rubin, K. S., Jones, P. M., & Mitchell, C. M. (1988). OFMspert: Inference of operator intentions in supervisory control using a blackboard architecture. *IEEE Transactions on Systems, Man and Cybernetics, 18*(4), 618–637.

Sanders, M. S., & McCormick, E. J. (1993). *Human factors in engineering and design.* New York: McGraw-Hill.

Sarter, N. B., & Woods, D. D. (1994). Pilot interaction with cockpit automation II: An experimental study of pilots' model and awareness of the flight management and guidance system. *International Journal of Aviation Psychology, 4*(1), 1–28.

Sebillote. S. (1995). Methodology guide to task analysis with the goal of extracting relevant characteristics for human-computer interfaces. *International Journal of Human-Computer Interaction, 7,* 341–363.

Shalin, V. L., & Geddes, N. D. (1994). Task dependent information management in a dynamic environment: Concept and measurement issues. In *Proceedings of the 1994 IEEE International Conference on Systems, Man and Cybernetics* (pp. 2102–2107). New York: IEEE.

Shalin, V. L., and Geddes, N. D. (1998). Pilot performance with mid-flight plan-based display changes. In *Proceedings of HICS '98: 4th Annual Symposium on Human Interaction with Complex Systems* (pp. 170–181). Los Alamitos, CA: IEEE Computer Society Press.

Shute, V. J., Torreano, L. A., & Wills, R. E. (2000). DNA: Providing the blueprint for instruction. In J. -M. Schraagen, S. Chipman, & V. L. Shalin (Eds.), *Cognitive task analysis* (pp. 71–86). Mahwah, NJ: Lawrence Erlbaum Associates.

Star, S. L. (1989). The structure of ill-structured solutions: Heterogeneous problem solving, boundary objects and distributed artificial intelligence. In M. Hahns and L. Gasser (Eds.), *Distributed artificial intelligence, 2* (pp. 37–54). Menlo Park, CA: Kaufmann.

Star, S. L., & Strauss, A. (1999). Layers of silence, arenas of voice: The ecology of visible and invisible work. *Computer Supported Cooperative Work, 8,* 9–30.

Suchman, L. A., & Trigg, R. H. (1993). Artificial intelligence as craftwork. In S. Chaiklin and J. Lave (Eds.), *Understanding practice: Perspectives on activity and context* (pp. 144–178). Cambridge, England: Cambridge University Press.

Symon, G., Long, K., & Ellis, J. (1996). The coordination of work activities: Cooperation and conflict in a hospital context. *Computer Supported Cooperative Work, 5,* 1–31.

Teege, G. (1995). Object-oriented activity support: A model for integrated CSCW systems. *Computer Supported Cooperative Work, 5,* 93–124.

Vaughn, M. W. (1998). Testing the boundaries of two user-centered design principles: Metaphors and memory load. *International Journal of Human-Computer Interaction, 10,* 265–282.

Watts-Perotti, J., & Woods, D. D. (1999). How experienced users avoid getting lost in large display networks. *International Journal of Human-Computer Interaction, 11,* 269–299.

Woods, D. D. (1993). Process-tracing methods for the study of cognition outside of the experimental psychology laboratory. In G. A. Klein, J. Orasanu, R. Calderwood, and C. Zsambok (Eds.). *Decision making in action: models and methods* (pp. 228–251). Norwood, NJ: Ablex.

Zachary, W. W., Ryder, J. M., & Hicinbothom, J. H. (2000). Building cognitive task analyses and models of a decision-making team in a complex real-time environment. In J.-M. Schraagen, S. Chipman, & V. L. Shalin (Eds.), *Cognitive task analysis* (pp. 365–383). Mahwah, NJ: Lawrence Erlbaum Associates.

Zhang, J. J. (1997). Distributed representation as a principle for the analysis of cockpit information displays. *International Journal of Aviation Psychology, 7,* 105–121.

Zhang, J. J., & Norman, D. A. (1994). Representations in distributed cognitive tasks. *Cognitive Science, 18,* 87–112.

30

Pilot's Understanding Process when Coping with Automation Surprises

Marielle Plat
Airbus, S. A. S., France

Abstract

This chapter is about the importance of considering the final user of a design, here the pilot, in the design cycle and to consider this user as he or she is. Designers must provide the pilot with an interface able to deal with rare situations, such as malfunctions of automated systems, and must therefore take into account how the pilot understands the situation. Industry must also revise procedure and training from the current "normative" point of view to provide an environment that is better adapted to pilots in their cockpits. This chapter presents an experimental study of pilot comprehension of software bugs in automated systems as the basis for a general discussion of what industry should take into account to ensure efficiency in unpredictable situations.

INTRODUCTION

In the field of human factors, it is a common idea that new technology such as the introduction of the glass-cockpit concept in aviation requires a period of adaptation to be effective and safe for users. Indeed, it takes years to become efficient in all the process domains such as users' adaptation, training, and design. To reach this point, the user has to regulate his or her own activity within this complex social system. As we will see, human factor specialists may help engineers to improve new design before in-service experience, by providing them with a "ready to use" reformulated user request, to design their systems. As we believe, the new glass-cockpit philosophy is now a more stable system for the users who have already gained experience and adopted habits. During the past 10 years, the automated systems introduced into aircraft cockpits have significantly increased safety and efficiency in the piloting task. These systems have been conceived by engineers who had already won their challenge to create a fully automated system, comparable to a computerised autonomous agent able to take over part of the pilot's activity with, sometimes, more precision. As Amalberti (2001) underlines, this glass-cockpit philosophy, as a system, has now revealed quite a good level in terms of safety. Consequently, it is hard to optimise it in terms of performance. Nevertheless, human–machine interaction could be improved to help users to cope with those systems, especially in crisis situations where they have to regulate the nonusual situation.

In a more general way, the aeronautical industry defined automated systems as a help for pilots during their flights. This means that these systems have been designed with an apparent logic, in the sense that in an ideal world, every type-qualified pilot would be able to deal with it: choosing the desired level of automated control and taking over flight control when necessary. However, we have to deal with a reality, which is not so ideal. Pilots in their normal activity are helped by automation, but they also have to deal with unusual complex situations, which automation is not always programmed for. Thus, it is still the pilot's task, in accordance with the airline philosophy, to choose his or her own level of automated control, or his or her level of help for each specific situation. Automated systems could be considered as a help in a sense that they partially take control of the piloting task and relieve pilots' workload. Nevertheless, in the event of an incident with these automated systems (which is still rare), in accordance with the generic procedure, pilots have to disconnect and revert to manual control. This is the expected behaviour with the normative way of thinking used in aviation and all domains, which is very suitable in most cases in terms of safety because it is used in today's safety models.

Amalberti holds a different point of view from this normative point of view. A lot of research work shows that people will not perform as they have been asked to, and we have to consider their activity with another idea than people behaving as required by a norm or a procedure. De Brito (2000) argues that pilots need to understand the procedure before acting. Writers all seem to agree that technology progress induces changes in pilots' operativeness and on the representation of their work. Some authors such as Corwin (1995), Sarter and Woods (1992, 1994), and Amalberti (1994) have listed the consequences of this transformation with the introduction of automated systems in cockpits. They emphasised that pilots prefer to develop an erroneous representation of what the system does to save some cognitive resources (Wioland & Amalberti, 1996).

To be more efficient when designing an interface, designers and human factor specialists have to improve the design of interfaces by understanding the human way of functioning as well as possible. This implies a need to update their knowledge about research in the domain as far as possible as well as in parallel fields, which can be transferred to the task the final user has to achieve. This is a basis, but this will never replace the work the designers have to do when analysing the user's task and his or her special operational needs to achieve it. This has to be done before the design. Thus, aware of the task and the constraints that users have to take into account every day, designers could create better systems and interfaces. However, designers also have constraints to take into account, such as dealing with providers and associated time constraints. Nevertheless, tests have to be done on paper and computer mock-up and then on research simulators to improve the system or interface design all along the design process. User involvement must be achieved all along this process with test pilots and, when possible, training instructors, at first to ensure the first design steps and the updating of the associated training programmes. Then the final users will help to finalise the design and associated procedures.

All through the user involvement process, human factors specialists have to help pilots to be aware of their needs and to communicate their requirements as clearly as possible. Indeed, some user behaviours are so much of a habit that they are difficult to be aware of and users may be unable to express them. Because of this, the human factors specialist intervenes by setting up appropriate systems and methods to allow the user to become aware of his or her real needs and manage to reformulate his or her request accordingly. Training development could also begin in the last part of the design process to be built accordingly. To ensure that design will be efficient in most cases, we have to take into account scenarios built for simulations of some rare cases that could occur in service. To reach this goal, in-service experience could help a designer to be aware of such cases. That is why human factor specialists and designers have to be aware of what could be done in real situations by having the ability to know these cases. Moreover, these cases are the examples of what a human could do in a situation with stress, anxiety, time constraints, and unpredictable events. This represents additive knowledge and operational illustrations of the possible human way of functioning. These cases may be reproduced when systems and interfaces are improved at the end of the design cycle.

From a practical point of view of what "cognitive task design" is, the aim of this chapter is to deal with the use of technological artefact as a complex process and also with the social artefact as a procedure in aviation. They both have a contextual implication for human cognitive activity management, especially during critical situation management, as we will show. From our particular point of view, we shall look at cases beyond the barrier of the acceptable safety rate 10^{-7} to learn more about pilots' cognitive behaviour in such rare situations. Crossing over this barrier is a new way of considering safety. We show an experimental study to illustrate a way of considering the real human behaviour when the human has to apply a procedure that will interrupt an action that is already underway.

BUILDING UP AN EXPERIMENTAL SITUATION TO IMPROVE THE RESPONSE TO AUTOMATION SURPRISES

The Experimental Study

The study (financed by the Human Factors Department of the French Ministry of Transport) began in late 1995 to prepare the first step: the modification of glass-cockpit full-flight simulator possibilities by a simulator manufacturer, the preparation of the experiment in coordination with a European airline training department, and the experiment itself.

Flight Simulator Modifications

The modification of flight simulator capacities was not easy, partly because we were using an existing simulator (Plat & Amalberti, 2000). This simulator was programmed with logic of implemented failures such as subsystem failures, with a list of failures corresponding to the manufacturer's predictions. The caution system for most of the failures is a warning correlated with the automated system that provides procedures to the crew. As a consequence, the current generation of full-flight simulators (FFSs) does not provide the possibility of installing computer bugs. The choice of faults that we were able to implement was driven by two main goals: being compatible with the current simulator certification status and being representative of computer- and human-induced malfunctions. We used in-service experience as a basis and programmed incidents similar to these experienced in line flights.

For facility reasons (simulator availability, ease of technical modifications), we chose a representative European glass-cockpit aircraft; for technical reasons, we focussed on bugs only programmed on the Flight Control Unit (FCU, responsible for immediate orders given to the autopilot: manual speed and altitude selection, preset functions of the same variables, and autopilot mode selection and autopilot descent selection modes). The list of incidents, failures, and so on was selectable from a menu in the FFS as usual FFS procedure for instructors.

TABLE 30.1
Automated Malfunction Descriptions

Nature of the Fault	Display	Effect on Flight or Aircraft
Speed corruption	The speed selected on the FCU is not consistent with the speed displayed on the Primary Flight Display (PFD).	The effective speed is the speed displayed on the PFD.
Mach corruption	The mach selected on the FCU is not consistent with the mach displayed on the PFD.	The effective mach is the one displayed on the PFD.
Autothrust corruption	A gong is heard that indicates that the Autothrottle (A/THR) disappears from the Flight Mode Annunciator. We provide no specific procedure on the electronic centralised aircraft monitoring to help the crew.	A/THR does not manage the thrust as usual.
Altitude preset ineffective	The preset function on the flight control unit is ineffective.	The altitude given to the preset function is immediately considered as active by the system.
Automatic reversion of the descent mode between heading–vertical speed (HDG–VS) and Track–flight-path angle (TRK–FPA) and vice versa	The selection and activation of one of the two available descent modes, HDG–VS and TRK–FPA, immediately leads to the engagement of the other mode. When asking for a heading in the FCU, you display a track on the Flight Mode Annunciator with a consistent display of the aircraft attitude in the middle of the PFD (with the FCU request) and vice versa with the other mode when requested.	The system reverts to the nonselected mode, which is displayed on the Flight Mode Annunciator (in terms of descent rate and profile), but continues to display consistent assistance with the selected mode.

Scenarios Chosen

We choose two line-oriented flight training (LOFT) scenarios with a list of automatic function faults (Table 30.1), and events according to the point of view of the operational instructors, the controller, and the experimental team. The LOFT scenarios were a Lyons–Madrid trans-European flight (Table 30.2). Each crew was programmed as for a real flight and had to prepare for it as usual.

The first LOFT was based mainly on computer malfunctions with no associated procedure and no way to recover them; each fault was programmed for a period of time before disappearing by itself or by engagement of another mode.

The second LOFT was based on usual faults with a computer malfunction common to the worst fault of the first LOFT; two of the traditional faults implemented were not sequential but cumulative with the computer malfunction common fault.

The important difference with the unusual faults that we programmed is that they were not signalled by warning or a system helping pilots to detect and manage it. Pilots then have to detect by themselves by checking displayed noncoherent information. Moreover, pilots do not have any training for such specific faults, so they will have to use all their competency to deal with such rare situations. Nevertheless, they have a generic procedure that they must use as a prescribed solution to avoid a problem-solving situation: when in doubt on automation behaviour, "go back to basics."

TABLE 30.2
Description of Scenarios

Scenario 1	*Scenario 2*

1. Departure Conditions
The crew begins with two items unavailable: Pack 2 inoperative, which has an implication on the max flight level of 310, and Spoilers 2 & 4, which become inoperative.

2. Preflight
(When checking the flight in the cockpit): event "pax baggage," a passenger is missing but his luggage is already on board.
Preflight time pressure

3. Takeoff: Standard International Departure

4. Climb
Bleed 1 fault 1500 ft Above Ground Level (AGL) (could be fixed)
Control asked for a "direct way point"
"Ground Protection Warning System" Fault: alarm functions without any known cause
Control requests the aircraft at level 210 because traffic
"Cab press sys 1 fault" (automated pressurisation management has failed)
Control requests a 330 and the aircraft is limited to 310 because Pack 2 faulty (since departure)

5. Cruise

Control requests to speed up to mach 0.81.	5 min after beginning of cruise, "Cab press sys 2 fault" (the
Erroneous selected mach occurs	last automated pressurisation management has failed; this
	obliges the crew to handle pressurisation manually for the
	rest of the flight).
Few minutes later, autothrust failure	

6. Descent

Preset altitude fault is activated	
Control requests a heading change	Radar regulation by Air Traffic Control
After 15 seconds mode reversion HDG–VS and TRK–FPA, which are the horizontal management modes (HDG & TRK) associated with VS & FPA	
"Speed corruption fault"	

7. Approach
ILS 18 (Instrument Landing System) runway 18
Go-around (external cause)
Rerouting runway 33 (instead of 18)

Flap fault (flaps jammed in position 1)	"Nose up not uplocked" (this fault could be handled by an emergency checklist)

8. Approach Number 2 on ILS 33

9. Landing and Parking

Experimental Situation Description

The experimental situation includes phone calls to the FFS room, where there is an instructor, a professional air traffic controller, and the crew observed. The situation was built so as to reproduce as far as possible the natural context of a real flight. The controller was there to do the control in real time with an audio record of real communications with other aircrafts to simulate a real outside context. The instructor's job was to make the fault input at the best time in agreement with the experimental team (an ergonomist and a pilot) outside the cockpit. Data were videotaped and sent outside the cockpit during the flight. The video recorded the main screens (PFD and Navigation Display) of the First Officer plus the crew and controller communications. It shows us aircraft parameters, automatic mode changes, and aircraft navigation. After the flight a debriefing was made as a way to know more about the way the crew felt about the situation and some details about their activity.

Pilots

10 crews were planned on this flight, as on a real one. The crew came on the simulator for half-day's training: flight preparation, a 2-hour flight, and the debriefing (after flight). Pilots were professional and experienced on this aircraft; their average experience was 3 ± 1.61 years. Six crews were flying the first scenario and 4 crews were flying the second. The number of crews initially planned was 12 (6 for each scenario), but there were two last-minute cancellations; because of heavy booking on the simulator, the second scenario was limited to 4 crews.

Hypothesis

Our first hypothesis is that pilots need a minimum of understanding to act and also that the depth of this comprehension process could change with complexity. This could lead pilots to act not exactly as immediately required by the procedure.

From a philosophical point of view, we assume that complexity is human related. However, conditions could be generated to facilitate its appearance. We assume that we try to create two different contexts of workload, with the goal of producing a different allocation of cognitive resources. We believe that because pilots need to allocate some cognitive resources to achieve the understanding process. If those resources are limited this is a factor that contributes to the generation of cognitive complexity accorded to Javaux (1997, 1998), Amalberti (1996), and Spérandio (1972, 1977). Moreover, complexity could increase with the growing number of incomprehensions waiting to be treated. To deal with incomprehension, pilots need to increase this reasoning abstraction level in accordance with Rasmussen's model (1986), and that is resource consuming. In accordance with these theoretical prerequisites, our prediction is that complexity increases, will have an impact on the importance of the comprehension process managed by pilots: limited in duration and in deep of the comprehension process itself.

To make a detailed analysis of this phenomenon, we focus on the common fault within two scenarios; the role of the other faults in the scenarios were to generate two different workloads. This should be sufficient to assess our prediction about the adaptation of the pilot's understanding process.

BUILDING UP A SPECIFIC METHODOLOGY TO SHOW THE UNDERSTANDING PROCESS DURING DIAGNOSIS

Different steps divided into two main lines will explain this methodology. The first one focuses on a global analysis of data to materialise and summarise pilots' activity during the fault, in terms of decision time, action, and choice within the cockpit (Steps 1 and 2). The second analyses more precisely the pilots' communications and actions while acting (Steps 3–7). We assume that operative languages explain a significant part of the common cognitive activity of a crew when in a collective work task. This could be explained by the fact that the task to be accomplished by the two crew members requires verbal communication, especially in the aeronautical field, which increments this process to have a better shared representation. We take the crew itself as an operator dealing with the automated systems process.

Step 1: Crew Performance Data

Explicit detection time will begin as soon as the fault appears until one of the crew members gives a sign that she or he has detected it, verbally or by action. The crew will consider this detection time as the beginning of the treatment of the fault.

Closure time corresponds to the total time the crew is concerned with the failure, either by discussing the consequences or the cause of it, or by conducting tests or recovery actions.

These data will be presented for both scenarios in order to compare them.

Step 2: "Coping Strategy" Diagram

The analysis of operative communications and actions is summarised in a diagram for the main fault we choose. We used domain knowledge to produce categories, which are considered with two main goals: change the external or the internal world of pilots. Considering these two main categories, and using domain knowledge, we divided the analysis into four categories:

- Object directed, such as "action on automated systems" (e.g., when disconnecting the autopilot) and "action on the aircraft as a mobile" (e.g., to increase speed);
- Cognition oriented, that is verbal acts concern the pilot himself or herself or the pilot's crew partner; action proposal, explicit detection, or distributed information process;
- "Internal cognitive-oriented communication" (e.g., "what is it doing to me now?"); and
- External cognitive-oriented communication" (e.g., "It seems that we did have a quick change in that column . . . [he shows the Flight Mode Annunciator display with his hand] when you engaged the autothrottle")

From these categories we build up a first diagram showing, for each fault, the crew treatment from which we summarised, in one diagram, all main treatments made by pilots to cope with the fault. This diagram is shown in the first part of the results.

Step 3: Operative Communication Analysis

In aviation, these communications are considered to be a very important part of the activity to allow the crew a common representation of the situation. Thus they are a complete part of the activity process and describe the crew activity as much as possible. We consider here that operative communications are built to make a change on action world objects (other agents and objects as defined earlier).

Verbal Protocol Reducing Method. We used the notion of the interaction unit developed by Rogalski and Samurçay (1993) and by Plat and Rogalski (2000) and Plat (2001). A unit is defined as communications transmitted by pilots related to the same local finality for these pilots. Crew dynamism will be materialised by a unit interaction succession. This unit in the aviation field is very often delimited by collating between the two crew members and even more so with the traffic controllers. An example of this is "when I need to know, I look at this. OK?" "Yes, OK!"

Nevertheless we can find exceptions to this delimitation. The exception is that the interaction could be finished by information or a situation evaluation with or without available medium, or action evaluation without the other pilot collating immediately.

With this method, we define reduced protocols, which could give us a representation of interaction dynamics during the chosen part of the flight. With these reduced protocols we will be able to see which pilot in the crew is initiating the unit or the action below. We also keep organisational marks of the interaction units as verbal acts. The number of messages dispensed by each pilot are also taken into account as a work-intensity marker.

From this method, we categorise interaction units into two main directions: external communication with Air Traffic Control (ATC) and the cabin crew and internal communication between the crew members. The most important thing is internal communication according to our purposes, which are the understanding process and action when coping with the diagnosis.

Unit Categorisation.   We maintain that comprehension and finalised action are completely dependent on one another in such a dynamic environment. To see the organisation between the two, we decided to categorise them separately: the first category is about comprehension and the second is about action.

Step 4: Comprehension Categorisation

The first phase of the understanding process is minimal; it is constituted by detection and first diagnosis of system status. This minimal process is consistent with the attempted behaviour to cope with such a faulty system symbolised by the "back to basics" generic procedure. Such a procedure is attempted as soon as the pilot has a doubt as to the way of functioning of an automated system. As the generic analyses will show, it seems that pilots do not stop this step and follow up a longer understanding process. We describe two categories of each level to give an overview of the method.

The level of analysis we describe mixes a finer analysis with interaction units coming from pilots' communication and action associated with interpretation issued from a domain expertise. Every category begins with SA, standing for situation awareness.

Detection Categorised As "SA.DETECT."   This is materialised by explicit fault detection or noncoherent information, such as "So, heading 260 and I have a track in front of me," and "FPA track . . . What's up on this . . . ."

Nevertheless, this detection could appear in different steps, as shown in the following example:

First step: "Careful, we are in track mode."
Second step: "Careful we are in heading mode [he shows the Flight Mode Annunciator Display] and we are in track mode in here [he shows the Flight Path Vector on the PFD] . . . ."

For this reason we decided to number our categories as follows: "SA.DETECT.1", "SA.DETECT.2," and through to "SA.DETECT.n."

First Diagnosis Of System State: Categorised As "SA.DIAG.IN."   The second phase of the understanding process goes deeper than the prescription of the procedure. We assume that pilots will try to identify more than what is not functioning. They will try to discover how it is malfunctioning and what could be trusted again. To attain that goal, they will combine action and comprehension processes with the following categories.

System State Evaluation: "SA.EVAL.ET.n."   This is more an incidental situation evolution status that the pilots are making, such as "No, see, it's different again." There are the

TABLE 30.3
Illustration of Mode Reversion Fault Coding for Crew 1

Aircraft Management	Fault Diagnosis	External Communication Management	Aircraft and Automated Systems Behaviour + Comments	Verbal Communications Account
		A.DEM CAP	TRK–FPA first fault input	ATC–FO
	SA.DETECT.1 + SA.DIAG.IN.1 + ACT.STAB.1		PA Off	CPT/FO (action)
•				FO
	SA.DIAG.IN.2 + SA.PROP.S.1 + ACT.OPER + ACT.SURV		Heading and mode change on FCU	CPT/FO (+ action)– CPT/FO

following:

- Shared representation tuning: "SA.AJUST.REP";
- Personal situation evaluation: "SA.EVAL.PS";
- Hypothesis generation: "SA.EME.HYP";
- Misunderstanding: "SA.INCOMP";
- Solution proposal: "SA.PROP.S.n"; and
- "Back to basics" diagnosis: "SA.DIAG.MANU."

Related to these comprehension categories, we defined action categories mostly associated with the previous ones. An exception must be underlined when the crew finds an appropriate solution and just has to repeat it each time it wants to operate the same way:

- Disconnection to stabilise the situation: "ACT.STAB.n";
- Final disconnection of all automated systems: "ACT.STAB.FIN";
- Information search action: "ACT.INF";
- Monitoring action: "ACT.SURV";
- Help to understand the action: "ACT.COMP";
- Action to operate according to a goal: "ACT.OPER";
- Action to demonstrate and validate an explanation to the other crew members: "ACT.AJUS";
- Action to restore situation: reset 1, that is, "ACT.RESET.1.deb" (disconnection) and "ACT.RESET.1.fin" (reconnection), and reset 2 (breakers), that is, "ACT.RESET.2.deb" and "ACT.RESET.2.fin."

Step 5: Verbal Protocol Reduction

This step is a table, which summarises verbal protocols according to previous step treatment. It also visualises the parallel or sequential management of different cognitive activity loops by a column materialisation. Table 30.3 allows us to look at resource allocations between different

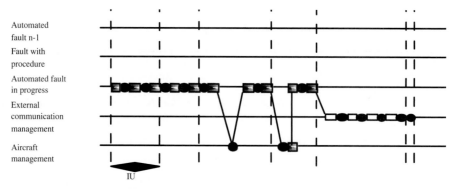

IU: Interaction Unit (vertical line separation)
Verbal exchanges:
● _CPT = PF
▣ FO = PNF
▢ ATC
○ Steward

FIG. 30.1.   Verbal reduced protocol activity profile of the crew one for an A/THR fault.

tasks that the pilots had to cope with during fault diagnosis treatment. The limitation of this methodology is that we just materialised the visible aspect of this cognitive treatment by using the pilots' communications and actions.

The aim of Table 30.3 is to materialise crew activity more quickly. This makes it possible to count the verbal exchanges in each interaction unit corresponding to each activity loop distribution (materialised by each line). This profile presents all verbal exchanges as having the same value as the others. This gives us a general fault treatment view according to the context changes (workload scenario difference and fault complexity variations) and crew resource distribution. Each line of the table will be like the previous table columns, except that we decided to add a line for the previous or simultaneous fault.

Step 6: Reduced Verbal Protocol Activity Profile

After the previous step (Table 30.3), we discovered that we had to deal with a new type of interaction unit—a mixed unit. These mixed units are interaction units divided into at least two activity loops in the same unit. This is, in our point of view, a way of materialising a parallel treatment of these activity loops. Even if we consider that these units are collective (two crew members), most of the time one crew member is still aware of what the other is doing, especially because these mixed units add fault treatment and ATC management or aircraft piloting tasks, which may be added to the diagnosis evolution, as shown in the result part. The crew member distribution is represented as detailed in the legend of Fig. 30.1. This allows us to see if the mixed units are mostly individual or collective.

A legend is specified in Fig. 30.1 with each interlocutor: CPT for captain (pilot flying, PF, or pilot not flying, PNF), FO for First Officer (PF or PNF), ATC for Air Traffic Controller. Each interaction unit is materialised by a vertical line. Arrows will be found at the end of each horizontal line when the profile continues on the following figure.

Step 7: Specific Profile for Understanding Process Visualisation

This profile illustrates the previous fault treatment column corpus with detailed categories. In the previous profile, time is considered to be constant; each communication is considered in the

TABLE 30.4
Detection Times (seconds) per Failure Type and Context (Workload)

Fault Types– Flight Phases	Ground Protection Warning System with Reflex Action to Operate	Standard Failures with Alerting System Active	Software Bugs with no Alerting System

Climb & Cruise
(medium & low
workload)
	n = 8	n = 14
		scenario 1:
		  A/THR fault
		scenario 2: Cab
		  Press Syst 1
		  fault; Cab Press
		  Syst 2 fault
			n = 9
			scenario 1: Mach
			  corruption; altitude
			  preset inop

	mean = 1.38 sec max = 2 s min = 1 s	mean = 1.57 s max = 4 s min = 1 s	mean = 2.22 s max = 10 s min = 1 s

Descent & final
approach
(medium & high
workload)
		n = 10
		scenario 1: flaps
		  jammed
		scenario 2: nose
		  wheel not
		  locked up
			n = 16
			scenario 1: speed
			  corruption;
			scenarios 1 & 2:
			  HDG,TRK–VS,
			  FPA reversion

		mean = 16.1 s max = 78 s min = 1 s	mean = 17 s max = 214 s min = 1 s

same way as another. Categories were already defined in Step 4. The aim of this profile is to see the distribution of categories between action and comprehension during a diagnosis accorded to situation dynamics. To facilitate the profile comprehension, the legend is as follows: each category is a small square unless this represents an action like disconnection (empty square) and connection (black square). The continuous line illustrates the continuous treatment of the fault and the dashed line illustrates the discontinuous treatment one. Each interaction unit end is materialised by a vertical dashed line. Each action that does not concern fault treatment is represented as an empty space up to three actions and, thereafter, like this: . . . // . . . . Whenever a crew is managing another fault we put an empty circle (not in these examples).

We identify a longer explicit detection time with computer malfunctions, which do not have an alert; the system doesn't recognise that it is faulty.

As predicted, the results show that detection time variations (Table 30.4) depend on workload increase. The main limitation is that we do not know if the pilots reacted as soon as they detected the fault, or if they delayed their reactions. However, heavy workload management could also generate this delay.

What surprised us in these results is more related to standard faults (with procedure)— detection time is also dependent on workload level.

Closure Time

Figures 30.2 and 30.3 illustrate that, for every fault, it takes more than a few minutes to control it and discuss consequences on the aircraft piloting task.

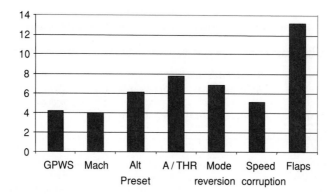

FIG. 30.2.   Fault scenario 1 closure time mean values.

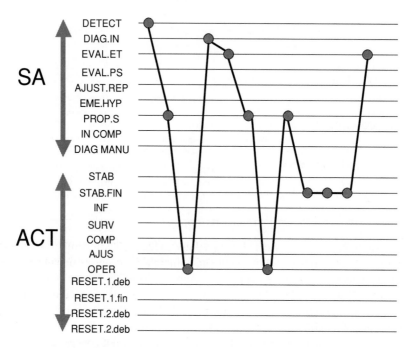

FIG. 30.3.   A/THR specific profile for understanding process visualisation, Crew 1.

We need to make it clear that our measurements do not consider that the pilots keep on thinking of the fault when they do not talk about it any longer. This limitation could be discussed in the sense that this could be on their minds without any possibility for us to identify it unless they do or say something. Nevertheless, we assume that with a high level of workload they might not have many resources to allocate to this thinking.

As we can see, the closer the fault appears to the end of the flight, the longer it takes to be treated. We suggest that this result is related to the workload increase, ATC request increase, and approach and landing preparation. We noticed that A/THR and reversion mode faults are more important concerns, with the second fault being slightly less than the first one (we voluntarily do not consider the flap fault, which continues until the end of the flight).

The fault common to the two scenarios shows us a considerable increase in closure time, which illustrates crew fault treatment (Fig. 30.2 & Fig. 30.4). We could underline that workload

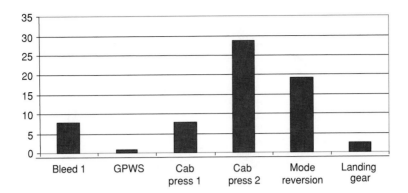

FIG. 30.4.   Fault scenario 2 closure time mean values.

and fault types seem to have an impact on closure time (times are much longer in the second scenario). Nevertheless, we note that the reversion mode fault time also decreases in this scenario.

From a global point of view, it is important to emphasise that all crews finished the flight in acceptable safety conditions. This type of diagram is a way of compacting crew strategies as an answer to a fault without any procedure to help them to deal with it (Fig. 30.4).

Coping Strategy Diagram

In the A/THR fault treatment, we could maintain that half of the crews who experienced the fault paid attention to transient A/THR information (1-s appearance) on the Flight Mode Annunciator (FMA) Display (Fig. 30.5). As we can see, this display seems to help them to generate hypotheses and consequently take actions on a more global automated system in the hierarchy. This display was a 2-second flash of A/THR information on the FMA (where active or prepared modes are indicated) when the crew was trying to reengage the automated mode. A crew who looks for clues to diagnose the fault picks up this sort of information. That small amount of information could, in some conditions, imply a different impact. Another result that we could emphasise is that most of all crews looked for breakers before making the decision to disconnect the automated system. They also did not go back to basics immediately as required, as all actions before this decision show us on the diagram.

The following fault, "mode reversion," is an even better illustration of this kind of strategy (Fig. 30.6).

No crews went back to basics immediately. Some crews did apply the generic procedure but went back to faulty automated management. During debriefing some who didn't apply the procedure were invoking excessive workload to cut of this kind of "helpful system."

This first level of analysis confirms our more global assumption, which is that pilots do not apply the procedure directly in the event of an automated system problem. We mean that they did not identify it and then immediately disconnect the system. This is confirmed by the time measurement that we have already presented.

In our next analysis presentation, we try to identify more precisely the treatment they engaged and the part of the understanding process involved in the global crew activity.

Verbal Reduced Protocol Activity Profile

The aim of Fig. 30.6 is to illustrate the fault treatment distribution throughout the crew activity. By the other materialisation used Step 6 (see Fig. 30.1) we moreover have access to the

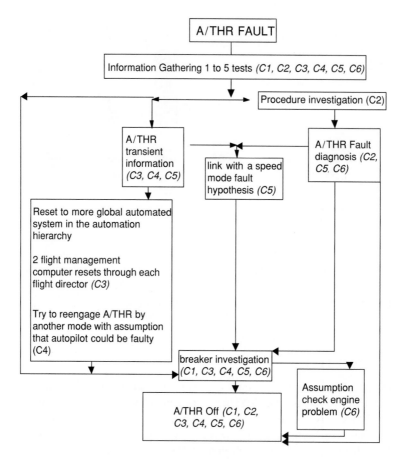

FIG. 30.5.   Coping strategy diagram for autothrust failure (scenario 1).

crew resource allocation during task sharing. Owing to the two previous analyses and materialisation, results show a significant difference in resource allocation management between the two scenarios (different workload in each) for the same fault treatment. All other profiles demonstrate the same global pattern (all profiles are available in the appendix of Plat, 2001).

Verbal exchanges are considered here as a work-density indicator. As we can see, there is a main difference between mode reversion treatment in scenarios 1 and 2. In the first one, the comprehension process is more sequential. The second scenario shows more changing between the different activity loops with a will to maintain the understanding process, even if the workload seems to oblige the pilot to attend to other tasks as well. As a consequence of this will, the global treatment is much longer in the second scenario. We could also compare the two fault issues from the same scenario: A/THR and mode reversion. These two faults show us that resource distribution seems to be quite similar.

To confirm these results we counted the verbal exchanges of all crews.

Fault Within the First Scenario Comparison: A/THR and Mode Reversion

Resource allocation distribution between these different activity loops seems to be quite similar, as shown in Table 30.5.

- Do we assume that a parallel management of loops is present, or do we show a sequential management?

TABLE 30.5
Verbal Exchange Distribution on Activity Loop for A/THR and Mode Reversion Faults
(Scenario 1)

Scenario 1	A/THR Fault (%)	Mode Reversion Fault (%)
Fault with procedure	0	0
Automated fault in progress	69	65
External communication management	12	19
Aircraft management	19	16

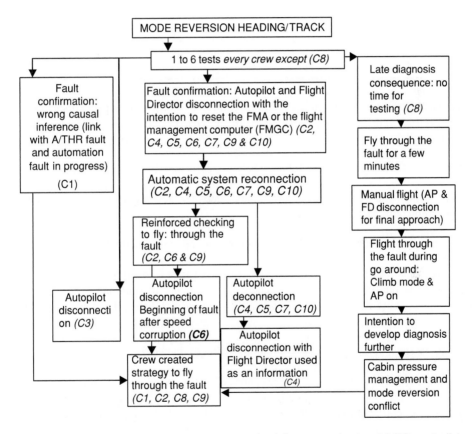

FIG. 30.6. Coping strategy diagram for mode reversion failure, scenarios 1 and 2 (AP = autopilot; FD = Flight Director).

We count a mean number of verbal exchanges within each interaction unit to highlight the mixed units with respect to the other unit distribution in the loop. We assume that these mixed units represent a parallel crew treatment of at least two or three loops at the same time (usually two).

Dominant units are units about fault treatment for both faults; for A/THR, fault treatment takes from 48% to 95% of crew activity compared with mode reversion, which shows 44% to 78% of crew activity within the same scenario. Another result is that units on the A/THR fault

TABLE 30.6

Verbal Exchange Initiator for A/THR and Mode Reversion Faults in the First Scenario

Verbal Exchange Initiator for A/THR Fault	C1	C2	C3	C4	C5	C6	Total
CPT	3	15	9	8	5	6	46
FO	1	2	5	4	0	1	13
Total	4	17	14	12	5	7	59
Verbal exchange initiator for mode reversion fault	C1	C2	C3	C4	C5	C6	Total
CPT	3	16	6	3	5	11	44
FO	2	7	4	6	5	1	37
Total	5	23	10	9	10	12	113

treatment are longer, especially when crews are gathering information on paper documentation. More mixed units are present in the A/THR treatment. That means to us that this fault is managed more in parallel than the other fault.

- Is there a balance in verbal exchange between crew members in the cockpit?

We observed that there is quite a good balance between crew members for every crew with a really slight dominance of Captains.

- Is this balance confirmed when we count interaction unit initiators?

As we can see, there is a considerable difference between Captain (CPT) and First Officer (FO) verbal exchange initiation (Table 30.6). Captains are mostly dominant. The A/THR fault shows that captains are dominant even when the First Officer is the pilot flying, which is not the case in the mode reversion fault.

- Do we show continuous or discontinuous verbal exchanges?

Results show that continuous verbal exchanges are more present in the A/THR fault than in the mode reversion one, which is already judged as continuous in both scenarios. One crew (C5) focussed on a continuous treatment without any interruption (60 successive verbal exchanges).

Same Fault in Both Scenario Comparisons: Different Workload Environment Effects

- Is there any verbal exchange density difference?

As we can see in Table 30.7, scenario 1 shows that the crews talk more than in the second scenario about an automated fault in progress. Crews in second scenarios concentrate less on this topic and they dispatch more to other parallel activities.

- Do we assume that a parallel management of loops is present, or do we show a sequential management?

TABLE 30.7
Verbal Exchange Mean for Crew Activity Loop in Both Scenarios

Verbal Exchange Mean for Each Crew Activity Loop	Scenario 1 (Intense but Sequential Workload Increase)	Scenario 2 (Heavy Workload)
Fault with procedure	0	25.3
Automated fault in progress	46.67	36.5
External communication management	13.3	51.5
Aircraft management	11.7	131.5

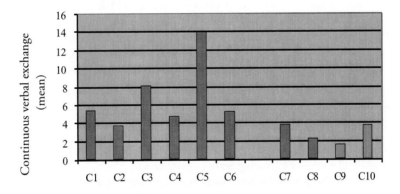

Crews C1 to C6 (Scenario 1) & C7 to C10 (scenario 2)

FIG. 30.7.   Continuous verbal exchange comparison during mode reversion fault treatment in the two scenarios.

When comparing the results of the two scenarios, we see that bigger variations are on the distribution of the way they manage the aircraft and the mixed interaction units. The variation is in favour of the second scenario, which illustrates a bigger parallel treatment and also a bigger consideration of aircraft management.

- Do we assume that a parallel management of loops is present, or do we show a sequential management?

We will show in Fig. 30.7 an account of continuous verbal exchange to illustrate by comparison the way to manage, resources with different workloads (Sc 1 = intense but sequential workload increase; Sc 2 = heavy workload).

Results show a quite different way of managing the fault in the two scenarios, which represents different workload environments for the same fault. In the first scenario, crews seems to focus more on fault treatment even if they still have the other activity loop to deal with. The second scenario illustrates shorter and more discontinuous verbal exchanges.

- Is there a balance in verbal exchange between crew members in the cockpit?

Results show that First Officers globally talk less than Captains, except for crew 1. The second scenario seems to confirm this tendency even better.

- Is this balance confirmed when we count the interaction unit initiator?

The difference is really more evident here in the second scenario, where mostly Captains initiate the interaction unit exchanges. This represents a bigger perception of complexity for us. Dörner (1980) and Kuklan (1988) showed that, in mediation situations, pilots took over control when the situation was perceived as complex because of its uncertainty or its intrinsic difficulties. This control taking by Captains could be analysed as showing a feeling of losing mastery of the situation and a will to keep control as a responsible member.

The same fault and the different treatments of the two scenarios show us that the activity loop distribution is different because of the workload variation with a will to keep an understanding process in progress. The heavier the workload, the more divided and shared the comprehension activity. However, this process will nevertheless be kept by the crew. This phenomenon is illustrated by longer profiles in the second scenario. Fault complexity implication in resource allocation variation is not clearly evident. Nevertheless, the workload increase related to flight phase evolution seems to produce the same phenomenon as the one observed through the two scenarios. Taking what we previously explained about this into account, Captain dominance on communication initiations seems to be predominant in A/THR fault and second scenario mode reversion fault treatments. Moreover, in the A/THR fault, Captains are dominant even if the First Officer is in charge of the flight. This fault, which seems to be more familiar than the other one, was trickier for the crews. This familiarity reinforced the solution as a procedure or a personal search for solution because the crews were thinking that it could be resolved.

A/THR Specific Profile for Understanding Process Visualisation (see Fig. 30.3)

The aim of these profiles is to illustrate with more dynamics the action–comprehension interaction process, with categories showing activity evolution.

Results show that there is no linearity such as comprehension and then an action process, but much more an interaction in between. We emphasise that an important process oriented to building a mental model is present but this process is also actions oriented in order to help building this comprehension process. These last actions seem to reinforce the comprehension process just as if, in a dynamic situation, actions were incorporated into the comprehension process.

Action–Comprehension Category Distribution

A/THR fault: As in the previous profile treatment, we counted category distribution.

These data confirm the importance given to comprehension for A/THR and mode reversion faults. The A/THR global intervention mean value is 19.67 (± 4). This mean is still smaller than other mean value of mode reversion in each scenario, that is, scenario 1 (26.5 ± 5) and scenario 2 (34.25 ± 6).

The SA category distribution in the A/THR fault results shows the following.

1. Every crew makes more than one solution proposal.
2. There is no personal situation evaluation.
3. Every crew detects the fault explicitly.
4. Every crew checks the system state, except crew 6.
5. Crews 3 and 5 are the only ones to make a manual diagnosis.
6. Crews 2 and 3 are the only ones to say they do not understand something.
7. Only crews 3 and 5 make assumptions.
8. Crews 2 and 5 update the common representation.
9. Crews 1 and 6 are the only ones to make an initial diagnosis.

The ACT category distribution in the A/THR fault results also shows the following.

1. Every crew operates quite often.
2. Every crew gathers information on paper and uses electronic documentation because they think there is a procedure to apply.
3. There is no act to help to understand.
4. There is no back to the manual procedure applied, as we have already noted (AP + FD + A/THR off).
5. Crew 3 only will try an electrical reset (level 2 for us).
6. Crew 3 reset an automated system by disconnecting and connecting it again (level 1) to reboot it.
7. One crew only (crew 5) tunes its representation with actions to support it.
8. Only one crew (crew 5) does a check action.
9. Most of crew will try to stabilise the situation to be able to find a solution.

The fact that most crews think they can find a solution maintains the comprehension. Thirty-three percent of the crew check the situation often (six or seven times). Information gathering is particularly intense for 40% of the crews. One third of the crew deals a lot with automated system reconnection as an operation.

The only crew that talked about going back to manual operation is the one that emitted a noncomprehension category. Every crew proposed a solution. There is no effective return to manual operation in this fault. That means that all crews tried to understand the situation and apply the solution prescribed. There is no action to help the understanding process, contrary to the mode reversion fault, which we judged as more complex. Results show no major transformation with the complexity variation as we programmed. In fact, a fault, which seems to be familiar but is not, is much trickier for pilots who activate the wrong resolution scheme (in the sense of Piaget, 1972) to resolve the problem. Even if we can underline a minimal common base in the understanding process, we have to explain that there are many individual variations in depth concerning crew problem solving.

Mode Reversion

Table 30.8 shows the global distribution of action (ACT) and comprehension (SA) categories for each crew during a mode reversion fault (scenarios 1 and 2). These data reinforce our thought that comprehension represents an important part of activity during fault treatment. As we already mentioned, global category mean values are higher. Scenario 2 global categories confirmed the increase of comprehension treatment with heavy workload. Three crews from scenario 1 use many more SA categories, three others (both scenarios) use more ACT categories, and the others are in between. Because of this action–comprehension category relationship, we just underline a difference in global strategy to cope with the situation. One is more thought based, and the other could be qualified as trial-and-error-based behaviour. With crews who are more thought based, one had flown through the fault without going back to manual operation (crew 1); the other two (crews 5 and 6) tried to do so as well, but they finally partially returned to manual flight without autopilot. Trial-and-error crews (crews 2, 8, and 9) all flew through the failure without going back to manual operation.

If we detail SA categories, we can underline that all crews from scenario 2 did a back to manual diagnosis but without action behind it. In comparison, scenario 1 crews never communicated this manual diagnosis (except crew 6). An important result is that crew 3, who is the only crew who went back to manual operation quite easily, is also the only one who does not have any solution proposal. Crews 1 and 4, who did not show any noncomprehension category, both pursued a long comprehension process (according to the coping strategy diagram). Three

TABLE 30.8

Number of Categories Concerning SA and ACT for Each Crew, A/THR, and Mode Reversion Fault Treatment

	A/THR Scenario 1						
	C1	C2	C3	C4	C5	C6	Total
SA	7	10	16	13	14	4	64
ACT	5	7	13	14	9	6	54
Total	12	17	29	27	23	10	118

	Mode Rev. Scenario 1						Mode Rev. Scenario 2				
	C1	C2	C3	C4	C5	C6	C7	C8	C9	C10	Total
SA	10	20	9	10	25	18	16	25	15	22	170
ACT	6	21	10	9	9	12	12	30	37	19	165
Total	16	41	19	19	34	30	28	55	52	41	335

crews (crews 4, 5, and 7) communicated assumptions; they all made a deeper situation analysis. In fact, they involved more global automated systems in the hierarchy of automation that the ones directly implied in the fault. They then went partially back to manual (AP off; FD and A/THR on). Without exception, all crews made a system state evaluation.

ACT category results show that there is no electrical reset (level 2 reset) action and some level 1 reset (automated system off–on). Operative action ACT.OPER seems to be much more present in scenario 2 crew activity than in scenario 1 crew activity. Comprehension-oriented actions are made to test the "bizarre" system behaviour. We think that such actions could help assumption emission. Action tuning, to illustrate comprehension tuning of the situation, is present in only half of the crews who emitted "SA.AJUST.REP" categories. There was no information gathering action during this fault treatment compared with A/THR. It seems to be clear that, for the crews, there was no specific procedure to help them as an existing solution (except a generic one to be specified), or the crew preferred to search for their own solution to apply. We observe that all crews took action to stabilise the situation; this phenomenon is amplified in scenario 2. We found final stabilisation actions with almost all crews, but they did go back to an automated flight.

DISCUSSION AND CONCLUSION

A Continuous Process

We demonstrate that every pilot was involved in a more or less important understanding process during these fault treatments. The procedure application, as required, would have produced a different activity: identifying the situation in a category, and then applying the associated prescribed solution without rebuilding the missing part of the occurring representation of a situation that the incident interrupted when it occurred. This is not the way that pilots choose to operate. They have at least to test the system before. Rasmussen (1986) described the "hypothetico-deductive" method that some pilots apply, which seems to be similar to what we observed. The pilots engaged themselves voluntarily in a problem solving process in order to apply the prescribed solution. We maintain that applying the prescribed solution is like accepting a change in point of view without running risk the of being stuck at an impasse.

We observed that pilots rather prefer to search for another solution than the one prescribed, because they think that they can find a better one according to their own goals. One of these goals is to stay within an acceptable framework of resource allocation. To attain this goal, they know that automated systems act as help, so they have a paradoxical situation to deal with. Nevertheless, they will try to find a solution to keep automated systems as help. That is why they keep the understanding process, even with heavy workload. They will adapt their behaviour, choosing to postpone a deeper diagnosis and deal with more than one activity loop at a time. This behaviour could be observed in situations other than aircraft control (e.g., control of cars or nuclear power stations), where a procedure acts as a short circuit in the user's activity. The designers must therefore take this behaviour into account as generic human behaviour, and not expect the procedure to be applied to the letter. The applicability of a procedure raises the problem of identification of the problem situation and the procedure category that refers to it. This problem is not simple to identify, even if many other elements come into play in the user's compromise.

The Depth of This Process Depends on Cognitive Resources

In the second scenario, activity during fault treatment is much more oriented to the comprehension process than in the first scenario. Rasmussen (1986) emphasised, in an emergency situation, a possibility for the pilot to postpone a deeper diagnosis process. We think that our emergency situation is related to fast action tempo and parallel management of different activities, which are resource consuming. This must have an effect on comprehension management. We judged it pertinent to study this from the point of view of cognitive complexity generated by the pilots. We should then be able to explain interaction between these variables, because we already know that complexity generates multiple goals. Moreover, the piloting activity we are studying is already identified as a multiple goal activity itself. As we already know, comprehension creates coherence from selected facts according to multiple goals and objectives related to temporal constraints. Cognitive costs increase when action planning begins to be difficult to manage. It is typically the situation we are looking at. That is why we thought that cognitive complexity was a pertinent variable to study.

Amalberti (1996) emphasised that the comprehension process depends on representation rebuilding, but also on a comprehension level choice. Moreover, Spérandio (1977) maintained that there is a workload effect on the operating regulation process when one is dealing with a dynamic environment. These effects are materialised by a lower performance objective chosen by the pilots themselves. Our results confirmed these views. Nevertheless, when people identify workload as a visible variable like "downgrading performance " or "error increase," we like to underline that this represents the normative point of view. It does not represent our point of view here, even if this is still a dominant way of looking at human resource management.

Our analysis shows us that, before accepting performance downgrading, pilots adapt their activity as much as they can (e.g., to keep an understanding process). Loss of situation mastery, as a feeling, could indicate to the pilot that he or she is close to reaching his or her own limit. Beyond this limit the pilot will not be able to manage his or her activity with an acceptable level of performance. The back to basics decision could be perceived by the pilot as activity downgrading acceptance, but is still a way to keep a situation under control by changing the performance goal level. We define this as activity downgrading, because almost all pilots are aware that their manual skills are weakened when they lack practice. Pilots jump from a cognitive compromise regulated by resource, risk, and comprehension management to another.

This description of activity shows that the normative point of view is too restrictive and represents a poor view compared with the complexity of real activity. In fact, performance degradation is just a result, but it does not say anything else about the pilots' activity before reaching this conclusion. Our situation illustrates a lower performance objective chosen by the pilot about the comprehension process, which means that a more superficial but sufficient process will be involved to manage the situation.

Hoc (1996) maintained that it is important to present causal structures to pilots for processes that are temporally marked, because such an illustration could help them make a diagnosis. By this, he underlines that comprehension activity could be better when one is analysing the cause of problems. But in fact, as we show, the pilots do not make a causal diagnosis as we could observe in a situation with less constraint of time. So, we maintain, it is not good to help pilots by a design to reinforce causal reasoning in an emergency situation. Our point of view is that this presentation will focus them on a deeper comprehension process that they cannot afford at this time. Nevertheless, this could be interesting afterwards when fault treatment is finished. In another way, Kantowitz (1994) demonstrated that, when pilots are conscious of their mental confusion, they generate more workload but also less confidence in automated systems. If this is still true in our situation, we could think that pilots would rather prefer to keep an understanding process, even if it is a superficial one. According to previous ideas, we define a dual-level time process. The first time level shows an economic adaptation of the operating process within optimised resource management. A second time level is marked by a lower performance objective choice, through internal risk management, which corresponds to noncomprehension acceptance.

We could underline that even if fault treatments are longer in the second scenario because of interrupted comprehension activity, this process seems to be less deep than in the first scenario. All crews propose solutions, which could illustrate that, as long as crew members think they could reach a solution, they will keep automated systems on. The mastery of the situation, described by Amalberti (1996), is illustrated here by this process of pursuit of a solution. As we already described, cognitive complexity illustrated by loss of the feeling of mastery of the situation acts as a stop indicator for the comprehension process. In fact, the more the cognitive resources are involved, the more the noncomprehension cognitive items are stocked. Stocked items could not be treated because of a lack of resource availability, and this also increases cognitive complexity. According to Rasmussen (1986), the transition between a low and high abstraction level consumes different resource costs. Surprisingly, the problem solving process, which is costly, is not considered by pilots as a constraint and a resource consumption, compared with the loss of the help of automated systems. Kantowitz (1994) seems to be right—this understanding process makes them feel as if they keep everything under control and gives them the feeling that the workload is slightly decreasing. That is why system interface and procedure designers must take the management of these resources into account by testing their products in situations close to the user's real activity in order to identify the impact of their creation on the user's activity as far as resource management is concerned.

A Process Depending on the Pilot's Own Procedure Building

The pilots show their will to maintain automated systems by attempting to restore the situation. We think this comes from a global cognitive scheme built by the global objective repetition of most procedures proposed to crews to deal with standard faults. This objective is more often to restore a system or to try at least to isolate the faulty system. This type of learned procedure could easily generate an organisational structure of cognitive activity, as a scheme and consequently be involved in pilot internal procedure creation when the pilot is dealing with similar situations. As we just explained, it could be important to anticipate such a way to operate by developing better knowledge of human behaviour and taking it into account when external procedures are created. These similar external procedures will be interpreted by pilots according to the situation and their own goals. We have to deal with the fact that the pilot will act in that way and that we have to take it into account by knowing such behaviour built by schemes internal to the pilot and operational constraints. Another point is that we have to consider pilot knowledge and information needs; the goal is not to encourage pilots to go deeper into the comprehension process but more to help them to identify the problem as being able to be solved, whether restored or not. This could be developed in training by operational knowledge on a simulator or in flight. This knowledge must be focussed more on automated system feedback and normal–abnormal behaviour of systems.

As we talked about causal diagnosis, we also have to take into account the consequences of a fault on resource management. This is important with regard to our results. The consequence of an A/THR fault is more problematic in terms of task management because the pilots will have to manage the thrust manually if this is not available, compared with the mode reversion fault, which is already a task that is not totally managed by automated systems (vertical and lateral navigation). In fact, the Air Traffic Controller asked the crew to make changes, which obliged them to manage the flight in the partially automatic mode, that is, the selected mode compared with the managed mode, which is considered a more automated way. Thus the gap in terms of resource allocation will be less important in terms of fault consequence task management. These kinds of things have to be taken into account by designers and training instructors while defining users' future activity elements.

A Collective Process

Cockpit crew communication is developed and regulated by procedure and crosscheck requirements compared with other activity domains. This requirement goal is materialised by an expression of the crew's own view, which develops a better common representation. This allows us to think that operational communication in the cockpit gives us an idea of pilots' representation. In another way, an activity analysis within our method shows resources allocated to different activity loops and a comprehension–action interactive structure during a diagnosis. This structure materialisation shows the importance of cognitive complexity within cognitive resources and time constraints of the pilot in the situation, and their impact on the comprehension process.

The collective point of view is a good way to see the common building of shared representation. In our view, a collective point of view on the activity is very important to be incorporated into cognitive methodologies and models. This is particularly true when users (such as pilots) are dealing with a multiagent environment. These models participate in building recommendations for design, training, and procedures, which will be used in a collective environment. Complexity acts as a regulator of depth of comprehension involvement, but it is still variable from one pilot to another. This is partly due to experience evolution. According to this variability, our invariant results take on an even more important dimension.

Experience evolution also depends on recurrent training, which is more and more restricted for financial reasons. Concerning this restriction, we have to be aware of the long-term consequences, especially as regard to manual ability to pilot the aircraft, which is also affected by this restriction. This could help in understanding the will of pilots to keep automated systems. The loss of manual ability must also question all aviation community players about the consequences of this loss and the growth of automated systems to help pilots during their everyday tasks. When automated systems are faulty, pilots have to use manual abilities. If they don't rely on manual flight skills frequently, they may not have complete confidence in these abilities. The cost of this can be high. We have to consider all aviation domain consequences of each decision when building systems, training, and procedures. Pilots have to consider it every day when they operate—they don't have the choice; they have to deal with it.

The industry is asking pilots to regulate the system when there is something that could not be managed by automated systems, for example, so pilots have to be more clever than a very good system and usually have to proceed quickly. Thus, they do need adapted training and understandable and easy to operate procedures and systems. That is why it is very important to take into account, as soon as possible, the pilots' natural way of acting and schemes internal to pilots and their communities when new systems are designed. This has to begin with a real need analysis and an iterative design process on research and integration simulators where new designs are tested with pilots. Moreover, instructors could provide a good feedback from difficulties to use some systems and the ease of learning about them.

As in every experimental study, the population of pilots has to be representative of different levels of experience, and different knowledge backgrounds must be representative of a real user population. It is important not to design only for really good pilots but also for every kind of user, because the understanding process and the way to operate systems depend on knowledge and the particular scheme. The aviation domain has an advantage compared with other domains (such as the automobile domain) in dealing with a professional population, which means that abilities are not too different from one limit to another and so could be taken into account in the design phase. Different workload increases and situations of stress have to be integrated in the tested scenarios to see how new designs could be considered as resistant to an activity increase as it appears in real incidental situations.

⁻Moreover, as an indispensable complement, training is a way to help pilots deal with every situation. This could be a good issue to improve system operation. That's why we underline that developing what the gestalt theory called "creative thinking" would help pilots to cope with rare situations. Today's training is more focussed on "reproductive thinking" based on procedure repetition. We are not saying that it is not good to use reproductive thinking, but rather that creative thinking will be a good complement, and that, in our point of view, they are not exclusive. The bad point of reproductive thinking is that it helps analogical transfer and restricts cognitive ability by the functional steadiness phenomenon. In contrast, creative thinking is an encouragement to focus the training more on the pilots' understanding of how the automated system functions, to allow them to deal with rare faults. This way helps pilots build their own procedure when necessary and encourages insightful behaviour (to reconsider situation elements). This could help pilots to get around an impasse or a situation that does not allow a solution to be reached. Creative thinking seems to be more suitable to new problem solving. In another way, to develop communication within the cockpit is a tool that helps pilots gain some distance from the problem-solving process, which avoids locking them into actions. That is why it is also important to develop shared representation in an operational training situation.

All of these conclusions enable us to say that close collaboration with human factors specialists during the design evolution cycle of a product (here, e.g., a cockpit) remains important in order to achieve the final product for the users and thus to minimise the changes to be made on already existing objects or along the object learning period by the users. Changes are called retrofits, which are usually expensive for the industry. However, this remark should be amended: a retrofit can also be the result of an evolution in technology and thus provide an improvement that could not have been envisaged at the time of initial design of the object. In contrast, an intuitive conception of objects permitting easy insertion into the user's professional activity remains a task to be performed in cooperation with designers, human factors specialists, and users.

REFERENCES

Amalberti, R. (2001). La maîtrise des situations dynamiques. *Psychologie Française, 2.*

Amalberti, R. (1996). *La conduite des systèmes à risque.* Paris: PUF.

Amalberti, R. (1994). *Briefings, a human factors course for professional pilots.* Paris: IFSA-DEDALE.

Brito, G. de (2000). *Analyse ergonomique du Suivi de Procedures Ecrites dans les Environnements Dynamiques (SPEED) appliquée à l'aéronautique.* Thèse de doctorat de psychologie cognitive spécialité ergonomie, Université Paris V René Descartes.

Corwin, W. H. (1995). Understanding mode annunciation: What is the pilot mental model. In *Proceedings of the 8th International Symposium on Aviation Psychology,* Columbus, OH, April 1995, Vol. 1, pp. 249–253. Ohio State University Aviation Section.

Dörner, D. (1980). On the difficulties people have when dealing with complexity, *Simulation and Games, 11,* 87–106.

Hoc, J. M. (1996). *Supervision et contrôle de processus, la cognition en situation dynamique.* Grenoble: Presse Universitaires.

Javaux, D. (1997). Measuring cognitive complexity in glass-cockpits. A generic framework and its application to autopilots and their modes. In R. Jenson & L. Rahovan (Eds.), *Proceedings of the Ninth Symposium on Aviation Psychology, OSU'97* (pp. 397–402). Columbus, OH. Ohio State University Aviation Section.

Javaux, D. (1998). The cognitive complexity of pilot-mode interaction; a possible explanation of Sarter and Woods' classical results. In *Proceedings of HCI-AERO'98* (pp. 49–54). Montreal: Polytechnique International.

Kantowitz, B. H. (1994). Pilot workload and fly deck automation. In R. Parasuraman & M. Mouloua (Eds.), *Human performance in automated systems: current research and trends* (pp. 212–223). Hillsdale, NJ: Lawrence Erlbaum Associates.

Kuklan, H. (1988). Perception and organizational crisis management. *Theory and Decision, 23*(3), 259–274.

Piaget, J. (1972). *L'épistémologie génétique.* Paris: PUF.

Plat, M. (2001). *Choisir de comprendre ou décider d'agir en environnement dynamique: le cas de l'activité de pilotage en situation incidentelle atypique.* Thèse de doctorat nouveau régime en psychologie cognitive. Université Paris VIII.

Plat, M., & Amalberti, R. (2000). Experimental crew training to automation surprises. In N. Sarter & R. Amalberti (Eds.), *Cognitive engineering in the aviation domain* (pp. 110–123). Mahwah, NJ: Lawrence Erlbaum Associates.

Plat, M., & Rogalski, J. (2000). Traitement de dysfonctionnement d'automatismes et modes de coopération dans le cockpit. In T. H. Benchekroun and A. Weill-Fassina (s/d), *Le travail collectif: perspectives actuelles en ergonomie* (pp. 135–163). Toulouse: Octarès Editions.

Rasmussen, J. (1986). *Information processing and human machine interaction.* Amsterdam: North-Holland.

Rogalski, J., & Samurçay, R. (1993). Analysing communication in complex distributed decision making. *Ergonomics, 36*(11), 1329–1343.

Sarter, N. B., & Woods, D. D. (1992). Pilot interaction with cockpit automation: Operational experiences with the Flight Management System. *International Journal of Aviation Psychology, 2*(4), 303–321.

Sarter, N. B., & Woods, D. D. (1994). Pilot interaction with cockpit automation II: An experimental study of pilots' model and awareness of the Flight Management and Guidance System. *International Journal of Aviation Psychology, 4*(1), 1–28.

Spérandio, J. C. (1977). La régulation des modes opératoires en fonction de la charge de travail chez les controleurs de trafic aérien. *Le Travail Humain, 40,* 249–256.

Spérandio, J. C. (1972). Charge de travail et régulation des processus opératoires. *Le Travail Humain, 35*(1), 85–98.

Wioland, L., & Amalberti, R. (1996). When errors serve safety: Toward a model of ecological safety, In *Proceedings of the first Asian Conference on Cognitive Systems Engineering in Process Control (CSEPC'96)* (pp. 184–191).

31

Team Design

Jan Maarten Schraagen and Peter Rasker
TNO Human Factors, The Netherlands

Abstract

Team design is about achieving a coordinated effort through the structuring of tasks, authority, and work flow. Just as with the design of an artefact or a technological system, team design affects the cognitive tasks of the members in the team. In this chapter, we provide an overview of criteria that are important to consider in team design. These criteria are derived from six large, multiyear research programs in the area of team design. On the basis of these criteria, we propose recommendations for team design and show how particular choices for a certain design affect the cognitive tasks of the team members. Seemingly disparate results may be reconciled by taking into account whether team members have to deal with unanticipated disturbances in the environment or not. Team structures that are better suited to deal with unanticipated disturbances are characterised by four Rs: redundancy, robustness, review, and reallocation. The four Rs are likely all to be mediated by the single cognitive construct of "shared mental models." These in turn are facilitated by certain antecedents, such as continuity of personnel, cross training, self-managed teams, consensual leadership, and organisational slack.

INTRODUCTION

Team design deals with the various ways in which teams can be designed, taking into account numerous factors moderating or mediating the resulting team performance. This is a relatively new and unexplored area of research. Therefore, there is a lack of psychologically motivated and empirically validated principles for team design. The purpose of the current chapter is to assess the state of the art in the area of team design, propose recommendations for team design, and show how particular choices for a certain design affect the cognitive tasks of the team members.

This chapter is structured as follows. The section following this one provides an overview of major schools of thought in the area of organisational design. The purpose of this section is to define the terminology and provide the reader with some background on different schools of thought that have emerged in the 20th century. We conclude that the dominant school of thought in the area of organisational design and team performance is the open systems school of thought. This school of thought places a large emphasis on the characteristics of the environment in which the team operates, and it views a team as a complex set of dynamically intertwined and interconnected elements that are highly responsive to their environment.

The next section takes as its starting point the open systems school of thought and describes a basic input–process–output model for team performance. This model is elaborated by incorporating a recently developed cognitive construct: shared mental models. We use shared mental model theory to illustrate the importance of shared cognitive structures in teamwork. We focus on two important cognitive team processes: communication and backup in high-workload situations.

Having set the stage theoretically, the next section provides a selective review of empirical studies on team design. The purpose of this review is to derive important criteria or elements in team design. The reader who is not interested in a detailed discussion of outcomes of empirical studies may skip this section and move directly to the penultimate section, where the major conclusions of the empirical studies are critically evaluated and recommendations for team design are provided. This section also shows how changing team organisational structures affects working conditions for people in the system, and how these changes then affect the cognitive tasks that are required. We illustrate these recommendations with case studies and examples. The final section provides a conclusion that ties the chapter together.

TEAM DESIGN: SCOPE OF THE AREA AND SCHOOLS OF THOUGHT

The area of team design is not as well defined in the literature as the area of organisational design. Szilagyi and Wallace (1990) define organisational design as ". . . the process of achieving a coordinated effort through the structuring of tasks, authority, and work flow" (p. 618). The same definition could be applied to the area of team design, with teams being a lower level unit in the organisation as a whole. In fact, Paley, Levchuk, Serfaty, and MacMillan (1999) described the team design process as an algorithm-based allocation among the mission tasks, system resources (e.g., information, raw materials, or equipment), and the human decision makers who will constitute the team. The result of the team design effort is a team structure that specifies both the structure and the strategy of the team, including who owns resources, who takes actions, who uses information, who coordinates with whom, the tasks about which they coordinate, who communicates with whom, who is responsible for what, and who shall provide backup to whom.

Current approaches to organisational design or team design mostly fall in a school of thought labelled as contingency theory. This theory views organisations or teams as open systems (Katz & Kahn, 1978). The open systems school views an organisation as a complex set of dynamically intertwined and interconnected elements, including its inputs, processes, outputs, feedback loops, and the environment in which it operates. The classical school, dominant in the first half of the 20th century, viewed organisations as static structures, basically ignoring the dependence of organisations on inputs from their environments. During the midcentury decades of rapid societal change, the inadequacies of closed system thinking about organisations became increasingly apparent. Subsequent large-scale empirical studies showed the following (Galbraith, 1973):

- there is no one best way to organise;
- all ways of organising are not equally effective.

These two conclusions form the basis for contingency theory. This school of thought states that we can observe a wide range of effective organisations, but that the form of an organisation does make a difference. For instance, organisations facing simple and stable environments should consider more functionally oriented structures (e.g., classical bureaucracies), whereas organisations facing complex and dynamic environments should consider structures that are more product oriented or divisionally oriented (Lawrence & Lorsch, 1969). The open systems school, with its emphasis on input, process, output, feedback loops, and the environment, is also the dominant school of thought in the area of models for team performance (e.g., Hackman, 1990; Tannenbaum, Beard, & Salas, 1992).

Galbraith (1973) is an excellent example of the systems–contingency perspective on organisations. In his view, organisations are faced with varying information processing requirements, dependent on the degree of task uncertainty. Task uncertainty is defined as the difference between the amount of information required to coordinate cooperative action and the amount of information actually possessed by the organisation. Hence, if task uncertainty increases, the information processing requirements increase. Suppose there is a large task in which the work is divided up on the basis of input skill specialisation. A design problem arises because the behaviours of all the employees contributing to the large task must be coordinated. According to Galbraith, the design problem is to create mechanisms that permit coordinated action across large numbers of interdependent roles. Most important, perceived variation in organisation form is hypothesised by Galbraith to be a variation in the capability of the organisation to process information about events that could not be anticipated in advance. In other words, what we perceive as different organisational structures are strategies that organisations must choose to cope with the increased information processing requirements caused by increased competition, technological change, or, in command and control, increases in time pressure, and volume and complexity of information.

According to Galbraith (1973), any organisation faced with increasing uncertainty must adopt at least one of the following four strategies (the set of alternatives is supposed to be exhaustive).

1. Creation of slack resources: this will result in a lower level of performance. For instance, allowing more time for a decision to be made will result in less time pressure. This could come at a cost to the organisation, however.
2. Creation of self-contained tasks: all resource groups will be self-contained, so there will be no need to process information about resource sharing among outputs, and, because of reduced division of labour, less need to coordinate roles.
3. Investment in vertical information systems: the capacity of hierarchical channels of communication can be expanded, and new ones can be created, by adding information systems (e.g., global databases) or by formalising information transfer.
4. Creation of lateral relations: communication channels across lines of authority are established, increasing the ability to process more information. Examples of lateral relations are direct contact between managers, liaison roles, teams, and the matrix organisation.

If the organisation does not consciously choose one of the four strategies, then slack is automatically generated, and performance will decrease.

The sociotechnical systems approach to organisational design is similar to the open systems approach in its emphasis on the importance of environmental factors, and flexibility and adaptability. The sociotechnical systems approach is based on the principle that any organisational system requires both a technology and a social system linking employees with the technology and to each other. The integration of the technical and social system leads to three sociotechnical design components.

1. Organisational factors: emphasis is placed on controlling variance in the work cycles.
2. Group factors: autonomous work groups are established. Three conditions enhance autonomy:
   - Task differentiation—the group's task should be as distinct and differentiable from the work of other organisational units as possible (groups should produce whole products, with a wide rather than limited scope);
   - Boundary control—group members should be able to affect transactions with their environment;
   - Task control—employees should have control over the actual process of performing the group's work.
3. Individual factors: emphasis is placed on providing some meaningful work (variety), with inherent responsibility (autonomy), knowledge of the task and results (feedback on one's performance is essential), and opportunities for growth (group members should develop multiple skills).

Empirical evidence has generally supported sociotechnical systems theory. Meta-analyses of interventions based on sociotechnical systems theory have indicated a high degree of success. Pasmore, Francis, Haldeman, and Shani (1982) analysed the results of 134 studies and found that the majority had positive effects on criteria of productivity, costs, employee withdrawal, employee attitudes, safety, grievances, and work quality. Goodman, Devadas, and Hughson (1988) concluded that the use of work teams according to sociotechnical principles can raise productivity as well as attitudes. In the most recent review, Cohen and Bailey (1997) concluded that autonomy is associated with higher performance for work teams, but not for project teams.

Recently, a different school of thought has emerged: the complex adaptive systems view (Duchon, Ashmos, & Nathan, 2000). This view is rooted in chaos theory and quantum theory. It states that the future behaviour of a system cannot be accurately predicted from its present state. Because the environment is inherently unpredictable, the organisation must continually make sense of its environment if it is to survive (Weick & Roberts, 1993). Sensemaking is not so much about imposing order as it is recognising an emergent order. Teams may be the appropriate unit for making sense. The surfacing of conflicting perspectives is essential for getting to some new collective understanding. From the traditional perspective, on one hand, teams function best when conflict is minimised. The complex adaptive systems view, on the other hand, embraces conflict and urges team leaders to learn to actively surface conflict, for instance by employing the devil's advocacy approach or the dialectical inquiry approach. As agents of disturbance, team leaders must resist the temptation to influence team members to align with each other or the leader. Second, this view puts more emphasis on the team's "connectedness" rather than its "cohesiveness." When groupthink (Janis, 1982) occurs, it can be said that the team is cohesive, but acts heedlessly, and does not allow for diversity in self-identities. Organisations existing in complex environments must match the external variety with internal variety. Teams will be effective to the extent that they develop for themselves complex relationships among their members, with the organisation as a whole, and with the environment. Therefore, the complex adaptive systems view values heterogeneity more than homogeneity (similar values and similar goals).

Conclusions

This brief review of major schools of thought in the area of organisational design has shown that the dividing line between organisations and teams is a thin one. Teams are often microcosms for the organisation, in the sense that all values, norms, goals, conflicts, and paradoxes that play an important role at the organisational level also play that role at the team level. This is reflected in the models that are developed for organisational and team performance: in both cases, the same open systems view is the dominant school of thought. However, organisations are composed of groups of teams, and those groups can be assembled, and have to be coordinated, in various ways. A team, being a lower level unit of aggregation, cannot be assembled with the same degrees of freedom as the organisation. For instance, teams mostly do not have their own staff, have shorter lines of communication, and are subject to reward systems defined at the organisational level. Furthermore, teams are often viewed as one type of coordination mechanism (Galbraith, 1973), with hierarchies, liaison roles, task forces, and integrating departments being alternatives to teams.

Team design is therefore more restricted than organisational design, even though teams share a lot of characteristics with the organisations they are part of. Exactly what has to be considered in team design is determined by one's theoretical framework. The dominant information processing framework focuses on the mission tasks, system resources (e.g., information, raw materials, or equipment), and the human decision makers who will constitute the team. The resulting team structure specifies who owns resources, who takes actions, who uses information, who coordinates with whom, the tasks about which they coordinate, who communicates with whom, who is responsible for what, and who shall provide backup to whom. The less dominant complex adaptive systems view focuses on sensemaking, and it encourages conflict, connectedness, and heterogeneity. The complex adaptive systems view is a response to increasing unpredictability and complexity in the environment.

SHARED MENTAL MODELS, TEAM PROCESSES, AND PERFORMANCE

Recent literature has advanced the cognitive construct of shared mental models among team members as an underlying mechanism of team processes and performance in teams (Cannon-Bowers, Salas, & Converse, 1993; Rouse, Cannon-Bowers, & Salas, 1992). This construct has emerged from the literature on individual mental models (Rouse & Morris, 1986; Wilson & Rutherford, 1989) that are organised knowledge structures that allow individuals to describe ("what is it?"), explain ("how does it work?"), and predict system functioning ("what is its future state?"). Bringing the mental model construct to a team level, shared mental models are organised knowledge structures that allow team members to describe, explain, and predict the teamwork demands. The knowledge that is shared comprises the internal team (e.g., knowledge about the tasks, roles, responsibilities, and informational needs of the team members, interdependencies in a team, and the characteristics of the team members) and the external situation (e.g., cues, patterns, and ongoing developments). The explanations and expectations generated by this knowledge allow team members to anticipate each other's task-related needs by providing each other information, resources, or other support in time (Cannon-Bowers et al., 1993). In turn, this has a positive impact on team processes, such as communication and backup behaviour that results in an improved performance.

The way shared mental models affect team processes and performance is illustrated in the basic input–process–output model depicted in Fig. 31.1. The input comprises antecedents that may facilitate shared mental models. Antecedents include specific team training methods such

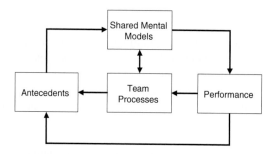

FIG. 31.1. Theoretical relationships among antecedents, shared mental models, team processes, and performance (adapted from Rasker, 2002).

as cross training, leader briefings, team planning, experience within the team, continuity of personnel, self-managed teams, consensual leadership, and organisational slack.

One of the most important team processes that is affected by shared mental models is communication. It is hypothesised that shared mental models allow team members to explain and predict the informational needs of teammates. Because team members rely on their shared mental models, communication can take place efficiently and effectively. Efficiently, because explicit and extensive communications to ask for information or to make arrangements concerning "who does what when" and "who provides which information when" are not needed. Effectively, because team members are able to provide each other with (a) the information needed to complete the tasks successfully, (b) without explicit communications, and (c) on the time in the task sequence of a teammate when this information is needed (Stout, Cannon-Bowers, & Salas, 1996). This is called implicit coordination, because team members exchange the necessary information and perform their tasks without the need for extensive communications to coordinate extensively (Kleinman & Serfaty, 1989). The result is the smooth team functioning of team members who are in sync with each other, and who know exactly when to talk and what to say.

Although shared mental models may result in efficient and effective communications, it is also hypothesised that communication is important for the development and maintenance of shared mental models (Orasanu, 1993; Stout et al., 1996). Communication during task execution refines team members' shared mental models with contextual cues. This may result in more accurate explanations and predictions of the teamwork demands (Stout et al., 1996). For maintenance purposes, communication is needed to keep the shared mental models up to date with the changes that occur during task execution. Especially in dynamic or novel situations, when teams have to deal with unanticipated disturbances, communication is needed to preserve an up-to-date shared mental model of the situation and to adjust strategies or develop new ones to deal with the situation (Orasanu, 1993). Shared mental models in changing and novel situations serve as an organising framework that enables team members to make suggestions, provide alternative explanations, employ their expertise, generate and test hypotheses, and offer information useful to determine strategies in that particular situation. In contrast to implicit coordination, which implies that mature teams are silent teams, this emphasises the need for explicit communication to arrive at a joint interpretation of the situation and the generation of strategies to deal with that situation.

It is also hypothesised that shared mental models have a positive impact on more generally formulated teamwork processes. The basic idea is that when team members have a common understanding of the team's goal and each other's roles, responsibilities, and tasks, teamwork is facilitated. Thorough knowledge of the teammate's tasks (presumably well structured and organised in a mental model) is needed to be able to monitor each other's performance, provide each other constructive feedback, and back each other up.

The shared mental model explains from a cognitive point of view what is required to design a robust and flexible team capable of handling unexpected disturbances in dynamic situations; team members must have sufficient information processing capacity available to be able to communicate extensively. These communications are needed to develop and maintain a up-to-date shared mental model of the ongoing developments in the team and situation. This is especially important in high-velocity environments containing uncertain and unforeseeable events. In such circumstances, team design must take into account that team members' individual tasks must not impose such excessive workload that team members cannot fulfill their teamwork responsibilities and communicate extensively about the developments in the ongoing situation.

The shared mental model also explains how the information processing capacity of team members can be broadened by limiting that type of communication that is not needed. Given the potential problems with communication in time-pressured situations, in that there is too little time to communicate and that it disrupts the individual task performance of team members (Hollenbeck, Ilgen, Tuttle, & Sego, 1995; Hutchins, 1992; Johnston & Briggs, 1968), this is highly important. Team members that can draw on their shared mental models are able to provide each other the necessary information in advance of requests. There is no need for explicit coordination, and team members can limit themselves to the necessary information exchange. In terms of team design, the necessary information exchange can be limited further by designing for a minimum of communication dependency among team members. This would leave team members free to perform their own tasks as well as they can. At the same time, this would leave as much spare information processing capacity available for that type of communication that is important for performance, namely for maintaining an up-to-date shared mental model of the situation and performing additional teamwork.

EMPIRICAL STUDIES ON TEAM DESIGN

We now turn to a selective review of empirical studies on team design. This review is selective in that we have focussed on large, integrated multi-year research efforts comprising multiple studies on team design in the area of command and control. The advantage of looking at larger research projects is that major themes and conclusions will emerge more clearly than when isolated experiments are looked at. Moreover, large research projects tend to employ the same research methods over a number of years, making results from various individual studies more easily comparable with each other. The empirical studies will be discussed in (rough) chronological order. The format for the discussion is as follows: background, method, major results, discussion, and implications for team design. In the discussion subsection of each paragraph, a critical look is taken at the research reviewed. We postpone an overall critical evaluation of the studies to the penultimate section (Integration of Empirical Studies on Team Design).

Briggs and Johnston

Background.   Briggs, Johnston, and coworkers at the Ohio State University carried out a 4-year program (1963–1967) of laboratory research on team training in a combat information centre (CIC) context (Briggs & Johnston, 1967). Their research was sponsored by the Naval Training Device Center, Orlando, Florida. The intent of the contract was to provide fundamental data by means of laboratory experimentation that would have implications for the training of CIC personnel. The research focussed not only on team training, but also on team structure and team communication.

Method.   An air traffic control task was used in which each of two radar controllers (RCs) guided aircraft into an approach gate by issuing verbal instructions via a simulated radio link to the aircraft pilots (played by well-trained experimental assistants). Each RC viewed a simulated radar display, representing the same airspace. The Alpha RC was assigned the northeast sector, whereas the Bravo RC was assigned the southeast sector. The two RCs were required to alternate in guiding aircraft through the approach gate, located at the centre of the airspace. That is, first an Alpha aircraft was to land, then a Bravo aircraft, then another Alpha aircraft, and so on. A successful approach occurred when an aircraft entered the approach gate at 200 knots on a heading of 280 degrees. RCs delivered only heading and speed commands to pilots over a voice communication channel, such as "Bravo one, speed 400 knots." An inter-RC communication link allowed the RCs to speak to each other; they could not see each other. After each landing, verbal feedback was provided by the experimenter (e.g., "40 seconds late").

In a second version of the task, each of the RCs' scopes displayed eight radar returns; four returns represented target aircraft and four represented corresponding interceptors. RCs directed interceptors by issuing heading and speed commands over a verbal channel to "pilots." Under coordination conditions, particular targets were to be intercepted simultaneously. Team coordination involved the coordinated responses of both RCs, so as to enable simultaneous interception of two targets. Individual coordination involved coordination within one RC, such that this person by himself or herself would intercept two targets simultaneously.

In their experiments, Briggs and Johnston varied numerous factors, such as team arrangement, workload, presence of inter-RC communication, training context (individual vs. team), and knowledge of results.

Major Results.   The following list summarises the major results of the Briggs and Johnston studies.

1. An hierarchical structure of teams appears to be a favourable arrangement provided that a strictly serial structure is not used. In the latter, team performance is determined in large measure by the weakest link in the chain, and overloading the input station, in particular, is a real possibility.

2. Independence of operators is highly desirable in that interaction between team members at either the input or the output level requires each operator to devote less than his or her full capacity to task-specific activities. Independence of assigned function permits each operator to utilise his or her entire information processing capacity to his or her specific job; if required to interact with other team members, an operator must share that capacity between specific demands of his or her job and the demands of the interaction process. Unfortunately, despite highly overlearned verbal behaviour, teams appear to require extensive experience before they learn efficient verbal interaction procedures or before they acquire interaction discipline.

3. Independence of team members one from another must not be achieved at the risk of overloading an individual operator. Both independence of function and load balancing result in superior team performance, but performance will suffer if either principle is ignored in system design and implementation.

4. Teams are able to adapt to higher loads by reassignment of function, development of new procedures, utilisation of shortcuts, and so on. Procedural flexibility is a desirable characteristic of systems in that it permits a team of human operators to devise work procedures, which fit task requirements that may not have been anticipated in the earlier design and implementation

of the system. However, flexibility in operating procedures leads to lower performance in emergencies, where more rigid rules would be better.

5. Team performance is superior following individual training in systems organised for relatively little interaction between team members.

6. Team coordination is evidenced by the presence of voluntary messages, which anticipate information needs of other teammates, rather than by the sheer volume of communications.

7. With training, teams exhibit progressively less volume of communications.

8. Time stress on a team will result in fewer communications. Communication discipline is important in this respect.

9. The opportunity for correcting or counteracting a teammate's error after it has been committed is an inverse function of system load: the opportunity is greater, the lower the system load. The higher the load, the more pressed a team member is to complete his or her own task, and the less time he or she has to compensate for a teammate's errors.

10. The opportunity to prevent a teammate's error is greater, the higher the system load, but only when team members are inexperienced. With experience, the fail-stop function loses its importance, because team members do not commit a great number of errors any more. Still, it would appear advisable to institute a compensatory arrangement under high-workload conditions in order to keep errors at a minimum.

11. In general, laboratory research on team communications indicates that the less such interoperator interaction, the better. Team communication hinders team performance the most when there is the least need and the least freedom to communicate. When there is enough time, and perhaps some reason to communicate, team communication neither retards nor enhances team performance.

Discussion.   The work by Briggs and Johnston has resulted in a large number of recommendations and concepts, some of which predate current concepts, such as "implicit coordination" or "team self-correction." Nevertheless, the extent to which their conclusions generalise to other task environments may be questioned. For instance, one of their most consistent findings is the negative effect of team communication on team performance. Their recommendation is therefore to limit team communication as much as possible, by assigning functions to operators such that a large degree of independence of operators is achieved. However, the negative effect of team communication was probably caused by the requirement to verbally recode visual information on the radar display. Performance was better when team members had visual access to each others' airspace compared with when they did not have visual access to their team member's airspace, and verbal communication was necessary. For this particular task, then, coordination is facilitated by the visual channel. This does not imply that the verbal channel is always less efficient. In particular, more recent formulations of multiple resource theory (Wickens, 1992) claim that the verbal channel is more efficient with auditory input rather than visual input.

A second issue concerns the division of attention between the individual task and the team task. Briggs and Johnston have clearly demonstrated that particular teamwork functions, such as correcting a teammate's error, will suffer as workload increases. That is, the higher the load, the more a team member will focus on his or her own task, to the neglect of team tasks. This reinforces the importance of the principle of independence of operators. However, three arguments can be brought against independence of operators. First, individual operators should not be overloaded. Second, a high level of independence leads to a lack of backup (errors made by teammates will not be noticed any more, let alone corrected). Third, Briggs and Johnston have never considered task switching in cases where the individual task does not demand constant attention. Frequently, operators have some freedom in scheduling their own activities, and they can switch between their own tasks and team tasks.

Implications. The work by Briggs and Johnston has shown that the following elements are important to consider designing teams are designed.

1. Task allocation: who should do what? The recommendation by Briggs and Johnston is to assign tasks to operators such that the required interaction is minimised.
2. Load balancing: assign tasks to operators such that workload is balanced across operators.
3. Communication: who talks to whom? The recommendation by Briggs and Johnston is to minimise interoperator interaction. This recommendation should probably be qualified as follows: minimise routine and standard communications, particularly when the task is highly visually demanding.
4. Procedural flexibility: who decides about what? The recommendation is to maximise procedural flexibility and have the team of human operators devise work procedures themselves, because particular task requirements may not have been anticipated in the earlier system design. The only exception are emergencies, where more rigid rules are better.
5. Backup (compensation and error prevention): The recommendation is that team members need to have the possibility to prevent, or compensate for, each other's errors. This can be achieved more easily, the more similar team members' tasks are to each other.
6. Coordination: who coordinates with whom? The recommendation is that team members need to be aware of each other's information needs, so they can send voluntary messages in anticipation of those needs (which reduces overt communication).
7. Structure: who is responsible for what; who owns resources? The recommendation is that a centralised hierarchy is preferred, with team performance being dependent on the best person on the team, rather than on all members including the poorest performer.

Research Into High-Reliability Organisations

Background. The High Reliability Organisation Project at the University of California, Berkeley, conducted field research from 1986 until 1991 in three very complex, technology-intensive organisations that are held to a failure-free standard: air traffic control, naval air operations at sea, and electric power systems. The goal of the project was to discover how these organisations attained their nearly failure-free performance, and what implications this has for other organisations striving toward that goal. Major references are LaPorte and Consolini (1991), Weick and Roberts (1993), Roberts (1990), and Rochlin, LaPorte, and Roberts (1987).

Method. Field observation notes were made and interviews were conducted while researchers were aboard Nimitz class carriers USS Carl Vinson and USS Enterprise, and while they observed the Federal Aviation Administration's air traffic control system. Researchers spent from 4 days to 3 weeks aboard the carriers at any one time. They usually made observations from different vantage points during the evolutions of various events. Observations were entered into computer systems and later compared across observers and across organisational members for clarity of meaning. Quarterly workshops were held with senior officers over a 2-year period to discuss findings.

Major Results. Three modes of operation were found, each with unique characteristics, but existing side by side in the same organisation, carried out by the same people facing different degrees of pressure.

In routine operations, under low to moderate demand, the hierarchy is the preferred structure. In the routine mode, there is high reliance on standard operating procedures (SOPs), discipline, and reliable performance. Feedback is not valued. Centralised, directive authority structure, with substantial information flows in the interests of coordination.

Under high demand, a different mode of operation takes over: the high-tempo mode. In this mode, technical expertise is more important than hierarchical rank; there is a professionalisation of the work teams: operators are granted more discretion. Feedback and negotiations increase in importance. Extra pairs of eyes quietly gather around the expert to help him or her out in case of overload: they give supportive assistance, sound alerts, and provide suggestions, usually in the form of questions rather than directives. When communication demands are heaviest, yet another person comes forward to take over part of the communications.

Under life-threatening situations, the emergency-response mode takes over. Predetermined allocation of duties takes over immediately. Teams know exactly what to do in cases of emergency, because these situations have been practised often.

The following characteristics of high-reliability organisations were found:

- Close interdependence;
- Close reciprocal coordination and information sharing, resulting in overlapping knowledge;
- High redundancy: multiple people observing the same event and sharing information;
- Broad definition of who belongs to the team;
- Teammates are included in the communication loops rather than excluded;
- Lots of error correction;
- High levels of situation comprehension: maintain constant awareness of the possibility of accidents;
- Dynamic reallocation of tasks depending on workload (flexibility in task sequencing);
- High levels of interpersonal skills;
- Very high level of technical competence: high levels of technological knowledge of the system and its components, and high levels of performance;
- High visibility for those activities that enhance reliability, including easy access by lower grades of personnel to report problems to senior levels of management;
- Maintainence of detailed records of past accidents and incidents that are closely examined with a view to learning from them;
- People are trained to perform many functions; high degree of staffing redundancy: organisational slack;
- Patterns of authority are changed to meet the demands of the events: organisational flexibility;
- Decision making is dispersed and need not be referred to a central authority;
- A constant search is undertaken for improvement in safety and reliability;
- The reporting of errors and faults is rewarded, not punished.

Discussion.   There are two limitations to the work of the high-reliability theorists. One is that, although good management and organisational design may reduce accidents in certain systems, they can never prevent them. As discussed by Vaughan (1996), NASA had incorporated many of the recommendations put forward by high-reliability theorists, yet this did not prevent the Challenger accident. A second point is that organisations are not stable structures, but have to deal with changing resource availability. Resource reduction may lead to organisational drift toward the safety border (Marcus & Nichols, 1999). Therefore, organisations have to maintain a distribution of resources that enables them to learn from unusual events in their routine functioning.

During much of the time, the hierarchical organisational structure governs the operations of the ship or the air traffic control tower. However, a very different organisational structure is adopted during complex operations, when workload and tempo increase. This organisational structure is informal in that it is not officially documented. Furthermore, the informal work organisation is flat and distributed rather than hierarchical and centralised. Workers who are at the bottom of the command chain will make important decisions without first obtaining approval from their superiors. The reason for this is that events can happen too quickly to allow for appeals through a chain of command. In addition, the people at the bottom of the command chain are those who have access to the information locally—information that is often vital for making correct decisions.

The switch from the formal to the informal organisation was a result of an exploratory process. It was not determined beforehand. Through experience, the workers have found that a flat, distributed organisation provides an effective structure for balancing the challenging demands of the job with the limited capacity of the resources available.

The conclusions from this project generalise well beyond aircraft carrier operations at sea or air traffic control (see, e.g., Vicente, 1999). The general principle is that if there are disturbances that cannot be anticipated up front by the designers, then a centralised organisational structure cannot possibly deal effectively with the nature of the demands. Instead, a distributed, adaptive organisational structure is required. This conclusion is not a new one. As already discussed, the contingency school of thought already came to the same conclusion. However, as Vicente (1999) pointed out, it is a conclusion with which information systems designers have not yet fully come to terms. How do we deliberately design to support adaptive self-organisation? In the implications that follow, we formulate some criteria on the basis of the high-reliability research that could guide designers.

Implications.   The following criteria are important when adaptive teams are designed.

1. Access to information or action means: who owns resources? The recommendation is that the actors who have the most immediate access to a particular information set are responsible for making decisions associated with that information.
2. Communication: who talks to whom? The recommendation is to facilitate communication within teams by including as many perspectives as needed for making a decision or for coordinating. Communication across teams should be minimised.
3. Workload sharing: the recommendation is to allow for dynamic reallocation of tasks depending on workload. Multiple people need to be available to share work (organisational slack).
4. Redundancy: the recommendation is to design for redundancy in case of emergencies. Two people are more reliable than one, because one individual can detect the mistakes of another.
5. Structure: who is responsible for what? The recommendation is that structure should be dictated by situational demands. Under low to moderate demand, the hierarchy is the preferred structure; under high demand, a flat, distributed organisational structure is preferred (organisational flexibility).
6. Procedural flexibility: who decides about what? The recommendation is that following procedures is dependent on the situational demand. Under low to moderate demand, and under life-threatening situations, procedures are followed closely. Under high demand, procedures are abandoned, and operators are granted more discretion.

Computational Modelling of Team and Organisational Behaviour

Background.   From 1990 onward, Carley and coworkers at Carnegie-Mellon University in Pittsburgh have been working on computational models to examine various aspects of organisational design (e.g., training, communication, and command structure, and resource access). Their main emphasis has been on organisational designs suited to high performance under stress. Recently, they have been working on organisational adaptation in a changing environment. Major references are Carley (1991), Carley and Lin (1995, 1997), Lin and Carley (1995, 1997), Lin (1998), and Carley and Lee (1998).

Method.   Carley's method has not involved any empirical studies involving human operators, but rather has consisted of developing computer models that simulate various aspects of organisational design. The basic task posed to the models was a stylised anti-air warfare radar task. The organisation must decide whether the single object in the airspace is friendly, neutral, or hostile. Each object is characterised by its score on nine factors, such as speed, direction, range, altitude, and angle. For each factor, the score for that factor is either 1, 2, or 3, resulting in $3^9 = 19.683$ possible unique problems. The meaning of these scores depends on the factor. For example, if the factor is speed, then a 1 would indicate a low speed, a 2 a moderate speed, and a 3 a high speed. Each object is uniquely characterized by a nine-digit number. Each object also has a true state, unknown a priori by the organisation or any agent within the organisation. The true state depends on the unweighed sum of the scores for the nine characteristics. For example, if the object is 123111333, then the sum is 18. Depending on the task environment, this object's true state in this case could be either neutral or hostile. This is because a different set of rules is used to determine an object's true state, depending on the task. For an unbiased task, the object is neutral if the sum is >16 and <20, whereas for a biased task, the object is neutral if the sum is >13 and <18. Hence, in a biased task, more of the objects are hostile.

The models consist of nine analysts at the lowest level who gather and evaluate information about the object in the airspace. These analysts pass on their decisions to their superiors, who integrate this information and then make their own decision, which, depending on the organisational structure, is either passed on to their superiors or is the organisation's decision. After the organisation makes a decision, a new problem is given to the organisation, and the process starts over with the analysts gathering information. In this way, all 19.683 problems are presented to the model.

Carley has varied various factors in her models that are important to organisational performance: the formal command and control structure (team, hierarchy, and matrix), the resource access structure (who has access to which resources or incoming information: is the view of the problem completely different for each analyst or overlapping to certain degrees), the procedures for training personnel (experientially or operationally), the task environment (biased or unbiased: the number of hostile objects in the environment), time pressure (how much time there is for making a decision about the object), number of review opportunities, and information distortion (missing information, incorrect information, communication breakdowns, and agent turnover).

Major Results.   The following list summarises the major results of the studies by Carley and associates.

1. Complex organisations (hierarchies and matrix structures) perform better in complex environments, whereas simple organisations (teams) perform better in simple environments (the complexity of the environment is defined here as whether the decision to be made about the target object is a binary, e.g., friendly or hostile, decision or a ternary, e.g., friendly, neutral, or hostile, decision). This confirms contingency theory: there is no single organisational structure that is optimal for all situations.

2. The organisation gets more robust if each piece of information is examined from multiple perspectives.

3. Erroneous or missing information is generally more debilitating than breakdowns.

4. Within hierarchies, the closer the communication breakdown is to the top of the pyramid, the more drastic the consequences.

5. In a biased environment, with an unequal distribution of hostile and friendly objects, it is important to take into account this unequal distribution, and have people learn from experience. If the environment is well known, then a SOP is better for making decisions.

6. Turnover degrades performance, particularly in organisations that heavily rely on experience of their decision makers. Organisations that rely on procedures are less affected by turnover, because procedures can be learned faster than experience.

7. The task environment is a more important determinant of performance than the organisational structure, the resource access structure, or the information distortion.

8. Generally, experience-based organisations perform better than procedure-based organisations. This is partly because experienced managers can ignore mistakes made by subordinates. For another part, this is because an experience-based organisation uses specific knowledge of the environment, and the task environment is the most important determinant of performance. A prerequisite is that the experience is drawn from the same environment the organisation is operating in. An organisation trained for war and having to operate in a peace-keeping operation will likely make mistakes.

9. Organisational structure has little effect on performance in situations of either high or low time pressure. Under moderate time pressure, organisational structure makes a difference, with the hierarchy resulting in the most correct decisions. This is because the hierarchy generates the most information, which increases the chances for a correct decision.

10. When organisations have more opportunities to review decisions, they will perform better, but only under moderate time pressure.

11. Time pressure and number of review opportunities are more important determinants of performance than organisational structure.

12. Procedural organisations working according to SOPs are much less dependent on a particular organisational structure than experience-based organisations. With organisational redesign, it is therefore more important to devote attention to the contents of SOPs than to the organisational structure.

Discussion.   It is important to be aware of the limitations of the approach taken by Carley and coworkers. As they are the first to admit, their models and results only apply to fully trained organisations and to organisations that routinely face the same type of problem. The agents in the model exhibit the same problem solving and information processing skills during crisis as they do under normal operating conditions. The results are most applicable to simple tasks, with only minimal choice options, or tasks where the work is relatively easy to compartmentalise. Moreover, their agents are mechanical entities that do not exhibit the same kind of behaviour that human teammates would exhibit, such as recognising signs of stress with teammates, providing backup when workload increases, providing information before being asked to do so, and changing the organisational structure as a response to changing demands in the environment. Their model is a purely bottom-up model, without any top-down influence on the analysts at the bottom. For instance, there is no priority setting, no elaboration of intentions, and no directions on sensor settings.

Of course, a computational approach also has a number of advantages compared with an experimental approach. First, assumptions have to be made explicit in order for a model to be run successfully. Second, with relatively little cost, the impact of large numbers of variables can be assessed. This allows us to prioritise variables, and it provides suggestions for hypotheses that might be tested in follow-on studies with human subjects.

In conclusion, it would be best to take Carley's work as a description of how a well-trained organisation operates under routine, and stable, conditions. Under these conditions, it appears that task environment, time pressure, and number of review opportunities are more important determinants of performance than organisational structure, information distortion, training, or resource access structure. This leaves us with a paradox: those factors that are beyond our control (task environment and time pressure) have a larger effect on performance than factors that we can control (structure and training). What should the organisation do? There are several options.

First, match the organisational structure to the environment. In general, the recommendation would be to increase the complexity of the organisation as the complexity of the environment increases. This can be achieved in two ways: either beforehand, by analysing the type of environment the organisation is likely to operate in, and then fix the structure for that type of most likely environment; or, second, by allowing the organisation to adaptively change its own structure in response to environmental demands (as is proposed by the high-reliability theorists). This recommendation is only valid under moderate time pressure; under low or high time pressure, organisational structure has little effect on performance.

Match the environment to the organisational structure. Although we have no complete control over the environment, to a certain extent we can determine our own environment, and, by that decision, determine how our organisation will fare. For instance, the Captain of the Vincennes had the choice to either engage or not engage the Iranian gunboats. By choosing to engage the gunboats, he created an environment of high-speed action and time pressure. By itself, this choice would not have had any repercussions. However, shortly thereafter, the Iranian Airbus entered the scene, which further increased time pressure. As the Vincennes' Captain's attention was largely devoted to the surface battle with the gunboats, too little time was left to deal with the Airbus, which was subsequently misclassified and shot down. Others at the scene, for instance the Captain of the USS Sides, have argued that the gunboats should not have been engaged (Collyer & Malecki, 1998). There was, therefore, a legitimate choice. By choosing to go down the path of increased time pressure, the Vincennes' environment, the primary determinant of its performance, changed dramatically.

Implications.   According to Carley, the following criteria are important to consider in organisational design.

1. Task environment: the recommendation is that the organisational structure should match the complexity of the environment—the more complex the environment, the more complex the structure has to be. Further, the organisation should expend a lot of effort acquiring knowledge of the task environment it will be operating in, as this is the most important factor determining performance.

2. Time pressure: the recommendation is that under low or (extremely) high time pressure, organisations should rely on procedures or hierarchies, whereas under moderate time pressure, a flatter organisation structure is recommended.

3. Number of review opportunities: the recommendation is to increase the number of review opportunities in organisations. This would result in more hierarchical rather than flat organisations, as hierarchies have more opportunities for review (see also Carzo & Yanouzas, 1969, for an empirical validation).

4. Structure: the recommendation is that teams are better for making simple decisions; hierarchies or matrices are better for making complex decisions.

5. Access to information: the recommendation is that people who share the same information should be in different teams or departments rather than blocked within a single team or department. A distributed information access structure will decrease competition or quarrelling between groups and will prevent catastrophic information loss. A second recommendation is that managers at the top of the pyramid should be available and their communication channels intact during crises. The reason is that, within hierarchies, the closer the communication breakdown is to the top of the pyramid, the more drastic the consequences.

6. Information distortions: the recommendation is that organisations should expend more effort in acquiring correct information than in setting up extra communication channels.

7. Training: the recommendation is that training to recognize the specific features of the environment helps in comparison with training in SOPs, when the environment is biased, and contains specific features. However, the wrong kind of training may be worse than no training at all. Because organisations that are trained to work according to SOPs are hardly dependent on a particular organisational structure, when such organisations are redesigned, it is more important to devote attention to the contents of the SOPs rather than to the organisational structure.

8. Redundancy: Redundant communication channels are costly and rarely improve the performance of the fully trained organisation. The recommendation is that the only organisations that may find redundant communication channels efficacious are those that expect frequent crises or for which the cost of a crisis, however unlikely, is extreme.

Multilevel Theory of Team Decision Making

Background. From 1990 onward, Hollenbeck and coworkers at Michigan State University have been working on a multilevel theory of team decision making. Their focus was on hierarchical teams with distributed expertise. It is interesting to compare their results to those of Carley and coworkers, as the two research programs complement each other: Carley and coworkers have used computational modelling as their research tool, whereas Hollenbeck and coworkers have tested large numbers of human subjects. Both groups of researchers, however, have employed basically the same naval classification task. Major references are Hollenbeck, Ilgen, Sego, Hedlund, Major, and Philips (1995); Hollenbeck, Ilgen, Tuttle, and Sego (1995); Hollenbeck, Sego, Ilgen, Major, Hedlund, and Phillips (1997); and Hollenbeck, Ilgen, LePine, Colquitt, and Hedlund (1998).

Method. Four-person hierarchical teams, consisting of a leader and three staff members, had to monitor the airspace surrounding a carrier. When an aircraft came into this airspace, each team member needed to gather some information about particular attributes of the aircraft (e.g., its speed, direction, angle, range, and size) and then arrive at a judgement about the appropriate response to make toward the aircraft (varying from ignore, review, monitor, warn, ready, and lock on to defend, the most aggressive response). Staff member judgements were forwarded over a computer network to the leader, who considered them along with the information on the attributes that were obtained at that level. The leader then made a final decision for the team. Although the aircraft was characterised by nine attributes, no team member was an expert on all nine attributes, but rather on six, at most. Some attributes were shared among team members, whereas other attributes were uniquely held by team members.

Three team-level constructs determine team decision-making accuracy. First, there is the level of team information: a team that is highly informed knows a great deal about the decision objects on average (i.e., it has sampled all relevant information). Second, there is staff validity: a team that is high in staff validity generates judgements about the decision object that predict the true state. Individual staff members do not necessarily have to generate correct judgements, provided their leader properly corrects for their bias. This leads to the third construct, which is hierarchical sensitivity: a team that is high in hierarchical sensitivity has a leader who uses the best possible weight for each staff member's opinion when combining these to arrive at the team's decision. These three core constructs together have explained 25–65% of the variance in team decision making in Hollenbeck's research. Noncore constructs such as cohesion, experience, familiarity of team members with each other, attrition, or redundancy hardly have a unique contribution to team decision making once the core constructs are controlled statistically.

Major Results.   The following statements summarise the major results of the studies by Hollenbeck and coworkers.

1. The leader of a familiar team, in comparison with the leader of an unfamiliar one, is better able to differentially weight the contributions of his or her staff. However, instability in staff membership disrupts the ability of leaders to develop an effective weighting scheme for their staff. Attrition is more disruptive to staff members in familiar teams than to staff members in unfamiliar teams.

2. Increased experience leads to enhanced capacity to use information to arrive at sound recommendations.

3. Teams that are low in cohesion and teams that contain an incompetent member perform worse than teams characterised in the opposite fashion. Redundancy has no effect on team decision accuracy. Furthermore, neither high cohesiveness nor high redundancy is able to buffer teams from the impact of the incompetent member.

4. Teams with highly redundant members are able to gather a great deal more information for each decision relative to teams that are low in redundancy. It is critical for teams to collect and distribute as much information as they can for each of the decision objects they encounter.

5. Teams staffed with members who are adept at translating raw data in their area of expertise into valid recommendations make better decisions, as long as those recommendations are appropriately weighted by the team leaders.

6. Teams that are high on hierarchical sensitivity, staff validity, and level of team information perform 1.5–2.0 standard deviation units above teams that are low on all three.

7. Most of the effects for noncore variables (e.g., experience, cohesiveness, and staff competency) are eliminated when the core constructs are controlled statistically.

8. High-performing teams send out more information voluntarily than poorly performing teams; poor teams send more requests for information than high-performing teams. Furthermore, high-performing teams translate raw data into a format that is understandable by other team members, whereas poor teams more often send raw, unfiltered, data to other team members.

9. Effective teams always close the communication loop, whereas ineffective teams fail to do so (they leave questions unanswered or fail to read responses to questions they originally asked).

10. Individual decision makers performing in a context requiring sustained attention exhibit a performance failure on both the critical and postcritical event. The same applies to decision-making teams. This is probably caused by social distraction in the context of a task that is perceived as boring. The exact same decision object is handled much better by a team when it is common than when it is rare, and rare events disrupt the team's ability to handle subsequent common decision objects.

11. On the postcritical trial, there is a major surge in message traffic flow dealing with the performance failure on the previous trial. Teams tried to comprehend what went wrong on the previous decision, leading attention away from the current decision. Also on the postcritical trial, team leaders relied more heavily on staff members, compared with the critical trials, even though the leaders were in the best position to get an accurate assessment of the aircraft as a whole.

12. Shifting from a low to a high time pressure environment (less time available for a decision) led to higher stress and lower accuracy, despite the fact that subjects were gaining experience. In addition, informational ambiguity (aircraft's attributes falling in "grey zones") leads to more perceptions of uncertainty, more time for staff members to make their judgements, and lower team decision accuracy.

13. High ambiguity–high time pressure conditions lead to more failures to register a decision. Ambiguity is especially harmful to teams at relatively low levels of experience.

14. Process feedback in terms of the core constructs of the multilevel theory leads to better team decision making than outcome feedback.

15. Experience results in better team decision making. This effect is mainly caused by an improved level of team information and staff validity.

16. Feedback results in improved hierarchical sensitivity. Process feedback results in better information distribution in teams: with process feedback, it becomes clearer who is still lacking what information. This helps inexperienced teams in particular.

17. Level of team information influences staff validity, particularly when teams receive feedback. Staff validity influences hierarchical sensitivity, particularly when teams are experienced. Learning to correctly weigh the judgements of staff members is a time-consuming process that is not improved merely by experience. Generally, team leaders weigh the judgements of staff members in a too uniform way, and they attach too much importance to those judgements.

Discussion. Hollenbeck's theory presupposes a static, bureaucratic organisation, particularly suitable for routine environments. As with Carley's computational agents, no mechanisms are invoked to change the information processing within the team: there is no workload sharing, no mutual consultation among team members, no top-down directions to shift priorities, and so on. This clearly limits the generalisability of the results. Nevertheless, for the type of task involved (classification), and the type of team considered (hierarchical teams with distributed expertise), the results of Hollenbeck and coworkers have clearly provided support for the multilevel theory of team decision making.

As already mentioned, it is appropriate to make a comparison between Hollenbeck's work and Carley's work, as these researchers have used more or less the same classification task. One of the variables studied by both researchers is attrition, or turnover in teams. Carley found that turnover degrades performance, particularly in experience-based organisations. Hollenbeck confirmed this empirically and attributed the performance degradation to a diminished ability of leaders to develop an effective weighting scheme for their staff. Hollenbeck also found that turnover was more disruptive to staff members in familiar teams than to staff members in unfamiliar teams; Carley found that organisations operating according to SOPs were less affected than organisations operating according to experience. It is not clear whether Hollenbeck's unfamiliar teams operated more according to SOPs, and whether his familiar teams operated more on the basis of their experience.

A second variable studied by both researchers is informational redundancy or the degree to which there is overlap in expertise among the staff members. Carley found that a distributed access structure, with staff members sharing a lot of information, led to better performance than a blocked access structure. Hollenbeck only found an effect of redundancy on level of team information, but not on overall team decision accuracy. In Hollenbeck's study, the core constructs explained far more of the variance than the redundancy construct. The positive effect of informational redundancy may therefore be restricted to level of team information: teams with highly redundant members were able to gather a great deal more information for each decision relative to teams that were low in redundancy. However, high redundancy did not buffer teams from the impact of an incompetent member, as might be expected. This confirms Carley's finding that incorrect information has a more debilitating effect than missing information.

Hollenbeck and Carley also found similar effects of time pressure and the task environment. Time pressure resulted in lower accuracy, higher stress, and more failures to register a decision. The nature of the task environment (the distribution of friendly and hostile objects) in both studies had a large impact on performance as well. Generally, rare events, for which teams are not trained, disrupt the team's ability to handle subsequent common decision objects, for which they are trained. Because Hollenbeck used human subjects, he could attribute the negative impact of rare events on the fact that teams tried to comprehend what went wrong on the previous decision, leading attention away from the current decision. This result could not have been obtained with the computer agents used by Carley.

In general, then, Hollenbeck's studies with humans have provided support for the computational studies by Carley. Hollenbeck has been able to attribute some results to unique human capabilities and limitations.

Given the general level of support for their theory, what practical implications does this have for team design? First, consider level of team information. This construct implies, first, that each team member needs to have direct access to all task relevant information; second, translate raw data into a format that is understandable by other team members; third, distribute as much information as they can for each of the decision objects they encounter; fourth, send out information voluntarily. Second, consider staff validity. This construct implies, first, that each team member correctly combines the raw data into a valid judgement; second, that this is carried out in a consistent rather than random manner, such that a team leader may be able to correct for a particular bias. Third, consider hierarchical sensitivity. This construct implies, first, that the team leader appropriately weighs the team member's judgements by knowing the team members well (as a result, for instance, of familiarity, stability in staff membership, or high physical proximity); second, it is important for the team leader to receive feedback about each team member's judgements.

Implications.   The following criteria are important when teams are designed.

1. Information access: the recommendation is that team members should have direct access to all task-relevant information.
2. Information distribution: the recommendation is to design direct communication links between team members to ensure proper information distribution.
3. Task work skills: the recommendation is that teams should be staffed with competent members who are adept at translating raw data in their area of expertise into valid recommendations.
4. Communication skills: the recommendation is that teams should be staffed with members who display proper communication skills (e.g., send out information voluntarily; close the communication loop).
5. Stability in staff membership: the recommendation is that stability in staff membership enables leaders to develop effective weighting schemes for their staff, because the leader knows the team members well.
6. Redundancy: the recommendation is that high redundancy is not able to buffer teams from the impact of an incompetent member. More effort should be expended in acquiring correct information rather than in extra communication channels.
7. Familiarity: the recommendation is that the leader of a familiar team is better able to differentially weigh the contributions of his or her staff. However, attrition is much more debilitating to familiar teams relative to unfamiliar teams. There is a trade-off, therefore: team members who know each other well perform better, but they are more vulnerable at the same time when a team member has to be replaced.

Team Architectural Design

Background.   Bowers and coworkers at the University of Central Florida have studied the effects of different team structures and workload on team performance. In their work, team structure referred to the information and capabilities assigned to individual team members for performing the team task. Bowers and coworkers have carried out three laboratory experiments, using the same test bed, although slightly varying the team structure in each experiment. This makes their work suitable for our purposes. The first experiment was reported in Bowers, Urban, and Morgan (1992); the second in Urban, Bowers, Monday, and Morgan (1995), and the third in Urban, Weaver, Bowers, and Rhodenizer (1996). A review chapter by Urban, Bowers, Cannon-Bowers, and Salas (1995) summarised the results of two of these experiments and also discussed member proximity and communication modality as elements of a team's architecture.

Method.   Bowers and coworkers have used the Team Performance Assessment Battery (TPAB) as their test bed. This is a five-person test bed, containing a number of individual and team tasks. Team members performed the individual tasks continuously and in conjunction with the team task. The team task was a resource management task that was designed to assess team skills such as communication, decision making, coordination, and resource allocation in a naval warfare simulation. The task was presented on two different displays. One of the two displays was a graphically simulated radar display, which displayed the distance of threats from a home base. The other display was a data table that provided information about approaching targets and amounts of resources available for use.

Team members were required to monitor the simulated radar display for incoming targets, by clicking on a pattern of dots. Once targets were clicked, team members needed to decide how to allocate and manage the resources among themselves. Information pertaining to the resources required and length of time available to prosecute the target appeared in the data table. Team members used the mouse to click on specific areas of the display to transfer resources to one another. Finally, team members used the resources allotted to them to engage enemy targets. Thus, there were three tasks that had to be allocated to team members: monitor targets, manage resources, and engage targets.

The structure manipulations in the various experiments were as follows. All studies used a nonhierarchical, or horizontal, structure as a control. In the control condition, each team member was presented with identical information and capabilities for performing the team task. That is, each team member could monitor targets, manage resources, and engage targets. The hierarchical, or vertical, structure was operationalised differently in the different studies.

In the 1992 study, three team members were Scope Watchers. Each of the Scope Watchers could identify a particular type of target and could send resources to the Resource Allocator. The Resource Allocator could not identify targets but received information about all three types of targets by the Scope Watchers. The Resource Allocator received resources from the Scope Watchers and passed these on to the Target Engager, who had to engage the enemy targets.

In the 1995 study, four out of five team members could only see one fourth of the radar display to monitor targets. The fifth team member did not see the radar display at all, but could distribute resources among the team members. The four team members with access to the radar display could engage targets, whereas the fifth team member without access to the radar display could not engage targets.

In the 1996 study, three out of five team members each had access to one third of the radar display to monitor targets. The fourth team member saw the complete radar display but could only perform resource allocation. The fifth team member could only engage targets. This is similar to the 1992 setup; the difference is that in the 1992 study the Scope Watchers had access to the complete radar display (but had to select a specific target), whereas in the 1996 study, access was limited to one third of the radar display.

Major Results.   The following list summarises the major results of the studies by Bowers and coworkers.

1. In both the 1992 and the 1995 studies, the nonhierarchical structure resulted in better team performance than the hierarchical structure. In the 1996 study, there was no difference between hierarchical and nonhierarchical structures.

2. In the 1992 study, nonhierarchical teams communicated more than hierarchical teams, whereas in the 1995 study, this was exactly the other way around. In the 1996 study, hierarchical teams communicated more about resource allocation, but nonhierarchical teams communicated more about resource availability and demand.

3. There was no interaction between structure and workload in any of the studies. There is some evidence that the poorer performance of the hierarchical teams in the 1992 study was caused by neglect of the team task relative to one of the individual tasks.

The following results are drawn from the 1995 literature review. First, large distances between team members are likely to interact with structure to exert negative effects in certain operational settings (in air traffic control, physical separation of controllers and radar operators reduced collective orientation; personnel did not know what effect their actions had on the other group, and therefore what information they needed). Second, the utility of introducing a new communication modality (e.g., e-mail) is likely to depend on the team structure that is in place (e.g., more communication resulting from e-mail could lead to positive effects on team performance if a team were organised nonhierarchically, but could lead to negative effects if a team were organised hierarchically). Third, the effect of member proximity is likely to depend on communication modality; specifically, perceived distance and ease of communication moderate the effect of physical distance on the amount of communication that occurs (also see Williges, Johnston, & Briggs, 1996, who demonstrated that team coordination was improved by having team members watch each other's radar scopes, as compared with limiting them to exchange information verbally only). Fourth, sometimes more communication will facilitate effective team performance, and at other times it will impair team performance, depending on the structure employed at a particular time.

Discussion.   Although the results from the three empirical studies may seem contradictory, they can be explained by the different ways tasks were allocated to team members.

First, consider the 1992 study. In this study, the Scope Watchers had to click on each target to decide whether it was their type of target or not. As they had access to the whole screen, they had to click on a large number of targets. It is inevitable that team members did overlapping work in the hierarchical structure.

Second, in the 1995 study, team members were forced to communicate a lot, because they had extreme task specialisation on the input side, and no team member had an overview of the whole display. In contrast, a lot of output coordination was required, because all team members could engage targets.

Third, in the 1996 study, one team member was provided with an overview, and there was less overlap in the tasks carried out by the Scope Watchers. Note that this was the only study in which the hierarchical structure performed at the same level as the nonhierarchical structure; in both previous studies, the nonhierarchical structure performed better.

Taken together, these results suggest the importance of the following factors. First, there should be a balance between task specialisation and load balancing: too little task specialisation results in duplication of work (as in the 1992 study), whereas extreme task specialisation results in lack of overview (as in the 1995 study). Only in the 1996 study was this balance achieved.

Second, communication patterns in a team are the result of particular task allocations. Extreme task specialisation results in a high volume of communications, because team members have to get missing information from someone else. As long as the team task does not suffer, high volumes of communication are not a problem. However, with increasing workload or time pressure, team members give priority to their individual tasks and neglect their team task. They will communicate less, and will therefore obtain less information relevant to the team task, which will subsequently suffer.

Third, nonhierarchical, autonomous, teams are very well able to arrive at an effective task allocation themselves. In none of the experiments did an experimenter-imposed structure lead to better results than a team-imposed structure (although Bowers and coworkers do not describe how the nonhierarchical teams carried out their task, i.e., what task structure they settled on; classic research on communication networks in groups has shown that groups themselves will move toward the simpler communication networks even though given the opportunity to utilise a more time-consuming network; see Guetzkow & Dill, 1957; Guetzkow & Simon, 1955).

Implications. The following elements are important to consider in team design, according to Bowers and coworkers.

1. Structure: the recommendation is that nonhierarchical teams outperformed hierarchical teams, but as it is not clear what structure the nonhierarchical teams chose, it remains unclear why this is the case. At least, a recommendation can be provided on what to avoid: (a) duplication of work and (b) lack of overview.
2. Communication: the recommendation is that communication is dependent on the structure employed and the workload imposed; hence, no general recommendation can be given.
3. Member proximity: the recommendation is that, in general, physical distance between team members should be minimised; if this is impossible or undesirable, perceived distance and ease of communication should be optimised.
4. Communication modality: the recommendation is that modality interacts with team structure. Some modalities result in higher information flows, which could overload a hierarchical structure, but not a nonhierarchical structure.

SC21/Manning Affordability Initiative

Background. One of the three major thrusts in the Manning Affordability Initiative is Human Performance Models (the other two are the Advanced Multi-Modal Watchstation and Human-Cantered Design Tools). The optimal design of command and control organisations onboard the next generation of surface combatants constitutes part of the human performance modelling thrust. Aptima has developed a quantitative methodology to design an organisation optimised for a typical mission for the Navy's next generation of surface combatants.

The algorithms that are fundamental to this team design method were originally developed for the Adaptive Architectures for Command and Control (A2/C2) program. Major references are Cannon-Bowers, Hamburger, Osga, Bost, Crisp, and Perry (1997); Paley et al. (1999); and MacMillan, Paley, Levchuk, and Serfaty (1999).

Method. The assumption behind the modelling approach taken by Aptima is based on contingency theory: an organisation operates best when its structure and processes fit, or match, the corresponding mission environment. Therefore, the critical information about the mission structure must be captured and quantified to establish a mathematical framework for application of optimisation techniques to the organisational design process.

We do not describe in detail here Aptima's optimisation techniques (the reader is referred to Levchuk, Pattipati, & Kleinman, 1999, for a detailed description). Rather, we focus on the underlying design objectives that deal with communication, coordination, task scheduling, and workload distribution. The design objectives are:

- Reduce team size by 50%;
- Maximise mission tempo to meet or beat mission tempo requirements;
- Maintain workload at or below the level of today's surface combatants;

- Balance workload;
- Minimise coordination overhead; and
- Minimise task-specific communication requirements.

The algorithmic process consists of five phases:

- Phase A: mission representation;
- Phase B: task scheduling;
- Phase C: clustering tasks into roles;
- Phase D: decomposition of role overlap;
- Phase E: organisational structuring.

For each phase, we describe how each objective has an impact on that phase.

For Phase A, mission representation, scenario events are described and the required responses (tasks) are mapped onto those events. Tasks are defined in terms of mean duration, standard deviation of task duration, and load imposed on watchstanders (visual, auditory, cognitive, and psychomotor). First, events are reduced to classes that require unique task responses (e.g., new air contact radar detection). Then, the tasks are matched to each class of events. In some cases, deadlines force some tasks to be performed in parallel. Finally, a family of scenarios is derived that define the "feasible scenario space." The goal is to define a mission space that samples difficult but plausible variants of the reference scenario. The algorithms will define a team that can handle all scenarios in the feasible scenario space. This creates a team that is robust to variation across scenarios, but that may be suboptimal for any given scenario.

For Phase B, task scheduling, the goal is to generate a schedule for task execution that completes the mission at criterion tempo or better within workload constraints assuming abundant resources. Later, this schedule is iteratively refined to account for resource constraints, watchstander constraints, workload constraints, and communication constraints. Tasks are scheduled as early as possible, and made parallel if necessary. Subsequently, the task execution schedule is refined to account for delays due to unavailable resources.

For Phase C, clustering tasks into roles, the algorithms should be given guidance concerning which groupings of tasks (into watchstander roles) are most feasible. For some tasks, watchstanders have primary responsibility; for others they act as primary or secondary backup. Whenever more than one watchstander has primary responsibility for a task, they must coordinate. The clustering is accomplished with the help of subject-matter experts. Tasks are clustered on multiple dimensions: circuits, resources, object types (team, system, friend, or commercial airliner), functions (identifying tracks, managing own air, managing own team, or dealing with threats), or expertise (task execution skills). One issue that should be dealt with when one is clustering on expertise is how much overlap of task execution skills between watchstanders is appropriate. On the one hand, greater overlap leads to greater redundancy, greater resiliency, and greater individual knowledge of team roles, but greater training cost or lower expertise. On the other hand, less overlap leads to greater expertise at lower training cost, but more brittleness (susceptible to overload) caused by lack of backup. This trade-off cannot be dealt with by the algorithms.

Once watchstander roles are defined, tasks are scheduled again, this time with constraints on resources and watchstanders taken into account. Workload data are now taken into account. Tasks are made parallel as much as possible, until the workload threshold for an individual watchstander is reached. Then, tasks are rescheduled into a queue.

Serialising tasks for one watchstander may result in the watchstander failing to meet deadlines. Therefore, a trade-off exists between making tasks serial and parallel tasks for one watchstander: either deadlines may not be met, or workload may exceed 100%. Another option is to make tasks parallel between watchstanders, but doing so introduces workload in the form of coordination overhead (caused by communication and communication queuing). This may push workload over 100%. Finally, workload is balanced by minimising the maximum workload per watchstander per unit time. Tasks are shifted off workload peaks, and troughs (low workload periods) are filled with tasks.

For Phase D, decomposition of role overlap, the goal is to uniquely assign tasks when possible to minimise routine communications between watchstanders. This is accomplished by decomposing shared tasks (with subject-matter experts) and splitting shared tasks so that individual watchstanders have independent pieces, where feasible. If a task must be split between watchstanders because it generates overload, then it is split at a point of minimal communication and maximal difference in information content. Minimising routine communications between watchstanders can be accomplished by serialising tasks to shorten communications queues, but at the expense of delaying task execution.

For Phase E, organisational structuring, the team is defined and evaluated. Roles are analysed qualitatively with subject-matter experts, the command hierarchy is defined, and team performance is analysed quantitatively (in terms of accuracy and workload). The next step is to refine the inputs and run the algorithms again.

In conclusion, this algorithmic design process is constantly balancing between maximising mission tempo and minimising workload. Maximising mission tempo implies making tasks parallel, but this could lead to unacceptably high workload levels. Therefore, a certain amount of serialisation is required for each watchstander. Another option besides serialisation is to make tasks parallel between watchstanders, so that two tasks can still be carried out at the same time. However, this introduces workload in the form of coordination overhead: watchstanders have to communicate with each other to know when to hand over information. This problem can at least be partially solved by splitting tasks at points of minimal communication and maximal difference in information content.

Major Results.   Within the A2/C2 program, the optimised command and control architectures generated by the algorithms have been evaluated empirically, and compared to nonoptimised architectures. Empirical results supported the model linking architecture type through team processes to performance, but only when teams were afforded sufficient training with the nontraditional optimised architecture (Entin, 1999). Initial reluctance by operators to use counterintuitive designs (optimised teams were markedly different from current operations) had to be overcome with training in order to obtain these results. Teams performing with a four-node nontraditional optimised command and control architecture performed as well as the six-node traditional nonoptimised command and control architecture. Performance of a six-node optimised architecture was higher than that of the six-node nonoptimised architecture. These findings were ascertained within the low-fidelity Distributed Dynamic Decision (DDD) environment; recently (Entin, 2000), the findings were replicated in a high-fidelity simulation environment (the Marine Air Ground Task Force Tactical Warfare Simulation environment).

Hollenbeck et al. (2002) have compared three different types of architectures within the DDD environment. In DDD, the subplatforms (resources) that subjects could use (AWACS, Jet, Helicopter, and Tank) were divided up differently according to the different type of architecture. In the functional structure, each of the four decision makers controlled one single type of subplatform. Another third of the teams were configured into a divisional structure in which each of the four decision makers controlled each of the four different subplatforms. Finally, a third set of teams was set up as a robust structure in which each decision maker controlled two different subplatforms, thus forming two separate North–South Divisions. There were two types of environment: predictable and unpredictable. In the predictable environment, all targets entered and went through the Demilitarised Zone in the same manner; in the unpredictable environment, each target changed direction as it approached the Demilitarised Zone. The results showed that there was no one best structure across all situations. However, for predictable environments, team performance was higher for teams with functional structures. For teams structured divisionally, performance was higher in unpredictable environments. Teams with robust structures performed relatively poorly in both environments.

Interestingly, the algorithms developed by Aptima often suggest roles for team members that require the use of many different types of resources (e.g., both tanks and aircraft controlled by the same individual). These structures resemble the divisional structures investigated by Hollenbeck et al. (2002).

Discussion.   The team-integrated design environment (TIDE) methodology has been applied to a wide range of projects. The majority of the projects have dealt with mission–organisational analysis and architecture design rather than with team performance improvement. This reflects the maturity of the work on mission analysis and architectures. The major advantage of TIDE is that it represents a principled, replicable design process. Important variables influencing team behaviour are represented (communication, coordination, backup behaviour, and workload balancing). However, the allocation process is currently primarily driven by the internal structure of the mission, rather than by team theory. This raises a set of questions dealing with the added value of team theory to the algorithms currently used: has TIDE left out important variables influencing team behaviour? Do unexpected mission requirements break the team? Can unexpectedly intense missions break the team? Can a team adaptively switch structure when faced with unexpectedly intense missions? Can we derive principles for team design in a post hoc fashion by running the algorithms across a wide variety of missions?

Implications.   The following elements are important to consider in team design, according to Aptima.

1. Communication: the recommendation is to minimise communication requirements.
2. Coordination: the recommendation is to minimise the need to coordinate between team members. Split tasks at points of minimal communication and maximal difference in information content.
3. Task scheduling: the recommendation is that those tasks that must be made parallel to meet a deadline are made parallel first; those tasks that can be made parallel within workload limits (to accelerate tempo) are handled second; the other tasks are serialised.
4. Task allocation: the recommendation is that tasks should be allocated in such a way that workload is distributed evenly among team members.
5. Redundancy: the recommendation is that high redundancy comes with higher training cost or lower expertise. In contrast, greater redundancy comes with greater resiliency and greater individual knowledge of team roles.
6. Task environment: the recommendation is that team structure is dependent on the task environment. There is no one best structure for all situations. Evidence suggests that functional structures are better in predictable environments, whereas divisional or product structures are more suited for unpredictable environments.
7. Training: the recommendation is that teams need sufficient training in the optimal non-traditional designs in order to show performance improvements relative to the traditional designs.

Empirical Studies on Team Design: Recommendations

The review of empirical studies on team design has brought up an extensive list of recommendations with regard to team design. In Table 31.1 we have summarised these recommendations.

TABLE 31.1
Recommendations for Team Design

Team Design Element	Recommendation
Task environment	If predictable, then a functional structure If unpredictable, then a divisional–product structure
Time pressure	If low, then hierarchy If moderate, then flat structure If very high, then hierarchy
Information distortions	If information distortions are likely, then expend a lot of effort in acquiring correct information
Training	Cross training leads to more knowledge of each other's tasks, greater resiliency, and more redundancy or backup in case of emergencies (associated with divisional–product structures) Cross training brings higher training costs or lower expertise with it, and potential for role confusion On-the-job training and mentoring requires redundancy in personnel Training is required in both taskwork and teamwork skills (e.g., communication and leadership)
Task allocation	Minimise required interaction: split tasks at points of minimal communication and maximal difference in information content Avoid duplication of work At least one member of the team should have a complete overview Distribute workload evenly
Task scheduling	First, make tasks parallel to meet deadlines Second, make tasks parallel within workload limits to accelerate tempo Third, serialise the rest of the tasks
Structure	If the environment is stable and predictable, and demand is low to moderate, then a hierarchy is recommended If the environment is dynamic and complex, and demand is high, then flatter structures are recommended
Access to information	Provide direct access to all task-relevant information, without assistants transmitting or relaying information Information needed for a decision should be available in one physical location, instead of being physically separated To enhance robustness, information about a particular object should be shared by multiple persons (creates overlap in who knows what) Those with the most immediate access to information should be responsible for making decisions about that information Managers at the top of hierarchies should be available during crises (this often implies reducing staff size)
Information distribution	Key decision makers should talk to other key decision makers without going through several intermediaries Eliminate excessive links in the information chain
Redundancy of information	Redundant communication channels are only worthwhile when frequent crises are expected, or the cost of a crisis is extreme
Number of review opportunities	Increase the number of review opportunities to ensure more correct team performance (this requires extra personnel)
Communication	Minimise routine, standard communication requirements Minimise communication across units that are not functionally related Maximise opportunities for extensive communication during nonroutine situations

(Continued)

TABLE 31.1
(Continued)

Team design element	Recommendation
	Facilitate communication within teams by minimising physical distance between team members, and by providing rich communication media (face to face)
Coordination	Enhance implicit coordination by cross training
	Minimise the need to coordinate between team members (if two people need to coordinate frequently, perhaps their positions should be merged)
Procedural flexibility	If demand is low to moderate or life threatening, then procedures should be followed closely
	If demand is high, then procedures may be abandoned
Workload sharing	Multiple people need to be available to dynamically reallocate tasks and share work (organisational slack)
Backup (staffing redundancy)	Design for redundancy in case of emergencies: allow for opportunities for error prevention and compensation
	Do not overload individual team members, such that they focus on their individual task to the exclusion of providing error checks
Team composition	Promote stability in team membership (this fosters teammate specific knowledge and enables team leaders to get to know their team members)
	Team members who know each other well, perform better

INTEGRATION OF EMPIRICAL STUDIES ON TEAM DESIGN

Here we attempt to integrate the diverse findings and recommendations from the various empirical studies on team design that were just discussed. First, however, we need to discuss the assumptions underlying the studies discussed. One could object that the studies discussed in the preceding section provide a biased account of team design because they are based on command and control tasks only, and have used overly simplified laboratory tasks to study teamwork. Other approaches have used naturalistic analyses of collaborative work in real world settings other than command and control (e.g., Heath & Luff, 1992; Hutchins, 1992; Rogalski & Samurçay, 1993). However, the difference between the two approaches is often not made clear enough. The goal of laboratory studies is to test theories. Theoretical constructs and their interrelationships are best studied, we believe, under controlled conditions (see also Driskell & Salas, 1992). The criticism that the typical laboratory study is not realistic enough is simply nonsensical. It is the theory that is applied to concrete settings, not the concrete features and results of any one laboratory study. The theory is then further refined and revised by the results of applied research in the real world setting. Therefore, the naturalistic analyses of collaborative work in real world settings are most often the appropriate setting for applying theory; the "artificial" laboratory environment is most often the appropriate setting for testing theory (Driskell & Salas, 1992, p. 109).

This being said, we may now turn to a critical evaluation of the research reviewed in the preceding section. The important theoretical constructs derived from this research will then be applied to real world settings through discussion of some naturalistic case studies. The research reviewed is actually split into two distinct groups. The first group is represented by the work of Briggs and Johnston, Carley and coworkers, Hollenbeck and colleagues, Bowers and associates (to a lesser extent), and the Manning Affordability Initiative. This group puts great emphasis on the efficiency and productivity of teams. These can be achieved, according to this group, by allocating tasks to team members in such a way that communication and coordination among team members are minimised.

The second group is represented by the work of high-reliability theorists. This group puts great emphasis on failure-free performance and safety. This can be achieved, according to this group, by redundancy in personnel (organisational slack), review opportunities, cross training, and redundant communication channels.

Both groups need not be antagonistic in their recommendations if we consider the role of the environment in team design. As emphasised by the open systems school and contingency theorists, the type of environment (e.g., stable vs. dynamic) plays a decisive role in team design. Relatively stable environments, with few unpredictable disturbances, and highly proceduralised tasks lend themselves more to the approach taken by the first group. In contrast, dynamic or open systems (Vicente, 1999) are subject to unpredictable disturbances that are external to the system. These dynamic environments lend themselves more to the approach taken by the second group. As extensively argued by Vicente (1999), systems are becoming more and more open (indeed, as we saw in the second section, this was one of the reasons for the emergence of contingency theory). The need to support worker flexibility and adaptation will only increase in the future (Lesgold, 2000; Vicente, 1999). Therefore, we may expect that theoretical constructs such as redundancy (both in people and in information channels), review (error correction), cross training, autonomy, and organisational flexibility will only increase in importance when teams are designed.

Minimising communication and coordination may still be important, however. First, communication can be inefficient and disrupt the work flow during high-workload periods or after critical, rare events (Hollenbeck, Ilgen, Tuttle, & Sego, 1995; Hutchins, 1992; Johnston & Briggs, 1968; Rasker, 2002). Second, people are not always flexible and adaptive in unpredictable situations, and they may need as much procedural support as they can get to minimise workload. Third, communication interdependencies among team members are particularly harmful in unpredictable situations, because in those situations team members tend to focus on their individual tasks to the neglect of their team tasks (Bowers et al., 1992). Fourth, our own research (Rasker, 2002) and previous research by Orasanu (1993) clearly showed the need for communication to be highest in novel situations. Communication is needed to develop a shared problem model that is necessary to ensure that all members are solving the same problem. In order to allow team members to communicate as fully as possible in novel situations, there should be as few communication interdependencies as possible in routine situations.

One way to reconcile the seemingly disparate results is, therefore, to take into account whether team members have to deal with unanticipated disturbances in the environment or not. As stated earlier, team structures that are better suited to deal with unanticipated disturbances are characterised by four Rs: redundancy, robustness, review, and reallocation. The four Rs are likely to be all mediated by the single cognitive construct of "shared mental models." These in turn are facilitated by certain antecedents, such as continuity of personnel, cross training, self-managed teams, consensual leadership, and organisational slack. When very few unanticipated disturbances are expected, task specialisation, functional structures, and hierarchical relationships may be suitable. These will lead to a lower level of shared mental models, caused by little task overlap and less communication. In this analysis, "shared mental models" is the key cognitive construct that is closely tied to communication. An organisational climate that fosters communication and collaboration will also result in higher levels of shared knowledge among team members. These shared knowledge structures have been shown to influence team processes (such as coordination, cooperation, and communication) that in turn influence team performance (Mathieu, Goodwin, Heffner, Salas, & Cannon-Bowers, 2000).

Communication therefore both causes shared mental model development and is in turn influenced by shared mental models (see Rasker, 2002, for empirical support of this bidirectional relationship). High levels of communication among team members are conducive for shared mental model development, because of the exchange of knowledge among team members. In contrast, once shared mental models are developed, there is less need for communication, because team members are well aware of each others' information requirements and will provide information in advance to fellow team members, particularly under high time pressure.

We now turn to some selected case studies illustrating the theoretical concepts described herein. We also show how team organisational structures affect the cognitive tasks team members have to carry out.

First, consider the studies by Rochlin et al. (1987) and Hutchins (1992) on aircraft operation and navigation at sea (see also Seifert & Hutchins, 1992). Typically, these military organisations have adopted a centralised hierarchical organisation, at least on paper. During much of the time, this hierarchical organisation does indeed govern the operations of the ship. However, during complex flight operations, Rochlin et al. found that a flat and distributed work organisation was adopted. This is because events on the flight deck can happen too quickly to allow for appeals through a chain of command. This remarkable degree of flexibility is essential for coping with the challenging demands of the job. A flat and distributed work organisation can only function effectively when there is a large amount of knowledge sharing. As Hutchins (1992) has shown, there is substantial sharing of knowledge between individuals, with the task knowledge of more expert performers completely subsuming the knowledge of those who are less experienced. This shared task knowledge in the navigation team also ensures that most errors will be detected by members who already have experience with the operations that cause error (Seifert & Hutchins, 1992). The detection of error requires access to erroneous performance, for instance, through open communication lines, and the correction of error requires a functionally redundant distribution of knowledge, for instance, by working one's way up through the positions (which implies low rates of turnover of personnel). It should be clear from these real-life examples that the constructs already identified in laboratory research all apply to this field setting. One may go even one step further: the laboratory research by, for instance, Briggs and Johnston (1967) specifies the conditions under which the error detection and correction process observed by Hutchins will be successful. As Seifert and Hutchins (1992) noted, each actor in the navigation task was not constantly occupied. Briggs and Johnston found that, as actors get more occupied, the less time they have to compensate for their teammate's errors. Therefore, a practical recommendation would be to provide organisational slack, and make sure that team members are not overloaded.

As a second case study, consider the work carried out by Marcus and Nichols (1999) on the effects of resource reduction on drift toward the safety border in two nuclear power plants. Through an in-depth study of two plants with contrasting reputations for safety, the authors identified the characteristics, behaviours, and capabilities that organisations on the edge exhibit. The characteristics, behaviours, and capabilities of the plant with the stronger safety reputation changed after a significant reduction of resources. Personnel were put in a position where they had to accomplish more with less. Workload duplication was eliminated, the number of senior reactor operators was cut, the amount of requalification training time was reduced, and equipment expenses were cut. Along with budget pressures, the plants were aging. Systems wore out and needed more maintenance. As stress levels grew, workloads increased. The plant started to react to unexpected problems after the fact, rather than anticipating these problems. Plants operating closer to the border of safety typically sought new knowledge and skills by dismissing old staff and hiring new people, buying advice from outside experts or consultants, and modelling, benchmarking, and copying other organisations' best practices. These organisations emphasise hierarchy and powerful headquarters staff, and their leaders tend to act as commanders and controllers rather than as catalysts and facilitators.

These results make sense in terms of the theoretical concepts derived from laboratory studies on team design. As a result of resource reductions, Marcus and Nichols (1999) observed poorer communication, less backup, less redundancy, higher turnover, fewer possibilities for review because of higher levels of workload, and, as a consequence, lower levels of shared mental model development. As less knowledge was shared among operators, communication started to become less informal. Before the resource cuts, employees talked and heard about problems in fairly closed groups. Every person in the group was responsible for solving its problems. Headquarters gave the people closest to the information the right to make decisions (cf. Rochlin et al. on board the aircraft carriers). Here, we see a clear example of a team organisational structure influencing working conditions, which, in its turn, influence cognitive tasks carried out by employees. Broad and deep delegation leads to motivated personnel that learns from its mistakes. Thickly articulated communication networks are developed as a result of the open climate. Together with stability and continuity of personnel (why leave if one is motivated?), this results in high levels of shared, tacit, understanding of operations. When unexpected problems occur, employees assist each other, provide error correction, and exhibit robust behaviour as a result. The cognitive tasks they have to carry out, monitoring, diagnosis, and repair, are guided by rich representations of the system. This allows for anticipatory rather than reactive behaviour. Lest the reader think this socially constructed knowledge is peculiar for nuclear power plants, he or she is referenced to Brown and Duguid (2000), who provide numerous case studies showing the importance of communication and collaboration in solving problems with computers, copiers, or other devices.

CONCLUSIONS

When one is designing a team, it is highly important to appreciate, first, that teamwork is actually work in the sense that team members must attend to it and that it, therefore, requires information processing capacity. The potential problem is that the attention needed to perform teamwork can be at the expense of the attention necessary to perform the individual tasks accurately and vice versa. Consider, for instance, the problems that occur when team members' taskwork is distorted as a result of communicating extensively with a teammate. Especially in high-workload or time-pressured conditions, these types of problems may occur. For team design, this implies that if teamwork is required or desired, a team member must have sufficient capacity available to be able to perform both the teamwork and the individual tasks properly. In other words, the demands in terms of cognitive workload of both the teamwork as well as the individual tasks must be assessed and taken into account when one is deciding on the assignment of (team)tasks to team members.

Given that teamwork requires attention and the team member's processing limits may be reached, the question is how environmental conditions affect teamwork and team design. We conclude that the choice for particular team designs depends on the type of situation (i.e., stable, routine vs. novel, unexpected situations) and the requirements to design for safety and failure-free performance.

When the situation is characterised as closed loop, relatively stable, routine, with few disturbances, and containing highly proceduralised tasks, one can design for a minimum of teamwork. Tasks must be assigned to team members such that there is a minimum of inter-action interdependency. Furthermore, it is recommendable to train team members thoroughly. Not only on their individual task skills, in order to cut back the need for teamwork such as performance monitoring and backing each other up, but also on their teamwork to facilitate team members' shared mental models. On the basis of these shared cognitions, team members can avoid extensive discussions to coordinate explicitly. Instead, team members can restrict themselves to the timely exchange of the necessary information. The common thread in this approach is efficiency; by limiting the teamwork as much as possible, team members have more capacity to attend to their own tasks.

Nevertheless, when the situation is characterised as open loop, novel, and with unpredictable disturbances, teamwork can be a great contributor to performance. In these circumstances, one must design teams according to the four Rs of redundancy, robustness, review, and reallocation. The consequence for team design is that one must bear in mind that team members' workload is not only determined by their individual tasks, but also by their teamwork. Team members must have sufficient capacity available to perform additional teamwork. This can be at the expense of the desired efficiency; additional team members may be needed and organisational slack must be permitted.

This being said, however, it is still important to aim at the minimisation of communication and explicit coordination in teams. Given the potential costs of communication, we recommend designing for a minimum of interaction interdependency among team members. Minimising this dependency would leave team members free to perform their own tasks as well as they can. At the same time, this would leave as much spare communication capacity available for that type of communication that facilitates the teamwork that is responsible for the four Rs—that is, for performance monitoring, backing each other up, and maintaining an up-to-date shared mental model of the developments in the situation.

In summary, the following steps must be taken in team design. First, determine the organisational and environmental context in which the future organisation will operate (e.g., predictability of the environment or expected levels of time pressure). This will suggest initial hypotheses about the structure the organisation should adopt. If the context is uncertain, multiple structures could be evaluated simultaneously.

Second, given a set of tasks the future organisation has to carry out, allocate tasks in such a way that the required interaction among team members is minimised. The principle here is that, for routine tasks, communication and coordination requirements should be minimised. Tasks require information to be carried out. This information should be accessible in one location, and it should require a minimal amount of time to obtain. A lot of effort should be expended to ensure the information is correct, as incorrect information has a devastating effect on the system.

Third, each team member now has a well-defined set of tasks, different from all other team members (there is no duplication of work). Now the handover points have to be determined, that is, those points where information has to be passed from one team member to another. The handover points should be as direct as possible: direct communication links are preferred; if the information to be passed along is of a routine nature, direct voice communication may be sufficient; however, if the information is rich and equivocal in nature, face-to-face interaction should be arranged for.

Fourth, design for the four Rs: redundancy, robustness, review, and reallocation. In case of emergencies, the structure requires backup possibilities. The four Rs provide those backup possibilities. These come with a cost associated with them: either increased training costs (cross training), increased personnel costs, or increased time to carry out tasks. A way to minimise these costs, yet maximise the four Rs, is to assign to each team member one other team member who can, in principle, either take over all the tasks of the first team member, or with whom work agreements can be made before task execution (Rasker & Willeboordse, 2001). In this way, the organisation will consist of dyads who each have their own tasks, but who can support the other team member when needed. In some cases, one may be forced to design for more redundancy, and include "dyads of dyads." In this case, if a complete dyad is eliminated, another dyad can take over.

Fifth, and finally, the resulting design is not the final answer, but should be viewed as a test bed, and an opportunity to test new technologies and new procedures. This requires an iterative approach to change.

REFERENCES

Bowers, C. A., Urban, J. M., & Morgan, B. B. (1992). *The study of crew coordination and performance in hierarchical team decision making* (Tech. Rep. No. 92-01). Orlando, FL: University of Central Florida, Team Performance Laboratory.

Briggs, G. E., & Johnston, W. A. (1967). *Team training* (Tech. Rep. No. NAVTRADEVCEN 1327-4). Orlando, FL: Naval Training Device Center.

Brown, J. S., & Duguid, P. (2000). *The social life of information.* Boston, MA: Harvard Business School Press.

Cannon-Bowers, J. A., Hamburger, T., Osga, G., Bost, R., Crisp, H., & Perry, A. (1997). *Achieving affordability through human systems integration.* Paper presented at the Third Annual Naval Aviation Systems Engineering Supportability Symposium, Arlington, VA.

Cannon-Bowers, J. A., Salas, E., & Converse, S. A. (1993). Shared mental models in expert team decision making. In N. J. Castellan, Jr. (Ed.), *Individual and group decision making: current issues* (pp. 221–246). Hillsdale, NJ: Lawrence Erlbaum Associates.

Carley, K. M. (1991). Designing organizational structures to cope with communication breakdowns: A simulation model. *Industrial Crisis Quarterly, 5,* 19–57.

Carley, K. M., & Lin, Z. (1995). Organizational designs suited to high performance under stress. *IEEE Transactions on Systems, Man, and Cybernetics, 25,* 221–230.

Carley, K. M., & Lin, Z. (1997). A theoretical study of organizational performance under information distortion. *Management Science, 43,* 976–997.

Carley, K. M., & Lee, J.-S. (1998). Dynamic organizations: Organizational adaptation in a changing environment. In J. Baum (Ed.), *Advances in strategic management* (Vol. 15, pp. 267–295). Stamford, CT: JAI.

Carzo, R., & Yanouzas, J. N. (1969). Effects of flat and tall organization structure. *Administrative Science Quarterly, 14,* 178–191.

Cohen, S. G., & Bailey, D. E. (1997). What makes teams work: Group effectiveness research from the shop floor to the executive suite. *Journal of Management, 23,* 239–290.

Collyer, S. C., & Malecki, G. S. (1998). TADMUS history, background, and paradigm. In J. A. Cannon-Bowers & E. Salas (Eds.), *Making decisions under stress: Implications for individual and team training* (pp. 3–16). Washington, DC: American Psychology Association.

Driskell, J. E., & Salas, E. (1992). Can you study real teams in contrived settings? The value of small group research to understanding teams. In R. W. Swezey & E. Salas (Eds.), *Teams: Their training and performance* (pp. 101–124). Norwood, NJ: Ablex.

Duchon, D., Ashmos, D. P., & Nathan, M. (2000). Complex systems and sensemaking teams: Conflict, connectedness, and leadership. In M. M. Beyerlein, D. A. Johnson, & S. T. Beyerlein (Eds.), *Team performance management* (Advances in interdisciplinary studies of work teams, Vol. 6, pp. 219–238). Stamford, CT: JAI.

Entin, E. E. (1999). Optimized command and control architectures for improved process and performance. In *Proceedings of the 1999 Command and Control Research & Technology Symposium,* Newport, RI: United States Naval War College.

Entin, E. E. (2000). Performance and process relationships in transitioning from a low to a high fidelity simulation environment. In *Proceedings of the IEA 2000/HFES 2000 Congress* (pp. 280–285). Santa Monica, CA: Human Factors and Ergonomics Society.

Galbraith, J. (1973). *Designing complex organizations.* Reading, MA: Addison-Wesley.

Goodman, P. S., Devadas, R., & Hughson, T. L. (1988). Groups and productivity: Analyzing the effectiveness of self-managing teams. In J. P. Campbell, R. J. Campbell, & Associates (Eds.), *Productivity in organizations* (pp. 295–327). San Francisco: Jossey-Bass.

Guetzkow, H., & Dill, W. R. (1957). Factors in the organizational development of task-oriented groups. *Sociometry, 20,* 175–204.

Guetzkow, H., & Simon, H. A. (1955). The impact of certain communication nets upon organization and performance in task-oriented groups. *Management Science, 1,* 233–250.

Hackman, J. R. (1990). *Groups that work (and those that don't). Creating conditions for effective teamwork.* San Francisco: Jossey-Bass.

Heath, C., & Luff, P. (1992). Collaboration and control: Crisis management and multimedia technology in London Underground line control rooms. *Computer Supported Cooperative Work (CSCW), 1,* 69–94.

Hollenbeck, J. R., Ilgen, D. R., Sego, D. J., Hedlund, J., Major, D. A., & Philips, J. (1995). Multilevel theory of team decision making: Decision performance in teams incorporating distributed expertise. *Journal of Applied Psychology, 80,* 292–316.

Hollenbeck, J. R., Ilgen, D. R., Tuttle, D. B., & Sego, D. J. (1995). Team performance on monitoring tasks: An examination of decision errors in contexts requiring sustained attention. *Journal of Applied Psychology, 80,* 685–696.

Hollenbeck, J. R., Sego, D. J., Ilgen, D. R., Major, D. A., Hedlund, J., & Phillips, J. (1997). Team decision-making accuracy under difficult conditions: Construct validation of potential manipulations using the TIDE2 simulation. In M. T. Brannick, E. Salas, & C. Prince (Eds.), *Team performance assessment and measurement: Theory, methods, and applications* (pp. 111–136). Mahwah, NJ: Lawrence Erlbaum Associates.

Hollenbeck, J. R., Ilgen, D. R., LePine, J. A., Colquitt, J. A., & Hedlund, J. (1998). Extending the multilevel theory of team decision making: Effects of feedback and experience in hierarchical teams. *Academy of Management Journal, 41,* 269–282.

Hollenbeck, J. R., Moon, H., Ellis, A. P. J., West, B. J., Ilgen, D. R., Sheppard, L., Porter, C. O. L. H., & Wagner, J. A. (2002). Structural contingency theory and individual differences: Examination of external and internal person-team fit. *Journal of Applied Psychology, 87,* 599–606.

Hutchins, E. L. (1992). The technology of team navigation. In J. Galegher, R. E. Kraut, & C. Egido (Eds.), *Intellectual teamwork: Social and technological foundations of cooperative work* (pp. 191–220). Hillsdale, NJ: Lawrence Erlbaum Associates.

Janis, I. L. (1982). *Groupthink: Psychological studies of policy decisions and fiascoes.* Boston: Houghton Mifflin.

Johnston, W. A., & Briggs, G. E. (1968). Team performance as a function of team arrangement and workload. *Journal of Applied Psychology, 52*(2), 89–94.

Katz, D., & Kahn, R. L. (1978). *The social psychology of organizations* (2nd ed.). New York: Wiley.

Kleinman, D. L., & Serfaty, D. (1989). Team performance assessment in distributed decision-making. In R. Gibson, J. P. Kincaid, & B. Goldiez (Eds.), *Proceedings of the Interactive Networked Simulation for Training Conference* (pp. 22–27). Orlando, FL: Naval Training System Center.

LaPorte, T. R., & Consolini, P. M. (1991). Working in practice but not in theory: Theoretical challenges of "High-reliability organizations." *Journal of Public Administration Research and Theory, 1,* 19–47.

Lawrence, P. R., & Lorsch, J. W. (1969). *Organization and environment.* Homewood, IL: Irwin.

Lesgold, A. (2000). On the future of cognitive task analysis. In J. M. Schraagen, S. F. Chipman, & V. L. Shalin (Eds.), *Cognitive task analysis* (pp. 451–465). Mahwah, NJ: Lawrence Erlbaum Associates.

Levchuk, Y. N., Pattipati, K. R., & Kleinman, D. L. (1999). Analytic model driven organizational design and experimentation in adaptive command and control. *Systems Engineering, 2,* 78–107.

Lin, Z. (1998). The choice between accuracy and errors: A contingency analysis of external conditions and organizational decision making performance. In M. J. Prietula, K. M. Carley, & L. G. Gasser (Eds.), *Simulating organizations: Computational models of institutions and groups* (pp. 67–87). Cambridge, MA: MIT Press.

Lin, Z., & Carley, K. M. (1995). DYCORP: A computational framework for examining organizational performance under dynamic conditions. *The Journal of Mathematical Sociology, 20,* 193–217.

Lin, Z., & Carley, K. M. (1997). Organizational response: The cost performance tradeoff. *Management Science, 43,* 217–234.

MacMillan, J., Paley, M. J., Levchuk, Y. N., & Serfaty, D. (1999). Model-based team design: Tools and techniques. In *Proceedings of the IEA 2000/HFES 2000 Congress* (pp. 741–746). Santa Monica, CA: Human Factors and Ergonomics Society.

Marcus, A. A., & Nichols, M. L. (1999). On the edge: Heeding the warnings of unusual events. *Organization Science, 10,* 482–499.

Mathieu, J. E., Goodwin, G. F., Heffner, T. S., Salas, E., & Cannon-Bowers, J. A. (2000). The influence of shared mental models on team process and performance. *Journal of Applied Psychology, 85,* 273–283.

Orasanu, J. (1993). Decision making in the cockpit. In E. L. Wiener, B. G. Kanki, & R. L. Helmreich (Eds.), *Cockpit resource management* (pp. 132–172). San Diego, CA: Academic Press.

Paley, M. J., Levchuk, Y. N., Serfaty, D., & MacMillan, J. (1999). Designing optimal organizational structures for combat information centers in the next generation of navy ships. In *Proceedings of the 1999 Command and Control Research and Technology Symposium.* Newport, RI: United States Naval War College.

Pasmore, W., Francis, C., Haldeman, J., & Shani, A. (1982). Sociotechnical systems: A North American reflection on empirical studies of the seventies. *Human Relations, 12,* 1179–1204.

Rasker, P. C. (2002). *Communication and performance in teams.* Unpublished doctoral dissertation, University of Amsterdam.

Rasker, P. C., & Willeboordse, E. W. (2001). *Werkafspraken in de commandocentrale [Work arrangements in the Combat Information Centre]* (Report No. TM-01-A002). Soesterberg: TNO Human Factors.

Roberts, K. H. (1990). Some characteristics of one type of high reliability organization. *Organization Science, 1,* 160–176.

Rochlin, G. I., LaPorte, T. R., & Roberts, K. H. (1987). The self-designing high-reliability organization: Aircraft carrier flight operations at sea. *Naval War College Review, 40,* 76–90.

Rogalski, J., & Samurçay, R. (1993). Analysing communication in complex distributed decision making. *Ergonomics, 36*(11), 1329–1343.

Rouse, W. B., Cannon-Bowers, J. A., & Salas, E. (1992). The role of mental models in team performance in complex systems. *IEEE Transactions on Systems, Man, and Cybernetics, 22*(6), 1296–1308.

Rouse, W. B., & Morris, N. M. (1986). On looking into the black box: Prospects and limits in the search for mental models. *Psychological Bulletin, 100*(3), 349–363.

Seifert, C. M., & Hutchins, E. L. (1992). Error as opportunity: Learning in a cooperative task. *Human-Computer Interaction, 7,* 409–435.

Stout, R. J., Cannon-Bowers, J. A., & Salas, E. (1996). The role of shared mental models in developing team situational awareness: Implications for team training. *Training Research Journal, 2,* 85–116.

Szilagyi, A. D., & Wallace, M. J. (1990). *Organizational behavior and performance* (5th ed.). New York: HarperCollins.

Tannenbaum, S. I., Beard, R. L., & Salas, E. (1992). Team building and its influence on team effectiveness: An examination of conceptual and empirical developments. In K. Kelley (Ed.), *Issues, theory, and research in industrial/ organizational psychology* (pp. 117–153). Amsterdam: Elsevier Science.

Urban, J. M., Bowers, C. A., Cannon-Bowers, J. A., & Salas, E. (1995). The importance of team architecture in understanding team processes. In M. M. Beyerlein, D. A. Johnson, & S. T. Beyerlein (Eds.), *Advances in interdisciplinary studies of work teams* (Vol. 2, pp. 205–228). Greenwich, CT: JAI.

Urban, J. M., Bowers, C. A., Monday, S. D., & Morgan, B. B. (1995). Workload, team structure, and communication in team performance. *Military Psychology, 7,* 123–139.

Urban, J. M., Weaver, J. L., Bowers, C. A., & Rhodenizer, L. (1996). Effects of workload and structure on team processes and performance: Implications for complex team decision making. *Human Factors, 38,* 300–310.

Vaughan, D. (1996). *The Challenger launch decision: Risky technology, culture, and deviance at NASA.* Chicago, IL: The University of Chicago Press.

Vicente, K. J. (1999). *Cognitive work analysis: Toward safe, productive, and healthy computer-based work.* Mahwah, NJ: Lawrence Erlbaum Associates.

Weick, K. E., & Roberts, K. H. (1993). Collective mind in organizations: Heedful interrelating on flight decks. *Administrative Science Quarterly, 38,* 357–381.

Wickens, C. D. (1992). *Engineering psychology and human performance* (2nd ed.). New York: HarperCollins.

Williges, R. C., Johnston, W. A., & Briggs, G. E. (1966). Role of verbal communication in teamwork. *Journal of Applied Psychology, 50,* 473–478.

Wilson, J. R., & Rutherford, R. (1989). Mental models: Theory and application in human factors. *Human Factors, 31*(6), 617–634.

Indexes

Author Index

Note: Reference pages are shown in *italics*.

Subject Index